Topological and Variational Methods
with Applications to Nonlinear Boundary
Value Problems

Dumitru Motreanu • Viorica Venera Motreanu
Nikolaos Papageorgiou

# Topological and Variational Methods with Applications to Nonlinear Boundary Value Problems

Dumitru Motreanu
Department of Mathematics
University of Perpignan
Perpignan, France

Viorica Venera Motreanu
Department of Mathematics
Ben-Gurion University of the Negev
Beer-Sheva, Israel

Nikolaos Papageorgiou
Department of Mathematics
National Technical University
   Zografou Campus
Athens, Greece

ISBN 978-1-4939-4474-3        ISBN 978-1-4614-9323-5 (eBook)
DOI 10.1007/978-1-4614-9323-5
Springer New York Heidelberg Dordrecht London


Mathematics Subject Classification (2010): 34Bxx; 34Cxx; 34Lxx; 35Bxx; 35Dxx; 35Gxx; 35Jxx; 35Pxx; 47Hxx; 47Jxx; 49Jxx; 49Rxx; 58Cxx; 58Exx; 58Jxx; 58Kxx

Printed on acid-free paper

Springer is part of Springer Science+Business Media (www.springer.com)

# Preface

This monograph presents fundamental methods and topics in nonlinear analysis and their efficient application to nonlinear boundary value problems for elliptic equations. The book is divided into 12 chapters, with 9 chapters covering the theoretical material – Sobolev spaces, nonlinear operators, nonsmooth analysis, degree theory, variational principles and critical point theory, Morse theory, bifurcation theory, regularity theorems and maximum principles, and spectrum of differential operators – followed by three chapters containing applications to ordinary differential equations and nonlinear elliptic equations with Dirichlet or Neumann boundary conditions. The last three chapters, but not only those, consist to a large extent of original results due to the authors, and many of these results appear here in a novel form, with significant improvements and developments. We emphasize that the first nine chapters devoted to general theories are not just a collection of relevant tools to study the nonlinear boundary value problems considered in the last three chapters. They offer broad and essential insight into powerful abstract theories. Major objectives for us have been to make a self-contained presentation for every treated subject and show that it applies to different types of problems.

This book originated in the collaboration of the three authors that gave rise during a period of about 10 years to a series of research papers studying nonlinear boundary value problems with Dirichlet and Neumann boundary conditions and having in the differential part Laplacian, $p$-Laplacian, or, more generally, even nonhomogeneous differential operators. These papers are reflected in our book, although the initial results are mostly rewritten, revised, and sharpened in the text here. A distinct feature of our work is that we combine various methods such as nonlinear operator theories, degree theory, lower and upper solutions, variational methods, Morse theory, regularity, maximum principles, and spectral theory. For instance, this can be seen in the study of multiple solutions, where every solution is usually obtained through a different approach and method.

The material in our book mainly addresses researchers in pure and applied mathematics, physics, mechanics, and engineering. It is also accessible to graduate students in mathematical and applied sciences, who will find updated information and a systematic exposition of important parts of modern mathematics.

The authors are deeply indebted to Dr. Lucas Fresse for his immense and generous help related to the whole work for the present book. We have decisively benefited from his brilliant ideas and insight. For instance, the version of the first deformation theorem and its application to the limit case in the minimax principle, Lemma 6.65, which is helpful in our presentation of Morse theory, the general version of the Moser iteration procedure, and the unified formulation of the local minimizer principle given in Theorem 12.18 are due to him. His outstanding contributions in improving every chapter of our book are gratefully acknowledged.

The authors express their gratitude to Springer Science+Business Media, LLC, New York, for its highly professional assistance, and first of all we thank our editors Vaishali Damle, Eve Mayer, and Marc Strauss for strong moral support and kind understanding.

The author Viorica Venera Motreanu acknowledges with thanks the support of Marie Curie Intra-European Fellowship for Career Development within the European Community's Seventh Framework Programme (Grant Agreement No. PIEF-GA-2010-274519).

Perpignan, France                                                    Dumitru Motreanu
Beer Sheva, Israel                                              Viorica Venera Motreanu
Athens, Greece                                                  Nikolaos Papageorgiou

# Contents

# Introduction

Nonlinear elliptic boundary value problems have proven to be an extremely fruitful area of application of topological and variational methods. Such methods usually exploit the special form of the nonlinearities entering the problem, for instance their symmetries, and offer complementary information. They are powerful tools to show the existence of multiple solutions and establish qualitative results on these solutions, for instance information regarding their location. The topological and variational approach provides not just the existence of a solution, usually several solutions, but allows one to acquire relevant knowledge about the behavior and properties of the solutions, which is extremely useful because generally the problems cannot be effectively solved, so the precise expression of the solutions is unknown. As a specific example of a property of a solution that we look for is the sign of the solution, for example, to be able to determine whether it is positive, negative, or nodal (i.e., sign changing). Such topics will be addressed in the present work.

The aim of this monograph is twofold: (1) to present, in a rigorous, modern, and coherent way, topological and variational methods from the point of view of nonlinear analysis; (2) to study nonlinear elliptic boundary value problems in order to infer qualitative properties of the solutions. These two major goals strongly interact. On the one hand, topological and variational methods enable us to discover important information about solutions, including their existence. On the other hand, investigation of the nonlinear elliptic boundary value problems illustrates and justifies in an ideal manner the power of the abstract techniques developed through topological and variational methods. Our book is based on the idea of capturing this close relationship and is designed to maintain a unifying treatment. Actually, the study of every topic considered here relies on previous results that can be found in the body of the book. In this sense, our work is self-contained. To show the unity of the book, we mention a few aspects that will be encountered in our text. The topological degree is used to investigate the linking properties related to critical point theory, also in bifurcation theory, in handling different nonlinear elliptic equations. Nonsmooth analysis is utilized to study boundary value problems with multivalued terms. Minimax results, such as the celebrated mountain pass and saddle point theorems, are extremely useful in handling nonlinear boundary value problems exhibiting a variational structure. The bifurcation theory permits us to deal with nonlinear equations depending on parameters. Various index theories, such as the genus, lead to multiplicity results for the solutions. Minimization with constraints, for instance on submanifolds, provides spectral information about nonlinear operators and is exploited for treating nonlinear boundary value problems subject to additional restrictions. There are many other situations that show the unifying character of our work.

The book consists of nine chapters devoted to abstract topological and variational methods and three chapters focusing on boundary value problems for nonlinear elliptic equations. It is worth mentioning that the first nine chapters are projected not just to provide the mathematical background to be applied in the last three chapters. They are of independent interest and give a comprehensive account, sometimes with traits of originality, of large areas of contemporary mathematics. Moreover, in many applications considered in our work, we make use of several methods for the same problem: minimization, variational principle, minimax methods, degree theory, lower and upper solutions, nonlinear operators, Morse theory, regularization, truncation, and maximum principle. Except in the first chapter, we provide complete proofs for almost all of the stated results, in many situations simplifying the arguments, sometimes correcting obscurities that exist in different references, or from being able to have hypotheses that are more general than those known from other works and even to improve the conclusions. Many examples are given in the text to illustrate the applicability of the abstract statements, especially in the parts focusing on applications to nonlinear boundary value problems. Every chapter has a final section where we indicate the relevance of the discussed topics, related references, and our specific contributions with respect to the available literature.

Finally, we briefly describe the contents of the chapters in the book. A more detailed description can be found in the abstracts appearing at the beginning of the chapters. Chapter 1 reports on the background of Sobolev spaces that is needed in the sequel. Chapter 2 discusses important classes of nonlinear operators: compact, maximal monotone, pseudomonotone, generalized pseudomonotone, $(S)_+$-maps, and Nemytskii operators. Chapter 3 has as its object convex analysis and subdifferentiability theory for locally Lipschitz functions. Chapter 4 presents degree theories: Brouwer's degree, Leray–Schauder degree, degree for $(S)_+$-maps, and degree for set-valued maps. Chapter 5 addresses variational principles and critical point theory, including minimax theorems formulated both for smooth and nonsmooth functions. Chapter 6 sets forth the basic facts of Morse theory emphasizing the study of critical groups. Chapter 7 highlights bifurcation results for parametric equations obtained through degree theory and the implicit function theorem. Chapter 8 consists of basic results in regularity theory and maximum principles for nonlinear elliptic equations. Chapter 9 is devoted to the spectral properties of some fundamental differential operators: Laplacian, $p$-Laplacian, and $p$-Laplacian plus an indefinite potential. Chapter 10 focuses on the periodic solutions of nonlinear ordinary differential equations. Chapter 11 examines nonlinear Dirichlet boundary value problems in a multitude of cases, such as sublinear, asymptotically linear, superlinear, coercive, noncoercive, parametric, resonant, and near resonant, and through various methods such as degree theory, variational methods, lower and upper solutions, Morse theory, and nonlinear operators techniques. Chapter 12 contains recent results on nonlinear elliptic equations with Neumann boundary conditions, pointing out advances in topics such as resonance from the left and from the right for Neumann problems depending on a parameter, Neumann equations whose differential part is expressed by means of a nonhomogeneous operator, and a unifying approach to sublinear and superlinear cases for semilinear Neumann problems. A list of symbols, references, and an index conclude the book.

# Chapter 1
# Sobolev Spaces

**Abstract** This chapter provides a comprehensive survey of the mathematical background of Sobolev spaces that is needed in the rest of the book. In addition to the standard notions, results, and calculus rules, various other useful topics, such as Green's identity, the Poincaré–Wirtinger inequality, and nodal domains, are also discussed. A careful distinction between various properties of Sobolev functions is made with respect to whether they are defined on a one-dimensional interval or a multidimensional domain. Bibliographical information and related comments can be found in the Remarks section.

## 1.1 Sobolev Spaces

In this chapter we gather some basic results from the theory of Sobolev spaces that we will need in the sequel. We simply state the results, and for their proofs we refer to one of the standard books on the subject mentioned in the final section of this chapter. Sobolev spaces are the main tool in the modern approach to the study of nonlinear boundary value problems.

We start by fixing our notation. For a measurable set $E \subset \mathbb{R}^N$ ($N \geq 1$) and $1 \leq p \leq +\infty$, we denote by $(L^p(E, \mathbb{R}^M), \|\cdot\|_p)$ the Banach space of measurable functions $u : E \to \mathbb{R}^M$ ($M \geq 1$) for which the quantity

$$
\|u\|_p := \begin{cases} \left( \int_E |u(x)|^p \, dx \right)^{\frac{1}{p}} & \text{if } 1 \leq p < +\infty, \\ \operatorname*{ess\,sup}_{E} |u(x)| & \text{if } p = +\infty \end{cases}
$$

is finite. Hereafter, $|\cdot|$ denotes the Euclidean norm of $\mathbb{R}^M$, which coincides with the absolute value if $M = 1$. We abbreviate $L^p(E) = L^p(E, \mathbb{R})$.

Let $\Omega \subset \mathbb{R}^N$ be an open set. Recall that $f : \Omega \to \mathbb{R}$ is *locally integrable* if, for every $K \subset \Omega$ compact, $f \in L^1(K)$. The space of locally integrable

D. Motreanu et al., *Topological and Variational Methods with Applications to Nonlinear Boundary Value Problems*, DOI 10.1007/978-1-4614-9323-5_1,
© Springer Science+Business Media, LLC 2014

functions on $\Omega$ is denoted by $L^1_{\text{loc}}(\Omega)$. Also, set $C^\infty_c(\Omega) = \{\vartheta \in C^\infty(\Omega) :$ $\vartheta$ has compact support in $\Omega\}$.

A *multi-index* is an $N$-tuple $\alpha = (\alpha_k)^N_{k=1} \in (\mathbb{N}_0)^N$, where $\mathbb{N}_0 = \mathbb{N} \cup \{0\}$. For an index $k \in \{1,\ldots,N\}$, let $D_k = \frac{\partial}{\partial x_k}$ denote the $k$th partial derivation operator for differentiable real functions on $\Omega$. A multi-index gives rise to a classical differential operator of higher order, $D^\alpha = D^{\alpha_1}_1 \cdots D^{\alpha_N}_N$, defined on smooth functions.

**Definition 1.1.** Let $\alpha = (\alpha_k)^N_{k=1}$ be a multi-index and $u,v \in L^1_{\text{loc}}(\Omega)$. We say that $v$ is the *weak* (or *distributional*) *derivative* of $u$, denoted by $D^\alpha u$, if

$$\int_\Omega u D^\alpha \vartheta \, dx = (-1)^{\alpha_1 + \cdots + \alpha_N} \int_\Omega v \vartheta \, dx \quad \text{for all} \ \vartheta \in C^\infty_c(\Omega).$$

If $\alpha = (0,\ldots,0)$, then we write $D^\alpha u = u$.

*Remark 1.2.* If $u$ is smooth enough to have a classical continuous derivative $D^\alpha u$, then we can integrate by parts and conclude that the classical derivative coincides with the weak one. Of course, the weak derivative may exist without having the existence of the classical derivative. To define the weak derivative $D^\alpha u$, we do not need the existence of derivatives of smaller order. Moreover, the weak derivative $D^\alpha u$, being an element of $L^1_{\text{loc}}(\Omega)$, is defined up to a Lebesgue-null set, and it is unique.

**Definition 1.3.** Let $\Omega \subset \mathbb{R}^N$ be an open set and $1 \leq p \leq +\infty$. The *Sobolev space* $W^{1,p}(\Omega)$ is defined by

$$W^{1,p}(\Omega) = \{u \in L^p(\Omega) : D_k u \in L^p(\Omega) \ \text{for all} \ k = 1,\ldots,N\}.$$

Also, we define

$$W^{1,p}_{\text{loc}}(\Omega) = \{u \in L^1_{\text{loc}}(\Omega) : u \in W^{1,p}(\Omega') \ \text{for all} \ \Omega' \subset\subset \Omega\}$$

(recall that the notation $\Omega' \subset\subset \Omega$ means $\Omega' \subset \Omega$ is bounded and $\overline{\Omega'} \subset \Omega$). Finally, for $u \in W^{1,p}(\Omega)$, we denote $\nabla u = (D_k u)^N_{k=1} \in L^p(\Omega,\mathbb{R}^N)$.

**Proposition 1.4.** *If* $\Omega \subset \mathbb{R}^N$ *is an open set and* $1 \leq p < +\infty$, *then*

(a) $W^{1,p}(\Omega)$ *is a Banach space for the norm (called Sobolev norm)*

$$\|u\| = (\|u\|^p_p + \|\nabla u\|^p_p)^{\frac{1}{p}} \ \text{for all} \ u \in W^{1,p}(\Omega).$$

*This Banach space is separable and, for* $1 < p < +\infty$, *uniformly convex (hence reflexive);*

(b) *The space* $H^1(\Omega) = W^{1,2}(\Omega)$ *is a Hilbert space with the inner product*

$$(u,v)_{H^1(\Omega)} = \int_\Omega uv \, dx + \int_\Omega (\nabla u, \nabla v)_{\mathbb{R}^N} \, dx \ \text{for all} \ u,v \in H^1(\Omega).$$

*Remark 1.5.* The uniform convexity of $W^{1,p}(\Omega)$ will be justified in Remark 2.47 (a). The Sobolev space $W^{1,\infty}(\Omega)$ is also a Banach space for the norm $u \mapsto \|u\|_\infty + \|\nabla u\|_\infty$, but it is neither reflexive nor separable. The foregoing proposition remains valid if we consider the equivalent norms $u \mapsto (\|u\|_p^r + \|\nabla u\|_p^r)^{\frac{1}{r}}$ and $u \mapsto \left(\|u\|_p^r + \sum_{k=1}^N \|D_k u\|_p^r\right)^{\frac{1}{r}}$ on $W^{1,p}(\Omega)$ with $1 < r < +\infty$. It remains valid for the equivalent norms $u \mapsto \|u\|_p + \|\nabla u\|_p$ and $u \mapsto \|u\|_p + \sum_{k=1}^N \|D_k u\|_p$, too, except the claim about uniform convexity.

The Sobolev spaces of higher order $m \geq 2$ are defined as follows.

**Definition 1.6.** Let $\Omega \subset \mathbb{R}^N$ be an open set, $m \in \mathbb{N}$, $m \geq 2$, and $1 \leq p \leq +\infty$. Inductively, we define the Sobolev space $W^{m,p}(\Omega)$ by

$$W^{m,p}(\Omega) = \{u \in L^p(\Omega) : \nabla u \in W^{m-1,p}(\Omega)^N\}.$$

*Remark 1.7.* The result in Proposition 1.4 remains valid for the higher-order Sobolev space $W^{m,p}(\Omega)$ furnished with the norm $u \mapsto \left(\sum \|D^\alpha u\|_p^p\right)^{\frac{1}{p}}$, where the sum is taken over all multi-indices $\alpha = (\alpha_N)_{k=1}^N$ such that $\alpha_1 + \cdots + \alpha_N \leq m$.

**Definition 1.8.** Let $\Omega \subset \mathbb{R}^N$ be an open set, $m \in \mathbb{N}$, $m \geq 1$, and $1 \leq p \leq +\infty$. The space $W_0^{m,p}(\Omega)$ is defined as the closure of $C_c^\infty(\Omega)$ in $W^{m,p}(\Omega)$. We write $H_0^m(\Omega)$ for $W_0^{m,2}(\Omega)$.

*Remark 1.9.* Evidently, $W_0^{m,p}(\Omega)$ is a subspace of $W^{m,p}(\Omega)$, and in general it is a strict subspace. However, if $\Omega = \mathbb{R}^N$, then $W^{m,p}(\mathbb{R}^N) = W_0^{m,p}(\mathbb{R}^N)$, i.e., $C_c^\infty(\mathbb{R}^N)$ is dense in $W^{m,p}(\mathbb{R}^N)$. Since uniform convergence preserves continuity, we have $W_0^{1,\infty}(\Omega) \subset C^1(\Omega)$.

**Proposition 1.10.** *If $\Omega \subset \mathbb{R}^N$ is open, $m \in \mathbb{N}$, $m \geq 1$, $1 \leq p \leq +\infty$, $u \in W_0^{m,p}(\Omega)$, and we define*

$$\tilde{u}(x) = \begin{cases} u(x) & \text{if } x \in \Omega, \\ 0 & \text{if } x \in \mathbb{R}^N \setminus \Omega, \end{cases}$$

*then $\tilde{u} \in W_0^{m,p}(\tilde{\Omega})$ for any open set $\tilde{\Omega}$ such that $\Omega \subset \tilde{\Omega}$. In particular, $\tilde{u} \in W^{m,p}(\mathbb{R}^N)$.*

*Remark 1.11.* If $u \in W^{m,p}(\Omega)$ and $\tilde{u}$ is defined as in the foregoing proposition, then in general $\tilde{u}$ does not have weak derivatives, and so $\tilde{u} \notin W^{m,p}(\mathbb{R}^N)$.

Combining Definitions 1.1 and 1.8, we see that the following integration-by-parts formula holds.

**Proposition 1.12.** *If $\Omega \subset \mathbb{R}^N$ is open, $m \in \mathbb{N}$, $m \geq 1$, $1 \leq p \leq +\infty$, $\frac{1}{p} + \frac{1}{p'} = 1$, and $u \in W^{m,p}(\Omega)$, $v \in W_0^{m,p'}(\Omega)$, then*

$$\int_\Omega (D^\alpha u) v \, dx = (-1)^{\alpha_1 + \cdots + \alpha_N} \int_\Omega u (D^\alpha v) \, dx$$

*for all multi-indices $\alpha = (\alpha_k)_{k=1}^N$ such that $\alpha_1 + \cdots + \alpha_N \leq m$.*

A Sobolev function can have a bad behavior.

*Example 1.13.* Let $\Omega = B_1 = \{x \in \mathbb{R}^N : |x| < 1\}$, and let $u(x) = \frac{1}{|x|^\theta}$ for $x \in \Omega \setminus \{0\}$, with $\theta > 0$. Then $u \in W^{1,p}(\Omega)$ if and only if $(\theta + 1)p < N$ [in particular, $u \notin W^{1,p}(\Omega)$ for $p \geq N$]. Similarly, if $\{x_n\}_{n\geq 1} \subset \Omega$ is dense and $u(x) = \sum_{n\geq 1} \frac{1}{2^n} \frac{1}{|x - x_n|^\theta}$ for $x \in \Omega \setminus \{x_n\}_{n\geq 1}$, with $\theta > 0$, then $u \in W^{1,p}(\Omega)$ if and only if $(\theta + 1)p < N$.

Nevertheless, we have the following approximation result known in the literature as the *Meyers–Serrin theorem*.

**Theorem 1.14.** *If $\Omega \subset \mathbb{R}^N$ is open and $1 \leq p < +\infty$, then $C^\infty(\Omega) \cap W^{1,p}(\Omega)$ is dense in $W^{1,p}(\Omega)$.*

*Remark 1.15.* The result is false for $W^{1,\infty}(\Omega)$. For instance, the function $u(x) = |x|$ for all $x \in (-1,1)$ belongs to $W^{1,\infty}(-1,1)$ but not to the closure of $C^\infty(-1,1) \cap W^{1,\infty}(-1,1)$. However, we have a weaker version that says that if $u \in W^{1,\infty}(\Omega)$, then we can find $\{u_n\}_{n\geq 1} \subset C^\infty(\Omega) \cap W^{1,\infty}(\Omega)$ such that $\|u_n - u\|_\infty \to 0$, $\|\nabla u_n\|_\infty \to \|\nabla u\|_\infty$, and $\nabla u_n(x) \to \nabla u(x)$ in $\mathbb{R}^N$ for a.a. $x \in \Omega$.

In Theorem 1.14, in general, we cannot replace $C^\infty(\Omega)$ by $C^\infty(\overline{\Omega})$. However, we have the following result, known in the literature as the *Friedrichs theorem*.

**Theorem 1.16.** *If $\Omega \subset \mathbb{R}^N$ is open, $1 \leq p < +\infty$, and $u \in W^{1,p}(\Omega)$, then there exists a sequence $\{u_n\}_{n\geq 1} \subset C_c^\infty(\mathbb{R}^N)$ such that $u_n|_\Omega \to u$ in $L^p(\Omega)$ and $\nabla u_n|_{\Omega'} \to \nabla u|_{\Omega'}$ in $L^p(\Omega', \mathbb{R}^N)$ for all $\Omega' \subset\subset \Omega$.*

To replace $C^\infty(\Omega)$ by $C^\infty(\overline{\Omega})$ in Theorem 1.14, we need to extend functions in $W^{1,p}(\Omega)$. We have seen that extension by zero preserves Sobolev functions in $W_0^{1,p}(\Omega)$ but not in $W^{1,p}(\Omega)$. To extend functions in $W^{1,p}(\Omega)$, we need to impose additional conditions on $\Omega$.

**Theorem 1.17.** *If $\Omega \subset \mathbb{R}^N$ is open and bounded, $\partial\Omega$ is Lipschitz, $1 \leq p < +\infty$, and $\Omega \subset\subset U$, then there exists a linear, continuous operator $P : W^{1,p}(\Omega) \to W^{1,p}(U)$ such that*

$$P(u)|_\Omega = u \quad \text{and} \quad \text{supp}(Pu) \subset U \quad \text{for all } u \in W^{1,p}(\Omega).$$

*Remark 1.18.* The requirement that $\partial\Omega$ be Lipschitz means that, near every $x \in \partial\Omega$, $\partial\Omega$ is the graph of a Lipschitz function. Of course, this is the case if $\partial\Omega$ is a

$C^1$-submanifold of $\mathbb{R}^N$. The operator $P$ obtained in the preceding theorem is known as an *extension operator*.

Relying on Theorem 1.17, we can prove the following theorem.

**Theorem 1.19.** *If $\Omega \subset \mathbb{R}^N$ is open, $\partial\Omega$ is Lipschitz, and $1 \le p < +\infty$, then $C_c^\infty(\mathbb{R}^N)|_\Omega$ is dense in $W^{1,p}(\Omega)$. In particular, $C^\infty(\overline{\Omega}) \cap W^{1,p}(\Omega)$ is dense in $W^{1,p}(\Omega)$.*

*Remark 1.20.* In fact, using Theorem 1.19, we can show that the piecewise affine functions are dense in $W^{1,p}(\Omega)$ ($1 \le p < +\infty$) when $\partial\Omega$ is Lipschitz.

For functions of one variable (i.e., $N = 1$), the Sobolev spaces exhibit stronger properties.

**Theorem 1.21.** *If $\Omega \subset \mathbb{R}$ is open and $1 \le p < +\infty$, then $u \in W^{1,p}(\Omega)$ if and only if it admits an absolutely continuous representative $\overline{u} : \Omega \to \mathbb{R}$ such that $\overline{u}$ and its classical derivative $\overline{u}'$ both belong to $L^p(\Omega)$. Moreover, if $p > 1$, then $\overline{u}$ is Hölder continuous with exponent $\frac{1}{p'}$ [i.e., $|\overline{u}(x) - \overline{u}(x')| \le c|x - x'|^{\frac{1}{p'}}$ for all $x, x' \in \Omega$ and with a constant $c > 0$].*

*Remark 1.22.* If $\Omega \subset \mathbb{R}$ is open and bounded, then by the fundamental theorem of Lebesgue calculus, $u \in W^{1,1}(\Omega)$ if and only if it admits an absolutely continuous representative $\overline{u} : \Omega \to \mathbb{R}$.

For the case where $p = +\infty$, we have the following theorem.

**Theorem 1.23.** *If $I \subset \mathbb{R}$ is an open interval, then $u \in W^{1,\infty}(I)$ if and only if it admits a bounded, Lipschitz continuous representative $\overline{u} : I \to \mathbb{R}$.*

Let us see how Theorems 1.21 and 1.23 can be extended to multivariable functions (i.e., $N > 1$). First we state the multivariable counterpart of Theorem 1.21. In what follows, $\lambda^{N-1}$ and $\lambda^N$ stand for the $(N-1)$-dimensional and $N$-dimensional Lebesgue measures, respectively.

**Theorem 1.24.** *If $\Omega \subset \mathbb{R}^N$ is open, $1 \le p < +\infty$, and $u \in L^p(\Omega)$, then $u \in W^{1,p}(\Omega)$ if and only if it has a representative $\overline{u}$ that is absolutely continuous on $\lambda^{N-1}$-a.a. line segments of $\Omega$ that are parallel to the coordinate axes and whose first-order (classical) partial derivatives belong to $L^p(\Omega)$. The first-order classical partial derivatives of $\overline{u}$ agree ($\lambda^N$ a.e. in $\Omega$) with the corresponding weak partial derivatives of $u$.*

The multivariable counterpart of Theorem 1.23 is the following result.

**Theorem 1.25.** *If $\Omega \subset \mathbb{R}^N$ is a domain (i.e., nonempty, open, connected) and $u \in W^{1,\infty}(\Omega)$, then $|u(x) - u(x')| \le \|\nabla u\|_\infty d_\Omega(x, x')$ for all $x, x' \in \Omega$, where $d_\Omega$ denotes the geodesic distance; if $\Omega$ is convex, then $|u(x) - u(x')| \le \|\nabla u\|_\infty |x - x'|$ for all $x, x' \in \Omega$.*

As a consequence of Theorem 1.24 and the properties of absolutely continuous functions, we have the following two results. The first is the *product rule* for Sobolev functions, and the second is the *chain rule* for Sobolev functions.

**Theorem 1.26.** *If $\Omega \subset \mathbb{R}^N$ is open, $1 \leq p \leq +\infty$, and $u, v \in W^{1,p}(\Omega) \cap L^\infty(\Omega)$, then $uv \in W^{1,p}(\Omega)$ and $D_k(uv)(x) = v(x)D_k u(x) + u(x)D_k v(x)$ for a.a. $x \in \Omega$ and all $k \in \{1, \ldots, N\}$. If, in addition, $u \in W_0^{1,p}(\Omega)$, then $uv \in W_0^{1,p}(\Omega)$.*

**Theorem 1.27.** *If $\Omega \subset \mathbb{R}^N$ is open, $1 \leq p < +\infty$, $u \in W^{1,p}(\Omega)$, and $f : \mathbb{R} \to \mathbb{R}$ is Lipschitz continuous with $f(0) = 0$ when $\Omega$ has infinite Lebesgue measure, then $f \circ u \in W^{1,p}(\Omega)$ and $D_k(f \circ u)(x) = f^*(u(x))D_k u(x)$ for a.a. $x \in \Omega$, $k \in \{1, \ldots, N\}$ and $f^* : \mathbb{R} \to \mathbb{R}$ is any measurable Lebesgue function such that $f^* = f'$ a.e.*

*Remark 1.28.* The preceding theorem remains true if we replace $W^{1,p}(\Omega)$ by $W_0^{1,p}(\Omega)$.

A direct consequence of the chain rule (Theorem 1.27) is the following useful proposition.

**Proposition 1.29.** *If $\Omega \subset \mathbb{R}^N$ is open, $1 \leq p < +\infty$, $X = W^{1,p}(\Omega)$ or $W_0^{1,p}(\Omega)$, and $u \in X$, then $|u|, u^+, u^- \in X$, we have*

$$\nabla u^+ = \begin{cases} \nabla u & \lambda^N \text{a.e. in } \{u > 0\}, \\ 0 & \lambda^N \text{a.e. in } \{u \leq 0\}, \end{cases}$$

$$\nabla u^- = \begin{cases} -\nabla u & \lambda^N \text{a.e. in } \{u < 0\}, \\ 0 & \lambda^N \text{a.e. in } \{u \geq 0\}, \end{cases}$$

*and*

$$\nabla |u| = \begin{cases} -\nabla u & \lambda^N \text{a.e. in } \{u < 0\}, \\ 0 & \lambda^N \text{a.e. in } \{u = 0\} \\ \nabla u & \lambda^N \text{a.e. in } \{u > 0\}. \end{cases}$$

*Remark 1.30.* As a consequence of Proposition 1.29, for $u \in W^{1,p}(\Omega)$ we have $\nabla u(x) = 0$ a.e. in $\{u = \mu\}$ for every $\mu \in \mathbb{R}$. We can say that $X = W^{1,p}(\Omega)$ or $W_0^{1,p}(\Omega)$ is a lattice, that is, if $u, v \in X$, then $\max\{u, v\} \in X$, $\min\{u, v\} \in X$, and we have

$$\nabla(\max\{u, v\}) = \begin{cases} \nabla u & \lambda^N \text{ a.e. in } \{u \geq v\}, \\ \nabla v & \lambda^N \text{ a.e. in } \{v \geq u\} \end{cases}$$

and

$$\nabla(\min\{u,v\}) = \begin{cases} \nabla u \; \lambda^N \text{ a.e. in } \{u \le v\}, \\ \nabla v \; \lambda^N \text{ a.e. in } \{v \le u\}. \end{cases}$$

It is an interesting aspect of the theory that the invariance under truncation of the spaces $W^{1,p}(\Omega)$, $W_0^{1,p}(\Omega)$ fails in the higher-order Sobolev spaces $W^{m,p}(\Omega)$, $W_0^{m,p}(\Omega)$, $m \in \mathbb{N}$, $m \ge 2$.

Next we describe the dual space of $W_0^{1,p}(\Omega)$, $1 \le p < +\infty$. Let $p'$ be the conjugate exponent of $p$, i.e., $\frac{1}{p} + \frac{1}{p'} = 1$. We denote by $W^{-1,p'}(\Omega)$ the dual of $W_0^{1,p}(\Omega)$. By $\langle \cdot, \cdot \rangle$ we denote the duality brackets for this pair of dual spaces.

**Theorem 1.31.** *If $\Omega \subset \mathbb{R}^N$ is open and $1 \le p < +\infty$, then $\xi \in W^{-1,p'}(\Omega) = W_0^{1,p}(\Omega)^*$ if and only if there exist functions $u_0, \dots, u_N \in L^{p'}(\Omega)$ such that*

$$\langle \xi, h \rangle = \int_\Omega u_0 h\, dx + \sum_{k=1}^N \int_\Omega u_k (D_k h)\, dx \text{ for all } h \in W_0^{1,p}(\Omega)$$

*and*

$$\|\xi\| = \max_{0 \le k \le N} \|u_k\|_{p'}.$$

*Moreover, if $\Omega$ is bounded, then we can take $u_0 = 0$.*

**Remark 1.32.** According to this theorem, $W^{-1,p'}(\Omega)$ can be identified with the subspace of distributions of the form $u_0 - \sum_{k=1}^N D_k u_k$, where $u_0, \dots, u_N \in L^{p'}(\Omega)$. Note that in general there is no analog of Theorem 1.31 for describing the dual of $W^{1,p}(\Omega)$.

For $\Omega \subset \mathbb{R}^N$ open, we have $\lambda^N(\partial\Omega) = 0$, and thus for $u \in W^{1,p}(\Omega)$ it is not a priori meaningful to talk about the values of $u$ on $\partial\Omega$, unless, say, $u$ is at least continuous on $\overline{\Omega}$. To give a meaning of $u|_{\partial\Omega}$, we introduce the notion of trace of $u$, which generalizes the concept of boundary values to Sobolev functions. The main result in this direction is the following theorem.

**Theorem 1.33.** *If $\Omega \subset \mathbb{R}^N$ is open and bounded, $\partial\Omega$ is Lipschitz, and $1 \le p < +\infty$, then:*

(a) *There exists a bounded linear operator $\gamma : W^{1,p}(\Omega) \to L^p(\partial\Omega, H^{N-1})$ [with $H^{N-1}$ being the $(N-1)$-dimensional Hausdorff measure on the topological submanifold $\partial\Omega$ of $\mathbb{R}^N$] such that $\gamma(v) = v$ on $\partial\Omega$ for all $v \in W^{1,p}(\Omega) \cap C(\overline{\Omega})$;*
(b) *For all $\vartheta \in C^1(\mathbb{R}^N, \mathbb{R}^N)$ and $v \in W^{1,p}(\Omega)$,*

$$\int_\Omega (\mathrm{div}\,\vartheta) v\, dx + \int_\Omega (\vartheta, \nabla v)_{\mathbb{R}^N}\, dx = \int_{\partial\Omega} (\vartheta, n)_{\mathbb{R}^N} \gamma(v)\, dH^{N-1},$$

*where $n(\cdot)$ denotes the outward unit normal on $\partial\Omega$ (which exists $H^{N-1}$ a.e. on*
*$\partial\Omega$ since, by hypothesis, the latter is Lipschitz);*
(c) $\ker\gamma = W_0^{1,p}(\Omega)$.

This theorem leads to the following definition.

**Definition 1.34.** For $u \in W^{1,p}(\Omega)$, the function $\gamma(u) \in L^p(\partial\Omega, H^{N-1})$ (uniquely
defined up to $H^{N-1}$-null subsets of $\partial\Omega$) is called the *trace* of $u$ on $\partial\Omega$ and is
interpreted as the *boundary values* of $u$ on $\partial\Omega$.

*Remark 1.35.* In addition to being $\mathbb{R}$-linear, the trace operator satisfies the following
properties:

$$\gamma(u^+) = \gamma(u)^+, \quad \gamma(u^-) = \gamma(u)^-, \quad \text{and so} \quad \gamma(|u|) = |\gamma(u)| \quad \text{for all } u \in W^{1,p}(\Omega)$$

(see, e.g., Carl et al. [72, p. 35]). In particular, we have $u^- \in W_0^{1,p}(\Omega)$ whenever
$u \in W^{1,p}(\Omega)$, $\gamma(u) \geq 0$, and $u^+ \in W_0^{1,p}(\Omega)$ whenever $u \in W^{1,p}(\Omega)$, $\gamma(u) \leq 0$.

*Remark 1.36.* If $\partial\Omega \subset \mathbb{R}^N$ is Lipschitz, then there exists a bounded linear operator
$\hat{\gamma}: W^{2,p}(\Omega) \to L^p(\partial\Omega, H^{N-1}) \times L^p(\partial\Omega, H^{N-1})$ such that if $u \in W^{2,p}(\Omega) \cap C^1(\overline{\Omega})$
and $\hat{\gamma} = (\gamma, \gamma_1)$, then $\gamma(u) = u|_{\partial\Omega}$ and $\gamma_1(u) = \frac{\partial u}{\partial n}|_{\partial\Omega}$ ($\frac{\partial u}{\partial n}$ is the normal derivative
of $u$). A similar property holds true for higher-order Sobolev spaces.

Using Theorem 1.33 (b) and the continuity of the trace operator, we obtain the
following theorem, known as *Green's identity*.

**Theorem 1.37.** *If $\Omega \subset \mathbb{R}^N$ is open and bounded, $\partial\Omega$ is Lipschitz, $u \in H^2(\Omega)$, and
$v \in H^1(\Omega)$, then*

$$\int_\Omega (\Delta u)v\,dx + \int_\Omega (\nabla u, \nabla v)_{\mathbb{R}^N}\,dx = \int_{\partial\Omega} \frac{\partial u}{\partial n}\gamma(v)\,dH^{N-1}.$$

We can extend Green's identity to the case where $p \neq 2$. To this end, let $\Omega \subset \mathbb{R}^N$ be
an open, bounded set with Lipschitz boundary $\partial\Omega$ and $1 < q < +\infty$. We introduce
the following space:

$$V^q(\Omega, \text{div}) = \{h \in L^q(\Omega, \mathbb{R}^N): \text{div}\,h \in L^q(\Omega)\}$$

[recall $\text{div}\,h = \sum_{k=1}^N D_k h_k$, where $h = (h_k)_{k=1}^N$]. We endow $V^q(\Omega, \text{div})$ with the norm

$$\|h\|_{V^q(\Omega, \text{div})} = (\|h\|_q^q + \|\text{div}\,h\|_q^q)^{\frac{1}{q}}.$$

It is easy to see that $V^q(\Omega, \text{div})$ with this norm becomes a separable, uniformly
convex (hence reflexive) Banach space, and $C^\infty(\overline{\Omega}, \mathbb{R}^N)$ is dense in it. The next
theorem extends the divergence formula in Theorem 1.33 (b).

**Theorem 1.38.** *If $\Omega \subset \mathbb{R}^N$ is open and bounded, $\partial\Omega$ is Lipschitz, $1 < p < +\infty$, and $\frac{1}{p} + \frac{1}{p'} = 1$, then there exists a unique bounded linear operator $\gamma_n : V^{p'}(\Omega, \mathrm{div}) \to W^{-\frac{1}{p'},p'}(\partial\Omega) = W^{\frac{1}{p'},p}(\partial\Omega)^*$, where $W^{\frac{1}{p'},p}(\partial\Omega)$ is the image of the trace operator $\gamma$ (Theorem 1.33), such that $\gamma_n(h) = (h, n)_{\mathbb{R}^N}$ for all $h \in C^\infty(\overline{\Omega}, \mathbb{R}^N)$, and for all $v \in W^{1,p}(\Omega)$ and all $h \in V^{p'}(\Omega, \mathrm{div})$ we have*

$$\int_\Omega (\mathrm{div}\, h) v \, dx + \int_\Omega (h, \nabla v)_{\mathbb{R}^N} \, dx = (\gamma_n(h), \gamma(v))_{W^{\frac{1}{p'},p}(\partial\Omega)}.$$

For $u \in W^{1,p}(\Omega)$ we set $\Delta_p u = \mathrm{div}(|\nabla u|^{p-2} \nabla u)$ (the $p$-Laplace differential operator). Specializing Theorem 1.38 to the $p$-Laplacian, we obtain the following version of Green's identity (Theorem 1.37).

**Theorem 1.39.** *If $\Omega \subset \mathbb{R}^N$ is open and bounded, $\partial\Omega$ is Lipschitz, $1 < p < +\infty$, $\frac{1}{p} + \frac{1}{p'} = 1$, $u \in W^{1,p}(\Omega)$, and $\Delta_p u \in L^{p'}(\Omega)$, then there exists a unique element $\frac{\partial u}{\partial n_p} := \gamma_n(|\nabla u|^{p-2} \nabla u)$ of $W^{-\frac{1}{p'},p'}(\partial\Omega)$ satisfying*

$$\int_\Omega (\Delta_p u) v \, dx + \int_\Omega |\nabla u|^{p-2} (\nabla u, \nabla v)_{\mathbb{R}^N} \, dx = \left( \frac{\partial u}{\partial n_p}, \gamma(v) \right)_{W^{\frac{1}{p'},p}(\partial\Omega)}$$

*for all $v \in W^{1,p}(\Omega)$.*

*Remark 1.40.* The element $\frac{\partial u}{\partial n_p} = \gamma_n(|\nabla u|^{p-2} \nabla u)$ can be seen as a generalized normal derivative. In the case where $p = 2$ and $u \in C^1(\overline{\Omega})$, we have $\gamma_n(\nabla u) \in C(\partial\Omega)$ and $\gamma_n(\nabla u)(x) = \frac{\partial u}{\partial n}(x) = (\nabla u(x), n(x))_{\mathbb{R}^N}$ on $\partial\Omega$ (see Kenmochi [193]).

Next we present some important inequalities for Sobolev functions. We start with the so-called *Poincaré inequality*, which is important in the study of Dirichlet boundary value problems.

**Theorem 1.41.** *If $\Omega \subset \mathbb{R}^N$ is open and bounded and $1 \leq p < +\infty$, then there exists a constant $c = c(\Omega, p) > 0$ such that*

$$\|u\|_p \leq c \|\nabla u\|_p \text{ for all } u \in W_0^{1,p}(\Omega).$$

*Remark 1.42.* In fact, this result is true if $\Omega \subset \mathbb{R}^N$ is unbounded but of finite width, that is, it is located between two parallel hyperplanes. However, the result generally fails for unbounded open sets $\Omega \subset \mathbb{R}^N$.

An immediate consequence of Theorem 1.41 is the following corollary.

**Corollary 1.43.** *If $\Omega \subset \mathbb{R}^N$ is open and bounded and $1 \leq p < +\infty$, then $\|\nabla u\|_p$ is a norm on $W_0^{1,p}(\Omega)$ equivalent to the usual Sobolev norm.*

Next, we state another inequality known as the *Poincaré–Wirtinger inequality*, which extends Theorem 1.41 (Poincaré's inequality) to the Sobolev space $W^{1,p}(\Omega)$.

**Theorem 1.44.** *If $\Omega \subset \mathbb{R}^N$ is a bounded domain with $\partial\Omega$ Lipschitz, $1 < p < +\infty$, and $D \subset \Omega$ is a measurable set such that $\lambda^N(D) > 0$, then there exists a constant $c = c(\Omega, p, D) > 0$ such that*

$$\|u - u_D\|_p \le c\|\nabla u\|_p \quad \text{for all } u \in W^{1,p}(\Omega),$$

*where $u_D = \frac{1}{\lambda^N(D)} \int_D u \, dx$.*

*Remark 1.45.* If $\Omega \subset \mathbb{R}^N$ is open, bounded, and convex, then there exists a constant $c = c(N, p) > 0$ such that

$$\|u - u_\Omega\|_p \le c\,(\operatorname{diam}\Omega)\|\nabla u\|_p \quad \text{for all } u \in W^{1,p}(\Omega).$$

If $\Omega \subset \mathbb{R}^N$ is only star-shaped with respect to $x_0 \in \Omega$ [i.e., $tx + (1-t)x_0 \in \Omega$ for all $t \in (0,1)$ and all $x \in \Omega$] and $C_{4r}(x_0) \subset \Omega \subset B_R(x_0)$ for some $r, R > 0$, where $C_{4r}(x_0)$ is the open cube with center $x_0$ and side $4r$ [i.e., $C_{4r}(x_0) = x_0 + (-2r, 2r)^N$] and $B_R(x_0)$ is an open ball with center $x_0$ and radius $R$, then

$$\|u - u_\Omega\|_p \le cR\left(\frac{R}{r}\right)^{\frac{N-1}{p}} \|\nabla u\|_p \quad \text{for all } u \in W^{1,p}(\Omega).$$

When $N = 1$ (functions of one variable), then from Theorem 1.21 we know that every Sobolev function is absolutely continuous, and then the Poincaré–Wirtinger inequality (Theorem 1.44) takes the following form.

**Theorem 1.46.** *If $I \subset \mathbb{R}$ is a bounded, open interval of length $b$, $1 \le p < +\infty$, and $u \in W^{1,p}(I)$ satisfies $\int_I u \, dt = 0$, then*

(a) $\|u\|_\infty \le b^{\frac{1}{p'}} \|u'\|_p$, where $\frac{1}{p} + \frac{1}{p'} = 1$;

(b) *If $p = 2$, then $\|u\|_\infty \le \sqrt{\frac{b}{12}} \|u'\|_2$ and $\|u\|_2 \le \frac{b}{2\pi} \|u'\|_2$ (these inequalities are sharp).*

The next result, known as the *Gagliardo–Nirenberg–Sobolev inequality*, is the first of a series of remarkable embedding theorems for Sobolev spaces that constitute a major tool in the study of boundary value problems.

**Theorem 1.47.** *If $\Omega \subset \mathbb{R}^N$ is open, $1 \le p < N$, and $p^* = \frac{Np}{N-p}$, then there exists a constant $c = c(N, p) > 0$ such that*

$$\|u\|_{p^*} \le c\|\nabla u\|_p \quad \text{for all } u \in W_0^{1,p}(\Omega).$$

*Remark 1.48.* If $\Omega = \mathbb{R}^N$, then we know that $W^{1,p}(\mathbb{R}^N) = W_0^{1,p}(\mathbb{R}^N)$ (Remark 1.9), and so the result applies to the Sobolev space $W^{1,p}(\mathbb{R}^N)$. In particular, Theorem 1.47 implies that $W^{1,p}(\mathbb{R}^N)$ is embedded continuously into $L^q(\mathbb{R}^N)$ for all $q \in [1, p^*]$.

When $\Omega$ is bounded, we have the so-called *Rellich–Kondrachov embedding theorem*. We denote by $\hookrightarrow$ a continuous embedding and by $\overset{c}{\hookrightarrow}$ a compact embedding.

**Theorem 1.49.** *If $\Omega \subset \mathbb{R}^N$ is open and bounded with $\partial\Omega$ Lipschitz, then:*

(a) *When $1 \leq p < N$, $W^{1,p}(\Omega) \hookrightarrow L^q(\Omega)$ for all $q \in [1, p^*]$, and $W^{1,p}(\Omega) \overset{c}{\hookrightarrow} L^q(\Omega)$ for all $q \in [1, p^*)$;*

(b) *When $p = N$, $W^{1,p}(\Omega) \overset{c}{\hookrightarrow} L^q(\Omega)$ for all $q \in [1, +\infty)$;*

(c) *When $p > N$, $W^{1,p}(\Omega) \overset{c}{\hookrightarrow} C^{0,\alpha}(\overline{\Omega})$ with $\alpha = 1 - \frac{N}{p} \in (0,1)$.*

*Remark 1.50.*

(a) If $W^{1,p}(\Omega)$ is replaced by $W_0^{1,p}(\Omega)$, then the preceding theorem holds for any bounded, open set $\Omega \subset \mathbb{R}^N$ with no condition on the boundary.

(b) Because of Theorem 1.49 (a), $p^* = \frac{Np}{N-p}$ ($p < N$) is often called the *Sobolev critical exponent*. If $\Omega = \mathbb{R}^N$, we have $W^{1,p}(\mathbb{R}^N) \hookrightarrow L^q(\mathbb{R}^N)$ for all $1 \leq q \leq p^*$ when $1 \leq p < N$ (see also Remark 1.48), $W^{1,p}(\mathbb{R}^N) \hookrightarrow L^q(\mathbb{R}^N)$ for all $1 \leq q < +\infty$ when $p = N$, and, finally, $W^{1,p}(\mathbb{R}^N) \hookrightarrow L^\infty(\mathbb{R}^N)$ when $p > N$. In this case, the result is known as the *Sobolev embedding theorem*.

(c) We extend the definition of the Sobolev critical exponent to all $p \in [1, +\infty)$ by setting $p^* = +\infty$ if $p \geq N$. Then, note that Theorem 1.49 implies that we have a compact embedding $W^{1,p}(\Omega) \overset{c}{\hookrightarrow} L^q(\Omega)$ for all $q \in [1, p^*)$, for all $p \in [1, +\infty)$.

For higher-order Sobolev spaces, Theorem 1.49 takes the following form.

**Theorem 1.51.** *If $\Omega \subset \mathbb{R}^N$ is open and bounded with $\partial\Omega$ Lipschitz, $m \in \mathbb{N}$, and $1 \leq p < +\infty$, then:*

(a) *When $mp < N$, $W^{m,p}(\Omega) \hookrightarrow L^q(\Omega)$ for all $1 \leq q \leq \frac{Np}{N-mp}$ and $W^{m,p}(\Omega) \overset{c}{\hookrightarrow} L^q(\Omega)$ for all $1 \leq q < \frac{Np}{N-mp}$;*

(b) *When $mp = N$, $W^{m,p}(\Omega) \overset{c}{\hookrightarrow} L^q(\Omega)$ for all $1 \leq q < +\infty$;*

(c) *When $mp > N$, $W^{m,p}(\Omega) \overset{c}{\hookrightarrow} C^{k,\alpha}(\overline{\Omega})$ with $k$ the integer part of $m - \frac{N}{p}$ and $\alpha = m - \frac{N}{p} - k \in [0,1)$.*

*Remark 1.52.* When we consider $W_0^{m,p}(\Omega)$, we can drop the Lipschitz condition on $\partial\Omega$. If $\Omega = \mathbb{R}^N$, then as before (Remark 1.50) we have only continuous embeddings.

Using Theorem 1.49 we can generate some useful equivalent norms for the Sobolev spaces.

**Proposition 1.53.** *If $\Omega \subset \mathbb{R}^N$ is open and bounded with $\partial\Omega$ Lipschitz, then $u \mapsto \|u\|_q + \|\nabla u\|_p$ is a norm on $W^{1,p}(\Omega)$ equivalent to the usual Sobolev norm when (a) $1 \leq q \leq p^*$ if $1 \leq p < N$; (b) $1 \leq q < +\infty$ if $p = N$; and (c) $1 \leq q \leq +\infty$ if $p > N$.*

*Remark 1.54.* As before, if we consider the Sobolev space $W_0^{1,p}(\Omega)$, then we can drop the Lipschitz requirement on $\partial\Omega$. For $\Omega \subset \mathbb{R}^N$ open and bounded with $\partial\Omega$ Lipschitz, some other norms equivalent to the usual Sobolev norm on $W^{1,p}(\Omega)$ ($1 \leq p < +\infty$) are given by the following expressions:

$$\left( \|\nabla u\|_p^p + |\int_\Omega u \, dx|^p \right)^{\frac{1}{p}},$$

$$\left( \|\nabla u\|_p^p + |\int_{\partial\Omega} u \, dH^{N-1}|^p \right)^{\frac{1}{p}},$$

$$\left( \|\nabla u\|_p^p + \int_{\partial\Omega} |u|^p \, dH^{N-1} \right)^{\frac{1}{p}}.$$

For $N = 1$, Theorem 1.49 takes the following particular form.

**Theorem 1.55.** *If $I \subset \mathbb{R}$ is an interval, then*

(a) $W^{1,p}(I) \hookrightarrow L^\infty(I)$ *for all* $1 \le p \le +\infty$;

(b) *For $I$ bounded, $W^{1,p}(I) \overset{c}{\hookrightarrow} C(\bar{I})$ for all* $1 < p \le +\infty$;

(c) *For $I$ bounded, $W^{1,1}(I) \overset{c}{\hookrightarrow} L^q(I)$ for all* $1 \le q < +\infty$.

*Remark 1.56.* $W^{1,1}(I) \hookrightarrow C(\bar{I})$, but never compactly. Also, for all $1 \le q \le +\infty$, $u \mapsto \|u\|_q + \|u'\|_p$ is a norm on $W^{1,p}(I)$ ($I$ bounded) equivalent to the usual Sobolev norm.

We conclude this section with some useful facts about the spaces $W^{1,p}(\Omega)$ and Nemytskii operators.

**Proposition 1.57.** *If $\Omega \subset \mathbb{R}^N$ is open, $f : \mathbb{R} \to \mathbb{R}$ is Lipschitz continuous with $f(0) = 0$ when $\Omega$ is unbounded, $1 \le p < +\infty$, and $N_f : W^{1,p}(\Omega) \to W^{1,p}(\Omega)$ is defined by $N_f(u) = f \circ u$ (Theorem 1.27), then $N_f$ is continuous.*

*Remark 1.58.* The result remains true if $W^{1,p}(\Omega)$ is replaced by $W_0^{1,p}(\Omega)$. The proposition implies that the maps $u \mapsto |u|$, $u \mapsto u^+$, $u \mapsto u^-$ are continuous from $W^{1,p}(\Omega)$ or $W_0^{1,p}(\Omega)$ into itself.

**Proposition 1.59.** *If $\Omega \subset \mathbb{R}^N$ is open and bounded with $\partial\Omega$ Lipschitz, $1 \le p < +\infty$, and $u \in W^{1,p}(\Omega) \cap C(\bar{\Omega})$, then $u \in W_0^{1,p}(\Omega)$ if and only if $u|_{\partial\Omega} = 0$.*

**Definition 1.60.** Let $\Omega \subset \mathbb{R}^N$ be open and bounded and $u \in C(\Omega)$. We set

$$Z(u) = \{x \in \Omega : u(x) = 0\} \quad \text{(the \textit{zero set} of } u\text{)}.$$

A connected component of $\Omega \setminus Z(u)$ is a *nodal domain* of $u$.

**Proposition 1.61.** *If $\Omega \subset \mathbb{R}^N$ is open and bounded, $1 \le p < +\infty$, $X = W^{1,p}(\Omega)$ or $W_0^{1,p}(\Omega)$, $u \in X \cap C(\Omega)$, and $\Omega_1 \subset \Omega$ is a nodal domain of $u$, then $u\chi_{\Omega_1} \in X$.*

Here, $\chi_{\Omega_1}$ denotes the characteristic function of the set $\Omega_1$, i.e.,

$$\chi_{\Omega_1}(x) = \begin{cases} 1 \text{ if } x \in \Omega_1 \\ 0 \text{ if } x \notin \Omega_1. \end{cases}$$

The last result is a Lusin-type property for Sobolev functions.

**Proposition 1.62.** *If $\Omega \subset \mathbb{R}^N$ is open, $1 \leq p < +\infty$, $u \in W^{1,p}(\Omega)$, and $\varepsilon > 0$, then we can find $v \in C^1(\overline{\Omega})$ and $U_\varepsilon \subset \Omega$ measurable such that $\lambda^N(\Omega \setminus U_\varepsilon) < \varepsilon$, $v|_{U_\varepsilon} = u|_{U_\varepsilon}$, and $\nabla v|_{U_\varepsilon} = \nabla u|_{U_\varepsilon}$.*

## 1.2 Remarks

**Section 1.1:** Sobolev spaces play a fundamental role in the theory of partial differential equations. They were introduced in the mid-1930s by Sobolev [363, 364]. Related spaces were also considered by Morrey [265, 266] and Deny–Lions [115]. The theory of Sobolev spaces was placed on a solid mathematical foundation with the advent of the theory of distributions due to Schwartz [357, 358], which generalized the theory of Radon measures. Sobolev spaces provided the tools to justify the Dirichlet principle originally formulated by Riemann (1851) and later criticized by Weierstrass (1870). Theorem 1.14 was proved by Meyers–Serrin [257] and ended much confusion about the relationship between the space $W^{m,p}(\Omega)$ and the space resulting from the completion of $C^\infty(\Omega)$ with respect to the Sobolev norm. The chain rule for Sobolev functions, presented in Theorem 1.27, is due to Marcus–Mizel [245]. Then, given a Lipschitz continuous function $f : \mathbb{R} \to \mathbb{R}$, we can define the Nemytskii (superposition) map $N_f : W^{1,p}(\Omega) \to W^{1,p}(\Omega)$ ($1 \leq p < +\infty$) by $N_f(u) = f \circ u$. Marcus–Mizel [247] proved that this map is continuous (Proposition 1.57). The nonlinear versions of Green's identity presented in Theorems 1.38 and 1.39 are due to Casas–Fernandez [73] and Kenmochi [193]. In Theorem 1.41 (Poincaré's inequality), the assumption that $\Omega \subset \mathbb{R}^N$ is bounded can be replaced by the weaker one that $\Omega \subset \mathbb{R}^N$ has finite width (that is, it lies between two parallel hyperplanes). Also, if the injection of a subspace $V$ of $W^{1,p}(\Omega)$ into $L^p(\Omega)$ is compact, then Poincaré's inequality (Theorem 1.41) holds on $V$ if and only if the constant function 1 does not belong to $V$. The embedding properties of the Sobolev spaces are an essential feature of these spaces. The first such results are due to Sobolev [363, 364], Rellich [339], and Kondrachov [201]. Refinements can be found in Morrey [265] and Gagliardo [145].

Comprehensive introductions to the theory of Sobolev spaces can be found in the books by Adams and Fournier [2], Brezis [52], Evans and Gariepy [131], Kufner [206], Kufner et al. [207], Maz'ja [255], Tartar [374], and Ziemer [396].

# Chapter 2
# Nonlinear Operators

**Abstract** This chapter focuses on important classes of nonlinear operators stating abstract results that offer powerful tools for establishing the existence of solutions to nonlinear equations. Specifically, they are useful in the study of nonlinear elliptic boundary value problems as demonstrated in the final three chapters of the present book. The first section of the chapter is devoted to compact operators and emphasizes the spectral properties, including the Fredholm alternative theorem. The second section treats nonlinear operators of monotone type, possibly set-valued, among which a prominent place is occupied by maximal monotone, pseudomonotone, generalized pseudomonotone, and $(S_+)$-operators. The cases of duality maps and $p$-Laplacian are of high interest in the sequel. The third section contains essential results on Nemytskii operators highlighting their main continuity and differentiability properties. Comments on the material of this chapter and related literature are given in a remarks section.

## 2.1 Compact Operators

In this section we present some results from nonlinear functional analysis that will be useful in the study of boundary value problems that follow. We start with compact maps, which historically are the first nonlinear operators studied in detail (primarily by Leray and Schauder in connection with their degree map). Compact operators, by definition, are close to operators in finite-dimensional spaces. In what follows, we will use the following basic definitions (slightly different versions of these notions exist in the literature).

**Definition 2.1.** Let $X$ and $Y$ be Banach spaces and $C$ a nonempty subset of $X$.

(a) We say that $f : C \to Y$ is *compact* if it is continuous and for every $B \subset C$ bounded, $\overline{f(B)}$ is compact in $Y$.

(b) We say that $f : C \to Y$ is *completely continuous* if for every sequence $\{x_n\}_{n\geq 1} \subset$
   $C$ such that $x_n \xrightarrow{w} x \in C$, we have $f(x_n) \to f(x)$ in $Y$.

In general, the two notions introduced above are not comparable. However, if we enrich the structure of $X$ or we restrict the map $f$, then we can compare them.

**Proposition 2.2.** *If $X$ is a reflexive Banach space, $Y$ is a Banach space, $C \subset X$ is nonempty, closed, and convex, and $f : C \to Y$ is completely continuous, then $f$ is compact.*

*Proof.* Evidently, $f$ is continuous. Let $B \subset C$ be a bounded set and let $\{y_n\}_{n\geq 1} \subset f(B)$. Then $y_n = f(x_n)$ with $\{x_n\}_{n\geq 1} \subset B$. The reflexivity of $X$ and the Eberlein–Šmulian theorem (e.g., Brezis [52, p. 70]) imply that, along a relabeled subsequence, $x_n \xrightarrow{w} x \in C$. Then $y_n = f(x_n) \to f(x)$ in $Y$, which proves the compactness of $f$.  □

For linear operators, compactness implies complete continuity.

**Proposition 2.3.** *If $X$ and $Y$ are Banach spaces and $L : X \to Y$ is a linear, compact operator, then $L$ is completely continuous.*

*Proof.* Let $x_n \xrightarrow{w} x$ in $X$. Then we can find $r > 0$ such that $\{x_n\}_{n\geq 1} \cup \{x\} \subset \overline{B_r(0)} := \{u \in X : \|u\| \leq r\}$. The linearity and continuity of $L$ imply that $L(x_n) \xrightarrow{w} L(x)$ in $Y$. But due to the compactness of $L$, $\overline{L(B_r(0))}$ is compact in the norm topology of $Y$, and so we conclude that $L(x_n) \to L(x)$ in $Y$.  □

**Corollary 2.4.** *If $X$ is a reflexive Banach space, $Y$ is a Banach space, and $L : X \to Y$ is linear, then $L$ is compact if and only if $L$ is completely continuous.*

**Proposition 2.5.** *If $X$ and $Y$ are Banach spaces, $C$ is a nonempty subset of $X$, $f_n : C \to Y$, $n \geq 1$, are compact maps, and $f : C \to Y$ is such that $f_n \to f$ uniformly on bounded subsets of $C$ as $n \to \infty$, then $f$ is compact.*

*Proof.* Let $\{x_n\}_{n\geq 1} \subset C$ be a bounded sequence. For every fixed $n \geq 1$, the sequence $\{f_n(x_k)\}_{k\geq 1} \subset Y$ has a strongly convergent subsequence $\{f_n(x_k^n)\}_{k\geq 1}$. By a standard diagonal process, we produce $\{u_k\}_{k\geq 1}$, the diagonal sequence of $\{x_k^n\}_{k,n\geq 1}$, and we have that $\{f_n(u_k)\}_{k\geq 1}$ strongly converges in $Y$ for every $n \geq 1$. By hypothesis, we have that $f_n(u_k) \to f(u_k)$ in $Y$ as $n \to \infty$ uniformly for $k \geq 1$. Then, for $k, m, n \geq 1$, we have

$$\|f(u_m) - f(u_k)\| \leq \|f(u_m) - f_n(u_m)\| + \|f_n(u_m) - f_n(u_k)\| + \|f_n(u_k) - f(u_k)\|,$$

which implies that $\{f(u_k)\}_{k\geq 1}$ is a Cauchy sequence. This proves the compactness of $f$.  □

**Definition 2.6.** Let $X$ and $Y$ be Banach spaces and $C$ a nonempty subset of $X$. A continuous, bounded (i.e., one that maps bounded sets to bounded sets) map $f : C \to Y$ is said to be of *finite rank* if the range of $f$ lies in a finite-dimensional subspace of $Y$.

The next theorem explains why many results for maps between finite-dimensional vector spaces can be extended to compact maps between infinite-dimensional Banach spaces. The result is the starting point of the Leray–Schauder degree theory (Sect. 4.2).

**Theorem 2.7.** *If $X$ and $Y$ are Banach spaces, $C \subset X$ is nonempty and bounded, and $f : C \to Y$ is continuous, then the following statements are equivalent:*

(a) *$f$ is a compact operator;*
(b) *For every $\varepsilon > 0$, there is a finite-rank map $f_\varepsilon : C \to Y$ such that*

$$\|f(x) - f_\varepsilon(x)\| < \varepsilon \text{ for all } x \in C.$$

*Proof.* (a) $\Rightarrow$ (b): Since $\overline{f(C)}$ is compact in $Y$, for every $\varepsilon > 0$ we can find a finite set $F_\varepsilon \subset f(C)$ such that $\min_{y \in F_\varepsilon} \|f(x) - y\| < \varepsilon$ for all $x \in C$. Then for every $y \in F_\varepsilon$ we set $\xi_y(x) = \max\{0, \varepsilon - \|f(x) - y\|\}$. Evidently, $\xi_y$ takes values in $\mathbb{R}_+ = [0, +\infty)$ and is continuous on $C$ with $\sum_{y \in F_\varepsilon} \xi_y(x) > 0$ for all $x \in C$. We introduce $f_\varepsilon : C \to Y$ defined by

$$f_\varepsilon(x) = \frac{\sum\limits_{y \in F_\varepsilon} \xi_y(x) y}{\sum\limits_{y \in F_\varepsilon} \xi_y(x)} \text{ for all } x \in C.$$

Clearly, $f_\varepsilon$ is a finite-rank operator and

$$\|f(x) - f_\varepsilon(x)\| = \frac{\left\|\sum\limits_{y \in F_\varepsilon} \xi_y(x)(f(x) - y)\right\|}{\sum\limits_{y \in F_\varepsilon} \xi_y(x)} < \frac{\sum\limits_{y \in F_\varepsilon} \xi_y(x) \varepsilon}{\sum\limits_{y \in F_\varepsilon} \xi_y(x)} = \varepsilon \text{ for all } x \in C.$$

(b) $\Rightarrow$ (a): Let $\varepsilon = \frac{1}{n}$, $n \geq 1$. Then by hypothesis we can find a finite-rank operator $f_n : C \to Y$ such that $\|f(x) - f_n(x)\| < \frac{1}{n}$ for all $n \geq 1$ and all $x \in C$. Proposition 2.5 implies that $f$ is a compact operator. $\square$

*Remark 2.8.* A careful reading of the first part of the foregoing proof reveals that the range of $f_\varepsilon$ is contained in the convex hull $\operatorname{conv} f(C)$ of $f(C)$.

It is useful to know whether a compact map can be extended in a compact fashion to the whole space. This can be done with the help of the following version of the Tietze extension theorem due to Dugundji [122].

**Theorem 2.9.** *If $(X, d)$ is a metric space, $Y$ is a normed space, $C$ is a nonempty, closed subset of $X$, and $f : C \to Y$ is a continuous map, then there exists a continuous map $\hat{f} : X \to Y$ of $f$ such that $\hat{f}|_C = f$ and $\hat{f}(X) \subset \operatorname{conv} f(C)$.*

Using this theorem and the fact that the convex hull of a relatively compact set in a Banach space is relatively compact, we have the following theorem.

**Theorem 2.10.** *If X and Y are Banach spaces, $C \subset X$ is nonempty, closed, and bounded, and $f : C \rightarrow Y$ is a compact operator, then there exists a compact operator $\hat{f} : X \rightarrow Y$ such that $\hat{f}|_C = f$ and $\hat{f}(X) \subset \operatorname{conv} f(C)$.*

In the sequel, we focus on compact linear operators that exhibit remarkable spectral properties (in particular in the context of Hilbert spaces). In what follows, for $X$ and $Y$ Banach spaces, by $\mathscr{L}(X,Y)$ we denote the Banach space of all bounded linear operators from $X$ into $Y$. By $\mathscr{L}_c(X,Y)$ we denote the closed subspace of $\mathscr{L}(X,Y)$ consisting of compact linear operators. We abbreviate $\mathscr{L}(X) = \mathscr{L}(X,X)$ and $\mathscr{L}_c(X) = \mathscr{L}_c(X,X)$. We start with a result due to Schauder, which says that compactness is preserved by taking the adjoint of an element of $\mathscr{L}_c(X,Y)$.

**Theorem 2.11.** *If X and Y are Banach spaces and $L \in \mathscr{L}(X,Y)$, then $L \in \mathscr{L}_c(X,Y)$ if and only if $L^* \in \mathscr{L}_c(Y^*,X^*)$.*

*Proof.* $\Rightarrow$: Let $\{y_n^*\}_{n \geq 1} \subset \overline{B}_{Y^*} := \{y^* \in Y^* : \|y^*\| \leq 1\}$. If $\overline{B}_X := \{x \in X : \|x\| \leq 1\}$, then, by hypothesis, $L(\overline{B}_X)$ is compact in $Y$. Consider the restrictions $y_n^*|_{\overline{L(\overline{B}_X)}}$, $n \geq 1$. Then this sequence is uniformly bounded and equicontinuous, and so, by the Arzelà–Ascoli theorem (e.g., Brezis [52, p. 111]), we can find a subsequence $\{y_{n_k}^*\}_{k \geq 1}$ of $\{y_n^*\}_{n \geq 1}$ such that

$$\sup\{|\langle y_{n_k}^*, L(x)\rangle - \langle y_{n_m}^*, L(x)\rangle| : x \in \overline{B}_X\} \rightarrow 0 \ \text{ as } \ k, m \rightarrow \infty,$$

and thus $\|L^*(y_{n_k}^*) - L^*(y_{n_m}^*)\| \rightarrow 0$ as $k, m \rightarrow \infty$. This proves that $\{L^*(y_{n_k}^*)\} \subset X^*$ is a Cauchy sequence in $X^*$, and so we conclude that $\overline{L^*(\overline{B}_{Y^*})}$ is compact in $X^*$. $\Leftarrow$: We note that $L^{**}|_X = L$. From the first part of the proof we have that $L^{**} \in \mathscr{L}_c(X^{**}, Y^{**})$, and so if $\overline{B}_{X^{**}} := \{x^{**} \in X^{**} : \|x^{**}\| \leq 1\}$, then $\overline{L^{**}(\overline{B}_{X^{**}})}$ is compact in $Y^{**}$. But $\overline{L^{**}(\overline{B}_X)}$ is a closed subset of $\overline{L^{**}(\overline{B}_{X^{**}})}$. Therefore, $\overline{L^{**}(\overline{B}_X)} = \overline{L(\overline{B}_X)}$ is compact in $Y$, hence $L \in \mathscr{L}_c(X,Y)$.                                                                    $\square$

We turn our attention to the rich spectral properties of compact operators. For this purpose, we introduce the following notions.

**Definition 2.12.** Let $X$ be a Banach space and $L \in \mathscr{L}(X)$.

(a) The *resolvent set* $\rho(L)$ of $L$ is defined by

$$\rho(L) = \{\lambda \in \mathbb{R} : \lambda \operatorname{id} - L \text{ is invertible}\}.$$

If $\lambda \in \rho(L)$, then $R_\lambda(L) := (\lambda \operatorname{id} - L)^{-1}$ is the *resolvent* of $L$ (corresponding to $\lambda$).
(b) The *spectrum* $\sigma(L)$ of $L$ is defined by $\sigma(L) = \mathbb{R} \setminus \rho(L)$.
(c) An element $\lambda \in \sigma(L)$ is an *eigenvalue* of $L$ if $\ker(\lambda \operatorname{id} - L) \neq \{0\}$. A vector $x \in \ker(\lambda \operatorname{id} - L) \setminus \{0\}$ is an *eigenvector* of $L$ corresponding to the eigenvalue $\lambda$, and $\ker(\lambda \operatorname{id} - L)$ is the *eigenspace* corresponding to the eigenvalue $\lambda$. We denote the set of eigenvalues of $L$ by $\sigma_p(L)$.

**Proposition 2.13.** *If $X$ is a Banach space, $L \in \mathcal{L}_c(X)$, and $\lambda \in \sigma_p(L) \setminus \{0\}$, then* $\dim \ker(\lambda \operatorname{id} - L) < +\infty$.

*Proof.* Let $\overline{B}$ be the closed unit ball of $\ker(\lambda \operatorname{id} - L)$. Then $L(\overline{B}) = \lambda \overline{B}$, and $L(\overline{B})$ is compact in $X$. Therefore, $\overline{B}$ is compact in $\ker(\lambda \operatorname{id} - L)$, and this implies that $\dim \ker(\lambda \operatorname{id} - L) < +\infty$. $\qquad\square$

**Proposition 2.14.** *If $X$ is a Banach space, $L \in \mathcal{L}_c(X)$, and $\lambda \neq 0$, then* $\operatorname{im}(\lambda \operatorname{id} - L)$ *is closed.*

*Proof.* Without loss of generality, we may assume that $\lambda = 1$. Let $K = \operatorname{id} - L$. Then, by virtue of Proposition 2.13, we have that $\dim \ker K < +\infty$. Hence $X = \ker K \oplus V$, with $V$ a closed subspace of $X$. Let $K_0 = K|_V$. Then $K(X) = K(V) = K_0(V)$ and $\ker K_0 = \ker K \cap V = \{0\}$. Hence $K_0$ is injective.

We claim that $\inf\{\|K_0(v)\| : v \in V, \|v\| = 1\} > 0$. Proceeding by contradiction, suppose that we can find $\{v_n\}_{n \geq 1} \subset V$, $\|v_n\| = 1$, for all $n \geq 1$, such that $\|K_0(v_n)\| \to 0$ as $n \to \infty$. Since $L \in \mathcal{L}_c(X)$, by passing to a subsequence if necessary, we may assume that $L(v_n) \to u$. Then $v_n = (K_0 + L)(v_n) \to u$, hence $\|u\| = 1$. Also, $K_0(v_n) \to K_0(u)$, and so $K_0(u) = 0$, which contradicts the injectivity of $K_0$.

Using the claim, we see that we can find $c > 0$ such that $c\|v\| \leq \|K_0(v)\|$ for all $v \in V$. This implies that $\operatorname{im} K_0 = \operatorname{im}(\operatorname{id} - L)$ is closed. To show this, let $\{x_n\}_{n \geq 1} \subset \operatorname{im} K_0$, and assume that $x_n \to x$. Then $x_n = K_0(u_n)$ with $u_n \in V$ for all $n \geq 1$. We have

$$c\|u_n - u_m\| \leq \|K_0(u_n) - K_0(u_m)\| \to 0 \quad \text{as } n, m \to \infty,$$

which yields that $\{u_n\}_{n \geq 1}$ is convergent to some $u \in V$, and so $K_0(u_n) \to K_0(u) = x$. Hence $x \in \operatorname{im} K_0$. $\qquad\square$

**Proposition 2.15.** *If $X$ is a Banach space, $L \in \mathcal{L}_c(X)$, and $\varepsilon > 0$, then $L$ has finitely many linearly independent eigenvectors corresponding to eigenvalues $\lambda$ such that* $|\lambda| > \varepsilon$.

*Proof.* Arguing by contradiction, suppose that $\{x_n\}_{n \geq 1}$ is a sequence of linearly independent eigenvectors corresponding to eigenvalues $\lambda$ such that $|\lambda| > \varepsilon$. For every $n \geq 1$, let $X_n = \operatorname{span}\{x_m\}_{m=1}^n$. Then $L(X_n) = X_n$. By virtue of Riesz's lemma (e.g., Brezis [52, p. 160]), we can find $u_n \in X_n$, $\|u_n\| = 1$ and $d(u_n, X_{n-1}) \geq \frac{1}{2}$ for all $n \geq 2$. Let $v_n = \frac{1}{\lambda_n} u_n$, with $\lambda_n$ being the eigenvalue corresponding to $x_n$. Then $\|v_n\| < \frac{1}{\varepsilon}$ and $L(v_n) \in X_n$. Moreover, if $u_n = \sum\limits_{m=1}^n \mu_m x_m$, then

$$L(v_n) - u_n = \sum_{m=1}^{n-1} \left( \frac{\lambda_m}{\lambda_n} - 1 \right) \mu_m x_m \in X_{n-1}.$$

If $m < n$, then $L(v_m) \in X_m \subset X_{n-1}$, and so we have

$$\|L(v_n) - L(v_m)\| = \|u_n - (L(v_m) + u_n - L(v_n))\| \geq d(u_n, X_{n-1}) \geq \frac{1}{2}.$$

Since $\{v_n\}_{n \geq 1}$ is bounded, we have a contradiction of the fact that $L \in \mathscr{L}_c(X)$.  □

**Lemma 2.16.** *If $X$ is a Banach space, $L \in \mathscr{L}_c(X)$, $K = \mathrm{id} - L$, and $V = \mathrm{im}\, K$ is a proper subspace of $X$, then for every $\varepsilon > 0$ we can find $x_\varepsilon \in X$ with $\|x_\varepsilon\| = 1$ such that $d(L(x_\varepsilon), L(V)) \geq 1 - \varepsilon$.*

*Proof.* By Riesz's lemma, we can find $x_\varepsilon \in X$, $\|x_\varepsilon\| = 1$ such that $d(x_\varepsilon, V) \geq 1 - \varepsilon$. We have $K(x_\varepsilon) \in V$ and $L(V) = (\mathrm{id} - K)(V) \subset V$. Therefore, $d(L(x_\varepsilon), L(V)) \geq d(x_\varepsilon - K(x_\varepsilon), V) = d(x_\varepsilon, V) \geq 1 - \varepsilon$.  □

Using this lemma we can show that every nonzero spectral element of an operator $L \in \mathscr{L}_c(X)$ is an eigenvalue.

**Proposition 2.17.** *If $X$ is a Banach space, $L \in \mathscr{L}_c(X)$, $\lambda \in \sigma(L)$, and $\lambda \neq 0$, then $\lambda \in \sigma_p(L)$.*

*Proof.* We will show that if $\lambda \notin \sigma_p(L)$ and $\lambda \neq 0$, then $\lambda \notin \sigma(L)$. Without any loss of generality, we may assume that $\lambda = 1$. We set $K = \mathrm{id} - L$. Since, by hypothesis, $1 \notin \sigma_p(L)$, then $\ker K = \{0\}$. To show that $1 \notin \sigma(L)$, we need to show that $K$ is invertible. By Banach's theorem (e.g., Brezis [52, Corollary 2.7]), it suffices to show that $K$ is surjective. Let $V_n = K^n(X)$, $n \geq 0$. From Proposition 2.14 we know that $V_n$ is a closed subspace of $X$ and

$$V_n = K^{n-1}(K(X)) \subset K^{n-1}(X) = V_{n-1} \quad \text{for all } n \geq 1.$$

We claim that for some $n \geq 0$, we have $V_n = V_{n+1}$. Arguing by contradiction, suppose that the claim is not true. Then according to Lemma 2.16, we can find $x_n \in V_n$ with $\|x_n\| = 1$ such that $d(L(x_n), L(V_{n+1})) \geq \frac{1}{2}$ for all $n \geq 0$. Then we have

$$\|L(x_n) - L(x_m)\| \geq \frac{1}{2} \quad \text{for all } n \neq m,$$

which contradicts the fact that $L \in \mathscr{L}_c(X)$. Hence the claim is true.

Next we show that $V_0 = V_1$ (note that $V_0 = X$). Again we proceed by contradiction. Suppose that $V_0 \neq V_1$. Let $m \geq 1$ be the smallest integer such that $V_{m-1} \neq V_m = V_{m+1}$. Choose $y \in V_{m-1} \setminus V_m$. We have $K(y) \in V_m = V_{m+1}$. Hence we can find $z \in V_m$ such that $K(y) = K(z)$ and $y \neq z$. Then $K(y - z) = 0$ with $y - z \neq 0$, which contradicts the fact that $\ker K = \{0\}$. Therefore, $K(X) = V_1 = X$, and so we conclude that $1 \notin \sigma(L)$.  □

We recall the following easy fact from elementary linear algebra.

**Lemma 2.18.** *If $X$ is a Banach space, $L \in \mathscr{L}(X)$, $\{\lambda_k\}_{k=1}^n$ are distinct eigenvalues of $L$, and, for every $k = 1, 2, \ldots, n$, $e_k$ is an eigenvector corresponding to $\lambda_k$, then $\{e_k\}_{k=1}^n$ are linearly independent.*

Combining Propositions 2.15 and 2.17 and Lemma 2.18, we obtain the following fundamental result concerning the spectrum of compact linear operators.

**Theorem 2.19.** *If $X$ is an infinite-dimensional Banach space and $L \in \mathscr{L}_c(X)$, then $\sigma(L) = \{0, \{\lambda_k\}_k\}$, with $\{\lambda_k\}_k$ being either finite (possibly empty) or converging to zero, and each $\lambda_k$ is an eigenvalue of $L$ and has a finite-dimensional eigenspace.*

The next result is known as the *Fredholm alternative theorem* and has applications to boundary value problems.

**Theorem 2.20.** *If $X$ is a Banach space, $L \in \mathscr{L}_c(X)$, and $\lambda \neq 0$, then one and only one of the following statements is true:*

(a) *For every $u \in X$, the equation $(\lambda \operatorname{id} - L)(x) = u$ has a unique solution $x$.*
(b) *The equation $L(x) = \lambda x$ has a nontrivial solution.*

*Proof.* As before, without any loss of generality, we assume that $\lambda = 1$. Suppose that (b) is not true. Hence $1 \notin \sigma_p(L)$, and so, according to Proposition 2.17, $1 \in \rho(L)$. Therefore, $(\operatorname{id} - L)^{-1} \in \mathscr{L}(X)$, which means that (a) is true.

Next, suppose that $\operatorname{im}(\operatorname{id} - L) = X$. We will show that $1 \notin \sigma_p(L)$. We proceed by contradiction and assume that $1 \in \sigma_p(L)$. Let $V_k = \ker((\operatorname{id} - L)^k)$ for all integers $k \geq 1$. Clearly, $V_k \subset V_{k+1}$ for all $k \geq 1$. Since $1 \in \sigma_p(L)$ and $\operatorname{im}(\operatorname{id} - L) = X$, we can construct a sequence $\{x_k\}_{k \geq 1} \subset X$ such that $x_1 \neq 0$, $(\operatorname{id} - L)(x_1) = 0$, and $(\operatorname{id} - L)(x_{k+1}) = x_k$ for $k \geq 1$. Then

$$(\operatorname{id} - L)^k (x_{k+1}) = (\operatorname{id} - L)^{k-1}(x_k) = \cdots = x_1 \neq 0$$

and $(\operatorname{id} - L)^{k+1}(x_{k+1}) = (\operatorname{id} - L)(x_1) = 0$. It follows that $x_k \in V_k \setminus V_{k-1}$ for all $k \geq 1$. Hence $V_k \neq V_{k-1}$ for all $k \geq 1$.

So, by Riesz's lemma (e.g., Brezis [52, p. 160]), we can find $u_k \in V_k$ with $\|u_k\| = 1$ such that $d(u_k, V_{k-1}) \geq \frac{1}{2}$. Then for $n > m$ we have

$$(\operatorname{id} - L)^{n-1}(u_m + (\operatorname{id} - L)(u_n) - (\operatorname{id} - L)(u_m))$$
$$= (\operatorname{id} - L)^{n-1}(u_m) + (\operatorname{id} - L)^n(u_n) - (\operatorname{id} - L)^n(u_m) = 0,$$

which implies that $u_m + (\operatorname{id} - L)(u_n) - (\operatorname{id} - L)(u_m) \in V_{n-1}$. Therefore,

$$\|L(u_n) - L(u_m)\| = \|u_n - (u_m + (\operatorname{id} - L)(u_n) - (\operatorname{id} - L)(u_m))\| \geq d(u_n, V_{n-1}) \geq \frac{1}{2},$$

which contradicts the fact that $L \in \mathscr{L}_c(X)$. $\qquad\square$

Next we focus on self-adjoint compact linear operators on a Hilbert space. Let $H$ be a Hilbert space, and assume that $L \in \mathscr{L}(H)$ is self-adjoint. Then

$$\lambda \in \sigma(L) \text{ if and only if } \inf\{\|(\lambda \operatorname{id} - L)(x)\| : \|x\| = 1\} = 0, \qquad (2.1)$$

and the eigenvectors corresponding to different eigenvalues are orthogonal.

**Proposition 2.21.** *If $H$ is a Hilbert space with inner product $(\cdot,\cdot)_H$, $L \in \mathscr{L}(H)$ is self-adjoint, and*

$$m = \inf\{(L(x),x)_H : \|x\| = 1\}, \quad M = \sup\{(L(x),x)_H : \|x\| = 1\},$$

*then $\sigma(L) \subset [m,M]$ and $m,M \in \sigma(L)$.*

*Proof.* Let $r > 0$ and set $\lambda = M + r$. We will show that $\lambda \notin \sigma(L)$. According to (2.1), it suffices to show that

$$\inf\{\|(\lambda\,\mathrm{id} - L)(x)\| : \|x\| = 1\} > 0. \tag{2.2}$$

For $x \in H$ with $\|x\| = 1$ we have

$$((\lambda\,\mathrm{id} - L)(x),x)_H = \lambda\|x\|^2 - (L(x),x)_H \geq (\lambda - M)\|x\|^2 = r > 0,$$

which implies that $\|(\lambda\,\mathrm{id} - L)(x)\| \geq r$. Thus, (2.2) holds, and this ensures that $\lambda \notin \sigma(L)$. Similarly, if $\lambda = m - r$, then $\lambda \notin \sigma(L)$. We infer that $\sigma(L) \subset [m,M]$.

Next, we show that $M \in \sigma(L)$. Note that for every $\theta \in \mathbb{R}$, $\sigma(L+\theta\,\mathrm{id}) = \sigma(L)+\theta$. Thus, without any loss of generality, up to translation, we may assume that $0 \leq m \leq M$. Then $M = \|L\|_{\mathscr{L}(H)}$. Let $\{x_n\}_{n\geq 1} \subset H$ be such that $\|x_n\| = 1$ for all $n \geq 1$ and $(L(x_n),x_n)_H \to M$. We have

$$\|Mx_n - L(x_n)\|^2 = M^2\|x_n\|^2 + \|L(x_n)\|^2 - 2M(L(x_n),x_n)_H$$
$$\leq 2M^2 - 2M(L(x_n),x_n)_H \to 0 \text{ as } n \to \infty.$$

Hence we have $\inf\{\|Mx - L(x)\| : \|x\| = 1\} = 0$, and so, by (2.1), we have $M \in \sigma(L)$. Similarly, we show that $m \in \sigma(L)$. $\qquad\square$

**Corollary 2.22.** *If $H$ is a Hilbert space, $L \in \mathscr{L}(H)$ is self-adjoint, and $\sigma(L) = \{0\}$, then $L \equiv 0$.*

From linear algebra we know that a real symmetric $N \times N$-matrix is diagonalizable. The next theorem is a generalization of this result to compact self-adjoint operators on a separable Hilbert space.

**Theorem 2.23.** *If $H$ is a separable Hilbert space and $L \in \mathscr{L}_c(H)$ is self-adjoint, then there exists an orthonormal basis $\{e_n\}_{n\geq 1}$ consisting of eigenvectors of $L$ such that for every $x \in H$ we have $L(x) = \sum_{n\geq 1} \lambda_n(x,e_n)_H e_n$, with $\{\lambda_n\}_{n\geq 1}$ being the nonzero eigenvalues.*

*Proof.* Let $\{\lambda_n\}_{n\geq 1}$ be the distinct eigenvalues of $L$, except 0, and set $\lambda_0 = 0$.

We define $V_0 = \ker L$ and $V_n = \ker(\lambda_n\,\mathrm{id} - L)$ for all $n \geq 1$. We know that $0 \leq \dim V_0 \leq +\infty$ and $0 < \dim V_n < +\infty$ for all $n \geq 1$ (Proposition 2.13). Moreover, the subspaces $\{V_n\}_{n\geq 1}$ are mutually orthogonal. We set $Y = \mathrm{span}\left\{\bigcup_{n\geq 0} V_n\right\}$, and we claim that $Y$ is dense in $H$. To this end, note that $L(Y) \subset Y$ and $L(Y^\perp) \subset Y^\perp$. Moreover, $L_0 := L|_{Y^\perp}$ is compact and self-adjoint. If $\lambda \in \sigma(L_0) \setminus \{0\}$, then $\lambda \in \sigma_p(L_0)$ (Proposition 2.17), and so there exists $u \in Y^\perp$, $u \neq 0$, such that $L_0(u) = \lambda u$. Hence $\lambda = \lambda_n$ for some $n \geq 1$ and $u \in Y^\perp \cap V_n = \{0\}$, a contradiction. This proves that $\sigma(L_0) = \{0\}$, and so, by virtue of Corollary 2.22, we have $L_0 = 0$. Therefore, $Y^\perp \subset \ker L \subset Y$, thus $Y^\perp = \{0\}$. It follows that $Y$ is dense in $H$.

Finally, due to the separability of $H$, we can choose an orthonormal basis in each $V_n$, $n \geq 0$. Considering their union denoted by $\{e_m\}_{m \geq 1}$ we get an orthonormal basis of $H$, and we see that the linear operator $\hat{L} : H \to H$ defined by

$$\hat{L}(x) = \sum_{m \geq 1} \mu_m (x, e_m)_H e_m,$$

where $\mu_m$ is the eigenvalue corresponding to $e_m$, is continuous and $\hat{L}(e_m) = L(e_m)$ for all $m \geq 1$. Therefore, we infer that $\hat{L} = L$.                    $\square$

*Remark 2.24.* If $H$ is a Hilbert space and $L \in \mathscr{L}_c(H)$ is self-adjoint, then the spectrum $\sigma(L)$ of $L$ consists of at most countably many eigenvalues $\{\lambda_n\}_{n \geq 1}$ with the only possible limit point the zero. These eigenvalues admit minimax (variational) characterizations. More precisely, let $\{\lambda_n^+\}_{n \geq 1}$ be the distinct positive eigenvalues ordered in decreasing order. Then we have

$$\lambda_n^+ = \inf_{V \in \mathscr{S}_{n-1}} \sup_{x \in V} \frac{(L(x), x)_H}{\|x\|^2} = \sup_{Z \in \mathscr{M}_n} \min_{x \in Z} \frac{(L(x), x)_H}{\|x\|^2},$$

with $\mathscr{S}_{n-1} = \{V \subset H : V \text{ is a vector subspace of } H \text{ with codim} V = n - 1\}$ and $\mathscr{M}_n = \{V \subset H : V \text{ is a vector subspace of } H \text{ of dimension } n\}$. Similar formulas hold for the distinct negative eigenvalues ordered in increasing order. That is,

$$\lambda_n^- = \sup_{V \in \mathscr{S}_{n-1}} \inf_{x \in V} \frac{(L(x), x)_H}{\|x\|^2} = \inf_{Z \in \mathscr{M}_n} \max_{x \in Z} \frac{(L(x), x)_H}{\|x\|^2}.$$

We emphasize that these minimax expressions do not depend on knowledge of the other eigenvalues.

## 2.2 Operators of Monotone Type

We pass to operators of monotone type, which were introduced in the early 1960s in an attempt to expand the class of nonlinear equations that could be treated using the theory of compact operators. The theory of maximal monotone operators, through the subdifferential of convex analysis, leads naturally to nonsmooth analysis, which provides the notions and analytical tools to handle problems with a nonsmooth potential. Recall that a function $f : \mathbb{R} \to \mathbb{R}$ is monotone increasing if and only if

$x \le y$ implies $f(x) \le f(y)$. The drawback of this definition is that it depends on the order structure of $\mathbb{R}$, and so it cannot be extended to maps between general Banach spaces. However, we can describe the monotonicity property in an order-free fashion, namely, that $f$ is monotone increasing if and only if $(f(x) - f(y))(x - y) \ge 0$ for all $x, y \in \mathbb{R}$. Thus, if in this definition we replace the product in $\mathbb{R}$ by the duality brackets of a pair $(X^*, X)$ (with $X$ a Banach space, $X^*$ its topological dual), then we have a notion of monotonicity for maps $f : D \subset X \to X^*$. To accommodate the fundamental notion of maximal monotone map, it is necessary to develop the theory in the framework of set-valued maps.

If $X$ and $Y$ are sets and $A : X \to 2^Y$ is a set-valued map (or multimap, or multifunction), then we denote $D(A) = \{x \in X : A(x) \ne \emptyset\}$ (the domain of $A$), $R(A) = \bigcup_{x \in X} A(x)$ (the range of $A$), $\mathrm{Gr}\,A = \{(x,y) \in X \times Y : y \in A(x)\}$ (the graph of $A$), and $A^{-1}(y) = \{x \in X : y \in A(x)\}$ for all $y \in Y$ (the inverse of $A$). Also, we say that $\hat{A} : X \to 2^Y$ is an *extension* of $A$ if $\mathrm{Gr}\,A \subset \mathrm{Gr}\,\hat{A}$.

In what follows, $X$ is a Banach space and $X^*$ its topological dual. By $\langle \cdot, \cdot \rangle$ we denote the duality brackets for the pair $(X^*, X)$.

**Definition 2.25.** Let $A : X \to 2^{X^*}$ with $D(A) \ne \emptyset$.

(a) We say that $A$ is *monotone* if

$$\langle x^* - y^*, x - y \rangle \ge 0 \text{ for all } (x, x^*), (y, y^*) \in \mathrm{Gr}\,A.$$

(b) We say that $A$ is *strictly monotone* if the inequality in (a) is strict when $x \ne y$.
(c) We say that $A$ is *strongly monotone* if there exists a constant $c > 0$ such that

$$\langle x^* - y^*, x - y \rangle \ge c\|x - y\|^2 \text{ for all } (x, x^*), (y, y^*) \in \mathrm{Gr}\,A.$$

(d) We say that $A$ is *uniformly monotone* if there exists an increasing, continuous function $\xi : \mathbb{R}_+ \to \mathbb{R}_+$ such that $\xi(0) = 0$, $\xi(r) \to +\infty$ as $r \to +\infty$, and

$$\langle x^* - y^*, x - y \rangle \ge \xi(\|x - y\|)\|x - y\| \text{ for all } (x, x^*), (y, y^*) \in \mathrm{Gr}\,A.$$

*Remark 2.26.* It is clear that we always have the following implications:

strongly monotone $\Rightarrow$ uniformly monotone $\Rightarrow$ strictly monotone $\Rightarrow$ monotone.

*Example 2.27.*

(a) If $H$ is a Hilbert space and $f : H \to H$ is nonexpansive [i.e., $\|f(x) - f(y)\| \le \|x - y\|$ for all $x, y \in H$], then $A = \mathrm{id} - f : H \to H$ is monotone.
(b) If $H$ is a Hilbert space and $C \subset H$ a nonempty, closed, convex set, we can define the metric projection map $p_C : H \to C$ by $p_C(x) = u$, with $u$ being the unique element of $C$ such that $\|x - u\| = d(x, C)$. Then $p_C$ is monotone and nonexpansive (Theorem 5.1).

(c) Let $\Omega \subset \mathbb{R}^N$, $N \geq 1$, be a bounded domain, $1 < p < +\infty$, and $A : W_0^{1,p}(\Omega) \to W^{-1,p'}(\Omega) = W_0^{1,p}(\Omega)^*$ $(\frac{1}{p} + \frac{1}{p'} = 1)$ be defined by

$$\langle A(u), v \rangle = \int_\Omega |\nabla u|^{p-2} (\nabla u, \nabla v)_{\mathbb{R}^N} \, dx \text{ for all } u, v \in W_0^{1,p}(\Omega)$$

[i.e., $A$ is the operator corresponding to the negative $p$-Laplacian

$$-\Delta_p u = -\operatorname{div}(|\nabla u|^{p-2} \nabla u)$$

for all $u \in W_0^{1,p}(\Omega)$]. For $1 < p < 2$, $A$ is strictly monotone; for $p = 2$, $A$ (which is linear) is strongly monotone; and for $2 < p < +\infty$, $A$ is uniformly monotone. These facts follow from the following elementary inequalities:

$$(|y|^{p-2} y - |h|^{p-2} h, y - h)_{\mathbb{R}^N} \geq \begin{cases} c_1(p)(|y| + |h|)^{p-2} |y - h|^2 & \text{if } 1 < p < 2, \\ c_2(p)|y - h|^p & \text{if } p \geq 2, \end{cases}$$

for all $y, h \in \mathbb{R}^N$, where $c_1(p), c_2(p) > 0$ are constants.

(d) Let $X$ be a Banach space and $\varphi : X \to \mathbb{R} \cup \{+\infty\}$ a convex function not identically $+\infty$. The *subdifferential* of $\varphi$ is the set-valued map $\partial \varphi : X \to 2^{X^*}$ defined by

$$\partial \varphi(x) = \{x^* \in X^* : \varphi(y) \geq \varphi(x) + \langle x^*, y - x \rangle \text{ for all } y \in X\}.$$

This is an extension of the notion of derivative, and we will return to it in Sect. 3.1. The multimap $\partial \varphi$ is monotone. Note that if we choose $X - W_0^{1,p}(\Omega)$ and $\varphi(u) = \frac{1}{p} \|\nabla u\|_p^p$ (which is of class $C^1$), then $\partial \varphi$ coincides with the operator $A$ of example (c).

The next proposition gives a useful property of monotone maps. For the proof of a more general result we refer to Gasiński–Papageorgiou [151, p. 306]. Recall that a map $A : X \to 2^{X^*}$ is *locally bounded* at $x \in D(A)$ if there exists a neighborhood $U$ of $x$ such that $A(U)$ is bounded in $X^*$.

**Proposition 2.28.** *If $A : X \to 2^{X^*}$ is a monotone map, then $A$ is locally bounded at each $x \in \operatorname{int} D(A)$.*

**Definition 2.29.** Let $C \subset X$ be a nonempty set and $A : C \to X^*$ a single-valued map.

(a) We say that $A$ is *hemicontinuous* at $x \in C$ if for every $h \in X$ and every sequence $t_n \downarrow 0$ with $x + t_n h \in C$ for all $n \geq 1$ we have $A(x + t_n h) \overset{w^*}{\to} A(x)$ in $X^*$. If this is true for every $x \in C$, then we say that $A$ is *hemicontinuous*.

(b) We say that $A$ is *demicontinuous at* $x \in C$ if for every sequence $\{x_n\}_{n \geq 1} \subset C$ such that $x_n \to x$ in $X$ we have $A(x_n) \overset{w^*}{\to} A(x)$ in $X^*$. If this is true for every $x \in C$, then we say that $A$ is *demicontinuous*.

*Remark 2.30.* It is clear from the preceding definitions that we always have

$$\text{continuity} \Rightarrow \text{demicontinuity} \Rightarrow \text{hemicontinuity}.$$

Moreover, a monotone map $A : C \to X^*$, with $C \subset X$ open, is hemicontinuous if and only if it is demicontinuous.

Among monotone operators, the following subclass is of special interest since it exhibits remarkable properties useful in applications.

**Definition 2.31.** A map $A : X \to 2^{X^*}$ is said to be *maximal monotone* if it is monotone and for $(x, x^*) \in X \times X^*$ the inequalities

$$\langle x^* - u^*, x - u \rangle \geq 0 \text{ for all } (u, u^*) \in \operatorname{Gr} A$$

imply that $(x, x^*) \in \operatorname{Gr} A$.

*Remark 2.32.* Definition 2.31 expresses that $A : X \to 2^{X^*}$ is maximal monotone if and only if $\operatorname{Gr} A$ is maximal with respect to inclusion among the graphs of all monotone maps from $X$ to $2^{X^*}$. Zorn's lemma (e.g., Brezis [52, p. 2]) implies that every monotone map admits a maximal monotone extension.

*Example 2.33.* An increasing function $f : \mathbb{R} \to \mathbb{R}$ is monotone but need not be maximal monotone, unless it is continuous. To produce a maximal monotone map out of $f$, we need to fill in the jump discontinuities. This example illustrates the need to consider set-valued maps.

From Remark 2.32 we see at once that the following proposition is true.

**Proposition 2.34.** *If $X$ is reflexive, then $A : X \to 2^{X^*}$ is maximal monotone if and only if $A^{-1} : X^* \to 2^X$ is.*

**Proposition 2.35.** *If $A : X \to 2^{X^*}$ is maximal monotone, then for every $x \in \mathrm{D}(A)$, $A(x)$ is convex and $w^*$-closed.*

*Proof.* Let $x^*, u^* \in A(x)$ and set $y_t^* = tx^* + (1-t)u^*, t \in [0,1]$. For all $(z, z^*) \in \operatorname{Gr} A$ we have

$$\langle y_t^* - z^*, x - z \rangle = t \langle x^* - z^*, x - z \rangle + (1-t) \langle u^* - z^*, x - z \rangle \geq 0.$$

The maximality of $A$ implies $y_t^* \in A(x)$, and so $A$ is convex-valued. In addition, let $\{x_\alpha^*\}_{\alpha \in J} \subset A(x)$ be a net such that $x_\alpha^* \overset{w^*}{\to} x^*$. Then

$$\langle x_\alpha^* - u^*, x - u \rangle \geq 0 \text{ for all } (u, u^*) \in \operatorname{Gr} A,$$

which yields $\langle x^* - u^*, x - u \rangle \geq 0$ for all $(u, u^*) \in \mathrm{Gr}\,A$. Once again the maximality of $A$ implies that $(x, x^*) \in \mathrm{Gr}\,A$, and so $A(x)$ is $w^*$-closed. $\qquad\square$

We next introduce notions replacing continuity for multivalued maps.

**Definition 2.36.** Let $Y$ and $V$ be Hausdorff topological spaces and $G : Y \to 2^V$ with nonempty values.

(a) We say that $G$ is *upper semicontinuous* (*u.s.c.*) if for all $C \subset V$ nonempty closed the set $G^-(C) := \{y \in Y : G(y) \cap C \neq \emptyset\}$ is closed.
(b) We say that $G$ is *lower semicontinuous* (*l.s.c.*) ) if for all $C \subset V$ nonempty open, the set $G^-(C)$ is open.

*Remark 2.37.*

(a) It is clear that if $G$ is single-valued, then the preceding definition coincides with the definition of the continuity of $G$.
(b) An equivalent definition of u.s.c. and l.s.c. multimaps can be given by using the set $G^+(C) := \{y \in Y : G(y) \subset C\}$ instead of $G^-(C)$. We have that $G$ is u.s.c. (resp. l.s.c.) if for all $C \subset V$ nonempty open (resp. closed) the set $G^+(C)$ is open (resp. closed).
(c) If $V$ is regular and $G$ has closed values, then the upper semicontinuity of $G$ implies that $\mathrm{Gr}\,G$ is closed in $Y \times V$ with the product topology (in fact, if $G$ has compact values, we can drop the regularity condition on $V$, and in this case $G$ maps compact sets in $Y$ to compact sets in $V$). Finally, if $G$ is locally compact [i.e., for every $y \in Y$ we can find a neighborhood $U$ of $y$ such that $\overline{G(U)}$ is compact in $V$] and $G(y)$ is closed in $V$ for all $y \in Y$, then $G$ is u.s.c. if and only if $\mathrm{Gr}\,G$ is closed in $Y \times V$.
(d) If $(V, d)$ is a metric space, then $G$ is l.s.c. if and only if for every $v \in V$ the function $y \mapsto d(v, G(y))$ is upper semicontinuous on $Y$. Moreover, if we assume that $G$ has compact values, then $G$ is l.s.c. if and only if for every $y_0 \in Y$ the function $y \mapsto \sup\{d(y^*, G(y_0)) : y^* \in G(y)\}$ is continuous on $Y$.

**Proposition 2.38.** *If $A : X \to 2^{X^*}$ is maximal monotone, then $A|_{\mathrm{int}\,D(A)}$ is u.s.c. from $X$ with the norm topology into $X^*$ with the $w^*$-topology.*

*Proof.* Let $C \subset X^*$ be nonempty and $w^*$-closed, and consider a sequence $\{x_n\}_{n \geq 1} \subset (A|_{\mathrm{int}\,D(A)})^-(C)$ such that $x_n \to x$ in $\mathrm{int}\,D(A)$. Then we can find $\{x_n^*\}_{n \geq 1} \subset C$ such that $x_n^* \in A(x_n)$ for all $n \geq 1$. Proposition 2.28 implies that $\{x_n^*\}_{n \geq 1} \subset X^*$ is bounded, and so, by Alaoglu's theorem (e.g., Brezis [52, p. 66]), there is a subnet $\{x_\alpha^*\}_{\alpha \in J}$ of $\{x_n^*\}_{n \geq 1}$ such that $x_\alpha^* \overset{w^*}{\to} x^* \in C$ in $X^*$. For every $(u, u^*) \in \mathrm{Gr}\,A$ we have $\langle x_\alpha^* - u^*, x_\alpha - u \rangle \geq 0$ [where $x_\alpha^* \in A(x_\alpha)$], and so $\langle x^* - u^*, x - u \rangle \geq 0$. Since $A$ is maximal monotone, we derive that $x^* \in A(x)$, hence $x \in (A|_{\mathrm{int}\,D(A)})^-(C)$. Therefore, $A|_{\mathrm{int}\,D(A)}$ is u.s.c., as claimed. $\qquad\square$

Directly from the definition of maximal monotone maps, we also have the following proposition.

**Proposition 2.39.** *If $A : X \to 2^{X^*}$ is a maximal monotone map, then $\mathrm{Gr}\,A$ is closed in $X \times X^*_{w^*}$ and in $X_w \times X^*$ (by $X^*_{w^*}$ we denote $X^*$ endowed with the $w^*$-topology and by $X_w$ the space $X$ endowed with the $w$-topology).*

We have a partial converse to Proposition 2.38.

**Proposition 2.40.** *If $A : X \to 2^{X^*}$ is a monotone map, $D(A) = X$, $A$ has convex, $w^*$-closed values and is u.s.c. from $X$ with the norm topology into $X^*_{w^*}$, then $A$ is maximal monotone.*

*Proof.* We argue by contradiction. Suppose that

$$\langle x^* - u^*, x - u \rangle \geq 0 \text{ for all } (u, u^*) \in \mathrm{Gr}\,A, \tag{2.3}$$

but $x^* \notin A(x)$. Then, by the strong separation theorem for convex sets (e.g., Brezis [52, Theorem 1.7]), we can find $v \in X \setminus \{0\}$ and $\varepsilon > 0$ such that

$$\langle z^*, v \rangle \leq \langle x^*, v \rangle - \varepsilon \text{ for all } z^* \in A(x). \tag{2.4}$$

Let $U = \{y^* \in X^* : \langle y^*, v \rangle < \langle x^*, v \rangle\}$. Clearly, $U \subset X^*$ is $w^*$-open and $A(x) \subset U$ [see (2.4)]. For $t > 0$, let $y_t = x + tv$ and $y_t^* \in A(y_t)$. The upper semicontinuity of $A$ [Remark 2.37 (b)] implies that for small $t > 0$ we have $A(y_t) \subset U$, that is,

$$\langle y_t^*, v \rangle < \langle x^*, v \rangle. \tag{2.5}$$

On the other hand, if in (2.3) we choose $u = y_t$ and $u^* = y_t^*$, then

$$\langle x^* - y_t^*, v \rangle \leq 0. \tag{2.6}$$

Comparing (2.5) and (2.6), we reach a contradiction. Therefore, $A$ is maximal monotone. ☐

*Remark 2.41.* In view of the preceding proof, we can see that Proposition 2.40 remains true if the assumption that $A$ is u.s.c. from $X$ to $X^*_{w^*}$ is replaced by the weaker assumption that, for every $x, v \in X$, the multimap $t \mapsto A(x + tv)$ is u.s.c from $[0, 1]$ to $X^*_{w^*}$, or, even more generally that, for each sequence $\{t_n\}_{n \geq 1}$ converging to $0$, given $U \subset X^*$ $w^*$-neighborhood of $A(x)$, we have $A(x + t_n v) \subset U$ whenever $n \geq 1$ is large enough.

In view of Remark 2.41, for a single-valued monotone operator $A : X \to X^*$, Proposition 2.40 becomes:

**Corollary 2.42.** *If $A : X \to X^*$ is monotone and hemicontinuous, then $A$ is maximal monotone.*

*Example 2.43.* The $p$-Laplace differential operator on $W_0^{1,p}(\Omega)$ [given in Example 2.27 (c)] is monotone and continuous, hence maximal monotone from $W_0^{1,p}(\Omega)$ into $W^{-1,p'}(\Omega)$, where $\frac{1}{p} + \frac{1}{p'} = 1$.

The map that we introduce next is a valuable tool in many parts of nonlinear analysis.

**Definition 2.44.** The map $\mathscr{F} : X \to 2^{X^*}$ defined by

$$\mathscr{F}(x) = \{x^* \in X^* : \langle x^*, x \rangle = \|x\|^2 = \|x^*\|^2\}$$

is called a *(normalized) duality map*.

*Remark 2.45.* If $\varphi(x) = \frac{1}{2}\|x\|^2$ for all $x \in X$, then $\mathscr{F}(x) = \partial\varphi(x)$ for all $x \in X$.

The duality map $\mathscr{F}$ is defined for any Banach space $(X, \|\cdot\|)$. However, its properties strongly depend on those of the Banach space. In this respect, we recall the following notions.

**Definition 2.46.**

(a) A Banach space $X$ is said to be *strictly convex* if the unit ball of $X$ is strictly convex, that is, we have

$$\|tx + (1-t)y\| < 1 \quad \text{for all } t \in (0,1), \text{ all } x, y \in X \text{ with } \|x\| \leq 1, \|y\| \leq 1, x \neq y.$$

(b) A Banach space $X$ is said to be *locally uniformly convex* if, for every $\varepsilon > 0$ and every $x \in X$ with $\|x\| \leq 1$, we can find $\delta > 0$ such that

$$y \in Y, \ \|y\| \leq 1, \ \|x - y\| > \varepsilon \ \Rightarrow \ \left\|\frac{x+y}{2}\right\| < 1 - \delta.$$

Moreover, if $\delta$ can be chosen independently of $x$, then we say that $X$ is *uniformly convex*.

*Remark 2.47.* We refer to Deville et al. [117] and Megginson [256] for a comprehensive introduction to strict, locally uniform, and uniform convexity. We point out the following aspects.

(a) It is clear that we have the following implications:

$$\text{uniformly convex} \Rightarrow \text{locally uniformly convex} \Rightarrow \text{strictly convex}.$$

It follows from the parallelogram law that a Hilbert space is always uniformly convex. It follows from Clarkson's inequalities (see [256, p. 450]) that the space $L^p(\Omega)$ is uniformly convex whenever $1 < p < +\infty$; more generally, $L^p(\Omega, \mathbb{R}^M)$ ($M \geq 1$) is uniformly convex (see Day [103]). For every $1 < p < +\infty$ the Sobolev space $W^{1,p}(\Omega)$ is uniformly convex for the Sobolev norm $\|u\| = (\|u\|_p^p + \|\nabla u\|_p^p)^{\frac{1}{p}}$, whereas the space $W_0^{1,p}(\Omega)$ is uniformly convex for the Sobolev norm and, when $\Omega$ is bounded, also for the equivalent norm $u \mapsto \|\nabla u\|_p$ [using [256, Proposition 5.2.7] and the isometrical embeddings $W^{1,p}(\Omega) \subset L^p(\Omega) \times L^p(\Omega, \mathbb{R}^N)$, $u \mapsto (u, \nabla u)$, and $(W_0^{1,p}(\Omega), \|\nabla \cdot\|_p) \subset L^p(\Omega, \mathbb{R}^N)$, $u \mapsto \nabla u$].

(b) The *Milman–Pettis theorem* asserts that a uniformly convex Banach space is always reflexive (e.g., [256, p. 452]). A kind of converse is provided by the *Troyanski renorming theorem* [377], which says that a reflexive Banach space can be equivalently renormed so that both $X$ and $X^*$ are locally uniformly convex (and thus also strictly convex) and with Fréchet differentiable norms (except at the origins).

(c) A locally uniformly convex Banach space always satisfies the so-called *Kadec–Klee property*, namely: given a sequence $\{x_n\}_{n \geq 1} \subset X$, if $x_n \xrightarrow{w} x$ in $X$ and $\|x_n\| \to \|x\|$ as $n \to \infty$, then $x_n \to x$ in $X$ as $n \to \infty$ (see [256, p. 453]).

(d) Note that strict, uniform, locally uniform convexity is sometimes called strict, uniform, locally uniform rotundity. The Kadec–Klee property is also called the Radon–Riesz property.

The following theorem summarizes some of the main properties of the duality map. For a proof we refer the reader to Zeidler [389, p. 861] and Gasiński–Papageorgiou [151, p. 313].

**Theorem 2.48.**

(a) *If $X$ is reflexive and $X^*$ is strictly convex, then $\mathscr{F}$ is single-valued, bounded, odd, demicontinuous, and maximal monotone.*

(b) *If $X$ is reflexive and both $X$ and $X^*$ are strictly convex, then $\mathscr{F} : X \to X^*$ is strictly monotone.*

(c) *If $X$ is reflexive and $X^*$ is locally uniformly convex, then $\mathscr{F} : X \to X^*$ is continuous.*

(d) *If $X^*$ is uniformly convex, then $\mathscr{F} : X \to X^*$ is uniformly continuous on bounded sets.*

*Example 2.49.*

(a) If $(\Omega, \Sigma, \mu)$ is a $\sigma$-finite measure space, $1 \leq p < +\infty$, and $X = L^p(\Omega)$, then

$$\mathscr{F}(u)(\cdot) = \frac{|u(\cdot)|^{p-2} u(\cdot)}{\|u\|_p^{p-2}} \text{ for all } u \in L^p(\Omega).$$

(b) If $\Omega \subset \mathbb{R}^N$ is a bounded domain, $1 < p < +\infty$, and $X = W_0^{1,p}(\Omega)$, then

$$\mathscr{F}(u) = -\frac{1}{\|\nabla u\|_p^{p-2}} \sum_{k=1}^{N} D_k(|D_k u|^{p-2} D_k u) \text{ for all } u \in W_0^{1,p}(\Omega).$$

In particular, if $p = 2$, then $\mathscr{F} = -\Delta$.

(c) If $H$ is a Hilbert space that is identified with its dual $H^*$ by virtue of the Riesz representation theorem (e.g., Brezis [52, p. 135]), then $\mathscr{F} = \mathrm{id}_H$.

What distinguishes maximal monotone maps is their remarkable surjectivity properties. Next, we present some basic results in this direction. We start with a definition.

**Definition 2.50.** Let $A : X \to 2^{X^*}$.

(a) We say that $A$ is *coercive* if either $D(A)$ is bounded or $D(A)$ is unbounded and $\inf\{\|x^*\| : x^* \in A(x)\} \to +\infty$ as $\|x\| \to +\infty$, $x \in D(A)$.

(b) We say that $A$ is *strongly coercive* if either $D(A)$ is bounded or $D(A)$ is unbounded and $\frac{1}{\|x\|} \inf\{\langle x^*, x \rangle : x^* \in A(x)\} \to +\infty$ as $\|x\| \to +\infty$, $x \in D(A)$.

The next result is a basic tool in the study of variational inequalities. For its proof we refer to Barbu [29, p. 34] and Papageorgiou–Kyritsi [318, p. 170].

**Theorem 2.51.** *If $X$ is reflexive, $C \subset X$ is nonempty, closed, and convex, $A : X \to 2^{X^*}$ is monotone with $D(A) \subset C$, and $B : C \to X^*$ is monotone, demicontinuous, bounded, and strongly coercive, then there exists $x_0 \in C$ such that $\langle u^* + B(x_0), u - x_0 \rangle \geq 0$ for all $(u, u^*) \in \mathrm{Gr}\,A$.*

We deduce from this theorem the following surjectivity result.

**Theorem 2.52.** *If $X$ is reflexive, $C \subset X$ is nonempty, closed, and convex, $A : X \to 2^{X^*}$ is maximal monotone with $D(A) \subset C$, and $B : C \to X^*$ is monotone, demicontinuous, bounded, and strongly coercive, then $A + B$ is surjective [i.e., $R(A + B) = X^*$].*

*Proof.* Given $y^* \in X^*$, we set $\hat{A}(x) = A(x) - y^*$. Evidently, $\hat{A}$ is maximal monotone. Hence, by Theorem 2.51, we can find $x_0 \in C$ such that $\langle \hat{u}^* + B(x_0), u - x_0 \rangle \geq 0$ for all $(u, \hat{u}^*) \in \mathrm{Gr}\,\hat{A}$, which implies that

$$\langle u^* - (y^* - B(x_0)), u - x_0 \rangle \geq 0 \text{ for all } (u, u^*) \in \mathrm{Gr}\,A.$$

From the maximal monotonicity of $A$ it follows that $(x_0, y^* - B(x_0)) \in \mathrm{Gr}\,A$, and this implies that $y^* \in A(x_0) + B(x_0)$. We conclude that $R(A + B) = X^*$. □

This theorem leads to a characterization of maximal monotonicity in terms of the duality map.

**Theorem 2.53.** *If $X$ is reflexive, $X$ and $X^*$ are both strictly convex, and $A : X \to 2^{X^*}$ is monotone, then $A$ is maximal monotone if and only if $R(A + \mathscr{F}) = X^*$.*

*Proof.* From Theorem 2.48 we know that $\mathscr{F} : X \to X^*$ is single-valued, strictly monotone, demicontinuous, and bounded.

$\Rightarrow$: Since $\mathscr{F}$ is strongly coercive (which is clear from Definition 2.44), we can apply Theorem 2.52 and obtain $R(A + \mathscr{F}) = X^*$.

$\Leftarrow$: Suppose that for a pair $(x, x^*) \in X \times X^*$ we have

$$\langle x^* - u^*, x - u \rangle \geq 0 \text{ for all } (u, u^*) \in \mathrm{Gr}\,A. \tag{2.7}$$

Since $R(A + \mathscr{F}) = X^*$, we can find $(y, y^*) \in \mathrm{Gr}\,A$ such that

$$x^* + \mathscr{F}(x) = y^* + \mathscr{F}(y). \tag{2.8}$$

Then from (2.7) we have $\langle \mathscr{F}(y) - \mathscr{F}(x), x - y \rangle \geq 0$. Since $\mathscr{F}$ is strictly monotone, we infer that $x = y \in D(A)$, and then, from (2.8), we conclude that $x^* = y^* \in A(y) = A(x)$. Therefore, $A$ is maximal monotone.                                                          $\square$

Now we have a second surjectivity result.

**Theorem 2.54.** *If $X$ is reflexive and $A : X \to 2^{X^*}$ is a maximal monotone map, then $A$ is surjective if and only if $A^{-1}$ is locally bounded.*

*Proof.* $\Rightarrow$: Since $A$ is maximal monotone, from Proposition 2.34 we know that $A^{-1}$ is maximal monotone. Moreover, since $A$ is surjective, we have $D(A^{-1}) = X^*$. Then from Proposition 2.28 we have that $A^{-1}$ is locally bounded.

$\Leftarrow$: We will show that $R(A)$ is both closed and open in $X^*$, hence $R(A) = X^*$. First we show that $R(A)$ is closed in $X^*$. To this end, let $\{x_n^*\}_{n \geq 1} \subset R(A)$, and assume that $x_n^* \to x^*$ in $X^*$. Let $\{x_n\}_{n \geq 1} \subset X$ be such that $x_n^* \in A(x_n)$. Because $A^{-1}$ is locally bounded, we have that $\{x_n\}_{n \geq 1}$ is bounded. Thus, we may assume that $x_n \xrightarrow{w} x$ in $X$. Then Proposition 2.39 implies that $x^* \in A(x)$, hence $x^* \in R(A)$, and so $R(A)$ is closed in $X^*$.

Next we prove that $R(A)$ is open in $X^*$. By virtue of the Troyanski renorming theorem [Remark 2.47(b)], we may assume that both $X$ and $X^*$ are locally uniformly convex. Let $x_0^* \in R(A)$, and let $r > 0$ be such that $A^{-1}(B_r(x_0^*))$ is bounded in $X$. We will show that $B_{\frac{r}{2}}(x_0^*) \subset R(A)$. Let $x_0 \in D(A)$ be such that $x_0^* \in A(x_0)$. In view of Theorem 2.48(a), we can apply Theorem 2.52, which yields that for every $\lambda > 0$ we have $R(A + \lambda \mathscr{F}(\cdot - x_0)) = X^*$. Thus, given $x^* \in B_{\frac{r}{2}}(x_0^*)$, we have $x^* = u_\lambda^* + \lambda \mathscr{F}(u_\lambda - x_0)$ with $(u_\lambda, u_\lambda^*) \in \mathrm{Gr}A$. Hence

$$\langle x^* - \lambda \mathscr{F}(u_\lambda - x_0) - x_0^*, u_\lambda - x_0 \rangle = \langle u_\lambda^* - x_0^*, u_\lambda - x_0 \rangle \geq 0,$$

that is, $\langle x^* - x_0^*, u_\lambda - x_0 \rangle \geq \lambda \|u_\lambda - x_0\|^2$, and so

$$\|x^* - u_\lambda^*\| = \lambda \|\mathscr{F}(u_\lambda - x_0)\| = \lambda \|u_\lambda - x_0\| \leq \|x^* - x_0^*\| < \frac{r}{2}.$$

Then we obtain

$$\|u_\lambda^* - x_0^*\| \leq \|u_\lambda^* - x^*\| + \|x^* - x_0^*\| < r.$$

Since $A^{-1}(B_r(x_0^*))$ is bounded, it follows that $\{u_\lambda\}_{\lambda > 0}$ is bounded in $X$, and so

$$\|x^* - u_\lambda^*\| = \lambda \|u_\lambda - x_0\| \to 0 \quad \text{as } \lambda \downarrow 0.$$

From the previous part of the proof we know that $R(A)$ is closed. Using that $u_\lambda^* \in R(A)$ for all $\lambda > 0$, we derive that $x^* \in R(A)$, and thus $B_{\frac{r}{2}}(x_0^*) \subset R(A)$, which proves the openness of $R(A)$.                                                          $\square$

From Definition 2.50 it is clear that the coercivity of $A$ implies that $A^{-1}$ is locally bounded. Thus, using Theorem 2.54 we have the following theorem.

**Theorem 2.55.** *If $X$ is reflexive and $A : X \to 2^{X^*}$ is maximal monotone and coercive, then $A$ is surjective.*

These surjectivity results lead to some useful single-valued approximations of a maximal monotone map. Let $X$ be reflexive and $A : X \to 2^{X^*}$ a maximal monotone map. We may assume that both $X$ and $X^*$ are locally uniformly convex [Remark 2.47(b)]. By virtue of Theorems 2.52 and 2.48(a), (b), for every $x \in X$ and every $\lambda > 0$ the inclusion $0 \in \lambda A(x_\lambda) + \mathscr{F}(x_\lambda - x)$ has a unique solution $x_\lambda \in D(A)$. Then we can define

$$J_\lambda^A(x) = x_\lambda \quad \text{and} \quad A_\lambda(x) = -\frac{1}{\lambda}\mathscr{F}(x_\lambda - x). \tag{2.9}$$

The maps $J_\lambda^A : X \to X$ and $A_\lambda : X \to X^*$ are called the *resolvent of $A$* and the *Yosida approximation of $A$*, respectively. The next proposition summarizes their main properties. For a proof, see Barbu [29, p. 49].

**Proposition 2.56.** *If $X$ is reflexive, $X$ and $X^*$ are both locally uniformly convex, $A : X \to 2^{X^*}$ is a maximal monotone map, and $\lambda > 0$, then:*

(a) $A_\lambda$ *is single-valued, demicontinuous, and monotone and* $D(A_\lambda) = X$ *(hence $A_\lambda$ is maximal monotone; see Corollary 2.42);*
(b) $A_\lambda(x) \in A(J_\lambda^A(x))$ *for all $x \in X$;*
(c) $A_\lambda$ *and $J_\lambda^A$ are bounded;*
(d) *If $x \in \overline{D(A)}$, $x_n \to x$ in $X$, and $\lambda_n \downarrow 0$, then $J_{\lambda_n}^A(x_n) \to x$ in $X$;*
(e) *If $(0,0) \in \mathrm{Gr}\,A$, then $J_\lambda^A(0) = 0$, $A_\lambda(0) = 0$, $\|J_\lambda^A(x)\| \leq 2\|x\|$, and $\|A_\lambda(x)\| \leq \frac{1}{\lambda}\|x\|$.*

If $H$ is a Hilbert space identified with its dual, then we have

$$J_\lambda^A = (\mathrm{id} + \lambda A)^{-1} \quad \text{and} \quad A_\lambda = \frac{1}{\lambda}(\mathrm{id} - J_\lambda^A).$$

In this case, Proposition 2.56 becomes (see Gasiński–Papageorgiou [151, p. 323]) the following proposition.

**Proposition 2.57.** *If $H$ is a Hilbert space, $A : H \to 2^H$ is maximal monotone, and $\lambda > 0$, then*

(a) $J_\lambda^A$ *is nonexpansive;*
(b) $A_\lambda(x) \in A(J_\lambda^A(x))$ *for all $x \in H$;*
(c) $A_\lambda$ *is monotone and Lipschitz continuous with Lipschitz constant $\frac{1}{\lambda}$;*
(d) $\|A_\lambda(x)\| \leq \|A^0(x)\|$ *for all $x \in D(A)$, where $A^0(x) \in A(x)$ is the unique element with smallest norm;*
(e) $\lim_{\lambda \downarrow 0} A_\lambda(x) = A^0(x)$ *for all $x \in D(A)$;*
(f) $\overline{D(A)}$ *is convex and $\lim_{\lambda \downarrow 0} J_\lambda^A(x) = p_{\overline{D(A)}}(x)$ for all $x \in H$.*

Now we introduce some important generalizations of maximal monotonicity.

**Definition 2.58.** Let $X$ be reflexive and $A : X \to 2^{X^*}$. We say that $A$ is *pseudomono-tone* if:

(a) $D(A) = X$ and for every $x \in X$, $A(x)$ is bounded, closed, and convex;
(b) $A$ is u.s.c. from every finite-dimensional subspace of $X$ into $X_w^*$;
(c) If $x_n \overset{w}{\to} x$ in $X$, $x_n^* \in A(x_n)$ for all $n \geq 1$, and $\limsup_{n \to \infty} \langle x_n^*, x_n - x \rangle \leq 0$, then for

each $u \in X$ we can find $y^*(u) \in A(x)$ such that

$$\langle y^*(u), x - u \rangle \leq \liminf_{n \to \infty} \langle x_n^*, x_n - u \rangle.$$

*Remark 2.59.* A completely continuous map $A : X \to X^*$ is pseudomonotone. Similarly, if $X$ is finite dimensional, then any continuous operator $A : X \to X^*$ is pseudomonotone.

**Proposition 2.60.** *If $X$ is reflexive and $A : X \to 2^{X^*}$ is maximal monotone with $D(A) = X$, then $A$ is pseudomonotone.*

*Proof.* Requirements (a) and (b) in Definition 2.58 follow from Proposi-tions 2.28, 2.35, and 2.38. Thus, we need to verify (c). To that end, let $x_n \overset{w}{\to} x$ in $X$ and $x_n^* \in A(x_n)$ for all $n \geq 1$, and suppose that $\limsup_{n \to \infty} \langle x_n^*, x_n - x \rangle \leq 0$. Since $D(A) = X$, there exists $x^* \in A(x)$. Then we have $\langle x^*, x_n - x \rangle \leq \langle x_n^*, x_n - x \rangle$ (since $A$ is monotone), hence

$$\lim_{n \to \infty} \langle x_n^*, x_n - x \rangle = 0. \tag{2.10}$$

Note that whenever $(y, y^*) \in \mathrm{Gr}\, A$, on the one hand, due to (2.10), we have

$$\liminf_{n \to \infty} \langle x_n^*, x_n - y \rangle = \liminf_{n \to \infty} \langle x_n^*, x - y \rangle, \tag{2.11}$$

and on the other hand, the monotonicity property of $A$ implies

$$\langle y^*, x - y \rangle \leq \liminf_{n \to \infty} \langle x_n^*, x - y \rangle. \tag{2.12}$$

Given $u \in X$ and $t > 0$, let $y_t = x + t(u - x)$ and $y_t^* \in A(y_t)$. In (2.12) we replace $(y, y^*)$ by $(y_t, y_t^*)$ and obtain

$$\langle y_t^*, x - u \rangle \leq \liminf_{n \to \infty} \langle x_n^*, x - u \rangle.$$

If $t_k \downarrow 0$, then $y_{t_k} \to x$ in $X$ and, by virtue of Proposition 2.28, we may assume that $y_{t_k}^* \overset{w}{\to} y^*(u)$ for some $y^*(u) \in X^*$. Proposition 2.39 implies that $y^*(u) \in A(x)$. Moreover, passing to the limit as $t_k \downarrow 0$ in the previous inequality, by (2.11), we have

$$\langle y^*(u), x-u \rangle \leq \liminf_{n \to \infty} \langle x_n^*, x-u \rangle = \liminf_{n \to \infty} \langle x_n^*, x_n - u \rangle.$$

Therefore, $A$ is pseudomonotone. □

The property of pseudomonotonicity is preserved by addition.

**Proposition 2.61.** *If $X$ is reflexive and $A_1, A_2 : X \to 2^{X^*}$ are pseudomonotone operators, then $A_1 + A_2$ is pseudomonotone.*

*Proof.* Clearly, $A_1 + A_2$ has nonempty, bounded, closed, and convex values and it is u.s.c. from any finite-dimensional subspace $Y$ of $X$ into $X_w^*$ (to see this, apply Hu–Papageorgiou [175, p. 59] with $Y = X_w^*$).

Let $x_n \xrightarrow{w} x$ in $X$ and $x_n^* \in (A_1 + A_2)(x_n)$, and suppose that $\limsup_{n \to \infty} \langle x_n^*, x_n - x \rangle \leq 0$. We have $x_n^* = v_n^* + z_n^*$ with $v_n^* \in A_1(x_n)$, $z_n^* \in A_2(x_n)$. We claim that

$$\limsup_{n \to \infty} \langle v_n^*, x_n - x \rangle \leq 0 \quad \text{and} \quad \limsup_{n \to \infty} \langle z_n^*, x_n - x \rangle \leq 0. \tag{2.13}$$

If (2.13) is not true, then by passing to a subsequence if necessary, we may assume that

$$\lim_{n \to \infty} \langle v_n^*, x_n - x \rangle > 0 \quad \text{and} \quad \lim_{n \to \infty} \langle z_n^*, x_n - x \rangle < 0. \tag{2.14}$$

Since $A_2$ is pseudomonotone, for each $u \in X$ there exists $y^*(u) \in X^*$ such that

$$\langle y^*(u), x-u \rangle \leq \liminf_{n \to \infty} \langle z_n^*, x_n - u \rangle.$$

Setting $u = x$, we find

$$0 \leq \liminf_{n \to \infty} \langle z_n^*, x_n - x \rangle. \tag{2.15}$$

Comparing (2.14) and (2.15), we reach a contradiction. Thus, (2.13) is valid.

By (2.13) and because $A_1$ and $A_2$ are pseudomonotone, for every $u \in X$ we can find $y_1^*(u) \in A_1(x)$ and $y_2^*(u) \in A_2(x)$ such that

$$\langle y_1^*(u), x-u \rangle \leq \liminf_{n \to \infty} \langle v_n^*, x_n - u \rangle \quad \text{and} \quad \langle y_2^*(u), x-u \rangle \leq \liminf_{n \to \infty} \langle z_n^*, x_n - u \rangle.$$

We infer that

$$\liminf_{n \to \infty} \langle x_n^*, x_n - u \rangle \geq \liminf_{n \to \infty} \langle v_n^*, x_n - u \rangle + \liminf_{n \to \infty} \langle z_n^*, x_n - u \rangle \geq \langle y_1^*(u) + y_2^*(u), x-u \rangle.$$

Since $y_1^*(u) + y_2^*(u) \in (A_1 + A_2)(x)$, we obtain that $A_1 + A_2$ is pseudomonotone. □

Pseudomonotone maps, much like maximal monotone ones, exhibit nice surjectivity properties. To show this, we will need the following proposition (see Papageorgiou and Kyritsi-Yiallourou [318, Proposition 3.2.59]).

**Proposition 2.62.** *If $X$ is finite dimensional and $A : X \to 2^{X^*}$ has nonempty, compact, convex values and is u.s.c. and strongly coercive, then $A$ is surjective, i.e., $R(A) = X^*$.*

Using Proposition 2.62 and a Galerkin approximation, we can prove the main surjectivity result for pseudomonotone maps.

**Theorem 2.63.** *If $X$ is reflexive and separable, and $A : X \to 2^{X^*}$ is a pseudomonotone, bounded (i.e., for every bounded subset $C$ of $X$, $\bigcup_{x \in C} A(x)$ is bounded in $X^*$), and strongly coercive map, then $A$ is surjective.*

*Proof.* Due to separability, there exists an increasing sequence $\{V_n\}_{n \geq 1}$ of finite-dimensional subspaces of $X$ such that $\overline{\bigcup_{n \geq 1} V_n} = X$. For $n \geq 1$, $i_n : V_n \to X$ denotes the inclusion map and $i_n^* : X^* \to V_n^*$ its adjoint. We set $A_n = i_n^* \circ A \circ i_n : V_n \to 2^{V_n^*}$. Clearly, $A_n$ is u.s.c. and strongly coercive. Moreover, $A_n(x)$ is nonempty, convex, and compact for every $x \in V_n$ [since $A_n(x) = i_n^*(A(x))$ and $A(x)$ is nonempty, convex, closed, bounded, and so weakly compact].

To prove the theorem, by translating things if necessary, it suffices to show that $0 \in R(A)$. Proposition 2.62 implies that for every $n \geq 1$ we can find $x_n \in V_n$ such that $0 \in A_n(x_n)$ [i.e., $i_n^*(x_n^*) = 0$ for some $x_n^* \in A(x_n)$]. Since $A$ is strongly coercive, the sequence $\{x_n\}_{n \geq 1}$ is bounded. Thanks to the reflexivity of $X$, by passing to a subsequence if necessary, we may assume that $x_n \xrightarrow{w} x$ in $X$.

We claim that

$$\lim_{n \to \infty} \langle x_n^*, x_n - u \rangle = 0 \text{ for all } u \in X. \tag{2.16}$$

To see this, first we note that $\bigcup_{n \geq 1} A(x_n)$ is bounded in $X^*$ (because $A$ is bounded), hence there is a constant $M > 0$ such that $\|x_n^*\| \leq M$ for all $n \geq 1$. Let $u \in X$ and $\varepsilon > 0$. Since $\overline{\bigcup_{n \geq 1} V_n} = X$, there exist $n_\varepsilon \geq 1$ and $u_\varepsilon \in V_{n_\varepsilon}$ such that $\|u_\varepsilon - u\| < \frac{\varepsilon}{M}$. Then for all $n \geq n_\varepsilon$ we have

$$|\langle x_n^*, x_n - u \rangle| \leq |\langle x_n^*, x_n - u_\varepsilon \rangle| + |\langle x_n^*, u_\varepsilon - u \rangle| \leq \|x_n^*\| \, \|u_\varepsilon - u\| < \varepsilon,$$

where we used that $i_n^* x_n^* = 0$. This proves (2.16).

Taking (2.16) into account, the pseudomonotonicity of $A$ ensures that we can find $y^*(u) \in A(x)$ such that

$$\langle y^*(u), x - u \rangle \leq \liminf_{n \to \infty} \langle x_n^*, x_n - u \rangle = 0. \tag{2.17}$$

If $0 \notin A(x)$, then by the strong separation theorem for convex sets (e.g., Brezis [52], Theorem 1.7]), we can find $z \in X$ with the property

$$0 < \inf\{\langle x^*, z \rangle : x^* \in A(x)\}. \tag{2.18}$$

Choosing $u = x - z$ in (2.17) leads to

$$\langle y^*(x-z), z \rangle \leq 0 \quad \text{with } y^*(x-z) \in A(x). \tag{2.19}$$

Comparing (2.18) and (2.19), we reach a contradiction. Hence $0 \in A(x) \subset R(A)$. $\square$

*Remark 2.64.* In fact, Theorem 2.63 holds without requiring that $X$ be separable and $A$ is bounded (see Gasiński and Papageorgiou [151, Theorem 3.2.52]).

In this context of surjectivity results, we cite from Zeidler [389, Problem 32.4] the following extension of Theorem 2.63.

**Theorem 2.65.** *Let $C$ be a nonempty, closed, convex subset of a real reflexive Banach space $X$, $A : C \to 2^{X^*}$ maximal monotone, $B : C \to 2^{X^*}$ pseudomonotone, bounded, with nonempty, closed, convex values and demicontinuous on simplices (i.e., convex hulls of finite subsets of $C$), and let $u_0^* \in X^*$. Assume that $B$ is $A$-coercive with respect to $u_0^*$ in the sense that there are $u_0 \in C \cap D(A)$ and $r > 0$ such that $\langle x^*, x - u_0 \rangle > \langle u_0^*, x - u_0 \rangle$ for all $(x, x^*) \in \mathrm{Gr}\, B$ with $\|x\| > r$. Then there exists $u \in C$ such that $u_0^* \in Au + Bu$.*

A related notion, which is in general easier to handle, is the following.

**Definition 2.66.** Let $X$ be reflexive and $A : X \to 2^{X^*}$. We say that $A$ is *generalized pseudomonotone* if for any sequence $\{(x_n, x_n^*)\}_{n \geq 1} \subset \mathrm{Gr}\, A$ with $x_n \xrightarrow{w} x$ in $X$, $x_n^* \xrightarrow{w} x^*$ in $X^*$ and

$$\limsup_{n \to \infty} \langle x_n^*, x_n - x \rangle \leq 0,$$

we have $(x, x^*) \in \mathrm{Gr}\, A$ and $\langle x_n^*, x_n \rangle \to \langle x^*, x \rangle$ as $n \to \infty$.

**Proposition 2.67.** *If $X$ is reflexive and $A : X \to 2^{X^*}$ is a pseudomonotone operator, then $A$ is generalized pseudomonotone.*

*Proof.* Let $\{(x_n, x_n^*)\}_{n \geq 1} \subset \mathrm{Gr}\, A$, and assume that $x_n \xrightarrow{w} x$ in $X$, $x_n^* \xrightarrow{w} x^*$ in $X^*$, and

$$\limsup_{n \to \infty} \langle x_n^*, x_n - x \rangle \leq 0. \tag{2.20}$$

The pseudomonotonicity of $A$ yields for every $u \in X$ an element $y^*(u) \in A(x)$ such that

$$\langle y^*(u), x - u \rangle \leq \liminf_{n \to \infty} \langle x_n^*, x_n - u \rangle. \tag{2.21}$$

By passing to a subsequence if necessary, we may assume that $\langle x_n^*, x_n \rangle \to \xi \in \mathbb{R}$. Thus, from (2.20) we have $\xi \leq \langle x^*, x \rangle$. Then (2.21) implies

$$\langle y^*(u), x - u \rangle \leq \xi - \langle x^*, u \rangle \leq \langle x^*, x - u \rangle \quad \text{for all } u \in X. \tag{2.22}$$

If $x^* \notin A(x)$, then from the strong separation theorem for convex sets (e.g., Brezis [52, Theorem 1.7]), we can find $v \in X \setminus \{0\}$ such that

$$\langle x^*, v \rangle < \inf\{\langle v^*, v \rangle : v^* \in A(x)\}. \tag{2.23}$$

In (2.22) we insert $u = x - v$; thus,

$$\langle y^*(u), v \rangle \leq \langle x^*, v \rangle. \tag{2.24}$$

Since $y^*(u) \in A(x)$, from (2.23) and (2.24) we reach a contradiction. Therefore, $x^* \in A(x)$. Moreover, if in (2.21) we let $u = x$, then

$$0 \leq \liminf_{n \to \infty} \langle x_n^*, x_n - x \rangle. \tag{2.25}$$

From (2.20) and (2.25) we conclude that $\langle x_n^*, x_n \rangle \to \langle x^*, x \rangle$; hence $A$ is generalized pseudomonotone. $\qquad \square$

Under some extra conditions on the values of $A$, the converse is also true.

**Proposition 2.68.** *If $X$ is reflexive, $A : X \to 2^{X^*}$ is bounded and generalized pseudomonotone, and for every $x \in X$, $A(x)$ is nonempty, closed, and convex, then $A$ is pseudomonotone.*

*Proof.* Since $X$ is reflexive and $A$ is bounded and generalized pseudomonotone, we can easily see that for every finite-dimensional subspace $Y$ of $X$, $A : Y \to 2^{X^*_w}$ is u.s.c.

Let $\{(x_n, x_n^*)\}_{n \geq 1} \subset \operatorname{Gr} A$ with $x_n \xrightarrow{w} x$ in $X$ and $\limsup_{n \to \infty} \langle x_n^*, x_n - x \rangle \leq 0$. Let $u \in X$, and let $\{(x_{n_k}, x_{n_k}^*)\}_{k \geq 1}$ be a subsequence of $\{(x_n, x_n^*)\}_{n \geq 1}$ such that

$$\liminf_{n \to \infty} \langle x_n^*, x_n - u \rangle = \lim_{k \to \infty} \langle x_{n_k}^*, x_{n_k} - u \rangle.$$

Due to the boundedness of $A$, we have that $\{x_{n_k}^*\}_{k \geq 1}$ is bounded in $X^*$; hence we may assume that $x_{n_k}^* \xrightarrow{w} y^*(u)$ in $X^*$ for some $y^*(u) \in X^*$. The generalized pseudomonotonicity of $A$ implies

$$y^*(u) \in A(x) \quad \text{and} \quad \langle x_{n_k}^*, x_{n_k} \rangle \to \langle y^*(u), x \rangle. \tag{2.26}$$

Note that (2.26) implies $\langle y^*(u), x - u \rangle = \lim_{k \to \infty} \langle x_{n_k}^*, x_{n_k} - u \rangle = \liminf_{n \to \infty} \langle x_n^*, x_n - u \rangle$. Thus, $A$ is pseudomonotone. $\qquad \square$

A notion closely related to pseudomonotonicity and generalized pseudomonotonicity is the notion of $(S)_+$-maps. It is very useful in the application of variational methods, specifically, in the verification of the compactness conditions of functionals (the Palais–Smale and the Cerami conditions).

**Definition 2.69.** Let $X$ be reflexive, $C \subset X$ be nonempty, and $A : C \to X^*$. We say that $A$ is an $(S)_+$-*map* if for every sequence $\{x_n\}_{n \geq 1} \subset C$ such that $x_n \xrightarrow{w} x$ in $X$ and $\limsup_{n \to \infty} \langle A(x_n), x_n - x \rangle \leq 0$ we have $x_n \to x$ in $X$.

The following results summarize some basic properties of $(S)_+$-maps.

**Proposition 2.70.**

(a) *Let $A : C \to X^*$ be an $(S)_+$-map and $\lambda > 0$. Then $\lambda A : C \to X^*$ is an $(S)_+$-map.*

(b) *Let $A : C \to X^*$ be an $(S)_+$-map and $B : C \to X^*$ be a demicontinuous $(S)_+$-map. Then $A + B : C \to X^*$ is an $(S)_+$-map.*

(c) *Let $A : C \to X^*$ be an $(S)_+$-map and $B : X \to X^*$ be monotone. Then $A + B : C \to X^*$ is an $(S)_+$-map.*

(d) *Let $A : C \to X^*$ be an $(S)_+$-map and $B : X \to X^*$ be completely continuous (i.e., if $x_n \xrightarrow{w} x$ in $X$, then $B(x_n) \to B(x)$ in $X^*$). Then $A + B : C \to X^*$ is an $(S)_+$-map.*

*Proof.* Part (a) of the proposition is clear.

(b) Let $\{x_n\}_{n \geq 1} \subset C$ be such that $x_n \xrightarrow{w} x$ in $X$ and $\limsup_{n \to \infty} \langle A(x_n) + B(x_n), x_n - x \rangle \leq 0$. We claim that

$$\limsup_{n \to \infty} \langle A(x_n), x_n - x \rangle \leq 0. \tag{2.27}$$

Note that (2.27), together with the assumption that $A$ is an $(S)_+$-map, implies that $x_n \to x$, and we are done. Thus, it is sufficient to establish (2.27). To do this, we argue by contradiction. Up to considering subsequences, we may assume that

$$\lim_{n \to \infty} \langle A(x_n), x_n - x \rangle > 0 \quad \text{and} \quad \lim_{n \to \infty} \langle B(x_n), x_n - x \rangle < 0.$$

Then, the fact that $B$ is an $(S)_+$-map yields $x_n \to x$, and the demicontinuity of $B$ then implies that $\lim_{n \to \infty} \langle B(x_n), x_n - x \rangle = 0$, which is contradictory. This establishes (2.27).

(c) Let $\{x_n\}_{n > 1} \subset C$ be such that $x_n \xrightarrow{w} x$ in $X$ and $\limsup_{n \to \infty} \langle A(x_n) + B(x_n), x_n - x \rangle \leq 0$. The monotonicity of $B$ yields

$$\langle A(x_n), x_n - x \rangle = \langle A(x_n) + B(x_n), x_n - x \rangle - \langle B(x_n) - B(x), x_n - x \rangle - \langle B(x), x_n - x \rangle$$
$$\leq \langle A(x_n) + B(x_n), x_n - x \rangle - \langle B(x), x_n - x \rangle$$

for all $n \geq 1$. This implies $\limsup_{n \to \infty} \langle A(x_n), x_n - x \rangle \leq 0$, whence $x_n \to x$ [since $A$ is an $(S)_+$-map].

(d) If $\{x_n\}_{n\geq 1} \subset C$ satisfies $x_n \xrightarrow{w} x$ in $X$ and $\limsup_{n\to\infty}\langle A(x_n)+B(x_n), x_n - x\rangle \leq 0$, then the assumption on $B$ yields $\lim_{n\to\infty}\langle B(x_n), x_n - x\rangle = 0$, whence $\limsup_{n\to\infty}\langle A(x_n), x_n -$ $x\rangle \leq 0$, so that, from the fact that $A$ is an $(S)_+$-map, we infer that $x_n \to x$.     □

**Proposition 2.71.**

(a) *If $X$ is reflexive and locally uniformly convex with $X^*$ strictly convex, then the duality map $\mathscr{F} : X \to X^*$ is an $(S)_+$-map.*
(b) *If $X$ is reflexive and $A : X \to X^*$ is demicontinuous and an $(S)_+$-map, then $A$ is pseudomonotone.*

*Proof.* (a) Note that $\mathscr{F}$ is single-valued by virtue of Theorem 2.48(a). To see that it is an $(S)_+$-map, let $\{x_n\}_{n\geq 1}$ be such that $x_n \xrightarrow{w} x$ and that $\limsup_{n\to\infty}\langle \mathscr{F}(x_n), x_n - x\rangle \leq 0$. Then $\limsup_{n\to\infty}\langle \mathscr{F}(x_n) - \mathscr{F}(x), x_n - x\rangle \leq 0$. We have

$$\langle \mathscr{F}(x_n) - \mathscr{F}(x), x_n - x\rangle = (\|x_n\| - \|x\|)^2 + (\|x_n\|\,\|x\| - \langle \mathscr{F}(x_n), x\rangle)$$
$$+ (\|x_n\|\,\|x\| - \langle \mathscr{F}(x), x_n\rangle)$$
$$\geq (\|x_n\| - \|x\|)^2,$$

whence $\|x_n\| \to \|x\|$. But since $X$ is a locally uniformly convex Banach space, it has the Kadec–Klee property [Remark 2.47(c)], so $x_n \to x$ in $X$.

(b) Let $x_n \xrightarrow{w} x$ in $X$, and assume $\limsup_{n\to\infty}\langle A(x_n), x_n - x\rangle \leq 0$. Since $A$ is an $(S)_+$-map, we get that $x_n \to x$ in $X$. Then, for all $u \in X$, due to the demicontinuity of $A$, we have

$$\langle A(x), x - u\rangle = \lim_{n\to\infty}\langle A(x_n), x_n - u\rangle,$$

and so $A$ is pseudomonotone.     □

Let $\Omega \subset \mathbb{R}^N$ be a bounded domain, $1 < p < +\infty$, and consider the map $A : W^{1,p}(\Omega) \to W^{1,p}(\Omega)^*$ defined by

$$\langle A(u), v\rangle = \int_\Omega |\nabla u|^{p-2}(\nabla u, \nabla v)_{\mathbb{R}^N}\, dx \text{ for all } u,v \in W^{1,p}(\Omega). \tag{2.28}$$

**Proposition 2.72.** *The map $A : W^{1,p}(\Omega) \to W^{1,p}(\Omega)^*$ defined by (2.28) is an $(S)_+$-map. In particular, the negative $p$-Laplacian Dirichlet operator $(-\Delta_p, W_0^{1,p}(\Omega))$ [Example 2.27(c)] is an $(S)_+$-map.*

*Proof.* We note that $A$ is monotone and continuous, hence it is maximal monotone (Corollary 2.42). Then Propositions 2.60 and 2.67 imply that $A$ is generalized pseudomonotone. Thus, if $u_n \xrightarrow{w} u$ in $W^{1,p}(\Omega)$ and $\limsup_{n\to\infty}\langle A(u_n), u_n - u\rangle \leq 0$, then $\|\nabla u_n\|_p^p = \langle A(u_n), u_n\rangle \to \langle A(u), u\rangle = \|\nabla u\|_p^p$. Since $u_n \to u$ in $L^p(\Omega)$ (Theorem 1.49), we have that $\|u_n\| \to \|u\|$ [where $\|\cdot\|$ denotes the Sobolev norm on $W^{1,p}(\Omega)$]. Because $W^{1,p}(\Omega)$ is uniformly convex and $u_n \xrightarrow{w} u$ in $W^{1,p}(\Omega)$, from the Kadec–Klee property we conclude that $u_n \to u$ in $W^{1,p}(\Omega)$ [Remark 2.47(a), (c)]. $\qquad\square$

## 2.3  Nemytskii Operators

In this section, we introduce a nonlinear map that arises naturally in the study of nonlinear problems. Let $(\Omega, \Sigma, \mu)$ be a $\sigma$-finite measure space and $f : \Omega \times \mathbb{R}^N \to \mathbb{R}$ a function, and consider the *Nemytskii* map $N_f(u)(\cdot) = f(\cdot, u(\cdot))$ defined on classes of measurable functions $u : \Omega \to \mathbb{R}^N$. The following notion is important in the study of the map $N_f$.

**Definition 2.73.** Let $(\Omega, \Sigma, \mu)$ be a measure space, $X$ a separable metric space, and $Y$ a metric space. A map $f : \Omega \times X \to Y$ is said to be a *Carathéodory map* if

(a) For all $\xi \in X$, $x \mapsto f(x, \xi)$ is $(\Sigma, \mathscr{B}(Y))$-measurable [$\mathscr{B}(Y)$ is the Borel $\sigma$-field of $Y$];
(b) For $\mu$-almost all $x \in \Omega$, $\xi \mapsto f(x, \xi)$ is continuous.

Carathéodory maps are jointly measurable (for a proof, see Denkowski et al. [113, p. 189]).

**Proposition 2.74.** *Let $f : \Omega \times X \to Y$ be a Carathéodory map. Then $f$ is $(\Sigma \times \mathscr{B}(X), \mathscr{B}(Y))$-measurable.*

**Corollary 2.75.** *If $f : \Omega \times X \to Y$ is a Carathéodory map and $u : \Omega \to X$ is $(\Sigma, \mathscr{B}(X))$-measurable, then $x \mapsto f(x, u(x))$ is $(\Sigma, \mathscr{B}(Y))$-measurable.*

Returning to the setting of the Nemytskii map $N_f$ for a Carathéodory function $f : \Omega \times \mathbb{R}^N \to \mathbb{R}$, by virtue of Corollary 2.75, we see that $N_f$ maps $\Sigma$-measurable functions to $\Sigma$-measurable ones. More precisely, we have the following theorem due to Krasnosel'skiĭ [202] (see also Gasiński and Papageorgiou [151, p. 407]).

**Theorem 2.76.** *If $\{p_k\}_{k=1}^N \subset [1, +\infty)$, $q \in [1, +\infty)$, and $f : \Omega \times \mathbb{R}^N \to \mathbb{R}$ is a Carathéodory function such that*

$$|f(x, \xi)| \leq a(x) + c \sum_{k=1}^N |\xi_k|^{\frac{p_k}{q}} \ \text{for } \mu\text{-a.a. } x \in \Omega \text{ and all } \xi = (\xi_k)_{k=1}^N \in \mathbb{R}^N,$$

*with $a \in L^q(\Omega)_+$, $c > 0$, then $N_f : \prod_{k=1}^N L^{p_k}(\Omega) \to L^q(\Omega)$, $N_f(u)(\cdot) = f(\cdot, u(\cdot))$, is well defined, bounded, and continuous.*

*Remark 2.77.* The result fails if $(x, \xi) \mapsto f(x, \xi)$ is jointly measurable and $\xi \mapsto$ $f(x, \xi)$ is either lower or upper semicontinuous.

The differentiability properties of the Nemytskii map $N_f$ are important. In this direction we have the following result.

**Proposition 2.78.** *If $\Omega \subset \mathbb{R}^N$ is a nonempty, open subset, $f : \Omega \times \mathbb{R} \to \mathbb{R}$ is a Carathéodory function such that $f(x, \cdot) \in C^1(\mathbb{R})$ for a.a. $x \in \Omega$, and $(x, s) \mapsto f'_s(x, s)$ is a Carathéodory function from $\Omega \times \mathbb{R}$ into $\mathbb{R}$ that satisfies the growth condition*

$$|f'_s(x, s)| \leq a_1(x) + c_1|s|^r \quad \text{for a.a. } x \in \Omega \text{ and all } s \in \mathbb{R},$$

$$\text{and } f(\cdot, 0) \in L^{\frac{r\theta}{r+1}}(\Omega),$$

*where $a_1 \in L^\theta(\Omega)$, $\theta \in (1, +\infty)$, $r \in [\frac{1}{\theta-1}, +\infty)$, $c_1 > 0$, then for $p = r\theta$ and $q = \frac{r\theta}{r+1}$ we have $N_f \in C^1(L^p(\Omega), L^q(\Omega))$ and $N'_f(u)(h) = N_{f'_s}(u)h$ for all $u, h \in L^p(\Omega)$.*

*Proof.* Integrating the growth condition for $f'_s$ and using Young's inequality (e.g., Brezis [52, p. 92]) we obtain

$$|f(x, s)| \leq |f(x, 0)| + \frac{r}{r+1} a_1(x)^{\frac{r+1}{r}} + \frac{1+c_1}{r+1} |s|^{r+1} \quad \text{for a.a. } x \in \Omega, \text{ all } s \in \mathbb{R}.$$

Since $|f(\cdot, 0)| + \frac{r}{r+1} a_1^{\frac{r+1}{r}} \in L^q(\Omega)$, we know from Theorem 2.76 that $N_f$ is continuous from $L^p(\Omega)$ into $L^q(\Omega)$.

Next we show that for all $u, h \in L^p(\Omega)$, $N_{f'_s}(u)h \in L^q(\Omega)$. To this end, using Hölder's inequality we may write

$$\int_\Omega |f'_s(x, u(x))h(x)|^q \, dx \leq \left( \int_\Omega |f'_s(x, u(x))|^{\frac{pq}{p-q}} \, dx \right)^{\frac{p-q}{p}} \left( \int_\Omega |h(x)|^p \, dx \right)^{\frac{q}{p}}. \quad (2.29)$$

Note that $\frac{pq}{p-q} = \theta$ and, by virtue of Theorem 2.76, $N_{f'_s}$ maps $L^p(\Omega)$ continuously into $L^\theta(\Omega)$. Thus, from (2.29) it follows that $N_{f'_s}(u)h \in L^q(\Omega)$.

Fix $u \in L^p(\Omega)$, and let

$$w(h) = N_f(u+h) - N_f(u) - N_{f'_s}(u)h \quad \text{for all } h \in L^p(\Omega). \quad (2.30)$$

Let $h \in L^p(\Omega)$. We have

$$f(x, u(x) + h(x)) - f(x, u(x)) = \int_0^1 \frac{d}{d\lambda} f(x, u(x) + \lambda h(x)) \, d\lambda$$

$$= \int_0^1 f'_s(x, u(x) + \lambda h(x)) h(x) \, d\lambda. \quad (2.31)$$

From (2.30) and (2.31) and using $\frac{q}{p} + \frac{q}{\theta} = 1$, Hölder's inequality, and Fubini's theorem, we obtain

$$\int_{\Omega} |w(h)(x)|^q \, dx$$

$$= \int_{\Omega} \left| \int_{0}^{1} \left( f'_s(x, u(x) + \lambda h(x)) - f'_s(x, u(x)) \right) h(x) \, d\lambda \right|^q dx$$

$$\leq \|h\|_p^q \left( \int_{\Omega} \left| \int_{0}^{1} \left( f'_s(x, u(x) + \lambda h(x)) - f'_s(x, u(x)) \right) d\lambda \right|^{\theta} dx \right)^{\frac{q}{\theta}}$$

$$\leq \|h\|_p^q \left( \int_{0}^{1} \int_{\Omega} \left| f'_s(x, u(x) + \lambda h(x)) - f'_s(x, u(x)) \right|^{\theta} dx d\lambda \right)^{\frac{q}{\theta}}. \qquad (2.32)$$

Since $N_{f'_s} : L^p(\Omega) \to L^{\theta}(\Omega)$ is continuous, from (2.32) it follows that

$$\frac{\|w(h)\|_q}{\|h\|_p} \to 0 \quad \text{as } \|h\|_p \to 0.$$

This shows that $N'_f(u)(h) = N_{f'_s}(u)h$. Then the continuity of the map $N'_f : L^p(\Omega) \to \mathscr{L}(L^p(\Omega), L^q(\Omega))$ easily follows from the continuity of $N_{f'_s} : L^p(\Omega) \to L^{\theta}(\Omega)$. Thus, $N_f \in C^1(L^p(\Omega), L^q(\Omega))$. □

*Remark 2.79.* Since $r > 0$, we have $q < p$. If $N_f : L^p(\Omega) \to L^p(\Omega)$ $(1 \leq p \leq +\infty)$, then $N_f$ is Fréchet differentiable if and only if $f(x,s) = a(x) + c(x)s$ for a.a. $x \in \Omega$ and all $s \in \mathbb{R}$, with $a \in L^p(\Omega)$, $c \in L^{\infty}(\Omega)$.

## 2.4 Remarks

**Section 2.1:** Due to their similarity to finite-dimensional maps, compact operators were the first class of maps used in the study of infinite-dimensional, nonlinear operator equations. Leray and Schauder [222] considered compact perturbations of the identity in order to extend Brouwer's degree to infinite-dimensional Banach spaces. We should mention that for linear operators, the notion of compactness can be traced earlier to the work of Riesz [341], while even earlier Hilbert [171] had introduced the notion of a completely continuous operator. The main property of compact maps is their uniform approximation on bounded sets by finite-rank maps (Theorem 2.7), and it is due to Schauder [350]. Theorem 2.9 is due to Dugundji [122] (see also Dugundji [123, p. 188]). Recall that a subset $E$ of the metric space $(X, d)$ is a *retract* of $X$ if there exists a continuous map $r : X \to E$ such that $r|_E = \mathrm{id}_E$ (i.e., the identity map on $E$ admits a continuous extension on $X$). Theorem 2.9 implies that every closed convex set of a normed space is a retract. This fact turns out to be useful in fixed point theory. Theorem 2.11 is due to Schauder [350], while Theorem 2.19 on the spectral properties of compact linear operators is due to Riesz [341]. Theorem 2.20 (the Fredholm alternative theorem) was proved in the context of linear integral equations by Fredholm [142].

Linear compact operators and their spectral properties are discussed in Akhiezer and Glazman [7], Dunford and Schwartz [124], Halmos [169], Kato [190], and Schechter [354].

**Section 2.2:** Monotone operators, and in particular maximal monotone ones, were introduced in an effort to go beyond compact maps and enlarge the class of problems that can be studied. Their origin can be traced to problems of the calculus of variations. Their systematic study started in the early 1960s and coincided with the advent of the so-called nonsmooth analysis (Chap. 3). The term *monotone operator* is due to Kačurovskiĭ [187], who proved that the subdifferential of a convex function on a Hilbert space is monotone (see also Kačurovskiĭ [188]). However, the real launching of monotone operator theory occurred with the work of Minty [258, 259], where monotone maps in Hilbert spaces were studied in detail and the significance of maximality was brought to light. Soon thereafter, Browder [57] initiated the extension of the theory to reflexive Banach spaces and their duals. The duality map (Definition 2.44) was introduced by Beurling and Livingstone [44] and is a basic tool in the study of evolution equations and the investigation of the geometric properties of Banach spaces. Further properties of the duality can be found in Browder [59], Ciorănescu [82], Gasiński and Papageorgiou [151], and Zeidler [388, 389]. Theorem 2.52 is due to Browder [58] and is useful in producing solutions for nonlinear boundary value problems. Theorem 2.53 was first proved for Hilbert spaces (where $\mathscr{F} = \mathrm{id}$) by Minty [258] and was extended to Banach spaces by Rockafellar [344]. Theorem 2.54 is due to Browder [58]. The notion of pseudomonotonicity (Definition 2.58) was first introduced by Brezis [50] using nets, and soon thereafter Browder [59] provided the sequential definition given here. The importance of this generalization of maximal monotonicity stems from Theorem 2.63, which is due to Browder and Hess [62]. Detailed investigations of pseudomonotone maps can be found in the papers of Browder and Hess [62] and Kenmochi [192, 193]. The notion of $(S)_+$-map was introduced by Browder [59].

Monotone operators and operators of a monotone type are discussed in Barbu [29], Brezis [51], Browder [59], Gasiński and Papageorgiou [151], Papageorgiou and Kyritsi-Yiallourou [318], Pascali and Sburlan [324], Phelps [328], and Zeidler [388, 389].

**Section 2.3:** The Nemytskii operator is the most common nonlinear map and in implicit form can be found in any calculus book under the name *composite function*. Other names used in the literature are *composition operator* and *superposition operator*. Theorem 2.76 is due to Krasnosel'skiĭ [202].

For a detailed study of the Nemytskii map on $L^p$ and Sobolev spaces, consult the books of Appell and Zabreĭko [22], Denkowski et al. [114], and Krasnosel'skiĭ et al. [203] and the papers of Marcus and Mizel [246, 247].

# Chapter 3
# Nonsmooth Analysis

**Abstract** This chapter offers a systematic presentation of nonsmooth analysis containing all that is necessary in this direction for the rest of the book. The first section of the chapter gathers significant results of convex analysis, especially related to the convex subdifferential such as its property of being a maximal monotone operator. The second section has as its main focus the subdifferentiability theory for locally Lipschitz functions. Further information and references are indicated in a remarks section.

## 3.1 Convex Analysis

In this section, we recall a few basic definitions and facts from *convex analysis*.

**Definition 3.1.** Let $X$ be a vector space and $\varphi : X \to \mathbb{R} \cup \{+\infty\}$. We say that $\varphi$ is *convex* if for all $x_1, x_2 \in X$ and for all $\lambda \in [0, 1]$ we have

$$\varphi(\lambda x_1 + (1 - \lambda)x_2) \leq \lambda \varphi(x_1) + (1 - \lambda)\varphi(x_2).$$

We will always consider functions that are not identical to $+\infty$. In the "nonsmooth" literature, such functions are called *proper*. We avoid the use of this name since in nonlinear analysis proper maps are the ones that return compact sets to compact ones.

By $\Gamma_0(X)$ we will denote the cone of convex, lower semicontinuous functions on a Banach space $X$ that are not identically $+\infty$. The *effective domain* of a function $\varphi : X \to \mathbb{R} \cup \{+\infty\}$ is defined by

$$\mathrm{dom}\, \varphi = \{x \in X : \varphi(x) < +\infty\}.$$

The next two theorems summarize the main continuity properties of convex functions. Their proofs can be found in Ekeland and Temam [129, p. 12] and Gasiński and Papageorgiou [151, p. 489].

D. Motreanu et al., *Topological and Variational Methods with Applications to Nonlinear Boundary Value Problems*, DOI 10.1007/978-1-4614-9323-5_3,
© Springer Science+Business Media, LLC 2014

**Theorem 3.2.** *Let $X$ be a Banach space and $\varphi : X \to \mathbb{R} \cup \{+\infty\}$ a convex function. For $x_0 \in \operatorname{dom} \varphi$, the following statements are equivalent:*

(a) *$\varphi$ is bounded above in a neighborhood of $x_0$;*
(b) *$\varphi$ is continuous at $x_0$.*

*Moreover, in each of these cases, $\operatorname{dom} \varphi$ has a nonempty interior and $\varphi$ is locally Lipschitz (Definition 3.20) on $\operatorname{int} \operatorname{dom} \varphi$.*

*Remark 3.3.* The *epigraph* of $\varphi : X \to \mathbb{R} \cup \{+\infty\}$ is defined by

$$\operatorname{epi} \varphi = \{(x, \lambda) \in X \times \mathbb{R} : \ \varphi(x) \leq \lambda\}.$$

If $X$ is a Banach space and $\varphi$ is a convex function, then it is easy to see that the interior of $\operatorname{epi} \varphi$ is nonempty if and only if $\varphi$ is bounded above on some open subset of $X$. Moreover, in this case, one can check (using Theorem 3.2) that $\operatorname{int} \operatorname{epi} \varphi = \{(x, \lambda) \in X \times \mathbb{R} : x \in \operatorname{int} \operatorname{dom} \varphi, \ \varphi(x) < \lambda\}$.

**Theorem 3.4.** *If $X$ is a Banach space and $\varphi \in \Gamma_0(X)$, then $\varphi|_{\operatorname{int} \operatorname{dom} \varphi}$ is locally Lipschitz.*

*Remark 3.5.* If $X$ is finite dimensional, then every convex function $\varphi : X \to \mathbb{R} \cup \{+\infty\}$ is locally Lipschitz on $\operatorname{int} \operatorname{dom} \varphi$.

Convex functions (even continuous ones) are not differentiable in general. For this reason we introduce the following multivalued replacement of the usual derivative.

**Definition 3.6.** Let $X$ be a Banach space and $\varphi : X \to \mathbb{R} \cup \{+\infty\}$ a convex function that is not identically $+\infty$. For $x \in \operatorname{dom} \varphi$ the *subdifferential* of $\varphi$ at $x$ is the set $\partial \varphi(x) \subset X^*$ defined by

$$\partial \varphi(x) = \{x^* \in X^* : \ \varphi(x+h) - \varphi(x) \geq \langle x^*, h \rangle \text{ for all } h \in X\}.$$

The elements of $\partial \varphi(x)$ are called *subgradients* of $\varphi$ at $x$. For $x \notin \operatorname{dom} \varphi$ we set $\partial \varphi(x) = \emptyset$.

*Remark 3.7.* Clearly, $\operatorname{D}(\partial \varphi) := \{x \in X : \ \partial \varphi(x) \neq \emptyset\} \subset \operatorname{dom} \varphi$, and the preceding definition of the subdifferential is equivalent to that given in Example 2.27(d). If $x \in \operatorname{D}(\partial \varphi)$, then we say that $\varphi$ is subdifferentiable at $x$.

It is clear from Definition 3.6 that the following proposition is true.

**Proposition 3.8.** *For every $x \in \operatorname{D}(\partial \varphi)$ the set $\partial \varphi(x)$ is nonempty, convex, and $w^*$-closed.*

**Proposition 3.9.** *If $\varphi : X \to \mathbb{R} \cup \{+\infty\}$ is convex and continuous at $x_0 \in \operatorname{dom} \varphi$, then $\partial \varphi(x_0)$ is nonempty and $w^*$-compact.*

*Proof.* From Theorem 3.2 and Remark 3.3 we know that $\operatorname{int}\operatorname{epi}\varphi \neq \emptyset$ and that $(x_0, \varphi(x_0))$ lies on the boundary of $\operatorname{epi}\varphi$. Then, by the weak separation theorem for convex sets (e.g., Brezis [52, Theorem 1.6]), we can find $(x^*, \xi) \in X^* \times \mathbb{R}$, $(x^*, \xi) \neq (0,0)$ such that

$$\langle x^*, x_0 \rangle + \xi \varphi(x_0) \leq \langle x^*, x \rangle + \xi \lambda \quad \text{for all } (x, \lambda) \in \operatorname{epi}\varphi. \tag{3.1}$$

Since $(x_0, \varphi(x_0) + 1) \in \operatorname{epi}\varphi$, from (3.1) we infer that $\xi \geq 0$. Suppose $\xi = 0$. Then

$$\langle x^*, x - x_0 \rangle \geq 0 \quad \text{for all } x \in \operatorname{dom}\varphi. \tag{3.2}$$

Because $\varphi$ is continuous at $x_0$, we have that $x_0 \in \operatorname{int}\operatorname{dom}\varphi$, and so from (3.2) it follows that for some $\delta > 0$ we have

$$\langle x^*, h \rangle \geq 0 \quad \text{for all } \|h\| \leq \delta,$$

hence $x^* = 0$, which contradicts the fact that $(x^*, \xi) \neq (0,0)$. Therefore $\xi > 0$, and so, from (3.1), we obtain

$$\left\langle \frac{1}{\xi} x^*, x_0 \right\rangle + \varphi(x_0) \leq \left\langle \frac{1}{\xi} x^*, x \right\rangle + \lambda \quad \text{for } (x, \lambda) \in \operatorname{epi}\varphi,$$

hence

$$\left\langle -\frac{1}{\xi} x^*, h \right\rangle \leq \varphi(x_0 + h) - \varphi(x_0) \quad \text{for all } h \in X,$$

and so $-\frac{1}{\xi} x^* \in \partial \varphi(x_0)$ (Definition 3.6). Therefore, we have $\partial \varphi(x_0) \neq \emptyset$. Moreover, since $\varphi$ is continuous at $x_0$, it is bounded above in a neighborhood of $x_0$, and so from Definition 3.6 we infer that $\partial \varphi(x_0)$ is bounded in $X^*$. Since $\partial \varphi(x_0)$ is also $w^*$-closed (Proposition 3.8), by Alaoglu's theorem (e.g., Brezis [52, p. 66]), it is $w^*$-compact. $\square$

Combining Theorem 3.4 and Proposition 3.9, we have the following corollary.

**Corollary 3.10.** *If* $\varphi \in \Gamma_0(X)$, *then* $\operatorname{int}\operatorname{dom}\varphi \subset D(\partial\varphi)$.

The following remark points out two noticeable properties of the convex subdifferential (see Ekeland and Temam [129, p. 26] and Gasiński and Papageorgiou [151, p. 528]).

*Remark 3.11.*

(a) For a convex function $\varphi : X \to \mathbb{R} \cup \{+\infty\}$ that is continuous at $x_0 \in \operatorname{dom}\varphi$, we have $\varphi'(x_0; h) = \sigma_{\partial\varphi(x_0)}(h)$ for all $h \in X$, where $\varphi'(x_0; h) = \lim_{\lambda \downarrow 0} \frac{\varphi(x_0 + \lambda h) - \varphi(x_0)}{\lambda}$ and $\sigma_{\partial\varphi(x_0)}(h) = \sup\{\langle x^*, h \rangle : x^* \in \partial\varphi(x_0)\}$.

(b) If $\varphi_1, \varphi_2 : X \to \mathbb{R} \cup \{+\infty\}$ are both convex, then $\partial(\varphi_1 + \varphi_2)(x) \subset \partial\varphi_1(x) + \partial\varphi_2(x)$ for every $x \in D(\partial\varphi_1) \cap D(\partial\varphi_2)$. Moreover, in the case where there exists a point in $\mathrm{dom}\,\varphi_1 \cap \mathrm{dom}\,\varphi_2$ where one of the two functions is continuous, then $\partial(\varphi_1 + \varphi_2)(x) = \partial\varphi_1(x) + \partial\varphi_2(x)$ for every $x \in D(\partial\varphi_1) \cap D(\partial\varphi_2)$.

Now we emphasize the link between convex subdifferentials and Gâteaux differentials.

**Proposition 3.12.** *If $\varphi : X \to \mathbb{R} \cup \{+\infty\}$ is convex, then:*

(a) *If $\varphi$ is Gâteaux differentiable at $x_0 \in \mathrm{int}\,\mathrm{dom}\,\varphi$, then $\partial\varphi(x_0) = \{\varphi'(x_0)\}$.*
(b) *If $\varphi$ is continuous at $x_0 \in \mathrm{dom}\,\varphi$ and $\partial\varphi(x_0)$ is a singleton, then $\varphi$ is Gâteaux differentiable at $x_0$ and $\partial\varphi(x_0) = \{\varphi'(x_0)\}$.*

*Proof.* (a) For every $0 < \lambda \leq 1$ and every $h \in X$ we have, by the convexity of $\varphi$, that

$$\frac{1}{\lambda}\left(\varphi(x_0 + \lambda h) - \varphi(x_0)\right) \leq \varphi(x_0 + h) - \varphi(x_0).$$

This entails

$$\langle \varphi'(x_0), h \rangle \leq \varphi(x_0 + h) - \varphi(x_0) \ \text{ for all } h \in X,$$

so $\varphi'(x_0) \in \partial\varphi(x_0)$ (Definition 3.6).

On the other hand, if $x^* \in \partial\varphi(x_0)$, then

$$\langle x^*, h \rangle \leq \frac{1}{\lambda}\left(\varphi(x_0 + \lambda h) - \varphi(x_0)\right) \ \text{ for all } \lambda > 0 \text{ and all } h \in X,$$

hence $\langle x^* - \varphi'(x_0), h \rangle \leq 0$ for all $h \in X$, that is, $x^* = \varphi'(x_0)$. Therefore, we obtain $\partial\varphi(x_0) = \{\varphi'(x_0)\}$.

(b) Since, by hypothesis, $\partial\varphi(x_0) = \{x^*\}$, from Remark 3.11(a) we have $\varphi'(x_0; h) = \langle x^*, h \rangle$ for all $h \in X$. Hence $\varphi$ is Gâteaux differentiable at $x_0$ with $\varphi'(x_0) = x^*$. $\square$

The next result is an easy consequence of Definition 3.6 and plays a central role in variational analysis (it is the nonsmooth version of Fermat's rule).

**Proposition 3.13.** *If $\varphi : X \to \mathbb{R} \cup \{+\infty\}$ is convex and not identically $+\infty$, then the global minimum of $\varphi$ on $X$ is attained at $x_0 \in X$ if and only if $0 \in \partial\varphi(x_0)$.*

We recall the following useful result.

**Lemma 3.14.** *If $Z$ is a locally convex topological vector space and $\varphi \in \Gamma_0(Z)$, then $\varphi$ is bounded below by an affine, continuous function.*

It is clear from Definition 3.6 that the subdifferential map $\partial\varphi : X \to 2^{X^*}$ is monotone. Furthermore, we have the following theorem.

**Theorem 3.15.** *If $X$ is a reflexive Banach space and $\varphi \in \Gamma_0(X)$, then $\partial\varphi : X \to 2^{X^*}$ is maximal monotone.*

*Proof.* We may assume that $X$ and $X^*$ are both strictly convex [Remark 2.47(b)]. Let $x^* \in X^*$, and let

$$\psi(x) = \frac{1}{2}\|x\|^2 + \varphi(x) - \langle x^*, x\rangle \quad \text{for all } x \in X.$$

Evidently, $\psi \in \Gamma_0(X)$. Consequently, $\psi$ is minorized by an affine, continuous function (Lemma 3.14), so it is coercive [i.e., $\psi(x) \to +\infty$ as $\|x\| \to +\infty$]. Thus, we can find $x_0 \in X$ such that

$$\psi(x_0) = \inf_X \psi.$$

Then Proposition 3.13 implies that $0 \in \partial\psi(x_0) = \mathscr{F}(x_0) + \partial\varphi(x_0) - x^*$ [Remark 3.11(b)], hence $x^* \in \partial\varphi(x_0) + \mathscr{F}(x_0)$. Since $x^* \in X^*$ is arbitrary, we infer that $R(\partial\varphi + \mathscr{F}) = X^*$, which by Theorem 2.53 ensures the maximal monotonicity of $\partial\varphi$. □

**Corollary 3.16.** *If $X$ is a reflexive Banach space and $\varphi \in \Gamma_0(X)$, then $D(\partial\varphi)$ is dense in $\operatorname{dom}\varphi$.*

*Proof.* We can assume that $X$ and $X^*$ are strictly convex [Remark 2.47(b)]. Let $x \in \operatorname{dom}\varphi$ and, for every $\lambda > 0$, let $x_\lambda \in D(\partial\varphi)$ be the unique solution of the operator inclusion

$$0 \in \lambda\partial\varphi(x_\lambda) + \mathscr{F}(x_\lambda - x)$$

[Theorems 2.53 and 2.48(b)]. Using Definition 3.6 we obtain

$$\|x_\lambda - x\|^2 + \lambda\varphi(x_\lambda) \le \lambda\varphi(x). \tag{3.3}$$

Since $\varphi \in \Gamma_0(X)$, it is minorized by an affine, continuous function (Lemma 3.14). Hence, from (3.3) we infer that $x_\lambda \to x$ as $\lambda \downarrow 0$. Because $x \in \operatorname{dom}\varphi$ is arbitrary, we conclude that $\operatorname{dom}\varphi \subset \overline{D(\partial\varphi)}$. □

*Remark 3.17.* Both Theorem 3.15 and Corollary 3.16 are true for any Banach space $X$ that is not necessarily reflexive. For proofs we refer the reader to Brønsted and Rockafellar [55] and Rockafellar [345].

Finally, let $H$ be a Hilbert space and $\varphi \in \Gamma_0(H)$. For each $\lambda > 0$ we define

$$\varphi_\lambda(x) = \inf_{y \in H}\left(\varphi(y) + \frac{1}{2\lambda}\|x - y\|^2\right) \quad \text{for all } x \in H. \tag{3.4}$$

The function $\varphi_\lambda$ is called the *Moreau–Yosida regularization of* $\varphi$. In the next theorem we summarize the main properties of this function (Brezis [51, p. 39]). We denote by $J_\lambda^{\partial\varphi}$ and $(\partial\varphi)_\lambda$ the resolvent of $\partial\varphi$ and the Yosida approximation of $\partial\varphi$, respectively (Sect. 2.2).

**Theorem 3.18.**

(a)  $\varphi_\lambda$ *is continuous, convex, and* $\mathbb{R}$*-valued (i.e.,* $\mathrm{dom}\,\varphi_\lambda = H$*).*

(b)  $\varphi_\lambda(x) = \varphi(J_\lambda^{\partial\varphi}(x)) + \frac{1}{2\lambda}\|x - J_\lambda^{\partial\varphi}(x)\|^2.$

(c)  $\partial\varphi_\lambda = (\partial\varphi)_\lambda$*, and so* $\varphi_\lambda$ *is Fréchet differentiable.*

(d)  $\varphi_\lambda(x) \leq \varphi(x)$ *for all* $x \in H$ *and* $\varphi_\lambda(x) \to \varphi(x)$ *as* $\lambda \downarrow 0$.

(e)  *If* $\lambda_n \downarrow 0$, $x_n \to x$ *in* $H$*, and* $\partial\varphi_{\lambda_n}(x_n) \xrightarrow{w} x^*$ *in* $H$*, then* $(x, x^*) \in \mathrm{Gr}\,\partial\varphi$.

(f)  *If* $\lambda_n \downarrow 0$ *and* $x_n \to x$ *in* $H$*, then* $\varphi(x) \leq \liminf\limits_{n\to\infty} \varphi_{\lambda_n}(x_n)$*; moreover, if in addition* $\{\partial\varphi_{\lambda_n}(x_n)\}_{n\geq 1}$ *is bounded in* $H$*, then* $\varphi_{\lambda_n}(x_n) \to \varphi(x)$.

(g)  $\varphi_\lambda(x) \leq \varphi_\lambda(y) + \|\partial\varphi_\lambda(x)\|\,\|x - y\|$ *for all* $x, y \in H$.

(h)  $|\varphi_\lambda(x) - \varphi_\lambda(y)| \leq \left(2\|\partial\varphi_\lambda(z)\| + \frac{\|x-z\|}{\lambda} + \frac{\|y-z\|}{\lambda}\right)\|x-y\|$ *for all* $x, y, z \in H$.

*Remark 3.19.* If $X$ is reflexive, $X, X^*$ are locally uniformly convex, and $\varphi \in \Gamma_0(X)$, then for every $\lambda > 0$ we can still define the Moreau–Yosida regularization $\varphi_\lambda$ of $\varphi$ by (3.4), and we have (Barbu [30, p. 65]) that

(a)  $\varphi_\lambda$ is continuous, convex, and $\mathbb{R}$-valued;

(b)  $\varphi_\lambda(x) = \varphi(J_\lambda^{\partial\varphi}(x)) + \frac{1}{2\lambda}\|x - J_\lambda^{\partial\varphi}(x)\|^2$;

(c)  $\partial\varphi_\lambda = (\partial\varphi)_\lambda$, and so $\varphi_\lambda$ is Gâteaux differentiable;

(d)  $\varphi_\lambda(x) \leq \varphi(x)$ for all $x \in X$ and $\varphi_\lambda(x) \to \varphi(x)$ as $\lambda \downarrow 0$.

## 3.2  Locally Lipschitz Functions

Now we want to extend the subdifferentiability theory of continuous, convex functions to nonconvex functions. Theorem 3.2 suggests that we should consider locally Lipschitz functions.

As before, the setting is a Banach space $X$, with $X^*$ its topological dual and $\langle \cdot, \cdot \rangle$ the duality brackets for the pair $(X^*, X)$.

**Definition 3.20.** A function $\varphi : X \to \mathbb{R}$ is *locally Lipschitz* if for every $x \in X$ we can find an open neighborhood $U(x)$ of $x$ and a constant $k(x) > 0$ such that

$$|\varphi(y) - \varphi(z)| \leq k(x)\|y - z\| \quad \text{for all } y, z \in U(x). \tag{3.5}$$

If (3.5) is satisfied for all $y, z \in X$ and for $k(x) = k > 0$ independent of $x \in X$, then we say that $\varphi$ is *Lipschitz continuous* (or simply *Lipschitz*).

*Remark 3.21.* Clearly, if $\varphi : X \to \mathbb{R}$ is Lipschitz on every bounded set in $X$, then $\varphi$ is locally Lipschitz. Moreover, if $X$ is finite dimensional, then the converse is also true. If $\varphi : X \to \mathbb{R}$ is continuous and convex, or $\varphi \in C^1(X, \mathbb{R})$, then $\varphi$ is locally Lipschitz.

**Definition 3.22.** Let $\varphi : X \to \mathbb{R}$ be a locally Lipschitz function. The *generalized directional derivative of $\varphi$ at $x \in X$ in the direction $h \in X$*, denoted by $\varphi^0(x; h)$, is defined by

$$\varphi^0(x; h) = \limsup_{\substack{x' \to x \\ \lambda \downarrow 0}} \frac{\varphi(x' + \lambda h) - \varphi(x')}{\lambda}.$$

In the next proposition we list the basic properties of $\varphi^0(x; h)$. They are direct consequences of the preceding definition.

**Proposition 3.23.**

(a) $\varphi^0(x; \cdot)$ *is sublinear (i.e., subadditive, positively homogeneous) and Lipschitz.*
(b) $(x, h) \mapsto \varphi^0(x; h)$ *is upper semicontinuous on $X \times X$.*
(c) $\varphi^0(x; -h) = (-\varphi)^0(x; h)$ *for all $x, h \in X$.*

The main definition of this section is as follows.

**Definition 3.24.** For a locally Lipschitz function $\varphi : X \to \mathbb{R}$, the *generalized subdifferential* of $\varphi$ at $x \in X$ is the set $\partial\varphi(x)$ defined by

$$\partial\varphi(x) = \{x^* \in X^* : \langle x^*, h \rangle \leq \varphi^0(x; h) \text{ for all } h \in X\}.$$

*Remark 3.25.*

(a) Part (a) of Proposition 3.23, together with the Hahn–Banach theorem (e.g., Brezis [52, p. 1]), implies that $\partial\varphi(x)$ is nonempty for every $x \in X$.
(b) Note that the multifunction $\partial\varphi : X \to 2^{X^*}$ is locally bounded. Indeed, given $x \in X$, we know that $\varphi$ is Lipschitz in a neighborhood $U(x)$ of $x$ with Lipschitz constant $k(x) > 0$. This clearly implies that $\varphi^0(y; h) \leq k(x)\|h\|$ for all $y \in U(x)$, all $h \in X$, whence $\|y^*\| \leq k(x)$ for all $y \in U(x)$, all $y^* \in \partial\varphi(y)$.
(c) It is clear from Definition 3.24 that $\partial\varphi(x)$ equals the convex subdifferential of $\varphi^0(x; \cdot)$ at $0$ (Definition 3.6). This observation leads to the following proposition.

**Proposition 3.26.** *If $\varphi : X \to \mathbb{R}$ is locally Lipschitz, then:*

(a) *For every $x \in X$, $\partial\varphi(x) \subset X^*$ is convex and w\*-compact;*
(b) *For every $x \in X$, $\varphi^0(x; \cdot) = \sigma_{\partial\varphi(x)}(\cdot)$, where $\sigma_{\partial\varphi(x)}(h) = \sup\{\langle x^*, h \rangle : x^* \in \partial\varphi(x)\}$ for all $h \in X$.*

If $\varphi, \psi : X \to \mathbb{R}$ are locally Lipschitz functions, then it is clear from Definition 3.20 that

$$(\varphi + \psi)^0(x; h) \leq \varphi^0(x; h) + \psi^0(x; h) \quad \text{for all } x, h \in X. \tag{3.6}$$

Definition 3.24 and (3.6) lead to the following result.

**Proposition 3.27.** *If $\varphi, \psi : X \to \mathbb{R}$ are locally Lipschitz, then*

(a) $\partial(\varphi + \psi)(x) \subset \partial\varphi(x) + \partial\psi(x)$ *for all $x \in X$, with equality if one of the sets $\partial\varphi(x)$, $\partial\psi(x)$ is a singleton;*
(b) $\partial(\lambda\varphi)(x) = \lambda\partial\varphi(x)$ *for all $\lambda \in \mathbb{R}$, all $x \in X$.*

**Proposition 3.28.** *If $\varphi : X \to \mathbb{R}$ is locally Lipschitz, then $\partial\varphi : X \to 2^{X^*}$ is u.s.c. from $X$ with the norm topology into $X^*$ with the $w^*$-topology (Definition 2.36).*

*Proof.* Let $C \subset X^*$ be nonempty and $w^*$-closed. Let $\{x_n\}_{n\geq 1} \subset (\partial\varphi)^-(C)$ such that $x_n \to x$ in $X$. Let $x_n^* \in \partial\varphi(x_n) \cap C$. Then, in view of Remark 3.25, $\{x_n^*\}_{n\geq 1}$ is bounded in $X^*$. By Alaoglu's theorem (e.g., Brezis [52, p. 66]), we can find a subnet $\{x_\alpha^*\}_{\alpha \in J}$ of $\{x_n^*\}_{n\geq 1}$ such that $x_\alpha^* \xrightarrow{w^*} x^*$ in $X^*$. We have

$$\langle x_\alpha^*, h \rangle \leq \varphi^0(x_\alpha; h) \text{ for all } h \in X, \text{ all } \alpha \in J.$$

By Proposition 3.23(b), we deduce that

$$\langle x^*, h \rangle \leq \limsup \varphi^0(x_\alpha; h) \leq \varphi^0(x; h) \text{ for all } h \in X.$$

Thus, $x^* \in \partial\varphi(x)$ (Definition 3.24) and, since $x^* \in C$, we get $x \in (\partial\varphi)^-(C)$. This proves that the set $(\partial\varphi)^-(C)$ is norm closed, and so $\partial\varphi(\cdot)$ is u.s.c., as claimed. $\quad\square$

Combining Propositions 3.26(a) and 3.28 with Remark 2.37, we deduce the following corollary.

**Corollary 3.29.** *If $\varphi : X \to \mathbb{R}$ is locally Lipschitz, then $\mathrm{Gr}\,\partial\varphi \subset X \times X^*$ is closed when $X$ is equipped with the norm topology and $X^*$ is equipped with the $w^*$-topology.*

It is natural to ask what the relation is between the convex and the generalized subdifferentials. From Remark 3.11(a) and Proposition 3.26(b), we infer the following proposition.

**Proposition 3.30.** *If $\varphi : X \to \mathbb{R}$ is continuous and convex (and hence locally Lipschitz, see Theorem 3.2), then the convex subdifferential (Definition 3.6) and the generalized subdifferential (Definition 3.24) coincide.*

The generalized subdifferential has a rich calculus, and many results of the smooth calculus extend to the generalized subdifferential. First we mention a *mean value theorem.*

**Proposition 3.31.** *If $\varphi : X \to \mathbb{R}$ is locally Lipschitz and $x, y \in X$, then we can find $u = x + t_0(y - x)$, with $t_0 \in (0, 1)$, and $u^* \in \partial\varphi(u)$ such that*

$$\varphi(y) - \varphi(x) = \langle u^*, y - x \rangle.$$

*Proof.* Let $\xi : \mathbb{R} \to \mathbb{R}$ be defined by $\xi(t) = \varphi(x + t(y - x))$. Clearly, $\xi$ is locally Lipschitz. We claim that

$$\partial \xi(t) \subset \{\langle x^*, y - x \rangle : x^* \in \partial \varphi(x + t(y - x))\} =: C_t \quad \text{for all } t \in [0, 1].$$

Since both sets are closed and convex in $\mathbb{R}$, to prove the claim, it suffices to show the corresponding inequality for the support functions, namely, that

$$\sigma_{\partial \xi(t)}(\pm 1) \leq \sigma_{C_t}(\pm 1). \tag{3.7}$$

To this end, we have

$$\xi^0(t; \pm 1) = \limsup_{\substack{s \to t \\ \lambda \downarrow 0}} \frac{\xi(s + \lambda(\pm 1)) - \xi(s)}{\lambda}$$

$$= \limsup_{\substack{s \to t \\ \lambda \downarrow 0}} \frac{\varphi(x + (s + \lambda(\pm 1))(y - x)) - \varphi(x + s(y - x))}{\lambda}$$

$$\leq \limsup_{\substack{v \to x + t(y - x) \\ \lambda \downarrow 0}} \frac{\varphi(v + \lambda(\pm 1)(y - x)) - \varphi(v)}{\lambda}$$

$$= \varphi^0(x + t(y - x); \pm(y - x)). \tag{3.8}$$

From (3.8) and Proposition 3.26(b) we infer that (3.7) holds, and this proves the claim.

Now let $\theta : \mathbb{R} \to \mathbb{R}$ be the locally Lipschitz function defined by $\theta(t) = \xi(t) + t(\varphi(x) - \varphi(y))$. Note that $\theta(0) = \theta(1) = \varphi(x)$, and so we can find $t_0 \in (0, 1)$ at which $\theta$ has a local extremum. Then it is clear from Definitions 3.22 and 3.24 that $0 \in \partial \theta(t_0)$. Hence we obtain

$$0 \in \partial \xi(t_0) + \varphi(x) - \varphi(y) \subset \{\langle x^*, y - x \rangle : x^* \in \partial \varphi(x + t_0(y - x))\} + \varphi(x) - \varphi(y)$$

(see Proposition 3.27 and the claim). Thus, we conclude that for some $u^* \in \partial \varphi(x + t_0(y - x))$ we have $\langle u^*, y - x \rangle = \varphi(y) - \varphi(x)$. □

From the calculus with generalized gradients, particularly useful are the chain rules. In the next proposition, we present two results in this direction. For the proofs we refer the reader to Clarke [85, pp. 42–45]. First, a definition.

**Definition 3.32.** Let $\varphi : X \to \mathbb{R}$ be locally Lipschitz. We say that $\varphi$ is *regular at* $x$ if $\varphi^0(x; \cdot) = \varphi'(x; \cdot)$, where $\varphi'(x; \cdot)$ is the usual one-sided directional derivative defined by

$$\varphi'(x; h) = \lim_{\lambda \downarrow 0} \frac{\varphi(x + \lambda h) - \varphi(x)}{\lambda}.$$

*Remark 3.33.* Continuous, convex functions and $C^1$-functions are regular at $x \in X$ (see also Proposition 3.45).

**Proposition 3.34.**

(a) *If $Y$ is another Banach space, $\varphi \in C^1(X,Y)$, $g : Y \to \mathbb{R}$ is a locally Lipschitz function, and $\psi = g \circ \varphi : X \to \mathbb{R}$, then $\psi$ is locally Lipschitz, and for all $x \in X$ we have*

$$\partial \psi(x) \subset \partial g(\varphi(x)) \circ \varphi'(x);$$

*equality holds if either $g$ (or $-g$) is regular at $\varphi(x)$ or $\varphi$ maps every neighborhood of $x$ to a set that is dense in a neighborhood of $\varphi(x)$.*

(b) *If $\varphi = (\varphi_1, \dots, \varphi_N) : X \to \mathbb{R}^N$ and $g : \mathbb{R}^N \to \mathbb{R}$ are locally Lipschitz and $\psi = g \circ \varphi$, then $\psi : X \to \mathbb{R}$ is locally Lipschitz, and for all $x \in X$ we have*

$$\partial \psi(x) \subset \overline{\mathrm{conv}}^{w^*} \big( \partial g(\varphi(x))(\partial \varphi_1(x) \times \cdots \times \partial \varphi_N(x)) \big).$$

*Moreover, in the case where $N = 1$, if $g \in C^1(\mathbb{R})$, or if $g$ (or $-g$) is regular at $\varphi(x)$ and $\varphi \in C^1(X,\mathbb{R})$, then the $\overline{\mathrm{conv}}^{w^*}$ is superfluous and equality holds.*

**Corollary 3.35.** *Let $\varphi_1, \varphi_2 : X \to \mathbb{R}$ be locally Lipschitz. Then $\varphi_1 \varphi_2 : X \to \mathbb{R}$ is locally Lipschitz, and we have*

$$\partial(\varphi_1 \varphi_2)(x) \subset \varphi_1(x) \partial \varphi_2(x) + \varphi_2(x) \partial \varphi_1(x) \ \text{for all } x \in X.$$

*Proof.* Apply Proposition 3.34(b) to the functions $\varphi : X \to \mathbb{R}^2$ given by $\varphi(x) = (\varphi_1(x), \varphi_2(x))$ and $g : \mathbb{R}^2 \to \mathbb{R}$ given by $g(s_1, s_2) = s_1 s_2$, taking Proposition 3.26(a) into account. $\quad\square$

Another useful consequence of Proposition 3.34(a) is the following corollary.

**Corollary 3.36.** *If $Y$ is another Banach space, $X$ is embedded continuously and densely in $Y$, $\varphi : Y \to \mathbb{R}$ is locally Lipschitz, and $\hat\varphi = \varphi|_X$, then $\partial \hat\varphi(x) = \partial \varphi(x)$ for all $x \in X$.*

*Proof.* Apply Proposition 3.34(a), with $g : X \to Y$ being the embedding of $X$ into $Y$. $\quad\square$

**Proposition 3.37.** *If $\eta \in C^1([0,1],X)$ and $\varphi : X \to \mathbb{R}$ is locally Lipschitz, then $\theta := \varphi \circ \eta : [0,1] \to \mathbb{R}$ is differentiable a.e., and for all $t \in [0,1]$ at which $\theta$ is differentiable we have*

$$\theta'(t) \leq \max \big\{ \langle x^*, \eta'(t) \rangle : x^* \in \partial \varphi(\eta(t)) \big\}.$$

*Proof.* Evidently, $\theta$ is locally Lipschitz and so differentiable a.e. Let $t \in (0,1)$ be a point of differentiability of $\theta$. Since $\varphi$ is locally Lipschitz, we have

$$\theta'(t) = \lim_{\lambda \downarrow 0} \frac{1}{\lambda} \big( \varphi(\eta(t+\lambda)) - \varphi(\eta(t)) \big)$$

$$= \lim_{\lambda \downarrow 0} \frac{1}{\lambda} \big( \varphi(\eta(t) + \lambda \eta'(t) + o(\lambda)) - \varphi(\eta(t)) \big)$$

$$= \lim_{\lambda \downarrow 0} \frac{1}{\lambda} \big( \varphi(\eta(t) + \lambda \eta'(t)) - \varphi(\eta(t)) \big)$$

$$\leq \limsup_{\substack{h \to 0 \\ \lambda \downarrow 0}} \frac{1}{\lambda} \big( \varphi(\eta(t) + h + \lambda \eta'(t)) - \varphi(\eta(t) + h) \big)$$

$$= \varphi^0(\eta(t); \eta'(t)) = \max\{ \langle x^*, \eta'(t) \rangle : x^* \in \partial \varphi(\eta(t)) \},$$

where $o(\lambda) \to 0$ as $\lambda \downarrow 0$ [Proposition 3.26(b)].                      □

**Definition 3.38.** Given a locally Lipschitz function $\varphi : X \to \mathbb{R}$, a point $x \in X$ is a *critical point* of $\varphi$ if $0 \in \partial \varphi(x)$.

*Remark 3.39.* Evidently, $x \in X$ is a critical point of $\varphi$ if and only if $\varphi^0(x;h) \geq 0$ for all $h \in X$.

From Definition 3.22 we infer the following result.

**Proposition 3.40.** *If $\varphi : X \to \mathbb{R}$ is locally Lipschitz and $x \in X$ is a local extremum of $\varphi$ (i.e., a local minimum or a local maximum of $\varphi$), then $x$ is a critical point of $\varphi$.*

If $\varphi : X \to \mathbb{R}$ is locally Lipschitz, then we define

$$m_\varphi(x) = \inf\{ \|x^*\| : x^* \in \partial \varphi(x) \}. \tag{3.9}$$

This quantity plays an important role in the critical point theory for locally Lipschitz functions.

**Proposition 3.41.** $m_\varphi : X \to \mathbb{R}$ *is lower semicontinuous.*

*Proof.* Since the norm functional on the dual of a Banach space is $w^*$ lower semicontinuous and the set $\partial \varphi(x) \subset X^*$ is $w^*$-compact [Proposition 3.26(a)], the infimum in (3.9) is attained.

Suppose $x_n \to x$ in $X$ and $m_\varphi(x_n) \leq \lambda$ for all $n \geq 1$ and some $\lambda \in \mathbb{R}_+$. Let $x_n^* \in \partial \varphi(x_n)$ such that $m_\varphi(x_n) = \|x_n^*\|$. By Alaoglu's theorem (e.g., Brezis [52, p. 66]), there is a subnet $\{x_\alpha^*\}_{\alpha \in J}$ of $\{x_n^*\}_{n \geq 1}$ such that $x_\alpha^* \xrightarrow{w^*} x^*$ in $X^*$. Corollary 3.29 implies that $x^* \in \partial \varphi(x)$. Also $\|x^*\| \leq \liminf \|x_\alpha^*\| = \liminf m_\varphi(x_\alpha) \leq \lambda$, hence $m_\varphi(x) \leq \lambda$. This proves the lower semicontinuity of $m_\varphi$.                      □

**Proposition 3.42.** *If $\varphi : X \to \mathbb{R}$ is locally Lipschitz, $x \in X$, and $\varepsilon > 0$, then the following statements are equivalent:*

(a) $0 \leq \varphi^0(x;h) + \varepsilon\|h\|$ for all $h \in X$;
(b) $0 \in \partial\varphi(x) + \varepsilon\overline{B_1}^*$, where $\overline{B_1}^* = \{x^* \in X^* : \|x^*\| \leq 1\}$;
(c) $m_\varphi(x) \leq \varepsilon$.

*Proof.* (a) $\Rightarrow$ (b): By virtue of Proposition 3.26(b), for all $h \in X$ we have

$$0 \leq \varphi^0(x;h) + \varepsilon\|h\| = \sigma_{\partial\varphi(x)}(h) + \varepsilon\sigma_{\overline{B_1}^*}(h)$$

$$= \sigma_{\partial\varphi(x)}(h) + \sigma_{\varepsilon\overline{B_1}^*}(h) = \sigma_{\partial\varphi(x)+\varepsilon\overline{B_1}^*}(h);$$

thus, $0 \in \partial\varphi(x) + \varepsilon\overline{B_1}^*$.

(b) $\Rightarrow$ (c): By hypothesis, we can find $x^* \in \partial\varphi(x)$ and $u^* \in \overline{B_1}^*$ such that $x^* + \varepsilon u^* = 0$, so $m_\varphi(x) \leq \|x^*\| \leq \varepsilon$ [see (3.9)].

(c) $\Rightarrow$ (a): Let $x^* \in \partial\varphi(x)$ be such that $m_\varphi(x) = \|x^*\|$. By hypothesis, we have $\|x^*\| \leq \varepsilon$, and so $0 \leq \langle x^*, h\rangle + \varepsilon\|h\| \leq \varphi^0(x;h) + \varepsilon\|h\|$ for all $h \in X$. $\qquad\square$

We know that for continuous, convex functions, convex and generalized subdifferentials coincide (Proposition 3.30). The next proposition provides a criterion for convexity in terms of the generalized subdifferential.

**Proposition 3.43.** *If $\varphi : X \to \mathbb{R}$ is locally Lipschitz, then $\varphi$ is convex if and only if the generalized subdifferential $\partial\varphi : X \to 2^{X^*}$ is monotone.*

*Proof.* $\Rightarrow$: This implication follows from Proposition 3.30 and Definition 3.6.

$\Leftarrow$: Let $\lambda \in (0,1)$ and $x,y \in X$. By virtue of Proposition 3.31, we can find $u^* \in \partial\varphi(u)$, with $u = \tau y + (1 - \tau)(\lambda x + (1 - \lambda)y)$, for some $\tau \in (0,1)$ such that

$$\varphi(y) - \varphi(\lambda x + (1 - \lambda)y) = \lambda\langle u^*, y - x\rangle. \tag{3.10}$$

Similarly, we can find $v^* \in \partial\varphi(v)$, with $v = \eta x + (1 - \eta)(\lambda x + (1 - \lambda)y)$, for some $\eta \in (0,1)$ such that

$$\varphi(x) - \varphi(\lambda x + (1 - \lambda)y) = (1 - \lambda)\langle v^*, x - y\rangle. \tag{3.11}$$

Note that

$$v - u = (\eta(1 - \lambda) + \tau\lambda)(x - y). \tag{3.12}$$

Using (3.10)–(3.12) and the monotonicity of $\partial\varphi$, we have

$$\lambda\varphi(x) + (1 - \lambda)\varphi(y) - \varphi(\lambda x + (1 - \lambda)y)$$

$$= \lambda(\varphi(x) - \varphi(\lambda x + (1 - \lambda)y)) + (1 - \lambda)(\varphi(y) - \varphi(\lambda x + (1 - \lambda)y))$$

$$= \lambda(1-\lambda)\langle v^* - u^*, x-y \rangle$$
$$= \lambda(1-\lambda)(\eta(1-\lambda) + \tau\lambda)^{-1}\langle v^* - u^*, v-u \rangle \geq 0,$$

so the convexity holds. □

In general, the Fréchet differentiability at $x \in X$ of a locally Lipschitz function does not imply that $\partial\varphi(x)$ is a singleton.

*Example 3.44.* Let $\varphi : \mathbb{R} \to \mathbb{R}$ be defined by $\varphi(x) = x^2 \sin\frac{1}{x}$. Then $\varphi$ is locally Lipschitz, differentiable at $x = 0$ with $\varphi'(0) = 0$, and $\varphi^0(0; h) = |h|$ for all $h \in \mathbb{R}$. Hence $\partial\varphi(0) = [-1, 1]$.

However, we have (see Clarke [85, pp. 32–33]) the following proposition.

**Proposition 3.45.** *If $\varphi \in C^1(X, \mathbb{R})$, then $\partial\varphi(x) = \{\varphi'(x)\}$ for all $x \in X$.*

In fact, in finite-dimensional spaces, we can strengthen the preceding result.

**Proposition 3.46.** *If $X$ is a finite-dimensional Banach space and $\varphi : X \to \mathbb{R}$ is locally Lipschitz, then $\partial\varphi$ is single-valued if and only if $\varphi \in C^1(X, \mathbb{R})$.*

*Remark 3.47.* We recall *Rademacher's theorem*, which says that if $\varphi : \mathbb{R}^m \to \mathbb{R}^k$ is a locally Lipschitz function, then $\varphi$ is differentiable almost everywhere (with respect to the Lebesgue measure). Thus, if $\varphi : \mathbb{R}^m \to \mathbb{R}$ is locally Lipschitz and $E_\varphi \subset \mathbb{R}^m$ is the Lebesgue-null set where $\varphi$ fails to be differentiable, then, using Rademacher's theorem, we have the following more intuitive (and geometric) characterization of the generalized subdifferential (see Clarke [85, p. 63]).

**Proposition 3.48.** *If $\varphi : \mathbb{R}^m \to \mathbb{R}$ is locally Lipschitz and $D \subset \mathbb{R}^m$ is a Lebesgue-null set, then $\partial\varphi(x) = \operatorname{conv}\{\lim\varphi'(x_n) : x_n \to x, x_n \notin D \cup E_\varphi, \lim\varphi'(x_n)$ exists$\}$.*

Finally, we conclude with a result on the generalized subdifferential of certain integral functionals. Let $(\Omega, \Sigma, \mu)$ be a finite measure space, $X$ a separable Banach space, and $f : \Omega \times X \to \mathbb{R}$ a function such that for all $\xi \in X$, $x \mapsto f(x, \xi)$ is $\Sigma$-measurable and, for a.a. $x \in \Omega$, $\xi \mapsto f(x, \xi)$ is locally Lipschitz. Let $p \in (1, +\infty]$, and we assume that either

(a)  $p \in (1, +\infty)$ and there is a constant $c_0 > 0$ such that

$$\|\xi^*\| \leq c_0(1 + \|\xi\|^{p-1}) \quad \text{for } \mu\text{-a.e. } x \in \Omega, \text{ all } \xi \in X, \text{ all } \xi^* \in \partial_\xi f(x, \xi)$$

or

(b)  $p = +\infty$ and there are $\alpha \in L^1(\Omega)_+$ and $\eta : [0, +\infty) \to [0, +\infty)$ nondecreasing such that

$$\|\xi^*\| \leq \alpha(x)(1 + \eta(\|\xi\|)) \quad \text{for } \mu\text{-a.e. } x \in \Omega, \text{ all } \xi \in X, \text{ all } \xi^* \in \partial_\xi f(x, \xi),$$

where $\partial_\xi f(x, \xi)$ stands for the generalized subdifferential of $\xi \mapsto f(x, \xi)$.

We consider the integral functional

$$I_f(u) = \int_\Omega f(x, u(x)) \, d\mu \quad \text{for all } u \in L^p(\Omega, X).$$

The assumption (together with Proposition 3.31) implies that this integral is well defined. We have (see Clarke [85, pp. 80, 83]) the following proposition.

**Proposition 3.49.** *Assume that f satisfies* (a) *or* (b) *given previously. The functional* $I_f : L^p(\Omega, X) \to \mathbb{R}$ *is Lipschitz continuous on bounded sets, and so locally Lipschitz, and*

$$\partial I_f(u) \subset \{u^* \in L^\theta(\Omega, X^*) : \ u^*(x) \in \partial_\xi f(x, u(x)) \ \mu\text{-a.e. in } \Omega\},$$

*where* $\theta = p' := \frac{p}{p-1}$ *if* $p \in (1, +\infty)$ *and* $\theta = 1$ *if* $p = +\infty$.

## 3.3 Remarks

**Section 3.1:** Among nonlinear functions, convex functions are the closest ones to linear functions, and in fact lower semicontinuous convex functions can be obtained as the upper envelope of all the affine, continuous minorants. Moreover, note that the linear functions are just those functions that are simultaneously convex and concave. An important convex function is one given by the distance to a convex set. More precisely, let $C \subset \mathbb{R}^N$ be a nonempty, closed set. We define the distance to $C$ by

$$d_C(x) = \inf\{|x - c| : \ c \in C\}.$$

A well-known result of Motzkin says that $C$ is convex if and only if $d_C$ is differentiable at every $x \in \mathbb{R}^N \setminus C$, if and only if, for all $x \in \mathbb{R}^N \setminus C$, there is a unique nearest point to $x$ in $C$. Theorem 3.15, which relates this section to Chap. 2, is due to Rockafeller [345]. Also, we recall that if $U \subset \mathbb{R}^N$ is open and convex and $\varphi \in C^2(U, \mathbb{R})$, then $\varphi$ is convex if and only if the Hessian of $\varphi$ is nonnegative definite at every $x \in U$, i.e.,

$$(\varphi''(x)h, h)_{\mathbb{R}^N} \geq 0 \quad \text{for all } h \in \mathbb{R}^N.$$

Moreover, if $\varphi''(x)$ is positive definite at every $x \in U$, then $\varphi$ is strictly convex. The converse is not in general true [e.g., $\varphi(x) = x^4$ for $x \in \mathbb{R}$ is strictly convex, but $\varphi''(0) = 0$].

There are many books on convex functions and their subdifferentiability and duality theories. We mention those by Borwein and Vanderwerff [49], Ekeland and Temam [129], Giles [159], Ioffe and Tihomirov [181], Laurent [217], Phelps [328], Roberts and Varberg [342], Rockafeller [343], and Rockafeller and Wets [346].

**Section 3.2:** The subdifferentiability theory of locally Lipschitz functions started with Clarke [84], who, using Rademacher's theorem, introduced the description of the subdifferential included in Proposition 3.48. Later, Clarke [85] extended the subdifferential to locally Lipschitz functions defined on a Banach space. Proposition 3.31, which is a useful tool in the nonsmooth critical point theory, is due to Lebourg [221]. The main source of information for the subdifferentiability theory of locally Lipschitz functions is the book by Clarke [85]. A subdifferentiability theory where the values of the subdifferential are not convex sets can be found in Mordukhovich [263]. We should mention that, using the notion of Haar-null sets, we can have a version of Rademacher's theorem for a locally Lipschitz map between Banach spaces (see Christensen [81] and Gasiński and Papageorgiou [151]).

# Chapter 4
# Degree Theory

**Abstract** This chapter provides the fundamental elements of degree theory used later in the book for showing abstract results of critical point theory or bifurcation theory as well as for the study of the existence and multiplicity of solutions to nonlinear problems. The first section of the chapter introduces Brouwer's degree and its important applications such as Brouwer's fixed point theorem, Borsuk's theorem, Borsuk–Ulam, and Lyusternik–Schnirelmann–Borsuk theorems. The second section sets forth the Leray–Schauder degree theory for compact perturbations of the identity. The third section amounts to a description of the degree for $(S)_+$-maps using Galerkin approximations and construction of the degree theory for multifunctions of the form $f + A$ with $f$ an $(S)_+$-map and $A$ a maximal monotone operator. Comments and historical notes are given in a remarks section.

## 4.1 Brouwer's Degree

In many situations we are led to a nonlinear equation of the form $\varphi(u) = y$, with $\varphi$ defined on a space $X$ and $y \in X$. We are interested in establishing the existence of solutions and "counting" the solutions and seeing how they behave with respect to some perturbations of $\varphi$. These issues are answered in part by the notion of *topological degree*. In this section we present a quick overview of the finite-dimensional degree theory (Brouwer's degree).

Brouwer's degree is defined on certain triples $(\varphi, U, y)$, with a continuous function $\varphi : \overline{U} \to \mathbb{R}^N$ ($N \geq 1$), $U \subset \mathbb{R}^N$ nonempty, bounded, and open, and $y \in \mathbb{R}^N$. Recall that if $\varphi \in C^1(U, \mathbb{R}^N)$, then the Jacobian of $\varphi$ at $x \in U$ is defined by

$$J_\varphi(x) = \det(\varphi'(x)), \quad \text{where } \varphi'(x) = \left( \frac{\partial \varphi_i}{\partial x_j}(x) \right)_{i,j=1}^N.$$

D. Motreanu et al., *Topological and Variational Methods with Applications to Nonlinear Boundary Value Problems*, DOI 10.1007/978-1-4614-9323-5_4,
© Springer Science+Business Media, LLC 2014

We say that $y \in \varphi(U)$ is a *regular value* of $\varphi$ if $J_\varphi(x) \neq 0$ for all $x \in \varphi^{-1}(y)$. The points $x \in U$ for which $J_\varphi(x) = 0$ are called *critical points* of $\varphi$. The set of critical points of $\varphi$ is denoted by $K_\varphi$ (or $K_\varphi(U)$).

The process of constructing Brouwer's degree $d(\varphi, U, y)$ is rather laborious and is done in a sequence of steps.

We start by defining the degree map $d(\varphi, U, y)$ in the "generic" case, that is, for $\varphi \in C^1(U, \mathbb{R}^N) \cap C(\overline{U}, \mathbb{R}^N)$ and $y \in \mathbb{R}^N \setminus \varphi(K_\varphi \cup \partial U)$ (where $\partial U$ denotes the boundary of $U$). In this case, $\varphi^{-1}(y) \neq \emptyset$ if and only if $y$ is a regular value, in which case $\varphi^{-1}(y)$ is compact and discrete [by virtue of the inverse function theorem, since $J_\varphi(x) \neq 0$ for all $x \in \varphi^{-1}(y)$]; hence $\varphi^{-1}(y)$ is finite. Therefore, the following definition makes sense.

**Definition 4.1.** If $U \subset \mathbb{R}^N$ is nonempty, bounded, and open, $\varphi \in C^1(U, \mathbb{R}^N) \cap C(\overline{U}, \mathbb{R}^N)$, and $y \in \mathbb{R}^N \setminus \varphi(K_\varphi \cup \partial U)$, then

$$d(\varphi, U, y) = \sum_{x \in \varphi^{-1}(y)} \mathrm{sgn}\, J_\varphi(x),$$

setting $d(\varphi, U, y) = 0$ if $\varphi^{-1}(y) = \emptyset$.

One fundamental property of the degree in Definition 4.1 is that $d(\varphi, U, y_1) = d(\varphi, U, y_2)$ whenever $y_1, y_2 \in \mathbb{R}^N \setminus \varphi(K_\varphi \cup \partial U)$ belong to the same connected component of $\mathbb{R}^N \setminus \varphi(\partial U)$. This property makes it possible to extend the degree to the case of $y \in \varphi(K_\varphi) \setminus \varphi(\partial U)$, i.e., the next definition makes sense.

**Definition 4.2.** Let $U \subset \mathbb{R}^N$ be nonempty, bounded, and open, $\varphi \in C^1(U, \mathbb{R}^N) \cap C(\overline{U}, \mathbb{R}^N)$, and $y \in \varphi(K_\varphi) \setminus \varphi(\partial U)$. We define $d(\varphi, U, y) = d(\varphi, U, y')$ whenever $y' \in \mathbb{R}^N \setminus \varphi(K_\varphi \cup \partial U)$ is such that $\|y' - y\| < d(y, \varphi(\partial U))$ [where $d(y, \varphi(\partial U))$ stands for the distance from $y$ to $\varphi(\partial U)$].

Here and subsequently in this section, $\|\cdot\|$ is a norm on $\mathbb{R}^N$ that is not necessarily the Euclidean norm. In the preceding definition, the existence of $y'$ is guaranteed by *Sard's theorem*, which implies that the set of critical values of $\varphi$ is Lebesgue-null in $\mathbb{R}^N$.

**Theorem 4.3.** *If $U \subset \mathbb{R}^N$ is nonempty and open and $\varphi : U \to \mathbb{R}^m$ is a $C^k$-map, then the set $\varphi(K_\varphi)$ is a Lebesgue-null set provided $k > \max\{0, N - m\}$.*

The degree defined in Definitions 4.1 and 4.2 for a function $\varphi \in C^1(U, \mathbb{R}^N) \cap C(\overline{U}, \mathbb{R}^N)$ and $y \in \mathbb{R}^N \setminus \varphi(\partial U)$ satisfies the following continuity property: if $\psi \in C^1(U, \mathbb{R}^N) \cap C(\overline{U}, \mathbb{R}^N)$ satisfies $\|\varphi - \psi\|_\infty < d(y, \varphi(\partial U))$ [and thus $y \notin \psi(\partial U)$], then $d(\varphi, U, y) = d(\psi, U, y)$. This leads to the following definition.

**Definition 4.4.** Let $U \subset \mathbb{R}^N$ be nonempty, bounded, and open, $\varphi \in C(\overline{U}, \mathbb{R}^N)$, and $y \notin \varphi(\partial U)$. If $\varphi \in C^1(U, \mathbb{R}^N)$, then Brouwer's degree $d(\varphi, U, y)$ is the one in Definitions 4.1 and 4.2. Otherwise, Brouwer's degree $d(\varphi, U, y)$ is defined by $d(\varphi, U, y) = d(\psi, U, y)$ whenever $\psi \in C^1(U, \mathbb{R}^N) \cap C(\overline{U}, \mathbb{R}^N)$ satisfies $\|\varphi - \psi\|_\infty < d(y, \varphi(\partial U))$.

In Definition 4.4, the existence of $\psi$ is guaranteed by the density of $C^1(U, \mathbb{R}^N) \cap C(\overline{U}, \mathbb{R}^N)$ in $C(\overline{U}, \mathbb{R}^N)$.

For details, the reader can consult any of the books we give in the relevant remarks in Sect. 4.4.

In the next theorem, we summarize the main properties of Brouwer's degree map. In what follows, we consider $U \subset \mathbb{R}^N$ nonempty, bounded, and open, $\varphi \in C(\overline{U}, \mathbb{R}^N)$, and $y \in \mathbb{R}^N \setminus \varphi(\partial U)$.

**Theorem 4.5.** *Brouwer's degree map* $(\varphi, U, y) \mapsto d(\varphi, U, y)$, *defined on triples* $(\varphi, U, y)$ *as previously, introduced in Definition 4.4, is the unique integer-valued map satisfying the following properties:*

(a) *Normalization:* $d(\mathrm{id}, U, y) = 1$ *for all* $y \in U$;

(b) *Domain additivity: if* $U_1, U_2 \subset U$ *are disjoint, nonempty, open sets and* $y \notin \varphi(\partial U_1) \cup \varphi(\partial U_2)$, *then*

$$d(\varphi, U_1 \cup U_2, y) = d(\varphi, U_1, y) + d(\varphi, U_2, y);$$

(c) *Homotopy invariance:* $d(h(t, \cdot), U, y(t))$ *is independent of* $t \in [0, 1]$ *whenever* $h : [0, 1] \times \overline{U} \to \mathbb{R}^N$ *and* $y : [0, 1] \to \mathbb{R}^N$ *are continuous and* $y(t) \notin h([0, 1] \times \partial U)$ *for all* $t \in [0, 1]$;

(d) *Excision: if* $C \subset \overline{U}$ *is closed and* $y \notin \varphi(C) \cup \varphi(\partial U)$, *then*

$$d(\varphi, U, y) = d(\varphi, U \setminus C, y);$$

(e) *Solution property: if* $d(\varphi, U, y) \neq 0$, *then there exists* $x \in U$ *such that* $\varphi(x) = y$;

(f) *Continuous dependence:* $d(\cdot, U, y)$ *is constant on*

$$\{\psi \in C(\overline{U}, \mathbb{R}^N) : \|\psi - \varphi\|_\infty < \rho\}$$

*and* $d(\varphi, U, \cdot)$ *is constant on*

$$B_\rho(y) := \{v \in \mathbb{R}^N : \|v - y\| < \rho\},$$

*where* $\rho = d(y, \varphi(\partial U))$; *moreover,* $d(\varphi, U, \cdot)$ *is constant on each connected component of* $\mathbb{R}^N \setminus \varphi(\partial U)$;

(g) *Boundary dependence: if* $\varphi, \psi \in C(\overline{U}, \mathbb{R}^N)$ *satisfy* $\varphi|_{\partial U} = \psi|_{\partial U}$, *then we have* $d(\varphi, U, y) = d(\psi, U, y)$ *whenever* $y \notin \varphi(\partial U)$;

(h) *Translation: if* $\varphi \in C(\overline{U}, \mathbb{R}^N)$ *and* $y \notin \varphi(\partial U)$, *then for all* $u \in \mathbb{R}^N$, *we have*

$$d(\varphi, U, y) = d(\varphi - u, U, y - u).$$

*Remark 4.6.* Properties (a)–(c) actually define Brouwer's degree. Property (c) (arguably the most important property) allows us to calculate $d(\varphi, U, y)$ by replacing it with a (hopefully) simpler calculation. The degree is independent of the coordinate system, i.e., if $h : \mathbb{R}^N \to \mathbb{R}^N$ is a diffeomorphism, then $d(\varphi, U, y) = d(h \circ \varphi \circ h^{-1}, h(U), h(y))$.

Next let us present some useful applications of Brouwer's degree. We start with the celebrated *Brouwer's fixed point theorem*.

**Theorem 4.7.** *If $C \subset \mathbb{R}^N$ is nonempty, compact, and convex and $\varphi : C \to C$ is continuous, then $\varphi$ admits a fixed point, i.e., there exists $x \in C$ such that $\varphi(x) = x$.*

*Proof.* First suppose $C = \overline{B_\rho(0)}$ for some $\rho > 0$. We may assume that $\varphi(x) \neq x$ for all $x \in \partial B_\rho(0)$, or otherwise we are done. Consider the homotopy $h : [0,1] \times \overline{B_\rho(0)} \to \mathbb{R}^N$ defined by $h(t,x) = x - t\varphi(x)$ for all $t \in [0,1]$, all $x \in \overline{B_\rho(0)}$. Then for all $t \in [0,1)$ and all $x \in \partial B_\rho(0)$ we have $\|h(t,x)\| \geq \|x\| - t\|\varphi(x)\| \geq (1-t)\rho > 0$, while $h(1,x) \neq 0$ for all $x \in \partial B_\rho(0)$ since $\varphi$ has no fixed points on $\partial B_\rho(0)$. Therefore, by the homotopy invariance and normalization properties in Theorem 4.5, we have

$$d(\mathrm{id} - \varphi, B_\rho(0), 0) = d(\mathrm{id}, B_\rho(0), 0) = 1.$$

Thus, by Theorem 4.5(e), we can find $x \in B_\rho(0)$ such that $\varphi(x) = x$.

Now let $C$ be an arbitrary nonempty, compact, convex set in $\mathbb{R}^N$. Let $p_C : \mathbb{R}^N \to C$ be the metric projection map on $C$, and let $\rho > 0$ be such that $C \subset \overline{B_\rho(0)}$. If $\hat{\varphi} = \varphi \circ p_C$, then $\hat{\varphi} : \overline{B_\rho(0)} \to \overline{B_\rho(0)}$ is continuous and $\hat{\varphi}(\overline{B_\rho(0)}) \subset C$. Hence, by the first part of the proof, there exists $x \in C$ such that $\varphi(x) = \hat{\varphi}(x) = x$.                                $\square$

Another important topological consequence of Brouwer's degree is the nonretraction property of the unit sphere in $\mathbb{R}^N$. First a definition.

**Definition 4.8.** Let $Y$ be a Hausdorff topological space and $C \subset Y$. We say that $C$ is a *retract* of $Y$ if there exists a continuous map $r : Y \to C$ such that $r|_C = \mathrm{id}_C$. The map $r$ is called a *retraction* of $Y$ on $C$.

*Remark 4.9.* By Dugundji's theorem (Theorem 2.9), every nonempty, closed, convex subset of a normed space is a retract. A retract is always closed.

**Theorem 4.10.** *If $B_1 = \{x \in \mathbb{R}^N : \|x\| < 1\}$ and $\partial B_1 = \{x \in \mathbb{R}^N : \|x\| = 1\}$, then $\partial B_1$ is not a retract of $\overline{B_1}$.*

*Proof.* Arguing indirectly, suppose that there is a retraction $r : \overline{B_1} \to \partial B_1$. Then from Theorem 4.5(g), (a) we have

$$d(r, B_1, 0) = d(\mathrm{id}, B_1, 0) = 1.$$

Then, by virtue of Theorem 4.5(e), we can find $x \in B_1$ such that $r(x) = 0$, a contradiction of the fact that the range of the retraction map $r$ is $\partial B_1$.         $\square$

*Remark 4.11.* This result is in sharp contrast to the infinite-dimensional case, where the unit sphere $\partial B_1$ is a retract of the closed unit ball (see also Example 4.20). Note that Theorems 4.7 and 4.10 are equivalent.

Another useful result is the so-called *Borsuk's theorem*.

**Theorem 4.12.** *If $U \subset \mathbb{R}^N$ is symmetric (i.e., $U = -U$), bounded, and open, with $0 \in U$, and $\varphi \in C(\overline{U}, \mathbb{R}^N)$ is odd [i.e., $\varphi(-x) = -\varphi(x)$ for all $x \in U$] and $0 \notin \varphi(\partial U)$, then $d(\varphi, U, 0)$ is an odd integer (in particular different from zero).*

*Proof.* First, we prove the theorem in the "generic" case, that is, for a map $\varphi \in C^1(U, \mathbb{R}^N) \cap C(\overline{U}, \mathbb{R}^N)$ satisfying $0 \notin \varphi(K_\varphi)$. Since $\varphi$ is odd, we have that $\varphi^{-1}(0) = \{0, -x_1, x_1, \ldots, -x_m, x_m\}$, for some $x_1, \ldots, x_m \in U \setminus \{0\}$ pairwise distinct, and that $J_\varphi(-x_i) = J_\varphi(x_i)$ for all $i$. Then, by Definition 4.1, we have

$$d(\varphi, U, 0) = \sum_{x \in \varphi^{-1}(0)} \operatorname{sgn} J_\varphi(x) = \operatorname{sgn} J_\varphi(0) + 2 \sum_{i=1}^{m} \operatorname{sgn} J_\varphi(x_i);$$

therefore, $d(\varphi, U, 0)$ is odd.

In the general case, we want to find $\tilde{\varphi} \in C^1(U, \mathbb{R}^N) \cap C(\overline{U}, \mathbb{R}^N)$ odd, with $0 \notin \tilde{\varphi}(K_{\tilde{\varphi}})$, such that $\|\varphi - \tilde{\varphi}\|_\infty < \rho := d(0, \varphi(\partial U))$. If this is done, then we get that $d(\varphi, U, 0) = d(\tilde{\varphi}, U, 0)$ is odd, by Definition 4.4 and the generic case treated previously. We construct $\tilde{\varphi}$ in two steps.

*Step 1*: There is $\hat{\varphi} \in C^1(U, \mathbb{R}^N) \cap C(\overline{U}, \mathbb{R}^N)$ odd, with $J_{\hat{\varphi}}(0) \neq 0$, such that $\|\varphi - \hat{\varphi}\|_\infty < \frac{\rho}{2}$.

Indeed, choose $\psi \in C^1(U, \mathbb{R}^N) \cap C(\overline{U}, \mathbb{R}^N)$ such that $\|\psi - \varphi\|_\infty < \frac{\rho}{8}$. Let $\psi_0$ be the odd part of $\psi$ defined by $\psi_0(x) = \frac{1}{2}(\psi(x) - \psi(-x))$, and choose $\delta \in (0, \frac{\rho}{4 \operatorname{diam} U}]$ (where $\operatorname{diam} U$ is the diameter of $U$) such that $\delta$ is not an eigenvalue of $\psi_0'(0)$. Set $\hat{\varphi} = \psi_0 - \delta \operatorname{id}$. Then $\hat{\varphi} \in C^1(U, \mathbb{R}^N) \cap C(\overline{U}, \mathbb{R}^N)$, $\hat{\varphi}$ is odd, and $J_{\hat{\varphi}}(0) \neq 0$. Moreover, for all $x \in \overline{U}$, we have

$$\|\varphi(x) - \hat{\varphi}(x)\| = \|\varphi(x) - \psi_0(x) + \delta x\|$$

$$\leq \|\varphi(x) - \frac{1}{2}(\psi(x) - \psi(-x))\| + \delta \operatorname{diam} U$$

$$\leq \frac{1}{2}\|\varphi(x) - \psi(x)\| + \frac{1}{2}\|\varphi(x) + \psi(-x))\| + \delta \operatorname{diam} U$$

$$= \frac{1}{2}\|\varphi(x) - \psi(x)\| + \frac{1}{2}\|\psi(-x) - \varphi(-x))\| + \delta \operatorname{diam} U$$

(since $\varphi$ is odd), which implies that $\|\varphi - \hat{\varphi}\|_\infty < \frac{\rho}{4} + \frac{\rho}{4} = \frac{\rho}{2}$.

*Step 2*: There is $\tilde{\varphi} \in C^1(U, \mathbb{R}^N) \cap C(\overline{U}, \mathbb{R}^N)$ odd, with $0 \notin \tilde{\varphi}(K_{\tilde{\varphi}})$, such that $\|\hat{\varphi} - \tilde{\varphi}\|_\infty \leq \frac{\rho}{2}$.

Set $U_0 = \emptyset$ and

$$U_k = \{x = (x_i)_{i=1}^N \in U : x_i \neq 0 \text{ for some } i \leq k\}.$$

For any $\psi \in C^1(U, \mathbb{R}^N)$ and $V \subset U$, we denote by $K_\psi(V)$ the set of critical points of $\psi$ contained in $V$.

*Claim 1*: For every $k \in \{0, 1, \ldots, N\}$ there is $\psi_k \in C^1(U, \mathbb{R}^N) \cap C(\overline{U}, \mathbb{R}^N)$ odd, with $\psi_k'(0) = \hat{\varphi}'(0)$ and $0 \notin \psi_k(K_{\psi_k}(U_k))$, such that $\|\hat{\varphi} - \psi_k\|_\infty \leq \frac{k\rho}{2N}$.

Once Claim 1 is proved, we have that the function $\tilde{\varphi} := \psi_N$ satisfies the requirements of Step 2. Hence it suffices to establish Claim 1.

We reason by induction on $k \in \{0, 1, \ldots, N\}$. Claim 1 is clearly satisfied at rank $k = 0$ by taking $\psi_0 = \hat{\varphi}$. Now, let $k \in \{1, \ldots, N\}$ such that Claim 1 is satisfied at rank $k - 1$. Fix a function $\theta \in C^1(\mathbb{R})$ odd, such that $\theta'(0) = 0$, $\theta(t) = 0$ if and only if $t = 0$, and $\|\theta\|_\infty \leq 1$ [e.g., $\theta(t) = \frac{t^3}{t^4+1}$]. Let $\hat{U}_k = \{x = (x_i)_{i=1}^N \in U : x_k \neq 0\}$, and let $\hat{\psi}_k \in C^1(\hat{U}_k, \mathbb{R}^N)$ be defined by $\hat{\psi}_k(x) = \frac{\psi_{k-1}(x)}{\theta(x_k)}$. By Sard's theorem (Theorem 4.3), we can find $y = y(k) \in \mathbb{R}^N \setminus \hat{\psi}_k(K_{\hat{\psi}_k}(\hat{U}_k))$, with $\|y\| \leq \frac{\rho}{2N}$. Then we define

$$\psi_k(x) = \psi_{k-1}(x) - \theta(x_k)y \quad \text{for all } x = (x_i)_{i=1}^N \in \overline{U}.$$

Clearly, $\psi_k \in C^1(U, \mathbb{R}^N) \cap C(\overline{U}, \mathbb{R}^N)$, $\psi_k$ is odd, and $\|\hat{\varphi} - \psi_k\|_\infty \leq \|\hat{\varphi} - \psi_{k-1}\|_\infty + \|\theta(x_k)y\|_\infty \leq \frac{(k-1)\rho}{2N} + \frac{\rho}{2N} = \frac{k\rho}{2N}$. Since $\theta'(0) = 0$, we have $\psi_k'(0) = \psi_{k-1}'(0) = \hat{\varphi}'(0)$. It remains to check that $0 \notin \psi_k(K_{\psi_k}(U_k))$. Arguing by contradiction, assume that there is $z = (z_i)_{i=1}^N \in U_k$ such that

$$\psi_k(z) = 0 \quad \text{and} \quad J_{\psi_k}(z) = 0.$$

We distinguish two cases.

*Case 1*: $z_k = 0$.

In this case, $z \in U_{k-1}$. Moreover, we have $\psi_{k-1}(z) = \psi_k(z) + \theta(0)y = 0$ and $J_{\psi_{k-1}}(z) = J_{\psi_k}(z) = 0$ [using that $\theta'(0) = 0$]. This contradicts the fact that $0 \notin \psi_{k-1}(K_{\psi_{k-1}}(U_{k-1}))$.

*Case 2*: $z_k \neq 0$.

In this case, $z \in \hat{U}_k$. Then we obtain a contradiction if we show that $z$ is a critical point of $\hat{\psi}_k$ such that $\hat{\psi}_k(z) = y$ [indeed, this contradicts the choice of $y \in \mathbb{R}^N \setminus \hat{\psi}_k(K_{\hat{\psi}_k}(\hat{U}_k))$]. To do this, we first note that the equality $\psi_k(z) = 0$ implies that $\hat{\psi}_k(z) = \frac{\psi_{k-1}(z)}{\theta(z_k)} = y$. Moreover, a straightforward calculation [based on the definition of $\hat{\psi}_k$ and the assumption that $\psi_k(z) = 0$] shows that $\hat{\psi}_k'(z) = \frac{1}{\theta(z_k)}\psi_k'(z)$, whence $J_{\hat{\psi}_k}(z) = (\frac{1}{\theta(z_k)})^N J_{\psi_k}(z) = 0$. This completes the proof of the theorem. $\quad\square$

**Corollary 4.13.** *The result in Theorem 4.12 still holds if we replace the assumption that $\varphi$ is odd by the weaker one that $\varphi|_{\partial U}$ is odd [in particular different from zero, and so there exists $x \in U$ such that $\varphi(x) = 0$].*

*Proof.* Let $\varphi_0$ be the odd part of $\varphi$ defined by $\varphi_0(x) = \frac{1}{2}(\varphi(x) - \varphi(-x))$ for all $x \in \overline{U}$. Then $\varphi|_{\partial U} = \varphi_0|_{\partial U}$ and so, by Theorem 4.5(g), we have that $d(\varphi, U, 0) = d(\varphi_0, U, 0)$. The corollary follows by applying Theorem 4.12 to the function $\varphi_0$. $\quad\square$

We mention further consequences of Borsuk's theorem. First we mention the following more refined version of Corollary 4.13.

**Proposition 4.14.** *If $U \subset \mathbb{R}^N$ is symmetric, bounded, and open, with $0 \in U$, $\varphi \in C(\overline{U}, \mathbb{R}^N)$, $0 \notin \varphi(\partial U)$, and*

$$\frac{\varphi(x)}{\|\varphi(x)\|} \neq \frac{\varphi(-x)}{\|\varphi(-x)\|} \quad \text{for all } x \in \partial U, \tag{4.1}$$

*then $d(\varphi, U, 0)$ is an odd integer.*

*Proof.* Let $h(t,x) = t\varphi(x) + (1-t)\frac{\varphi(x) - \varphi(-x)}{2}$ for all $t \in [0,1]$, all $x \in \overline{U}$. Then $h \in C([0,1] \times \overline{U}, \mathbb{R}^N)$, and because of (4.1), we have $0 \notin h([0,1] \times \partial U)$. Theorem 4.5(c) implies that

$$d(h(0,\cdot), U, 0) = d(h(1,\cdot), U, 0). \tag{4.2}$$

Note that $h(0,x) = \frac{\varphi(x) - \varphi(-x)}{2}$ for all $x \in \overline{U}$, which is an odd function. Thus, Theorem 4.12 ensures that $d(h(0,\cdot), U, 0)$ is an odd integer. Noting that $h(1,\cdot) = \varphi$, from (4.2) we conclude that $d(\varphi, U, 0)$ is an odd integer. $\qquad\square$

The next byproduct of Theorem 4.12 is the following result, known in the literature as the *Borsuk–Ulam theorem*.

**Theorem 4.15.** *If $U \subset \mathbb{R}^N$ is symmetric, bounded, and open, with $0 \in U$, and $\varphi : \partial U \to \mathbb{R}^N$ is continuous and such that $\varphi(\partial U) \subset H$, where $H \subset \mathbb{R}^N$ is an affine subspace with $\dim H < N$, then there exists $x \in \partial U$ such that $\varphi(x) = \varphi(-x)$. In particular if, in addition, $\varphi$ is odd, then $\varphi(x) = 0$.*

*Proof.* By virtue of Theorem 2.9, we can extend $\varphi$ to a function $\hat{\varphi} \in C(U, \mathbb{R}^N)$ such that $\hat{\varphi}(\overline{U}) \subset H$. Let $\hat{\varphi}_0 \in C(\overline{U}, \mathbb{R}^N)$ be the odd part of $\hat{\varphi}$, that is, $\hat{\varphi}_0(x) = \frac{1}{2}(\hat{\varphi}(x) - \hat{\varphi}(-x))$ for all $x \in \overline{U}$. Thus, $\hat{\varphi}_0(\overline{U}) \subset V$, where $V \subset \mathbb{R}^N$ is the direction of $H$, so a vector subspace with $\dim V < N$.

To prove the theorem, we assume by contradiction that $\varphi(x) \neq \varphi(-x)$ for all $x \in \partial U$. This implies that $0 \notin \hat{\varphi}_0(\partial U)$. Hence the degree $d(\hat{\varphi}_0, U, 0)$ is well defined, and Theorem 4.12 guarantees that it is nonzero. Moreover, letting $\rho = d(0, \hat{\varphi}_0(\partial U))$, we have that $d(\hat{\varphi}_0, U, y) = d(\hat{\varphi}_0, U, 0) \neq 0$ for every $y \in B_\rho(0) := \{x \in \mathbb{R}^N : \|x\| < \rho\}$ [Theorem 4.5(f)]. Then, by virtue of Theorem 4.5(e), we get $B_\rho(0) \subset \hat{\varphi}_0(U)$. Since $\hat{\varphi}_0(U) \subset V$ and $\dim V < N$, we reach a contradiction. $\qquad\square$

A direct application of Theorem 4.15 leads to the following corollary.

**Corollary 4.16.** *If $N > m \geq 1$ are integers, $B_1^N = \{x \in \mathbb{R}^N : \|x\| < 1\}$, and $B_1^m = \{x \in \mathbb{R}^m : \|x\| < 1\}$, then there is no continuous, odd map $\varphi : \partial B_1^N \to \partial B_1^m$.*

We state another important consequence of Borsuk's theorem, which is the so-called *Lyusternik–Schnirelmann–Borsuk theorem*.

**Theorem 4.17.** *If* $U \subset \mathbb{R}^N$ *is symmetric, bounded, and open, with* $0 \in U$, *and* $\{C_k\}_{k=1}^M$ *is a closed covering of* $\partial U$ *satisfying* $C_k \cap (-C_k) = \emptyset$ *for all* $k = 1, \ldots, M$, *then* $M \geq N+1$.

*Proof.* We have $\overset{M}{\underset{k=1}{\cap}} C_k = \emptyset$. Indeed, if $x_0 \in \overset{M}{\underset{k=1}{\cap}} C_k$, then because $\partial U = \overset{M}{\underset{k=1}{\cup}} (-C_k)$, we must have $x_0 \in C_k \cap (-C_k)$ for some $k \in \{1, \ldots, M\}$, a contradiction of the hypothesis.

Next, we argue by contradiction. Thus, suppose that $M \leq N$. Set $d_k(x) = d(x, C_k)$ for $k \in \{1, \ldots, M\}$, all $x \in \partial U$. We consider $\varphi \in C(\partial U, \mathbb{R}^{M-1})$ defined by

$$\varphi(x) = (d_k(x))_{k=1}^{M-1} .$$

Invoking Theorem 4.15, we can find $x \in \partial U$ such that $\varphi(x) = \varphi(-x)$. Since $\{C_k\}_{k=1}^M$ is a covering of $\partial U$, we can find $k_0 \in \{1, \ldots, M\}$ minimal such that $x \in C_{k_0}$. There are two possibilities:

(a) $k_0 \leq M - 1$: then $d_{k_0}(-x) = d_{k_0}(x) = 0$, hence $x \in C_{k_0} \cap (-C_{k_0})$, a contradiction.

(b) $k_0 = M$: then $x \notin \overset{M-1}{\underset{k=1}{\cup}} C_k$, and so $d_k(-x) = d_k(x) > 0$ for all $k \in \{1, \ldots, M - 1\}$, hence $x \notin \overset{M-1}{\underset{k=1}{\cup}} (-C_k)$. Since $\{C_k\}_{k=1}^M$ and $\{-C_k\}_{k=1}^M$ are coverings of $\partial U$, it follows that $x \in C_M \cap (-C_M)$, again a contradiction. $\qquad\square$

The next result is useful in the definition of the Leray–Schauder degree, which is the infinite-dimensional extension of Brouwer's degree (Definition 4.22).

**Proposition 4.18.** *Let* $U \subset \mathbb{R}^N$ *be nonempty, bounded, and open, and let* $\varphi \in C(\overline{U}, \mathbb{R}^N)$ *be such that* $0 \notin \varphi(\partial U)$. *Let* $V \subset \mathbb{R}^N$ *be a vector subspace such that* $(\mathrm{id} - \varphi)(U) \subset V$. *If* $U \cap V = \emptyset$, *then* $d(\varphi, U, 0) = 0$. *Otherwise, we have*

$$d(\varphi, U, 0) = d(\varphi|_{\overline{U \cap V}}, U \cap V, 0),$$

*where* $d(\varphi|_{\overline{U \cap V}}, U \cap V, 0)$ *stands for the Brouwer degree of the map* $\varphi|_{\overline{U \cap V}} \in C(\overline{U \cap V}, V)$.

*Proof.* If $d(\varphi, U, 0) \neq 0$, then Theorem 4.5(e) yields $x \in U$ such that $\varphi(x) = 0$. Then $x = x - \varphi(x) \in V$, which implies that $U \cap V \neq \emptyset$. This proves the first claim of the proposition.

Next, we assume that $U \cap V \neq \emptyset$. Set $\hat{\varphi} = \varphi|_{\overline{U \cap V}}$. The assumption that $(\mathrm{id} - \varphi)(U) \subset V$ guarantees that $\hat{\varphi}(\overline{U \cap V}) \subset V$, whence $\hat{\varphi} \in C(\overline{U \cap V}, V)$. Denoting by $\partial^V(U \cap V)$ the boundary of $U \cap V$ in $V$, we clearly have $\partial^V(U \cap V) \subset \partial U$, whence $0 \notin \hat{\varphi}(\partial^V(U \cap V))$. This shows that the degree $d(\hat{\varphi}, U \cap V, 0)$ is well defined.

Let $\rho = d(0, \varphi(\partial U)) > 0$. Using the Weierstrass approximation theorem, we find $\psi \in C(\overline{U}, V) \cap C^1(U, V)$ such that $\|(\varphi - \mathrm{id}) - \psi\|_\infty < \rho$. Up to dealing with $\psi + \mathrm{id}$ instead of $\varphi$ [by virtue of Theorem 4.5(f)], we may assume that $\varphi \in C(\overline{U}, \mathbb{R}^N) \cap C^1(U, \mathbb{R}^N)$ [and so $\hat{\varphi} \in C(\overline{U \cap V}, V) \cap C^1(U \cap V, V)$]. Moreover, Sard's theorem (Theorem 4.3) implies that there is $y \in V$ with $\|y\| < \rho$, which is not a critical value of $\hat{\varphi}$. Up to dealing with the translation $\varphi - y$ instead of $\varphi$, and so $\hat{\varphi} - y$ instead of $\hat{\varphi}$, we may assume in what follows that $0 \notin \hat{\varphi}(K_{\hat{\varphi}})$.

We claim that

$$\varphi^{-1}(0) = \hat{\varphi}^{-1}(0) \subset V \quad \text{and} \quad J_{\varphi}(x) = J_{\hat{\varphi}}(x) \text{ for all } x \in U \cap V. \qquad (4.3)$$

In particular, from (4.3) we have that $0 \notin \varphi(K_{\varphi})$, whence, by Definition 4.1,

$$d(\varphi, U, 0) = \sum_{x \in \varphi^{-1}(0)} \operatorname{sgn} J_{\varphi}(x) = \sum_{x \in \hat{\varphi}^{-1}(0)} \operatorname{sgn} J_{\hat{\varphi}}(x) = d(\hat{\varphi}, U \cap V, 0).$$

Thus, it suffices to prove (4.3). The inclusion $\hat{\varphi}^{-1}(0) \subset \varphi^{-1}(0)$ is immediate. Next, every $x \in \varphi^{-1}(0)$ satisfies $x = x - \varphi(x) \in U \cap V$, so that $x \in \hat{\varphi}^{-1}(0)$, whence the first relation in (4.3).

By virtue of Remark 4.6, we may assume that $V = \{(x_i)_{i=1}^{N} : x_{m+1} = \ldots = x_N = 0\}$ for some $m \leq N$. Set $\varphi(x) = (\varphi_i(x))_{i=1}^{N}$. The assumption that $(\mathrm{id} - \varphi)(U) \subset V$ ensures that $\varphi_k(x) = x_k$ for all $k \in \{m+1, \ldots, N\}$, all $x = (x_i)_{i=1}^{N} \in U$. Then, for every $x \in U \cap V$ we get

$$J_{\varphi}(x) = \det(\varphi'(x)) = \det\begin{pmatrix} \hat{\varphi}'(x) & * \\ 0 & I_{N-m} \end{pmatrix} = \det(\hat{\varphi}'(x)) = J_{\hat{\varphi}}(x),$$

where $I_{N-m}$ stands for the identity matrix and where we identify $\hat{\varphi}'(x)$ with its matrix in the natural basis. This proves the second relation in (4.3). $\qquad\square$

We conclude our discussion on Brouwer's degree with the following useful result concerning potential operators due to Rabinowitz [335] (for $C^2$-functions) and Amann [11] (for $C^1$-functions).

**Proposition 4.19.** *If $U \subset \mathbb{R}^N$ is bounded and open, $\varphi \in C^1(U, \mathbb{R})$, and $x_0 \in U$ is an isolated critical point of $\varphi$ at which $\varphi$ has a local minimum, then we can find $r > 0$ such that $d(\varphi', B_r(x_0), 0) = 1$.*

Since applications to boundary value problems require an infinite-dimensional ambient space, it is natural to ask whether we can extend Brouwer's theory to infinite-dimensional spaces. The answer is no, without any additional restrictions on the map. The next example illustrates this.

*Example 4.20.* Brouwer's fixed point theorem (Theorem 4.7) fails in infinite dimensions. Consider the Hilbert space $(\ell^2, \|\cdot\|)$ and its unit closed ball $\overline{B_1(0)}$, and let $\varphi : \overline{B_1(0)} \to \overline{B_1(0)}$ be defined by

$$\varphi(x) = (\sqrt{1 - \|x\|^2}, x_1, x_2, \ldots, x_n, \ldots) \text{ for all } x = (x_n)_{n \geq 1} \in \overline{B_1(0)}.$$

Clearly, $\varphi$ is continuous and $\varphi(\overline{B_1(0)}) \subset \partial B_1(0) = \{x \in \ell^2 : \|x\| = 1\}$. Suppose $\varphi$ has a fixed point $x = (x_n)_{n\geq 1} \in \overline{B_1(0)}$, i.e., $\varphi(x) = x$. Then $x_{n+1} = x_n$ for all $n \geq 1$ and $x_1 = \sqrt{1 - \|x\|^2}$. Since $\|x\| = \|\varphi(x)\| = 1$, we have $x_1 = 0$, and so $x_n = 0$ for all $n \geq 1$, hence $x = 0$, a contradiction. This proves that $\varphi$ is fixed-point-free.

## 4.2  Leray–Schauder Degree

We pass to the infinite-dimensional theory, focusing on the Leray–Schauder degree theory for compact perturbations of the identity. As seen in Example 4.20, we need to restrict the family of maps. Since compact maps can be approximated by finite-rank ones, we look at maps that are compact perturbations of the identity. Then the Leray–Schauder theory follows from Brouwer's degree theory. The key step is provided by the next lemma.

**Lemma 4.21.** *Let $X$ be a Banach space, $U \subset X$ nonempty, bounded, and open, and $K : \overline{U} \to X$ a compact map with $0 \notin (\mathrm{id} - K)(\partial U)$, so that $\rho := d(0, (\mathrm{id} - K)(\partial U)) > 0$. If $K_1, K_2 : \overline{U} \to X$ are finite rank maps such that*

$$\|K_i - K\|_\infty < \rho \quad and \quad K_i(\overline{U}) \subset Z \quad for \, i \in \{1,2\},$$

*where $Z \subset X$ is a finite-dimensional vector subspace intersecting $U$, then*

$$d((\mathrm{id} - K_1)|_{\overline{U \cap Z}}, U \cap Z, 0) = d((\mathrm{id} - K_2)|_{\overline{U \cap Z}}, U \cap Z, 0),$$

*where $d((\mathrm{id} - K_i)|_{\overline{U \cap Z}}, U \cap Z, 0)$ stands for the Brouwer degree of the map $(\mathrm{id} - K_i)|_{\overline{U \cap Z}} \in C(\overline{U \cap Z}, Z)$, $i \in \{1,2\}$.*

*Proof.* Since $K$ is a compact map, the assumption that $0 \notin (\mathrm{id} - K)(\partial U)$ guarantees that $d(0, (\mathrm{id} - K)(\partial U)) > 0$. Let $\partial^Z (U \cap Z)$ be the boundary of $U \cap Z$ in $Z$. Since $\partial^Z (U \cap Z) \subset \partial U$, we have $0 \notin (\mathrm{id} - K_i)(\partial^Z (U \cap Z))$, hence the degree $d((\mathrm{id} - K_i)|_{\overline{U \cap Z}}, U \cap Z, 0)$ is well defined for $i \in \{1,2\}$. We consider $h : [0,1] \times \overline{U \cap Z} \to Z$ defined by

$$h(t,x) = (1-t)K_1(x) + tK_2(x).$$

Thus $\|h(t,x) - K(x)\| < \rho$ for all $t \in [0,1]$, all $x \in \overline{U \cap Z}$. This implies that

$$\|x - h(t,x)\| \geq \|x - K(x)\| - \|K(x) - h(t,x)\| > 0$$

for all $t \in [0,1]$, all $x \in \partial^Z(U \cap Z)$. Then, from Theorem 4.5(c) we derive

$$d(\mathrm{id} - h(0,\cdot), U \cap Z, 0) = d(\mathrm{id} - h(1,\cdot), U \cap Z, 0),$$

and the conclusion follows.                                                                                   $\square$

Now we give the definition of the Leray–Schauder degree.

**Definition 4.22.** Let $X$ be a Banach space, $U \subset X$ a nonempty, bounded, and open set, and $\varphi : \overline{U} \to X$ a compact perturbation of the identity, that is, $\varphi = \mathrm{id} - K$, where $K : \overline{U} \to X$ is a compact map.

(a) If $0 \notin \varphi(\partial U)$, then the Leray–Schauder degree of the triple $(\varphi, U, 0)$ is defined by

$$d_{\mathrm{LS}}(\varphi, U, 0) = d((\mathrm{id} - \tilde{K})|_{\overline{U \cap Z}}, U \cap Z, 0), \qquad (4.4)$$

where $Z \subset X$ is a finite-dimensional vector subspace intersecting $U$ and $\tilde{K} : \overline{U} \to X$ is a finite-rank map such that $\|K - \tilde{K}\|_\infty < d(0, \varphi(\partial U))$ and $\tilde{K}(\overline{U}) \subset Z$.
(b) If $y \in X \setminus \varphi(\partial U)$, $y \neq 0$, then the Leray–Schauder degree of the triple $(\varphi, U, y)$ is defined by

$$d_{\mathrm{LS}}(\varphi, U, y) = d_{\mathrm{LS}}(\varphi - y, U, 0).$$

*Remark 4.23.*

(a) The existence of the pair $(\tilde{K}, Z)$ in Definition 4.22(a) is guaranteed by Theorem 2.7.
(b) To show that Definition 4.22(a) makes sense, we need to check that it is independent of the pair $(\tilde{K}, Z)$. Thus, let $Z_1, Z_2 \subset X$ be finite-dimensional subspaces intersecting $U$, and let $\tilde{K}_1, \tilde{K}_2 : \overline{U} \to X$ be finite-rank maps such that $\|K - \tilde{K}_i\|_\infty < d(0, \varphi(\partial U))$ and $\tilde{K}_i(\overline{U}) \subset Z_i$ for $i \in \{1, 2\}$. Set $Z = Z_1 + Z_2$. Then the equalities

$$d((\mathrm{id} - \tilde{K}_1)|_{\overline{U \cap Z_1}}, U \cap Z_1, 0) = d((\mathrm{id} - \tilde{K}_1)|_{\overline{U \cap Z}}, U \cap Z, 0)$$
$$= d((\mathrm{id} - \tilde{K}_2)|_{\overline{U \cap Z}}, U \cap Z, 0)$$
$$= d((\mathrm{id} - \tilde{K}_2)|_{\overline{U \cap Z_2}}, U \cap Z_2, 0)$$

are implied by Proposition 4.18 and Lemma 4.21.
(c) If $X = \mathbb{R}^N$ ($N \geq 1$) and $\varphi \in C(\overline{U}, \mathbb{R}^N)$, then $\varphi$ is a compact perturbation of the identity and $d_{\mathrm{LS}}(\varphi, U, y) = d(\varphi, U, y)$ for all $y \in \mathbb{R}^N \setminus \varphi(\partial U)$.

By virtue of (4.4), the main properties of the Leray–Schauder degree follow from the corresponding properties of Brouwer's degree. We consider triples $(\varphi, U, y)$ such that $U \subset X$ nonempty, bounded, and open, $\varphi = \mathrm{id} - K$, with $K : \overline{U} \to X$ compact, and $y \notin \varphi(\partial U)$.

**Theorem 4.24.** *The Leray–Schauder degree map* $(\varphi, U, y) \mapsto d_{\mathrm{LS}}(\varphi, U, y)$, *defined on triples* $(\varphi, U, y)$ *as previously, introduced in Definition 4.22, is the unique integer-valued map satisfying the following properties:*

(a) *Normalization:* $d_{LS}(\text{id}, U, y) = 1$ *if* $y \in U$;
(b) *Domain additivity: if* $U_1, U_2 \subset U$ *are disjoint, nonempty, open sets and* $y \notin \varphi(\partial U_1) \cup \varphi(\partial U_2)$, *then*

$$d_{LS}(\varphi, U_1 \cup U_2, y) = d_{LS}(\varphi, U_1, y) + d_{LS}(\varphi, U_2, y);$$

(c) *Homotopy invariance: if* $h : [0,1] \times \overline{U} \to X$ *is compact and for all* $t \in [0,1]$, *letting* $\varphi_t = \text{id} - h(t, \cdot)$, *we have* $y \notin \varphi_t(\partial U)$, *then* $d_{LS}(\varphi_t, U, y)$ *is independent of* $t \in [0,1]$;
(d) *Excision: if* $C \subset \overline{U}$ *is closed and* $y \notin \varphi(C) \cup \varphi(\partial U)$, *then*

$$d_{LS}(\varphi, U, y) = d_{LS}(\varphi, U \setminus C, y);$$

(e) *Solution property: if* $d_{LS}(\varphi, U, y) \neq 0$, *then there exists* $x \in U$ *such that* $\varphi(x) = y$;
(f) *Continuous dependence: if* $K, G : \overline{U} \to X$ *are compact maps,* $y \notin (\text{id} - K)(\partial U)$, *and* $\|K - G\|_\infty < d(y, (\text{id} - K)(\partial U))$, *then* $y \notin (\text{id} - G)(\partial U)$ *and*

$$d_{LS}(\text{id} - K, U, y) = d_{LS}(\text{id} - G, U, y);$$

*moreover,* $d_{LS}(\text{id} - K, U, \cdot)$ *is constant on each connected component of* $X \setminus (\text{id} - K)(\partial U)$;
(g) *Boundary dependence: if* $K, G : \overline{U} \to X$ *are compact maps,* $K|_{\partial U} = G|_{\partial U}$, *and* $y \notin (\text{id} - K)(\partial U)$, *then* $d_{LS}(\text{id} - K, U, y) = d_{LS}(\text{id} - G, U, y)$;
(h) *Translation:* $d_{LS}(\varphi, U, y) = d_{LS}(\varphi - u, U, y - u)$ *for all* $u \in X$.

As was already mentioned in Remark 4.11, in an infinite-dimensional normed space, $\partial B_1$ is a retract of $\overline{B}_1$. However, the retraction cannot be chosen among the compact perturbations of the identity. The proof of the following result is the same as the proof of Theorem 4.10, using Theorem 4.24 instead of Theorem 4.5.

**Proposition 4.25.** *If* $X$ *is a Banach space and* $B_1$ *is its open unit ball, then we cannot find* $\varphi \in C(\overline{B}_1, \partial B_1)$ *such that* $\varphi = \text{id} - K$, *with* $K : \overline{B}_1 \to X$ *compact, and* $\varphi|_{\partial B_1} = \text{id}_{\partial B_1}$.

The next result is the infinite-dimensional version of Brouwer's fixed point theorem (Theorem 4.7), known as *Schauder's fixed point theorem*.

**Theorem 4.26.** *If* $X$ *is a Banach space,* $C \subset X$ *is nonempty, bounded, closed, and convex, and* $K : C \to C$ *is a compact map, then* $K$ *has a fixed point, i.e., there exists* $x \in C$ *such that* $K(x) = x$.

*Proof.* From Theorem 2.7 and Remark 2.8, and since $C$ is convex, we can find $K_n : C \to C$, $n \geq 1$, finite-rank maps with range in a finite-dimensional subspace $X_n$ of $X$ such that $\|K_n(x) - K(x)\| \leq \frac{1}{n}$ for all $n \geq 1$, all $x \in C$. Then $K_n : C \cap X_n \to C \cap X_n$ satisfies the requirements of Theorem 4.7, and so we can find $x_n \in C \cap X_n$ such that $K_n(x_n) = x_n$. By the compactness of $K$, we may assume that $K(x_n) \to x$ in $X$ for some $x \in X$. Hence $x_n \to x$ in $X$, and so $K(x_n) \to K(x)$ in $X$. Therefore, $K(x) = x$, with $x \in C$.                                                                          $\square$

The next result has numerous applications to boundary value problems and is known as *Schaefer's alternative*.

**Theorem 4.27.** *If $X$ is a Banach space and $K : X \to X$ is a compact map, then either $x = tK(x)$ has a solution for every $t \in [0,1]$ or $S = \{x \in X : x = tK(x) \text{ for some } t \in [0,1)\}$ is unbounded.*

*Proof.* Assume that for some $t_0 \in (0,1]$ the equation $x = t_0 K(x)$ has no solution. Let $\rho > 0$ and let $r : X \to \overline{B_\rho}(0)$ be the $\rho$-radial retraction of $X$ onto $\overline{B_\rho}(0)$, that is,

$$r(x) = \begin{cases} x & \text{if } x \in \overline{B_\rho}(0) \\ \frac{\rho x}{\|x\|} & \text{otherwise.} \end{cases}$$

Let $K_0 = t_0 K$. The map $r \circ K_0 : \overline{B_\rho}(0) \to \overline{B_\rho}(0)$ is compact (since $K$ is) and so, by virtue of Theorem 4.26, it has a fixed point $x_\rho \in \overline{B_\rho}(0)$, i.e., $x_\rho = r(K_0(x_\rho))$. Since $K_0(x_\rho) = t_0 K(x_\rho) \neq x_\rho$ (by assumption), we must have $\|K_0(x_\rho)\| > \rho$, and so $x_\rho = tK(x_\rho)$ with $0 < t := \frac{\rho t_0}{\|K_0(x_\rho)\|} < 1$ (hence $x_\rho \in S$) and $\|x_\rho\| = \rho$. Since $\rho > 0$ is arbitrary, we conclude that $S$ is unbounded.          □

*Remark 4.28.* According to Theorem 4.27, if $K : X \to X$ is compact and the set $S = \{x \in X : x = tK(x) \text{ for some } t \in [0,1)\}$ is bounded, then $K$ has a fixed point. Theorem 4.27 illustrates the importance of a priori bounds in the study of boundary value problems.

Borsuk's theorem (Theorem 4.12) has an analog in the case of the Leray–Schauder degree.

**Theorem 4.29.** *If $X$ is a Banach space, $U \subset X$ is symmetric, bounded, and open, with $0 \in U$, and $K : \overline{U} \to X$ is a compact, odd map such that $0 \notin (\mathrm{id} - K)(\partial U)$, then $d_{\mathrm{LS}}(\mathrm{id} - K, U, 0)$ is an odd integer (in particular different from zero).*

*Proof.* By Theorem 2.7, we can find $\tilde{K} : \overline{U} \to X$, a finite-rank map, such that

$$\|\tilde{K}(x) - K(x)\| < d(0, (\mathrm{id} - K)(\partial U)) \quad \text{for all } x \in \overline{U}.$$

Up to dealing with $x \mapsto \frac{1}{2}(\tilde{K}(x) - \tilde{K}(-x))$, we may assume that $\tilde{K}$ is odd. Let $V$ be a finite-dimensional subspace of $X$ such that $U \cap V \neq \emptyset$ and $\tilde{K}(\overline{U}) \subset V$. Then $(\mathrm{id} - \tilde{K})(\overline{U} \cap V) \subset V$, and, by virtue of Definition 4.22, we have

$$d_{\mathrm{LS}}(\mathrm{id} - K, U, 0) = d((\mathrm{id} - \tilde{K})|_{\overline{U \cap V}}, U \cap V, 0).$$

By Theorem 4.12, $d((\mathrm{id} - \tilde{K})|_{\overline{U \cap V}}, U \cap V, 0)$ is odd, which completes the proof.   □

*Remark 4.30.* Because of Theorem 4.24(g), it suffices to assume that $(\mathrm{id} - K)|_{\partial U}$ is odd.

In a similar fashion, we obtain an infinite-dimensional analog of Proposition 4.14.

**Proposition 4.31.** *If $X$ is a Banach space, $U \subset X$ is symmetric, bounded, and open, with $0 \in U$, and $\varphi = \mathrm{id} - K$, with $K : \overline{U} \to X$ a compact map, is such that $0 \notin \varphi(\partial U)$ and*

$$\frac{\varphi(x)}{\|\varphi(x)\|} \neq \frac{\varphi(-x)}{\|\varphi(-x)\|} \quad \text{for all } x \in \partial U,$$

*then $d_{\mathrm{LS}}(\varphi, U, 0)$ is an odd integer.*

An application of Theorem 4.29 is the following result, known as the *invariance of domain theorem*, which gives sufficient conditions for a map to be open (that is, to map open sets to open ones).

**Theorem 4.32.** *Let $X$ be a Banach space and $U \subset X$ be nonempty and open. Let $\varphi : U \to X$ be a compact perturbation of the identity (i.e., $\varphi = \mathrm{id} - K$, with $K : U \to X$ compact) and locally one-to-one [i.e., for every $x \in U$ there exists a neighborhood $V(x)$ of $x$ such that $\varphi|_{V(x)}$ is one-to-one]. Then $\varphi(U)$ is open.*

*Proof.* Given any $x \in U$, we need to find a neighborhood of $\varphi(x)$ contained in $\varphi(U)$. Up to dealing with $\varphi(\cdot + x) - \varphi(x)$ (defined on $U - x$) instead of $\varphi$, we may assume that $x = 0$ and $\varphi(x) = 0$. Let $r > 0$ be such that $\hat{\varphi} := \varphi|_{\overline{B_r(0)}}$ is one-to-one. Note that $0 \notin \hat{\varphi}(\partial B_r(0))$.

We claim that

$$d_{\mathrm{LS}}(\hat{\varphi}, B_r(0), 0) \neq 0. \tag{4.5}$$

If this is true, then Theorem 4.24(f) yields $d_{\mathrm{LS}}(\hat{\varphi}, B_r(0), y) = d_{\mathrm{LS}}(\hat{\varphi}, B_r(0), 0) \neq 0$ for all $y \in B_\rho(0)$, where $\rho = d(0, \hat{\varphi}(\partial B_r(0)))$, so that Theorem 4.24(e) implies that $B_\rho(0) \subset \varphi(B_r(0)) \subset \varphi(U)$, and we are done.

To prove (4.5), consider the map $h : [0,1] \times \overline{B_r(0)} \to X$ defined by

$$h(t,x) = K\left(\frac{x}{1+t}\right) - K\left(-\frac{tx}{1+t}\right) \quad \text{for all } t \in [0,1], \text{ all } x \in \overline{B_r(0)}.$$

Since $K|_{\overline{B_r(0)}}$ is compact, it is clear that $h$ is compact. Let $\varphi_t = \mathrm{id} - h(t, \cdot)$ for all $t \in [0,1]$. We see that $0 \notin \varphi_t(\partial B_r(0))$ for all $t \in [0,1]$. Indeed, if for some $(t,x) \in [0,1] \times \partial B_r(0)$ we have

$$0 = x - h(t,x) = \hat{\varphi}\left(\frac{x}{1+t}\right) - \hat{\varphi}\left(-\frac{tx}{1+t}\right),$$

then $\frac{x}{1+t} = -\frac{tx}{1+t}$ (since $\hat{\varphi}$ is one-to-one), which implies that $x = 0$, a contradiction of the fact that $x \in \partial B_r(0)$. Now, applying Theorems 4.24(c) and 4.29 (since $\varphi_1$ is odd), we obtain $d_{\mathrm{LS}}(\hat{\varphi}, B_r(0), 0) = d_{\mathrm{LS}}(\varphi_0, B_r(0), 0) = d_{\mathrm{LS}}(\varphi_1, B_r(0), 0) \neq 0$. This proves (4.5), which completes the proof.                                                                                   $\square$

It is clear from the definition of Brouwer's degree that if $L \in \mathcal{L}(\mathbb{R}^N)$ is invertible and $U \subset \mathbb{R}^N$ is nonempty, bounded, and open, with $0 \notin \partial U$, then $d(L,U,0) = \operatorname{sgn} \det L$ if $0 \in U$, and $d(L,U,0) = 0$ if $0 \notin U$. This result has the following counterpart in the Leray–Schauder theory.

**Proposition 4.33.** *If $X$ is a Banach space, $U \subset X$ is nonempty, bounded, and open, with $0 \notin \partial U$, $\varphi = \mathrm{id} - L$ is one-to-one, with $L \in \mathcal{L}_c(X)$, then $d_{\mathrm{LS}}(\varphi, U, 0) = (-1)^m$ if $0 \in U$, with $m$ being the sum of the multiplicities of the eigenvalues of $L$ that belong to $(1, +\infty)$, and $d_{\mathrm{LS}}(\varphi, U, 0) = 0$ if $0 \notin U$.*

The next proposition relates the degree of a compact perturbation of the identity to the degree of its differential.

**Proposition 4.34.** *Let $X$ be a Banach space, $U \subset X$ nonempty and open, and $K \in C^1(U, X)$ compact. Then:*

(a) *For every $x \in U$ we have $K'(x) \in \mathcal{L}_c(X)$.*
(b) *Moreover, assume $x \in U$ such that $\mathrm{id} - K'(x)$ is one-to-one. Let $y = x - K(x)$. Then there is $r_0 > 0$ such that, for every $r \in (0, r_0)$, the degree $d_{\mathrm{LS}}(\mathrm{id} - K, B_r(x), y)$ is well defined and*

$$d_{\mathrm{LS}}(\mathrm{id} - K, B_r(x), y) = d_{\mathrm{LS}}(\mathrm{id} - K'(x), B_r(0), 0) = (-1)^{m(x)},$$

*where $m(x)$ denotes the sum of the multiplicities of the eigenvalues of $K'(x)$ that belong to $(1, +\infty)$.*

*Proof.* (a) Let $B \subset X$ be bounded. Let $n_0 \geq 1$ be an integer such that $x + \frac{1}{n_0} \overline{B} \subset U$. Since $K$ is of class $C^1$, the differential $K'(x)$ is the uniform limit on $\overline{B}$ of the compact maps $K_n : \overline{B} \to X$ $(n \geq n_0)$ given by $K_n(h) = n(K(x + \frac{1}{n}h) - K(x))$. By Proposition 2.5, we obtain that $K'(x)(B)$ is relatively compact in $X$, whence $K'(x)$ is compact.

(b) Set $\varphi = \mathrm{id} - K$. The map $\psi : V_x := \{h \in X : x + h \in U\} \to X$ given by

$$\psi(h) = \varphi(x + h) - y - \varphi'(x)(h) \quad \text{for all } h \in V_x$$

is compact and satisfies $\frac{\psi(h)}{\|h\|} \to 0$ as $h \to 0$. Since $\varphi'(x)$ is injective, we have $m := \min_{x \in \partial B_1(0)} \|\varphi'(x)\| > 0$. Choose $r_0 > 0$ such that $B_{r_0}(0) \subset V_x$ and $\|\psi(h)\| < m\|h\|$ for all $h \in B_{r_0}(0)$. Then we have $\varphi'(x)(h) + t\psi(h) \neq 0$ for all $h \in B_{r_0}(0)$, all $t \in [0, 1]$. Invoking Theorem 4.24(h) and Proposition 4.33, for every $r \in (0, r_0)$ we obtain

$$d_{\mathrm{LS}}(\varphi, B_r(x), y) = d_{\mathrm{LS}}(\varphi(x + \cdot) - y, B_r(0), 0) = d_{\mathrm{LS}}(\varphi'(x), B_r(0), 0) = (-1)^{m(x)}.$$

The proof is now complete.

$\square$

We conclude with the infinite-dimensional generalization of Proposition 4.19. Again, for details consult Rabinowitz [335] and Amann [11].

**Proposition 4.35.** *If $H$ is a Hilbert space, $U \subset H$ is nonempty, bounded, and open, $\varphi \in C^1(U, \mathbb{R})$, $\varphi' = \mathrm{id} - K$ with $K : \overline{U} \to H$ compact, and $x_0 \in U$ is an isolated critical point of $\varphi$ at which $\varphi$ has a local minimum, then we can find $r > 0$ such that $d_{\mathrm{LS}}(\varphi', B_r(x_0), 0) = 1$.*

## 4.3  Degree for Operators of Monotone Type

In this section, we focus on Browder–Skrypnik degree theory for operators of monotone type. We will consider maps from $X$ into its topological dual $X^*$. For a certain class of such maps, which are of monotone type and arise naturally in the study of nonlinear boundary value problems, we define a degree map that exhibits the usual properties.

### Degree for $(S)_+$-Maps

Let $X$ be a reflexive Banach space and $X^*$ its topological dual. By $\langle \cdot, \cdot \rangle$ we denote the duality brackets for the pair $(X^*, X)$. Let $U \subset X$ be nonempty, bounded, and open. In what follows, the purpose is to define a notion of degree for a demicontinuous $(S)_+$-map $f : \overline{U} \to X^*$.

Henceforth, we fix an equivalent norm $\| \cdot \|$ on $X$ fulfilling the Troyanski renorming theorem [Remark 2.47(b)], in particular such that both $X$ and $X^*$ are locally uniformly convex. This norm gives rise to a duality map $\mathscr{F} : X \to X^*$ defined by

$$\mathscr{F}(x) = \{ x^* \in X^* : \langle x^*, x \rangle = \|x\|^2 = \|x^*\|^2 \}$$

(Definition 2.44) that is single-valued, strictly monotone, a homeomorphism (its inverse is the duality $X^* \to X$), and an $(S)_+$-map (Theorem 2.48 and Proposition 2.71). Then the restriction $\mathscr{F} : \overline{U} \to X^*$ is a demicontinuous $(S)_+$-map, and the degree that we aim to construct will be normalized with respect to $\mathscr{F}$ (in place of the identity map for Brouwer's and Leray–Schauder degrees).

In the case where $\dim X < +\infty$, the assumption on $f$ is equivalent to saying that $f : \overline{U} \to X^*$ is continuous. Then we can consider Brouwer's degree $d(f, U, y^*)$ [for $y^* \in X^* \setminus f(\partial U)$] by identifying $f$ with the composition $\mathscr{F}^{-1} \circ f : \overline{U} \to X$, so that

$$d(f, U, y^*) = d(\mathscr{F}^{-1} \circ f, U, \mathscr{F}^{-1}(y^*)).$$

It will follow from Definition 4.39 that $d_{(S)_+}(f, U, y^*) = d(f, U, y^*)$.

The main tool for the construction of the degree is provided by Proposition 4.38. Before stating it, we give two preliminary lemmas. Recall that the Galerkin approximation of $f : \overline{U} \to X^*$ with respect to a finite-dimensional subspace $Y \subset X$ is the map $f_Y : \overline{U \cap Y} \to Y^*$ defined by

$$\langle f_Y(x), y \rangle_Y = \langle f(x), y \rangle \quad \text{for all } x \in \overline{U \cap Y}, \text{ all } y \in Y,$$

where by $\langle \cdot, \cdot \rangle_Y$ we denote the duality brackets for the pair $(Y^*, Y)$. By $\partial^Y(V)$ we will denote the boundary of $V \subset Y$ with respect to the topology of $Y$. The next result follows from Browder [60, Proposition 11].

**Lemma 4.36.** *Assume that* $\dim X < +\infty$. *Let* $U \subset X$ *be a bounded, open subset and* $Y \subset X$ *a linear subspace intersecting* $U$. *Let* $f : \overline{U} \to X^*$ *be a continuous map, and let* $f_Y : \overline{U \cap Y} \to Y^*$ *be its Galerkin approximation. Assume that one of the following conditions is fulfilled:*

(i) $0 \in f(\partial U) \cup f_Y(\partial^Y(U \cap Y))$ *or*
(ii) $0 \notin f(\partial U) \cup f_Y(\partial^Y(U \cap Y))$ *and* $d(f, U, 0) \neq d(f_Y, U \cap Y, 0)$.

*Then there exists* $x \in \partial U$ *such that*

$$\langle f(x), x \rangle \leq 0 \quad \text{and} \quad \langle f(x), y \rangle = 0 \quad \text{for all } y \in Y.$$

The following result from the theory of Banach spaces can be found in Floret [140, p. 30].

**Lemma 4.37.** *If* $X$ *is a reflexive Banach space,* $D \subset X$ *is bounded, and* $x \in \overline{D}^w$, *then we can find a sequence* $\{x_n\}_{n \geq 1} \subset D$ *such that* $x_n \overset{w}{\to} x$ *in* $X$.

Now let us return to the situation of a reflexive Banach space $X$ and a demicontinuous $(S)_+$-map $f : \overline{U} \to X$. Let $\{X_\alpha\}_{\alpha \in J}$ be the family of all finite-dimensional subspaces of $X$ such that $U_\alpha := U \cap X_\alpha \neq \emptyset$. Note that the restriction of $\|\cdot\|$ to $X_\alpha$ makes $X_\alpha$ and $X_\alpha^*$ locally uniformly convex. For $\alpha, \beta \in J$, we set $\alpha \leq \beta$ if $X_\alpha \subset X_\beta$, so that $J$ is a partially ordered set. Let $f_\alpha := f_{X_\alpha} : \overline{U_\alpha} \to X_\alpha^*$ be the Galerkin approximation of $f$ with respect to $X_\alpha$. The following proposition shows that Brouwer's degree of the Galerkin approximations eventually stabilizes.

**Proposition 4.38.** *If* $X$ *is reflexive,* $U \subset X$ *is nonempty, bounded, and open, and* $f : \overline{U} \to X^*$ *is a demicontinuous* $(S)_+$-*map with* $0 \notin f(\partial U)$, *then there exists* $\alpha_0 \in J$ *such that for all* $\alpha \in J$, *with* $\alpha \geq \alpha_0$ (*i.e.,* $X_{\alpha_0} \subset X_\alpha$), *we have*

$$0 \notin f_\alpha(\partial^{X_\alpha}(U_\alpha)) \quad \text{and} \quad d(f_\alpha, U_\alpha, 0) = d(f_{\alpha_0}, U_{\alpha_0}, 0).$$

*Proof.* We proceed by contradiction. Suppose that for each $\alpha \in J$ we can find $\beta \geq \alpha$ (i.e., $X_\beta \supset X_\alpha$) such that either $0 \in f_\alpha(\partial^{X_\alpha}(U_\alpha)) \cup f_\beta(\partial^{X_\beta}(U_\beta))$ or

$$0 \notin f_\alpha(\partial^{X_\alpha}(U_\alpha)) \cup f_\beta(\partial^{X_\beta}(U_\beta)) \quad \text{and} \quad d(f_\beta, U_\beta, 0) \neq d(f_\alpha, U_\alpha, 0).$$

Note that $f_\alpha$ is the Galerkin approximation of $f_\beta$ with respect to $X_\alpha$. Applying Lemma 4.36, we can find $x \in \partial^{X_\beta}(U_\beta) \subset \partial U$ such that

$$\langle f(x),x \rangle = \langle f_\beta(x),x \rangle_{X_\beta} \leq 0 \quad \text{and} \quad \langle f(x),y \rangle = \langle f_\beta(x),y \rangle_{X_\beta} = 0 \quad \text{for all } y \in X_\alpha.$$

We have shown that the set

$$D_\alpha := \{x \in \partial U : \langle f(x),x \rangle \leq 0 \text{ and } \langle f(x),y \rangle = 0 \text{ for all } y \in X_\alpha\}$$

is nonempty for all $\alpha \in J$. Moreover, if $\alpha_1,\dots,\alpha_k \in J$, then $\bigcap_{i=1}^{k} D_{\alpha_i} \neq \emptyset$ (indeed, letting $\gamma \in J$ be such that $X_\gamma = X_{\alpha_1} + \cdots + X_{\alpha_k}$, we see that $\bigcap_{i=1}^{k} D_{\alpha_i} = D_\gamma \neq \emptyset$). Thus, $\{\overline{D_\alpha}^w\}_{\alpha \in J}$ is a family of weakly closed subsets of $\overline{\partial U}^w$ that has the finite intersection property. By Alaoglu's theorem (e.g., Brezis [52, p. 66]), $\overline{\partial U}^w$ is weakly compact. Therefore,

$$\bigcap_{\alpha \in J} \overline{D_\alpha}^w \neq \emptyset.$$

Let $x_0 \in \bigcap_{\alpha \in J} \overline{D_\alpha}^w$ and $u \in X$. We choose a finite-dimensional subspace $X_\alpha$ of $X$ such that $\{x_0,u\} \subset X_\alpha$. By virtue of Lemma 4.37, we can find $\{y_n\}_{n \geq 1} \subset D_\alpha$ such that $y_n \overset{w}{\to} x_0$ in $X$. From the definition of $D_\alpha$ we have

$$\langle f(y_n),y_n \rangle \leq 0 \quad \text{and} \quad \langle f(y_n),y \rangle = 0 \quad \text{for all } y \in X_\alpha, \text{ all } n \geq 1. \tag{4.6}$$

From (4.6) it follows that

$$\limsup_{n\to\infty} \langle f(y_n),y_n - x_0 \rangle \leq 0. \tag{4.7}$$

Because $f$ is an $(S)_+$-map, from (4.7) we infer that $y_n \to x_0 \in \partial U$ in $X$. Then the demicontinuity of $f$ ensures that $f(y_n) \overset{w}{\to} f(x_0)$ in $X^*$, and so $\langle f(x_0),u \rangle = 0$ [see (4.6)]. Since $u \in X$ is arbitrary, we obtain that $f(x_0) = 0$, a contradiction of the hypothesis that $0 \notin f(\partial U)$.                                                                 $\square$

Since Brouwer's degree of the Galerkin approximations eventually stabilizes, as in the case of the Leray–Schauder degree, we are led to the following definition.

**Definition 4.39.** Let $U \subset X$ be a nonempty, bounded, open subset and $f : \overline{U} \to X^*$ a demicontinuous $(S)_+$-map such that $0 \notin f(\partial U)$. We define

$$d_{(S)_+}(f,U,0) = d(f_\alpha,U \cap X_\alpha,0) \quad \text{whenever } \alpha \in J \text{ with } \alpha \geq \alpha_0,$$

where $\alpha_0 \in J$ is as in Proposition 4.38. For $y^* \in X^* \setminus f(\partial U)$, $y^* \neq 0$, we define

$$d_{(S)_+}(f,U,y^*) = d_{(S)_+}(f-y^*,U,0).$$

As was the case with $d_{LS}$, the properties of $d_{(S)_+}$ follow from the corresponding properties of the Brouwer degree of the Galerkin approximations. First we need to introduce the admissible homotopies.

**Definition 4.40.** Let $U \subset X$ be nonempty, bounded, and open and $h : [0,1] \times \overline{U} \to X^*$. We say that $h$ is a *homotopy of class* $(S)_+$ if the following condition holds: if $t_n \to t$ in $[0,1]$, $x_n \overset{w}{\to} x$ in $X$ and $\limsup_{n \to \infty} \langle h(t_n, x_n), x_n - x \rangle \leq 0$, then $x_n \to x$ in $X$ and $h(t_n, x_n) \overset{w}{\to} h(t,x)$ in $X^*$.

**Proposition 4.41.** *Let $f_0, f_1 : \overline{U} \to X^*$ be demicontinuous $(S)_+$-maps. Then the homotopy $h : [0,1] \times \overline{U} \to X^*$ given by $h(t,x) = (1-t)f_0(x) + tf_1(x)$ for all $t \in [0,1]$, all $x \in \overline{U}$, is of class $(S)_+$.*

*Proof.* Assume that $t_n \to t$ in $[0,1]$, $x_n \overset{w}{\to} x$ in $X$, and $\limsup_{n \to \infty} \langle h(t_n, x_n), x_n - x \rangle \leq 0$, and let us check that $x_n \to x$ in $X$ [which, together with the demicontinuity of $f_0$ and $f_1$, guarantees that $h(t_n, x_n) \overset{w}{\to} h(t,x)$, and so that $h$ is a homotopy of class $(S)_+$]. We claim that

$$\limsup_{n \to \infty} \langle (1 - t_n)f_0(x_n), x_n - x \rangle \leq 0 \quad \text{and} \quad \limsup_{n \to \infty} \langle t_n f_1(x_n), x_n - x \rangle \leq 0. \quad (4.8)$$

Arguing by contradiction, assume that (4.8) is not true. Up to considering subsequences, we may assume that, say,

$$\lim_{n \to \infty} \langle (1 - t_n)f_0(x_n), x_n - x \rangle > 0 \quad \text{and} \quad \lim_{n \to \infty} \langle t_n f_1(x_n), x_n - x \rangle < 0.$$

The second inequality implies that $t_n \neq 0$ for $n$ large enough. Then $\lim_{n \to \infty} \langle f_1(x_n), x_n - x \rangle = \lim_{n \to \infty} \frac{1}{t_n} \langle t_n f_1(x_n), x_n - x \rangle < 0$, whence $x_n \to x$ [since $f_1$ is an $(S)_+$-map], and so $\lim_{n \to \infty} \langle f_1(x_n), x_n - x \rangle = 0$ (since $f_1$ is demicontinuous), which is contradictory. This proves (4.8).

Now, to show that (4.8) implies that $x_n \to x$, we distinguish two cases.

*Case 1:* $t \in (0,1]$. Then $t_n \neq 0$ for $n$ large enough, and the second inequality in (4.8) implies that $\limsup_{n \to \infty} \langle f_1(x_n), x_n - x \rangle = \limsup_{n \to \infty} \frac{1}{t_n} \langle t_n f_1(x_n), x_n - x \rangle \leq 0$, whence $x_n \to x$ since $f_1$ is an $(S)_+$-map.

*Case 2:* $t = 0$. Then $t_n \neq 1$ for $n$ large enough, and the first inequality in (4.8) implies that $\limsup_{n \to \infty} \langle f_0(x_n), x_n - x \rangle = \limsup_{n \to \infty} \frac{1}{1-t_n} \langle (1-t_n)f_0(x_n), x_n - x \rangle \leq 0$, whence $x_n \to x$ since $f_0$ is an $(S)_+$-map. $\qquad \square$

The next theorem summarizes the main properties of the degree $d_{(S)_+}$. Recall that the standing assumptions are as follows: $X$ is a reflexive Banach space endowed with an equivalent norm $\| \cdot \|$, making $X$ and $X^*$ locally uniformly convex, $U \subset X$ is nonempty, bounded, and open, and $f : \overline{U} \to X^*$ is a demicontinuous $(S)_+$-map.

**Theorem 4.42.**

(a) *Normalization:* $d_{(S)_+}(\mathscr{F},U,y^*) = 1$ *if* $y^* \in \mathscr{F}(U)$, *where* $\mathscr{F} : X \to X^*$ *is the duality map corresponding to the norm* $\|\cdot\|$;
(b) *Domain additivity: if* $U_1,U_2 \subset U$ *are nonempty, disjoint, open sets and* $y^* \notin f(\partial U_1) \cup f(\partial U_2)$, *then*

$$d_{(S)_+}(f,U_1 \cup U_2,y^*) = d_{(S)_+}(f,U_1,y^*) + d_{(S)_+}(f,U_2,y^*);$$

(c) *Homotopy invariance: if* $h : [0,1] \times \overline{U} \to X^*$ *is a homotopy of class* $(S)_+$, $y^* : [0,1] \to X^*$ *is a continuous map, and* $y^*(t) \notin h(t,\partial U)$ *for all* $t \in [0,1]$, *then*

$$d_{(S)_+}(h(t,\cdot),U,y^*(t)) \text{ is independent of } t \in [0,1];$$

(d) *Excision: if* $C \subset \overline{U}$ *is closed and* $y^* \notin f(C) \cup f(\partial U)$, *then*

$$d_{(S)_+}(f,U,y^*) = d_{(S)_+}(f,U \setminus C,y^*);$$

(e) *Solution property: if* $d_{(S)_+}(f,U,y^*) \neq 0$, *then there exists* $x \in U$ *such that* $f(x) = y^*$;
(f) *Boundary dependence: if* $f,g : \overline{U} \to X^*$ *are both demicontinuous* $(S)_+$-*maps such that* $f|_{\partial U} = g|_{\partial U}$ *and* $y^* \notin f(\partial U)$, *then*

$$d_{(S)_+}(f,U,y^*) = d_{(S)_+}(g,U,y^*).$$

Next we will compute the degree for certain potential maps that are $(S)_+$. These results are useful in the degree theoretic methods for the study of nonlinear boundary value problems. We start with the following finite-dimensional result due to Amann [11]. For $\lambda,\mu \in \mathbb{R}$ we denote $\{\varphi < \mu\} = \{x \in U : \varphi(x) < \mu\}$, $\{\varphi \leq \lambda\} = \{x \in U : \varphi(x) \leq \lambda\}$, $\{\lambda \leq \varphi \leq \mu\} = \{x \in U : \lambda \leq \varphi(x) \leq \mu\}$ (whenever $\lambda \leq \mu$).

**Proposition 4.43.** *Let* $U \subset \mathbb{R}^N$ *be nonempty and open. Let* $\varphi \in C^1(U,\mathbb{R})$ *be such that there exist* $\lambda,\mu \in \mathbb{R}$, $\lambda < \mu$, $r > 0$, *and* $x_0 \in U$ *satisfying the following properties:*

(i) $V := \{\varphi < \mu\}$ *is bounded and such that* $\overline{V} \subset U$;
(ii) $\{\varphi \leq \lambda\} \subset \overline{B_r(x_0)} \subset V$;
(iii) $\varphi'(x) \neq 0$ *for all* $x \in \{\lambda \leq \varphi \leq \mu\}$.
    *Then* $d(\varphi',V,0) = 1$.

*Remark 4.44.* A careful reading of part (iv) of the proof of the theorem in Amann [11] reveals that hypothesis (ii) in Proposition 4.43 can be replaced by the following weaker one:

(ii)$'$ $x \in \{\varphi \leq \lambda\} \Rightarrow \{tx + (1-t)x_0 : t \in [0,1]\} \subset V$.

Using Galerkin approximations, we will extend this result to $(S)_+$-maps on reflexive Banach spaces.

**Theorem 4.45.** *Let $X$ be reflexive and $U \subset X$ be nonempty and open. Let $\varphi : U \to \mathbb{R}$ be Gâteaux differentiable and continuous such that $\varphi'$ is a demicontinuous $(S)_+$-map and for which there exist $\lambda, \mu \in \mathbb{R}$, $\lambda < \mu$, $x_0 \in X$, satisfying the following properties:*

(i) $V := \{\varphi < \mu\}$ *is bounded and* $\overline{V} \subset U$;
(ii) $x \in \{\varphi \le \lambda\} \Rightarrow \{tx + (1-t)x_0 : t \in [0,1]\} \subset V$;
(iii) $\varphi'(x) \ne 0$ *for all* $x \in \{\lambda \le \varphi \le \mu\}$.

*Then* $d_{(S)_+}(\varphi', V, 0) = 1$.

*Proof.* Since $\partial V \subset \varphi^{-1}(\mu)$, from hypothesis (iii), we see that $0 \notin \varphi'(\partial V)$. Therefore, $d_{(S)_+}(\varphi', V, 0)$ is well defined. We introduce

$$\mathscr{S} = \{Y \subset X : Y \text{ is a finite-dimensional subspace of } X, \ V \cap Y \ne \emptyset\}.$$

For $Y \in \mathscr{S}$, let $\varphi_Y = \varphi|_{U \cap Y} : U \cap Y \to \mathbb{R}$.

*Claim 1:* There exists $Y_0 \in \mathscr{S}$ such that, for all $Y \supset Y_0$, we have $(\varphi_Y)'(x) \ne 0$ for all $x \in U \cap Y$ such that $\lambda \le \varphi(x) \le \mu$.

Suppose that Claim 1 is not true. Then for all $Y \in \mathscr{S}$ we can find $\tilde{Y} \in \mathscr{S}$ with $\tilde{Y} \supset Y$ and $x_Y \in U \cap \tilde{Y}$ such that $\lambda \le \varphi(x_Y) \le \mu$ and $(\varphi_{\tilde{Y}})'(x_Y) = 0$, whence $\langle \varphi'(x_Y), u \rangle = 0$ for all $u \in \tilde{Y}$. In particular, $x_Y$ belongs to the set

$$D_Y := \{x \in \{\lambda \le \varphi \le \mu\} : \langle \varphi'(x), x \rangle = 0 \text{ and } \langle \varphi'(x), u \rangle = 0 \text{ for all } u \in Y\}.$$

Note that $\overset{k}{\underset{i=1}{\cap}} D_{Y_i} = D_{Y_1 + \ldots + Y_k} \ne \emptyset$ for all $Y_1, \ldots, Y_k \in \mathscr{S}$, hence the family $\{\overline{D_Y}^{\mathrm{w}}\}_{Y \in \mathscr{S}}$ has the finite intersection property. We see that $\{\lambda \le \varphi \le \mu\} \subset \overline{V}$ [indeed, otherwise, we find $x \in \{\lambda \le \varphi \le \mu\} \setminus \overline{V}$, which is then a minimizer of $\varphi|_{U \setminus \overline{V}}$, whence $\varphi'(x) = 0$; this contradicts hypothesis (iii)]. Thus $\overline{D_Y}^{\mathrm{w}} \subset \overline{V}^{\mathrm{w}}$ for all $Y \in \mathscr{S}$. Since $V$ is bounded [hypothesis (i)], we know that $\overline{V}^{\mathrm{w}}$ is weakly compact. Therefore,

$$D_0 := \underset{Y \in \mathscr{S}}{\cap} \overline{D_Y}^{\mathrm{w}} \ne \emptyset.$$

Let $x \in D_0$, $u \in X$, and consider $Y \in \mathscr{S}$ such that $\{x, u\} \subset Y$. Since $x \in \overline{D_Y}^{\mathrm{w}}$, according to Lemma 4.37, we can find $\{x_n\}_{n \ge 1} \subset D_Y$ such that $x_n \overset{\mathrm{w}}{\to} x$ in $X$ as $n \to \infty$. The fact that $x_n \in D_Y$ implies that $\langle \varphi'(x_n), x_n - x \rangle = 0$ for all $n \ge 1$. Since $\varphi'$ is an $(S)_+$-map, we infer that $x_n \to x$ in $X$, and so

$$\langle \varphi'(x), u \rangle = \lim_{n \to \infty} \langle \varphi'(x_n), u \rangle = 0$$

(using that $x_n \in D_Y$ and $u \in Y$). Since $u \in X$ is arbitrary, we obtain that $\varphi'(x) = 0$. Moreover $x = \lim_{n \to \infty} x_n \in \{\lambda \le \varphi \le \mu\}$. This contradicts hypothesis (iii). Thus Claim 1 is proved.

According to Claim 1, for every $Y \in \mathscr{S}$ with $Y \supset Y_0$, $\varphi_Y$ satisfies all the hypotheses of Proposition 4.43 and Remark 4.44. It follows that

$$d((\varphi_Y)', V \cap Y, 0) = 1 \text{ for all } Y \in \mathscr{S}, Y \supset Y_0.$$

Finally, Definition 4.39 implies that $d_{(S)_+}(\varphi', V, 0) = 1$ [where we use that $(\varphi_Y)'$ coincides with the Galerkin approximation of $\varphi'$ with respect to $Y \in \mathscr{S}$].          □

A first consequence of Theorem 4.45 is the following corollary.

**Corollary 4.46.** *Let $X$ be reflexive, and let $\varphi : X \to \mathbb{R}$ be a Gâteaux differentiable and continuous function, with $\varphi' : X \to X^*$ a demicontinuous $(S)_+$-map. Assume that $\varphi(x) \to +\infty$ as $\|x\| \to +\infty$ and that there is $r_0 > 0$ such that $\varphi'(x) \neq 0$ for all $\|x\| \geq r_0$. Then there exists $r_1 \geq r_0$ such that*

$$d_{(S)_+}(\varphi', B_r(0), 0) = 1 \text{ for all } r \in [r_1, +\infty).$$

*Proof.* Take $\lambda > \sup\{\varphi(x) : x \in B_{r_0}(0)\}$, and let $r_1 = \sup\{\|x\| : x \in \{\varphi \leq \lambda\}\}$. Given $r \geq r_1$, let $\mu > \max\{\lambda, \sup\{\varphi(x) : x \in B_r(0)\}\}$. The corollary follows by applying Theorem 4.45 with $x_0 = 0$, $U = X$, $V = \{\varphi < \mu\}$, and then by invoking the excision property [Theorem 4.42(d)], with $C = \overline{V} \setminus B_r(0)$.          □

Before stating another useful consequence of Theorem 4.45, we need two preliminary lemmas. We state them under a general form that will be needed later in the book.

**Lemma 4.47.** *Let $X$ be a reflexive Banach space, $U \subset X$ a nonempty, convex, and open set, and $\varphi \in C^1(U, \mathbb{R})$ be such that $\varphi' : U \to X^*$ is an $(S)_+$-map. Then $\varphi$ is sequentially weakly lower semicontinuous (l.s.c.).*

*Proof.* Arguing by contradiction, assume that we can find $\{x_n\}_{n \geq 1} \subset U$ and $x \in U$ such that $x_n \xrightarrow{w} x$ in $X$ and

$$\lim_{n \to \infty} \varphi(x_n) < \varphi(x). \tag{4.9}$$

By the mean value theorem, we can find $t_n \in (0, 1)$ such that

$$\varphi(x_n) - \varphi(x) = \langle \varphi'(x + t_n(x_n - x)), x_n - x \rangle. \tag{4.10}$$

Using (4.9) and (4.10), we get

$$\limsup_{n \to \infty} \langle \varphi'(x + t_n(x_n - x)), t_n(x_n - x) \rangle \leq 0.$$

Since $\varphi'$ is an $(S)_+$-map, we derive that $x + t_n(x_n - x) \to x$ in $X$, so $\varphi(x_n) \to \varphi(x)$ [using (4.10) and the continuity of $\varphi'$], a contradiction of (4.9). This proves the lemma.          □

**Lemma 4.48.** *Let $X$ be a reflexive Banach space, $U \subset X$ a nonempty, open set, $\varphi :$ $U \to \mathbb{R}$ a sequentially weakly l.s.c., Gâteaux differentiable map such that $\varphi' : U \to \mathbb{R}$ is an $(S)_+$-map, $\psi : X \to \mathbb{R} \cup \{+\infty\}$ a sequentially weakly l.s.c. map, $x_0 \in U$ a strict local minimizer of $\varphi + \psi$, and $r_0 > 0$ be such that $\overline{B}_{r_0}(x_0) \subset U$ and*

$$\varphi(x) + \psi(x) > \varphi(x_0) + \psi(x_0) \quad \text{for all } x \in \overline{B}_{r_0}(x_0) \setminus \{x_0\}.$$

*Then, for all $r \in (0, r_0]$, we have*

$$\inf \left\{ \varphi(x) + \psi(x) : x \in \overline{B}_{r_0}(x_0) \setminus B_r(x_0) \right\} > \varphi(x_0) + \psi(x_0).$$

*Proof.* Arguing by contradiction, suppose that we can find $r \in (0, r_0]$ and $\{x_n\}_{n \geq 1} \subset \overline{B}_{r_0}(x_0) \setminus B_r(x_0)$ such that

$$\lim_{n \to \infty} (\varphi(x_n) + \psi(x_n)) = \varphi(x_0) + \psi(x_0). \tag{4.11}$$

Since $X$ is reflexive and $\|x_n\| \leq r_0$ for all $n \geq 1$, we may assume that $x_n \xrightarrow{w} x$ in $X$ for some $x \in \overline{B}_{r_0}(x_0)$. Since $\varphi$ and $\psi$ are sequentially weakly l.s.c., we have

$$\varphi(x) + \psi(x) \leq \liminf_{n \to \infty}(\varphi(x_n) + \psi(x_n)) = \lim_{n \to \infty}(\varphi(x_n) + \psi(x_n)) = \varphi(x_0) + \psi(x_0).$$

By the assumption on $x_0$, it follows that $x = x_0$. The mean value theorem yields $t_n \in (0,1)$ such that

$$\varphi(x_n) - \varphi\left(\frac{x_n + x_0}{2}\right) = \left\langle \varphi'(z_n), \frac{x_n - x_0}{2} \right\rangle,$$

with $z_n = t_n x_n + (1 - t_n)\frac{x_n + x_0}{2}$. Then, due to the fact that $\frac{x_n + x_0}{2} \xrightarrow{w} x_0$ in $X$, (4.11), and the fact that $\varphi$ and $\psi$ are sequentially weakly l.s.c., we obtain

$$\limsup_{n \to \infty}\left\langle \varphi'(z_n), \frac{x_n - x_0}{2} \right\rangle$$

$$= \lim_{n \to \infty} (\varphi(x_n) + \psi(x_n)) - \liminf_{n \to \infty}\left( \varphi\left(\frac{x_n + x_0}{2}\right) + \psi(x_n) \right) \leq 0,$$

whence

$$\limsup_{n \to \infty}\langle \varphi'(z_n), z_n - x_0 \rangle = \limsup_{n \to \infty}(1 + t_n)\left\langle \varphi'(z_n), \frac{x_n - x_0}{2} \right\rangle \leq 0.$$

Since $\varphi'$ is an $(S)_+$-map and $z_n \xrightarrow{w} x_0$, this implies that $z_n \to x_0$ in $X$. However,

$$\|z_n - x_0\| = \frac{1 + t_n}{2}\|x_n - x_0\| \geq \frac{r}{2} \quad \text{for all } n \geq 1,$$

which is contradictory. This proves the lemma. □

**Corollary 4.49.** *Let $X$ be reflexive, $U \subset X$ be nonempty and open, and $\varphi \in C^1(U, \mathbb{R})$ be such that $\varphi' : U \to X^*$ is an $(S)_+$-map. Assume that $\varphi$ has a local minimum at $x_0 \in U$ and that $x_0$ is an isolated critical point of $\varphi$. Then we can find $\tilde{r}_1 > 0$ such that*

$$d_{(S)_+}(\varphi', B_r(x_0), 0) = 1 \text{ for all } r \in (0, \tilde{r}_1].$$

*Proof.* Let $\tilde{\varphi}(x) = \varphi(x) - \varphi(x_0)$ for all $x \in U$. We take $\tilde{U} \subset U$ a convex, open neighborhood of $x_0$, so that, by Lemma 4.47, $\tilde{\varphi}|_{\tilde{U}}$ is sequentially weakly l.s.c. Because $x_0$ is a local minimizer and an isolated critical point of $\tilde{\varphi}$, we can find $r_0 > 0$ such that $\overline{B_{r_0}(x_0)} \subset \tilde{U}$ and

$$0 = \tilde{\varphi}(x_0) < \tilde{\varphi}(x) \text{ and } \tilde{\varphi}'(x) \neq 0 \text{ for all } x \in \overline{B_{r_0}(x_0)} \setminus \{x_0\}. \tag{4.12}$$

Applying Lemma 4.48 with $\tilde{\varphi}$ and $\psi = 0$, we have

$$\mu_r := \inf\{\tilde{\varphi}(x) : x \in \overline{B_{r_0}(x_0)} \setminus B_r(x_0)\} > 0 \quad \text{for all } r \in (0, r_0].$$

Let

$$V = \{x \in B_{r_0}(x_0) : \tilde{\varphi}(x) < \mu_{\frac{r_0}{2}}\}.$$

Clearly, $V$ is nonempty and open and $V \subset B_{\frac{r_0}{2}}(x_0)$. Fix $\tilde{r}_1 \in (0, \frac{r_0}{2})$ such that $\overline{B_{\tilde{r}_1}(x_0)} \subset V$. Let $r \in (0, \tilde{r}_1]$, and choose $\lambda \in (0, \mu_r)$. Note that

$$\{x \in B_{r_0}(x_0) : \tilde{\varphi}(x) \leq \lambda\} \subset B_r(x_0) \subset \overline{B_r(x_0)} \subset V. \tag{4.13}$$

On the basis of (4.12) and (4.13), we can apply Theorem 4.45 to the function $\tilde{\varphi}|_{B_{r_0}(x_0)}$ and the numbers $\lambda < \mu_{\frac{r_0}{2}}$, which yields $d_{(S)_+}(\varphi', V, 0) = 1$. Since $0 \notin \varphi'(C)$ for $C = \overline{V} \setminus B_r(x_0)$, the excision property implies $d_{(S)_+}(\varphi', B_r(x_0), 0) = 1$.  $\square$

**Degree for Operators of Monotone Type**

Our next purpose is to develop a degree theory for multifunctions of the form $f + A$, where $f$ is a bounded, demicontinuous $(S)_+$-map and $A$ is a maximal monotone map with $(0, 0) \in \mathrm{Gr}\, A$. Such maps arise in the study of variational inequalities. For $A = 0$ this degree theory will coincide with the degree theory for $(S)_+$-maps developed previously.

The mathematical setting is the following: $X$ is a reflexive Banach space furnished with a norm such that both $X$ and $X^*$ are locally uniformly convex; the Troyanski renorming theorem [Remark 2.47(b)] ensures the existence of such a norm. Also, $U \subset X$ is nonempty, bounded, and open, $f : \overline{U} \to X^*$ is a bounded, demicontinuous $(S)_+$-map, and $A : D(A) \subset X \to 2^{X^*}$ is a maximal monotone map such that $(0,0) \in \mathrm{Gr}\,A$. We assume that $y^* \notin (f+A)(\partial U)$ [if $T$ (here $T = f+A$) is a multifunction and $C \subset X$ a subset, then we denote $T(C) = \bigcup_{x \in C} T(x)$]. We will define a degree on such triples $(f+A, U, y^*)$.

For all $\lambda > 0$ we consider the map $x \mapsto (f+A_\lambda)(x)$, where $A_\lambda$ is the Yosida approximation of $A$ [see (2.9)]. From Propositions 2.56(a), (c) and 2.70(b) we have that $x \mapsto (f+A_\lambda)(x)$ is a bounded, demicontinuous $(S)_+$-map. The following proposition is the basis for defining the degree on triples $(f+A, U, y^*)$, as previously. Its proof can be found in Browder [60, Theorem 8].

**Proposition 4.50.** *If $(f+A, U, y^*)$ is a triple as previously, then there exists $\lambda_0 > 0$ such that*

(a)  $y^* \notin (f+A_\lambda)(\partial U)$ *whenever* $0 < \lambda \leq \lambda_0$ *[hence the degree* $d_{(S)_+}(f+A_\lambda, U, y^*)$ *is well defined];*
(b)  $d_{(S)_+}(f+A_\lambda, U, y^*)$ *is independent of* $\lambda \in (0, \lambda_0]$.

According to Proposition 4.50, the degree $d_{(S)_+}(f+A_\lambda, U, 0)$ stabilizes for $\lambda > 0$ small, which leads to the following definition.

**Definition 4.51.** Let $U \subset X$ be a nonempty, bounded, open subset, $f : \overline{U} \to X^*$ a bounded, demicontinuous $(S)_+$-map, $A : X \to 2^{X^*}$ a maximal monotone map with $(0,0) \in \mathrm{Gr}\,A$, and $y^* \in X^* \setminus (f+A)(\partial U)$. We define

$$d_M(f+A, U, y^*) = d_{(S)_+}(f+A_\lambda, U, y^*) \quad \text{whenever } \lambda \in (0, \lambda_0],$$

where $\lambda_0 > 0$ is as in Proposition 4.50.

By Definition 4.51, it is clear that in the case of $A = 0$ we get $d_M(f, U, y^*) = d_{(S)_+}(f, U, y^*)$. The admissible homotopies for the degree $d_M$ will be obtained by combining the class of homotopies of class $(S)_+$ introduced in Definition 4.40 and a second class of homotopies, introduced in the next definition.

**Definition 4.52.** Let $\{A^{(t)}\}_{t \in [0,1]}$ be a family of maximal monotone maps from $X$ into $2^{X^*}$ such that $(0,0) \in \mathrm{Gr}\,A^{(t)}$ for all $t \in [0,1]$. We say that $\{A^{(t)}\}_{t \in [0,1]}$ is a *pseudomonotone homotopy* if it satisfies the following mutually equivalent properties:

(a)  If $t_n \to t$ in $[0,1]$, $(x_n, x_n^*) \in \mathrm{Gr}\,A^{(t_n)}$ for all $n \geq 1$, $x_n \xrightarrow{w} x$ in $X$, $x_n^* \xrightarrow{w} x^*$ in $X^*$, and $\limsup_{n \to \infty} \langle x_n^*, x_n \rangle \leq \langle x^*, x \rangle$, then $(x, x^*) \in \mathrm{Gr}\,A^{(t)}$ and $\langle x_n^*, x_n \rangle \to \langle x^*, x \rangle$.
(b)  The map $h : [0,1] \times X^* \to X$ defined by $h(t, x^*) = (A^{(t)} + \mathscr{F})^{-1}(x^*)$ for all $t \in [0,1]$, all $x^* \in X^*$, is continuous when both $X^*$ and $X$ are furnished with their norm topologies.
(c)  For every $x^* \in X^*$, the map $h_{x^*} : [0,1] \to X$ defined by $h_{x^*}(t) = (A^{(t)} + \mathscr{F})^{-1}(x^*)$ for all $t \in [0,1]$ is continuous into $X$ furnished with the norm topology.
(d)  If $t_n \to t$ in $[0,1]$ and $(x, x^*) \in \mathrm{Gr}\,A^{(t)}$, then there exists $(x_n, x_n^*) \in \mathrm{Gr}\,A^{(t_n)}$, $n \geq 1$, such that $x_n \to x$ in $X$ and $x_n^* \to x^*$ in $X^*$.

The next theorem summarizes the basic properties of the degree map $d_M$ and is a consequence of Definition 4.51 and Theorem 4.42.

**Theorem 4.53.**

(a) *Normalization:* $d_M(\mathscr{F}, U, y^*) = 1$ *if* $y^* \in \mathscr{F}(U)$, *where* $\mathscr{F} : X \to X^*$ *is the duality map corresponding to the considered norm;*

(b) *Domain additivity: if* $U_1, U_2 \subset U$ *are nonempty, disjoint, open sets and* $y^* \notin (f+A)(\partial U_1) \cup (f+A)(\partial U_2)$, *then*

$$d_M(f+A, U_1 \cup U_2, y^*) = d_M(f+A, U_1, y^*) + d_M(f+A, U_2, y^*);$$

(c) *Homotopy invariance: if* $h : [0,1] \times \overline{U} \to X^*$ *is a homotopy of class* $(S)_+$ *such that* $h(t, \cdot)$ *is bounded for every* $t \in [0,1]$, $\{A^{(t)}\}_{t \in [0,1]}$ *is a pseudomonotone homotopy,* $y^* : [0,1] \to X^*$ *is a continuous map and* $y^*(t) \notin (h(t, \cdot) + A^{(t)})(\partial U)$ *for all* $t \in [0,1]$, *then* $d_M(h(t, \cdot) + A^{(t)}, U, y^*(t))$ *is independent of* $t \in [0,1]$;

(d) *Excision: if* $C \subset \overline{U}$ *is closed and* $y^* \notin (f+A)(C)$, *then*

$$d_M(f+A, U, y^*) = d_M(f+A, U \setminus C, y^*).$$

(e) *Solution property: if* $d_M(f+A, U, y^*) \neq 0$, *then there exists* $x \in U$ *such that* $y^* \in f(x) + A(x)$.

While for the previous degrees affine homotopies were admissible and a useful tool, in contrast, for $d_M$ such homotopies may fail to be admissible.

**Proposition 4.54.** *Let* $A : D(A) \subset X \to 2^{X^*}$ *and* $T : D(T) \subset X \to 2^{X^*}$ *be two maximal monotone maps such that* $(0,0) \in \operatorname{Gr} A \cap \operatorname{Gr} T$, *and let* $V : X \to X^*$ *be a continuous, monotone map (and so* $V$ *is maximal monotone; see Corollary 2.42).*

(a) *If* $\overline{D(A)} \neq \overline{D(T)}$, *then* $G^{(t)} = (1-t)A + tT$, $t \in [0,1]$, *is not a pseudomonotone homotopy.*

(b) *If* $\overline{D(A)} = X$, *then* $G^{(t)} = (1-t)V + tA$, $t \in [0,1]$, *is a pseudomonotone homotopy.*

*Proof.* (a) Assume without any loss of generality that $\overline{D(A)} \not\subset \overline{D(T)}$, whence there is $x_0 \in D(A) \setminus \overline{D(T)}$. Let $x_0^* \in A(x_0)$, and let $t_n = \frac{1}{n+1}$ for $n \geq 1$. If $\{G^{(t)}\}_{t \in [0,1]}$ were a pseudomonotone homotopy, then, by virtue of Definition 4.52(d), we could find $\{(x_n, x_n^*)\}_{n \geq 1} \subset \operatorname{Gr} G^{(t_n)}$ such that $x_n \to x_0$ in $X$ and $x_n^* \to x_0^*$ in $X^*$. Then $x_n \in D(G^{(t_n)}) = D(A) \cap D(T)$ for all $n \geq 1$, hence $x_0 \in \overline{D(T)}$, a contradiction. This shows that $\{G^{(t)}\}_{t \in [0,1]}$ is not a pseudomonotone homotopy.

(b) Let $\{t_n\}_{n \geq 1}$ be a sequence such that $t_n \to t \in [0,1]$, and let $(x, x^*) \in \operatorname{Gr} G^{(t)}$.

  *Case 1:* $t \in (0,1]$. Then $x^* = (1-t)V(x) + tu^*$ with $u^* \in A(x)$. We set $x_n = x$ and $x_n^* = (1-t_n)V(x) + t_n u^*$ for all $n \geq 1$. Then $(x_n, x_n^*) \in \operatorname{Gr} G^{(t_n)}$ for all $n \geq 1$, $x_n \to x$, and $x_n^* \to x^*$ in $X^*$.

*Case 2:* $t = 0$. We construct the sequences $\{x_n\}_{n\geq 1}$, $\{x_n^*\}_{n\geq 1}$ as follows. For $n \geq 1$ such that $t_n = 0$ we set $x_n = x$ and $x_n^* = x^*$; and for $n \geq 1$ such that $t_n > 0$ we set $x_n = J_{t_n}^A(x)$ and $x_n^* = (1 - t_n)V(x_n) + t_n A_{t_n}(x)$. Then $(x_n, x_n^*) \in \mathrm{Gr}\, G^{(t_n)}$ for all $n \geq 1$. From Proposition 2.56(d) we know $x_n \to x$ in $X$ [recall that by hypothesis $\overline{D(A)} = X$]. Hence $V(x_n) \to V(x)$ in $X^*$. For $n \geq 1$ such that $t_n > 0$ we see that

$$\|t_n A_{t_n}(x)\| = \|\mathscr{F}(x_n - x)\| = \|x_n - x\| \to 0 \text{ as } n \to \infty$$

[see (2.9)]. Therefore, we have $x_n^* \to V(x) = x^*$ in $X^*$, and so $\{G^{(t)}\}_{t \in [0,1]}$ is a pseudomonotone homotopy.

$\square$

Next we will prove an extended version of the normalization property, which is convenient when $D(A)$ is not dense in $X$. First a lemma.

**Lemma 4.55.** *Let* $C \subset X$ *be a nonempty, closed set,* $f : C \to X^*$ *a bounded, demicontinuous* $(S)_+$-map, *and* $A : D(A) \subset X \to 2^{X^*}$ *a maximal monotone map such that* $(0,0) \in \mathrm{Gr}\, A$. *Let* $\{\lambda_n\}_{n\geq 1} \subset (0, +\infty)$ *and* $\{x_n\}_{n\geq 1} \subset \overline{U}$ *such that* $\lambda_n \downarrow 0$, $x_n \xrightarrow{w} x$ *in* $X$, $A_{\lambda_n}(x_n) \xrightarrow{w} x^*$ *in* $X^*$, *and* $f(x_n) + A_{\lambda_n}(x_n) \to y^*$ *in* $X^*$. *Then* $x_n \to x$ *in* $X$, $(x, x^*) \in \mathrm{Gr}\, A$, *and* $f(x) + x^* = y^*$.

*Proof.* We claim that

$$\liminf_{n\to\infty} \langle f(x_n), x_n - x \rangle \geq 0. \tag{4.14}$$

Suppose that (4.14) is not true. Then we can find a subsequence of $\{x_n\}_{n\geq 1}$ (still denoted for notational simplicity by the same index) such that

$$\lim_{n\to\infty} \langle f(x_n), x_n - x \rangle < 0. \tag{4.15}$$

Since $f$ is an $(S)_+$-map, from (4.15) it follows that $x_n \to x$ in $X$, hence $\langle f(x_n), x_n - x \rangle \to 0$, a contradiction of (4.15). This proves (4.14).

We have

$$\|J_{\lambda_n}^A(x_n) - x_n\| = \lambda_n \|A_{\lambda_n}(x_n)\| \to 0 \text{ as } n \to \infty \tag{4.16}$$

[see (2.9)]. Then, by (4.16), since $x_n \xrightarrow{w} x$ in $X$, and using (4.14), we have

$$\limsup_{n\to\infty} \langle A_{\lambda_n}(x_n) - x^*, J_{\lambda_n}^A(x_n) - x \rangle = \limsup_{n\to\infty} \langle A_{\lambda_n}(x_n) - x^*, x_n - x \rangle$$

$$= \limsup_{n\to\infty} \langle A_{\lambda_n}(x_n), x_n - x \rangle = \lim_{n\to\infty} \langle f(x_n) + A_{\lambda_n}(x_n), x_n - x \rangle$$

$$- \liminf_{n\to\infty} \langle f(x_n), x_n - x \rangle \leq 0,$$

which implies that

$$\limsup_{n\to\infty}\langle A_{\lambda_n}(x_n),J^A_{\lambda_n}(x_n)\rangle \le \langle x^*,x\rangle. \tag{4.17}$$

Recall that $(J^A_{\lambda_n}(x_n),A_{\lambda_n}(x_n)) \in \mathrm{Gr}A$ for all $n \ge 1$. Hence

$$\langle A_{\lambda_n}(x_n)-A_{\lambda_m}(x_m),J^A_{\lambda_n}(x_n)-J^A_{\lambda_m}(x_m)\rangle \ge 0 \text{ for all } n,m \ge 1. \tag{4.18}$$

Let $\mu = \liminf_{m\to\infty}\langle A_{\lambda_m}(x_m),J^A_{\lambda_m}(x_m)\rangle$. In (4.18) we fix $n \ge 1$ and let $m \to \infty$. We obtain

$$\langle A_{\lambda_n}(x_n),J^A_{\lambda_n}(x_n)\rangle - \langle x^*,J^A_{\lambda_n}(x_n)\rangle - \langle A_{\lambda_n}(x_n),x\rangle + \mu \ge 0,$$

which yields

$$\mu = \liminf_{n\to\infty}\langle A_{\lambda_n}(x_n),J^A_{\lambda_n}(x_n)\rangle \ge \langle x^*,x\rangle. \tag{4.19}$$

Combining (4.17) and (4.19), we infer that

$$\lim_{n\to\infty}\langle A_{\lambda_n}(x_n),J^A_{\lambda_n}(x_n)\rangle = \langle x^*,x\rangle. \tag{4.20}$$

From the monotonicity of $A$ we have

$$\langle A_{\lambda_n}(x_n)-v^*,J^A_{\lambda_n}(x_n)-v\rangle \ge 0 \text{ for all } n \ge 1, \text{ all } (v,v^*) \in \mathrm{Gr}A,$$

which, by (4.20), gives $\langle x^*-v^*,x-v\rangle \ge 0$ for all $(v,v^*) \in \mathrm{Gr}A$. The maximal monotonicity of $A$ implies $(x,x^*) \in \mathrm{Gr}A$. Also, we have $\lim_{n\to\infty}\langle A_{\lambda_n}(x_n),x_n\rangle = \langle x^*,x\rangle$ [see (4.16) and (4.20)]. Therefore,

$$\lim_{n\to\infty}\langle f(x_n),x_n-x\rangle = \lim_{n\to\infty}\langle f(x_n)+A_{\lambda_n}(x_n),x_n-x\rangle - \lim_{n\to\infty}\langle A_{\lambda_n}(x_n),x_n-x\rangle = 0,$$

which implies that $x_n \to x$ in $X$ [since $f$ is an $(S)_+$-map]. Finally, since by hypothesis $f(x_n)+A_{\lambda_n}(x_n) \to y^*$ in $X^*$ and $f$ is demicontinuous, we have $f(x)+x^* = y^*$.  □

Using this lemma, we can have the following extended normalization property for $d_M$.

**Theorem 4.56.** *If $U \subset X$ is a nonempty, bounded, open set, $A : D(A) \subset X \to 2^{X^*}$ is a maximal monotone map, $(0,0) \in \mathrm{Gr}A$, and $y^* \in (\mathscr{F}+A)(U)$, then $d_M(\mathscr{F}+A,U,y^*) = 1$.*

*Proof.* Let $x_0,x_\lambda \in X$ be the unique solutions of

$$y^* \in \mathscr{F}(x_0)+A(x_0) \text{ and } y^* = \mathscr{F}(x_\lambda)+A_\lambda(x_\lambda), \ \lambda > 0. \tag{4.21}$$

The hypothesis that $y^* \in (\mathscr{F} + A)(U)$ yields $x_0 \in U$ and ensures that $y^* \notin (\mathscr{F} + A)(\partial U)$; hence $d_M(\mathscr{F} + A, U, y^*)$ is well defined.

*Claim 1*: $x_\lambda \to x_0$ in $X$ as $\lambda \downarrow 0$.

For all $\lambda > 0$ we have

$$\langle y^*, x_\lambda \rangle = \langle \mathscr{F}(x_\lambda), x_\lambda \rangle + \langle A_\lambda(x_\lambda), x_\lambda \rangle$$

[see (4.21)], which implies that $\|x_\lambda\|^2 \le \|y^*\| \|x_\lambda\|$ [since $\langle A_\lambda(x_\lambda), x_\lambda \rangle \ge 0$], so $\{x_\lambda\}_{\lambda > 0}$ is bounded in $X$. Thus, by (4.21), $\{A_\lambda(x_\lambda)\}_{\lambda > 0}$ is bounded in $X^*$.

We consider a sequence $\{\lambda_n\}_{n \ge 1} \subset (0, +\infty)$ such that

$$\lambda_n \to 0, \quad x_{\lambda_n} \xrightarrow{w} x \text{ in } X \text{ and } A_{\lambda_n}(x_{\lambda_n}) \xrightarrow{w} x^* \text{ in } X^* \text{ as } n \to \infty. \tag{4.22}$$

Since $y^* = \mathscr{F}(x_{\lambda_n}) + A_{\lambda_n}(x_{\lambda_n})$ for all $n \ge 1$, from (4.22) and Lemma 4.55 we have $x_{\lambda_n} \to x$ in $X$, $(x, x^*) \in \mathrm{Gr}\, A$, and $y^* = \mathscr{F}(x) + x^*$, hence $y^* \in \mathscr{F}(x) + A(x)$. Due to the uniqueness of $x_0$ in (4.21), we have $x = x_0$. Then $x_\lambda \to x_0$ in $X$ as $\lambda \downarrow 0$. This proves Claim 1.

Since $x_0 \in U$, from Claim 1 and Definition 4.51 we have

$$x_\lambda \in U \text{ and } d_M(\mathscr{F} + A, U, y^*) = d_{(S)_+}(\mathscr{F} + A_\lambda, U, y^*) \text{ for } \lambda > 0 \text{ small.} \tag{4.23}$$

Because $U$ is bounded, we can find $r > 0$ such that $U \subset B_r(0)$. Since $x_\lambda \in U$ is the unique solution of $y^* = (\mathscr{F} + A_\lambda)(x_\lambda)$, we have $y^* \notin (\mathscr{F} + A_\lambda)(\overline{B_r(0)} \setminus U)$. Hence, by virtue of the excision property of the degree map $d_{(S)_+}$, we have

$$d_{(S)_+}(\mathscr{F} + A_\lambda, U, y^*) = d_{(S)_+}(\mathscr{F} + A_\lambda, B_r(0), y^*) \text{ for } \lambda > 0 \text{ small.} \tag{4.24}$$

*Claim 2*: $d_{(S)_+}(\mathscr{F} + A_\lambda, B_r(0), y^*) = 1$ for $\lambda > 0$ small.

Let $h(t, x) = \mathscr{F}(x) + t A_\lambda(x)$ for all $(t, x) \in [0, 1] \times \overline{B_r(0)}$. Reasoning as in Proposition 2.70(c) [using that $\mathscr{F}$ is an $(S)_+$-map and $A_\lambda$ is monotone] and noting that $\mathscr{F}$ and $A_\lambda$ are demicontinuous, we see that $h$ is a homotopy of class $(S)_+$ (Definition 4.40). Moreover, let us check that $t y^* \notin h(t, \partial B_r(0))$ for all $t \in [0, 1]$. Arguing by contradiction, suppose that we can find $t \in [0, 1]$ and $x \in \partial B_r(0)$ such that

$$t y^* = \mathscr{F}(x) + t A_\lambda(x). \tag{4.25}$$

Since $y^* \notin (\mathscr{F} + A_\lambda)(\overline{B_r(0)} \setminus U)$, we necessarily have that $t \in [0, 1)$. Using (4.21) and (4.25), we have

$$t(\mathscr{F}(x_\lambda) + A_\lambda(x_\lambda)) - t(\mathscr{F}(x) + A_\lambda(x)) = (1 - t)\mathscr{F}(x).$$

Then the monotonicity of $\mathscr{F}+A_\lambda$ implies that

$$0 \le (1-t)\langle \mathscr{F}(x), x_\lambda - x \rangle \le (1-t)(\|x\|\,\|x_\lambda\| - \|x\|^2) = (1-t)(r\|x_\lambda\| - r^2) < 0$$

[since $x_\lambda \in U \subset B_r(0)$ for $\lambda > 0$ small], a contradiction. By the homotopy invariance property, we now obtain

$$d_{(S)_+}(\mathscr{F}+A_\lambda, B_r(0), y^*) = d_{(S)_+}(\mathscr{F}, B_r(0), 0) = 1.$$

This proves Claim 2.

Then from (4.23), (4.24), and Claim 2 we get that $d_M(\mathscr{F}+A, U, y^*) = 1$. □

**Corollary 4.57.** *If $U \subset X$ is a nonempty, bounded, open set, $f : \overline{U} \to X^*$ is a bounded, demicontinuous, monotone $(S)_+$-map, $A : D(A) \subset X \to 2^{X^*}$ is a maximal monotone map such that $(0,0) \in \mathrm{Gr}\,A$, and $y^* \in (f+A)(U) \setminus (f+A)(\partial U)$, then $d_M(f+A, U, y^*) = 1$.*

*Proof.* Let $x_0 \in U$ such that $y^* = f(x_0) + x_0^*$ with $x_0^* \in A(x_0)$, and set $\hat{y}^* = \mathscr{F}(x_0) + x_0^*$. Define $h(t,x) = (1-t)f(x) + t\mathscr{F}(x) + A(x)$ for all $(t,x) \in [0,1] \times \overline{U}$. We show that $(1-t)y^* + t\hat{y}^* \notin h(t, \partial U)$ for all $t \in [0,1]$. Arguing by contradiction, suppose that we can find $t \in [0,1]$ and $x \in \partial U$ (hence $x \ne x_0$) such that

$$(1-t)y^* + t\hat{y}^* = (1-t)f(x) + t\mathscr{F}(x) + x^*,$$

with $x^* \in A(x)$. If $t = 0$, then we get $y^* = f(x) + x^* \in (f+A)(\partial U)$, which contradicts the hypothesis. Thus, $t \in (0,1]$. We have

$$(1-t)f(x_0) + t\mathscr{F}(x_0) + x_0^* - (1-t)f(x) - t\mathscr{F}(x) - x^* = 0. \qquad (4.26)$$

Acting on (4.26) with $x_0 - x$ and using the monotonicity of $f$ and $A$ and the strict monotonicity of $\mathscr{F}$, we reach a contradiction. Therefore, the homotopy $h$ satisfies the requirements of Theorem 4.53(c), with $y^*(t) = (1-t)y^* + t\hat{y}^*$. Then we have $d_M(f+A, U, y^*) = d_M(\mathscr{F}+A, U, \hat{y}^*) = 1$ (Theorem 4.56). □

Next we will extend Corollary 4.49 to the case of the degree $d_M$ and maps of the form $f+A$, as previously.

**Proposition 4.58.** *Let $U \subset X$ be a nonempty, open set, $\varphi \in C^1(U, \mathbb{R})$ be such that $\varphi' : U \to X^*$ is a bounded $(S)_+$-map, and $\psi \in \Gamma_0(X)$ be such $\psi(x) \ge \psi(0)$ for all $x \in X$. Assume that $x_0 \in U$ is a local minimizer and an isolated critical point of $\varphi + \psi$ [i.e., $0 \in \varphi'(x_0) + \partial\psi(x_0)$ and $0 \notin \varphi'(x) + \partial\psi(x)$ for $x$ near $x_0$]. Then we can find $\tilde{r}_1 > 0$ such that*

$$d_M(\varphi' + \partial\psi, B_r(x_0), 0) = 1 \quad \text{for all } r \in (0, \tilde{r}_1].$$

*Proof.* Up to dealing with $\varphi - (\varphi(x_0) + \psi(x_0))$ instead of $\varphi$, we may assume that $\varphi(x_0) + \psi(x_0) = 0$. Note that the assumption implies that $x_0$ is a strict local minimizer of $\varphi + \psi$ (otherwise, there would be other local minimizers, hence other critical points, in any neighborhood of $x_0$). Then we can find $r_0 > 0$ such that $\overline{B_{r_0}(x_0)} \subset U$ and

$$\varphi(x) + \psi(x) > 0 \quad \text{and} \quad 0 \notin \varphi'(x) + \partial\psi(x) \quad \text{for all } x \in \overline{B_{r_0}(x_0)} \setminus \{x_0\}. \tag{4.27}$$

From Proposition 2.56(e) we know that we can find $\lambda_0 > 0$ and $r_1 \in (0, r_0)$ such that

$$J_\lambda^{\partial\psi}(x) \in B_{r_0}(x_0) \quad \text{for all } \lambda \in (0, \lambda_0) \text{ and all } x \in B_{r_1}(x_0). \tag{4.28}$$

Let $\beta = \sup\{\|\varphi'(x)\| : x \in B_{r_0}(x_0)\}$. Since $\varphi'$ is bounded, $\beta \geq 0$ is finite. By the mean value theorem, we have

$$\varphi(y) - \beta\|x - y\| \leq \varphi(x) \leq \varphi(y) + \beta\|x - y\| \quad \text{for all } x, y \in B_{r_0}(x_0). \tag{4.29}$$

By Lemma 4.48 (see also Lemma 4.47), we have

$$\mu_r := \inf\{\varphi(x) + \psi(x) : x \in \overline{B_{r_0}(x_0)} \setminus B_r(x_0)\} > 0 \quad \text{for all } r \in (0, r_0]. \tag{4.30}$$

*Claim 1*: Let $r \in (0, r_1)$, and let $\xi \in (0, \mu_{\frac{r}{2}})$. Then there exists $\lambda_{r,\xi} > 0$ such that for $\lambda \in (0, \lambda_{r,\xi})$ we have $\{\varphi + \psi_\lambda \leq \xi\} \cap B_{r_1}(x_0) \subset B_r(x_0)$.

From (4.30) we have $\{\varphi + \psi < \mu_{\frac{r}{2}}\} \cap B_{r_0}(x_0) \subset B_{\frac{r}{2}}(x_0)$. Therefore, choosing $\varepsilon > 0$ small enough and denoting $\{\varphi + \psi < \mu_{\frac{r}{2}}\}_\varepsilon = \{x \in X : d(x, \{\varphi + \psi < \mu_{\frac{r}{2}}\}) < \varepsilon\}$, there holds

$$\{\varphi + \psi < \mu_{\frac{r}{2}}\}_\varepsilon \cap B_{r_1}(x_0) \subset B_r(x_0) \tag{4.31}$$

(since $r_1 < r_0$). We can find $\lambda_{r,\xi} \in (0, \lambda_0)$ such that for $\lambda \in (0, \lambda_{r,\xi})$ we have

$$\xi < \frac{s^2}{2\lambda} - \beta s \text{ for all } s \in [\varepsilon, +\infty) \text{ and } \xi < \frac{s^2}{2\lambda} - \beta s + \mu_{\frac{r}{2}} \text{ for all } s \in [0, +\infty). \tag{4.32}$$

Let $x \in B_{r_1}(x_0) \setminus B_r(x_0)$. Thus, by (4.31), we have $x \notin \{\varphi + \psi < \mu_{\frac{r}{2}}\}_\varepsilon$. By Remark 3.19(b), we see that

$$\varphi(x) + \psi_\lambda(x) = \varphi(x) + \psi(J_\lambda^{\partial\psi}(x)) + \frac{1}{2\lambda}\|x - J_\lambda^{\partial\psi}(x)\|^2$$

$$\geq \varphi(J_\lambda^{\partial\psi}(x)) - \beta\|x - J_\lambda^{\partial\psi}(x)\| + \psi(J_\lambda^{\partial\psi}(x))$$

$$+ \frac{1}{2\lambda}\|x - J_\lambda^{\partial\psi}(x)\|^2 \tag{4.33}$$

[see (4.28) and (4.29)]. If $J_\lambda^{\partial\psi}(x) \in \{\varphi + \psi < \mu_{\frac{r}{2}}\}$, then $\|x - J_\lambda^{\partial\psi}(x)\| \geq \varepsilon$ (since $x \notin \{\varphi + \psi < \mu_{\frac{r}{2}}\}_\varepsilon$), and from (4.33), the first inequality in (4.32), (4.28), and (4.27), we have

$$\varphi(x) + \psi_\lambda(x) > \varphi(J_\lambda^{\partial\psi}(x)) + \psi(J_\lambda^{\partial\psi}(x)) + \xi \geq \xi,$$

which implies $x \notin \{\varphi + \psi_\lambda \leq \xi\}$. If $J_\lambda^{\partial\psi}(x) \notin \{\varphi + \psi < \mu_{\frac{\xi}{2}}\}$, then from (4.33), using this time the second inequality in (4.32), we see that $x \notin \{\varphi + \psi_\lambda \leq \xi\}$ again. This proves Claim 1.

Now we fix $\tilde{r}_1 \in (0, r_1)$ and let $r \in (0, \tilde{r}_1]$. We fix $\xi \in (0, \mu_{\frac{r}{2}})$. Hence, by Claim 1, we have

$$\{\varphi + \psi_\lambda \leq \xi\} \cap B_{r_1}(x_0) \subset B_r(x_0) \text{ for } \lambda \in (0, \lambda_{r,\xi}). \tag{4.34}$$

Moreover, we fix $r_2 \in (0, r_1)$ such that $4\beta r_2 \leq \xi$, and we choose $\eta \in (0, \mu_{\frac{r_2}{2}})$ such that $\eta < \frac{\xi}{2}$. Again, by Claim 1, we have

$$\{\varphi + \psi_\lambda \leq \eta\} \cap B_{r_1}(x_0) \subset B_{r_2}(x_0) \text{ for } \lambda \in (0, \lambda_{r_2,\eta}). \tag{4.35}$$

*Claim 2*: For $\lambda \in (0, \lambda_{r_2,\eta})$, if $x \in \{\varphi + \psi_\lambda \leq \eta\} \cap B_{r_1}(x_0)$, then $(1-t)x_0 + tx \in \{\varphi + \psi_\lambda < \xi\}$ for all $t \in [0,1]$.

For all $x \in B_{r_1}(x_0)$ and all $t \in [0,1]$ we have

$$\varphi((1-t)x_0 + tx) = (1-t)\varphi(x_0 + t(x - x_0)) + t\varphi(x + (1-t)(x_0 - x))$$
$$\leq (1-t)\varphi(x_0) + t\varphi(x) + 2t(1-t)\beta \|x - x_0\| \tag{4.36}$$

[see (4.29)]. Also, from the convexity of $\psi_\lambda$ and using Remark 3.19(d), for all $x \in B_{r_1}(x_0)$ and all $t \in [0,1]$ we see that

$$\psi_\lambda((1-t)x_0 + tx) \leq (1-t)\psi_\lambda(x_0) + t\psi_\lambda(x) \leq (1-t)\psi(x_0) + t\psi_\lambda(x). \tag{4.37}$$

If $x \in \{\varphi + \psi_\lambda \leq \eta\} \cap B_{r_1}(x_0)$, then from (4.35), (4.36), and (4.37) we obtain

$$\varphi((1-t)x_0 + tx) + \psi_\lambda((1-t)x_0 + tx) \leq t\eta + 2\beta r_2 < \xi.$$

This proves Claim 2.

*Claim 3*: There is $\tilde{\lambda} > 0$ such that, for $\lambda \in (0, \tilde{\lambda})$, if $x \in \{\eta \leq \varphi + \psi_\lambda\} \cap B_{r_1}(x_0)$, then $\varphi'(x) + \psi_\lambda'(x) \neq 0$.

Arguing by contradiction, suppose that Claim 3 is not true. We can find $\{\lambda_n\}_{n\geq 1} \subset (0,1)$ and $\{x_n\}_{n\geq 1} \subset \{\eta \leq \varphi + \psi_{\lambda_n}\} \cap B_{r_1}(x_0)$ such that

$$\lambda_n \to 0 \text{ and } \varphi'(x_n) + \psi_{\lambda_n}'(x_n) = 0 \text{ for all } n \geq 1. \tag{4.38}$$

Since, by hypothesis, $\varphi'$ is bounded, $\{\varphi'(x_n)\}_{n\geq 1}$ is bounded in $X^*$. Hence, so is $\{\psi_{\lambda_n}'(x_n)\}_{n\geq 1}$ [see (4.38)]. Therefore, we may assume that

$$x_n \xrightarrow{w} x \text{ in } X \text{ and } \psi_{\lambda_n}'(x_n) \xrightarrow{w} x^* \text{ in } X^* \text{ as } n \to \infty. \tag{4.39}$$

Because of (4.38) and (4.39) [and since $(0,0) \in \mathrm{Gr}\,\partial\psi$] we can use Lemma 4.55, which implies that

$$x_n \to x \text{ in } X, \ (x,x^*) \in \mathrm{Gr}\,\partial\psi \ \text{ and } \ \varphi'(x) + x^* = 0. \tag{4.40}$$

Also, from Remark 3.19(d), the convexity of $\psi_{\lambda_n}$, (4.39), and (4.40) we have

$$\psi(x) - \limsup_{n\to\infty} \psi_{\lambda_n}(x_n) = \liminf_{n\to\infty}(\psi_{\lambda_n}(x) - \psi_{\lambda_n}(x_n))$$

$$\geq \lim_{n\to\infty} \langle \psi'_{\lambda_n}(x_n), x - x_n \rangle = 0.$$

Passing to the lim sup in the inequality $\eta \leq \varphi(x_n) + \psi_{\lambda_n}(x_n)$ and using (4.40), it turns out that $\eta \leq \varphi(x) + \psi(x)$, which implies that $x \neq x_0$. This, combined with (4.40), contradicts (4.27). This proves Claim 3.

Set $\hat{\lambda} = \min\{\lambda_{r,\xi}, \lambda_{r_2,\eta}, \tilde{\lambda}\}$, and let $\lambda \in (0,\hat{\lambda})$. We will apply Theorem 4.45 to the function $(\varphi + \psi_\lambda)|_{B_{r_1}(x_0)}$ and the numbers $\eta < \xi$. By (4.34), we have $V_\lambda := \{\varphi + \psi_\lambda < \xi\} \cap B_{r_1}(x_0) \subset B_r(x_0) \subset \overline{B_r(x_0)} \subset B_{r_1}(x_0)$, so condition (i) in Theorem 4.45 is satisfied. Moreover, Claims 2 and 3 ensure that conditions (ii) and (iii) of Theorem 4.45 are also satisfied. We therefore obtain $d_{(S)_+}(\varphi' + \psi'_\lambda, V_\lambda, 0) = 1$. Then, applying Theorem 4.42(d) with $C = \overline{B_r(x_0)} \setminus V_\lambda$, we obtain

$$d_{(S)_+}(\varphi' + \psi'_\lambda, B_r(x_0), 0) = 1 \ \text{ for all } \lambda \in (0,\hat{\lambda}).$$

By Definition 4.51, we finally deduce that $d_M(\varphi' + \partial\psi, B_r(x_0), 0) = 1$. $\square$

Finally, we focus on the case where $A = \partial\psi$, with $\psi \in \Gamma_0(X)$, which is important in the study of obstacle problems.

**Definition 4.59.** Let $\psi_n : X \to \mathbb{R} \cup \{+\infty\}$ $(n \geq 1)$ and $\psi : X \to \mathbb{R} \cup \{+\infty\}$. We say that the sequence $\{\psi_n\}_{n\geq 1}$ *converges to* $\psi$ *in the Mosco sense*, denoted by $\psi_n \xrightarrow{M} \psi$, if the following two conditions hold:

(a) For every $x \in X$ and every $x_n \xrightarrow{w} x$ in $X$ we have

$$\psi(x) \leq \liminf_{n\to\infty} \psi_n(x_n).$$

(b) For every $x \in X$ there exists a sequence $x_n \to x$ in $X$ such that

$$\psi_n(x_n) \to \psi(x).$$

*Remark 4.60.* Note that we get a definition equivalent to Definition 4.59 if we require condition (b) only for $x \in \mathrm{dom}\,\psi$. Indeed, (a) guarantees that condition (b) is satisfied whenever $x \in X \setminus \mathrm{dom}\,\psi$ (by any sequence such that $x_n \to x$).

The next two results show that the Mosco convergence is a useful tool for constructing pseudomonotone homotopies.

**Proposition 4.61.** *Let $\{\psi^{(t)}\}_{t\in[0,1]} \subset \Gamma_0(X)$ be such that $\psi^{(t)}(x) \geq \psi^{(t)}(0)$ for all $t \in [0,1]$, all $x \in X$, and suppose that $\psi^{(t_n)} \xrightarrow{M} \psi^{(t)}$ whenever $t_n \to t$ in $[0,1]$. Then $\{\partial \psi^{(t)}\}_{t\in[0,1]}$ is a pseudomonotone homotopy.*

*Proof.* We check (a) in Definition 4.52. Assume that $t_n \to t$ in $[0,1]$, $(x_n, x_n^*) \in \mathrm{Gr}\,\partial\psi^{(t_n)}$, $x_n \xrightarrow{w} x$ in $X$, $x_n^* \xrightarrow{w} x^*$ in $X^*$, and

$$\limsup_{n\to\infty}\langle x_n^*, x_n \rangle \leq \langle x^*, x \rangle. \tag{4.41}$$

Let $u \in \mathrm{dom}\,\psi^{(t)}$. By virtue of Definition 4.59(b), we can find a sequence $\{u_n\}_{n\geq 1} \subset X$ such that $u_n \to u$ in $X$ and $\psi^{(t_n)}(u_n) \to \psi^{(t)}(u)$. Because $(x_n, x_n^*) \in \mathrm{Gr}\,\partial\psi^{(t_n)}$ for all $n \geq 1$ we have

$$\psi^{(t_n)}(u_n) - \psi^{(t_n)}(x_n) \geq \langle x_n^*, u_n - x_n \rangle \quad \text{for all } n \geq 1.$$

Passing to the lim sup as $n \to \infty$ and using Definition 4.59(a) and (4.41), we obtain

$$\psi^{(t)}(u) - \psi^{(t)}(x) \geq \limsup_{n\to\infty}\langle x_n^*, u_n - x_n \rangle = \langle x^*, u \rangle - \liminf_{n\to\infty}\langle x_n^*, x_n \rangle$$

$$\geq \langle x^*, u \rangle - \limsup_{n\to\infty}\langle x_n^*, x_n \rangle \geq \langle x^*, u - x \rangle. \tag{4.42}$$

Because $u \in \mathrm{dom}\,\psi^{(t)}$ is arbitrary, from (4.42) we infer that $(x, x^*) \in \mathrm{Gr}\,\partial\psi^{(t)}$. Moreover, if we take $u = x$ (hence $u_n \to x$ in $X$), then (4.42) becomes

$$0 \geq \langle x^*, x \rangle - \liminf_{n\to\infty}\langle x_n^*, x_n \rangle \geq \langle x^*, x \rangle - \limsup_{n\to\infty}\langle x_n^*, x_n \rangle \geq 0,$$

which yields $\langle x_n^*, x_n \rangle \to \langle x^*, x \rangle$, and thus $\{\partial\psi^{(t)}\}_{t\in[0,1]}$ is a pseudomonotone homotopy. $\qquad\square$

**Proposition 4.62.** *Let $\psi^{(0)}, \psi^{(1)} \in \Gamma_0(X)$ be such that $\psi^{(0)}(x) \geq \psi^{(0)}(0)$, $\psi^{(1)}(x) \geq \psi^{(1)}(0)$ for all $x \in X$, and $\mathrm{dom}\,\psi^{(0)} = \mathrm{dom}\,\psi^{(1)}$. Suppose that there are functions $c_k : (0,1] \to [0,+\infty)$ $(k = 1, 2)$ such that $\lim_{\lambda\downarrow 0} c_k(\lambda) = 0$ and*

$$\psi^{(0)}(J_\lambda^{\partial\psi^{(1)}}(x)) \leq (1 + c_1(\lambda))\psi^{(0)}(x) + c_2(\lambda) \quad \text{for all } \lambda > 0, \text{ all } x \in \mathrm{dom}\,\psi^{(0)}, \tag{4.43}$$

$$\psi^{(1)}(J_\lambda^{\partial\psi^{(0)}}(x)) \leq (1 + c_1(\lambda))\psi^{(1)}(x) + c_2(\lambda) \quad \text{for all } \lambda > 0, \text{ all } x \in \mathrm{dom}\,\psi^{(1)}. \tag{4.44}$$

*Then, denoting $\psi^{(t)} = (1-t)\psi^{(0)} + t\psi^{(1)}$ $(t \in [0,1])$, the family $\{\partial\psi^{(t)}\}_{t\in[0,1]}$ is a pseudomonotone homotopy.*

*Proof.* We will apply Proposition 4.61. It is clear that $\psi^{(t)} \in \Gamma_0(X)$ for all $t \in [0,1]$. Now we fix a sequence $t_n \to t$ in $[0,1]$ and we check that $\psi^{(t_n)} \xrightarrow{M} \psi^{(t)}$ as $n \to \infty$.

Note that if $x_n \xrightarrow{w} x$ in $X$, then, using that $\psi^{(0)}, \psi^{(1)} \in \Gamma_0(X)$, we get

$$\psi^{(t)}(x) \leq \liminf_{n\to\infty} \psi^{(t_n)}(x_n).$$

Thus, part (a) of Definition 4.59 holds.

Next, let $x \in \text{dom}\,\psi^{(t)}$. If $t \in (0,1)$, then $\text{dom}\,\psi^{(t)} = \text{dom}\,\psi^{(0)} \cap \text{dom}\,\psi^{(1)}$, and so the sequence $x_n = x$ $(n \geq 1)$ satisfies part (b) of Definition 4.59.

Next we assume $t = 0$. In this case, for $n \geq 1$ we let

$$x_n = \begin{cases} J_{t_n}^{\partial\psi^{(1)}}(x) & \text{if } t_n > 0, \\ x & \text{if } t_n = 0. \end{cases}$$

Note that $x \in \text{dom}\,\psi^{(0)} \subset \overline{\text{dom}\,\psi^{(1)}} = \overline{D(\partial\psi^{(1)})}$ (by hypothesis and Corollary 3.16). Then Proposition 2.56(d) yields $x_n \to x$ in $X$. From (4.43) we have

$$\psi^{(0)}(x_n) \leq (1 + c_1(t_n))\psi^{(0)}(x) + c_2(t_n) \quad \text{whenever } t_n > 0,$$

which implies that $\limsup_{n\to\infty} \psi^{(0)}(x_n) \leq \psi^{(0)}(x)$. Due to the lower semicontinuity of $\psi^{(0)}$, we then derive

$$\lim_{n\to\infty} \psi^{(0)}(x_n) = \psi^{(0)}(x). \tag{4.45}$$

On the other hand, by hypothesis and Remark 3.19(b), (d), we have

$$\psi^{(1)}(0) \leq \psi^{(1)}(x_n) \leq \psi^{(1)}(x) \quad \text{for all } n \geq 1,$$

hence the sequence $\{\psi^{(1)}(x_n)\}_{n\geq1}$ is bounded. This fact, together with (4.45), yields

$$\psi^{(t_n)}(x_n) = (1-t_n)\psi^{(0)}(x_n) + t_n\psi^{(1)}(x_n) \to \psi^{(0)}(x) \quad \text{as } n \to \infty. \tag{4.46}$$

Similarly, in the case where $t = 1$, interchanging the roles of $\psi^{(0)}$ and $\psi^{(1)}$ and using this time (4.44) instead of (4.43) we can produce a sequence $\{x_n\}_{n\geq1} \subset X$ such that $x_n \to x$ in $X$ and $\psi^{(t_n)}(x_n) \to \psi^{(1)}(x)$ as $n \to \infty$.

In all the cases, we have checked part (b) of Definition 4.59, so $\psi^{(t_n)} \xrightarrow{M} \psi^{(t)}$ as $n \to \infty$. We can apply Proposition 4.61, and we conclude that $\{\partial\psi^{(t)}\}_{t\in[0,1]}$ is a pseudomonotone homotopy. $\square$

*Remark 4.63.* In Proposition 4.62, the hypothesis that $\overline{\operatorname{dom}\psi^{(0)}} = \overline{\operatorname{dom}\psi^{(1)}}$ implies that $\overline{D(\partial\psi^{(0)})} = \overline{D(\partial\psi^{(1)})}$ (Corollary 3.16). Actually, we know from Proposition 4.54 that this condition is necessary for $\{\partial\psi^{(t)}\}_{t\in[0,1]}$ to be a pseudomonotone homotopy.

## 4.4  Remarks

**Section 4.1:** The classical degree theory for functions defined on $\mathbb{R}^N$ started with Brouwer [56], who introduced the degree map described in Theorem 4.5. Here, we present the analytical approach to the construction of Brouwer's degree, due to Heinz [170] and Nagumo [306]. For the uniqueness of Brouwer's degree, we refer to Führer [144] and Amann and Weiss [12].

Detailed presentations of the construction and properties of Brouwer's degree can be found in the books of Deimling [108], Denkowski et al. [114], Fonseca and Gangbo [141], Krawcewicz and Wu [205], Lloyd [237], and Nirenberg [309].

**Section 4.2:** The extension of Brouwer's degree to maps defined on an infinite-dimensional Banach space and having the form $\mathrm{id} - K$ with $K$ compact (compact perturbations of the identity) is due to Leray–Schauder [222]. Of course, the uniqueness of Brouwer's degree leads to the uniqueness of the Leray–Schauder degree. The books mentioned previously for Brouwer's degree also contain extended discussions of the Leray–Schauder degree. Nussbaum [311] extended the Leray–Schauder degree to maps of the form $\mathrm{id} - K$, with $K$ being $\gamma$-condensing ($\gamma$ being a measure of noncompactness). Also, some of the results of the Leray–Schauder degree can be cast in the language of essential and inessential maps (see Granas and Dugundji [165]). Finally, we should also mention the equivariant degree, a basic tool of equivariant analysis, which deals with the impact of symmetries (represented by a certain group $G$ and translated as the equivariance of the corresponding operators) on the existence, multiplicity, stability, and topological structure of the solutions to nonlinear operator equations. Equivariant degree theory is discussed in the books of Balanov et al. [28] and Ize and Vignoli [182].

**Section 4.3:** The degree theory for operators of the monotone type started with Skrypnik [360] and was extended to broader classes of maps by Browder [60, 61]. The key ingredients are the renorming theorems for Banach spaces and Lemma 4.37. For the renorming theory, we refer the reader to the book by Deville et al. [117], while Lemma 4.37 is a property of the so-called angelic spaces. For more information on them, consult Floret [140, p. 30].

Our presentation here is based on the works of Browder [60, 61], Kobayashi and Ôtani [197, 198], and Motreanu et al. [291]. Additional information and results can be found in the papers of Aizicovici et al. [3, 4], Hu and Papageorgiou [174, 175], Kartsatos and Skrypnik [189], Kien et al. [196], and Kobayashi and Ôtani [199].

# Chapter 5
# Variational Principles and Critical Point Theory

**Abstract** This chapter addresses variational principles and critical point theory
that will be applied later in the book for setting up variational methods in the
case of nonlinear elliptic boundary value problems. The first section of the chapter
illustrates the connection between the variational principles of Ekeland and Zhong
and compactness-type conditions such as the Palais–Smale and Cerami conditions.
The second section contains the deformation theorems that form the basis of the
critical point and Morse theories. These results are proved in the setting of Banach
spaces relying on the construction of a pseudogradient vector field and by using
the Cerami condition. The third section focuses on important minimax theorems
encompassing various linking situations: mountain pass, saddle point, generalized
mountain pass, and local linking. The fourth section studies critical points for
functionals with symmetries providing minimax values corresponding to index
theories whose prototype is the Krasnosel'skiĭ genus. The fifth section is devoted
to generalizations: critical point theory on Banach manifolds and nonsmooth
critical point theories. Comments and related references are available in a remarks
section.

## 5.1 Ekeland Variational Principle

Many of the partial differential equations in physics and mechanics arise from a
variational principle. That is, we have a set of admissible solutions and an energy
functional defined on them. We look for the minimizers of the energy functional
over the set of admissible solutions (least energy solutions). Then such a minimizer
is a solution to the corresponding Euler equation, which is the partial differential
equation we started with. In the first part of this section, we discuss some abstract
variational principles that are useful tools in the study of boundary value problems.

First we recall a classical result concerning the minimization of the distance of a
point to a closed, convex set in a Hilbert space.

D. Motreanu et al., *Topological and Variational Methods with Applications to Nonlinear
Boundary Value Problems*, DOI 10.1007/978-1-4614-9323-5_5,
© Springer Science+Business Media, LLC 2014

**Theorem 5.1.** *If $H$ is a Hilbert space with inner product $(\cdot,\cdot)$ and induced norm $\|\cdot\|$, $C \subset H$ is nonempty, closed, and convex, and $u \in H \setminus C$, then there exists a unique $p_C(u) \in C$ such that $\|u - p_C(u)\| = \min\{\|u - x\| : x \in C\}$, and this element is characterized by $(u - p_C(u), x - p_C(u)) \leq 0$ for all $x \in C$. Moreover, the map $p_C : H \to C$ is nonexpansive, i.e.,*

$$\|p_C(u) - p_C(v)\| \leq \|u - v\| \ \text{for all } u, v \in H.$$

*Remark 5.2.* The map $p_C : H \to C$ is called the *metric projection onto C*.

Recall that the Riesz representation theorem says that every continuous linear functional on $H$ (i.e., an element of $H^*$) can be identified with an element of $H$ (e.g., Brezis [52, p. 135]). Next we will see that this remarkable result can be generalized from the inner product to general bilinear forms on $H$. This is the context of the well-known Lax–Milgram theorem, which we will produce here as a consequence of a more general result for variational inequalities due to Stampacchia [366].

**Definition 5.3.** Let $H$ be a Hilbert space and $a : H \times H \to \mathbb{R}$ a bilinear form. We say that:

(a)  *a is continuous* if there is $M > 0$ such that $|a(u,y)| \leq M\|u\|\,\|y\|$ for all $u, y \in H$;
(b)  *a is coercive* if there is $c > 0$ such that $a(u,u) \geq c\|u\|^2$ for all $u \in H$;
(c)  *a is symmetric* if $a(u,y) = a(y,u)$ for all $u, y \in H$.

**Theorem 5.4.** *If $H$ is a Hilbert space with inner product $(\cdot,\cdot)$, $a : H \times H \to \mathbb{R}$ is a continuous, coercive, bilinear form on $H$, $C \subset H$ is nonempty, closed, and convex, and $h \in H$, then there exists a unique $u \in C$ such that*

$$a(u, x - u) \geq (h, x - u) \ \text{for all } x \in C. \tag{5.1}$$

*Moreover, if $a(\cdot,\cdot)$ is also symmetric, then $u \in C$ is the unique minimizer of*

$$\varphi(y) = \frac{1}{2}a(y,y) - (h,y) \ \text{on } C.$$

*Proof.* Fix $u \in H$ and consider the function $y \mapsto a(u,y)$. This is a continuous, linear functional on $H$, and so by the Riesz representation theorem there is a unique $A(u) \in H$ such that $(A(u), y) = a(u,y)$ for all $y \in H$. Then $A \in \mathscr{L}(H)$ and

$$(A(u), u) \geq c\|u\|^2 \ \text{for all } u \in H. \tag{5.2}$$

Let $i_C$ denote the indicator function of $C$, that is,

$$i_C(y) = \begin{cases} 0 & \text{if } y \subset C, \\ +\infty & \text{if } y \notin C. \end{cases}$$

Since $C \subset H$ is nonempty, closed, and convex, we have $i_C \in \Gamma_0(H)$, and, identifying $H^*$ with $H$, for all $u \in C$ we have

$$\partial i_C(u) = N_C(u) := \{v \in H : (v, x - u) \le 0 \text{ for all } x \in C\}$$

(the normal cone to $C$ at $u$). Thus, we see that $u \in C$ satisfies (5.1) if and only if $0 \in A(u) + \partial i_C(u) - h$. Theorem 2.52 (using Theorem 3.15) guarantees the existence of such $u \in C$. Moreover, since $A + \partial i_C$ is strictly monotone [by (5.2) and the linearity of $A$], we obtain that $u$ is unique.

If $a(\cdot, \cdot)$ is symmetric, since $u \in C$ is the solution to (5.1), then, exploiting the symmetry of $a$, for all $y \in C$ we have

$$\varphi(y) = \varphi(u + y - u) = \frac{1}{2} a(u + y - u, u + y - u) - (h, u + (y - u))$$

$$= \frac{1}{2} a(u, u) - (h, u) + a(u, y - u) - (h, y - u) + \frac{1}{2} a(y - u, y - u)$$

$$\ge \varphi(u) + \frac{c}{2} \|y - u\|^2,$$

hence $u$ is the unique minimizer of $\varphi$ on $C$.                                         □

*Remark 5.5.* The proof of the theorem just given differs from the original proof in [366]. If $C = Y$ is a closed subspace of $H$, then the variational inequality (5.1) becomes $a(u, x) \ge (h, x)$ for all $x \in Y$. Replacing $x$ by $-x \in Y$, we infer that $a(u, x) = (h, x)$ for all $x \in Y$. In particular, this is true if $Y = H$, and so we have the *Lax–Milgram theorem*, which we state in what follows.

**Corollary 5.6.** *If $H$ is a Hilbert space with inner product $(\cdot, \cdot)$, $a : H \times H \to \mathbb{R}$ is a continuous, coercive, bilinear form on $H$, and $h \in H$, then there exists a unique $u \in H$ such that $a(u, y) = (h, y)$ for all $y \in H$. Moreover, if $a(\cdot, \cdot)$ is also symmetric, then $u \in H$ is the unique minimizer of $\varphi(y) = \frac{1}{2} a(y, y) - (h, y)$, $y \subset H$.*

Next we present the so-called *Ekeland variational principle*, which has proven to be a very powerful tool in nonlinear analysis with remarkable applications in optimization and optimal control. It asserts the existence of a minimizing sequence of a particular kind. Along this sequence we reach the infimal value of the minimization problem, and the first-order optimality condition is satisfied up to any desired approximation.

**Theorem 5.7.** *If $(X, d)$ is a complete metric space, $\varphi : X \to \mathbb{R} \cup \{+\infty\}$ is l.s.c., $\varphi \not\equiv +\infty$ and it is bounded below, then given $\varepsilon > 0$, $u \in X$ such that $\varphi(u) \le \inf_X \varphi + \varepsilon$, and $\lambda > 0$, we can find $v \in X$ such that*

$$\varphi(v) \le \varphi(u), \quad d(u, v) \le \lambda$$

*and*

$$\varphi(v) \le \varphi(y) + \frac{\varepsilon}{\lambda} d(v, y) \text{ for all } y \in X.$$

*Proof.* Without loss of generality, we may assume that $\lambda = 1$ (otherwise, we replace $d$ by the equivalent metric $d_\lambda = \frac{1}{\lambda} d$). We introduce the relation $\leq$ on $X$ defined by

$$y \leq x \text{ if and only if } \varphi(y) + \varepsilon d(x,y) \leq \varphi(x). \tag{5.3}$$

It is easy to see that $\leq$ is a partial order on $X$ (i.e., $\leq$ is reflexive, antisymmetric, and transitive). Using induction, we construct a sequence $\{u_n\}_{n \geq 0} \subset X$ as follows. Let $u_0 = u$, and suppose that $u_n$ has been defined. Let $L_n = \{h \in X : h \leq u_n\}$, and choose $u_{n+1} \in L_n$ such that

$$\varphi(u_{n+1}) \leq \inf_{L_n} \varphi + \frac{1}{n+1}. \tag{5.4}$$

The lower semicontinuity of $\varphi$ implies that $L_n$ is closed, and we have that $L_{n+1} \subset L_n$ (because $u_{n+1} \leq u_n$).

*Claim 1*: diam $L_n \to 0$ as $n \to \infty$.

Let $h \in L_{n+1}$. Then $h \leq u_{n+1} \leq u_n$, and so from (5.3) and (5.4) we have

$$\varepsilon d(u_{n+1}, h) \leq \varphi(u_{n+1}) - \varphi(h) \leq \inf_{L_n} \varphi + \frac{1}{n+1} - \varphi(h) \leq \frac{1}{n+1},$$

so diam $L_{n+1} \leq \frac{2}{\varepsilon(n+1)} \to 0$ as $n \to \infty$.

Since $X$ is complete, invoking the Cantor intersection theorem (e.g., Dugundji [123, p. 296]) in light of Claim 1, we have $\bigcap_{n \geq 0} L_n = \{v\}$ for some $v \in X$. In particular, $v \in L_0$, and so $v \leq u_0 = u$, which implies that

$$\varphi(v) \leq \varphi(u) - \varepsilon d(u,v) \leq \varphi(u)$$

[see (5.3)]. Also, we have

$$d(u,v) \leq \frac{1}{\varepsilon}(\varphi(u) - \varphi(v)) \leq \frac{1}{\varepsilon}\left(\inf_X \varphi + \varepsilon - \inf_X \varphi\right) = 1.$$

Note that $v$ is minimal for the order $\leq$. Indeed, if $z \leq v$, then $z \leq u_n$ for all $n \geq 0$, and so $z \in \bigcap_{n \geq 0} L_n = \{v\}$, hence $z = v$. We conclude that $\varphi(v) \leq \varphi(y) + \varepsilon d(v,y)$ for all $y \in X$. $\square$

*Remark 5.8.* In the conclusion of Theorem 5.7, the relations $d(u,v) \leq \lambda$ and $\varphi(v) \leq \varphi(y) + \frac{\varepsilon}{\lambda} d(v,y)$ for all $y \in X$ are complementary. The choice of $\lambda > 0$ allows us to strike a balance between them. If $\lambda > 0$ is large, then the inequality $d(u,v) \leq \lambda$ gives little information on the whereabouts of $v$, while the inequality $\varphi(v) \leq \varphi(y) + \frac{\varepsilon}{\lambda} d(v,y)$ for all $y \in X$ becomes sharper and implies that $v$ is close to being a global minimizer of $\varphi$ since the perturbation $\frac{\varepsilon}{\lambda} d(v,\cdot)$ is very small. The situation is reversed if $\lambda > 0$ is small. Two important special cases are when $\lambda = 1$ (which means that we are unconcerned with the whereabouts of $v$) and when $\lambda = \sqrt{\varepsilon}$ (which means that we need to have information from both inequalities). Below we state both cases as corollaries of Theorem 5.7. The inequality $\varphi(v) \leq \varphi(y) + \varepsilon d(v,y)$ for all $y \in X$ has a clear interpretation when $\varphi$ is Gâteaux differentiable on a Banach space. This is illustrated subsequently in Corollaries 5.11 and 5.12.

**Corollary 5.9.** *If $(X,d)$ and $\varphi : X \to \mathbb{R} \cup \{+\infty\}$ are as in Theorem 5.7, then for any $\varepsilon > 0$ we can find $v_\varepsilon \in X$ such that*

$$\varphi(v_\varepsilon) \leq \inf_X \varphi + \varepsilon \ \text{ and } \ \varphi(v_\varepsilon) \leq \varphi(y) + \varepsilon d(v_\varepsilon, y) \ \text{ for all } y \in X.$$

**Corollary 5.10.** *If $(X,d)$ and $\varphi : X \to \mathbb{R} \cup \{+\infty\}$ are as in Theorem 5.7 and $\varepsilon > 0$ and $u_\varepsilon \in X$ satisfy*

$$\varphi(u_\varepsilon) \leq \inf_X \varphi + \varepsilon,$$

*then we can find $v_\varepsilon \in X$ such that*

$$\varphi(v_\varepsilon) \leq \varphi(u_\varepsilon), \ \ d(u_\varepsilon, v_\varepsilon) \leq \sqrt{\varepsilon} \ \text{ and } \ \varphi(v_\varepsilon) \leq \varphi(y) + \sqrt{\varepsilon} d(v_\varepsilon, y) \ \text{ for all } y \in X.$$

**Corollary 5.11.** *If $X$ is a Banach space, $\varphi : X \to \mathbb{R}$ is l.s.c., bounded below, and Gâteaux differentiable, then for every $\varepsilon > 0$ we can find $v_\varepsilon \in X$ such that*

$$\varphi(v_\varepsilon) \leq \inf_X \varphi + \varepsilon \ \text{ and } \ \|\varphi'(v_\varepsilon)\| \leq \varepsilon.$$

**Corollary 5.12.** *If $X$ and $\varphi$ are as in Corollary 5.11 and $\varepsilon > 0$ and $u_\varepsilon \in X$ satisfy*

$$\varphi(u_\varepsilon) \leq \inf_X \varphi + \varepsilon,$$

*then we can find $v_\varepsilon \in X$ such that*

$$\varphi(v_\varepsilon) \leq \varphi(u_\varepsilon), \ \ \|u_\varepsilon - v_\varepsilon\| \leq \sqrt{\varepsilon} \ \text{ and } \ \|\varphi'(v_\varepsilon)\| \leq \sqrt{\varepsilon}.$$

*Remark 5.13.* The preceding corollary guarantees the existence of a minimizing sequence $\{x_n\}_{n \geq 1} \subset X$ such that $\varphi'(x_n) \to 0$ in $X^*$. This leads to the introduction of the following compactness condition for functions $\varphi \in C^1(X, \mathbb{R})$, which brings us to the doorsteps of critical point theory.

**Definition 5.14.** Let $(X, \|\cdot\|)$ be a Banach space and $\varphi \in C^1(X, \mathbb{R})$.

(a) $\varphi$ satisfies the *Palais–Smale condition at the level* $c \in \mathbb{R}$ ($(\mathrm{PS})_c$-*condition*) if every sequence $\{x_n\}_{n \geq 1} \subset X$ such that

$$\varphi(x_n) \to c \ \text{ in } \mathbb{R} \ \text{ and } \ \varphi'(x_n) \to 0 \ \text{ in } X^*$$

admits a strongly convergent subsequence. We say that $\varphi$ satisfies the *Palais–Smale condition* ((PS)-*condition*) if it satisfies the (PS)$_c$-condition at every level $c \in \mathbb{R}$.

(b) $\varphi$ satisfies the *Cerami condition at the level* $c \in \mathbb{R}$ ((C)$_c$-*condition*) if every sequence $\{x_n\}_{n \geq 1} \subset X$ such that

$$\varphi(x_n) \to c \text{ in } \mathbb{R} \text{ and } (1 + \|x_n\|)\varphi'(x_n) \to 0 \text{ in } X^*$$

admits a strongly convergent subsequence. We say that $\varphi$ satisfies the *Cerami condition* ((C)-*condition*) if it satisfies the (C)$_c$-condition at every level $c \in \mathbb{R}$.

*Remark 5.15.* The (C)$_c$-condition is weaker than the (PS)$_c$-condition.

The next extension of Theorem 5.7, due to Zhong [395], fits well with the (C)-condition.

**Theorem 5.16.** *Let $\xi : \mathbb{R}_+ \to \mathbb{R}_+$ be a continuous, nondecreasing function such that $\int_0^{+\infty} \frac{1}{1+\xi(s)} ds = +\infty$. Let $(X,d)$ be a complete metric space, $x_0 \in X$ be fixed, $\varphi : X \to \mathbb{R} \cup \{+\infty\}$ be l.s.c., and $\varphi \not\equiv +\infty$ and bounded below. Let $\varepsilon > 0$, $u \in X$ be such that $\varphi(u) \leq \inf_X \varphi + \varepsilon$, and $\lambda > 0$. Set $r_0 = d(x_0,u)$ and fix $\bar{r} > 0$ such that $\int_{r_0}^{r_0+\bar{r}} \frac{1}{1+\xi(s)} ds \geq \lambda$. Then there exists $v \in X$ such that*

$$\varphi(v) \leq \varphi(u), \ d(v,x_0) \leq r_0 + \bar{r}$$

*and*

$$\varphi(v) \leq \varphi(y) + \frac{\varepsilon}{\lambda(1+\xi(d(x_0,v)))} d(v,y) \text{ for all } y \in X. \tag{5.5}$$

*Remark 5.17.* If $\xi = 0$, $x_0 = u$, and $\bar{r} = \lambda$, then Theorem 5.16 reduces to Theorem 5.7.

**Corollary 5.18.** *Let $\xi : \mathbb{R}_+ \to \mathbb{R}_+$ be a continuous, nondecreasing function such that $\int_0^{+\infty} \frac{1}{1+\xi(s)} ds = +\infty$. Let $X$ be a Banach space, and let $\varphi : X \to \mathbb{R}$ be l.s.c., bounded below, and Gâteaux differentiable. Then, given $\varepsilon > 0$, $u \in X$ such that $\varphi(u) \leq \inf_X \varphi + \varepsilon$, $\lambda > 0$, and $\bar{r} > 0$ such that $\int_0^{\bar{r}} \frac{1}{1+\xi(s)} ds \geq \lambda$, we can find $v \in X$ such that*

$$\varphi(v) \leq \varphi(u), \ \|v - u\| \leq \bar{r} \text{ and } \|\varphi'(v)\| \leq \frac{\varepsilon}{\lambda(1+\xi(\|v\|))}.$$

**Corollary 5.19.** *If $\xi$, $X$, and $\varphi$ are as in Corollary 5.18, then for every $\varepsilon > 0$ we can find $v_\varepsilon \in X$ such that*

$$\varphi(v_\varepsilon) \leq \inf_X \varphi + \varepsilon \text{ and } \|\varphi'(v_\varepsilon)\| \leq \frac{\varepsilon}{1+\xi(\|v_\varepsilon\|)}.$$

**Corollary 5.20.** *If $\xi$, $X$, and $\varphi$ are as in Corollary 5.18, then $\varphi$ admits a minimizing sequence $\{x_n\}_{n\geq 1} \subset X$ such that*

$$(1 + \xi(\|x_n\|))\varphi'(x_n) \to 0 \text{ in } X^*.$$

Combining this corollary with the $(C)_c$-condition for $c = \inf_X \varphi$, we have the following corollary.

**Corollary 5.21.** *If $X$ is a Banach space, $\varphi \in C^1(X,\mathbb{R})$, $\varphi$ is bounded below, and $\varphi$ satisfies the $(C)_c$-condition with $c = \inf_X \varphi$, then there is $x \in X$ such that $\varphi(x) = \inf_X \varphi$.*

*Proof.* By virtue of Corollary 5.20 [with $\xi(r) = r$], we find a sequence $\{x_n\}_{n\geq 1} \subset X$ such that

$$\varphi(x_n) \to c = \inf_X \varphi \quad \text{and} \quad (1 + \|x_n\|)\varphi'(x_n) \to 0 \text{ in } X^*.$$

Since $\varphi$ satisfies the $(C)_c$-condition, by passing to a subsequence if necessary, we may assume that $x_n \to x$ in $X$ for some $x \in X$. Then $\varphi(x) = \inf_X \varphi$. $\square$

The next proposition establishes an interesting connection between the $(C)$-condition and the coercivity of $\varphi$.

**Proposition 5.22.** *If $X$ is a Banach space and $\varphi \in C^1(X,\mathbb{R})$ is bounded below and satisfies the $(C)$-condition, then $\varphi$ is coercive, i.e., $\varphi(x) \to +\infty$ as $\|x\| \to +\infty$.*

*Proof.* Arguing by contradiction, suppose that $\varphi$ is not coercive. Then we can find $c \in \mathbb{R}$ and a sequence $\{u_n\}_{n\geq 1} \subset X$ such that

$$\varphi(u_n) \leq c + \frac{1}{n} \quad \text{and} \quad \|u_n\| \geq 2(e^n - 1). \tag{5.6}$$

We apply Corollary 5.18 with $\xi(r) = r$, $\varepsilon = c + \frac{1}{n} - \inf_X \varphi$, $\lambda = n$, $\bar{r} = e^n - 1$ and obtain a sequence $\{v_n\}_{n\geq 1} \subset X$ such that

$$\varphi(v_n) \leq \varphi(u_n), \ \|v_n - u_n\| \leq e^n - 1 \ \text{and} \ \|\varphi'(v_n)\| \leq \frac{c + \frac{1}{n} - \inf_X \varphi}{n(1 + \|v_n\|)} \ \text{for all } n \geq 1. \tag{5.7}$$

By (5.6) and (5.7), note that

$$\|v_n\| \geq \|u_n\| - \|v_n - u_n\| \geq e^n - 1,$$

which yields $\|v_n\| \to +\infty$, and so $\limsup_{n\to\infty} \varphi(v_n) \leq c$, and $(1 + \|v_n\|)\varphi'(v_n) \to 0$ in $X^*$ [see (5.7)]. This contradicts the fact that $\varphi$ satisfies the $(C)$-condition. $\square$

This leads to the following comparison of the two compactness conditions in Definition 5.14.

**Proposition 5.23.** *If $X$ is a Banach space and $\varphi \in C^1(X,\mathbb{R})$ is bounded below, then the* (PS)- *and* (C)-*conditions are equivalent.*

*Proof.* We only need to show that the (C)-condition implies the (PS)-condition (Remark 5.15). Let $\{x_n\}_{n\geq 1} \subset X$ be a sequence such that $\{\varphi(x_n)\}_{n\geq 1}$ is bounded and $\varphi'(x_n) \to 0$ in $X^*$. Then, by virtue of Proposition 5.22, $\{x_n\}_{n\geq 1}$ is bounded, and so $(1 + \|x_n\|)\varphi'(x_n) \to 0$ in $X^*$. Since $\varphi$ satisfies the (C)-condition, we can find a strongly convergent subsequence of $\{x_n\}_{n\geq 1}$. Therefore, $\varphi$ satisfies the (PS)-condition.                                                                                    □

Finally, we mention a consequence of the (PS)-condition on functionals that are bounded below.

**Proposition 5.24.** *Let $X$ be a Banach space and $\varphi \in C^1(X,\mathbb{R})$ be bounded below and satisfy the* (PS)$_c$-*condition with* $c = \inf_X \varphi$. *Then, every minimizing sequence* $\{u_n\}_{n\geq 1}$ *of $\varphi$ admits a convergent subsequence whose limit is a global minimizer of $\varphi$.*

*Proof.* Up to considering a subsequence, we may assume that $\varphi(u_n) \leq \inf_X \varphi + \frac{1}{n^2}$ for all $n \geq 1$. By Corollary 5.12, we can find a sequence $\{v_n\}_{n\geq 1}$ such that

$$\lim_{n\to\infty} \varphi(v_n) = \inf_X \varphi, \quad \lim_{n\to\infty} \|\varphi'(v_n)\| = 0, \text{ and } \|u_n - v_n\| \leq \frac{1}{n} \text{ for all } n \geq 1. \quad (5.8)$$

Since $\varphi$ satisfies the (PS)$_c$-condition, $\{v_n\}_{n\geq 1}$ admits a convergent subsequence $\{v_{n_k}\}_{k\geq 1}$. Then the third relation in (5.8) implies that the subsequence $\{u_{n_k}\}_{k\geq 1}$ is convergent. By the continuity of $\varphi$, the limit is a global minimizer of $\varphi$.       □

## 5.2   Critical Points and Deformation Theorems

The (PS) and (C)-conditions are compactness-type conditions on the energy functional to compensate for the lack of local compactness of the ambient space $X$, which in general is an infinite-dimensional Banach space. In this section we see some significant consequences of these conditions related to the critical points.

Let $X$ be a Banach space and $\varphi \in C^1(X,\mathbb{R})$. Let

$$K_\varphi = \{x \in X : \varphi'(x) = 0\} \text{ and } K_\varphi^C = \{x \in K_\varphi : \varphi(x) \in C\}$$

denote the critical set of $\varphi$ and the critical set of $\varphi$ with critical values in $C \subset \mathbb{R}$, respectively. Moreover, we abbreviate $K_\varphi^c := K_\varphi^{\{c\}}$ with $c \in \mathbb{R}$.

The next result follows at once from Definition 5.14.

**Proposition 5.25.** *If $C \subset \mathbb{R}$ is compact and $\varphi$ satisfies the* $(C)_c$-*condition for all* $c \in C$, *then* $K_\varphi^C$ *is compact.*

The compactness conditions yield various properties of the sublevel sets of $\varphi$. Locating critical points of a smooth functional $\varphi$ essentially reduces to capturing the changes in the topology of the sublevel sets

$$\varphi^\lambda := \{x \in X : \varphi(x) \leq \lambda\}$$

as $\lambda$ varies in $\mathbb{R}$. A way to do this is through the deformation theorems. These results exploit the fact that near a regular value $c \in \mathbb{R}$, for $\varepsilon > 0$ small, the sublevel sets $\varphi^{c+\varepsilon}$ and $\varphi^{c-\varepsilon}$ are topologically the same (i.e., roughly speaking there is a suitable deformation transforming $\varphi^{c+\varepsilon}$ into $\varphi^{c-\varepsilon}$ or vice versa). This is no longer true the moment we cross a critical value $c \in \mathbb{R}$. Then $\varphi^{c+\varepsilon}$ and $\varphi^{c-\varepsilon}$ exhibit distinct structures, as the following examples illustrate.

*Example 5.26.*

(a) Let $\varphi : \mathbb{R}^2 \to \mathbb{R}$ be defined by $\varphi(x,y) = x^2 - y^2$ for all $(x,y) \in \mathbb{R}^2$. Then $K_\varphi = \{(0,0)\}$. Then we can see that for $\varepsilon > 0$ small, $\varphi^\varepsilon$ is connected, while $\varphi^{-\varepsilon}$ is disconnected with two components.

(b) Let $\varphi : \mathbb{R}^2 \to \mathbb{R}$ be defined by $\varphi(u) = |u|^4 - 2|u|^2$ for all $u \in \mathbb{R}^2$. Then $K_\varphi = \{0\} \cup \partial B_1(0)$, and so the only critical values of $\varphi$ are 0 and $-1$. For $c < -1$, $\varphi^c = \emptyset$; for $c \in (-1,0)$, $\varphi^c$ is an annulus (i.e., a set of the form $\{u \in \mathbb{R}^2 : r^2 \leq |u|^2 \leq R^2\}$); and for $c > 0$, $\varphi^c$ is a ball $\overline{B_\rho(0)}$. Note that the ball is simply connected while the annulus is not. Of course, both are connected.

Suppose that $H$ is a Hilbert space with the inner product $(\cdot,\cdot)$. Then we can define the gradient of $\varphi \in C^1(H,\mathbb{R})$, denoted by $\nabla \varphi$, via the relation

$$(\nabla \varphi(x),y) = \langle \varphi'(x),y \rangle \quad \text{for all } x,y \in H.$$

If $\varphi \in C^2(H,\mathbb{R})$, then we can use the steepest descent method to study the transformations of the sublevel sets. However, if $\varphi$ is only $C^1$ or the ambient space is not a Hilbert space, then the steepest descent method fails. For this reason, we introduce the notion of a pseudogradient vector field.

**Definition 5.27.** *Let $X$ be a Banach space, $\varphi \in C^1(X,\mathbb{R})$, and set $X_0 = \{x \in X : \varphi'(x) \neq 0\} = X \setminus K_\varphi$. Then a pseudogradient vector field for $\varphi$ is a locally Lipschitz map $v : X_0 \to X$ such that*

$$\|v(x)\| \leq 2\|\varphi'(x)\| \quad \text{and} \quad \langle \varphi'(x),v(x) \rangle \geq \|\varphi'(x)\|^2 \quad \text{for all } x \in X_0.$$

*Remark 5.28.*

(a) If $H$ is a Hilbert space and $\varphi \in C^1(H,\mathbb{R})$ has a locally Lipschitz derivative, then $\nabla \varphi$ is a pseudogradient vector field for $\varphi$.

(b) The pseudogradient vector field is not in general unique (for instance, for $\varphi$ as in (a), $\lambda \nabla \varphi$ is a pseudogradient vector field for $\varphi$ for all $\lambda \in [1,2]$).

(c) Any convex combination of pseudogradient vector fields for $\varphi$ is a pseudogradient vector field for $\varphi$.

**Proposition 5.29.** *If $\varphi \in C^1(X,\mathbb{R})$, then there exists a pseudogradient vector field for $\varphi$.*

*Proof.* For each $x \in X_0$ we claim that we can find an element $h_x \in X$ and a neighborhood $V_x$ of $x$ such that

$$\|h_x\| \leq 2\|\varphi'(z)\| \quad \text{and} \quad \langle \varphi'(z), h_x \rangle \geq \|\varphi'(z)\|^2 \quad \text{for all } z \in V_x. \tag{5.9}$$

Indeed, take $y \in X$ such that $\|y\| = 1$ and $\langle \varphi'(x), y \rangle > \frac{2}{3}\|\varphi'(x)\|$; then $h_x := \frac{3}{2}\|\varphi'(x)\|y$ satisfies $\|h_x\| = \frac{3}{2}\|\varphi'(x)\|$ and $\langle \varphi'(x), h_x \rangle > \|\varphi'(x)\|^2$, and the continuity of $\varphi'$ allows us to find a neighborhood $V_x$ satisfying (5.9).

The family $\mathscr{S} = \{V_x : x \in X_0\}$ is an open covering of $X_0$. Note that $X_0$, being a metric space, is paracompact (e.g., Dugundji [123, p. 186]). So there is $\mathscr{S}' = \{W_i\}_{i \in I}$, a locally finite open refinement of $\mathscr{S}$. Thus, for all $i \in I$ we can find $x_i \in X_0$ such that $W_i \subset V_{x_i}$. We set $\xi_i(z) = d(z, X \setminus W_i)$ and

$$v(z) = \sum_{i \in I} \frac{\xi_i(z)}{\sum_{j \in I} \xi_j(z)} h_{x_i} \quad \text{for all } z \in X_0.$$

The map $v : X_0 \to X$ is well defined because the covering $\{W_i\}_{i \in I}$ is locally finite. We claim that $v$ is locally Lipschitz. Indeed, given $z \in X_0$ and letting $i \in I$ such that $z \in W_i$, one can find $r < \frac{1}{2}d(z, X \setminus W_i)$ positive small enough such that $B_r(z)$ intersects a finite number of sets $W_j$ $(j \in I)$. Due to the choice of $r$, one has $\sum_{j \in I} \xi_j(y) \geq \xi_i(y) > \frac{1}{2}d(z, X \setminus W_i)$ for all $y \in B_r(z)$. Since each $\xi_j$ $(j \in I)$ is Lipschitz and since only a finite number of functions $\xi_j$ $(j \in I)$ are nonzero on $B_r(z)$, we deduce that $v|_{B_r(z)}$ is Lipschitz. This demonstrates our claim.

It is readily seen [using (5.9)] that $\|v(z)\| \leq 2\|\varphi'(z)\|$ and $\langle \varphi'(z), v(z) \rangle \geq \|\varphi'(z)\|^2$ for all $z \in X_0$. Hence $v$ is a pseudogradient vector field for $\varphi$. $\qquad\square$

We introduce the following weaker version of the Cerami condition.

**Definition 5.30.** Let $X$ be a Banach space, $\varphi \in C^1(X,\mathbb{R})$, $Z \subset X$, and $c \in \mathbb{R}$. We say that $\varphi$ satisfies the $(C)_{Z,c}$-condition if every sequence $\{x_n\}_{n \geq 1} \subset Z$ such that $\varphi(x_n) \to c$ and $(1 + \|x_n\|)\varphi'(x_n) \to 0$ in $X^*$ as $n \to \infty$ admits a strongly convergent subsequence.

*Remark 5.31.* Note that the $(C)_{Z,c}$-condition is implied by the $(C)_c$-condition.

Now we can prove the *first deformation theorem*. For $A, B \subset X$ we denote $d(A,B) := \inf_{(x,y) \in A \times B} \|x - y\|$.

**Theorem 5.32.** *Let $X$ be a Banach space, $\varphi \in C^1(X, \mathbb{R})$, and $c \in \mathbb{R}$. Fix $\varepsilon_0 > 0$, $\theta > 0$, and an open subset $U \subset X$ (possibly empty) such that $K_\varphi^c \subset U$. Assume that $\varphi$ satisfies the $(C)_{X \setminus S, c}$-condition for some bounded subset $S \subset U$ (possibly empty) such that $d(S, X \setminus U) > 0$. Then there exist $\varepsilon \in (0, \varepsilon_0)$ and a continuous map $h : [0,1] \times X \to X$ such that, for every $(t,x) \in [0,1] \times X$, we have:*

(a) *$\|h(t,x) - x\| \leq \theta(1 + \|x\|)t$;*

(b) *$\varphi(h(t,x)) \leq \varphi(x)$;*

(c) *$h(t,x) \neq x \Rightarrow \varphi(h(t,x)) < \varphi(x)$;*

(d) *$|\varphi(x) - c| \geq \varepsilon_0 \Rightarrow h(t,x) = x$;*

(e) *$h(1, \varphi^{c+\varepsilon}) \subset \varphi^{c-\varepsilon} \cup U$ and $h(1, \varphi^{c+\varepsilon} \setminus U) \subset \varphi^{c-\varepsilon}$.*

*Proof.* From Proposition 5.25 we know that $K_\varphi^c$ is compact (possibly empty). Moreover, by hypothesis, we have that $d(S, X \setminus U) > 0$. Thus, we can find $\rho > 0$ such that

$$(S \cup K_\varphi^c)_{3\rho} := \{x \in X : d(x, S \cup K_\varphi^c) < 3\rho\} \subset U$$

(take any $\rho > 0$ if $S \cup K_\varphi^c = \emptyset$).

*Claim 1*: There exist $\bar{\varepsilon} \in (0, \frac{\varepsilon_0}{2})$ and $\mu > 0$ such that we have $(1 + \|x\|)\|\varphi'(x)\| \geq \mu$ whenever $x \in A := \{x \in X : |\varphi(x) - c| \leq 2\bar{\varepsilon}$ and $x \notin (S \cup K_\varphi^c)_\rho\}$.

Supposing that this is not true, we can find a sequence $\{x_n\}_{n \geq 1} \subset X$ such that

$$\varphi(x_n) \to c, \quad x_n \notin (S \cup K_\varphi^c)_\rho \quad \text{and} \quad (1 + \|x_n\|)\varphi'(x_n) \to 0 \text{ in } X^*.$$

Because $\varphi$ satisfies the $(C)_{X \setminus S, c}$-condition, we may assume that $x_n \to x$ in $X$, and then we have $\varphi(x) = c$, $\varphi'(x) = 0$, a contradiction of the fact that $d(x, K_\varphi^c) \geq \rho$. This proves Claim 1.

Note that Claim 1 yields $A \subset X_0 = X \setminus K_\varphi$. We fix some notation. Let

$$B = \{x \in X : |\varphi(x) - c| \geq 2\bar{\varepsilon} \text{ or } x \in \overline{(S \cup K_\varphi^c)_\rho}\},$$

$$C = \{x \in X : |\varphi(x) - c| \leq \bar{\varepsilon} \text{ and } x \notin (S \cup K_\varphi^c)_{2\rho}\}.$$

Thus, $B \supset X \setminus A$ and $C \subset \text{int} A$. The sets $B, C$ are disjoint and closed, so the map $\gamma : X \to [0,1]$ defined by

$$\gamma(x) = \frac{d(x, B)}{d(x, B) + d(x, C)}$$

is locally Lipschitz and satisfies $\gamma|_B = 0$, $\gamma|_C = 1$. Furthermore, we choose $\eta > 0$ such that $e^\eta \leq \theta + 1$, and we fix $v : X_0 \to X$ as a pseudogradient vector field for $\varphi$ (Proposition 5.29). Finally, we define

$$g(x) = \begin{cases} -\mu\eta\gamma(x)\frac{v(x)}{\|v(x)\|^2} & \text{if } x \in A, \\ 0 & \text{if } x \in X \setminus A. \end{cases} \qquad (5.10)$$

*Claim 2*: The map $g : X \to X$ is locally Lipschitz and satisfies

$$\|g(x)\| \leq \eta(1 + \|x\|) \quad \text{and} \quad \langle\varphi'(x), g(x)\rangle \leq -\frac{1}{4}\mu\eta\gamma(x) \quad \text{for all } x \in X.$$

Clearly, if $x \in X \setminus A$, then $g$ is Lipschitz in the neighborhood of $x$, and both claimed relations are satisfied [since $\gamma(x) = 0$]. Next, let $x \in A$ (so $x \in X_0$). Note that we have $g(y) = -\mu\eta\gamma(y)\frac{v(y)}{\|v(y)\|^2}$ for all $y \in X_0$. Since $\gamma$, $v$, and $\|v\|^2$ are locally Lipschitz and since $\|v(x)\| \geq \frac{\mu}{1+\|x\|} > 0$ (by Claim 1 and Definition 5.27), we obtain that $g$ is Lipschitz in the neighborhood of $x$. The previous relation also implies that $\|g(x)\| \leq \eta(1 + \|x\|)$, and, moreover, Definition 5.27 yields

$$\langle\varphi'(x), g(x)\rangle \leq -\mu\eta\gamma(x)\frac{\|\varphi'(x)\|^2}{\|v(x)\|^2} \leq -\frac{1}{4}\mu\eta\gamma(x).$$

This proves Claim 2.

Given $x \in X$, we consider the Cauchy problem

$$\begin{cases} \xi'(t) = g(\xi(t)) & \text{in } [0,1], \\ \xi(0) = x. \end{cases} \qquad (5.11)$$

By Claim 2, the map $g$ is locally Lipschitz and of sublinear growth, hence problem (5.11) admits a unique global solution $\xi_x : [0,1] \to X$. We set

$$h(t,x) = \xi_x(t) \quad \text{for all } t \in [0,1], \text{ all } x \in X.$$

We must now check that the map $h : [0,1] \times X \to X$ so obtained satisfies the conditions of the theorem.

First, since $h(\cdot, x)$ is continuous for all $x \in X$ and since the map $X \to C([0,1], X)$, $x \mapsto h(\cdot, x)$ is continuous, we obtain that the map $h$ is continuous.

Second, integrating (5.11) and invoking Claim 2, we have

$$\|h(t,x) - x\| \leq \int_0^t \|g(h(s,x))\| \, ds \leq \eta \int_0^t (1 + \|h(s,x)\|) \, ds$$

$$\leq \eta \int_0^t \|h(s,x) - x\| \, ds + \eta(1 + \|x\|)t.$$

By Gronwall's inequality, we obtain

$$\|h(t,x) - x\| \leq \eta(1 + \|x\|)t + \int_0^t \eta^2(1 + \|x\|)s e^{\eta(t-s)} \, ds = (1 + \|x\|)(e^{\eta t} - 1).$$

Moreover, the fact that $t \in [0,1]$ and the choice of $\eta$ imply

$$e^{\eta t} - 1 = \sum_{k=1}^{\infty} \frac{(\eta t)^k}{k!} \leq t \sum_{k=1}^{\infty} \frac{\eta^k}{k!} = t(e^{\eta} - 1) \leq \theta t,$$

whence $\|h(t,x) - x\| \leq \theta(1 + \|x\|)t$ for all $t \in [0,1]$, all $x \in X$. This proves part (a) of the theorem.

Next, by the chain rule and Claim 2, we have

$$\frac{d}{dt}\varphi(h(t,x)) = \langle \varphi'(h(t,x)), g(h(t,x)) \rangle \leq -\frac{1}{4}\eta\mu\gamma(h(t,x)) \leq 0 \text{ in } [0,1], \quad (5.12)$$

hence $\varphi(h(t,x)) \leq \varphi(x)$ for all $t \in [0,1]$, which proves part (b) of the theorem.

To deduce part (c) of the theorem from (5.12), it is sufficient to check that one has $\gamma(h(t,x)) > 0$ whenever $h(t,x) \neq x$. Arguing indirectly, we assume that $x_0 := h(t,x)$ is different from $x$ and satisfies $\gamma(x_0) = 0$, so $g(x_0) = 0$. Then the functions $\xi_1 \equiv x_0$ and $\xi_2(s) = h(s,x)$ are solutions of the Cauchy problem

$$\begin{cases} \xi'(s) = g(\xi(s)) & \text{for all } s \in [0,1], \\ \xi(t) = x_0. \end{cases}$$

By the uniqueness of the solution, this yields $\xi_1 = \xi_2$, hence $h(s,x) = x_0$ for all $s \in [0,1]$. In particular, we deduce that $x = h(t,x)$, and the proof of (c) is complete.

Note that if $x \in X$ is such that $|\varphi(x) - c| \geq \varepsilon_0$, then $x \in B$, which implies that $\gamma(x) = 0$, and so $g(x) = 0$. In this case, the function $\xi \equiv x$ is a solution to problem (5.11). Hence $h(t,x) = x$ for all $t \in [0,1]$. This proves part (d) of the theorem.

We now fix the constant $\varepsilon$ needed in part (e) of the theorem. Fix $R > 0$ and $\varepsilon \in (0,\bar{\varepsilon})$ such that

$$(S \cup K_{\varphi}^c)_{3\rho} \subset B_R(0), \quad 8\varepsilon \leq \mu\eta \quad \text{and} \quad 8\theta(1+R)\varepsilon \leq \mu\eta\rho. \quad (5.13)$$

It remains to check that the map $h$ and the constant $\varepsilon$ satisfy part (e) of the theorem. Arguing indirectly, suppose that we can find $x \in \varphi^{c+\varepsilon}$ such that $\varphi(h(1,x)) > c - \varepsilon$ and $[x \notin U \text{ or } h(1,x) \notin U]$. By (5.12), we have

$$c - \varepsilon < \varphi(h(1,x)) \leq \varphi(h(t,x)) \leq \varphi(x) \leq c + \varepsilon \quad \text{for all } t \in [0,1]. \quad (5.14)$$

We claim that $h([0,1],x) \cap (S \cup K_{\varphi}^c)_{2\rho} \neq \emptyset$. Indeed, otherwise, (5.14) guarantees that $h(t,x) \in C$ for all $t \in [0,1]$, hence $\gamma(h(t,x)) = 1$ for all $t \in [0,1]$; then (5.12) and (5.14) imply

$$\frac{1}{4}\mu\eta \leq \varphi(x) - \varphi(h(1,x)) < 2\varepsilon,$$

a contradiction of the second relation in (5.13), so the claim is justified. This claim, together with the assumption that $(x = h(0,x) \notin U$ or $h(1,x) \notin U)$ and $(S \cup K_\varphi^c)_{3\rho} \subset U$, allows us to find $t_1, t_2 \in [0,1]$, $t_1 \neq t_2$, such that

$$d(h(t_1,x), S \cup K_\varphi^c) = 2\rho, \quad d(h(t_2,x), S \cup K_\varphi^c) = 3\rho, \tag{5.15}$$

$$\text{and } 2\rho < d(h(t,x), S \cup K_\varphi^c) < 3\rho \text{ for all } t \in (t_1', t_2'), \tag{5.16}$$

where $t_1' = \min\{t_1,t_2\}$ and $t_2' = \max\{t_1,t_2\}$. Relations (5.14) and (5.16) imply that $h(t,x) \in C$ [so $\gamma(h(t,x)) = 1$] for all $t \in (t_1', t_2')$. By (5.12) and (5.14), we deduce that

$$t_2' - t_1' \leq \frac{4}{\eta\mu}(\varphi(h(t_1',x)) - \varphi(h(t_2',x))) \leq \frac{8\varepsilon}{\eta\mu}. \tag{5.17}$$

Note that $h(t_1' + t,x) = h(t, h(t_1',x))$ for all $t \in (0, 1 - t_1')$ [since both functions are solutions of the Cauchy problem (5.11) with initial condition $\xi(0) = h(t_1',x)$]. By this observation, combined with (5.13), (5.15), (5.17), and part (a) of the theorem, we obtain

$$\rho \leq \|h(t_2',x) - h(t_1',x))\| \leq \theta(1 + \|h(t_1',x\|)(t_2' - t_1') < \theta(1+R)\frac{8\varepsilon}{\eta\mu} \leq \rho,$$

a contradiction. This proves part (e) of the theorem. $\qquad\square$

Let us recall the following topological notions, which are important in critical point theory and in Morse theory.

**Definition 5.33.** Let $Y$ be a Hausdorff topological space.

(a) A continuous map $h : [0,1] \times Y \to Y$ is a *deformation* of $Y$ if $h(0,\cdot) = \text{id}_Y$. Moreover, if $h(1,Y) \subset A \subset Y$, then we say that $h$ is a *deformation of $Y$ into $A$*.
(b) A closed set $A \subset Y$ is a (resp. *strong*) *deformation retract of $Y$* if there exists a deformation $h : [0,1] \times Y \to Y$ of $Y$ into $A$ such that $h(1,\cdot)|_A = \text{id}_A$ (resp. such that $h(t,\cdot)|_A = \text{id}_A$ for all $t \in [0,1]$).

The next result is known in the literature as the *second deformation lemma*. In the statement, we allow $b = +\infty$, in which case $\varphi^b \setminus K_\varphi^b = X$.

**Theorem 5.34.** *Let $X$ be a Banach space, $\varphi \in C^1(X)$, $a \in \mathbb{R}$, and $b \in (a, +\infty]$. Assume that $\varphi$ satisfies the $(C)_c$-condition for every $c \in [a,b)$, that $\varphi$ has no critical values in $(a,b)$, and that $\varphi^{-1}(a)$ contains at most a finite number of critical points of $\varphi$. Then there exists a deformation $h : [0,1] \times (\varphi^b \setminus K_\varphi^b) \to \varphi^b \setminus K_\varphi^b$ of $\varphi^b \setminus K_\varphi^b$ into $\varphi^a$ such that:*

(a) *If $x \in \varphi^a$, then $h(t,x) = x$ for all $t \in [0,1]$ (hence, $\varphi^a$ is a strong deformation retract of $\varphi^b \setminus K_\varphi^b$);*
(b) *If $x \in \varphi^b \setminus (\varphi^a \cup K_\varphi^b)$, then $a \leq \varphi(h(t,x)) < \varphi(h(s,x))$ for all $s,t \in [0,1]$, $s < t$.*

*Proof.* Choose $v : X_0 = X \setminus K_\varphi \to X$ a pseudogradient vector field for $\varphi$ (Definition 5.27). We write

$$\varphi^b \setminus K_\varphi^b = A \cup \varphi^a, \quad \text{where } A = \varphi^b \setminus (\varphi^a \cup K_\varphi^b).$$

By hypothesis, $A \subset X_0$; hence, for $x \in A$, the Cauchy problem

$$\begin{cases} \xi'(t) = -\dfrac{v(\xi(t))}{(1 + \|\xi(t)\|)\|v(\xi(t))\|^2} & \text{in } \mathbb{R}_+, \\ \xi(0) = x, \end{cases} \tag{5.18}$$

has a unique solution $\xi_x$ defined on a maximal interval $[0, \eta_+(x))$. Moreover, using the properties of the pseudogradient vector field $v$, we have

$$\frac{d}{dt}\varphi(\xi_x(t)) = \langle \varphi'(\xi_x(t)), \xi_x'(t) \rangle \leq -\frac{1}{4(1 + \|\xi_x(t)\|)} \quad \text{for all } t \in [0, \eta_+(x)). \tag{5.19}$$

In particular, the map $t \mapsto \varphi(\xi_x(t))$ is decreasing on $[0, \eta_+(x))$. We set

$$t(x) = \sup\{t \in [0, \eta_+(x)) : \varphi(\xi_x(t)) > a\}.$$

In this way, we obtain a map $A \to (0, +\infty]$, $x \mapsto t(x)$. The construction of the deformation $h$ of the statement is based on a series of claims. First, we emphasize two preliminary facts (Claims 1 and 2).

*Claim 1*: For every $\delta > 0$ and $b' \in (a, b)$ there is $\mu = \mu(\delta, b') > 0$ such that

$$x \in A, \quad \varphi(x) \leq b', \quad \text{and} \quad d(x, K_\varphi^a) > \delta \implies (1 + \|x\|)\|\varphi'(x)\| \geq \mu.$$

This easily follows from the assumptions that $\varphi$ satisfies the $(C)_c$-condition for every $c \in [a, b]$ and that the interval $(a, b)$ contains no critical value.

*Claim 2*: Let $\{x_n\}_{n \geq 1} \subset A$, and let $\{t_n\}_{n \geq 1}$, $\{s_n\}_{n \geq 1}$ be real sequences such that $0 \leq s_n < t_n < t(x_n)$ for all $n \geq 1$. Assume that the limit $\lim_{n \to \infty} \xi_{x_n}(s_n) =: \hat{x}$ exists and belongs to $\varphi^{-1}(a)$. Then we also have $\lim_{n \to \infty} \xi_{x_n}(t_n) = \hat{x}$.

This comparison principle will be a useful ingredient in reasonings later on in the proof. We argue by contradiction: assume that $\xi_{x_n}(t_n) \not\to \hat{x}$. Up to dealing with subsequences, we can find $\rho > 0$ such that $\|\xi_{x_n}(t_n) - \hat{x}\| \geq 2\rho$ for all $n \geq 1$. Since $K_\varphi^a$ is a finite set (by assumption) and $\varphi(\hat{x}) = a$, up to taking $\rho > 0$ smaller if necessary, we find $\delta > 0$ and $b' \in (a, b)$ such that

$$x \in A, \quad \rho \leq \|x - \hat{x}\| \leq 2\rho \implies \varphi(x) \leq b' \text{ and } d(x, K_\varphi^a) > \delta.$$

Since $\lim_{n \to \infty} \xi_{x_n}(s_n) = \hat{x}$, there is $n_0 \geq 1$ such that $\|\xi_{x_n}(s_n) - \hat{x}\| \leq \rho$ for all $n \geq n_0$. Then, for all $n \geq n_0$, we can find numbers $s_n', t_n'$ satisfying

$$s_n \le s_n' < t_n' \le t_n, \quad \|\xi_{x_n}(s_n') - \hat{x}\| = \rho, \quad \|\xi_{x_n}(t_n') - \hat{x}\| = 2\rho,$$

$$\text{and } \rho \le \|\xi_{x_n}(t) - \hat{x}\| \le 2\rho \text{ for all } t \in [s_n', t_n'].$$

Let $\mu = \mu(\delta, b') > 0$ be as in Claim 1 with $\delta$ and $b'$ as previously. Using Claim 1, we have

$$\|\xi_{x_n}'(t)\| \le \frac{1}{(1 + \|\xi_{x_n}(t)\|)\|v(\xi_{x_n}(t))\|} \le \frac{1}{(1 + \|\xi_{x_n}(t)\|)\|\varphi'(\xi_{x_n}(t))\|} \le \frac{1}{\mu}$$

for all $t \in [s_n', t_n']$, all $n \ge n_0$, thus

$$\rho \le \|\xi_{x_n}(t_n') - \xi_{x_n}(s_n')\| \le \frac{t_n' - s_n'}{\mu}. \tag{5.20}$$

Next, by (5.19), we have

$$\frac{d}{dt}\varphi(\xi_{x_n}(t)) \le -\frac{1}{4(1 + \|\hat{x}\| + 2\rho)} \text{ for all } t \in [s_n', t_n'], \text{ all } n \ge n_0.$$

Integrating this relation and invoking (5.20), we obtain

$$\varphi(\xi_{x_n}(t_n')) - \varphi(\xi_{x_n}(s_n')) \le -\frac{t_n' - s_n'}{4(1 + \|\hat{x}\| + 2\rho)} \le -\frac{\mu\rho}{4(1 + \|\hat{x}\| + 2\rho)}$$

for all $n \ge n_0$. Note that $\varphi(\xi_{x_n}(t_n')) > a$ [since $t_n' < t(x_n)$] and $\varphi(\xi_{x_n}(s_n')) \le \varphi(\xi_{x_n}(s_n))$ [by (5.19)]. Thus,

$$a + \frac{\mu\rho}{4(1 + \|\hat{x}\| + 2\rho)} < \varphi(\xi_{x_n}(s_n)) \text{ for all } n \ge n_0.$$

This is contradictory because $\lim_{n \to \infty} \varphi(\xi_{x_n}(s_n)) = \varphi(\hat{x}) = a$. Claim 2 ensues.

*Claim 3*: Let $x \in A$. Then $t(x) \in (0, +\infty)$ and $\hat{\xi}(x) := \lim_{t \to t(x)} \xi_x(t)$ exists and belongs to $\varphi^{-1}(a)$. More precisely, there are two possible situations:

(a) If $\liminf_{t \to t(x)} d(\xi_x(t), K_\varphi^a) > 0$, then $t(x) < \eta_+(x)$ and $\hat{\xi}(x) \in \varphi^{-1}(a) \setminus K_\varphi^a$;

(b) If $\liminf_{t \to t(x)} d(\xi_x(t), K_\varphi^a) = 0$, then $t(x) = \eta_+(x) < +\infty$ and $\hat{\xi}(x) \in K_\varphi^a$.

First we establish (a). The assumption in (a) implies that there are $\gamma \in (0, t(x))$ and $\delta > 0$ such that

$$d(\xi_x(t), K_\varphi^a) > \delta \text{ for all } t \in [\gamma, t(x)).$$

By (5.19), we have $\varphi(\xi_x(t)) \le \varphi(\xi_x(\gamma)) =: b' < b$ for all $t \in [\gamma, t(x))$. Then, letting $\mu = \mu(\delta, b') > 0$ be as in Claim 1, we have

$$(1+\|\xi_x(t)\|)\|v(\xi_x(t))\| \geq (1+\|\xi_x(t)\|)\|\varphi'(\xi_x(t))\| \geq \mu \text{ for all } t \in [\gamma,t(x)),$$

whence

$$\|\xi_x'(t)\| \leq \frac{1}{\mu}, \text{ and so } \|\xi_x(t)\| \leq \|\xi_x(\gamma)\| + \frac{t-\gamma}{\mu} \text{ for all } t \in [\gamma,t(x)). \quad (5.21)$$

Now, integrating (5.19), we obtain

$$a - \varphi(\xi_x(\gamma)) < \varphi(\xi_x(t)) - \varphi(\xi_x(\gamma))$$

$$\leq -\int_\gamma^t \frac{\mu}{4(\theta+s-\gamma)} \, ds = -\frac{\mu}{4}\left(\ln(\theta+t-\gamma) - \ln\theta\right)$$

for all $t \in [\gamma,t(x))$, where $\theta = \mu(1+\|\xi_x(\gamma)\|)$, which guarantees that $t(x) < +\infty$. Then the first part of (5.21) yields

$$\int_\gamma^{t(x)} \|\xi_x'(t)\| \, dt \leq \frac{t(x)-\gamma}{\mu} < +\infty.$$

This implies that the limit $\hat{\xi}(x) = \lim_{t \to t(x)} \xi_x(t)$ exists; moreover, $\hat{\xi}(x) \in \varphi^{-1}([a,b'])$ and, clearly, $\hat{\xi}(x) \notin K_\varphi^a$. Thus, $\hat{\xi}(x) \in X_0$. Since $\xi_x$ is the maximal solution of (5.18), it must be defined beyond the interval $[0,t(x))$. In this way, we have $t(x) < \eta_+(x)$, and the definition of $t(x)$ yields $\varphi(\hat{\xi}(x)) = \varphi(\xi_x(t(x))) = a$. This proves (a).

Next, we establish (b). The assumption in (b) yields sequences $\{s_n\}_{n\geq 1} \subset [0,t(x))$ and $\{x_n\}_{n\geq 1} \subset K_\varphi^a$ such that $s_n \to t(x)$ and $\|\xi_x(s_n) - x_n\| \to 0$ as $n \to \infty$. Since $K_\varphi^a$ is compact (here it is a finite set), up to extracting a subsequence, we may assume that $\{\xi_x(s_n)\}_{n\geq 1}$ converges to some $\hat{x} \in K_\varphi^a$. We can see that

$$\xi_x(t) \to \hat{x} \text{ in } X \text{ as } t \to t(x). \quad (5.22)$$

Indeed, if this is not the case, then there is a sequence $\{t_n\}_{n\geq 1}$ such that

$$s_n < t_n < t(x) \text{ for all } n \geq 1 \text{ and } \xi_x(t_n) \nrightarrow \hat{x},$$

but this contradicts Claim 2, so (5.22) holds. Thus, $\hat{\xi}(x) = \lim_{t \to t(x)} \xi_x(t) = \hat{x}$ exists and belongs to $K_\varphi^a$. In particular, since $\hat{\xi}(x) \notin X_0$, the solution $t \mapsto \xi_x(t)$ to the Cauchy problem (5.18) is not defined at $t = t(x)$. This yields $t(x) = \eta_+(x)$. To complete the proof of (b), we need to check that $t(x) < +\infty$. To do this, note that the set $\{\xi_x(t) : t \in [0,t(x))\} \cup \{\hat{\xi}(x)\}$ is compact, and hence we can find $M > 0$ such that $\|\xi_x(t)\| \leq M$ for all $t \in [0,t(x))$. Then (5.19) implies

$$a - \varphi(x) < \varphi(\xi_x(t)) - \varphi(\xi_x(0)) \leq -\frac{t}{4(1+M)} \text{ for all } t \in [0,t(x)),$$

whence $t(x) < +\infty$. This proves (b), which completes the proof of Claim 3.

*Claim 4*: Let $x \in A$ with $\hat{\xi}(x) = \lim_{t \to t(x)} \xi_x(t) \in \varphi^{-1}(a)$ as in Claim 3. Let $\{x_n\}_{n \geq 1} \subset A$ and $\{t_n\}_{n \geq 1}$, with $0 \leq t_n < t(x_n)$, be sequences satisfying $\lim_{n \to \infty} x_n = x$ and $\limsup_{n \to \infty} t_n \geq t(x)$. Then we have $\xi_{x_n}(t_n) \to \hat{\xi}(x)$ as $n \to \infty$.

It is sufficient to show that any subsequence of $\{\xi_{x_n}(t_n)\}_{n \geq 1}$ admits a subsequence that converges to $\hat{\xi}(x)$. Fix a subsequence that, for simplicity, we still denote by $\{\xi_{x_n}(t_n)\}_{n \geq 1}$. Fix an increasing sequence $\{s_k\}_{k \geq 1} \subset [0, t(x))$ such that

$$s_k \to t(x) \text{ as } k \to \infty \text{ and } \|\xi_x(s_k) - \hat{\xi}(x)\| < \frac{1}{k} \text{ for all } k \geq 1.$$

Given $k \geq 1$, due to the continuous dependence of the solution to (5.18) on the initial condition, we have $\|\xi_{x_n}(s_k) - \hat{\xi}(x)\| < \frac{2}{k}$ whenever $n$ is large enough. Moreover, since $\limsup_{n \to \infty} t_n \geq t(x)$, we have $t_n > s_k$ along a subsequence of $\{t_n\}_{n \geq 1}$. In this way, we can construct an increasing map $\mathbb{N} \to \mathbb{N}$, $k \mapsto n_k$ such that

$$s_k < t_{n_k} \text{ for all } k \geq 1 \text{ and } \xi_{x_{n_k}}(s_k) \to \hat{\xi}(x) \text{ as } k \to \infty.$$

By Claim 2, we obtain that $\xi_{x_{n_k}}(t_{n_k}) \to \hat{\xi}(x)$ as $k \to \infty$. This achieves the proof of Claim 4.

*Claim 5*: The map $A \to (0, +\infty)$, $x \mapsto t(x)$ is continuous.

Let $x \in A$. Arguing by contradiction, assume that we can find a sequence $\{x_n\}_{n \geq 1} \subset A$ such that $x_n \to x$ in $X$ but $t(x_n) \not\to t(x)$. For every $\varepsilon \in (0, t(x))$, due to the continuous dependence of the solution to (5.18) on the initial condition, we have $\eta_+(x_n) > \eta_+(x) - \varepsilon$ and $a < \varphi(\xi_{x_n}(t(x) - \varepsilon))$ whenever $n$ is large enough, hence $t(x_n) \geq t(x) - \varepsilon$; this yields $\liminf_{n \to \infty} t(x_n) \geq t(x)$. Thus, $\limsup_{n \to \infty} t(x_n) > t(x)$, so that, up to considering a subsequence, we may assume that

$$t(x_n) > t(x) + \gamma \text{ for all } n \geq 1 \tag{5.23}$$

for some $\gamma > 0$. It easily follows from Claim 4 that we can find $\varepsilon > 0$, $M > 0$, and $n_0 \geq 1$ such that

$$\|\xi_{x_n}(t)\| \leq M \text{ for all } n \geq n_0, \text{ all } t \in (t(x) - \varepsilon, t(x_n)). \tag{5.24}$$

Fix $\delta \in (0, \varepsilon)$. Due to the continuous dependence of the solution of (5.18) on the initial condition and due to the continuity of $\varphi$, we have

$$\varphi(\xi_x(t(x) - \delta)) = \lim_{n \to \infty} \varphi(\xi_{x_n}(t(x) - \delta)). \tag{5.25}$$

Taking $n \geq n_0$ and $\theta \in (0, \gamma)$, by (5.19) and (5.24), we have

$$\varphi(\xi_{x_n}(t(x_n) - \theta)) - \varphi(\xi_{x_n}(t(x) - \delta)) \leq -\frac{t(x_n) - \theta - t(x) + \delta}{4(1 + M)}.$$

By definition of $t(x_n)$, we have $\varphi(\xi_{x_n}(t(x_n) - \theta)) > a$. Letting $\theta \to 0$, through (5.23) we obtain

$$\varphi(\xi_{x_n}(t(x) - \delta)) > a + \frac{\gamma + \delta}{4(1 + M)} \quad \text{for all } n \geq n_0.$$

Letting $n \to \infty$ [using (5.25)] and then letting $\delta \to 0$, by invoking Claim 3, we obtain

$$a = \lim_{\delta \to 0} \varphi(\xi_x(t(x) - \delta)) \geq a + \frac{\gamma}{4(1 + M)} > a,$$

which is contradictory. This completes the proof of Claim 5.

We now define a map $h : [0, 1] \times \varphi^b \setminus K_\varphi^b \to \varphi^b \setminus K_\varphi^b$ by letting

$$h(t, x) = \begin{cases} \xi_x(tt(x)) & \text{if } x \in A \text{ and } t \in [0, 1), \\ \hat{\xi}(x) & \text{if } x \in A \text{ and } t = 1, \\ x & \text{if } x \in \varphi^a, \end{cases}$$

for all $t \in [0, 1]$, all $x \in \varphi^b \setminus K_\varphi^b$, with $\hat{\xi}(x)$ as in Claim 3. It is easy to see that $h$ satisfies conditions (a) and (b), of the statement of the theorem. To complete the proof of the theorem, it remains to check that $h$ is continuous. Clearly, $h|_{[0,1] \times \varphi^a}$ is continuous, and it follows from Claim 5 that $h$ is continuous on $[0, 1) \times A$. It is then clear that the proof of the theorem will be complete once we prove the following claim.

*Claim 6*: Let $(t, x) \in (\{1\} \times A) \cup ([0, 1] \times \varphi^{-1}(a))$, and let the sequences $\{x_n\}_{n \geq 1} \subset A$ and $\{t_n\}_{n \geq 1} \subset [0, 1]$ such that $(t_n, x_n) \to (t, x)$ as $n \to \infty$. Then $h(t_n, x_n) \to h(t, x)$ as $n \to \infty$.

For all $n \geq 1$, since $h(\cdot, x_n)$ is clearly continuous on $[0, 1]$, we can find $\hat{t}_n \in (0, 1)$ such that $\|h(t_n, x_n) - h(\hat{t}_n, x_n)\| \leq \frac{1}{n}$ and $|t_n - \hat{t}_n| \leq \frac{1}{n}$. Up to dealing with $\hat{t}_n$ instead of $t_n$, we may assume that $t_n \in (0, 1)$ for all $n \geq 1$. Thus, $h(t_n, x_n) = \xi_{x_n}(t_n t(x_n))$ for all $n \geq 1$. We distinguish two cases depending on whether we have ($x \in A$ and $t = 1$) or $x \in \varphi^{-1}(a)$. In the first case, $\lim_{n \to \infty} t_n t(x_n) = t(x)$, and, by Claim 4, we get $\xi_{x_n}(t_n t(x_n)) \to \hat{\xi}(x) = h(1, x)$ as $n \to \infty$. In the second case, set $s_n = 0$, so that $s_n < t_n t(x_n) < t(x_n)$ for all $n \geq 1$ and $\xi_{x_n}(s_n) = x_n \to x$ as $n \to \infty$; applying Claim 2, we deduce that $\xi_{x_n}(t_n t(x_n)) \to x = h(t, x)$ as $n \to \infty$. In both cases, we obtain that $h(t_n, x_n) = \xi_{x_n}(t_n t(x_n)) \to h(t, x)$ as $n \to \infty$. This establishes Claim 6. The proof of the theorem is now complete. $\qquad \square$

As a direct consequence of Theorem 5.34, we have the following corollary.

**Corollary 5.35.** *If $\varphi \in C^1(X, \mathbb{R})$, $a \in \mathbb{R}$, $a < b \leq +\infty$, $\varphi$ satisfies the $(C)_c$-condition at every $c \in [a, b)$, and $\varphi$ has no critical values in $[a, b]$, then $\varphi^a$ is a strong deformation retract of $\varphi^b$.*

## 5.3  Minimax Theorems for Critical Points

The next notion in different forms is central in critical point theory.

**Definition 5.36.** Let $Y$ be a Hausdorff topological space, $E_0 \subset E$ and $D$ be nonempty sets in $Y$, and $\gamma^* \in C(E_0, Y)$. We say that the pair $\{E_0, E\}$ *links* $D$ in $Y$ *via* $\gamma^*$ if the following conditions hold:

(a) $\gamma^*(E_0) \cap D = \emptyset$.
(b) For any $\gamma \in C(E, Y)$ with $\gamma|_{E_0} = \gamma^*$ we have $\gamma(E) \cap D \neq \emptyset$.

*Remark 5.37.* The sets $\{E_0, E, D\}$ are said to be *linking sets in* $Y$ *via* $\gamma^*$. If $\gamma^* = \mathrm{id}_{E_0}$ (which is usually the case), then we simply say that $\{E_0, E, D\}$ are *linking sets*.

In what follows, $(X, \|\cdot\|)$ is a Banach space.

*Example 5.38.*

(a) Let $E_0 = \{x_0, x_1\} \subset X$, $E = \{(1-t)x_0 + tx_1 : t \in [0,1]\}$, and $D = \partial\Omega$, where $\Omega$ is an open neighborhood of $x_0$ such that $x_1 \notin \overline{\Omega}$. Then, by connectedness, we see that the sets $\{E_0, E, D\}$ are linking in $X$.
(b) Suppose $X = Y \oplus V$ with $\dim Y < +\infty$. Let $E_0 = \{x \in Y : \|x\| = R\} = \partial B_R(0) \cap Y$, $E = \{x \in Y : \|x\| \leq R\} = \overline{B_R(0)} \cap Y$ and $D = V$. We claim that the sets $\{E_0, E, D\}$ are linking. To this end, let $p_Y \in \mathcal{L}(X)$ be the projection onto $Y$ (it exists since $Y$ is finite dimensional), and let $\gamma \in C(E, X)$ be such that $\gamma|_{E_0} = \mathrm{id}_{E_0}$. We will show that $0 \in p_Y(\gamma(E))$, which implies that $\gamma(E) \cap D \neq \emptyset$. To do this, we use Brouwer's degree theory. We consider the homotopy

$$h(t, x) = t p_Y(\gamma(x)) + (1-t)x \quad \text{for all } (t, x) \in [0,1] \times Y.$$

It is clear that $0 \notin h([0,1] \times E_0)$. From the homotopy invariance and normalization properties of Brouwer's degree [Theorem 4.5(c), (a)] we have

$$d(p_Y \circ \gamma, B_R(0) \cap Y, 0) = d(\mathrm{id}, B_R(0) \cap Y, 0) = 1,$$

which yields $0 \in p_Y(\gamma(x))$ for some $x \in B_R(0) \cap Y$ [by Theorem 4.5(e)].
(c) Let $X = Y \oplus V$, with $\dim Y < +\infty$, $R_1 > r > 0$, $R_2 > 0$, and let $v_0 \in V$, with $\|v_0\| = 1$. We set

$$E_0 = \{y + \lambda v_0 : y \in Y, (0 < \lambda < R_1, \|y\| = R_2) \text{ or } (\lambda \in \{0, R_1\}, \|y\| \leq R_2)\},$$

$$E = \{y + \lambda v_0 : y \in Y, 0 \leq \lambda \leq R_1, \|y\| \leq R_2\}, \quad D = \partial B_r(0) \cap V.$$

Clearly, $E$ is a cylinder with basis $\overline{B_{R_2}(0)} \cap Y$ and height $R_1$, and $E_0$ is the boundary of the cylinder (lateral surface and bottom and top bases). We claim that the sets $\{E_0, E, D\}$ are linking in $X$. Let $\gamma \in C(E, X)$ with $\gamma|_{E_0} = \mathrm{id}_{E_0}$. We will show that there exists $x \in E$ such that $\|\gamma(x)\| = r$ and $p_Y(\gamma(x)) = 0$. To this end, we consider the homotopy

$$h(t,x) = tp_Y(\gamma(x)) + (1-t)y + (t\|\gamma(x) - p_Y(\gamma(x))\| + (1-t)\lambda - r)v_0$$

for all $t \in [0,1]$, all $x = y + \lambda v_0 \in E$. It is easy to check that $0 \notin h([0,1] \times E_0)$. From the homotopy invariance property of Brouwer's degree [Theorem 4.5(c)] we get

$$d(h(0,\cdot),\mathrm{int}\,E,0) = d(h(1,\cdot),\mathrm{int}\,E,0).$$

We have $h(0,x) = y + (\lambda - r)v_0$ for all $x = y + \lambda v_0 \in E$, and so, in view of Theorem 4.5(h), (a), we obtain that $d(h(0,\cdot),\mathrm{int}\,E,0) = d(\mathrm{id},\mathrm{int}\,E,rv_0) = 1$, hence $d(h(1,\cdot),\mathrm{int}\,E,0) = 1$. Therefore, we can find $x = y + \lambda v_0$ such that $h(1,(y,\lambda)) = 0$ [Theorem 4.5(e)], hence $p_Y(\gamma(x)) = 0$ and $\|\gamma(x)\| = \|\gamma(x) - p_Y(\gamma(x))\| = r$.

(d) Let $X = Y \oplus V$, with $\dim Y < +\infty$, $0 < r < R$, and let $v_0 \in V$, with $\|v_0\| = 1$. Set

$$E_0 = \{x = y + \lambda v_0 : y \in Y, (\lambda \geq 0, \|x\| = R) \text{ or } (\lambda = 0, \|y\| \leq R)\},$$

$$E = \{x = y + \lambda v_0 : y \in Y, \lambda \geq 0, \|x\| \leq R\}, \quad D = \partial B_r(0) \cap V.$$

Thus, $E$ is the upper-half $R$-ball, $E_0$ is the northern hemisphere plus the equator disk, and $D$ is the $r$-sphere in $V$. As in (c), we can show that $\{E_0,E,D\}$ are linking in $X$.

Using the notion of linking sets, we have the following general minimax principle. Recall that for $A,B \subset X$ we denote $d(A,B) := \inf\limits_{(x,y)\in A\times B} \|x - y\|$.

**Theorem 5.39.** *Let $\{E_0,E,D\}$ be linking sets in $X$ via $\gamma^*$, with $\gamma^*(E_0)$ bounded, $D$ closed, and $d(\gamma^*(E_0),D) > 0$. Let $\Gamma = \{\gamma \in C(E,X) : \gamma|_{E_0} = \gamma^*\}$. Let $\varphi \in C^1(X,\mathbb{R})$ be such that*

$$a := \sup_{\gamma^*(E_0)} \varphi \leq \inf_D \varphi =: b, \tag{5.26}$$

*let $c = \inf\limits_{\gamma\in\Gamma} \sup\limits_{x\in E} \varphi(\gamma(x))$, and assume that $\varphi$ satisfies the $(C)_c$-condition. Then we have $c \geq b$, and $c$ is a critical value of $\varphi$ (i.e., $K_\varphi^c \neq \emptyset$). Moreover, if $c = b$, then $K_\varphi^c \cap D \neq \emptyset$.*

*Proof.* From Definition 5.36 we see that if $\gamma \in \Gamma$, then $\gamma(E) \cap D \neq \emptyset$, and so $c \geq b$.

First, suppose that $c > b$, and assume that $K_\varphi^c = \emptyset$ (i.e., $c$ is not a critical value of $\varphi$). Let $\varepsilon_0 \in (0,c-a]$. We apply Theorem 5.32 with the preceding $\varepsilon_0$ and $U = \emptyset$ and obtain $\varepsilon \in (0,\varepsilon_0)$ and a homotopy $h : [0,1] \times X \to X$ satisfying conditions (a)–(e) in Theorem 5.32. We choose $\gamma \in \Gamma$ such that

$$\varphi(\gamma(x)) \leq c + \varepsilon \quad \text{for all } x \in E.$$

Let $\gamma_0(x) = h(1,\gamma(x))$ for all $x \in E$. Then $\gamma_0 \in C(E,X)$, and if $x \in E_0$, then $\varphi(\gamma(x)) = \varphi(\gamma^*(x)) \leq a \leq c - \varepsilon_0$, hence $\gamma_0|_{E_0} = \gamma^*$ [by Theorem 5.32(d)], and so $\gamma_0 \in \Gamma$.

Also, we have $\varphi(\gamma_0(x)) \leq c - \varepsilon$ for all $x \in E$ [Theorem 5.32(e)], a contradiction of the definition of $c$.

Now suppose that $c = b$. Since $\gamma^*(E_0)$ is bounded and $d(\gamma^*(E_0), D) > 0$, we can find $S \subset X$, a closed, bounded neighborhood of $\gamma^*(E_0)$ such that $d(S, D) > 0$ and $d(\gamma^*(E_0), X \setminus S) > 0$. In fact, when $c = b$, we prove the theorem under the weaker assumption that $\varphi$ satisfies the $(C)_{X \setminus S, c}$-condition [instead of the $(C)_c$-condition]. We distinguish two cases, as follows.

*Case 1:* $b = c > a$. Arguing indirectly, suppose that $K_\varphi^c \cap D = \emptyset$. Since $D$ is closed, $U := X \setminus D$ is an open neighborhood of $K_\varphi^c$. Using Theorem 5.32 with $\varepsilon_0 \in (0, c-a]$ and the sets $S \subset U$, we find $\varepsilon > 0$ and $h : [0,1] \times X \to X$, a homotopy satisfying conditions (a)–(e) in Theorem 5.32. We choose $\gamma \in \Gamma$ such that $\varphi(\gamma(x)) \leq c + \varepsilon$ for all $x \in E$ and set $\gamma_0(x) = h(1, \gamma(x))$ for all $x \in E$. As before, we see that $\gamma_0 \in \Gamma$. Moreover, by Theorem 5.32(e), for all $x \in E$ we have

$$\varphi(\gamma_0(x)) \leq c - \varepsilon < \inf_D \varphi \quad \text{or} \quad \gamma_0(x) \in U.$$

Both cases imply that $\gamma_0(x) \notin D$ for all $x \in E$, and so $\gamma_0(E) \cap D = \emptyset$, a contradiction of the fact that $\{E_0, E, D\}$ are linking sets.

*Case 2:* $b = c = a$. Up to dealing with $\varphi - (a-1)$ instead of $\varphi$, we may assume that $a > 0$. Let $\zeta \in C^1(X, [0,1])$ be a function satisfying $\zeta|_{\gamma^*(E_0)} \equiv 0$ and $\zeta|_{X \setminus S} \equiv 1$. Set $\hat{\varphi} = \zeta \varphi \in C^1(X, \mathbb{R})$ and $\hat{c} = \inf_{\gamma \in \Gamma} \sup_{x \in E} \hat{\varphi}(\gamma(x))$. Observe that $\hat{\varphi}$ satisfies the $(C)_{X \setminus S, c}$-condition (since $\hat{\varphi} = \varphi$ on $X \setminus S$) and $\sup_{\gamma^*(E_0)} \hat{\varphi} = 0 < b = \inf_D \hat{\varphi}$; hence, applying Case 1 to the function $\hat{\varphi}$ we have $K_{\hat{\varphi}}^{\hat{c}} \cap D \neq \emptyset$. To complete the proof, it is sufficient to check that $K_{\hat{\varphi}}^{\hat{c}} \cap D \subset K_\varphi^c \cap D$. The inclusion $K_{\hat{\varphi}}^{\hat{c}} \cap D \subset K_\varphi^{\hat{c}} \cap D$ follows from the fact that $\varphi = \hat{\varphi}$ on $X \setminus S \supset D$. Thus, it remains to check that $\hat{c} = c$. On the one hand, for $\gamma \in \Gamma$ and $x \in E$ we always have $\hat{\varphi}(\gamma(x)) \leq \sup_E (\varphi \circ \gamma)$ [since $\zeta(\gamma(x)) \in [0,1]$ and $\sup_E (\varphi \circ \gamma) \geq c > 0$], whence $\hat{c} \leq c$ (by the definition of $c$ and $\hat{c}$). On the other hand, for $\gamma \in \Gamma$, the linking assumption yields $x \in E$ such that $\gamma(x) \in D$, so that $\hat{\varphi}(\gamma(x)) = \varphi(\gamma(x)) \geq b = c$, and thus $\hat{c} \geq c$. The proof is now complete. $\quad\square$

With suitable choices of the linking sets, we produce some well-known minimax theorems. We start with the so-called *mountain pass theorem*.

**Theorem 5.40.** *If $\varphi \in C^1(X, \mathbb{R})$, $x_0, x_1 \in X$, $\|x_1 - x_0\| > r$,*

$$\max\{\varphi(x_0), \varphi(x_1)\} \leq \inf\{\varphi(x) : \|x - x_0\| = r\} =: b,$$

$\Gamma = \{\gamma \in C([0,1], X) : \gamma(0) = x_0, \ \gamma(1) = x_1\}$, $c = \inf_{\gamma \in \Gamma} \sup_{t \in [0,1]} \varphi(\gamma(t))$, *and $\varphi$ satisfies the $(C)_c$-condition, then $c \geq b$, $c$ is a critical value of $\varphi$, and if $c = b$, then $K_\varphi^c \cap \partial B_r(x_0) \neq \emptyset$.*

*Proof.* Apply Theorem 5.39 with the sets $\{E_0, E, D\}$ in Example 5.38(a) [where $\Omega = B_r(x_0)$]. □

The next result is the so-called *saddle point theorem.*

**Theorem 5.41.** *If $X = Y \oplus V$ with $\dim Y < +\infty$, $\varphi \in C^1(X, \mathbb{R})$, there exists $R > 0$ such that*

$$\max\{\varphi(x) : x \in \partial B_R(0) \cap Y\} \leq \inf\{\varphi(x) : x \in V\} =: b,$$

$$\Gamma = \{\gamma \in C(\overline{B_R(0)} \cap Y, X) : \gamma|_{\partial B_R(0) \cap Y} = \mathrm{id}_{\partial B_R(0) \cap Y}\}, \quad c = \inf_{\gamma \in \Gamma} \sup_{x \in \overline{B_R(0)} \cap Y} \varphi(\gamma(x)),$$

*and $\varphi$ satisfies the $(C)_c$-condition, then $c \geq b$, $c$ is a critical value of $\varphi$, and if $c = b$, then $K_\varphi^c \cap V \neq \emptyset$.*

*Proof.* Apply Theorem 5.39 with the sets $\{E_0, E, D\}$ in Example 5.38(b). □

Similarly, Theorem 5.39 can be applied to the linking sets from Example 5.38(c) or (d) (this leads to the so-called *generalized mountain pass theorem*). We furthermore mention the following useful consequence of the mountain pass theorem.

**Proposition 5.42.** *Let $X$ be a Banach space, $\varphi \in C^1(X, \mathbb{R})$, and assume that $\varphi$ satisfies the $(C)$-condition. Let $x_0, x_1 \in X$, $x_0 \neq x_1$, satisfy $\varphi(x_0) \leq \varphi(x_1)$, and assume that $x_1$ is a strict local minimizer of $\varphi$. Then $\varphi$ admits a critical point $x_2$ such that*

$$\varphi(x_1) < \varphi(x_2) = \inf_{\gamma \in \Gamma} \sup_{t \in [0,1]} \varphi(\gamma(t)),$$

*where $\Gamma = \{\gamma \in C([0,1], X) : \gamma(0) = x_0, \ \gamma(1) = x_1\}$.*

*Proof.* Since $x_1$ is a strict local minimizer of $\varphi$, we can find $r_0 > 0$ such that

$$\varphi(x) > \varphi(x_1) \quad \text{for all } x \in B_{r_0}(x_1) \setminus \{x_1\}. \tag{5.27}$$

We claim that

$$\eta_r := \inf\{\varphi(x) : \|x - x_1\| = r\} > \varphi(x_1) \quad \text{for all } r \in (0, r_0). \tag{5.28}$$

Arguing by contradiction, assume that $\eta_r = \varphi(x_1)$ for some $r \in (0, r_0)$. Thus, we can find a sequence $\{u_n\}_{n \geq 1} \subset X$ such that $\|u_n - x_1\| = r$ and $\varphi(u_n) \leq \varphi(x_1) + \frac{1}{n^2}$ for all $n \geq 1$. By Corollary 5.10 on $\overline{B_{r_0}}(x_1)$, there is a sequence $\{v_n\}_{n \geq 1}$ such that

$$\varphi(v_n) \leq \varphi(u_n), \quad \|v_n - u_n\| \leq \frac{1}{n} \quad \text{and (eventually) } \|\varphi'(v_n)\| \leq \frac{1}{n} \text{ for } n \geq 1. \tag{5.29}$$

The second relation in (5.29) implies that the sequence $\{v_n\}_{n \geq 1}$ is bounded in $X$. Thus $(1 + \|v_n\|)\varphi'(v_n) \to 0$ in $X^*$ as $n \to \infty$. Moreover, for $n > \frac{1}{r_0 - r}$ we have

$\|v_n - x_1\| \leq \|u_n - x_1\| + \frac{1}{n} < r_0$, hence $\varphi(x_1) \leq \varphi(v_n) \leq \varphi(u_n) \leq \varphi(x_1) + \frac{1}{n^2}$, whence $\varphi(v_n) \to \varphi(x_1)$ as $n \to \infty$. Now, since $\varphi$ satisfies the (C)-condition, we obtain that the sequence $\{v_n\}_{n\geq 1}$ admits a strongly convergent subsequence $\{v_{n_k}\}_{k\geq 1}$ whose limit, denoted by $v_0$, satisfies $\varphi(v_0) = \varphi(x_1)$. Moreover, $u_{n_k} \to v_0$ as $k \to \infty$ [see (5.29)], hence $\|v_0 - x_1\| = r$. This contradicts (5.27). Thus, we have established (5.28).

Now we fix $r \in (0, r_0)$ such that $r < \|x_0 - x_1\|$. Then

$$\max\{\varphi(x_0), \varphi(x_1)\} = \varphi(x_1) < \inf_{\partial B_r(x_1)} \varphi = \eta_r.$$

This inequality permits the use of the mountain pass theorem (Theorem 5.40), which implies that $\varphi$ admits a critical point $x_2 \in X$ such that $\varphi(x_2) = \inf_{\gamma \in \Gamma} \sup_{t \in [0,1]} \varphi(\gamma(t)) \geq \eta_r > \varphi(x_1)$. This completes the proof. $\qquad\square$

Next we will present a general principle due to Ghoussoub and Preiss [157], which includes as a special case the mountain pass theorem. First we need to introduce some relevant notions.

**Definition 5.43.** The *Cerami metric* between two points $x_1, x_2 \in X$ is defined by

$$\delta(x_1, x_2) = \inf\{\ell(\gamma) : \gamma \in C^1([0,1], X), \ \gamma(0) = x_1, \ \gamma(1) = x_2\}, \qquad (5.30)$$

where

$$\ell(\gamma) = \int_0^1 \frac{\|\gamma'(t)\|}{1 + \|\gamma(t)\|} dt.$$

Moreover, let $\delta(x_1, C) = \inf_{x_2 \in C} \delta(x_1, x_2)$.

*Remark 5.44.* By considering $\gamma(t) = (1 - t)x_1 + tx_2$, $t \in [0, 1]$, we see that $\delta(x_1, x_2) \leq \|x_1 - x_2\|$. On the other hand, for any bounded set $C \subset X$ we can find $\gamma_C > 0$ such that $\delta(x_1, x_2) \geq \gamma_C \|x_1 - x_2\|$ for all $x_1, x_2 \in C$. Thus, on bounded sets the Cerami metric $\delta$ and the distance induced by the norm are equivalent. When $x_1 = 0$, the infimum in (5.30) is attained at the line segment from 0 to $x_2$, and so

$$\delta(0, x) = \int_0^1 \frac{\|x\|}{1 + t\|x\|} dt = \ln(1 + \|x\|) \quad \text{for all } x \in X.$$

Hence bounded and $\delta$-bounded sets in $X$ coincide. Moreover, every sequence $\{x_n\}_{n\geq 1} \subset X$ is $\delta$-Cauchy if and only if it is norm Cauchy. Therefore, $(X, \delta)$ is a complete metric space.

**Definition 5.45.** A closed set $C$ *separates* two points $x_0, x_1 \in X$ if $x_0$ and $x_1$ belong to disjoint connected components of $X \setminus C$.

The following theorem is a generalized version of the Ghoussoub–Preiss principle since we guarantee the existence of a $(C)_c$-sequence [instead of a $(PS)_c$-sequence]. For its proof consult Ekeland [128, p. 140].

**Theorem 5.46.** *If $X$ is a Banach space, $\varphi : X \to \mathbb{R}$ is continuous and Gâteaux differentiable, $\varphi' : X \to X^*$ is continuous from the norm topology to the weak$^*$ topology, $x_0, x_1 \in X$,*

$$\Gamma = \{\gamma \in C([0,1], X) : \ \gamma(0) = x_0, \ \gamma(1) = x_1\}, \quad c = \inf_{\gamma \in \Gamma} \ \sup_{t \in [0,1]} \ \varphi(\gamma(t)),$$

*and there exists a closed set $C$ such that $\{x \in C : \ \varphi(x) \geq c\}$ separates $x_0$ and $x_1$; then there exists a sequence $\{x_n\}_{n \geq 1} \subset X$ such that $\delta(x_n, C) \to 0$, $\varphi(x_n) \to c$, and $(1 + \|x_n\|)\varphi'(x_n) \to 0$ in $X^*$.*

*Remark 5.47.*

(a) Note that we recover the mountain pass theorem (Theorem 5.40) from Theorem 5.46. Indeed, under the hypotheses of Theorem 5.40, taking either $C = X$ [if $c > \inf_{\partial B_r(x_0)} \varphi$] or $C = \partial B_r(x_0)$ [if $c = \inf_{\partial B_r(x_0)} \varphi$], we have that the set $\{x \in C : \varphi(x) \geq c\}$ separates the points $x_0, x_1$, so that [due to the $(C)_c$-condition] Theorem 5.46 yields a critical point $x \in K_\varphi^c$ [which lies in $\partial B_r(x_0)$ in the case $c = \inf_{\partial B_r(x_0)} \varphi$].

(b) The original result in Ghoussoub and Preiss [157] (under the same hypotheses as in Theorem 5.46) guarantees the existence of a $(PS)_c$-type sequence, i.e., a sequence $\{x_n\}_{n \geq 1} \subset X$ such that $\varphi(x_n) \to c$, $\|\varphi'(x_n)\| \to 0$, and $d(x_n, C) \to 0$ (the norm distance).

Now we look for abstract theorems guaranteeing the existence of multiple critical points. First we consider the case in which the energy functional $\varphi \in C^1(X, \mathbb{R})$ has no symmetry properties. In this direction, the following notion will be helpful.

**Definition 5.48.** Let $X$ be a Banach space, $X = Y \oplus V$, and $\varphi \in C^1(X, \mathbb{R})$. We say that $\varphi$ has a *local linking at* $0$ [with respect to the pair $(Y, V)$] if there is $r > 0$ such that

$$\begin{cases} \varphi(x) \leq 0 & \text{if } x \in Y, \ \|x\| \leq r, \\ \varphi(x) \geq 0 & \text{if } x \in V, \ \|x\| \leq r. \end{cases}$$

*Remark 5.49.* Clearly, $0$ is a critical point of $\varphi$. If $\varphi \in C^2(\mathbb{R}^N, \mathbb{R})$, $N \geq 1$, $\varphi(0) = 0$, and $0$ is a critical point of $\varphi$ with $\varphi''(0)$ invertible (i.e., $0$ is a nondegenerate critical point of $\varphi$), then $\varphi$ has a local linking at $0$.

**Lemma 5.50.** *If $X$ is a Banach space, $\varphi \in C^1(X, \mathbb{R})$ and satisfies the $(PS)$-condition, $x_0 \in X$ is the unique global minimizer of $\varphi$ on $X$, $y \in X$ is such that $\varphi'(y) \neq 0$, and $\varphi$ has no critical value in the interval $(\varphi(x_0), \varphi(y))$, then for $v : X_0 = X \setminus K_\varphi \to X$ being a pseudogradient vector field of $\varphi$, the solution to the problem*

$$\begin{cases} \xi'(t) = -\dfrac{v(\xi(t))}{\|v(\xi(t))\|^2} & \text{in } \mathbb{R}_+, \\ \xi(0) = y \end{cases} \tag{5.31}$$

is defined on a maximal interval $[0, T(y))$, with $T(y) < +\infty$. Moreover, $t \mapsto \varphi(\xi(t))$ is decreasing on $[0, T(y))$, and we have $\lim_{t \to T(y)} \xi(t) = x_0$.

*Proof.* By the chain rule, the maximal solution $\xi$ of (5.31) satisfies

$$\frac{d\varphi(\xi(t))}{dt} = \langle \varphi'(\xi(t)), \xi'(t) \rangle = \left\langle \varphi'(\xi(t)), -\frac{v(\xi(t))}{\|v(\xi(t))\|^2} \right\rangle \le -\frac{\|\varphi'(\xi(t))\|^2}{\|v(\xi(t))\|^2} \le -\frac{1}{4} \tag{5.32}$$

for all $t \in [0, T(y))$ (Definition 5.27), which implies that $t \mapsto \varphi(\xi(t))$ is decreasing, as stated. From (5.32) and the assumption that $x_0$ is the unique global minimizer of $\varphi$ we have

$$\varphi(x_0) < \varphi(\xi(t)) \le -\frac{1}{4}t + \varphi(y) < \varphi(y) \quad \text{for all } t \in (0, T(y)) \tag{5.33}$$

and

$$T(y) \le 4(\varphi(y) - \varphi(x_0)) < +\infty.$$

The maximality of $T(y)$ yields

$$+\infty = \int_0^{T(y)} \|\xi'(t)\| \, dt = \int_0^{T(y)} \frac{1}{\|v(\xi(t))\|} \, dt \le \int_0^{T(y)} \frac{1}{\|\varphi'(\xi(t))\|} \, dt.$$

Since $T(y) < +\infty$, we can find a sequence $\{t_n\}_{n \ge 1} \subset (0, T(y))$ such that $t_n \to T(y)$, $\|\varphi'(\xi(t_n))\| \to 0$, and $\varphi(\xi(t_n)) \to c \in [\varphi(x_0), \varphi(y))$ as $n \to \infty$ [see (5.33)]. Since $\varphi$ satisfies the (PS)-condition, we may assume that $\xi(t_n) \to \hat{x}_0$ in $X$, with $\hat{x}_0 \in K_\varphi^c$. Because, by hypothesis, $\varphi$ has no critical values in $(\varphi(x_0), \varphi(y))$, we see that $\varphi(\hat{x}_0) = \varphi(x_0)$, and so $\hat{x}_0 = x_0$. Finally, we have found $\{t_n\}_{n \ge 1}$, with $\lim_{n \to \infty} t_n = T(y)$, such that $\lim_{n \to \infty} \xi(t_n) = x_0$. Since $t \mapsto \varphi(\xi(t))$ is decreasing [see (5.32)], we derive $\lim_{t \to T(y)} \varphi(\xi(t)) = \varphi(x_0)$. Then, from Proposition 5.24 and the fact that $x_0$ is the unique global minimizer of $\varphi$, we obtain $\lim_{t \to T(y)} \xi(t) = x_0$. $\qquad\square$

Using this lemma, we can prove the following multiplicity theorem, known as the *local linking theorem*.

**Theorem 5.51.** *Let $X$ be a Banach space, $X = Y \oplus V$ with $\dim Y < +\infty$, and let $\varphi \in C^1(X, \mathbb{R})$ be bounded below with $\inf_X \varphi < 0$ and satisfy the (PS)-condition. Assume that $\varphi$ has a local linking at $0$ with respect to the pair $(Y, V)$. Then $\varphi$ has at least two nontrivial critical points.*

*Proof.* From Corollary 5.21 we know that there exists $x_0 \in X$ such that $\varphi(x_0) = \inf_X \varphi < 0 = \varphi(0)$, and so $x_0 \neq 0$. Arguing indirectly, suppose that $0, x_0$ are the only critical points of $\varphi$.

*Case 1*: $Y \neq 0$ and $V \neq 0$.

Without any loss of generality, we may assume that $r < \|x_0\|$. Due to the continuity of $\varphi$, we can find $\mu > 0$ such that $\varphi(x) < 0$ for all $x \in X$, $\|x - x_0\| < \mu$.

*Claim 1*: We can find $\delta > 0$ such that

$$\{x \in X : \varphi(x) \leq \varphi(x_0) + \delta\} \subset \{x \in X : \|x - x_0\| < \mu\}.$$

If this is not the case, we can find $\{x_n\}_{n \geq 1} \subset X$, a minimizing sequence for $\varphi$, such that $\|x_n - x_0\| \geq \mu$ for all $n \geq 1$. Invoking Proposition 5.24, we can find a subsequence $\{x_{n_k}\}_{k \geq 1}$ that converges to some $\hat{x}_0 \in X$. Thus, $\varphi(\hat{x}_0) = \inf_X \varphi$, which yields $\hat{x}_0 \in K_\varphi$ and $\hat{x}_0 \neq 0$. Moreover, we see that $\|\hat{x}_0 - x_0\| \geq \mu$, whence $\hat{x}_0 \neq x_0$. So we have found a critical point different from $0, x_0$, which contradicts our assumption. This proves Claim 1.

Hereafter we fix a number $\delta > 0$ satisfying Claim 1 such that $\delta < \min_{Y \cap \partial B_r(0)} \varphi - \varphi(x_0)$ [in particular $\varphi(x_0) + \delta < 0$].

For every $y \in Y$ with $\|y\| = r < \|x_0\|$ we have $\varphi'(y) \neq 0$, and $\varphi$ has no critical value in $(\varphi(x_0), \varphi(y))$ [since $\varphi(y) \leq 0$], so we can apply Lemma 5.50 and have a maximal solution $\xi_y$ of (5.31) on a maximal interval $[0, T(y))$, with $T(y)$ finite and $\xi_y(t) \to x_0$ as $t \to T(y)$. Since $\varphi(x_0) + \delta < \varphi(y)$ (by the choice of $\delta$) and since $t \mapsto \varphi(\xi_y(t))$ is decreasing and $\lim_{t \to T(y)} \varphi(\xi_y(t)) = \varphi(x_0)$ (by Lemma 5.50), there is $t(y) \in (0, T(y))$ unique such that $\varphi(\xi_y(t(y))) = \varphi(x_0) + \delta$.

*Claim 2*: The map $y \mapsto t(y)$ is continuous on $Y \cap \partial B_r(0)$.

Let $\varepsilon > 0$ such that $\varepsilon < \min\{t(y), T(y) - t(y)\}$. Since $t \mapsto \varphi(\xi_y(t))$ is decreasing, we have

$$\varphi(\xi_y(t(y) + \varepsilon)) < \varphi(\xi_y(t(y))) = \varphi(x_0) + \delta < \varphi(\xi_y(t(y) - \varepsilon)).$$

By the continuous dependence of the solution of (5.31) on the initial condition and by the continuity of $\varphi$, for $\|y - z\|$ small enough we obtain

$$\varphi(\xi_z(t(y) + \varepsilon)) < \varphi(x_0) + \delta < \varphi(\xi_z(t(y) - \varepsilon)),$$

which yields $t(y) - \varepsilon < t(z) < t(y) + \varepsilon$ (since $\varphi \circ \xi_z$ is decreasing), i.e., $|t(y) - t(z)| < \varepsilon$. This proves Claim 2.

Fix $v_0 \in V$ with $\|v_0\| = r$. We introduce the following sets:

$$E = \{y + \lambda v_0 : y \in Y, \|y\| \leq r, \lambda \in [0, 1]\},$$

$$E_0 = \{y \in Y : \|y\| \leq r\} \cup \{(1-\lambda)y + \lambda v_0 : y \in Y, \|y\| = r, \lambda \in [0,1]\}.$$

We consider $\gamma^* \in C(E_0, X)$ defined by $\gamma^*(y) = y$ for $y \in Y$, $\|y\| \leq r$, and by

$$\gamma^*((1-\lambda)y + \lambda v_0) = \begin{cases} x_0 & \text{if } \lambda = 1, \\ (2\lambda - 1)x_0 + (2 - 2\lambda)\xi_y(t(y)) & \text{if } \lambda \in [\frac{1}{2}, 1), \\ \xi_y(2\lambda t(y)) & \text{if } \lambda \in [0, \frac{1}{2}], \end{cases}$$

for $y \in Y$, $\|y\| = r$, and $\lambda \in [0,1]$. Clearly, $\gamma^*$ is well defined. The continuity of $\gamma^*$ on $E_0 \setminus \{v_0\}$ is clear [by the continuity of $y \mapsto t(y)$ and $y \mapsto \xi_y$]. The continuity of $\gamma^*$ at $v_0$ can be checked as follows: let $v_n = (1 - \lambda_n)y_n + \lambda_n v_0$, with $y_n \in Y$, $\|y_n\| = r$, and $\lambda_n \in [\frac{1}{2}, 1)$, such that $v_n \to v_0$ as $n \to \infty$. The fact that $\{y_n\}_{n \geq 1}$ does not converge to $v_0$ (since $Y \subset X$ is closed) yields $\lambda_n \to 1$. Moreover, by Claim 1, the sequence $\{\xi_{y_n}(t(y_n))\}_{n \geq 1}$ is bounded. Then the definition of $\gamma^*$ yields $\gamma^*(v_n) \to x_0 = \gamma^*(v_0)$ as $n \to \infty$.

*Claim 3*: We have $\varphi(\gamma^*(x)) \leq 0$ for all $x \in E_0 \cap Y$ and $\varphi(\gamma^*(x)) < 0$ for all $x \in E_0 \setminus Y$.

Indeed, for $x \in E_0 \cap Y$ we have $\varphi(\gamma^*(x)) = \varphi(x) \leq 0$ by the hypothesis of local linking, whereas for $x \in E_0 \setminus Y$, i.e., $x = (1 - \lambda)y + \lambda v_0$, with $y \in Y$, $\|y\| = r$, and $\lambda \in (0, 1]$, we can see that:

- If $\lambda = 1$, then $\varphi(\gamma^*(x)) = \varphi(x_0) < 0$ by assumption;
- If $\lambda \in [\frac{1}{2}, 1)$, then we have $\|\gamma^*(x) - x_0\| = 2(1 - \lambda)\|\xi_y(t(y)) - x_0\| < \mu$ (by Claim 1), and so $\varphi(\gamma^*(x)) < 0$ (by the choice of $\mu$);
- If $\lambda \in (0, \frac{1}{2}]$, then $\varphi(\gamma^*(x)) = \varphi(\xi_y(2\lambda t(y))) < \varphi(\xi_y(0)) = \varphi(y) \leq 0$ [since $t \mapsto \varphi(\xi_y(t))$ is decreasing, and using the hypothesis of local linking to get that $\varphi(y) \leq 0$].

This proves Claim 3.

Let $\pi_Y : X \to Y$ denote the linear continuous projection according to the topological direct sum decomposition $X = Y \oplus V$, and let $g : X \to Y_0 := Y \oplus \mathbb{R}v_0$ be the map defined by

$$g(x) = \pi_Y(x) + \|x - \pi_Y(x)\|v_0 \quad \text{for all } x \in X. \tag{5.34}$$

Write $E_0 = E_{0,1} \cup E_{0,2}$, with $E_{0,1} = E_0 \cap Y = \{y \in Y : \|y\| \leq r\}$ and $E_{0,2} = \overline{E_0 \setminus Y} = \{(1 - \lambda)y + \lambda v_0 : y \in Y, \|y\| = r, \lambda \in [0,1]\}$.

*Claim 4*: There exists a continuous map $h : [0, 1] \times E_{0,2} \to Y_0$ such that

(a) $h(0, x) = g(\gamma^*(x))$, $h(1, x) = x$, and $h(t, x) \neq 0$ for all $t \in [0, 1]$, all $x \in E_{0,2}$;
(b) $h(t, x) = x$ for all $t \in [0, 1]$, all $x \in E_{0,1} \cap E_{0,2} = \{y \in Y : \|y\| = r\}$.

A preliminary observation (due to Claim 3) is that $\gamma^*(x) \neq 0$ whenever $x \in E_{0,2}$. This easily implies that $g(\gamma^*(x)) \neq 0$ for all $x \in E_{0,2}$. This also implies that the map $h_1 : [0, 1] \times E_{0,2} \to Y_0$ given by

$$h_1(t,x) = \left( \frac{rt}{r\|\gamma^*(x) - \pi_Y(\gamma^*(x))\| + \|\pi_Y(\gamma^*(x))\|} + (1-t) \right) g(\gamma^*(x))$$

is well defined and continuous and satisfies $h_1(t,x) \neq 0$ for all $t \in [0,1]$, all $x \in E_{0,2}$. As $\gamma^*|_{E_{0,1}} = \mathrm{id}_{E_{0,1}}$ and $g|_Y = \mathrm{id}_Y$, we obtain that $h_1(t,x) = x$ whenever $x \in E_{0,1} \cap E_{0,2}$.

Moreover, let us check that $h_1(1, E_{0,2}) \subset E_{0,2}$. Let $x \in E_{0,2}$. If $\pi_Y(\gamma^*(x)) = 0$, then we have $g(\gamma^*(x)) = \|\gamma^*(x)\|v_0$, and so $h_1(1,x) = v_0 \in E_{0,2}$. If $\pi_Y(\gamma^*(x)) \neq 0$, then letting

$$\lambda = \frac{r\|\gamma^*(x) - \pi_Y(\gamma^*(x))\|}{r\|\gamma^*(x) - \pi_Y(\gamma^*(x))\| + \|\pi_Y(\gamma^*(x))\|} \quad \text{and} \quad y = \frac{r\pi_Y(\gamma^*(x))}{\|\pi_Y(\gamma^*(x))\|}$$

(which satisfy $\lambda \in [0,1]$, $y \in Y$, $\|y\| = r$), from (5.34) we obtain $h_1(1,x) = (1 - \lambda)y + \lambda v_0 \in E_{0,2}$.

Note that the restriction of the linear projection $\pi_Y$ defines a homeomorphism $\hat{\pi} : E_{0,2} \to \{y \in Y : \|y\| \leq r\}$. Setting

$$h_2(t, \cdot) = \hat{\pi}^{-1} \circ (t \, \mathrm{id} + (1-t)(\hat{\pi} \circ h_1(1, \cdot) \circ \hat{\pi}^{-1})) \circ \hat{\pi}$$

provides a continuous map $h_2 : [0,1] \times E_{0,2} \to E_{0,2}$ such that $h_2(0, \cdot) = h_1(1, \cdot)$, $h_2(1, \cdot) = \mathrm{id}_{E_{0,2}}$, and $h_2(t,x) = x$ for all $t \in [0,1]$, all $x \in E_{0,1} \cap E_{0,2}$.

Finally, define $h : [0,1] \times E_{0,2} \to Y_0$ by

$$h(t,x) = \begin{cases} h_1(2t,x) & \text{if } t \in [0, \frac{1}{2}], \\ h_2(2t - 1, x) & \text{if } t \in [\frac{1}{2}, 1], \end{cases}$$

for all $t \in [0,1]$, all $x \in E_{0,2}$. The map $h$ satisfies the requirements of Claim 4.

*Claim 5*: There exists $\rho \in (0,r)$ such that the pair $\{E_0, E\}$ links $D := \{v \in V : \|v\| = \rho\}$ in $X$ via $\gamma^*$.

Let $h$ be as in Claim 4. Take any $\rho \in (0,r)$ such that

$$\|\rho v_0\| < \inf_{(t,x) \in [0,1] \times E_{0,2}} \|h(t,x)\| \tag{5.35}$$

(this infimum is positive because $h(t,x) \neq 0$ for all $t \in [0,1]$, all $x \in E_{0,2}$, and the set $[0,1] \times E_{0,2}$ is compact). In particular, since $\rho < r$, we have that $\varphi$ is nonnegative on $D$ (by the hypothesis of local linking), which, by Claim 3, yields $\gamma^*(E_0 \setminus Y) \cap D = \emptyset$. On the other hand, we have $\gamma^*(E_0 \cap Y) \cap D \subset Y \cap D = \emptyset$, whence $\gamma^*(E_0) \cap D = \emptyset$, so the triple $\{E_0, E, D\}$ satisfies condition (a) in Definition 5.36.

To check condition (b) in Definition 5.36, we need to show that $\gamma(E) \cap D \neq \emptyset$ whenever $\gamma : E \to X$ is a continuous extension of $\gamma^*$. Clearly, it suffices to find $\hat{x} \in E$ such that $g(\gamma(\hat{x})) = \rho v_0$ [see (5.34)]. To do this, we use Brouwer's degree theory. Note that $E$ is a closed subset of $Y_0 = Y \oplus \mathbb{R}v_0$ of boundary $E_0$. First, we extend the map $h$ from Claim 4 to a continuous map $\hat{h} : [0,1] \times E_0 \to Y_0$ by letting

$$\hat{h}(t,x) = \begin{cases} x & \text{if } x \in E_{0,1}, \\ h(t,x) & \text{if } x \in E_{0,2}, \end{cases}$$

for all $t \in [0,1]$ (the map $\hat{h}$ is well defined and continuous by part (b) of Claim 4). Next, by Theorem 2.9, we know that $\hat{h}$ can be extended to a continuous map $\overline{h} : [0,1] \times E \to Y_0$. Due to the choice of $\rho$ [see (5.35)], we have $\rho v_0 \notin \overline{h}([0,1] \times E_0)$. Note that $\overline{h}(0,\cdot)|_{E_0} = g \circ \gamma$ and $\overline{h}(1,\cdot)|_{E_0} = \mathrm{id}_{E_0}$ (Claim 4). Then, by Theorem 4.5(a), (c), and (g), we have

$$d(g \circ \gamma, E \setminus E_0, \rho v_0) = d(\overline{h}(0,\cdot), E \setminus E_0, \rho v_0) = d(\overline{h}(1,\cdot), E \setminus E_0, \rho v_0)$$
$$= d(\mathrm{id}, E \setminus E_0, \rho v_0) = 1.$$

Then, by Theorem 4.5(e), $\rho v_0 \in g(\gamma(E))$. This completes the proof of Claim 5.

Let $\Gamma = \{\gamma \in C(E,X) : \gamma|_{E_0} = \gamma^*\}$ and $c = \inf_{\gamma \in \Gamma} \max_{x \in E} \varphi(\gamma(x))$. Due to Claim 3, the hypothesis of local linking, and Claim 5, we have

$$\sup_{x \in E_0} \varphi(\gamma^*(x)) = 0 \leq \inf_{x \in D} \varphi(x) \leq c.$$

If $c > 0$, then Theorem 5.39 provides a critical point $x_1$ of $\varphi$ such that $\varphi(x_1) = c$; then $x_1$ is different from $0, x_0$ since $\varphi(x_1) > 0 = \varphi(0) > \varphi(x_0)$. If $c = 0$, then, by Theorem 5.39, there is $x_2 \in K_\varphi^c \cap D$; again, $x_2$ is different from $0, x_0$ since $\varphi(x_2) > \varphi(x_0)$ and $x_2 \in D$, whereas $0 \notin D$. In both cases, this contradicts the assumption that $0$ and $x_0$ are the only critical points of $\varphi$. This completes the proof in Case 1.

*Case 2*: $Y = 0$ and $V \neq 0$.

The local linking hypothesis implies that $0$ is a local minimizer of $\varphi$. Recall that we have assumed that $0, x_0$ are the only critical points of $\varphi$. In particular, $0$ is a strict local minimizer of $\varphi$. Moreover, we have $\varphi(x_0) < \varphi(0) = 0$. Therefore, we can apply Proposition 5.42, which yields a critical point $x_2 \in X$ different from $0$ and $x_0$, a contradiction. The proof is thus complete in Case 2.

*Case 3*: $Y \neq 0$ and $V = 0$.

In this case, due to the local linking hypothesis (and since we assume that $0, x_0$ are the only critical points of $\varphi$), $0$ is a strict local minimizer of $-\varphi$. Proposition 5.22 implies that $\varphi$ is coercive, and so we can find $y = -tx_0$, with $t > 0$, such that $-\varphi(y) < 0 = -\varphi(0)$. Since $-\varphi$ satisfies the (PS)-condition, by Proposition 5.42, there is a critical point $x_2$ of $\varphi$ such that

$$-\varphi(0) < -\varphi(x_2) = \inf_{\gamma \in \Gamma} \sup_{t \in [0,1]} (-\varphi(\gamma(t))),$$

with $\Gamma = \{\gamma \in C([0,1],X) : \gamma(0) = 0, \ \gamma(1) = y\}$. Thus, $x_2 \neq 0$. Moreover, consider $\gamma_0 \in \Gamma$ defined by $\gamma_0(s) = sy$ for all $s \in [0,1]$. Since $\gamma_0(s) \neq x_0$ for all $s \in [0,1]$ and since $x_0$ is the unique global minimizer of $\varphi$, we get

$$-\varphi(x_2) \le \sup_{t \in [0,1]} (-\varphi(\gamma_0(t))) < -\varphi(x_0),$$

whence $x_2 \neq x_0$, a contradiction of the fact that $0, x_0$ are the only critical points of $\varphi$. This proves the theorem in Case 3. Note that in this case we did not use that $\dim Y < +\infty$. $\square$

In the local linking theorem (Theorem 5.51), $\varphi$ is bounded below. What can be said if $\varphi$ is indefinite? In this direction, we have a result due to Li and Willem [224]. Let $X$ be a Banach space with a direct sum decomposition $X = Y \oplus V$. Consider two sequences of subspaces

$$Y_1 \subset Y_2 \subset \cdots \subset Y_k \subset \cdots \subset Y \text{ and } V_1 \subset V_2 \subset \cdots \subset V_k \subset \cdots \subset V$$

such that $Y = \overline{\bigcup_{k \ge 1} Y_k}$ and $V = \overline{\bigcup_{k \ge 1} V_k}$. For all $\alpha = (i,j) \in \mathbb{N}^2$ we let $X_\alpha = Y_i \oplus V_j$. Moreover, we say that a sequence $\{\alpha_n\}_{n \ge 1} \subset \mathbb{N}^2$ is *admissible* if $\alpha_n = (i_n, j_n)$, with $i_n \to +\infty$ and $j_n \to +\infty$ as $n \to \infty$.

**Definition 5.52.** We say that $\varphi \in C^1(X, \mathbb{R})$ satisfies the (PS)*-*condition* if every sequence $\{x_{\alpha_n}\}_{n \ge 1}$ such that $\{\alpha_n\}_{n \ge 1}$ is admissible and

$$x_{\alpha_n} \in X_{\alpha_n}, \ \{\varphi(x_{\alpha_n})\}_{n \ge 1} \text{ is bounded, and } \varphi'(x_{\alpha_n}) \to 0 \text{ in } X^*$$

admits a strongly convergent subsequence.

The result of Li and Willem [224, Theorem 2] is the following theorem.

**Theorem 5.53.** *If $X = Y \oplus V$, with $V \neq 0$, $\varphi \in C^1(X, \mathbb{R})$, and*

(i) *$\varphi$ has a local linking at 0 with respect to the pair $(Y,V)$;*
(ii) *$\varphi$ satisfies the (PS)*-condition;*
(iii) *$\varphi$ maps bounded sets to bounded ones;*
(iv) *For every $k \in \mathbb{N}$, $\varphi(x) \to -\infty$ as $\|x\| \to +\infty$, $x \in Y \oplus V_k$,*

*then $\varphi$ has at least one nontrivial critical point.*

## 5.4 Critical Points for Functionals with Symmetries

A favorable situation for obtaining multiplicity results is the presence of symmetries, i.e., when there exists some topological group $G$ acting continuously on $X$ and the functional $\varphi$ is invariant under this group action.

Let $X$ be a Banach space and $G$ be a topological group.

**Definition 5.54.** A *representation of $G$ over $X$* is a family $\{S(g)\}_{g \in G} \subset \mathscr{L}(X)$ with

(a) $S(e) = \mathrm{id}_X$ ($e$ is the identity element of $G$);
(b) $S(g_1 g_2) = S(g_1) S(g_2)$ for all $g_1, g_2 \in G$;
(c) $(g, x) \mapsto S(g)x$ is continuous from $G \times X$ into $X$.

We say that the representation is *isometric* if

$$\|S(g)x\| = \|x\| \quad \text{for all } g \in G \text{ and all } x \in X$$

[i.e., every $S(g) \in \mathscr{L}(X)$ is an isometry]. A set $C \subset X$ is *invariant* (or *G-invariant*) if $S(g)(C) \subset C$ for all $g \in G$. Similarly, a functional $\varphi : X \to \mathbb{R}$ is *invariant* if $\varphi \circ S(g) = \varphi$ for all $g \in G$. A map $h : X \to X$ is *equivariant* if $S(g) \circ h = h \circ S(g)$ for all $g \in G$. Finally, the set of *invariant points* of $X$ (or *fixed points* of $X$) is the set $X^G = \{x \in X : S(g)x = x \text{ for all } g \in G\}$.

*Remark 5.55.* Very often we identify $S(g)$ with $g$ and we speak about the *linear action* of $G$ on $X$. We will follow this custom here, too.

In what follows, we assume that we have the linear action of a topological group $G$ on a Banach space $X$.

**Proposition 5.56.** *If $\varphi \in C^1(X, \mathbb{R})$ is invariant, then we have:*

(a) $\langle \varphi'(gx), y \rangle = \langle \varphi'(x), g^{-1}y \rangle$ *for all $g \in G$, all $x, y \in X$;*
(b) *If the action of $G$ on $X$ is isometric, then $\|\varphi'(gx)\| = \|\varphi'(x)\|$ for all $g \in G$, all $x \in X$.*

*Proof.*

(a) Note that $S(g)^{-1}$ exists and is equal to $S(g^{-1})$ (Definition 5.54). Then

$$\langle \varphi'(gx), y \rangle = \lim_{\lambda \to 0} \frac{1}{\lambda} \big( \varphi(g(x + \lambda g^{-1}y)) - \varphi(gx) \big)$$

$$= \lim_{\lambda \to 0} \frac{1}{\lambda} \big( \varphi(x + \lambda g^{-1}y) - \varphi(x) \big) = \langle \varphi'(x), g^{-1}y \rangle.$$

(b) This is an immediate consequence of (a).

$\square$

**Proposition 5.57.** *If $\varphi \in C^1(X, \mathbb{R})$ is invariant, $G$ is compact, and the linear action of $G$ on $X$ is isometric, then there is an equivariant pseudogradient vector field $\hat{v} : X_0 \to X$ for $\varphi$, where $X_0 = X \setminus K_\varphi$.*

*Proof.* Recall that one feature of a compact topological group $G$ is that it admits a unique $G$-invariant measure $\mu = \mu_G$ such that $\mu(G) = 1$, called the *Haar measure*. From Proposition 5.29 we know that there exists a pseudogradient vector field $v : X_0 \to X$. Let

$$\hat{v}(x) = \int_G g v(g^{-1}x) \, d\mu(g) \quad \text{for all } x \in X_0.$$

We note that $\hat{v}$ is equivariant since, for all $g' \in G$, we have

$$\hat{v}(g'x) = \int_G gv(g^{-1}g'x)\,d\mu(g) = \int_G g'(g')^{-1}gv(((g')^{-1}g)^{-1}x)\,d\mu(g) = g'\hat{v}(x).$$

Now we check that $\hat{v} : X_0 \to X$ is a pseudogradient vector field for $\varphi$. We see that

$$\|\hat{v}(x)\| \le \int_G \|v(g^{-1}(x))\|\,d\mu(g) \le \int_G 2\|\varphi'(g^{-1}x)\|\,d\mu(g) = 2\|\varphi'(x)\|$$

[see Proposition 5.56(b)]. Also, using Proposition 5.56(a), (b), we have

$$\langle \varphi'(x), \hat{v}(x) \rangle = \int_G \langle \varphi'(x), gv(g^{-1}x) \rangle\,d\mu(g) = \int_G \langle \varphi'(g^{-1}x), v(g^{-1}x) \rangle\,d\mu(g)$$

$$\ge \int_G \|\varphi'(g^{-1}x)\|^2\,d\mu(g) = \int_G \|\varphi'(x)\|^2\,d\mu(g) = \|\varphi'(x)\|^2.$$

It remains to show that $\hat{v}$ is locally Lipschitz. For $x \in X_0$ let $O(x) = \{gx : g \in G\}$ be the orbit of $x$. The compactness of $G$ implies that $O(x)$ is compact. Hence we can find $\delta > 0$ such that $v$ is Lipschitz continuous on $O(x)_\delta := \{y \in X_0 : d(y, O(x)) \le \delta\}$ with Lipschitz constant $\theta > 0$. Thus, since $O(x)_\delta$ is invariant, for all $y, z \in B_\delta(x)$ we have

$$\|\hat{v}(y) - \hat{v}(z)\| \le \int_G \|g(v(g^{-1}y) - v(g^{-1}z))\|\,d\mu(g) = \int_G \|v(g^{-1}y) - v(g^{-1}z)\|\,d\mu(g)$$

$$\le \theta \int_G \|g^{-1}y - g^{-1}z\|\,d\mu(g) = \theta\|y - z\|.$$

This completes the proof.                                                                        □

Using this proposition, we obtain the following equivariant version of the first deformation theorem (Theorem 5.32).

**Theorem 5.58.** *Let $G$ be a compact topological group with isometric linear action on $X$, $\varphi \in C^1(X, \mathbb{R})$, and $c \in \mathbb{R}$. Fix $\varepsilon_0 > 0$, $\theta > 0$, and an invariant open subset $U \subset X$ (possibly empty) such that $K_\varphi^c \subset U$. Assume that $\varphi$ satisfies the $(C)_{X\setminus S, c}$-condition, for some invariant bounded subset $S \subset U$ (possibly empty) such that $d(S, X \setminus U) > 0$. Then there exists a continuous map $h : [0,1] \times X \to X$ that satisfies (a)–(e) of Theorem 5.32 and, in addition, the following property:*
*(f) for every $t \in [0,1]$, $h(t, \cdot)$ is equivariant.*

Let $\mathscr{S} = \{A \subset X : A \text{ is closed and } G\text{-invariant}\}$.

**Definition 5.59.** An *index* (or *G-index*) on $X$ is a map $i : \mathscr{S} \to \mathbb{N} \cup \{+\infty\}$ such that

(a) $i(A) = 0$ if and only if $A = \emptyset$;
(b) If $h : A \to C$ $(A, C \in \mathscr{S})$ is equivariant, then $i(A) \le i(C)$ (monotonicity);

(c) $i(A \cup C) \leq i(A) + i(C)$ for all $A, C \in \mathscr{S}$ (subadditivity);
(d) If $A \in \mathscr{S}$ is compact, then there exists an invariant, open neighborhood $U$ of $A$ such that $i(\overline{U}) = i(A)$ (continuity).

Suppose that we dispose of a $G$-index $i : \mathscr{S} \to \mathbb{N} \cup \{+\infty\}$. Given $k \in \mathbb{N}$, we set

$$\mathscr{S}_k = \{A \subset X : A \text{ is compact and invariant and } i(A) \geq k\}.$$

Let $\varphi \in C^1(X, \mathbb{R})$, and define the values $c_k$, $k \in \mathbb{N}$, by the formula

$$c_k = \inf_{A \in \mathscr{S}_k} \max_{x \in A} \varphi(x). \tag{5.36}$$

Clearly, $-\infty \leq c_1 \leq c_2 \leq \dots$.

**Theorem 5.60.** *If $G$ is a compact topological group with isometric linear action on $X$, $\varphi \in C^1(X, \mathbb{R})$ is invariant and satisfies the (C)-condition and $c_k > -\infty$ for some $k \in \mathbb{N}$, then $c_k$ is a critical value of $\varphi$. More precisely, given any $m \in \mathbb{N}$, $m \leq k$ such that $c_m = c_k =: c > -\infty$, we have $i(K_\varphi^c) \geq k - m + 1$.*

*Proof.* From Propositions 5.25 and 5.56 it follows that $K_\varphi^c$ is compact and invariant. Let $U$ be an invariant, open neighborhood of $K_\varphi^c$ such that $i(\overline{U}) = i(K_\varphi^c)$ [see Definition 5.59(d)]. We apply Theorem 5.58 and obtain a continuous deformation $h : [0, 1] \times X \to X$ satisfying (a)–(f). From (5.36) we see that we can find $A \in \mathscr{S}_k$ such that $\max_A \varphi \leq c + \varepsilon$. Let $C = A \setminus U$. Then

$$k \leq i(A) \leq i(C) + i(A \cap \overline{U}) \leq i(C) + i(\overline{U}) = i(C) + i(K_\varphi^c) \tag{5.37}$$

[see Definition 5.59(b), (c)]. Note that $C \subset \varphi^{c+\varepsilon} \setminus U$, and so from Theorem 5.58(e) we have $D = h(1, C) \subset \varphi^{c-\varepsilon}$. Since $h(1, \cdot)$ is equivariant [Theorem 5.58(f)] and $C$ is compact and invariant, we infer that $D$ is compact and invariant and $\max_D \varphi \leq c - \varepsilon$. From the definition of $c_m = c$ it follows that $i(D) \leq m - 1$. Then from Definition 5.59(b) we have $i(C) \leq i(D) \leq m - 1$, which implies that $i(K_\varphi^c) \geq k - m + 1$ [see (5.37)], and thus $K_\varphi^c \neq \emptyset$ [Definition 5.59(a)]. $\qquad\square$

Let us consider the linear action of the group $G = \mathbb{Z}_2 = \{\mathrm{id}, -\mathrm{id}\}$. Then $\mathscr{S}$ is the set of subsets $A \subset X$ that are closed and symmetric (i.e., $A = -A$). In this case, there exists a $G$-index on $X$ called the Krasnosel'skiĭ genus that is defined as follows (see Coffman [86]).

**Definition 5.61.** For $A \in \mathscr{S}$, the *Krasnosel'skiĭ genus* gen$A$ of $A$ is defined by gen $\emptyset = 0$, gen$A = \inf\{m \geq 1 : \text{there is } h \in C(A, \mathbb{R}^m \setminus \{0\}) \text{ odd}\}$ if $A$ is nonempty, with gen$A = +\infty$ by convention if no such $h$ exists.

*Remark 5.62.* Some useful properties of the genus include the following:

(a) If $U$ is any bounded, symmetric neighborhood of the origin in $\mathbb{R}^m$, then gen$\partial U = m$; in particular, gen $S^{m-1} = m$, where $S^{m-1} = \{x \in \mathbb{R}^m : |x| = 1\}$.

(b) If $C \subset X$ is nonempty and closed and $C \cap (-C) = \emptyset$, then $\text{gen}(C \cup (-C)) = 1$; in particular, if $A \in \mathscr{S}$, $0 \notin A$, and $\text{gen}\, A \geq 2$, then $A$ is infinite.
(c) If $A, B \in \mathscr{S}$ and there exists an odd homeomorphism $h : A \to B$, then $\text{gen}\, A = \text{gen}\, B$.

**Theorem 5.63.** *If $\varphi \in C^1(X, \mathbb{R})$ is even and bounded below and satisfies the* (C)-*condition, and there exists a compact, symmetric set $C$ such that* $\text{gen}\, C = m$ *and* $\sup_C \varphi < \varphi(0)$, *then $\varphi$ has at least $m$ distinct pairs $\{-x_k, x_k\}_{k=1}^m$ of critical points with* $\varphi(-x_k) = \varphi(x_k) < \varphi(0)$.

*Proof.* Let $\{c_k\}_{k \in \mathbb{N}}$ be as in (5.36). Using that $\varphi$ is bounded below, that $\text{gen}\, C = m$, and the definition of $c_m$, we have

$$-\infty < \inf_X \varphi \leq c_1 \leq c_2 \leq \cdots \leq c_m \leq \max_C \varphi < \varphi(0).$$

For each $k \in \{1, \ldots, m\}$ the set $K_\varphi^{c_k}$ is nonempty (by Theorem 5.60) and symmetric and $0 \notin K_\varphi^{c_k}$ (hence $K_\varphi^{c_k}$ has at least two elements). If the numbers $\{c_k\}_{k=1}^m$ are pairwise distinct, then the sets $K_\varphi^{c_k}$ ($k \in \{1, \ldots, m\}$) are pairwise disjoint, and we reach the desired conclusion. If there is $k \in \{2, \ldots, m\}$ with $c_{k-1} = c_k$, then Theorem 5.60 yields $\text{gen}\, K_\varphi^{c_k} \geq 2$, so $K_\varphi^{c_k}$ is infinite (by Remark 5.62), and again we reach the desired conclusion. $\qquad \square$

A related multiplicity result is the *symmetric mountain pass theorem* due to Ambrosetti and Rabinowitz [17] (see Rabinowitz [336, Theorem 9.12]).

**Theorem 5.64.** *Let $X$ be an infinite-dimensional Banach space decomposing as $X = Y \oplus V$ with $\dim Y < +\infty$, and let $\varphi \in C^1(X, \mathbb{R})$ be even, with $\varphi(0) = 0$, satisfying the* (PS)-*condition, and*

(i) *There exist $\eta, \rho > 0$ such that $\varphi \geq \eta$ on $\{x \in V : \|x\| = \rho\}$,*
(ii) *For each finite-dimensional subspace $W$ of $X$ with $W \supset Y$ there exists $R = R(W) > 0$ such that $\varphi \leq 0$ on $\{x \in W : \|x\| \geq R\}$.*

*Then $\varphi$ admits an unbounded sequence of critical values.*

The next theorem is known as the *symmetric criticality theorem* and is useful in problems exhibiting symmetry. Recall that $H^G = \{x \in H : gx = x \text{ for all } g \in G\}$.

**Theorem 5.65.** *If $H$ is a Hilbert space, $G$ is a topological group with isometric linear action on $H$, $\varphi \in C^1(H, \mathbb{R})$ is invariant, and $x_0$ is a critical point of $\varphi|_{H^G}$, then $x_0 \in K_\varphi$.*

*Proof.* Let $(\cdot, \cdot)$ be the inner product of $H$. Since the action of $G$ is isometric (hence it preserves inner products) and by virtue of Proposition 5.56(a), we have

$$(\varphi'(gx), y) = (\varphi'(x), g^{-1}y) = (g\varphi'(x), y) \quad \text{for all } x, y \in H, \text{ all } g \in G,$$

hence $\varphi'(gx) = g\varphi'(x)$, i.e., $\varphi'$ is equivariant.

If $x_0 \in H^G$, we see that $g\varphi'(x_0) = \varphi'(gx_0) = \varphi'(x_0)$, hence $\varphi'(x_0) \in H^G$. On the other hand, assuming that $x_0$ is a critical point of $\varphi|_{H^G}$, we have $\varphi'(x_0) \in (H^G)^\perp$. Therefore, $\varphi'(x_0) \in H^G \cap (H^G)^\perp = \{0\}$, which forces $\varphi'(x_0) = 0$.                □

We conclude our review of the critical point theory under a linear group action with the so-called *fountain theorem*. This result depends on the notion of admissible action of a compact group $G$.

**Definition 5.66.** Let $Y$ be a finite-dimensional space, and let $G$ be a compact topological group with linear action on $Y$. For $k \geq 2$, set $Y^k = Y \times \cdots \times Y$ ($k$-times) and let $G$ act on $Y^k$ diagonally, i.e., $g(y_1, \ldots, y_k) = (gy_1, \ldots, gy_k)$. We say that the action is *admissible* if, for each $k \geq 2$, each $U$ invariant, open, and bounded neighborhood of $0$ in $Y^k$, we have that every continuous, equivariant map $h : \partial U \to Y^{k-1}$ has a zero.

*Remark 5.67.* By the Borsuk–Ulam theorem (Theorem 4.15), the linear action of $G = \mathbb{Z}_2 = \{\mathrm{id}, -\mathrm{id}\}$ on $Y = \mathbb{R}$ is admissible.

The setting is as follows. We consider a compact topological group $G$, a finite-dimensional space $Y$ equipped with an admissible linear action of $G$, and a Banach space $X$ equipped with an isometric linear action of $G$, and we suppose that $X = \overline{\underset{k \geq 1}{\oplus} X_k}$, where for all $k \geq 1$ the space $X_k$ is invariant and there is an equivariant isomorphism $Y \to X_k$.

We use the following notation:

$$Z_m = \overset{m}{\underset{k=1}{\oplus}} X_k \text{ and } V_m = \overline{\underset{k \geq m}{\oplus} X_k}.$$

Then, using the equivariant deformation theorem (Theorem 5.58), we can have the *fountain theorem* due to Bartsch [35] (see Willem [382, p. 58]).

**Theorem 5.68.** *If the preceding setting holds, $\varphi \in C^1(X, \mathbb{R})$ is invariant and satisfies the* $(PS)_c$*-condition at every $c > 0$, and for every $m \geq 1$ there are $\rho_m > r_m > 0$ such that*

(i)  $\max\{\varphi(x) : x \in Z_m, \|x\| = \rho_m\} \leq 0,$
(ii) $\inf\{\varphi(x) : x \in V_m, \|x\| = r_m\} \to +\infty$ *as $m \to \infty$;*

*then $\varphi$ has an unbounded sequence of critical values.*

## 5.5  Generalizations

### Smooth Critical Point Theory on Banach Manifolds

Next, we turn our attention to the problem of finding the critical points of a smooth functional over a constraint set $M \subset X$, which is a smooth Banach manifold. First,

we recall the following generalization of the notion of critical point for a map into a
Banach space.

**Definition 5.69.** Let $X$, $Y$ be Banach spaces, let $U \subset X$ and $V \subset Y$ be nonempty,
open sets, and let $\varphi : U \to V$ be a Fréchet differentiable map.

(a) We say that $x \in U$ is a *critical point* of $\varphi$ if $\varphi'(x) \in \mathscr{L}(X,Y)$ is not surjective.
(b) We say that $x \in U$ is a *regular point* of $\varphi$ if $\varphi'(x) \in \mathscr{L}(X,Y)$ is surjective.

*Remark 5.70.* From Definition 5.69 we retrieve that, for $Y = \mathbb{R}$, a point $x \in U$ is a
critical (resp. regular) point of $\varphi$ if $\varphi'(x) = 0$ [resp. $\varphi'(x) \neq 0$].

In the next definition, we recall the notion of Banach submanifold.

**Definition 5.71.** Let $X$ be a Banach space and $M \subset X$ a subset.

(a) For $x \in M$, a *tangent vector* to $M$ at $x$ is the derivative $\gamma'(0)$ of a curve $\gamma \in$
$C^1(I,M)$ defined on an open interval $I \subset \mathbb{R}$ containing 0 and such that $\gamma(0) = x$.
We denote by $T_x M$ the set of all tangent vectors to $M$ at $x$.
(b) We say that $M$ is a $C^k$-*Banach submanifold* of $X$ ($k \geq 1$) if the following
conditions are satisfied:

   (i) The set $T_x M$ is a closed vector subspace of $X$ for all $x \in M$ (then it is called
   the *tangent space to $M$ at $x$*);
   (ii) For each $x \in M$ there are open subsets $U_x \subset M$ containing $x$, $V_x \subset T_x M$
   containing 0, and a homeomorphism $g_x : V_x \to U_x$;
   (iii) Given any $x,y \in M$ such that $U_x \cap U_y \neq \emptyset$, the chart change $g_y^{-1} \circ g_x :$
   $g_x^{-1}(U_x \cap U_y) \to g_y^{-1}(U_x \cap U_y)$ is of class $C^k$.

(c) We say that $M$ is a Banach submanifold of $X$ of *dimension $m$* (resp. of
*codimension $m$*) if, for every $x \in M$, the tangent space $T_x M$ is a subspace of
$X$ of dimension $m$ (resp. of codimension $m$).

The next theorem, known in the literature as *Lyusternik's theorem*, identifies
an important class of constraint sets as smooth Banach submanifolds. For a
proof, we refer the reader to Zeidler [387, p. 288] or Papageorgiou and Kyritsi
[318, p. 74].

**Theorem 5.72.** *If $X$, $Y$ are Banach spaces, $U \subset X$ is nonempty and open, $\varphi \in$
$C^k(U,Y)$, $k \geq 1$, $M = \{x \in U : \varphi(x) = 0\}$, and every $x \in M$ is a regular point of $\varphi$,
then $M$ is a $C^k$-Banach submanifold of $X$ and*

$$T_x M = \ker \varphi'(x) = \{y \in X : \langle \varphi'(x), y \rangle = 0\} \text{ for all } x \in M.$$

*Moreover, if $Y = \mathbb{R}^m$ and $M \neq \emptyset$, then $M$ is a $C^k$-Banach submanifold of $X$ of
codimension $m$.*

In general, Banach submanifolds can be interpreted as constraint sets for studying
the critical points of maps $\varphi \in C^1(X,Y)$. This idea leads to the notion of constrained
critical point.

**Definition 5.73.** Let $X$ be a Banach space, $\varphi \in C^1(X, \mathbb{R})$, and $M \subset X$ a $C^1$-Banach submanifold. A *constrained critical point* of $\varphi$ on $M$ is a point $x \in M$ such that

$$\langle \varphi'(x), y \rangle = 0 \quad \text{for all } y \in T_x M.$$

We denote by $K_{\varphi|_M}$ the set of constrained critical points of $\varphi$ on $M$.

Clearly, if $x \in M$ is a critical point of $\varphi$, then $x$ is also a constrained critical point of $\varphi$ on $M$. Hence we have $K_\varphi \cap M \subset K_{\varphi|_M}$. The inverse inclusion does not hold in general. For instance, for $\varphi$ and $M$ as in Theorem 5.72 we have $K_\varphi \cap M = \emptyset$ and $K_{\varphi|_M} = M$. This leads to the following definition.

**Definition 5.74.** Let $X$ be a Banach space and $\varphi \in C^1(X, \mathbb{R})$. A $C^1$-submanifold $M \subset X$ is a *natural constraint* for $\varphi$ if $K_\varphi \cap M = K_{\varphi|_M}$.

A natural constraint can then be used as a tool for obtaining existence results for critical points of $\varphi$. We present an important example of natural constraint, the *Nehari manifold*, and this will conclude our short overview.

**Proposition 5.75.** *Let $X$ be a Banach space, $\varphi \in C^2(X, \mathbb{R})$, and set*

$$\mathcal{N}_\varphi = \{x \in X \setminus \{0\} : \langle \varphi'(x), x \rangle = 0\}.$$

*Assume that $\mathcal{N}_\varphi$ is nonempty, $\langle \varphi''(x)x, x \rangle \neq 0$ for all $x \in \mathcal{N}_\varphi$, and there exists $r > 0$ such that $B_r(0) \cap \mathcal{N}_\varphi = \emptyset$. Then $\mathcal{N}_\varphi$ is a complete $C^1$-Banach submanifold of $X$ of codimension 1, and it is a natural constraint for $\varphi$.*

*Proof.* Let $\psi(x) = \langle \varphi'(x), x \rangle$. Then $\psi \in C^1(X, \mathbb{R})$, and we have $\mathcal{N}_\varphi = \{x \in X \setminus \{0\} : \psi(x) = 0\}$. Note that

$$\langle \psi'(x), x \rangle = \langle \varphi''(x)x, x \rangle + \langle \varphi'(x), x \rangle = \langle \varphi''(x)x, x \rangle \neq 0 \quad \text{for all } x \in \mathcal{N}_\varphi, \quad (5.38)$$

so every $x \in \mathcal{N}_\varphi$ is a regular point of $\psi$. Theorem 5.72 implies that $\mathcal{N}_\varphi$ is a $C^1$-Banach submanifold of $X$ of codimension 1. By assumption, we have $\mathcal{N}_\varphi = \psi^{-1}(0) \cap (X \setminus B_r(0))$, the intersection of two closed subsets of $X$, hence $\mathcal{N}_\varphi$ is complete.

It remains to check the inclusion $K_{\varphi|_{\mathcal{N}_\varphi}} \subset K_\varphi$. Let $x \in K_{\varphi|_{\mathcal{N}_\varphi}}$. This means that $\langle \varphi'(x), y \rangle = 0$ for all $y \in T_x \mathcal{N}_\varphi$, that is, $T_x \mathcal{N}_\varphi \subset \ker \varphi'(x)$. By Theorem 5.72, we have $T_x \mathcal{N}_\varphi = \ker \psi'(x)$. In this way, we obtain the inclusion $\ker \psi'(x) \subset \ker \varphi'(x)$, which implies that there is $\lambda \in \mathbb{R}$ such that

$$\varphi'(x) = \lambda \psi'(x). \quad (5.39)$$

Since $\mathcal{N}_\varphi \subset \psi^{-1}(0)$, this yields

$$0 = \psi(x) = \langle \varphi'(x), x \rangle = \lambda \langle \psi'(x), x \rangle. \quad (5.40)$$

From (5.38) and (5.40) we get $\lambda = 0$, and so $\varphi'(x) = 0$ [see (5.39)], i.e., $x \in K_\varphi$.  □

**Remark 5.76.** The submanifold $\mathcal{N}_\varphi$ is called the *Nehari manifold* of $\varphi$. It can be used to produce critical points of $\varphi$ on $X$. For instance, if $\varphi$ is bounded below on $\mathcal{N}_\varphi$ and $x \in \mathcal{N}_\varphi$ satisfies $\varphi(x) = \inf_{\mathcal{N}_\varphi} \varphi$, then we obtain that $x$ is a critical point of $\varphi$ different from 0. A critical point $x \in K_\varphi$ that is a global minimizer of $\varphi$ on $\mathcal{N}_\varphi$ is called a *ground state* (or *least energy*) critical point.

## Nonsmooth Critical Point Theory on Banach Spaces

We conclude this chapter with an overview of some nonsmooth critical point theories that arise naturally in the study of the so-called variational, hemivariational, and variational–hemivariational inequalities. In Sects. 3.1 and 3.2, we dealt with generalized gradients for nonsmooth functionals of the form $\varphi : X \to \mathbb{R}$ locally Lipschitz or $\psi : X \to \mathbb{R} \cup \{+\infty\}$ l.s.c., convex, and not identically $+\infty$ [i.e., $\psi \in \Gamma_0(X)$]. In what follows, for the convenience of the exposition, we fix a functional $\Xi : X \to \mathbb{R} \cup \{+\infty\}$ of the following form, which unifies these two approaches:

$$\Xi = \varphi + \psi \text{ with } \varphi : X \to \mathbb{R} \text{ locally Lipschitz and } \psi \in \Gamma_0(X). \tag{5.41}$$

**Definition 5.77.** We say that $x \in X$ is a *critical point* of the functional $\Xi$ in (5.41) if $0 \in \partial\varphi(x) + \partial\psi(x)$, where $\partial\varphi(x)$ and $\partial\psi(x)$ stand for the generalized subdifferential and the convex subdifferential of the functionals $\varphi$ and $\psi$, respectively.

**Remark 5.78.**

(a) In the case where $\Xi : X \to \mathbb{R}$ is a smooth functional [i.e., $\varphi \in C^1(X, \mathbb{R})$ and $\psi = 0$], we recover the usual notion of critical point. In particular, in this case, the critical points coincide with the solutions of the Euler problem $\Xi'(x) = 0$.
(b) If $\varphi \in C^1(X, \mathbb{R})$ and $\psi \in \Gamma_0(X)$, then the critical points of $\Xi$ in the sense of Definition 5.77 coincide with the solutions of the *variational inequality*

$$\varphi'(x; y - x) + \psi(y) - \psi(x) \geq 0 \text{ for all } y \in X. \tag{5.42}$$

In particular, if $\psi = i_C$ is the indicator function of a convex set $C \subset X$ [i.e., $i_C(x) = 0$ if $x \in C$ and $i_C(x) = +\infty$ if $x \in X \setminus C$], then (5.42) becomes the *problem with variational constraint*

$$\begin{cases} \varphi'(x; y - x) \geq 0 \text{ for all } y \in C, \\ x \in C. \end{cases}$$

The critical point theory for functionals of the form $\Xi = \varphi + \psi$ with $\varphi \in C^1(X, \mathbb{R})$ and $\psi \in \Gamma_0(X)$ was developed by Szulkin [369].

(c) In the case where $\psi = 0$, i.e., $\Xi = \varphi$ is locally Lipschitz, the critical points of $\Xi$ coincide with the solutions of the problem

$$\varphi^0(x;y) \geq 0 \text{ for all } y \in X.$$

For example, if $X \subset L^p(\Omega,\mathbb{R})$ ($\Omega \subset \mathbb{R}^N$ bounded domain and $p \in [1,+\infty]$) and $\varphi = \varphi_1 + \varphi_2$, with $\varphi_1 \in C^1(X,\mathbb{R})$ and $\varphi_2 : X \to \mathbb{R}$ locally Lipschitz, then, by Proposition 3.49, the critical points of $\Xi = \varphi$ are particular solutions of the integral inequality

$$\varphi_1'(u;v) + \int_\Omega \varphi_2^0(u(x);v(x)) \, dx \geq 0 \text{ for all } v \in X.$$

Problems of this type are called *hemivariational inequalities*, and their study was initiated by Panagiotopoulos [317] (see also Naniewicz and Panagiotopoulos [307]), relying on the critical point theory for locally Lipschitz functionals developed by Chang [77].

(d) As a natural extension of (b) and (c), the critical points of a general functional $\Xi = \varphi + \psi$ as in (5.41) are solutions of the so-called *variational–hemivariational inequalities*, introduced in Motreanu and Panagiotopoulos [274].

A first type of critical points in the sense of Definition 5.77 are local minimizers.

**Proposition 5.79.** *Let $\Xi : X \to \mathbb{R} \cup \{+\infty\}$ be a functional as in (5.41). Then every local minimizer of $\Xi$ is a critical point of $\Xi$.*

For further existence results of critical points, we need to introduce suitable compactness conditions.

**Definition 5.80.** Let $\Xi = \varphi + \psi : X \to \mathbb{R} \cup \{+\infty\}$ be a functional as in (5.41).

(a) We say that $\Xi$ satisfies the *Palais–Smale condition at level $c \in \mathbb{R}$ ((PS)$_c$-condition)* if every sequence $\{x_n\}_{n\geq 1} \subset X$ satisfying

   (i) $\Xi(x_n) \to c$ as $n \to \infty$ and
   (ii) there exists $\{x_n^*\}_{n\geq 1} \subset X^*$ with $x_n^* \to 0$ in $X^*$ as $n \to \infty$ such that

$$\varphi^0(x_n;y-x_n) + \psi(y) - \psi(x_n) \geq \langle x_n^*, y - x_n \rangle \text{ for all } n \geq 1, \text{ all } y \in X$$

   admits a strongly convergent subsequence.

(b) We say that $\Xi$ satisfies the *Palais–Smale condition ((PS)-condition)* if it satisfies the (PS)$_c$-condition at every level $c \in \mathbb{R}$.

*Remark 5.81.*

(a) If $\Xi \in C^1(X,\mathbb{R})$, then we recover the usual Palais–Smale condition in the smooth case (Definition 5.14).

(b) When $\Xi = \varphi : X \to \mathbb{R}$ is locally Lipschitz, it satisfies the (PS)$_c$-condition if and only if every sequence $\{x_n\}_{n\geq 1} \subset X$ such that

$$\varphi(x_n) \to c \text{ and } m_\varphi(x_n) \to 0 \text{ as } n \to \infty$$

admits a strongly convergent subsequence [with $m_\varphi$ from (3.9)].

A basic minimization result involving the (PS)-condition, relying on the Ekeland variational principle (Theorem 5.7), is the following proposition (see Motreanu and Rădulescu [278, p. 48]).

**Proposition 5.82.** *Let* $\Xi : X \to \mathbb{R} \cup \{+\infty\}$ *be a functional as in (5.41). Assume that* $\Xi$ *is bounded below and satisfies the* (PS)$_c$*-condition for* $c = \inf_X \Xi$. *Then* $\Xi$ *admits a critical point* $x_0 \in X$ *such that* $\Xi(x_0) = c$.

Suitable deformation results for different types of nonsmooth functionals fitting (5.41) have been obtained by Chang [77, Theorem 3.1], Szulkin [369, Proposition 2.3], Motreanu and Panagiotopoulos [274, Theorem 3.1], Marano and Motreanu [242, Theorem 3.1], and Gasiński and Papageorgiou [150, Theorem 2.1.1]. These results lead to minimax principles; here we state the general minimax principle obtained in Motreanu and Panagiotopoulos [274, Theorem 3.2].

**Theorem 5.83.** *Let* $\{E_0, E, D\}$ *be linking sets in the sense of Remark 5.37, with* $D$ *closed,* $E$ *compact, and* $E_0 = \partial E$. *Let* $\Gamma = \{\gamma \in C(E, X) : \gamma|_{E_0} = \mathrm{id}_{E_0}\}$. *Let* $\Xi : X \to \mathbb{R} \cup \{+\infty\}$ *be a functional as in (5.41) such that*

$$\sup_E \Xi \in \mathbb{R}, \quad b := \inf_D \Xi \in \mathbb{R} \text{ and } a := \sup_{E_0} \Xi < b.$$

*Let* $c = \inf_{\gamma \in \Gamma} \sup_{x \in E} \varphi(\gamma(x))$, *and assume that* $\varphi$ *satisfies the* (PS)$_c$*-condition. Then* $c \geq b$ *and* $\Xi$ *admits a critical point* $x_0$ *such that* $\Xi(x_0) = c$.

**Remark 5.84.** An extension of Theorem 5.83 in the limiting case $a = b$ can be found in Marano and Motreanu [242, 243] (see also Motreanu and Motreanu [273]).

Choosing the linking sets $E_0, E, D$ of the theorem as in Example 5.38(a)–(d), we obtain nonsmooth versions of the mountain pass theorem, the saddle point theorem, and the generalized mountain pass theorem.

Multiplicity results can also be obtained for nonsmooth functionals. We conclude our overview with the following nonsmooth version of the local linking theorem (see Gasiński and Papageorgiou [150, p. 178]).

**Theorem 5.85.** *If* $X$ *is a reflexive Banach space,* $X = Y \oplus V$, *with* $\dim Y < +\infty$, $\Xi = \varphi : X \to \mathbb{R}$ *is locally Lipschitz and bounded below and satisfies the* (PS)*-condition,* $\inf_X \varphi < 0 = \varphi(0)$, *and there exists* $r > 0$ *such that*

$$\begin{cases} \varphi(x) \leq 0 & \text{if } x \in Y, \ \|x\| \leq r, \\ \varphi(x) \geq 0 & \text{if } x \in V, \ \|x\| \leq r, \end{cases}$$

*then* $\varphi$ *has at least two nontrivial critical points.*

## 5.6   Remarks

**Section 5.1:** The Lax–Milgram theorem (Corollary 5.6) can be found in Lax and Milgram [218], while Stampacchia's theorem (Theorem 5.4) is proved in Stampacchia [366]. Both are simple and efficient tools for solving linear elliptic equations and inequalities, respectively. Theorem 5.7 (the so-called Ekeland variational principle) was proved by Ekeland [126]. More about the consequences and applications of this variational principle can be found in Ekeland [127, 128]. The extension formulated in Theorem 5.16 can be found in Zhong [395]. The (PS)- and (C)-conditions, which are basic tools in critical point theory, are compactness-type conditions that compensate for the fact that the ambient space $X$ is not in general locally compact (being an infinite-dimensional Banach space). They were introduced by Palais and Smale [316], Palais [314], and Cerami [75]. The relation between these two conditions and between them and the coercivity of the functional is discussed in Čaklović et al. [64], Costa and Silva [91], Motreanu and Motreanu [272], Motreanu et al. [285, 298], and Zhong [395].

**Section 5.2:** The literature contains two approaches to critical point theory. The first one is based on the deformation properties of the negative gradient flow or of a suitable substitute of it, namely, the pseudogradient flow. This is the approach that we have adopted. The second approach uses the Ekeland variational principle and can be found in the works of Cuesta [94] (for the critical point theory of functions defined on a $C^1$-Banach manifold), Ekeland [128], and de Figueiredo [105]. For the deformation approach, the notion of pseudogradient field (Definition 5.27) was introduced by Palais [314] in order to extend the classical Lyusternik–Schnirelmann theory to infinite-dimensional Banach manifolds. A first version of the deformation theorem (Theorem 5.32) was proved by Clark [83]. Usually, the deformation theorem is stated in terms of the (PS)-condition. The first to formulate the result in terms of the weaker (C)-condition were Bartolo et al. [33]. The second deformation theorem (Theorem 5.34) is due to Rothe [348], Marino and Prodi [249], Chang [78] (under the (PS)-condition), and Silva and Teixeira [361] (under the (C)-condition). The proof that we propose here combines original arguments with ideas from Chang [78] and Silva and Teixeira [361].

The deformation approach to critical point theory can be found in the books of Ambrosetti and Malchiodi [15], Chang [78], Costa [89], Gasiński and Papageorgiou [151], Ghoussoub [156], Jabri [183], Kavian [191], and Willem [382].

**Section 5.3:** The notion of linking sets (Definition 5.36) is crucial in critical point theory and was first introduced by Benci and Rabinowitz [40]. Various other versions of the notion of linking sets can be found in Corvellec et al. [88] and Schechter [352, 353]. Versions of Theorem 5.39 can be found in Ekeland [128], Mawhin and Willem [253], and Struwe [367]. The proof proposed here to treat the limit case $a = b$ is original. The mountain pass theorem (Theorem 5.40) is due to Ambrosetti and Rabinowitz [17], while the saddle point theorem (Theorem 5.41) and the generalized mountain pass theorem are due to Rabinowitz [336, 338].

The notion of local linking (Definition 5.48) is originally due to Liu and Li [232] under the stronger conditions that $\dim Y < +\infty$ and $\varphi(x) \geq r > 0$ for all $x \in V$ with $\|x\| = r$. Theorem 5.51 is due to Brezis and Nirenberg [53], while Theorem 5.53 is due to Li and Willem [224].

**Section 5.4:** Symmetry is a very helpful assumption for the existence of multiple critical points. The best-known examples are when the acting group is $G = \mathbb{Z}_2$ or $G = S^1$. In the first case we have the Krasnosel'skiĭ genus (Krasnosel'skiĭ [202] and Coffman [86]), and in the second case we have the cohomological index due to Fadell and Rabinowitz [133] (see also Benci [39] and Fadell et al. [134]). The symmetric criticality theorem (Theorem 5.65) is due to Palais [315] (in fact, as Palais points out in [315], an earlier implicit use of this principle can be found in Pauli [325] and Weyl [381] on problems of mathematical physics). This principle is not valid in general, and Palais [315] provided some counterexamples to this effect (see also Kobayashi and Ôtani [200]). An extension of the fountain theorem (Theorem 5.68) can be found in Zou [397]. Further discussions and results on the theme of multiplicity versus symmetry can be found in the book by Bartsch [34].

**Section 5.5:** Given $\varphi \in C^1(X, \mathbb{R})$ ($X$ a Banach space) and $M \subset X$ a submanifold, we say that $M$ is a *natural constraint* of $\varphi$ if the critical set of $\varphi$ coincides with the critical set of $\varphi|_M$ (Definition 5.73). Proposition 5.75 provides an example of a natural constraint for $\varphi$, the Nehari manifold. Critical points for constrained functionals and applications to elliptic problems and Hamiltonian systems can be found in Ambrosetti and Rabinowitz [17], Badiale and Serra [27], Blanchard and Brüning [47], Costa [89], and Zeidler [386]. The nonsmooth critical point theory started with Chang [77]. Since then, there have been various extensions to the theory due to Degiovanni and Marzocchi [107], Gasiński and Papageorgiou [150], Motreanu and Panagiotopoulos [274], Marano and Motreanu [242, 243], and Szulkin [369]. There are also nonsmooth extensions of the principle of symmetric criticality (Theorem 5.65), for which we refer readers to the papers of Krawcewicz and Marzantowicz [204] and Kobayashi and Ôtani [200]. For other developments, consult Candito et al. [66, 67].

# Chapter 6
# Morse Theory

**Abstract** This chapter represents a self-contained presentation of basic results and techniques of Morse theory that are useful for studying the multiplicity of solutions of nonlinear elliptic boundary value problems with a variational structure. The first section of the chapter contains the needed preliminaries of algebraic topology. The second section focuses on the Morse lemma and the splitting and shifting theorems. The third section is devoted to the Morse relations, including the Poincaré–Hopf formula, which involve the critical groups and critical groups at infinity. The fourth section sets forth efficient results for the computation of critical groups that are powerful tools in the study of multiple solutions. Here an original approach is developed, and improvements of known results are shown. Notes on related literature and comments are provided in a remarks section.

## 6.1 Elements of Algebraic Topology

This section provides the preliminaries necessary for the study of critical groups.

**Definition 6.1.** A *topological pair* is a pair $(X, A)$ formed by a Hausdorff topological space $X$ and a subset $A \subset X$.

**Definition 6.2.** Let $(X, A)$, $(Y, B)$, and $(V, C)$ be topological pairs.

(a) A *map of pairs* $f : (X, A) \to (Y, B)$ is a continuous map $f : X \to Y$ such that $f(A) \subset B$. We denote by $C((X, A), (Y, B))$ the set of maps of pairs from $(X, A)$ to $(Y, B)$ and by $\mathrm{id}_{(X,A)} : (X, A) \to (X, A)$ the identity map seen as a map of pairs.
(b) If $g : (Y, B) \to (V, C)$ is another map of pairs, then the composition $g \circ f : (X, A) \to (V, C)$ is a map of pairs.
(c) Two topological pairs $(X, A)$, $(Y, B)$ are *homeomorphic* if there is a homeomorphism $f : X \to Y$ such that $f(A) = B$.

D. Motreanu et al., *Topological and Variational Methods with Applications to Nonlinear Boundary Value Problems*, DOI 10.1007/978-1-4614-9323-5_6,
© Springer Science+Business Media, LLC 2014

*Remark 6.3.*

(a) Note that the space $X$ can be regarded as the topological pair $(X, \emptyset)$.
(b) If $A = \{x_0\}$, then the pair $(X, \{x_0\})$ is denoted $(X, x_0)$ and called a *pointed space*.

**Definition 6.4.** Let $(X, A)$ and $(Y, B)$ be two topological pairs and $f, g : (X, A) \to (Y, B)$ two maps of pairs. We say that $f$ *is homotopic to* $g$, denoted by $f \sim g$, if there is a map of pairs $h : ([0,1] \times X, [0,1] \times A) \to (Y, B)$ such that $h(0, x) = f(x)$ and $h(1, x) = g(x)$ for all $x \in X$.

*Remark 6.5.* Clearly, $\sim$ is an equivalence relation on $C((X, A), (Y, B))$. Compositions of homotopic maps remain homotopic, i.e., if $f, g : (X, A) \to (Y, B)$, $\vartheta, \eta : (Y, B) \to (V, C)$ and $f \sim g$, $\vartheta \sim \eta$, then $\vartheta \circ f \sim \eta \circ g$. Hence the composition is well defined on the equivalence classes and it is associative.

**Definition 6.6.** Two topological pairs $(X, A)$ and $(Y, B)$ are said to be *homotopy equivalent*, denoted by $(X, A) \sim (Y, B)$, if there are maps of pairs $f : (X, A) \to (Y, B)$ and $g : (Y, B) \to (X, A)$ such that $g \circ f \sim \mathrm{id}_{(X, A)}$ and $f \circ g \sim \mathrm{id}_{(Y, B)}$. The map $f$ is called a *homotopy equivalence*.

*Remark 6.7.* Homotopy equivalent topological pairs can be viewed as being continuously deformable into each other. Two homeomorphic pairs are always homotopy equivalent, but the converse is false. The following two examples are helpful in visualizing the notion of homotopy equivalence.

*Example 6.8.* Let $S^1, B^2 \subset \mathbb{R}^2$ denote the unit circle and the open unit ball.

(a) The solid torus $B^2 \times S^1$ is homotopy equivalent to the circle $S^1$. Also, the pairs $(B^2 \times S^1, \{0\} \times S^1)$, $(S^1, S^1)$ are homotopy equivalent. Moreover, whenever $x_0 \in S^1$, the pairs $(B^2 \times S^1, B^2 \times \{x_0\})$, $(S^1, x_0)$ are homotopy equivalent.
(b) The circle $S^1$ is not homotopy equivalent to a point.

In introducing homology groups, we will use the axiomatic approach (naive homology theory). We recall that a (finite or infinite) chain of group homomorphisms

$$\cdots \longrightarrow G_{k+1} \xrightarrow{\alpha_{k+1}} G_k \xrightarrow{\alpha_k} G_{k-1} \longrightarrow \cdots$$

is called an *exact sequence* if we have im $\alpha_{k+1} = \ker \alpha_k$ for all $k$. The basic definition is as follows.

**Definition 6.9.** A *homology theory* on a collection of topological pairs $(X, A)$ consists of:

(a) A sequence of abelian groups $H_k(X, A)$ (for $k \in \mathbb{N}_0 = \mathbb{N} \cup \{0\}$), called *homology groups*, attached to every topological pair $(X, A)$; we write $H_k(X) = H_k(X, \emptyset)$.
(b) A sequence of group homomorphisms $f_* : H_k(X, A) \to H_k(Y, B)$ (for $k \in \mathbb{N}_0$) attached to every map of pairs $f : (X, A) \to (Y, B)$.

(c) A sequence of group homomorphisms $\partial : H_k(X,A) \to H_{k-1}(A)$ [for $k \in \mathbb{N}_0$, with $H_{-1}(A) := 0$] attached to every topological pair $(X,A)$.

These data satisfy the following axioms.

*Axiom 1*: If $f = \mathrm{id}_{(X,A)}$, then $f_* = \mathrm{id}_{H_k(X,A)}$.

*Axiom 2*: If $f : (X,A) \to (Y,B)$ and $g : (Y,B) \to (V,C)$, then $(g \circ f)_* = g_* \circ f_*$.

*Axiom 3*: If $f : (X,A) \to (Y,B)$, then $\partial \circ f_* = (f|_A)_* \circ \partial$.

*Axiom 4*: If $i : A \to X$ and $j : (X,\emptyset) \to (X,A)$ are the inclusion maps, then the following sequence is exact:

$$\ldots \xrightarrow{\partial} H_k(A) \xrightarrow{i_*} H_k(X) \xrightarrow{j_*} H_k(X,A) \xrightarrow{\partial} H_{k-1}(A) \to \ldots \, , \ k \in \mathbb{N}_0.$$

*Axiom 5*: If $f,g : (X,A) \to (Y,B)$ are homotopic maps of pairs, then $f_* = g_*$.

*Axiom 6* (Excision): If $A,B \subset X$ are subsets such that $X = \mathrm{int}\, A \cup \mathrm{int}\, B$, then the inclusion map $e : (A, A \cap B) \to (X,A)$ induces an isomorphism $e_* : H_k(A, A \cap B) \to H_k(X,A)$ for all $k \in \mathbb{N}_0$.

*Axiom 7*: If $X$ is a singleton, then $H_k(X) = 0$ for all $k \in \mathbb{N}$.

*Remark 6.10.* A group $G$ such that $H_0(X) \simeq G$ for every singleton $X$ is called the *group of coefficients* of the homology theory. For $k < 0$ we set $H_k(X,A) = 0$.

In what follows, we consider a collection of topological pairs for which we dispose of a homology theory and we derive some useful consequences.

**Proposition 6.11.** *If the topological pairs $(X,A)$ and $(Y,B)$ are homotopy equivalent, then $H_k(X,A) = H_k(Y,B)$ for all $k \in \mathbb{N}_0$ (hereafter, the symbol "=" indicates that the groups are isomorphic).*

*Proof.* Let $f : (X,A) \to (Y,B)$ be a homotopy equivalence with homotopy inverse $g : (Y,B) \to (X,A)$ (Definition 6.6). Then $g \circ f \sim \mathrm{id}_{(X,A)}$, and so $g_* \circ f_* = \mathrm{id}_{H_k(X,A)}$ (Axioms 1, 2, and 5). Similarly, $f_* \circ g_* = \mathrm{id}_{H_k(Y,B)}$. Therefore, $f_* : H_k(X,A) \to H_k(Y,B)$ is a group isomorphism with inverse $g_*$. $\square$

**Proposition 6.12.** *If $A \subset X$ is a deformation retract [Definition 5.33(b)], then $H_k(X,A) = 0$ for all $k \in \mathbb{N}_0$.*

*Proof.* The assumption implies that the inclusion map $i : (A,\emptyset) \to (X,\emptyset)$ is a homotopy equivalence; hence, as in the proof of Proposition 6.11, $i_* : H_k(A) \to H_k(X)$ is a group isomorphism for all $k \in \mathbb{N}_0$. From the exact sequence in Axiom 4 we infer that $H_k(X,A) = 0$ for all $k \in \mathbb{N}_0$. $\square$

**Corollary 6.13.** $H_k(X,X) = 0$ *for all $k \in \mathbb{N}_0$.*

The following proposition generalizes the long exact sequence of Axiom 4.

**Proposition 6.14.** *Let $C \subset A \subset X$. The inclusion maps $i : (A,C) \to (X,C)$, $j : (X,C) \to (X,A)$, and $j_2 : (A,\emptyset) \to (A,C)$ and the homomorphism $\partial_1 : H_k(X,A) \to H_{k-1}(A)$ induce an exact sequence*

$$\ldots \xrightarrow{\ j_{2_*} \circ \partial_1\ } H_k(A,C) \xrightarrow{\ i_*\ } H_k(X,C) \xrightarrow{\ j_*\ } H_k(X,A) \xrightarrow{\ j_{2_*} \circ \partial_1\ } H_{k-1}(A,C) \xrightarrow{\quad} \ldots, \quad k \in \mathbb{N}_0.$$

*Proof.* First, we must check the inclusions

$$\operatorname{im} j_{2_*} \circ \partial_1 \subset \ker i_*, \quad \operatorname{im} i_* \subset \ker j_*, \quad \text{and} \quad \operatorname{im} j_* \subset \ker j_{2_*} \circ \partial_1. \tag{6.1}$$

By Axiom 2, the following diagram of homomorphisms is commutative:

$$
\begin{array}{ccc}
H_{k+1}(X,A) & \xrightarrow{\ \partial_1\ } H_k(A) & \xrightarrow{\ j_{2_*}\ } H_k(A,C) \\[2mm]
 & \downarrow{\scriptstyle i_{1_*}} & \downarrow{\scriptstyle i_*} \\[2mm]
 & H_k(X) & \xrightarrow{\ j_{3_*}\ } H_k(X,C),
\end{array}
$$

where $i_1 : A \to X$ and $j_3 : (X,\emptyset) \to (X,C)$ are the inclusion maps. As $i_{1_*} \circ \partial_1 = 0$ (Axiom 4), we obtain $i_* \circ (j_{2_*} \circ \partial_1) = 0$, whence the first inclusion in (6.1). The other two inclusions can be checked in a similar way. Second, we must show that

$$\ker i_* \subset \operatorname{im} j_{2_*} \circ \partial_1, \quad \ker j_* \subset \operatorname{im} i_*, \quad \text{and} \quad \ker j_{2_*} \circ \partial_1 \subset \operatorname{im} j_*. \tag{6.2}$$

Take $z \in \ker j_*$ and let us show that $z \in \operatorname{im} i_*$. We consider the following commutative diagram of homomorphisms (Axioms 2 and 3):

$$
\begin{array}{ccccc}
H_k(A) & \xrightarrow{\ i_{1_*}\ } & H_k(X) & & \\[2mm]
\downarrow{\scriptstyle j_{2_*}} & & \downarrow{\scriptstyle j_{3_*}} & \searrow{\scriptstyle j_{1_*}} & \\[2mm]
H_k(A,C) & \xrightarrow[\ \ \partial_2\ \ ]{i_*} & H_k(X,C) & \xrightarrow{\ j_*\ } & H_k(X,A) \\[2mm]
 & \searrow & \downarrow{\scriptstyle \partial_3} & & \downarrow{\scriptstyle \partial_1} \\[2mm]
 & & H_{k-1}(C) & \xrightarrow[\ i_{2_*}\ ]{} & H_{k-1}(A),
\end{array}
$$

where $j_1 : (X,\emptyset) \to (X,A)$ and $i_2 : C \to A$ are the inclusion maps and $\partial_2, \partial_3$ are the homomorphisms obtained from Definition 6.9. The argument involves chasing diagram. We reason in four steps, as follows.

1. Because $j_*(z) = 0$, we have $i_{2_*}(\partial_3(z)) = \partial_1(j_*(z)) = 0$. Since $\ker i_{2_*} = \operatorname{im} \partial_2$ (Axiom 4), we find $y \in H_k(A,C)$ such that $\partial_3(z) = \partial_2(y)$.
2. We have $\partial_3(i_*(y) - z) = 0$. Since $\ker \partial_3 = \operatorname{im} j_{3_*}$ (by Axiom 4), we find $x \in H_k(X)$, with $i_*(y) - z = j_{3_*}(x)$.
3. Note that $j_{1_*}(x) = j_*(i_*(y) - z) = 0$ (since $j_* \circ i_* = 0$ [see (6.1)] and $j_*(z) = 0$). As $\ker j_{1_*} = \operatorname{im} i_{1_*}$ (Axiom 4), there is $w \in H_k(A)$ satisfying $x = i_{1_*}(w)$.

4. All together we obtain $i_*(y) - z = j_{3*}(i_{1*}(w)) = i_*(j_{2*}(w))$, whence $z \in \operatorname{im} i_*$.

We have shown the second inclusion in (6.2). The other two inclusions can be obtained in the same way. □

**Corollary 6.15.** *Let $C \subset A \subset X$.*

(a) *If $C$ is a deformation retract of $A$, then $H_k(X,A) = H_k(X,C)$ for all $k \in \mathbb{N}_0$.*
(b) *If $A$ is a deformation retract of $X$, then $H_k(X,C) = H_k(A,C)$ for all $k \in \mathbb{N}_0$.*

*Proof.* This is an immediate consequence of Propositions 6.12 and 6.14. □

When $A$ is a retract of $X$ (Definition 4.8), the homology has the property to decompose as follows.

**Proposition 6.16.** *If $A$ is a retract of $X$, then $H_k(X) = H_k(A) \oplus H_k(X,A)$ for all $k \in \mathbb{N}_0$.*

*Proof.* Let $r : X \to A$ be a retraction and $i : A \to X$ be the inclusion map, so $r \circ i = \operatorname{id}_A$. This yields $r_* \circ i_* = \operatorname{id}_{H_k(A)}$ (Axioms 1 and 2), which implies that $i_* : H_k(A) \to H_k(X)$ is injective and $r_* : H_k(X) \to H_k(A)$ is surjective. Then it is readily seen that

$$H_k(X) = \operatorname{im} i_* \oplus \ker r_* \simeq H_k(A) \oplus \ker r_*.$$

Since $i_*$ is injective, Axiom 4 yields a short exact sequence:

$$0 \longrightarrow H_k(A) \overset{i_*}{\longrightarrow} H_k(X) \overset{j_*}{\longrightarrow} H_k(X,A) \longrightarrow 0 \ ,$$

where $j : (X,\emptyset) \to (X,A)$ is the inclusion map. By the exactness of the sequence, we obtain that $j_* : \ker r_* \to H_k(X,A)$ is an isomorphism, whence $H_k(X) \simeq H_k(A) \oplus H_k(X,A)$. □

We will need the following result of abstract algebra, called the *five lemma* (see Spanier [365, p. 185]).

**Lemma 6.17.** *If the commutative diagram of abelian groups and homomorphisms*

$$
\begin{array}{ccccccccc}
A & \longrightarrow & B & \longrightarrow & C & \longrightarrow & D & \longrightarrow & E \\
\downarrow{\scriptstyle \alpha} & & \downarrow{\scriptstyle \beta} & & \downarrow{\scriptstyle \gamma} & & \downarrow{\scriptstyle \delta} & & \downarrow{\scriptstyle \eta} \\
\hat{A} & \longrightarrow & \hat{B} & \longrightarrow & \hat{C} & \longrightarrow & \hat{D} & \longrightarrow & \hat{E}
\end{array}
$$

*has exact rows and $\alpha, \beta, \delta, \eta$ are isomorphisms, then $\gamma$ is an isomorphism.*

**Proposition 6.18.** *If $(X,A) = \overset{n}{\underset{i=1}{\cup}} (X_i, A_i)$, where $\{X_i\}_{i=1}^n$ are nonempty, closed, and pairwise disjoint, then $H_k(X,A) = \overset{n}{\underset{i=1}{\oplus}} H_k(X_i, A_i)$ for all $k \in \mathbb{N}_0$.*

*Proof.* We do the proof for $n = 2$, with the general case following by induction.

We claim that the inclusion maps $i_1 : X_1 \to X$ and $i_2 : X_2 \to X$ yield an isomorphism $i_{1*} \oplus i_{2*} : H_k(X_1) \oplus H_k(X_2) \to H_k(X)$ for all $k \in \mathbb{N}_0$. For this we need to check that $i_{1*}, i_{2*}$ are injective and satisfy $H_k(X) = \operatorname{im} i_{1*} \oplus \operatorname{im} i_{2*}$. Let $j_1 : (X, \emptyset) \to (X, X_1)$ be the inclusion map. Then the composition $j_1 \circ i_2 : (X_2, \emptyset) \to (X, X_1)$ is the inclusion map, so that, by the excision property (Axiom 6), $(j_1 \circ i_2)_* = j_{1*} \circ i_{2*}$ is an isomorphism. This implies that $i_{2*}$ is injective, and this easily implies that $H_k(X) = \ker j_{1*} \oplus \operatorname{im} i_{2*}$. Similarly, we prove that $i_{1*}$ is injective. Moreover, the exactness in Axiom 4 yields $\ker j_{1*} = \operatorname{im} i_{1*}$, so $H_k(X) = \operatorname{im} i_{1*} \oplus \operatorname{im} i_{2*}$, as required.

Similarly, we obtain that the inclusion maps $i_1^A : A_1 \to A$ and $i_2^A : A_2 \to A$ yield an isomorphism $i_{1*}^A \oplus i_{2*}^A : H_k(A_1) \oplus H_k(A_2) \to H_k(A)$ for all $k \in \mathbb{N}_0$. Then, for all $k \in \mathbb{N}_0$, Axiom 4 furnishes a commutative diagram:

$$
\begin{array}{ccccccccc}
\bigoplus_{i=1}^{2} H_k(A_i) & \to & \bigoplus_{i=1}^{2} H_k(X_i) & \to & \bigoplus_{i=1}^{2} H_k(X_i, A_i) & \to & \bigoplus_{i=1}^{2} H_{k-1}(A_i) & \to & \bigoplus_{i=1}^{2} H_{k-1}(X_i) \\
\downarrow{\sim} & & \downarrow{\sim} & & \downarrow & & \downarrow{\sim} & & \downarrow{\sim} \\
H_k(A) & \longrightarrow & H_k(X) & \longrightarrow & H_k(X, A) & \longrightarrow & H_{k-1}(A) & \longrightarrow & H_{k-1}(X),
\end{array}
$$

where the rows are exact and the arrows marked by "$\sim$" are isomorphisms. By Lemma 6.17, we obtain that $H_k(X, A) \simeq H_k(X_1, A_1) \oplus H_k(X_2, A_2)$ for all $k \in \mathbb{N}_0$.  $\square$

Next we turn our attention to the homology groups of the form $H_k(X, *)$, where $*$ denotes a point in $X$.

**Proposition 6.19.**

(a) $H_k(X, *)$ *is isomorphic to the kernel of* $r_*$ *whenever* $r : X \to Y$ *is the map to a singleton* $Y = \{y\}$.
(b) *We have* $H_k(X) = H_k(X, *) \oplus H_k(*)$ *for all* $k \in \mathbb{N}_0$.

*Proof.* Part (a) follows from the proof of Proposition 6.16. Part (b) is an immediate consequence of Proposition 6.16.  $\square$

*Remark 6.20.* The groups $H_k(X, *)$, $k \in \mathbb{N}_0$, are easier to work with and are called the *reduced homology groups of* $X$. Proposition 6.19(a) shows that they are essentially independent of the choice of $* \in X$.

**Proposition 6.21.** *If* $(X, A)$ *is a topological pair and* $* \in A$, *then there is an exact sequence*

$$
\ldots \to H_k(A, *) \to H_k(X, *) \to H_k(X, A) \to H_{k-1}(A, *) \to \ldots, \quad k \in \mathbb{N}_0.
$$

*Proof.* This is a particular case of Proposition 6.14.  $\square$

We recall the following basic notion from topology.

**Definition 6.22.** A Hausdorff topological space $X$ is said to be *contractible* if the identity map $\mathrm{id}_X : X \to X$ is homotopic to a constant map $f : X \to *$ (i.e., there is $h : [0,1] \times X \to X$ continuous such that $h(0,x) = x$ and $h(1,x) = *$ for all $x \in X$).

*Remark 6.23.* Clearly, $X$ is contractible if and only if some point of $X$ is a deformation retract of $X$ [Definition 5.33(b)] or if and only if $X$ is homotopy equivalent to a singleton [Definition 6.6]. Any convex subset or, more generally, any star-shaped subset of a topological vector space is contractible (recall that a subset $X$ of a vector space is star shaped if there is $x_0 \in X$ such that, for all $x \in X$, $[x_0,x] := \{(1-\lambda)x_0 + \lambda x : \lambda \in [0,1]\} \subset X$).

By virtue of Proposition 6.12, we have the following proposition.

**Proposition 6.24.** *If $X$ is contractible and $* \in X$, then $H_k(X,*) = 0$ for all $k \in \mathbb{N}_0$.*

Combining Propositions 6.21 and 6.24, we have the following proposition.

**Proposition 6.25.** *If $A \subset X$ is contractible and $* \in A$, then $H_k(X,A) = H_k(X,*)$ for all $k \in \mathbb{N}_0$.*

We recall the following result from homological algebra, known in the literature as the *Whitehead–Barratt lemma* (see Granas and Dugundji [165, p. 610]).

**Lemma 6.26.** *If the commutative diagram of abelian groups and homomorphisms*

$$\cdots \xrightarrow{\quad} A_k \xrightarrow{\ f_k\ } B_k \xrightarrow{\ g_k\ } C_k \xrightarrow{\ h_k\ } A_{k-1} \xrightarrow{\ f_{k-1}\ } \cdots$$
$$\Big\downarrow \alpha_k \qquad \Big\downarrow \beta_k \qquad \Big\downarrow \gamma_k \qquad \Big\downarrow \alpha_{k-1}$$
$$\cdots \xrightarrow{\quad} \hat{A}_k \xrightarrow{\ \hat{f}_k\ } \hat{B}_k \xrightarrow{\ \hat{g}_k\ } \hat{C}_k \xrightarrow{\ \hat{h}_k\ } \hat{A}_{k-1} \xrightarrow{\ \hat{f}_{k-1}\ } \cdots$$

*has exact rows and the $\gamma_k$ are isomorphisms, then the sequence*

$$\cdots \xrightarrow{\quad} A_k \xrightarrow{\ (\alpha_k,-f_k)\ } \hat{A}_k \oplus B_k \xrightarrow{\ \hat{f}_k \oplus \beta_k\ } \hat{B}_k \xrightarrow{\ h_k \gamma_k^{-1} \hat{g}_k\ } A_{k-1} \xrightarrow{\quad} \cdots$$

*is exact.*

Using this lemma, we can have the following result, known as the *Mayer–Vietoris theorem.*

**Theorem 6.27.** *If $A, B \subset X$ are two subsets whose interiors cover $X$ and $* \in A \cap B$, then there is an exact sequence*

$$\cdots \to H_k(A \cap B, *) \to H_k(A,*) \oplus H_k(B,*) \to H_k(A \cup B, *) \to H_{k-1}(A \cap B, *) \to \cdots$$

*(where $k \in \mathbb{N}_0$).*

*Proof.* By Proposition 6.21 (or Proposition 6.14), we have the following commutative diagram with exact rows:

$$\cdots \to H_k(A \cap B, *) \xrightarrow{i_{1*}} H_k(A, *) \xrightarrow{j_{1*}} H_k(A, A \cap B) \xrightarrow{\partial_1} H_{k-1}(A \cap B, *) \to \cdots$$
$$\downarrow \alpha_* \qquad\qquad \downarrow \beta_* \qquad\qquad \downarrow e_* \qquad\qquad \downarrow \alpha_*$$
$$\cdots \to H_k(B, *) \xrightarrow{i_{2*}} H_k(A \cup B, *) \xrightarrow{j_{2*}} H_k(A \cup B, B) \xrightarrow{\partial_2} H_{k-1}(B, *) \to \cdots,$$

where the maps $i_{1*}, i_{2*}, j_{1*}, j_{2*}, \alpha_*, \beta_*, e_*$ are those induced by the natural inclusion maps. Moreover, the excision property (Axiom 6) implies that $e_*$ is an isomorphism. Then the conclusion follows from Lemma 6.26. □

*Example 6.28.* We compute the homology groups for any homology theory of

$$\overline{B^n} = \{x \in \mathbb{R}^n : \|x\| \le 1\} \quad \text{and} \quad S^n = \{x \in \mathbb{R}^{n+1} : \|x\| = 1\},$$

where $\|\cdot\|$ is a given norm on $\mathbb{R}^n$, resp. on $\mathbb{R}^{n+1}$.

(a) From Proposition 6.24 we have $H_k(\overline{B^n}, *) = 0$ for all $k \in \mathbb{N}_0$.
(b) Since the homology groups $H_k(S^n, *)$ only depend on the homotopy type of $S^n$ (Proposition 6.11), we may assume that $\|\cdot\|$ is the Euclidean norm of $\mathbb{R}^{n+1}$. We compute the homology groups $H_k(S^n, *)$ by induction on $n \ge 0$. If $n = 0$, then we have

$$H_k(S^0, *) = H_k(*) \oplus H_k(*, *) = H_k(*) \quad \text{for all } k \in \mathbb{N}_0$$

(by Corollary 6.13 and Proposition 6.18). Now assume that $n \ge 1$. Let $x_N$ and $x_S$ be respectively the northern and southern poles of $S^n$. Let $S_1^n = S^n \setminus \{x_N\}$ and $S_2^n = S^n \setminus \{x_S\}$. The sets $S_1^n$ and $S_2^n$ form an open covering of $S^n$; hence, by virtue of Theorem 6.27, there is an exact sequence

$$\bigoplus_{i=1}^{2} H_k(S_i^n, *) \to H_k(S^n, *) \to H_{k-1}(S_1^n \cap S_2^n, *) \to \bigoplus_{i=1}^{2} H_{k-1}(S_i^n, *) .$$

Note that the spaces $S_1^n$ and $S_2^n$ are contractible, hence $H_k(S_1^n, *) = H_k(S_2^n, *) = 0$ for all $k \in \mathbb{N}_0$. Moreover, it is clear that the pair $(S_1^n \cap S_2^n, *)$ is homotopically equivalent to $(S^{n-1}, *)$, so $H_k(S_1^n \cap S_2^n, *) = H_k(S^{n-1}, *)$ for all $k \in \mathbb{N}_0$ (Proposition 6.11). This yields

$$H_k(S^n, *) = H_{k-1}(S^{n-1}, *) \quad \text{for all } k \in \mathbb{N}_0,$$

which easily implies (by induction)

$$H_k(S^n, *) = \begin{cases} H_0(*) & \text{if } k = n \\ 0 & \text{if } k \neq n. \end{cases}$$

Hence, for any homology theory, $H_n(S^n, *)$ is the only reduced homology group of $S^n$ that is nonzero, and it coincides with $H_0(*)$ (the group of coefficients of the homology theory).

(c) Combining (a) and (b), and then invoking Proposition 6.21, we obtain

$$H_k(\overline{B^n}, S^{n-1}) = H_{k-1}(S^{n-1}, *) = \begin{cases} H_0(*) & \text{if } k = n, \\ 0 & \text{if } k \neq n. \end{cases}$$

Now we present the construction of a homology theory to be used in the sequel, namely, the singular homology with coefficients in a ring. Hereafter, $\mathscr{R}$ denotes a commutative ring with unit. Let $\Delta^k$ be the *standard k-simplex*, defined by

$$\Delta^k = \left\{ (\lambda_0, \dots, \lambda_k) \in \mathbb{R}^{k+1} : \sum_{i=0}^{k} \lambda_i = 1, \ \lambda_i \geq 0 \right\}.$$

For $i \in \{0, \dots, k\}$ set $e_i = (0, \dots, 0, 1, 0, \dots, 0)$ [with 1 at the $(i+1)$th entry].

**Definition 6.29.** Let $X$ be a Hausdorff topological space. A *singular k-simplex* in $X$ is a continuous map $\sigma : \Delta^k \to X$.

*Remark 6.30.* Evidently, a singular 0-simplex is a map from the singleton $\Delta^0$ into $X$. Thus, we can identify it with a point of $X$. A singular 1-simplex is a continuous map $\sigma : \Delta^1 \simeq [0, 1] \to X$, hence it is a path in $X$. The word *singular* is used here to reflect the fact that $\sigma$ need not be a homeomorphism, and so its image may not look at all like a simplex.

**Definition 6.31.** The *singular chain group in dimension k* is the free $\mathscr{R}$-module $C_k(X; \mathscr{R})$ generated by the set of all singular $k$-simplices. An element of $C_k(X; \mathscr{R})$ is called a *singular k-chain* in $X$ and is a formal linear combination of singular $k$-simplices with coefficients in $\mathscr{R}$. Let $C_k(X; \mathscr{R}) = 0$ by convention if $k < 0$.

Let $C \subset \mathbb{R}^n$ be a convex set. For any $k+1$ points $u_0, \dots, u_k \in C$, let $a(u_0, \dots, u_k)$ : $\Delta^k \to \mathbb{R}^n$ denote the restriction of the unique affine map that takes $e_i$ to $u_i$ for $i = 0, \dots, k$. Evidently, the image of $a(u_0, \dots, u_k)$ lies in $C$, and so it is a singular $k$-simplex in $C$. It is called an *affine singular simplex*. A singular chain in which every singular simplex is affine is called an *affine singular chain*. For each $i \in \{0, \dots, k\}$ we let

$$\varphi_{i,k} = a(e_0, \dots, \hat{e}_i, \dots, e_k) : \Delta^{k-1} \to \Delta^k,$$

where the circumflex indicates that $e_i$ is omitted. We call $\varphi_{i,k}$ the *ith face map in dimension k*.

**Definition 6.32.** For every singular simplex $\sigma : \Delta^k \to X$, we define the *boundary of* $\sigma$ to be the singular $(k-1)$-chain $\partial\sigma$ given by

$$\partial\sigma = \sum_{i=0}^{k}(-1)^i\sigma\circ\varphi_{i,k}.$$

This extends uniquely to a homomorphism $\partial : C_k(X;\mathscr{R}) \to C_{k-1}(X;\mathscr{R})$, called the *boundary operator*.

*Remark 6.33.* Sometimes we write $\partial_k$ instead of $\partial$ to indicate the chain group on which the boundary operator is acting. The boundary of any 0-chain is defined as zero.

**Definition 6.34.**

(a) A singular $k$-chain $c$ is said to be a *$k$-cycle* if $\partial c = 0$.
(b) A singular $k$-chain $c$ is said to be a *$k$-boundary* if there exists a singular $(k+1)$-chain $b$ such that $\partial b = c$.
(c) By $Z_k(X;\mathscr{R})$ we denote the set of all $k$-cycles and by $B_k(X;\mathscr{R})$ the set of all $k$-boundaries. Both are abelian subgroups (in fact, submodules) of $C_k(X;\mathscr{R})$.

*Example 6.35.*

(a) Recall that a singular 1-simplex is a path $\sigma : [0,1] \to X$ and $\partial\sigma$ corresponds to the formal difference $\sigma(1) - \sigma(0)$. Therefore, a 1-cycle is a formal $\mathscr{R}$-linear combination of paths with the property that the set of initial points counted with multiplicities (in the ring $\mathscr{R}$) is exactly the same as the set of terminal points with multiplicities.
(b) The boundary of a singular 2-simplex $\sigma : \Delta^2 \to X$ is the sum of three paths with signs. For example, if $i_{\Delta^2} = a(e_0,e_1,e_2) : \Delta^2 \to \mathbb{R}^3$ is the inclusion map, then

$$\partial(i_{\Delta^2}) = \partial a(e_0,e_1,e_2) = a(e_1,e_2) - a(e_0,e_2) + a(e_0,e_1).$$

Hence $\partial(i_{\Delta^2})$ is associated with the sum of the 1-simplices in the boundary of $\Delta^2$ with appropriate signs.

The next proposition gives the most important feature of the boundary operator. Its proof is straightforward, but it involves tedious calculations and so it is omitted.

**Proposition 6.36.** $\partial\circ\partial = 0$ *(i.e., $\partial_k \circ \partial_{k+1} = 0$ for all $k \in \mathbb{N}_0$).*

This proposition implies that $B_k(X;\mathscr{R})$ is a subgroup of $Z_k(X;\mathscr{R})$. This leads to the following definition.

**Definition 6.37.** The *$k$th singular homology group of $X$ with coefficients in the ring $\mathscr{R}$* is defined as the quotient group

$$H_k(X;\mathscr{R}) = Z_k(X;\mathscr{R})/B_k(X;\mathscr{R}) = \ker\partial_k/\operatorname{im}\partial_{k+1}.$$

From the definition of singular homology groups we have (see Spanier [365, p. 173]) the following proposition.

**Proposition 6.38.** *If X is a singleton, then*

$$H_k(X;\mathscr{R}) = \begin{cases} \mathscr{R} & \text{if } k = 0 \\ 0 & \text{if } k \in \mathbb{N}. \end{cases}$$

Now we can introduce relative singular homology groups. These are based on the idea of ignoring the singular chains in a subspace $A$ of $X$.

Let $(X,A)$ be a topological pair, and for $k \in \mathbb{N}_0$ let

$$C_k(X,A;\mathscr{R}) = C_k(X;\mathscr{R})/C_k(A;\mathscr{R}).$$

Therefore, chains in $A$ are trivial in $C_k(X,A;\mathscr{R})$. Note that the boundary map $\partial : C_k(X;\mathscr{R}) \to C_{k-1}(X;\mathscr{R})$ takes $C_k(A;\mathscr{R})$ to $C_{k-1}(A;\mathscr{R})$, and so it induces a quotient boundary map $\partial : C_k(X,A;\mathscr{R}) \to C_{k-1}(X,A;\mathscr{R})$. Thus, we have a sequence of boundary maps

$$\cdots \to C_k(X,A;\mathscr{R}) \xrightarrow{\partial} C_{k-1}(X,A;\mathscr{R}) \to \ldots,$$

and the relation $\partial \circ \partial = 0$ still holds. Therefore, the next definition makes sense.

**Definition 6.39.** Let $(X,A)$ be a topological pair. Let $\partial_k$ denote the boundary map $C_k(X,A;\mathscr{R}) \to C_{k-1}(X,A;\mathscr{R})$ as previously. The *kth relative singular homology group* of the pair $(X,A)$ is defined by

$$H_k(X,A;\mathscr{R}) = \ker \partial_k / \operatorname{im} \partial_{k+1}.$$

**Theorem 6.40.** *The relative singular homology is a homology theory with coefficients in $\mathscr{R}$ on the topological pairs, in the sense of Definition 6.9 and Remark 6.10.*

This basic theorem (whose proof can be found in Spanier [365, Sect. 4]) implies that all the consequences of Axioms 1–7 in Definition 6.9 that we have presented in this section apply to singular homology groups.

*Remark 6.41.*

(a) Actually, in our construction, in addition to being abelian groups, the homology groups $H_k(X,A;\mathscr{R})$ carry a structure of $\mathscr{R}$-modules. The boundary homomorphism $\partial$, as well as every homomorphism $f_*$ induced by a map of pairs $f : (X,A) \to (Y,B)$, in addition to being a group homomorphism, is a morphism of $\mathscr{R}$-modules.

(b) The singular homology with integer coefficients, i.e., for $\mathscr{R} = \mathbb{Z}$ being the ring of integers, is the most standard singular homology. It is universal in the sense that if the ring $\mathscr{R}$ is without torsion as $\mathbb{Z}$-module, then there is a natural isomorphism

$$H_k(X,A;\mathscr{R}) = \mathscr{R} \otimes_{\mathbb{Z}} H_k(X,A;\mathbb{Z}) \quad \text{for all } k \in \mathbb{N}_0.$$

This holds, for instance, in the case where $\mathscr{R}$ is a field of characteristic zero.

(c) Singular homology groups with integers coefficients may have torsion, i.e., we may have rank $H_k(X,A;\mathbb{Z}) = 0$, although $H_k(X,A;\mathbb{Z}) \neq 0$.

For our purposes in this book, it is more convenient to avoid torsion phenomena. For this reason, in the following sections and chapters, we deal with *singular homologies with coefficients in a field*. We fix $\mathscr{R} = \mathbb{F}$ as a field of characteristic zero. We abbreviate

$$H_k(X) := H_k(X;\mathbb{F}) \quad \text{and} \quad H_k(X,A) := H_k(X,A;\mathbb{F}),$$

the singular homology groups and relative singular homology groups with coefficients in $\mathbb{F}$. In this manner, $H_k(X)$ and $H_k(X,A)$ are $\mathbb{F}$-vector spaces, and we denote by $\dim H_k(X)$ and $\dim H_k(X,A)$ their dimensions. The boundary homomorphism $\partial$ and the homomorphisms $f_*$ induced by maps of pairs are $\mathbb{F}$-linear.

*Example 6.42.*

(a) Specializing Example 6.28 to the case of singular homology with coefficients in $\mathbb{F}$, we obtain the following formulas for the reduced homology groups of the closed unit ball $\overline{B^n} \subset \mathbb{R}^n$ and the unit sphere $S^n \subset \mathbb{R}^{n+1}$:

$$H_k(\overline{B^n},*) = 0 \quad \text{and} \quad H_k(S^n,*) = H_k(\overline{B^n},S^{n-1}) = \delta_{k,n}\mathbb{F} \quad \text{for all } k \in \mathbb{N}_0,$$

where $\delta_{k,n}$ stands for the Kronecker symbol.

(b) From the definition of the singular homology groups it easily follows that, for a nonempty Hausdorff topological space $X$, we have that $\dim H_0(X)$ coincides with the number of path-connected components of $X$ (Remark 6.30 and Example 6.35). More generally, if $A \subset X$, then $\dim H_0(X,A)$ coincides with the number of path-connected components $C \subset X$ that do not intersect $A$. In particular, if each $x \in X$ can be connected to an element of $A$ by a path contained in $X$, then we have $H_0(X,A) = 0$.

## 6.2  Critical Groups

Now we are ready to introduce the notion of critical groups, which describe the local behavior of a $C^1$-function on a Banach space $X$ (or, more generally, on a Banach manifold $M$). Critical groups help to distinguish between different types of critical points and are extremely useful in producing multiple critical points for a functional. Recall that we denote by $H_\bullet$ the singular homology with coefficients in a fixed field $\mathbb{F}$ of characteristic zero.

**Definition 6.43.** Let $X$ be a Banach space, $\varphi \in C^1(X, \mathbb{R})$, and $x \in X$ an isolated critical point of $\varphi$. The *critical groups* of $\varphi$ at $x$ are defined by

$$C_k(\varphi, x) = H_k(\varphi^c \cap U, \varphi^c \cap U \setminus \{x\}) \quad \text{for all } k \in \mathbb{N}_0,$$

where $c = \varphi(x)$ and $U$ is a neighborhood of $x$ such that $K_\varphi \cap \varphi^c \cap U = \{x\}$ [recall that $\varphi^c = \{y \in X : \varphi(y) \le c\}$ and $K_\varphi = \{x \in X : \varphi'(x) = 0\}$]. Moreover, we set by convention $C_k(\varphi, x) = 0$ if $k \in \mathbb{Z}, k < 0$.

*Remark 6.44.*

(a) The excision property of the singular homology implies that the preceding definition of critical groups is independent of the particular choice of the neighborhood $U$.

(b) The critical groups $C_k(\varphi, x)$ are actually $\mathbb{F}$-vector spaces.

(c) The critical groups $C_k(\varphi, x)$ depend only on the behavior of $\varphi$ near $x$. In particular, they are also defined when $\varphi$ is defined only in a neighborhood of $x$.

*Example 6.45.*

(a) Let $X$ be a Banach space, $\varphi \in C^1(X, \mathbb{R})$, and $x \in X$ a local minimizer of $\varphi$ that is an isolated critical point. Then we can find a neighborhood $U$ of $x$ such that $K_\varphi \cap U = \{x\}$ and $c = \varphi(x) < \varphi(y)$ for all $y \in U \setminus \{x\}$. Hence

$$C_k(\varphi, x) = H_k(\{x\}, \emptyset) = H_k(\{x\}) = \delta_{k,0}\mathbb{F} \quad \text{for all } k \in \mathbb{N}_0,$$

with $\delta_{k,0}$ being the Kronecker $\delta$-symbol.

(b) Let $\varphi \in C^1(X, \mathbb{R})$ and $x$ be a local maximizer of $\varphi$ that is an isolated critical point. Then we can find $\rho > 0$ small such that $K_\varphi \cap B_\rho(x) = \{x\}$ and $c = \varphi(x) > \varphi(y)$ for all $y \in \overline{B_\rho(x)} \setminus \{x\}$. Thus,

$$C_k(\varphi, x) = H_k(\overline{B_\rho(x)}, \overline{B_\rho(x)} \setminus \{x\}) \quad \text{for all } k \in \mathbb{N}_0.$$

If $\dim X$ is infinite, then both $\overline{B_\rho(x)}$ and $\overline{B_\rho(x)} \setminus \{x\}$ are contractible [see Benyamini–Sternfeld [42] (see also Bessaga [43])], and we get $C_k(\varphi, x) = 0$ for all $k \in \mathbb{N}_0$. Next, assume that $m := \dim X < +\infty$. Then $\overline{B_\rho(x)} \setminus \{x\}$ is homotopy equivalent to the sphere $S^{m-1}$, so that $C_k(\varphi, x) = H_{k-1}(\overline{B^m}, S^{m-1}) = \delta_{k,m}\mathbb{F}$ for all $k \in \mathbb{N}_0$ [Example 6.42(a)]. In this way, in all cases, we have shown

$$C_k(\varphi, x) = \begin{cases} \mathbb{F} & \text{if } k = \dim X, \\ 0 & \text{otherwise,} \end{cases} \quad \text{for all } k \in \mathbb{N}_0.$$

(c) Critical groups for functions defined on $\mathbb{R}$ can be completely determined: let $\varphi \in C^1(\mathbb{R}, \mathbb{R})$ and let $x \in K_\varphi$ isolated. Then only three situations can occur:

(i) If $x$ is a local minimizer of $\varphi$, then $C_k(\varphi,x) = \delta_{k,0}\mathbb{F}$ for all $k \in \mathbb{N}_0$.
(ii) If $x$ is a local maximizer of $\varphi$, then $C_k(\varphi,x) = \delta_{k,1}\mathbb{F}$ for all $k \in \mathbb{N}_0$.
(iii) Otherwise, $C_k(\varphi,x) = 0$ for all $k \in \mathbb{N}_0$.

Indeed, claims (i) and (ii) are implied by (a) and (b) above. Claim (iii) can be checked as follows. Let $\delta > 0$ be such that $[x - \delta, x + \delta] \cap K_\varphi = \{x\}$. Since $\varphi$ has no local extremum at $x$, $\varphi$ is either increasing or decreasing on $[x - \delta, x + \delta]$. Say that it is increasing (the other case is solved in the same way). Hence $[x - \delta, x + \delta] \cap \varphi^{\varphi(x)} = [x - \delta, x]$. Thus, for all $k \in \mathbb{N}_0$ we have $C_k(\varphi,x) = H_k([x - \delta, x], [x - \delta, x)) = 0$ (Propositions 6.24 and 6.25).

A useful tool in the computation of critical groups is the so-called *Morse lemma* (Theorem 6.48 below). Here, we consider a Hilbert space $H$ with inner product $(\cdot,\cdot)_H$. Let $U \subset H$ be an open set and $\varphi \in C^2(U,\mathbb{R})$. For each $x \in U$, $\varphi''(x)$ can be seen as a symmetric bilinear form on $H$, and there is a unique $L_x \in \mathscr{L}(H)$ such that

$$(L_x(y),z)_H = \varphi''(x)(y,z) \quad \text{for all } y,z \in H. \tag{6.3}$$

In particular, $L_x$ is self-adjoint, so we have the orthogonal decomposition $H = \ker L_x \oplus \operatorname{im} L_x$. We can identify $L_x$ with $\varphi''(x)$.

**Definition 6.46.** Let $\varphi \in C^2(U,\mathbb{R})$ be as above and $x \in U$ be a critical point of $\varphi$.

(a) The *Morse index* of $x$ is defined as the supremum of the dimensions of the vector subspaces of $H$ on which $\varphi''(x)$ is negative definite.
(b) The *nullity* of $x$ is the dimension of $\ker \varphi''(x) = \ker L_x$.
(c) We say that $x \in K_\varphi$ is *nondegenerate* if $\varphi''(x)$ is nondegenerate (i.e., $L_x$ is invertible).

*Remark 6.47.*

(a) By the inverse function theorem, a nondegenerate critical point is always isolated.
(b) Note that if the nullity of $x$ is finite, then $L_x$ is a Fredholm operator of index 0 (i.e., $\dim \ker L_x = \operatorname{codim} \operatorname{im} L_x < +\infty$ in this case).

The Morse lemma describes the local behavior of $\varphi$ near a nondegenerate critical point.

**Theorem 6.48.** *If $H$ is a Hilbert space, $U \subset H$ is open, $\varphi \in C^2(U,\mathbb{R})$, and $x \in K_\varphi$ is nondegenerate, then there is a diffeomorphism $h$ from a neighborhood $V \subset H$ of $0$ into $H$ such that $h(0) = x$ and*

$$\varphi(h(y)) = \varphi(x) + \frac{1}{2}\varphi''(x)(y,y) \quad \text{for all } y \in V.$$

In fact, this result is a special case of a more general one (Theorem 6.49 below), known as the *generalized Morse lemma* or the *splitting theorem*, which permits the consideration of degenerate critical points. The proof can be found in Chang [78,

p. 44] and Mawhin and Willem [253, p. 184]. In the next statement, $L_x \in \mathcal{L}(H)$ is the self-adjoint operator given in (6.3).

**Theorem 6.49.** *Let $H$ be a Hilbert space, $U \subset H$ an open set, $\varphi \in C^2(U,\mathbb{R})$, and $x \in K_\varphi$ an isolated critical point of finite nullity. Then there exist a diffeomorphism $h$ from a neighborhood $V \subset H$ of $0$ into $H$ and a map $\hat{\varphi} \in C^2(W,\mathbb{R})$ defined on a neighborhood $W$ of $0$ in $\ker L_x$ such that $h(0) = x$, $\hat{\varphi}(0) = 0$, $\hat{\varphi}'(0) = 0$, $\hat{\varphi}''(0) = 0$, and*

$$\varphi(h(v)) = \varphi(x) + \frac{1}{2}\varphi''(x)(y,y) + \hat{\varphi}(z)$$

*for all $v \in V$, where $v = z + y$, with $z \in \ker L_x$ and $y \in (\ker L_x)^\perp = \operatorname{im} L_x$.*

*Remark 6.50.* Theorem 6.49 implies that if $z \in V \cap W$ is a critical point of $\hat{\varphi}$, then $h(z)$ is a critical point of $\varphi$. In particular, since $x = h(0)$ is an isolated critical point of $\varphi$, we obtain that $0$ is an isolated critical point of $\hat{\varphi}$, so that the critical groups $C_k(\hat{\varphi},0)$, for $k \in \mathbb{N}_0$, are well defined.

The critical groups of a nondegenerate critical point depend only on its Morse index.

**Theorem 6.51.** *If $H$ is a Hilbert space, $U \subset H$ is open, $\varphi \in C^2(U,\mathbb{R})$, and $x \in K_\varphi$ is nondegenerate with Morse index $m$ (possibly $+\infty$), then for every $k \in \mathbb{N}_0$ we have*

$$C_k(\varphi,x) = \begin{cases} \mathbb{F} & \text{if } k = m, \\ 0 & \text{otherwise.} \end{cases}$$

*Proof.* Without any loss of generality, we may assume that $x = 0$ and $c = \varphi(x) = \varphi(0) = 0$. Then, by virtue of Theorem 6.48, there exist $V \subset H$ a neighborhood of $0$ and a diffeomorphism $h$ from $V$ into $H$, with $h(0) = 0$, such that

$$\psi(y) := \varphi(h(y)) = \frac{1}{2}\varphi''(0)(y,y) \text{ for all } y \in V. \tag{6.4}$$

Let $B \subset V$ be a closed ball centered at $0$ such that $K_\varphi \cap h(B) = \{0\}$. Using Definition 6.43 and since $h$ is a homeomorphism, for all $k \in \mathbb{N}_0$ we have

$$C_k(\varphi,0) = H_k(\varphi^0 \cap h(B), \varphi^0 \cap h(B) \setminus \{0\}) = H_k(\psi^0 \cap B, \psi^0 \cap B \setminus \{0\}).$$

Because $0 \in K_\varphi$ is nondegenerate, we have the orthogonal (with respect to both $(\cdot,\cdot)_H$ and $\varphi''(0)(\cdot,\cdot)$) direct sum decomposition $H = H_- \oplus H_+$, with $\psi|_{H_-}$ (resp. $\psi|_{H_+}$) being negative (resp. positive) definite. So every $y \in H$ can be written in a unique way as $y = y_- + y_+$, with $y_- \in H_-$ and $y_+ \in H_+$. We consider the deformation $\eta : [0,1] \times B \to B$ of $B$ defined by

$$\eta(t,y) = y_- + (1-t)y_+.$$

By (6.4), we have

$$\psi(\eta(t,y)) = \psi(y_-) + (1-t)^2\psi(y_+) \le \psi(y) \quad \text{for all } t \in [0,1], \text{ all } y \in B,$$

hence $\eta$ restricts to a well-defined homotopy $[0,1] \times \psi^0 \cap B \to \psi^0 \cap B$ between $\mathrm{id}_{\psi^0 \cap B}$ and the retraction $\eta(1,\cdot): \psi^0 \cap B \to H_- \cap B$. Therefore, we obtain that the topological pairs $(\psi^0 \cap B, \psi^0 \cap B \setminus \{0\})$ and $(H_- \cap B, H_- \cap B \setminus \{0\})$ are homotopy equivalent. Hence, by virtue of Proposition 6.11, we have

$$H_k(\psi^0 \cap B, \psi^0 \cap B \setminus \{0\}) = H_k(H_- \cap B, H_- \cap B \setminus \{0\}) \quad \text{for all } k \in \mathbb{N}_0.$$

Recall that $\dim H_- = m$ (Definition 6.46). Arguing as in Example 6.45(b), we obtain

$$C_k(\varphi,0) = H_k(H_- \cap B, H_- \cap B \setminus \{0\}) = \begin{cases} \mathbb{F} & \text{if } k = m, \\ 0 & \text{otherwise.} \end{cases}$$

The proof is now complete.  $\square$

For the degenerate case, using Theorem 6.49, we have the so-called *shifting theorem*, which says that for a degenerate critical point the critical groups depend on the Morse index and on the "degenerate part" of the functional. Thus, the computation of the critical groups is reduced to a finite-dimensional problem. For a proof we refer the reader to Chang [78, p. 50] and Mawhin and Willem [253, p. 190].

**Theorem 6.52.** *Let $H$ be a Hilbert space, $U \subset H$ be open, $\varphi \in C^2(U,\mathbb{R})$, and $x \in K_\varphi$ be an isolated critical point of finite nullity and finite Morse index $m$. Then $C_k(\varphi,x) = C_{k-m}(\hat{\varphi},0)$ for all $k \in \mathbb{N}_0$, with $\hat{\varphi}$ as in Theorem 6.49.*

## 6.3 Morse Relations

We return to the situation of a Banach space $X$. The basic observation is that the singular homology groups make it possible to detect the presence of critical points.

**Proposition 6.53.** *Let $X$ be a Banach space, and let $\varphi \in C^1(X,\mathbb{R})$ satisfy the (C)-condition. Assume that there are $k \in \mathbb{N}_0$ and $a,b \in \mathbb{R}$, with $a < b$, such that $H_k(\varphi^b, \varphi^a) \ne 0$. Then $K_\varphi \cap \varphi^{-1}([a,b]) \ne \emptyset$.*

*Proof.* Arguing by contradiction, suppose that $K_\varphi \cap \varphi^{-1}([a,b]) = \emptyset$. Then Corollary 5.35 implies that $\varphi^a$ is a strong deformation retract of $\varphi^b$, which, in view of Proposition 6.12, implies that $H_k(\varphi^b, \varphi^a) = 0$, a contradiction.  $\square$

The *Morse relations* connect more precisely the dimensions of the homology group $H_k(\varphi^b, \varphi^a)$ and of the critical groups $C_k(\varphi,x)$, for $x \in K_\varphi \cap \varphi^{-1}([a,b])$. The next definition introduces algebraic quantities that play a role in this respect.

**Definition 6.54.** Let $X$ be a Banach space, $\varphi \in C^1(X, \mathbb{R})$, and $a, b \in \mathbb{R}$, $a < b$. Assume that $a, b$ are not critical values of $\varphi$ and that $\varphi^{-1}((a, b))$ contains a finite number of critical points $\{x_i\}_{i=1}^n \subset X$.

(a) The *Morse-type numbers* of $\varphi$ for $(a, b)$ are defined by

$$M_k(a, b) = \sum_{i=1}^n \dim C_k(\varphi, x_i), \quad k \in \mathbb{N}_0.$$

If $M_k(a, b)$ is finite for every $k \in \mathbb{N}_0$ and vanishes for all $k \in \mathbb{N}_0$ large, then $M(a, b)(t) := \sum_{k \geq 0} M_k(a, b) t^k$ is called the *Morse polynomial* of $\varphi$ for $(a, b)$.

(b) The *Betti-type numbers* of $\varphi$ for $(a, b)$ are defined by

$$\beta_k(a, b) = \dim H_k(\varphi^b, \varphi^a), \quad k \in \mathbb{N}_0.$$

If $\beta_k(a, b)$ is finite for every $k \in \mathbb{N}_0$ and vanishes for all $k \in \mathbb{N}_0$ large, then $P(a, b)(t) := \sum_{k \geq 0} \beta_k(a, b) t^k$ is called the *Poincaré polynomial* of $\varphi$ for $(a, b)$.

**Lemma 6.55.** *Let $X$ be a Banach space, $\varphi \in C^1(X, \mathbb{R})$, and $a, b \in \mathbb{R}$, with $a < b$. Assume that $\varphi$ satisfies the* $(C)_{c'}$*-condition at every level $c' \in [a, b)$, that the only critical value of $\varphi$ in $[a, b]$ is $c \notin \{a, b\}$, and that $K_\varphi^c = \{x_i\}_{i=1}^n$ is finite. Then*

$$H_k(\varphi^b, \varphi^a) = \bigoplus_{i=1}^n C_k(\varphi, x_i) \quad \text{for all } k \in \mathbb{N}_0.$$

*In particular, $M_k(a, b) = \beta_k(a, b)$ for all $k \in \mathbb{N}_0$.*

*Proof.* We have the inclusions $\varphi^a \subset \varphi^c \setminus K_\varphi^c \subset \varphi^c \subset \varphi^b$. By Theorem 5.34, $\varphi^a$ is a deformation retract of $\varphi^c \setminus K_\varphi^c$ and $\varphi^c$ is a deformation retract of $\varphi^b$. By Corollary 6.15, we obtain

$$H_k(\varphi^b, \varphi^a) = H_k(\varphi^c, \varphi^a) = H_k(\varphi^c, \varphi^c \setminus K_\varphi^c) \quad \text{for all } k \in \mathbb{N}_0. \tag{6.5}$$

Let $\{U_i\}_{i=1}^n$ be pairwise disjoint open neighborhoods of $\{x_i\}_{i=1}^n$ such that

$$C := \bigcup_{i=1}^n U_i \subset \varphi^{-1}([a, b]).$$

In this way, $U_i \cap K_\varphi = \{x_i\}$ for all $i$, so that $H_k(\varphi^c \cap U_i, \varphi^c \cap U_i \setminus \{x_i\}) = C_k(\varphi, x_i)$. From the properties of the singular homology (excision property and Proposition 6.18) we have

$$H_k(\varphi^c, \varphi^c \setminus K_\varphi^c) = H_k(\varphi^c \cap C, (\varphi^c \setminus K_\varphi^c) \cap C) = \bigoplus_{i=1}^n C_k(\varphi, x_i) \tag{6.6}$$

for all $k \in \mathbb{N}_0$. The lemma is obtained by combining (6.5) and (6.6). $\qquad\square$

We need the following property of singular homology groups.

**Lemma 6.56.** *Let $A_0 \subset A_1 \subset \ldots \subset A_m$ ($m \geq 2$) be Hausdorff topological spaces.*

(a) *For all $k \in \mathbb{N}_0$ we have*

$$\dim H_k(A_m, A_0) \leq \sum_{i=1}^{m} \dim H_k(A_i, A_{i-1}).$$

(b) *Assume that $H_k(A_i, A_{i-1})$ has finite dimension for all $k \in \mathbb{N}_0$, all $i \in \{1, \ldots, m\}$, and vanishes for all $k \in \mathbb{N}_0$ large. Then so does $H_k(A_m, A_0)$, and there is a polynomial $Q(t)$ with nonnegative integer coefficients such that*

$$\sum_{i=1}^{m} \left( \sum_{k \in \mathbb{N}_0} \dim H_k(A_i, A_{i-1}) t^k \right) = \sum_{k \in \mathbb{N}_0} \dim H_k(A_m, A_0) t^k + (1+t) Q(t).$$

*Proof.* We show the lemma for $m = 2$. The proof in the general case follows by easy induction. By Proposition 6.14, for every $k \in \mathbb{N}_0$ we have an exact sequence

$$H_{k+1}(A_2, A_1) \xrightarrow{\partial_k} H_k(A_1, A_0) \xrightarrow{i_*} H_k(A_2, A_0) \xrightarrow{j_*} H_k(A_2, A_1) \xrightarrow{\partial_{k-1}} H_{k-1}(A_1, A_0).$$

The rank formula and the exactness of the sequence yield

$$\dim H_k(A_2, A_0) = \dim \ker j_* + \dim \operatorname{im} j_* = \dim \operatorname{im} i_* + \dim \operatorname{im} j_*$$

$$\leq \dim H_k(A_1, A_0) + \dim H_k(A_2, A_1). \tag{6.7}$$

This proves (a). Now let $r_k = \dim \operatorname{im} \partial_k$. From (6.7) and the exactness of the above sequence we derive

$$\dim H_k(A_2, A_0) + r_k + r_{k-1} = (r_k + \dim \operatorname{im} i_*) + (r_{k-1} + \dim \operatorname{im} j_*)$$

$$= (\dim \ker i_* + \dim \operatorname{im} i_*) + (r_{k-1} + \dim \ker \partial_{k-1})$$

$$= \dim H_k(A_1, A_0) + \dim H_k(A_2, A_1). \tag{6.8}$$

Under the assumption in (b), it is clear that $\dim H_k(A_2, A_0)$ and $r_k$ are finite for all $k \in \mathbb{N}_0$ and vanish for all $k \in \mathbb{N}_0$ large. Thus, $Q(t) := \sum_{k \in \mathbb{N}_0} r_k t^k$ is a polynomial with nonnegative integer coefficients and, by (6.8), the formula in (b) is satisfied.  □

Now we are ready for the so-called *Morse relation*.

**Theorem 6.57.** *Let $X$ be a Banach space, $\varphi \in C^1(X, \mathbb{R})$, and $a, b \in \mathbb{R} \setminus \varphi(K_\varphi)$, $a < b$. Assume that $\varphi$ satisfies the $(C)_c$-condition for every $c \in [a, b)$ and that $\varphi^{-1}([a, b])$ contains a finite number of critical points of $\varphi$.*

(a) *For all $k \in \mathbb{N}_0$ we have $M_k(a,b) \geq \beta_k(a,b)$.*
(b) *If the Morse-type numbers $M_k(a,b)$ are finite and vanish for $k$ large, then so do
the Betti-type numbers $\beta_k(a,b)$, and we have*

$$\sum_{k \geq 0} M_k(a,b)t^k = \sum_{k \geq 0} \beta_k(a,b)t^k + (1+t)Q(t),$$

*where $Q(t)$ is a polynomial with nonnegative integer coefficients.*

*Proof.* Write $\varphi(K_\varphi) \cap [a,b] = \{c_i\}_{i=1}^m$. Let $\{d_i\}_{i=0}^m \subset [a,b] \setminus \varphi(K_\varphi)$ such that

$$a = d_0 < c_1 < d_1 < \cdots < d_{i-1} < c_i < d_i < \cdots < c_m < d_m = b.$$

By Definition 6.54 and Lemmas 6.55 and 6.56, we have

$$M_k(a,b) = \sum_{i=1}^m M_k(d_{i-1},d_i) = \sum_{i=1}^m \beta_k(d_{i-1},d_i) \geq \beta_k(a,b). \tag{6.9}$$

This proves (a). Now, assume that $M_k(a,b)$ is finite for all $k \in \mathbb{N}_0$ and vanishes for
all $k \in \mathbb{N}_0$ large. Then so do $\beta_k(d_{i-1},d_i)$ and $\beta_k(a,b)$, and by Lemma 6.56, we find
a polynomial $Q(t)$ with nonnegative integer coefficients such that

$$\sum_{k \in \mathbb{N}_0} \left( \sum_{i=1}^m \beta_k(d_{i-1},d_i) \right) t^k = \sum_{k \in \mathbb{N}_0} \beta_k(a,b)t^k + (1+t)Q(t).$$

Combining this with (6.9), the desired formula ensues.                                □

*Remark 6.58.* Choosing $t = -1$ in Theorem 6.57, we obtain

$$\sum_{k \geq 0} (-1)^k M_k(a,b) = \sum_{k \geq 0} (-1)^k \beta_k(a,b).$$

This equality is known as the *Poincaré–Hopf formula.*

The critical groups at infinity are useful tools for dealing with Morse relations.
They are defined for functionals whose set of critical values is bounded below.

**Definition 6.59.** Let $\varphi \in C^1(X,\mathbb{R})$ be a map such that $\inf \varphi(K_\varphi) > -\infty$ and
satisfying the (C)-condition. The *critical groups of $\varphi$ at infinity* are defined by

$$C_k(\varphi,\infty) = H_k(X,\varphi^a) \quad \text{for all } k \in \mathbb{N}_0$$

for any $a < \inf \varphi(K_\varphi)$.

*Remark 6.60.* From the second deformation theorem (Theorem 5.34) we know
that if $a' < a < \inf \varphi(K_\varphi)$, then $\varphi^{a'}$ is a strong deformation retract of $\varphi^a$. By
Corollary 6.15(a), this yields $H_k(X,\varphi^a) = H_k(X,\varphi^{a'})$ for all $k \in \mathbb{N}_0$. Therefore,
Definition 6.59 is independent of the choice of the level $a < \inf \varphi(K_\varphi)$.

The next proposition is an easy consequence of the second deformation theorem (Theorem 5.34) and of the properties of singular homology groups.

**Proposition 6.61.** *Let* $\varphi \in C^1(X, \mathbb{R})$ *satisfy the* (C)-*condition and* $\inf \varphi(K_\varphi) > -\infty$.

(a) *If* $a < \inf \varphi(K_\varphi) \leq \sup \varphi(K_\varphi) < b$, *then* $C_k(\varphi, \infty) = H_k(\varphi^b, \varphi^a)$ *for all* $k \in \mathbb{N}_0$.
(b) *If* $K_\varphi = \emptyset$, *then* $C_k(\varphi, \infty) = 0$ *for all* $k \geq 0$.
(c) *If* $K_\varphi = \{x_0\}$, *then* $C_k(\varphi, \infty) = C_k(\varphi, x_0)$ *for all* $k \geq 0$.

Combining Definition 6.54, Theorem 6.57, and Proposition 6.61(a), we have the following theorem.

**Theorem 6.62.** *Let* $\varphi \in C^1(X, \mathbb{R})$ *satisfy the* (C)-*condition and admit finitely many critical points.*

(a) *For all* $k \in \mathbb{N}_0$ *we have*

$$\sum_{x \in K_\varphi} \dim C_k(\varphi, x) \geq \dim C_k(\varphi, \infty).$$

(b) *Assume that* $C_k(\varphi, x)$ *has finite dimension for all* $k \in \mathbb{N}_0$, *all* $x \in K_\varphi$, *and vanishes for all* $k \in \mathbb{N}_0$ *large. Then there exists a polynomial* $Q(t)$ *with nonnegative integer coefficients such that*

$$\sum_{x \in K_\varphi} \left( \sum_{k \in \mathbb{N}_0} \dim C_k(\varphi, x) t^k \right) = \sum_{k \in \mathbb{N}_0} \dim C_k(\varphi, \infty) t^k + (1+t) Q(t).$$

Next, we present some computations of critical groups of $\varphi$ at infinity.

**Proposition 6.63.** *Let* $\varphi \in C^1(X, \mathbb{R})$ *satisfy the* (C)-*condition and* $\inf \varphi(K_\varphi) > -\infty$. *Assume that* $X = Y \oplus V$ *with* $\dim Y < +\infty$, $\varphi|_V$ *bounded below, and* $\varphi|_Y$ *anticoercive* [*i.e.,* $\varphi(y) \to -\infty$ *as* $\|y\| \to +\infty$, $y \in Y$ ]. *Then* $C_{\dim Y}(\varphi, \infty) \neq 0$.

*Proof.* Fix $a < \min\{\inf \varphi|_V, \inf \varphi(K_\varphi)\}$. The anticoercivity of $\varphi|_Y$ implies that, taking $r > 0$ large, we have $S_r^Y := \{y \in Y : \|y\| = r\} \subset \varphi^a$. We thus have inclusions $S_r^Y \subset \varphi^a \subset X \setminus V \subset X$. Considering the map $h : [0,1] \times X \setminus V \to X \setminus V$ given by

$$h(t, y+v) = (1-t)(y+v) + rt \frac{y}{\|y\|} \quad \text{for all } y+v \in X \setminus V \ (y \in Y, \ v \in V),$$

we see that $S_r^Y$ is a strong deformation retract of $X \setminus V$, hence $H_k(X \setminus V, S_r^Y) = 0$ for all $k \in \mathbb{N}_0$. We have the commutative diagram of homomorphisms (for $k \in \mathbb{N}_0$)

$$H_k(\varphi^a, S_r^Y) \xrightarrow{i_*} H_k(X, S_r^Y) \xrightarrow{j_*} H_k(X, \varphi^a) = C_k(\varphi, \infty)$$

$$\searrow \eta_* \qquad \qquad \uparrow v_*$$

$$H_k(X \setminus V, S_r^Y)$$

induced by the inclusion maps $i, j, \eta, \nu$ and whose first row is exact (Proposition 6.14). Since $i_*$ factors through $H_k(X \setminus V, S_r^Y) = 0$, we get $i_* = 0$. Thus, $j_*$ is injective for all $k \in \mathbb{N}_0$. Then, because $H_{\dim Y}(X, S_r^Y) = H_{\dim Y - 1}(S_r^Y, *) = \mathbb{F}$ [by Propositions 6.21 and 6.24, and Example 6.42(a)], we get $C_{\dim Y}(\varphi, \infty) \neq 0$. □

**Proposition 6.64.** *Let* $\varphi \in C^1(X, \mathbb{R})$ *satisfy the* (C)*-condition and* $\inf \varphi(K_\varphi) > -\infty$.

(a) *If* $\varphi$ *is bounded below, then* $C_k(\varphi, \infty) = \delta_{k,0} \mathbb{F}$ *for all* $k \in \mathbb{N}_0$.
(b) *If* $\varphi$ *is not bounded below, then* $C_k(\varphi, \infty) = H_{k-1}(\varphi^a, *)$ *for all* $k \in \mathbb{N}_0$, *all* $a < \inf \varphi(K_\varphi)$; *in particular,* $C_0(\varphi, \infty) = 0$.

*Proof.* (a) Let $a < \inf_X \varphi$. Then $C_k(\varphi, \infty) = H_k(X, \varphi^a) = H_k(X, \emptyset)$. As $X$ is contractible, from Propositions 6.19 and 6.24 we get $C_k(\varphi, \infty) = \delta_{k,0} \mathbb{F}$ for all $k \in \mathbb{N}_0$.

(b) Let $a < \inf \varphi(K_\varphi)$. By assumption, $\varphi^a \neq \emptyset$. Combining Propositions 6.21 and 6.24, we get $C_k(\varphi, \infty) = H_k(X, \varphi^a) = H_{k-1}(\varphi^a, *)$ for all $k \in \mathbb{N}_0$. □

Now we state three technical lemmas from which we will derive further properties of the critical groups, in light of the Morse relations. Lemma 6.65 will be used in Sect. 6.4, too.

**Lemma 6.65.** *Let* $X$ *be a reflexive Banach space,* $\varphi \in C^1(X, \mathbb{R})$ *satisfy the* (C)*-condition, and* $x_0 \in K_\varphi$ *be isolated with* $c := \varphi(x_0)$ *isolated in* $\varphi(K_\varphi)$. *Then we can find* $\breve{\varphi} \in C^1(X, \mathbb{R})$, *an open subset* $\breve{U} \subset X$, *with* $x_0 \in \breve{U}$, *and* $\delta > 0$ *such that*

(a) $\breve{\varphi}$ *satisfies the* (C)*-condition;*
(b) $\varphi \leq \breve{\varphi}$ *on* $X$ *and* $\varphi = \breve{\varphi}$ *on* $\breve{U}$;
(c) $K_{\breve{\varphi}} = K_\varphi$;
(d) $K_{\breve{\varphi}} \cap \breve{\varphi}^{-1}([c - \delta, c + \delta]) = \{x_0\}$;
(e) *If* $X = H$ *is a Hilbert space and* $\varphi \in C^k(H; \mathbb{R})$ *for* $k \geq 2$, *then* $\breve{\varphi}$ *can be taken in* $C^k(H, \mathbb{R})$.

*Proof.* Since $X$ is reflexive, using the Troyanski renorming theorem [Remark 2.47 (b)], we may assume that $X$ and $X^*$ are locally uniformly convex with Fréchet differentiable norms (except at the origins), so that $g : X \to [0, +\infty)$ given by $g(x) = \|x\|^2$ is of class $C^1$ in $X$ with $g'(x) = 2\mathscr{F}(x)$ for all $x$ (Proposition 3.12, Remark 2.45, and Theorem 2.48). Then it is easy to construct a map $\zeta \in C^1(X, \mathbb{R})$ such that

$$\zeta(x) = 0 \text{ if } \|x\| \leq \rho_1, \quad \zeta(x) = 1 \text{ if } \|x\| \geq \rho_2,$$

$$0 \leq \zeta \leq 1 \quad \text{and} \quad M := \sup_{x \in X} \|\zeta'(x)\| < +\infty$$

for given constants $\rho_2 > \rho_1 > 0$ chosen such that

$$\overline{B_{\rho_2}(x_0)} \cap K_\varphi = \{x_0\} \quad \text{and} \quad \varphi, \, \varphi' \text{ are bounded on } \overline{B_{\rho_2}(x_0)}.$$

Let $\check{U} = B_{\rho_1}(x_0)$. Since $\varphi$ satisfies the (C)-condition, there is $\mu > 0$ such that

$$\|\varphi'(x)\| \geq \mu \text{ for all } x \in X \text{ with } \rho_1 \leq \|x\| \leq \rho_2. \tag{6.10}$$

Since $c$ is isolated in $\varphi(K_\varphi)$, there are $c_0 \in (c - \frac{\mu}{2M}, c)$ and $\delta > 0$ such that

$$[c_0 - \delta, c_0 + \delta] \subset \mathbb{R} \setminus \varphi(K_\varphi).$$

We define $\check{\varphi} \in C^1(X, \mathbb{R})$ by letting

$$\check{\varphi}(x) = \varphi(x) + (c - c_0)\zeta(x) \text{ for all } x \in X.$$

Let us check that $\check{U}$, $\check{\varphi}$, and $\delta$ satisfy conditions (a)–(e) of the statement.

(a) Let $\{x_n\}_{n\geq 1} \subset X$ be a sequence such that $\{\check{\varphi}(x_n)\}_{n\geq 1}$ is bounded and $(1 + \|x_n\|)\|\check{\varphi}'(x_n)\| \to 0$ as $n \to \infty$. Note that

$$\|\check{\varphi}'(x)\| \geq \|\varphi'(x)\| - |c - c_0|\|\zeta'(x)\| > \mu - \frac{\mu}{2M}M = \frac{\mu}{2} \text{ for } \rho_1 \leq \|x\| \leq \rho_2; \tag{6.11}$$

hence there is $n_0 \geq 1$ such that $\|x_n\| \notin [\rho_1, \rho_2]$ for all $n \geq n_0$. Thus, $\check{\varphi}'(x_n) = \varphi'(x_n)$ for all $n \geq n_0$. Clearly, $\{\varphi(x_n)\}_{n\geq n_0}$ is bounded. Since $\varphi$ satisfies the (C)-condition, we infer that $\{x_n\}_{n\geq 1}$ has a strongly convergent subsequence. This proves (a).

(b) Follows from the fact that $c > c_0$, $\zeta \geq 0$ on $X$, and $\zeta = 0$ on $\check{U}$.

(c) Follows by noting that we have $\varphi'(x) = \check{\varphi}'(x)$ whenever $\|x\| \notin [\rho_1, \rho_2]$ and $\varphi'(x) \neq 0$, $\check{\varphi}'(x) \neq 0$ whenever $\|x\| \in [\rho_1, \rho_2]$ [see (6.10) and (6.11)].

(d) Let $x \in K_{\check{\varphi}} \setminus \{x_0\}$. Then (c) yields $x \in K_\varphi$, and so in particular $\|x\| > \rho_2$, whence $\check{\varphi}(x) = \varphi(x) + (c - c_0)$. From the choice of $c_0$ and $\delta$ we have $\varphi(x) \notin [c_0 - \delta, c_0 + \delta]$. This implies that $\check{\varphi}(x) \notin [c - \delta, c + \delta]$. Thus (d) is proven.

(e) If $X = H$ is a Hilbert space, then the map $g(x) = \|x\|^2$ is of class $C^\infty$, hence the map $\zeta$ used in the construction of $\check{\varphi}$ can be chosen in $C^\infty(H, \mathbb{R})$. Then, clearly, $\check{\varphi}$ belongs to $C^k(H, \mathbb{R})$ whenever $\varphi$ does. The proof is now complete.     □

*Remark 6.66.* Since $\varphi$ satisfies the (C)-condition, the set $(c - \gamma, c) \setminus \varphi(K_\varphi)$ is open for all $\gamma > 0$ (Proposition 5.25). Taking this observation into account, a careful reading of the foregoing proof indicates that Lemma 6.65 still holds if the assumption that $c$ is isolated in $\varphi(K_\varphi)$ is replaced by the weaker assumption that there is a sequence $\{c_n\}_{n\geq 1} \subset \mathbb{R} \setminus \varphi(K_\varphi)$, with $c_n < c$, such that $c_n \to c$ as $n \to \infty$.

**Lemma 6.67.** *Let $\varphi \in C^2(\mathbb{R}^N, \mathbb{R})$, and let $U \subset \mathbb{R}^N$ be open bounded and $K \subset U$ compact such that $K_\varphi \cap \overline{U} \setminus K = \emptyset$. Then, for every $\varepsilon > 0$, we can find $\check{\varphi} \in C^2(\mathbb{R}^N, \mathbb{R})$ such that*

(a) *$|\varphi(x) - \check{\varphi}(x)| + \|\varphi'(x) - \check{\varphi}'(x)\| \leq \varepsilon$ for all $x \in \mathbb{R}^N$;*

(b) *$\varphi(x) = \check{\varphi}(x)$ for all $x \notin U$;*

(c) *$\check{\varphi}$ has a finite number of critical points in $\overline{U}$, all nondegenerate.*

*Proof.* Let $\mu = \inf\{\|\varphi'(x)\| : x \in \overline{U \setminus K}\} > 0$. Fix $\xi \in C^\infty(\mathbb{R}^N, \mathbb{R})$ satisfying

$$\xi(x) = \begin{cases} 1 & \text{if } x \in K, \\ 0 & \text{if } x \notin U, \end{cases}$$

$R > 0$ with $U \subset B_R(0)$, and $\theta > 0$ such that

$$\theta R \|\xi\|_\infty \leq \frac{\varepsilon}{2} \quad \text{and} \quad \theta \|\xi\|_\infty + \theta R \|\xi'\|_\infty \leq \frac{1}{2}\min\{\mu, \varepsilon\}.$$

By Sard's theorem (Theorem 4.3), we can find $e \in \mathbb{R}^N$, with $|e| \leq \theta$, such that $-e$ is not a critical value of $\varphi'$ [i.e., $\varphi''(x)$ is nondegenerate whenever $\varphi'(x) = -e$]. Then we define $\tilde{\varphi} \in C^2(\mathbb{R}^N, \mathbb{R})$ by

$$\tilde{\varphi}(x) = \varphi(x) + \xi(x)(x, e)_{\mathbb{R}^N} \quad \text{for all } x \in \mathbb{R}^N.$$

Note that

$$\tilde{\varphi}'(x) = \varphi'(x) + \xi(x)e + \xi'(x)(x, e)_{\mathbb{R}^N} \quad \text{for all } x \in \mathbb{R}^N.$$

By the choices of $\xi$ and $|e| \leq \theta$, the map $\tilde{\varphi}$ satsifies conditions (a) and (b) of the statement, and

$$\inf\{\|\tilde{\varphi}'(x)\| : x \in \overline{U \setminus K}\} \geq \frac{\mu}{2}. \tag{6.12}$$

If $x \in \overline{U}$ is a critical point of $\tilde{\varphi}$, then (6.12) implies that $x \in \operatorname{int} K$. By the definition of $\tilde{\varphi}$, this yields $0 = \tilde{\varphi}'(x) = \varphi'(x) + e$. Since $-e$ is not a critical value of $\varphi'$, we infer that $\tilde{\varphi}''(x) = \varphi''(x)$ is nondegenerate. Thus, the critical points of $\tilde{\varphi}$ in $\overline{U}$ are nondegenerate, hence isolated [Remark 6.47(a)], and located in the compact set $K$, hence they are finitely many. This establishes (c). $\qquad \square$

**Lemma 6.68.** *Let* $\varphi \in C^2(\mathbb{R}^N, \mathbb{R})$, *with* $x_0 \in K_\varphi$ *isolated, and* $c = \varphi(x_0)$. *Then we can find* $\psi \in C^2(\mathbb{R}^N, \mathbb{R})$ *such that*

(a) $\psi = \varphi$ *in a neighborhood of* $x_0$;
(b) $K_\psi$ *is finite*;
(c) $K_\psi^c = \{x_0\}$;
(d) $\psi$ *is coercive [and so satisfies the* (PS)*-condition].*

*Proof.* Up to modifying $\varphi$ outside a ball centered in $x_0$, we may assume that $\varphi$ is coercive and that there is $R > 0$ such that $K_\varphi \cap (\mathbb{R}^N \setminus B_R(x_0)) = \emptyset$. Let $\rho \in (0, R)$ such that $K_\varphi \cap \overline{B_\rho(x_0)} = \{x_0\}$. Let $\tilde{\varphi} \in C^2(\mathbb{R}^N, \mathbb{R})$ be the function provided by Lemma 6.67 for the choices $U = B_{2R}(x_0) \setminus \overline{B_{\frac{\rho}{2}}(x_0)}$, $K = \overline{B_R(x_0)} \setminus B_\rho(x_0)$, and any $\varepsilon > 0$. Thus, $\tilde{\varphi}$ satisfies conditions (a), (b), and (d) of the lemma. Finally, apply

Lemma 6.65 to $\tilde{\varphi}$ and denote by $\psi = \check{\varphi} \in C^2(\mathbb{R}^N, \mathbb{R})$ the function so obtained. Then $\psi$ satisfies conditions (a)–(d) of the statement. □

A first consequence of these lemmas is the following vanishing property of the critical groups for functions defined on $\mathbb{R}^N$.

**Proposition 6.69.** *Let $\varphi \in C^2(\mathbb{R}^N, \mathbb{R})$, and let $x_0 \in K_\varphi$ isolated. Then $\dim C_k(\varphi, x_0)$ is finite for all $k \in \mathbb{N}_0$ and $C_k(\varphi, x_0) = 0$ whenever $k \notin \{0, 1, \ldots, N\}$.*

*Proof.* Recall that the critical groups $C_k(\varphi, x_0)$ only depend on $\varphi$ in a neighborhood of $x_0$. Hence, by Lemma 6.68, we may assume that $\varphi$ is coercive [so satisfying the (PS)-condition], $K_\varphi$ is finite, and $x_0$ is the only critical point for the value $\varphi(x_0)$. Hence there are $a, b \in \mathbb{R}$ with $a < \varphi(x_0) < b$ such that $K_\varphi \cap \varphi^{-1}([a, b]) = \{x_0\}$. By Lemma 6.55, this yields

$$C_k(\varphi, x_0) = H_k(\varphi^b, \varphi^a) \text{ for all } k \in \mathbb{N}_0.$$

Fix $\rho > 0$ such that $\overline{B_\rho(x_0)} \subset \{x \in \mathbb{R}^N : a < \varphi(x) < b\}$. Let $U = B_\rho(x_0)$ and $K = \overline{B_{\frac{\rho}{2}}(x_0)}$. Thus, $c := \inf_U \varphi > a$ and $d := \sup_U \varphi < b$. Let $\varepsilon > 0$ be such that $\varepsilon < \min\{c - a, b - d\}$. Finally, let $\tilde{\varphi} \in C^2(\mathbb{R}^N, \mathbb{R})$ be the map provided by Lemma 6.67 for our choices of $K$, $U$, and $\varepsilon$. From parts (a) and (b) in Lemma 6.67 and the choice of $\varepsilon$ we can see that $\tilde{\varphi}^b = \varphi^b$ and $\tilde{\varphi}^a = \varphi^a$. Then Theorem 6.57(a) yields

$$\dim C_k(\varphi, x_0) = \dim H_k(\tilde{\varphi}^b, \tilde{\varphi}^a) \leq \sum_{i=1}^n \dim C_k(\tilde{\varphi}, x_i) \text{ for all } k \in \mathbb{N}_0, \tag{6.13}$$

where $\{x_i\}_{i=1}^n = K_{\tilde{\varphi}} \cap U$. By Lemma 6.67(c), each $x_i$ is a nondegenerate critical point of $\tilde{\varphi}$, and hence from Theorem 6.51 we obtain that $\dim C_k(\tilde{\varphi}, x_i) \in \{0, 1\}$ for all $k \in \mathbb{N}_0$, all $i \in \{1, \ldots, n\}$, and $\dim C_k(\tilde{\varphi}, x_i) = 0$ whenever $k \notin \{0, 1, \ldots, N\}$. In view of (6.13), the proposition ensues. □

Combining this result with the shifting theorem (Theorem 6.52), we obtain the following corollary.

**Corollary 6.70.** *Let $H$ be a Hilbert space, $\varphi \in C^2(H, \mathbb{R})$, and $x_0 \in K_\varphi$ be an isolated critical point of finite nullity $\nu$ and finite Morse index $m$. Then $\dim C_k(\varphi, x_0)$ is finite for all $k \in \mathbb{N}_0$, and we have $C_k(\varphi, x_0) = 0$ whenever $k \notin \{m, m+1, \ldots, m+\nu\}$.*

The next results provide similar information for the critical groups at infinity. Their proofs can be found in Bartsch and Li [36]. Let $H$ be a Hilbert space and $\varphi \in C^1(H, \mathbb{R})$ satisfying the following condition:

$(A_\infty)$ $\quad \varphi(x) = \frac{1}{2}(A(x), x)_H + \psi(x)$, where $A \in \mathscr{L}(H)$ is self-adjoint, 0 is isolated in the spectrum of $A$, $\psi \in C^1(H, \mathbb{R})$, and $\lim_{\|x\| \to +\infty} \frac{\psi(x)}{\|x\|^2} = 0$. Moreover, both $\psi$ and $\psi'$ are bounded, and $\varphi$ is bounded below and satisfies the (C)-condition.

If $(A_\infty)$ holds, then we set $Y = \ker A$ and $Z = Y^\perp$. The space $Z$ admits the orthogonal direct sum decomposition $Z = Z_+ \oplus Z_-$, where both subspaces are $A$-invariant, $A|_{Z_+} > 0$ and $A|_{Z_-} < 0$. Then there exists $c_0 > 0$ such that

$$\pm \frac{1}{2}(A(x),x)_H \geq c_0\|x\|^2 \text{ for all } x \in Z_\pm .$$

We set $m = \dim Z_-$ (the *Morse index of* $\varphi$ *at infinity*) and $v = \dim Y$ (the *nullity of* $\varphi$ *at infinity*). The next result is an analog of Corollary 6.70 for the critical groups at infinity.

**Theorem 6.71.** *If $H$ is a Hilbert space and $\varphi \in C^1(H,\mathbb{R})$ satisfies the condition $(A_\infty)$, then $C_k(\varphi,\infty) = 0$ for all $k \notin \{m, m+1, \ldots, m+v\}$.*

*Remark 6.72.* The result does not require that $m$ and $v$ be finite. If $m < +\infty$ and $v = 0$, then $C_m(\varphi,\infty) = \mathbb{F}$.

By imposing in addition some "angle condition" on $\varphi$, we can say more.

**Theorem 6.73.** *Let $H$ be a Hilbert space and $\varphi \in C^1(H,\mathbb{R})$ satisfy hypothesis $(A_\infty)$. Assume that $m, v$ are finite. Then:*

(a) $C_k(\varphi,\infty) = \delta_{k,m}\mathbb{F}$ *for all $k \in \mathbb{N}_0$, provided the following angle condition at infinity holds:*

$(A_\infty^+)$   *there exist $M > 0$ and $\alpha \in (0,1)$ such that $\langle \varphi'(x),y \rangle \geq 0$ for $x = y+z$, $y \in Y$, $z \in Z$, $\|x\| \geq M$ and $\|z\| \leq \alpha\|x\|$.*

(b) $C_k(\varphi,\infty) = \delta_{k,m+v}\mathbb{F}$ *for all $k \in \mathbb{N}_0$, provided the following angle condition at infinity holds:*

$(A_\infty^-)$   *there exist $M > 0$ and $\alpha \in (0,1)$ such that $-\langle \varphi'(x),y \rangle \geq 0$ for $x = y+z$, $y \in Y$, $z \in Z$, $\|x\| \geq M$, and $\|z\| \leq \alpha\|x\|$.*

We conclude this section with a final result relating the topological degree and the critical groups. To do this, we introduce the notion of *Brouwer index*.

**Definition 6.74.** Let $f \in C(\mathbb{R}^N,\mathbb{R}^N)$, and let $x_0 \in X$ be an isolated solution of the equation $f(x) = 0$. Take $r > 0$ be such that $x_0$ is the only solution of the equation in $\overline{B_r(x_0)}$. The *Brouwer index of $f$ at $x_0$* is defined by

$$i(f,x_0) = d(f,B_r(x_0),0).$$

The excision property of the Brouwer degree [Theorem 4.5(d)] guarantees that the definition of $i(f,x_0)$ is independent of the choice of $r$.

Both $i(\varphi',x_0)$ and $C_k(\varphi,x_0)$ are topological invariants describing the local behavior at an isolated critical point $x_0 \in K_\varphi$. Thus, it is natural to expect some relation between them. Such a relation is provided by the following theorem.

**Theorem 6.75.** *Let $\varphi \in C^2(\mathbb{R}^N, \mathbb{R})$, and let $x_0 \in \mathbb{R}^N$ be an isolated critical point of $\varphi$. Then we have*

$$i(\varphi', x_0) = \sum_{k \geq 0} (-1)^k \dim C_k(\varphi, x_0).$$

*Proof.* By Lemma 6.68, we may assume that $\varphi$ is coercive and there are $a, b \in \mathbb{R}$, $a < \varphi(x_0) < b$, such that $K_\varphi^{[a,b]} = \{x_0\}$. Then, by Lemma 6.55, we have

$$C_k(\varphi, x_0) = H_k(\varphi^b, \varphi^a) \quad \text{for all } k \in \mathbb{N}_0.$$

Let $\rho > 0$ be small so that $i(\varphi', x_0) = d(\varphi', B_\rho(x_0), 0)$ and $\overline{B_\rho(x_0)} \subset \{x \in \mathbb{R}^N : a < \varphi(x) < b\}$. Let $U = B_\rho(x_0)$, $K = \overline{B_{\frac{\rho}{2}}(x_0)}$, and $\varepsilon > 0$ such that

$$\varepsilon < \min\{\inf_U \varphi - a, \ b - \sup_U \varphi, \ d(0, \varphi'(\partial U))\}.$$

Let $\tilde{\varphi} \in C^2(\mathbb{R}^N, \mathbb{R})$ be the map provided by Lemma 6.67 for our choices of $U$, $K$, and $\varepsilon$. By Lemma 6.67(a), we have $\|\varphi'(x) - \tilde{\varphi}'(x)\| < d(0, \varphi'(\partial U))$ for all $x \in \mathbb{R}^N$, hence

$$i(\varphi', x_0) = d(\varphi', U, 0) = d(\tilde{\varphi}', U, 0) \tag{6.14}$$

[Theorem 4.5(f)]. Moreover, from Lemma 6.67(a), (b) and the choice of $\varepsilon$ we get $\tilde{\varphi}^b = \varphi^b$ and $\tilde{\varphi}^a = \varphi^a$. This observation, together with Remark 6.58(b), yields

$$\sum_{k \geq 0} (-1)^k \dim C_k(\varphi, x_0) = \sum_{k \geq 0} (-1)^k \dim H_k(\tilde{\varphi}^b, \tilde{\varphi}^a)$$

$$= \sum_{i=1}^{n} \sum_{k \geq 0} (-1)^k \dim C_k(\tilde{\varphi}, x_i), \tag{6.15}$$

where $\{x_i\}_{i=1}^n := \overline{U} \cap K_{\tilde{\varphi}}$ [Lemma 6.67(c)].

By Lemma 6.67(c), each $x_i$ is a nondegenerate critical point of $\tilde{\varphi}$. By Theorem 6.51, we have $C_k(\tilde{\varphi}, x_i) = \delta_{k, m_i} \mathbb{F}$ for all $k \in \mathbb{N}_0$, where $m_i$ denotes the Morse index of $\tilde{\varphi}$ at $x_i$. Thus,

$$\sum_{i=1}^{n} \sum_{k \geq 0} (-1)^k \dim C_k(\tilde{\varphi}, x_i) = \sum_{i=1}^{n} (-1)^{m_i}. \tag{6.16}$$

For each $i \in \{1, \ldots, n\}$ choose $U_i$ as an open neighborhood of $x_i$ such that $\tilde{\varphi}''(x)$ is invertible for all $x \in U_i$ and $\{\overline{U_i}\}_{i=1}^n$ are pairwise disjoint. Note that $m_i$ is the sum of the multiplicities of the negative eigenvalues of $\tilde{\varphi}''(x_i)$, so $\operatorname{sgn} \det \tilde{\varphi}''(x_i) = (-1)^{m_i}$. Combining this observation with Definition 4.1 and Theorem 4.5(b), (d), we get

$$d(\tilde{\varphi}',U,0) = d\left(\tilde{\varphi}', \overset{n}{\underset{i=1}{\cup}} U_i, 0\right) = \sum_{i=1}^{n} d(\tilde{\varphi}',U_i,0) = \sum_{i=1}^{n} (-1)^{m_i}. \qquad (6.17)$$

The theorem is now obtained by combining (6.14)–(6.17). $\qquad\qquad\qquad\qquad\square$

*Remark 6.76.* This theorem actually shows that, for potential vector fields in $\mathbb{R}^N$, the critical groups carry more information than the index.

## 6.4 Computation of Critical Groups

In this section, we apply Morse theory to obtain existence and multiplicity results for critical points. In Definition 5.36, we introduced the notion of linking sets, which is crucial in the minimax theory of critical points. Here we introduce a similar notion, which is useful in obtaining pairs of sublevel sets with nontrivial homology groups and, in this way, in detecting critical points with nontrivial critical groups.

**Definition 6.77.** Let $X$ be a Banach space and $E_0, E$ and $D$ be nonempty subsets of $X$ such that $E_0 \subset E$ and $E_0 \cap D = \emptyset$. We say that the pair $\{E_0, E\}$ *homologically links* $D$ in dimension $m$ if the homomorphism $i_* : H_m(E,E_0) \to H_m(X,X \setminus D)$ induced by the inclusion is nontrivial.

*Remark 6.78.*

(a) In the literature, to distinguish between homological linking and the linking notion from Definition 5.36, the latter is often called *homotopical linking*.

(b) For every $m \in \mathbb{N}_0$ and $* \in E_0$ we have a commutative diagram of homomorphisms:

$$
\begin{array}{ccc}
H_m(E,E_0) & \xrightarrow{\partial_1} & H_{m-1}(E_0,*) \\
\downarrow{\scriptstyle i_*} & & \downarrow{\scriptstyle j_*} \\
H_m(X,X \setminus D) & \xrightarrow{\partial_2} & H_{m-1}(X \setminus D,*),
\end{array}
$$

where $j_*$ is induced by the natural inclusion map. When $E$ is contractible, it follows from the long exact sequence of Proposition 6.14 that $\partial_1, \partial_2$ are isomorphisms. In this case, we thus have that $\{E_0, E\}$ homologically links $D$ in dimension $m$ if and only if the homomorphism $j_*$ is nontrivial. In particular, the notion of homological linking is essentially independent of the choice of $E \supset E_0$ contractible.

*Example 6.79.*

(a) Let $E_0 = \{x_0, x_1\} \subset X$, $E = \{tx_0 + (1-t)x_1 : t \in [0,1]\}$, and $D = \partial\Omega$, with $\Omega$ an open neighborhood of $x_0$, such that $x_1 \notin \overline{\Omega}$. Let $j : (E_0, \{x_1\}) \to (X \setminus D, \{x_1\})$

be the inclusion map and $r : (X \setminus D, \{x_1\}) \to (E_0, \{x_1\})$ be given by

$$r(x) = \begin{cases} x_0 & \text{if } x \in \Omega, \\ x_1 & \text{if } x \in X \setminus \overline{\Omega}. \end{cases}$$

Then $r \circ j = \mathrm{id}_{(E_0,\{x_1\})}$. Hence $j_* : H_0(E_0, \{x_1\}) \to H_0(X \setminus D, \{x_1\})$ is injective. Since $H_0(E_0, \{x_1\}) = \mathbb{F}$ (see Example 6.42), we get that $j_*$ is nontrivial. By Remark 6.78(b), we get that $\{E_0, E\}$ homologically links $D$ in dimension 1.

(b) Let $X = Y \oplus V$, with $\dim Y = m < +\infty$, $E_0 = \{x \in Y : \|x\| = R\}$, $E = \{x \in Y : \|x\| \le R\}$, and $D = V$. As noted in the proof of Proposition 6.63, $E_0$ is a strong deformation retract of $X \setminus D$, so that $H_k(X \setminus D, E_0) = 0$ for all $k \in \mathbb{N}_0$. By Proposition 6.14, for $* \in E_0$ the injection $j : (E_0, *) \to (X \setminus D, *)$ induces an isomorphism $j_* : H_{m-1}(E_0, *) = \mathbb{F} \to H_{m-1}(X \setminus D, *)$ (in particular nontrivial). Thus, by Remark 6.78(b), the pair $\{E_0, E\}$ homologically links $D$ in dimension $m$.

(c) Similarly, if $E_0, E, D$ are as in Example 5.38(c), (d), then we can check that the sets $\{E_0, E\}$ homologically link $D$.

We present a first useful consequence of the notion of homological linking.

**Proposition 6.80.** *Let $X$ be a Banach space. Assume that the pair $\{E_0, E\}$ homologically links $D$ in dimension $m$. Let $\varphi \in C^1(X, \mathbb{R})$ and $a < b \le +\infty$ such that*

$$\varphi|_{E_0} \le a < \varphi|_D \quad \text{and} \quad \sup_{x \in E} \varphi(x) \le b.$$

(a) *We have $H_m(\varphi^b, \varphi^a) \ne 0$.*

(b) *Moreover, assume that $\varphi$ satisfies the (C)-condition, $a, b \notin \varphi(K_\varphi)$, and $K_\varphi \cap \varphi^{-1}((a,b))$ is finite. Then there exists $x \in K_\varphi \cap \varphi^{-1}((a,b))$ such that $C_m(\varphi, x) \ne 0$.*

*Proof.* (a) The assumptions yield the inclusion maps of topological pairs

$$(E, E_0) \xrightarrow{j} (\varphi^b, \varphi^a) \xrightarrow{\ell} (X, X \setminus D).$$

Since $i_* = \ell_* \circ j_* : H_m(E, E_0) \to H_m(X, X \setminus D)$ is nontrivial, we obtain that $j_* \ne 0$, $\ell_* \ne 0$, whence $H_m(\varphi^b, \varphi^a) \ne 0$.

(b) Follows from Theorem 6.57(a). $\qquad\square$

**Corollary 6.81.** *Let $X$ be a Banach space, and let $\varphi \in C^1(X, \mathbb{R})$ satisfy the (C)-condition and admit a finite number of critical points. Assume that $x_0, x_1 \in X$ and $r > 0$, with $r < \|x_0 - x_1\|$, are such that*

$$c := \max\{\varphi(x_0), \varphi(x_1)\} < \inf\{\varphi(x) : \|x - x_0\| = r\} =: d.$$

*Then there exists $x \in K_\varphi$ with $\varphi(x) \ge d$ and $C_1(\varphi, x) \ne 0$.*

*Proof.* We fix $a \in (c,d)$ such that $[a,d)$ contains no critical value. Let $E_0 = \{x_0, x_1\}$, $E = \{(1-t)x_0 + tx_1 : 0 \leq t \leq 1\}$, and $D = \partial B_r(x_0)$. By Example 6.79(a), $\{E_0, E\}$ and $D$ homologically link in dimension 1. Thus, by virtue of Proposition 6.80(b) applied with $b = +\infty$, we can find $x \in K_\varphi$ such that $\varphi(x) > a$ and $C_1(\varphi, x) \neq 0$. Since, by assumption, $[a,d)$ contains no critical value, we obtain $\varphi(x) \geq d$. $\qquad\square$

In Definition 5.48 we introduced the notion of *local linking at* 0 of a functional $\varphi$, which plays a central role in many variational problems. The local linking condition implies that the origin is a critical point of the functional $\varphi$, and so it is natural to ask what the critical groups of $\varphi$ are at the origin. An answer to this question will be given in Corollary 6.88. Actually, we deal with an extension of the notion of local linking, called *homological local linking*, introduced in the following definition.

**Definition 6.82.** Let $X$ be a Banach space, $\varphi \in C^1(X, \mathbb{R})$, and 0 be an isolated critical point of $\varphi$ with $\varphi(0) = 0$. Let $m, n \geq 1$ be integers. We say that $\varphi$ has a *local $(m,n)$-linking* near the origin if there are a neighborhood $U$ of 0 and nonempty subsets $E_0, E \subset U$ and $D \subset X$ such that $0 \notin E_0 \subset E$, $E_0 \cap D = \emptyset$, and

(a) 0 is the only critical point of $\varphi$ in $\varphi^0 \cap U$;
(b) $\dim \operatorname{im} i_{m-1} - \dim \operatorname{im} j_{m-1} \geq n$, where $i_{m-1} : H_{m-1}(E_0) \to H_{m-1}(X \setminus D)$ and $j_{m-1} : H_{m-1}(E_0) \to H_{m-1}(E)$ are the homomorphisms induced by the inclusion maps $i : E_0 \to X \setminus D$ and $j : E_0 \to E$;
(c) $\varphi|_E \leq 0 < \varphi|_{U \cap D \setminus \{0\}}$.

*Remark 6.83.* Definition 6.82 is slightly more general than Perera [326, Definition 1.1].

The next proposition provides a relation between the notions of local linking and homological local linking.

**Proposition 6.84.** *Let $X = Y \oplus V$, with $m = \dim Y < +\infty$ and $\varphi \in C^1(X, \mathbb{R})$, which has a local linking at 0 with respect to the pair $(Y,V)$, i.e., there is $r > 0$ such that*

$$\begin{cases} \varphi(x) \leq 0 & \text{if } x \in Y, \ \|x\| \leq r, \\ \varphi(x) \geq 0 & \text{if } x \in V, \ \|x\| \leq r. \end{cases}$$

*Moreover, assume that 0 is an isolated critical point of $\varphi$ and one of the next two conditions holds:*

(i) *0 is a strict local minimizer of $\varphi|_V$, or*
(ii) *$X = H$ is a Hilbert space and $\varphi'$ is Lipschitz near 0.*

*Then $\varphi$ has a local $(m, 1)$-linking at 0.*

The proof (given below) of this proposition relies on a deformation lemma.

**Lemma 6.85.** *Let $X, Y, V, \varphi, r$ be as in Proposition 6.84. Then there exist $\rho \in (0, r)$ and a homeomorphism $h : X \to X$ with $h(0) = 0$ such that*

(a) $h(\overline{B_\rho(0)}) \subset B_r(0)$;
(b) $h(x) = x$ for all $x \in \overline{B_\rho(0)} \cap Y$;
(c) $\varphi(x) > 0$ for all $x \in h(V \cap \overline{B_\rho(0)})$, $x \neq 0$.

*Proof (of Lemma 6.85).* If condition (i) of Proposition 6.84 holds, then we can take any $\rho \in (0, r)$ such that $\varphi(x) > 0$ for all $x \in V \cap \overline{B_\rho(0)} \setminus \{0\}$, and $h = \mathrm{id}_X$. Thus, we may assume that condition (ii) of Proposition 6.84 is satisfied. Let $B_1 = B_{\rho_1}(0), B_2 = B_{\rho_2}(0)$ (with $0 < \rho_1 < \rho_2 < r$) such that 0 is the only critical point of $\varphi$ in $B_1$, and $\varphi'$ is Lipschitz continuous in $B_2$. Take $\rho \in (0, \rho_1)$, and let $B_0 = B_\rho(0)$. The sets $\overline{B_0}$ and $H \setminus B_1$ are disjoint and closed. Hence the function $f : H \to [0,1]$ such that

$$f(x) = \frac{d(x, H \setminus B_1)}{d(x, \overline{B_0}) + d(x, H \setminus B_1)} \quad \text{for all } x \in H$$

is locally Lipschitz, and we have $f(x) = 1$ if $x \in \overline{B_0}$ and $f(x) = 0$ if $x \in H \setminus B_1$. Define $g : H \to H$ by

$$g(x) = f(x) \|p_V(x)\| \varphi'(x) \quad \text{for all } x \in H,$$

where $p_V$ is the linear continuous projection onto $V$ with respect to the topological decomposition $H = Y \oplus V$. Evidently, $g$ is locally Lipschitz and bounded. Thus, for all $t_0 \in [0, +\infty)$, all $x \in H$, the Cauchy problem

$$\frac{d\xi}{dt} = g(\xi) \text{ on } [0, +\infty), \quad \xi(t_0) = x, \tag{6.18}$$

has a unique global solution $\xi_{t_0, x} : [0, +\infty) \to H$. We define $h, k : H \to H$ by letting $h(x) = \xi_{0,x}(1)$ and $k(x) = \xi_{1,x}(0)$. By the continuous dependence of the solution of (6.18) with respect to the initial condition, $h, k$ are continuous, and clearly $h \circ k = k \circ h = \mathrm{id}_H$, hence $h$ is a homeomorphism. Since $g(0) = 0$, we have $h(0) = 0$. Let us check that $h$ satisfies conditions (a)–(c) of the statement.

(a) Note that if $x \in H \setminus B_1$, then we have $g(x) = 0$, and so $h(x) = x$; thus, $h(H \setminus B_1) = H \setminus B_1$. This implies that $h(\overline{B_0}) \subset h(B_1) \subset B_1 \subset B_r(0)$.
(b) Similarly, if $x \in Y$, then we have $g(x) = 0$, whence $h(x) = x$.
(c) Finally, let $x \in \overline{B_0} \cap V \setminus \{0\}$. We have

$$\varphi(h(x)) = \varphi(x) + \int_0^1 I(t)\, dt,$$

with $I(t) = f(\xi_{0,x}(t)) \|p_V \xi_{0,x}(t)\| \|\varphi'(\xi_{0,x}(t))\|^2$. Clearly, $I(t) \geq 0$ for all $t \in [0,1]$. Moreover, since $f(x) = 1$, $p_V(x) = x$, and $x \notin K_\varphi$, we have

$$I(0) = \|x\| \|\varphi'(x)\|^2 > 0,$$

whence $\varphi(h(x)) > \varphi(x) \geq 0$.                                                                  $\square$

*Proof (of Proposition 6.84).* Taking $r > 0$ smaller if necessary, we may assume that $B_r(0) \cap K_\varphi = \{0\}$. Let $\rho \in (0, r)$ and $h : X \to X$ provided by Lemma 6.85. Set

$$U = h(\overline{B_\rho(0)}), \quad E_0 = Y \cap \partial B_\rho(0), \quad E = Y \cap \overline{B_\rho(0)}, \quad \text{and} \quad D = h(V).$$

Conditions (a) and (c) of Definition 6.82 are clearly satisfied. It remains to check condition (b). Note that $E_0$ is a strong deformation retract of $X \setminus V$ (see the proof of Proposition 6.63), and hence $E_0 = h(E_0)$ is a strong deformation retract of $X \setminus D = h(X \setminus V)$. Thus, the homomorphism $i_{m-1} : H_{m-1}(E_0) \to H_{m-1}(X \setminus D)$ is bijective. By Example 6.42, we get

$$\dim \operatorname{im} i_{m-1} = \dim H_{m-1}(E_0) = \begin{cases} 1 & \text{if } m \geq 2, \\ 2 & \text{if } m = 1. \end{cases}$$

Since $E$ is contractible, combining Proposition 6.21 and Example 6.42, we see that $H_{m-1}(E, E_0) = H_{m-2}(E_0, *) = 0$. Hence, by Axiom 4 of Definition 6.9, the homomorphism $j_{m-1} : H_{m-1}(E_0) \to H_{m-1}(E)$ is surjective. Thus, Example 6.42 yields

$$\dim \operatorname{im} j_{m-1} = \dim H_{m-1}(E) = \begin{cases} 0 & \text{if } m \geq 2, \\ 1 & \text{if } m = 1. \end{cases}$$

All together, we obtain $\dim \operatorname{im} i_{m-1} - \dim \operatorname{im} j_{m-1} = 1$, whence we have proven condition (b) in Definition 6.82. $\qquad\square$

Here is a different example of homological local linking.

*Example 6.86.* The function $\varphi(x, y) = x^3 - 3xy^2$ has a local $(1, 2)$-linking near the origin. It does not have a local linking at 0 in the sense of Definition 5.48.

The next result estimates the critical groups at the origin for a functional that has a homological local linking.

**Theorem 6.87.** *If $X$ is a Banach space, $\varphi \in C^1(X, \mathbb{R})$, 0 is an isolated critical point of $\varphi$, and $\varphi$ has a local $(m, n)$-linking near the origin, then $\dim C_m(\varphi, 0) \geq n$.*

*Proof.* Let $U, E_0, E, D$ be as in Definition 6.82. Thus, $C_m(\varphi, 0) = H_m(\varphi^0 \cap U, \varphi^0 \cap U \setminus \{0\})$. We then have an exact sequence

$$C_m(\varphi, 0) \xrightarrow{\partial} H_{m-1}(\varphi^0 \cap U \setminus \{0\}) \xrightarrow{\ell_*} H_{m-1}(\varphi^0 \cap U), \qquad (6.19)$$

with $\ell_*$ induced by the inclusion map $\ell : \varphi^0 \cap U \setminus \{0\} \to \varphi^0 \cap U$ (Axiom 4 in Definition 6.9). From (6.19) and the rank formula we have

$$\dim \ker \ell_* = \dim \operatorname{im} \partial \leq \dim C_m(\varphi, 0). \qquad (6.20)$$

By Definition 6.82, we have the following commutative diagram induced by the corresponding inclusion maps:

$$
\begin{array}{ccc}
H_{m-1}(X \setminus D) & \xleftarrow{\hspace{1cm}} H_{m-1}(E_0) & \xrightarrow{\hspace{0.3cm} j_{m-1} \hspace{0.3cm}} H_{m-1}(E) \\
\end{array}
$$

$$
\begin{array}{ccc}
& \xleftarrow{i_{m-1}} & \quad \downarrow k_* \qquad\qquad \downarrow \eta_* \\
& v_* & H_{m-1}(\varphi^0 \cap U \setminus \{0\}) \xrightarrow{\ell_*} H_{m-1}(\varphi^0 \cap U).
\end{array}
$$

Note that $\dim \operatorname{im} i_{m-1} = \dim \operatorname{im} v_* \circ k_* \leq \dim \operatorname{im} k_*$. Similarly, using the rank formula, we see that $\dim \operatorname{im} j_{m-1} \geq \dim \operatorname{im} \ell_* \circ k_* = \dim \operatorname{im} k_* - \dim \ker \ell_*|_{\operatorname{im} k_*} \geq \dim \operatorname{im} k_* - \dim \ker \ell_*$. All together, we obtain

$$
\dim C_m(\varphi, 0) \geq \dim \ker \ell_* \geq \dim \operatorname{im} i_{m-1} - \dim \operatorname{im} j_{m-1} \geq n
$$

[see (6.20) and Definition 6.82(b)]. The proof is complete.                          $\square$

**Corollary 6.88.** *Let $X = Y \oplus V$, with $Y \neq 0$, $\dim Y = m < +\infty$, and let $\varphi \in C^1(X, \mathbb{R})$. Assume that $\varphi$ has a local linking at 0, that 0 is an isolated critical point of $\varphi$, and that either condition (i) or (ii) in Proposition 6.84 is satisfied. Then $C_m(\varphi, 0) \neq 0$.*

In many problems, it is clear from the hypotheses that 0 is a critical point of the considered functional $\varphi$ and we are interested in finding other critical points. Hereafter, we provide existence and multiplicity results that are helpful in this direction.

**Proposition 6.89.** *Let $X$ be a Banach space, and let $\varphi \in C^1(X, \mathbb{R})$ satisfy the (C)-condition and such that $0 \in K_\varphi$ and $K_\varphi$ is finite. Assume that, for some integer $k \geq 0$, we have*

$$
C_k(\varphi, 0) = 0 \quad and \quad C_k(\varphi, \infty) \neq 0.
$$

*Then there exists $x \in K_\varphi$, $x \neq 0$, such that $C_k(\varphi, x) \neq 0$.*

*Proof.* This easily follows from Theorem 6.62(a).                          $\square$

For the next existence results, we will need the following topological lemma.

**Lemma 6.90.** *If $X_1 \subset X_2 \subset X_3 \subset X_4$ are Hausdorff topological spaces, then*

$$
\dim H_k(X_3, X_2) \leq \dim H_{k-1}(X_2, X_1) + \dim H_{k+1}(X_4, X_3) + \dim H_k(X_4, X_1)
$$

*for all $k \in \mathbb{N}_0$. In particular, if we have $H_k(X_3, X_2) \neq 0$ and $H_k(X_4, X_1) = 0$, then $H_{k-1}(X_2, X_1) \neq 0$ or $H_{k+1}(X_4, X_3) \neq 0$.*

*Proof.* Applying Proposition 6.14 to the triples $(X_3, X_2, X_1)$ and $(X_4, X_3, X_1)$, we obtain exact sequences

$$H_k(X_3,X_1) \longrightarrow H_k(X_3,X_2) \longrightarrow H_{k-1}(X_2,X_1)$$

and $\quad H_{k+1}(X_4,X_3) \longrightarrow H_k(X_3,X_1) \longrightarrow H_k(X_4,X_1).$

Arguing on the basis of the rank formula, as in the proof of Lemma 6.56, we obtain

$$\dim H_k(X_3,X_2) \le \dim H_k(X_3,X_1) + \dim H_{k-1}(X_2,X_1) \qquad (6.21)$$

and $\quad \dim H_k(X_3,X_1) \le \dim H_{k+1}(X_4,X_3) + \dim H_k(X_4,X_1).$ $\qquad (6.22)$

Adding (6.21) and (6.22), we obtain the desired inequality. $\qquad\qquad\square$

Using this lemma, we prove the following existence result.

**Proposition 6.91.** *Let $X$ be a Banach space, and let $\varphi \in C^1(X,\mathbb{R})$ satisfy the (C)-condition, $x \in X$, and $a,b,c \in \mathbb{R}$, with $a < c < b$. Assume that $K_\varphi$ is finite, $K_\varphi^c = \{x\}$, $a,b \notin \varphi(K_\varphi)$, and we have $C_k(\varphi,x) \ne 0$ and $H_k(\varphi^b,\varphi^a) = 0$ for some $k \in \mathbb{N}_0$. Then there exists $y \in K_\varphi$ such that*

$$\textit{either} \quad a < \varphi(y) < c \ \textit{and} \ C_{k-1}(\varphi,y) \ne 0$$
$$\textit{or} \quad c < \varphi(y) < b \ \textit{and} \ C_{k+1}(\varphi,y) \ne 0.$$

*Proof.* Let $\varepsilon > 0$ small enough so that $K_\varphi \cap \varphi^{-1}([c-\varepsilon,c+\varepsilon]) = \{x\}$ and $a < c-\varepsilon < c+\varepsilon < b$. By assumption, we have

$$H_k(\varphi^{c+\varepsilon},\varphi^{c-\varepsilon}) = C_k(\varphi,x) \ne 0 \quad \text{and} \quad H_k(\varphi^b,\varphi^a) = 0$$

(Lemma 6.55). Applying Lemma 6.90 to the sets $\varphi^a \subset \varphi^{c-\varepsilon} \subset \varphi^{c+\varepsilon} \subset \varphi^b$, we obtain $H_{k-1}(\varphi^{c-\varepsilon},\varphi^a) \ne 0$ or $H_{k+1}(\varphi^b,\varphi^{c+\varepsilon}) \ne 0$. In both cases, using Theorem 6.57, we find $y \in K_\varphi$ with either $\varphi(y) \in (a,c-\varepsilon)$ and $C_{k-1}(\varphi,y) \ne 0$, or $\varphi(y) \in (c+\varepsilon,b)$ and $C_{k+1}(\varphi,y) \ne 0$. $\qquad\square$

**Corollary 6.92.** *Let $X$ be a Banach space, and let $\varphi \in C^1(X,\mathbb{R})$ satisfy the (C)-condition. Assume that $K_\varphi$ is finite, $K_\varphi^c = \{x\}$, and we have $C_k(\varphi,x) \ne 0$ and $C_k(\varphi,\infty) = 0$ for some $k \in \mathbb{N}_0$. Then there exists $y \in K_\varphi$ such that*

$$\textit{either} \quad \varphi(y) < \varphi(x) \ \textit{and} \ C_{k-1}(\varphi,y) \ne 0,$$
$$\textit{or} \quad \varphi(y) > \varphi(x) \ \textit{and} \ C_{k+1}(\varphi,y) \ne 0.$$

*Proof.* Take $a,b \in \mathbb{R}$ with $a < \inf\varphi(K_\varphi)$ and $b > \sup\varphi(K_\varphi)$. Then Proposition 6.61 yields $H_k(\varphi^b,\varphi^a) = C_k(\varphi,\infty) = 0$, so that we can apply Proposition 6.91. $\qquad\square$

Another abstract multiplicity result is the next one due to Liu and Su [234]. The proof that we give is somewhat different.

**Proposition 6.93.** *Let $X$ be a Banach space, and let $\varphi \in C^1(X, \mathbb{R})$ be bounded below and satisfy the (PS)-condition. Assume that there exists $x \in K_\varphi$ isolated, not a global minimizer of $\varphi$, satisfying $C_m(\varphi, x) \neq 0$ for some $m \in \mathbb{N}_0$. Then $\varphi$ has at least three critical points.*

*Proof.* By Corollary 5.21, there exists $x_0 \in K_\varphi$, a global minimizer of $\varphi$. By assumption, $x$ is not a global minimizer of $\varphi$; thus, $x \neq x_0$ and $\varphi(x_0) < \varphi(x)$. Arguing indirectly, assume that $K_\varphi = \{x_0, x\}$, and let us show that

$$C_k(\varphi, x) = 0 \quad \text{for all } k \in \mathbb{N}_0. \tag{6.23}$$

Fix $a, b \in \mathbb{R}$ with $\varphi(x_0) < a < \varphi(x) < b$. Then $\varphi^b$ is a strong deformation retract of $X$ and $\{x_0\}$ is a strong deformation retract of $\varphi^a$ (Theorem 5.34). Combining Corollary 6.15, the fact that $X$ is contractible, and Proposition 6.12, we obtain

$$H_k(\varphi^b, \{x_0\}) = H_k(X, \{x_0\}) = 0 \quad \text{and} \quad H_k(\varphi^a, \{x_0\}) = 0 \quad \text{for all } k \in \mathbb{N}_0.$$

The exact sequence of Proposition 6.21 yields $H_k(\varphi^b, \varphi^a) = 0$ for all $k \in \mathbb{N}_0$. Finally, (6.23) is derived by applying Lemma 6.55.                                                       □

Combining Theorem 6.87 and Proposition 6.93, we obtain the following multiplicity result, which yields an alternative proof of Theorem 5.51 in certain cases (Proposition 6.84).

**Corollary 6.94.** *Let $X$ be a Banach space, and let $\varphi \in C^1(X, \mathbb{R})$ be bounded below and satisfy the (PS)-condition. Assume that $\varphi$ has a local $(m, n)$-linking at 0, with $m, n \geq 1$, and 0 is not a global minimizer of $\varphi$. Then $\varphi$ has at least three critical points.*

Now we focus on two important special types of critical points of $\varphi$ whose critical groups can be completely determined: local minimizers and critical points of mountain pass type. We start with the local minimizers.

**Proposition 6.95.** *Let $X$ be a reflexive Banach space, $\varphi \in C^1(X, \mathbb{R})$ satisfy the (C)-condition, and $x_0 \in K_\varphi$ isolated with $c := \varphi(x_0)$ isolated in $\varphi(K_\varphi)$. Then the following statements are equivalent:*

   (i) *$x_0$ is a local minimizer of $\varphi$;*
  (ii) *$C_k(\varphi, x_0) = \delta_{k,0} \mathbb{F}$ for all $k \in \mathbb{N}_0$;*
 (iii) *$C_0(\varphi, x_0) \neq 0$.*

*Proof.* The implication (i)⇒(ii) is provided by Example 6.45(a), whereas (ii)⇒(iii) is immediate. It remains to show (iii)⇒(i). Thus, assume that $x_0$ is not a local minimizer of $\varphi$, and let us check that $C_0(\varphi, x_0) = 0$. By Lemma 6.65, we may assume that there are $a, b \in \mathbb{R}$ with $a < c < b$ such that $K_\varphi^{[a,b]} = \{x_0\}$. By Lemma 6.55 and Theorem 5.34, we have

$$C_0(\varphi, x_0) = H_0(\varphi^b, \varphi^a) = H_0(\varphi^b, \varphi^c \setminus \{x_0\}).$$

By Example 6.42(b), to show that $C_0(\varphi, x_0) = 0$, it is now sufficient to check that each $x \in \varphi^b$ can be connected to an element of $\varphi^c \setminus \{x_0\}$ by a path contained in $\varphi^b$. We reason in two steps.

- First, we show that each $x \in \varphi^b$ can be connected inside $\varphi^b$ to an element in $\varphi^c$. To do this, let $h : [0,1] \times \varphi^b \to \varphi^b$ be the deformation into $\varphi^c$ provided by Theorem 5.34. Then $h(\cdot, x)$ is a path in $\varphi^b$ connecting $x$ and $h(1,x) \in \varphi^c$.
- Second, we show that each $x \in \varphi^c$ can be connected inside $\varphi^b$ to an element in $\varphi^c \setminus \{x_0\}$. It suffices to take $x = x_0$. Let $r > 0$ small so that $\varphi(y) < b$ for all $y \in B_r(x_0)$. Since $x_0$ is not a local minimizer, there is $x_1 \in B_r(x_0)$ such that $\varphi(x_1) < c$. Then $\gamma(t) = (1-t)x_0 + tx_1$ is a path in $\varphi^b$ connecting $x_0$ and $x_1$.

The proof is now complete. ☐

Combining this proposition with Lemma 6.68, we obtain the following corollary.

**Corollary 6.96.** *Let* $\varphi \in C^2(\mathbb{R}^N, \mathbb{R})$, *and let* $x_0 \in K_\varphi$ *isolated. Then the following conditions are equivalent:*

(i) $x_0$ *is a local minimizer of* $\varphi$;
(ii) $C_k(\varphi, x_0) = \delta_{k,0} \mathbb{F}$ *for all* $k \in \mathbb{N}_0$;
(iii) $C_0(\varphi, x_0) \neq 0$.

In fact, a similar result can be obtained for local maximizers of functions defined on $\mathbb{R}^N$ (see Mawhin and Willem [253, p. 193]).

**Proposition 6.97.** *Let* $\varphi \in C^2(\mathbb{R}^N, \mathbb{R})$, *and let* $x_0 \in K_\varphi$ *isolated. Then the following conditions are equivalent:*

(i) $x_0$ *is a local maximizer of* $\varphi$;
(ii) $C_k(\varphi, x_0) = \delta_{k,N} \mathbb{F}$ *for all* $k \in \mathbb{N}_0$;
(iii) $C_N(\varphi, x_0) \neq 0$.

Next we pass to critical points of mountain pass type.

**Definition 6.98.** Let $X$ be a Banach space, $\varphi \in C^1(X, \mathbb{R})$, and $x \in K_\varphi$. We say that $x$ is of *mountain pass type* if, for any open neighborhood $U$ of $x$, the set $\{y \in U : \varphi(y) < \varphi(x)\}$ is nonempty and not path-connected.

The following result, due to Hofer [173], is a variant of the mountain pass theorem (Theorem 5.40) and establishes the existence of critical points of mountain pass type.

**Theorem 6.99.** *If* $X$ *is a Banach space,* $\varphi \in C^1(X, \mathbb{R})$ *satisfies the* (C)-*condition,* $x_0, x_1 \in X$, $\Gamma := \{\gamma \in C([0,1], X) : \gamma(0) = x_0, \gamma(1) = x_1\}$, $c := \inf_{\gamma \in \Gamma} \max_{t \in [0,1]} \varphi(\gamma(t))$, *and* $c > \max\{\varphi(x_0), \varphi(x_1)\}$, *then* $K_\varphi^c \neq \emptyset$, *and, moreover, if* $K_\varphi^c$ *is discrete, then we can find* $x \in K_\varphi^c$, *which is of mountain pass type.*

We now describe the critical groups for critical points of mountain pass type.

**Proposition 6.100.** *Let $X$ be a reflexive Banach space, $\varphi \in C^1(X, \mathbb{R})$, and $x_0 \in K_\varphi$ isolated with $c := \varphi(x_0)$ isolated in $\varphi(K_\varphi)$. If $x_0$ is of mountain pass type, then $C_1(\varphi, x_0) \neq 0$.*

*Proof.* Let $\check{\varphi} \in C^1(X, \mathbb{R})$ be the function provided by Lemma 6.65. In particular, $\varphi \leq \check{\varphi}$ on $X$, whereas $\varphi$ and $\check{\varphi}$ coincide on an open neighborhood $\check{U}$ of $x_0$. We claim that $x_0$ is a critical point of mountain pass type for $\check{\varphi}$. Indeed, letting $U \subset X$ be an open neighborhood of $x_0$ and setting $U_1 := \{x \in U : \check{\varphi}(x) < c\} \cup (\check{U} \cap U)$, we clearly have

$$\{x \in U : \check{\varphi}(x) < c\} = \{x \in U_1 : \varphi(x) < c\}.$$

Since $x_0$ is of mountain pass type for $\varphi$, it follows that $\{x \in U : \check{\varphi}(x) < c\}$ is nonempty and not path-connected, whence $x_0$ is of mountain pass type for $\check{\varphi}$. This proves our claim. Thus, up to dealing with $\check{\varphi}$ instead of $\varphi$, we may assume that there are $a, b \in \mathbb{R}$, with $a < c < b$, such that

$$K_\varphi \cap \varphi^{-1}([a, b]) = \{x_0\}.$$

Let $C$ be the connected component of $U := \{x \in X : a < \varphi(x) < b\}$ that contains $x_0$. In particular, $C$ is an open, path-connected neighborhood of $x_0$ and $C \cap K_\varphi = \{x_0\}$. Thus,

$$C_1(\varphi, x_0) = H_1(C \cap \varphi^c, C \cap \varphi^c \setminus \{x_0\}).$$

Let $h_1 : [0, 1] \times \varphi^b \to \varphi^b$ be the deformation into $\varphi^c$ provided by Theorem 5.34. We clearly have $h_1([0, 1] \times U) = U$, and so $h_1([0, 1] \times C) = C$ (because $h_1([0, 1] \times C)$ is connected and contains $C$). Thus, $h_1$ restricts to a deformation $[0, 1] \times C \to C$ into $C \cap \varphi^c$, which implies that $C \cap \varphi^c$ is a strong deformation retract of $C$. From this,

$$C_1(\varphi, x_0) = H_1(C, C \cap \varphi^c \setminus \{x_0\})$$

(Corollary 6.15). Then, by Axiom 4 of Definition 6.9 and Example 6.42(b), we get an exact sequence

$$C_1(\varphi, x_0) \longrightarrow H_0(C \cap \varphi^c \setminus \{x_0\}) \longrightarrow H_0(C) = \mathbb{F}. \tag{6.24}$$

Let $C^c = \{x \in C : \varphi(x) < c\}$. Let $d \in (a, c)$, and let $h_2 : [0, 1] \times \varphi^c \setminus \{x_0\} \to \varphi^c \setminus \{x_0\}$ be the deformation into $\varphi^d$ provided by Theorem 5.34. We claim that

$$h_2([0, 1] \times C \cap \varphi^c \setminus \{x_0\}) \subset C \cap \varphi^c \setminus \{x_0\} \quad \text{and} \quad h_2([0, 1] \times C^c) \subset C^c. \tag{6.25}$$

To see this, note that if $x \in C \cap \varphi^c \setminus \{x_0\}$, then clearly $h_2([0, 1] \times \{x\})$ lies in $U$, is connected, and intersects $C$ (because it contains $x$), so we have $h_2([0, 1] \times \{x\}) \subset C$ (by the definition of $C$); this proves the first part in (6.25). To check the second part, it is enough to note that if $x \in C^c$, then $\varphi(h_2(t,x)) \leq \varphi(x) < c$ for all $t \in [0,1]$ [by Theorem 5.34(b)]. This establishes (6.25). Thus, $h_2$ restricts to deformations $[0,1] \times (C \cap \varphi^c \setminus \{x_0\}) \to (C \cap \varphi^c \setminus \{x_0\})$ and $[0,1] \times C^c \to C^c$ into $C \cap \varphi^d$. Hence $C \cap \varphi^d$ is a strong deformation retract of both $C \cap \varphi^c \setminus \{x_0\}$ and $C^c$. This easily implies that $H_0(C \cap \varphi^c \setminus \{x_0\}) = H_0(C^c)$. By (6.24), we get an exact sequence

$$C_1(\varphi, x_0) \xrightarrow{\alpha} H_0(C^c) \xrightarrow{\beta} \mathbb{F}. \tag{6.26}$$

By the assumption that $x_0$ is of mountain pass type, $C^c$ is nonempty and not path-connected, whence $\dim H_0(C^c) > 1$ [Example 6.42(b)]. Hence the homomorphism $\beta$ of (6.26) cannot be injective. This implies that $C_1(\varphi, x_0) \neq 0$.  $\square$

By strengthening the assumption on $\varphi$, we can deduce a complete description of the critical groups at a critical point of mountain pass type. We need the next result.

**Proposition 6.101.** *Let $H$ be a Hilbert space, $\varphi \in C^2(H, \mathbb{R})$, and $x_0 \in K_\varphi$ an isolated critical point with finite nullity $v_0$ and finite Morse index $m_0$. Moreover, in the case where $m_0 = 0$, we assume that $v_0 \in \{0, 1\}$. Under these circumstances, if $C_1(\varphi, x_0) \neq 0$, then $C_k(\varphi, x_0) = \delta_{k,1}\mathbb{F}$ for all $k \in \mathbb{N}_0$.*

*Proof.* Let $\hat{\varphi} \in C^2(W, \mathbb{R})$, with $W \subset \ker \varphi''(x_0)$ a neighborhood of 0, be the map provided by Theorem 6.49, so that we have $C_k(\varphi, x_0) = C_{k-m_0}(\hat{\varphi}, 0)$ for all $k \in \mathbb{N}_0$ (Theorem 6.52). Because $C_1(\varphi, x_0) \neq 0$, this imposes $m_0 \in \{0, 1\}$. We distinguish two cases.

- If $m_0 = 1$, then we have $C_0(\hat{\varphi}, 0) \neq 0$, so that Corollary 6.96 yields $C_k(\varphi, x_0) = C_{k-1}(\hat{\varphi}, 0) = \delta_{k,1}\mathbb{F}$ for all $k \in \mathbb{N}_0$.
- If $m_0 = 0$, then we have $C_1(\hat{\varphi}, 0) \neq 0$. In this case, by Proposition 6.69 and the assumption, $\ker \varphi''(x_0)$ is one-dimensional. Invoking Example 6.45(c), we easily derive $C_k(\varphi, x_0) = C_k(\hat{\varphi}, 0) = \delta_{k,1}\mathbb{F}$ for all $k \in \mathbb{N}_0$.

The proof is now complete.  $\square$

Combining Propositions 6.100 and 6.101, we finally obtain the following corollary.

**Corollary 6.102.** *Let $H$ be a Hilbert space, $\varphi \in C^2(H, \mathbb{R})$ satisfy the (C)-condition, and $x_0 \in K_\varphi$ be isolated with $\varphi(x_0)$ isolated in $\varphi(K_\varphi)$, with finite nullity $v_0$ and finite Morse index $m_0$, and such that $v_0 \in \{0, 1\}$ whenever $m_0 = 0$. Under these assumptions, if $x_0$ is of mountain pass type, then $C_k(\varphi, x_0) = \delta_{k,1}\mathbb{F}$ for all $k \in \mathbb{N}_0$.*

## 6.5   Remarks

**Section 6.1:** The material from algebraic topology is standard and can be found in most books on the subject. In preparing this section, we consulted the books of Dold [119], Eilenberg and Steenrod [125], and Spanier [365].

**Section 6.2:** Critical groups (Definition 6.43) are a powerful tool that helps us distinguish between different kinds of critical points of a functional $\varphi \in C^1(X, \mathbb{R})$ and prove multiplicity theorems for various elliptic equations. Proofs of the Morse lemmas (Theorems 6.48 and 6.49) can be found in Mawhin and Willem [253, pp. 185–187]. This result was first proved by Morse [267] for functions defined on $\mathbb{R}^N$. Later, Morse [268] extended his theory to compact, smooth, finite-dimensional manifolds. In Morse [268], we encounter for the first time Theorem 6.51. The theory of Morse was extended to Hilbert spaces by Rothe [348] and to infinite dimensional Hilbert manifolds by Palais [313], Palais and Smale [316], and Smale [362]. As in the Leray–Schauder degree theory and in the deformation theory, the local compactness of the underlying space is replaced by the "compactness" of the functional. Concerning the Morse lemma, we should also mention the papers of Cambini [65] and Kuiper [208]. The shifting theorem (Theorem 6.52) is due to Gromoll and Meyer [166].

**Section 6.3:** The Morse relation (Theorem 6.57) can be found in Marino and Prodi [249]. The critical groups of $\varphi \in C^1(X, \mathbb{R})$ at infinity were introduced by Bartsch and Li [36], who also obtained Theorem 6.62. Theorem 6.75 is due to Rothe [348]. Theorem 6.73 is due to Bartsch and Li [36, Proposition 3.10], but the condition $(A_\infty)$ is slightly weaker. In Bartsch and Liu [37], it is assumed that $\varphi''(u) \to 0$ as $\|u\| \to +\infty$. However, a careful inspection of the proof in Bartsch and Liu [37] reveals that their asymptotic condition can be replaced by the weaker one $\|\psi'(u)\| = o(\|u\|)$ as $\|u\| \to +\infty$. This fact was first observed by Su and Zhao [368, Proposition 2.1]. Theorem 6.75 is due to Rothe [348].

**Section 6.4:** The notion of homological linking (Definition 6.77) goes back to the work of Liu [231], who also established the nontriviality of the critical group $C_1(\varphi, u_0)$, where $u_0 \in K_\varphi$ is of mountain pass type (Corollary 6.81). The notion of local $(m, n)$-linking near the origin (Definition 6.82) is due to Perera [326], who proved Theorem 6.87. Corollary 6.88 was earlier obtained by Liu [231] (see also Bartsch and Li [36, Proposition 2.3]). Definition 6.98 and Theorem 6.99 are both due to Hofer [173]. The structure of the critical set was investigated by Hofer [172, 173], Manes and Micheletti [240], and Pucci and Serrin [330, 331]. Results like Propositions 6.95 and 6.100, as well as Theorem 6.75 from Sect. 6.3, are usually proved through an argument based on the second deformation lemma (Theorem 5.34) and on the existence of neighborhoods that are stable by the pseudogradient vector flow in Theorem 5.34. Here, we have followed a different approach based on Lemma 6.65. For this reason, we formulate Propositions 6.95 and 6.100 under the mild assumption that the critical value of the critical point is isolated (see also Remark 6.66), but in counterpart we obtain these results for $C^1$-maps in reflexive Banach spaces, whereas they are usually stated for $C^2$-maps in Hilbert spaces.

# Chapter 7
# Bifurcation Theory

**Abstract** This chapter examines the bifurcation points of parametric equations, that is, values of a parameter from which the set of solutions splits into several branches. The deep connection between bifurcation points and the spectrum of linear operators involved in problems is pointed out. The presentation consists of two parts regarding the used approach: degree theory and implicit function theorem. In the latter, the theory of Fredholm operators is utilized in conjunction with the Lyapunov–Schmidt reduction method. Applications to ordinary differential equations are given. The proofs of the results presented in the chapter are complete, and novel ideas are incorporated. The basic references are mentioned in a remarks section.

## 7.1 Bifurcation Theory

Bifurcation theory deals with parametric equations of the form

$$\varphi(\lambda, x) = 0, \tag{7.1}$$

where $\lambda \in T$ is a parameter. The following phenomenon often occurs. There is a branch of solutions $u = u(\lambda)$ of (7.1) that at some critical parameter value $\lambda^* \in T$ may disappear or may split into several branches. This kind of phenomenon is called a bifurcation.

The mathematical setting of the bifurcation problem is as follows. Let $\varphi : T \times X \to Y$ be a map, where $X, Y$ are Banach spaces and $T$ is a Hausdorff topological space. We will always assume that

$$\varphi(\lambda, 0) = 0 \ \text{ for all } \lambda \in T,$$

i.e., 0 is a solution of (7.1) (the *trivial solution*). We introduce

$$S_\varphi = \{(\lambda, x) \in T \times X : \ \varphi(\lambda, x) = 0, \ x \neq 0\},$$

the set of *nontrivial solutions* of (7.1).

D. Motreanu et al., *Topological and Variational Methods with Applications to Nonlinear Boundary Value Problems*, DOI 10.1007/978-1-4614-9323-5_7,
© Springer Science+Business Media, LLC 2014

**Definition 7.1.** We say that $\lambda^* \in T$ is a *bifurcation point* of (7.1) if $(\lambda^*, 0) \in \overline{S_\varphi}$, i.e., every neighborhood $U$ of $(\lambda^*, 0)$ contains a point $(\lambda, x) \in S_\varphi$.

*Remark 7.2.* If $T$ is first countable and if $\lambda^* \in T$ is a bifurcation point of (7.1), then we can find a sequence $\{(\lambda_n, x_n)\}_{n \geq 1} \subset S_\varphi$ such that $(\lambda_n, x_n) \to (\lambda^*, 0)$ in $T \times X$.

The aim of bifurcation theory is to derive conditions for locating bifurcation points and to study the topological structure of the solution set $S_\varphi$. When $\varphi$ is sufficiently smooth, we can base the bifurcation theory on the calculus in Banach spaces, and the key tool is the *implicit function theorem*, which we recall next (for a proof, see, for example, Gasiński and Papageorgiou [151, p. 481]). Henceforth, by $\varphi'_x(\lambda_0, x_0)$ we denote the differential at $x_0$ of the map $\varphi(\lambda_0, \cdot)$ and by $\varphi'_\lambda(\lambda_0, x_0)$ the differential at $\lambda_0$ of the map $\varphi(\cdot, x_0)$.

**Theorem 7.3.** *If $T, X, Y$ are Banach spaces, $U \subset T \times X$ is an open set, $(\lambda_0, x_0) \in U$, $\varphi \in C^k(U, Y)$ with $k \geq 1$, $\varphi(\lambda_0, x_0) = 0$, and $\varphi'_x(\lambda_0, x_0) \in \mathcal{L}(X, Y)$ is an isomorphism, then there exist $T_0$, $X_0$ open neighborhoods of $\lambda_0$, $x_0$, respectively, with $T_0 \times X_0 \subset U$, and a map $\vartheta \in C^k(T_0, X_0)$ such that for all $(\lambda, x) \in T_0 \times X_0$*

$$\varphi(\lambda, x) = 0 \quad \text{if and only if} \quad x = \vartheta(\lambda).$$

*Moreover, $\vartheta'(\lambda) = -\varphi'_x(\lambda, \vartheta(\lambda))^{-1} \circ \varphi'_\lambda(\lambda, \vartheta(\lambda))$ for all $\lambda \in T_0$.*

The following statements are necessary conditions for $\lambda^* \in T$ to be a bifurcation point of (7.1), which we derive from Theorem 7.3. In what follows, we suppose that $T, X, Y$ are Banach spaces. Recall that we have fixed a map $\varphi : T \times X \to Y$ such that $\varphi(\lambda, 0) = 0$ for all $\lambda \in T$.

**Proposition 7.4.** *Suppose that $T, X, Y$ are Banach spaces and $\varphi \in C^1(T \times X, Y)$. If $\lambda^* \in T$ is a bifurcation point of (7.1), then $\varphi'_x(\lambda^*, 0) \in \mathcal{L}(X, Y)$ is not invertible.*

*Proof.* If $\varphi'_x(\lambda^*, 0)$ is invertible, then we can apply Theorem 7.3 and obtain a neighborhood $T_0 \times X_0$ of $(\lambda^*, 0)$ such that for all $\lambda \in T_0$ the equation $\varphi(\lambda, x) = 0$ has a unique solution $\vartheta(\lambda)$ in $X_0$. Since $\varphi(\lambda, 0) = 0$, this solution is necessarily the trivial one $\vartheta(\lambda) = 0$. This implies that $(T_0 \times X_0) \cap S_\varphi = \emptyset$. Then, by Definition 7.1, $\lambda^*$ is not a bifurcation point for (7.1). $\square$

Now we take $X = Y$ and $T \subset \mathbb{R}$, an open interval.

**Corollary 7.5.** *Let $\varphi(\lambda, x) = \lambda x - f(x)$, with $f \in C^1(X, X)$. If $\lambda^* \in T \subset \mathbb{R}$ is a bifurcation point of (7.1), then $\lambda^*$ belongs to the spectrum of $f'(0) \in \mathcal{L}(X)$.*

Moreover, from Definition 7.1 we have the following proposition.

**Proposition 7.6.** *Let $\varphi(\lambda, x) = \lambda x - L(x)$, with $L \in \mathcal{L}(X)$. Then $\lambda^* \in T$ is a bifurcation point of (7.1) if and only if $\lambda^*$ belongs to the closure of the set of eigenvalues of $L$.*

A closely related result is the following proposition.

**Proposition 7.7.** Let $\varphi(\lambda,x) = x - \lambda L(x) + f(\lambda,x)$, with $L \in \mathscr{L}(X)$ and $f \in C(\mathbb{R} \times X, X)$, such that $\frac{f(\lambda,x)}{\|x\|} \to 0$ as $x \to 0$ uniformly for all $\lambda \in T$. If $\lambda^* \in T$ is a bifurcation point of (7.1), then $\lambda^* \neq 0$ and $\frac{1}{\lambda^*}$ belongs to the spectrum of $L$.

*Proof.* If $\lambda^* = 0$ or $\frac{1}{\lambda^*}$ is not in the spectrum of $L$, then $\mathrm{id} - \lambda^* L$ is bijective, and there is $M > 0$ such that

$$\|x\| \leq M\|x - \lambda^* L(x)\| \leq M\|x - \lambda L(x)\| + M\|L(x)\|\,|\lambda - \lambda^*| \qquad (7.2)$$

for all $\lambda \in T$, all $x \in X$. Since $\lambda^*$ is a bifurcation point of (7.1), by Remark 7.2, we find sequences $\lambda_n \to \lambda^*$, $x_n \to 0$, $x_n \neq 0$, such that $\varphi(\lambda_n, x_n) = 0$ for all $n \geq 1$. By (7.2), this yields

$$1 \leq M \frac{\|f(\lambda_n, x_n)\|}{\|x_n\|} + M\|L\|\,|\lambda_n - \lambda^*| \quad \text{for all } n \geq 1.$$

Letting $n \to \infty$ we reach a contradiction. $\qquad\square$

The necessary conditions of bifurcation that we have pointed out here are not in general sufficient. The next example shows that the converse of Corollary 7.5 (as well as that of Proposition 7.7) is not valid.

*Example 7.8.* Let $X = \mathbb{R}^2$ and $\varphi(\lambda,x) = \lambda x - f(x)$, with

$$f(x) = \begin{pmatrix} -x_1 + x_2^3 \\ -x_2 - x_1^3 \end{pmatrix}, \text{ for all } \lambda \in \mathbb{R}, \text{ all } x = (x_1,x_2) \in \mathbb{R}^2.$$

It is straightforward to see that for all $\lambda \in \mathbb{R}$ the only solution of the equation $\varphi(\lambda,x) = 0$ is $x = 0$. Therefore, there is no bifurcation point for this equation. In particular, $\lambda^* := -1$ (the only eigenvalue of $f'(0) = -\mathrm{id}_{\mathbb{R}^2}$) is not a bifurcation point.

We are now looking for sufficient conditions for $\lambda^* \in T$ to be a bifurcation point of (7.1). Two approaches are possible for this purpose: a topological approach based on topological degree and an analytic approach based on the implicit function theorem and differential calculus in Banach spaces. The rest of the section is divided into two parts accordingly.

**Degree-Theoretic Approach**

We start with the following partial converse to Corollary 7.5 and Proposition 7.7 due to Krasnosel'skiĭ.

**Theorem 7.9.** Let $X$ be a Banach space, $T \subset \mathbb{R}$ an open interval, $\lambda^* \in T \setminus \{0\}$, $U \subset T \times X$ an open neighborhood of $(\lambda^*, 0)$, and $\varphi : T \times X \to X$ a map satisfying

$$\varphi(\lambda,x) = x - \lambda L(x) + f(\lambda,x) \ \text{ for all } (\lambda,x) \in U,$$

*where $L \in \mathscr{L}_c(X)$, and $f : \overline{U} \to X$ is a compact map such that $\lim_{x\to 0} \frac{f(\lambda,x)}{\|x\|} = 0$ for all $\lambda \in T$ in a neighborhood of $\lambda^*$. If $\frac{1}{\lambda^*}$ is an eigenvalue of $L$ of odd multiplicity, then $\lambda^*$ is a bifurcation point of (7.1).*

*Proof.* Arguing by contradiction, suppose that $\lambda^*$ is not a bifurcation point. Then, there are open bounded connected neighborhoods $T_0 \subset T$ of $\lambda^*$ and $X_0 \subset X$ of 0 such that $(\overline{T_0} \times \overline{X_0}) \cap S_\varphi = \emptyset$, that is,

$$\varphi(\lambda,x) \neq 0 \ \text{ for all } (\lambda,x) \in \overline{T_0} \times \overline{X_0}, x \neq 0. \tag{7.3}$$

We may assume that the neighborhoods $T_0, X_0$ are chosen so that

$$\overline{T_0} \times \overline{X_0} \subset U, \ \lim_{x\to 0} \frac{f(\lambda,x)}{\|x\|} = 0 \text{ for all } \lambda \in T_0, \tag{7.4}$$

$$0 \notin T_0, \text{ and } \tfrac{1}{\lambda} \text{ is not in the spectrum of } L \text{ for all } \lambda \in T_0 \setminus \{\lambda^*\}. \tag{7.5}$$

By (7.3), we have $0 \notin \varphi(\lambda, \partial X_0)$ for all $\lambda \in T_0$; hence the Leray–Schauder degree $d_{\mathrm{LS}}(\varphi(\lambda,\cdot), X_0, 0)$ is well defined. Since the map $(\lambda,x) \mapsto \lambda L(x) + f(\lambda,x)$ is compact on $\overline{T_0} \times \overline{X_0}$, it follows from Theorem 4.24(c) that

$$d_{\mathrm{LS}}(\varphi(\lambda,\cdot), X_0, 0) \text{ is independent of } \lambda \in T_0. \tag{7.6}$$

We fix elements $\lambda_1, \lambda_2 \in T_0$ such that $\lambda_1 < \lambda^* < \lambda_2$. By (7.6), we have

$$d_{\mathrm{LS}}(\varphi(\lambda_1,\cdot), X_0, 0) = d_{\mathrm{LS}}(\varphi(\lambda_2,\cdot), X_0, 0). \tag{7.7}$$

Fix $i \in \{1,2\}$ and let us compute independently $d_{\mathrm{LS}}(\varphi(\lambda_i,\cdot), X_0, 0)$. By (7.5), $\frac{1}{\lambda_i}$ is not in the spectrum of $L$, hence the endomorphism $\mathrm{id} - \lambda_i L$ is invertible. By (7.4), we can find $r_i > 0$ such that

$$\frac{\|f(\lambda_i,x)\|}{\|x\|} < \frac{1}{\|(\mathrm{id} - \lambda_i L)^{-1}\|} \text{ for all } x \in B_{r_i}(0) \subset X_0, x \neq 0. \tag{7.8}$$

On the one hand, by (7.3), the excision property of the Leray–Schauder degree gives

$$d_{\mathrm{LS}}(\varphi(\lambda_i,\cdot), X_0, 0) = d_{\mathrm{LS}}(\varphi(\lambda_i,\cdot), B_{r_i}(0), 0). \tag{7.9}$$

On the other hand, (7.8) implies that

$$\|x - \lambda_i L(x) + \varepsilon f(\lambda_i,x)\| \geq \frac{\|x\|}{\|(\mathrm{id} - \lambda_i L)^{-1}\|} - \|f(\lambda_i,x)\| > 0 \tag{7.10}$$

for all $x \in \partial B_{r_i}(0)$, all $\varepsilon \in [0,1]$. By assumption, the map $(\varepsilon,x) \mapsto \lambda_i L(x) - \varepsilon f(\lambda_i,x)$ is compact on $[0,1] \times \overline{B_{r_i}(0)}$. Combining this fact with (7.10) and invoking Theorem 4.24(c) and Proposition 4.33, we obtain

$$d_{LS}(\varphi(\lambda_i,\cdot),B_{r_i}(0),0) = d_{LS}(\mathrm{id} - \lambda_i L, B_{r_i}(0),0) = (-1)^{m_i},$$

where $m_i \in \mathbb{N}_0$ denotes the sum of the multiplicities of the eigenvalues of $L$ contained in $(\frac{1}{\lambda_i},+\infty)$. On the one hand, in view of (7.7) and (7.9), we should have

$$(-1)^{m_1} = (-1)^{m_2}, \quad \text{hence} \quad m_2 - m_1 \in 2\mathbb{Z}. \tag{7.11}$$

On the other hand, by (7.5), we have $m_2 = m_1 \pm m^*$, where $m^*$ denotes the multiplicity of $\frac{1}{\lambda^*}$ as an eigenvalue of $L$. By assumption, $m^*$ is odd, which contradicts (7.11). The proof of the theorem is then complete. $\square$

Theorem 7.9 is of a local nature in the sense that it says nothing about the global structure of the set $S_\varphi$. We next prove a global bifurcation result.

**Theorem 7.10.** *Let $X$ be a Banach space and $\varphi : \mathbb{R} \times X \to X$ a map of the form*

$$\varphi(\lambda,x) = x - \lambda L(x) + f(\lambda,x) \ \text{for all} \ (\lambda,x) \in \mathbb{R} \times X, \tag{7.12}$$

*where $L \in \mathscr{L}_c(X)$ and $f : \mathbb{R} \times X \to X$ is a compact map such that $\lim\limits_{x \to 0} \frac{f(\lambda,x)}{\|x\|} = 0$ uniformly for $\lambda$ on bounded subsets of $\mathbb{R}$. Then every connected component $C \subset \overline{S_\varphi}$ satisfies at least one of the following properties:*

(a) *$C$ is not compact;*
(b) *$C$ contains an even number of points of the form $(\lambda,0)$, with $\lambda \in \mathbb{R} \setminus \{0\}$, such that $\frac{1}{\lambda}$ is an eigenvalue of $L$ of odd multiplicity.*

In the proof, we need the following topological lemma.

**Lemma 7.11.** *If $(E,d)$ is a compact metric space, $C \subset E$ is a connected component, and $D \subset E$ is a closed set such that $C \cap D = \emptyset$, then there exist compact sets $K_1 \supset C$ and $K_2 \supset D$ such that $E = K_1 \cup K_2$ and $K_1 \cap K_2 = \emptyset$.*

*Proof.* Given $\varepsilon > 0$, we say that $x,y \in E$ are $\varepsilon$-chainable if there is a finite sequence $\{z_i\}_{i=1}^m \subset E$ (called a $\varepsilon$-chain) such that $z_1 = x$, $z_m = y$, and $d(z_i,z_{i+1}) < \varepsilon$ for all $i \in \{1,\dots,m-1\}$. We set

$$S_{\varepsilon(C)} = \{x \in E : \text{ there is } y \in C \text{ such that } x,y \text{ are } \varepsilon\text{-chainable}\}.$$

Then $C \subset S_{\varepsilon(C)}$, and clearly $S_{\varepsilon(C)}$ is both open and closed in $E$. We prove

$$\bigcap_{\varepsilon > 0} S_{\varepsilon(C)} \ \text{is connected}. \tag{7.13}$$

Arguing by contradiction, assume that there are $S_1, S_2 \subset E$ nonempty and closed (hence compact) subsets such that

$$\bigcap_{\varepsilon > 0} S_{\varepsilon(C)} = S_1 \cup S_2, \quad S_1 \cap S_2 = \emptyset.$$

Since $S_1, S_2$ are compact and disjoint, there is $\delta > 0$ such that $d(x, y) > 3\delta$ for all $x \in S_1$, all $y \in S_2$. Since $C$ is connected, it is contained in one of the subsets $S_1, S_2$, say, $C \subset S_1$. Fix $y \in S_2$. Let $\{\varepsilon_n\}_{n \geq 1}$ be a sequence such that $0 < \varepsilon_n < \frac{\delta}{n}$. For every $n \geq 1$ there is an $\varepsilon_n$-chain $\{z_i\}_{i=1}^{m_n} \subset E$ such that $z_1 \in C$ and $z_{m_n} = y$, and we can find an element $x_n$ belonging to this chain such that

$$d(x_n, S_1) > \delta, \quad d(x_n, S_2) > \delta, \quad x_n \in S_{\varepsilon_n}(C).$$

Since $E$ is compact, up to considering a subsequence, we may assume that $x_n \to x \in E$ as $n \to \infty$. Clearly, $x \in \bigcap_{\varepsilon > 0} S_{\varepsilon(C)}$. However, we have $d(x, S_1) \geq \delta$ and $d(x, S_2) \geq \delta$, thus $x \notin S_1 \cup S_2$, a contradiction. This proves (7.13).

Since $C$ is a connected component of $E$, (7.13) forces $C = \bigcap_{\varepsilon > 0} S_{\varepsilon(C)}$, whence $D \cap \left( \bigcap_{\varepsilon > 0} S_{\varepsilon(C)} \right) = \emptyset$. Because $D$ is compact and $\{S_{\varepsilon(C)}\}_{\varepsilon > 0}$ is a family of closed subsets totally ordered by inclusion, there is $\varepsilon > 0$ such that $D \cap S_{\varepsilon(C)} = \emptyset$. Thus,

$$K_1 := S_{\varepsilon(C)} \quad \text{and} \quad K_2 := E \setminus S_{\varepsilon(C)}$$

are open and closed (hence compact) subsets of $E$ that satisfy the requirements of the lemma.                                                                    □

*Proof (of Theorem 7.10).* Suppose that $C$ is compact, and let us show that situation (b) of the theorem occurs.

Since $L \in \mathscr{L}_c(X)$, the nonzero eigenvalues of $L$ form either a finite sequence or a sequence converging to 0 (Theorem 2.19). Hence, the set

$$\sigma(L)^{-1} := \{\lambda \in \mathbb{R} \setminus \{0\} : \tfrac{1}{\lambda} \text{ is an eigenvalue of } L\}$$

is discrete. Since $C$ is compact, we deduce that $\sigma(L)^{-1} \cap C$ is finite. We may assume that $\sigma(L)^{-1} \cap C$ is nonempty (otherwise, we are done), and we write

$$\sigma(L)^{-1} \cap C = \{\lambda_1, \lambda_2, \ldots, \lambda_p\}.$$

*Claim 1:* There is an open, bounded subset $U \subset \mathbb{R} \times X$ with $U \supset C$ such that

(i)  $\varphi(\lambda, x) \neq 0$ for all $(\lambda, x) \in \partial U$ with $x \neq 0$;
(ii) the only elements $\lambda \in \sigma(L)^{-1}$ such that $(\lambda, 0) \in \overline{U}$ are $\lambda_1, \lambda_2, \ldots, \lambda_p$.

Since $C$ is compact and $\sigma(L)^{-1}$ is discrete, we can find an open, bounded neighborhood $V \subset \mathbb{R} \times X$ of $C$ such that

$$\{\lambda \in \sigma(L)^{-1} : (\lambda,0) \in \overline{V}\} = \{\lambda_1,\ldots,\lambda_p\}. \tag{7.14}$$

The intersection $E := \overline{V} \cap \overline{S_\varphi}$ is compact. Applying Lemma 7.11 to the connected component $C \subset E$ and the closed subset $D := (\partial V) \cap \overline{S_\varphi}$, we find $K_1, K_2 \subset E$ compact such that

$$C \subset K_1, \quad D \subset K_2, \quad E = K_1 \cup K_2, \quad \text{and} \quad K_1 \cap K_2 = \emptyset.$$

In particular, $K_1 \subset V$. Since $K_1, K_2$ are compact and disjoint, we can find an open, bounded neighborhood $U \subset V$ of $K_1$ (hence of $C$) such that $\overline{U} \cap K_2 = \emptyset$. We have $(\partial U) \cap \overline{S_\varphi} = \emptyset$ [since $(\partial U) \cap K_1 = (\partial U) \cap K_2 = \emptyset$], hence $U$ satisfies property (i) of the claim. Moreover, by (7.14) and the fact that $\overline{U} \subset \overline{V}$, the set $U$ also satisfies property (ii) of the claim. The proof of Claim 1 is complete.

For $r \in (0,+\infty)$ we consider the map $\varphi_r : \mathbb{R} \times X \to \mathbb{R} \times X$ defined by

$$\varphi_r(\lambda,x) = (\|x\|^2 - r^2, \varphi(\lambda,x)) \quad \text{for all } (\lambda,x) \in \mathbb{R} \times X.$$

The assumptions on $L$ and $f$ imply that the map $(r,\lambda,x) \mapsto \varphi_r(\lambda,x) - (\lambda,x)$ is compact on $(0,+\infty) \times \overline{U}$. Moreover, part (i) of Claim 1 implies that for every $r \in (0,+\infty)$ the Leray–Schauder degree $d_{LS}(\varphi_r,\overline{U},0)$ is well defined. By Theorem 4.24(c), we have

$$d_{LS}(\varphi_r,\overline{U},0) \text{ is independent of } r \in (0,+\infty). \tag{7.15}$$

Our purpose in what follows is to compute the degree $d_{LS}(\varphi_r,\overline{U},0)$ in two different ways. From the comparison between the two results obtained, we will deduce that $C$ satisfies property (b) of the statement of the theorem.

*Claim 2:* For $r > 0$ large enough, we have $d_{LS}(\varphi_r,\overline{U},0) = 0$.

Since the set $U$ is bounded, we can find $R > 0$ such that $\|x\| < R$ for all $(\lambda,x) \in U$. Let $r \in [R,+\infty)$. We then have $\|x\|^2 - r^2 \neq 0$ for all $(\lambda,x) \in U$, hence the equation $\varphi_r(\lambda,x) = 0$ has no solution in $U$. By Theorem 4.24(e), it follows that $d_{LS}(\varphi_r,\overline{U},0) = 0$.

Before leading a second calculation of $d_{LS}(\varphi_r,\overline{U},0)$, we need a preparatory claim. For $\delta > 0$ and $i \in \{1,\ldots,p\}$ we write

$$V_i(\delta) = \{(\lambda,x) \in \mathbb{R} \times X : |\lambda - \lambda_i| + \|x\| < \delta\}.$$

Choose $\delta > 0$ small such that $V_i(2\delta) \subset U$ for all $i \in \{1,\ldots,p\}$.

*Claim 3:* There is $r \in (0,\delta)$ such that

$$\varphi_r^{(t)}(\lambda,x) := \left(\|x\|^2 - r^2, x - \lambda L(x) + t f(\lambda,x)\right) \neq (0,0)$$

for all $(\lambda,x) \in U \setminus \bigcup_{i=1}^{p} V_i(r+\delta)$, all $t \in [0,1]$.

Arguing by contradiction, assume that we can find sequences $\{t_n\}_{n\geq 1} \subset [0,1]$, $\{r_n\}_{n\geq 1} \subset (0,\delta)$ with $r_n \downarrow 0$, and $\{(\mu_n, x_n)\}_{n\geq 1} \subset U \setminus \overset{p}{\underset{i=1}{\cup}} V_i(r_n + \delta)$ such that $\varphi_{r_n}^{(t_n)}(\mu_n, x_n) = (0,0)$. Hence

$$\|x_n\| = r_n, \quad |\mu_n - \lambda_i| \geq \delta \quad \text{for all } i \in \{1,\dots,p\}, \tag{7.16}$$

and, letting $y_n = \frac{x_n}{\|x_n\|}$, we have

$$y_n - \mu_n L(y_n) = -t_n \frac{f(\mu_n, x_n)}{\|x_n\|} \quad \text{for all } n \geq 1. \tag{7.17}$$

Since $U$ is bounded, the sequence $\{\mu_n\}_{n\geq 1}$ is bounded and we may assume that $\mu_n \to \mu$ as $n \to \infty$, for some $\mu \in \mathbb{R}$. By (7.16), we have $\mu \notin \{\lambda_1,\dots,\lambda_p\}$. By (7.16), it also follows that $(\mu_n, x_n) \to (\mu, 0) \in \overline{U}$, hence

$$\mu \notin \sigma(L)^{-1} \tag{7.18}$$

[see Claim 1(ii)]. The assumption on $f$ implies that the right-hand side in (7.17) converges to 0. Since the operator $L$ is compact, we may assume that the sequence $\{L(y_n)\}_{n\geq 1}$ is convergent, and hence by (7.17), the sequence $\{y_n\}_{n\geq 1}$ is also convergent to some $y \in X$ such that $\|y\| = 1$, and we have

$$y - \mu L(y) = 0,$$

a contradiction of (7.18). This proves Claim 3.

We now fix $\delta > 0$ small enough so that the sets $\{V_i(2\delta)\}_{i=1}^{p}$ lie in $U$, do not contain $(0,0)$, and are pairwise disjoint. We fix $r \in (0,\delta)$ satsifying the conditions in Claim 3. Then, invoking Claim 3 (with $t = 1$) and Theorem 4.24(b), (d), we have

$$d_{\mathrm{LS}}(\varphi_r, U, 0) = \sum_{i=1}^{p} d_{\mathrm{LS}}(\varphi_r, V_i(\delta + r), 0). \tag{7.19}$$

Thus, to deduce a second calculation of $d_{\mathrm{LS}}(\varphi_r, U, 0)$, it suffices to compute $d_{\mathrm{LS}}(\varphi_r, V_i(\delta + r), 0)$ for every $i \in \{1,\dots,p\}$.

*Claim 4:* For $\delta, r$ as above and for every $i \in \{1,\dots,p\}$, we have

$$d_{\mathrm{LS}}(\varphi_r, V_i(\delta + r), 0) = \varepsilon_i(1 - (-1)^{m_i})$$

for some $\varepsilon_i \in \{-1,1\}$, where $m_i$ denotes the multiplicity of $\frac{1}{\lambda_i}$ as an eigenvalue of $L$.

Consider the homotopy $h_r : [0,1] \times \overline{U} \to \mathbb{R} \times X$ defined by

$$h_r(t,\lambda,x) = \left(t(\|x\|^2 - r^2) + (1-t)(\delta^2 - (\lambda - \lambda_i)^2), \, x - \lambda L(x) + t f(\lambda,x)\right).$$

Clearly, $(t, \lambda, x) \mapsto (\lambda, x) - h_r(t, \lambda, x)$ is compact on $[0, 1] \times \overline{U}$. We claim that

$$0 \notin h_r(t, \partial V_i(\delta + r)) \quad \text{for all } t \in [0, 1]. \tag{7.20}$$

Arguing by contradiction, suppose that there exist $t \in [0, 1]$ and $(\lambda, x) \in \partial V_i(\delta + r)$ such that $h_r(t, \lambda, x) = 0$. The fact that $(\lambda, x) \in \partial V_i(\delta + r)$ yields

$$\|x\| + |\lambda - \lambda_i| = r + \delta.$$

Then the relation $t(\|x\|^2 - r^2) + (1 - t)(\delta^2 - (\lambda - \lambda_i)^2) = 0$ forces $|\lambda - \lambda_i| = \delta$ and $\|x\| = r$. Using the notation of Claim 3, we have

$$(0, 0) = h_r(t, \lambda, x) = \varphi_r^{(t)}(\lambda, x).$$

Since $(\lambda, x) \in \partial V_i(\delta + r) \subset U \setminus \overset{p}{\underset{j=1}{\bigcup}} V_j(\delta + r)$, this contradicts the fact that $r$ satisfies Claim 3. We have established (7.20).

By virtue of (7.20), we can invoke Theorem 4.24(c), which yields

$$d_{\mathrm{LS}}(\varphi_r, V_i(\delta + r), 0) = d_{\mathrm{LS}}(h_r(0, \cdot, \cdot), V_i(\delta + r), 0). \tag{7.21}$$

We write $\psi_r(\lambda, x) := h_r(0, \lambda, x) = (\delta^2 - (\lambda - \lambda_i)^2, x - \lambda L(x))$ and we study the map $\psi_r : V_i(\delta + r) \to \mathbb{R} \times X$ thus obtained. Since $\lambda_i \pm \delta \notin \sigma(L)^{-1}$, the only zeros of $\psi_r$ in $V_i(\delta + r)$ are $(\lambda_i + \delta, 0)$ and $(\lambda_i - \delta, 0)$. Hence

$$d_{\mathrm{LS}}(\psi_r, V_i(\delta + r), 0) = d_{\mathrm{LS}}(\psi_r, W_+, 0) + d_{\mathrm{LS}}(\psi_r, W_-, 0) \tag{7.22}$$

whenever $W_+, W_- \subset V_i(\delta + r)$ are disjoint, open neighborhoods of $(\lambda_i + \delta, 0)$ and $(\lambda_i - \delta, 0)$, respectively [Theorem 4.24(b), (d)]. Note that $\psi_r$ is of class $C^1$ and

$$(\psi_r)'(\lambda_i \pm \delta, 0)(\lambda, x) = \left(-2(\pm\delta)\lambda, x - (\lambda_i \pm \delta)L(x)\right) = (\lambda, x) - \tilde{L}_\pm(\lambda, x),$$

with $\tilde{L}_\pm(\lambda, x) = ((1 \pm 2\delta)\lambda, (\lambda_i \pm \delta)L(x))$. In particular, $(\psi_r)'(\lambda_i \pm \delta, 0)$ is injective. By Proposition 4.34, we can find $W_+, W_- \subset V_i(\delta + r)$ disjoint, open neighborhoods of $(\lambda_i + \delta, 0)$ and $(\lambda_i - \delta, 0)$ such that

$$d_{\mathrm{LS}}(\psi_r, W_+, 0) = (-1)^{\tilde{m}_+} \quad \text{and} \quad d_{\mathrm{LS}}(\psi_r, W_-, 0) = (-1)^{\tilde{m}_-}, \tag{7.23}$$

where $\tilde{m}_\pm$ is the sum of the multiplicities of the eigenvalues of $\tilde{L}_\pm$ that are bigger than 1. In view of the form of $\tilde{L}_\pm$, we have $\tilde{m}_+ = 1 + m_+$ and $\tilde{m}_- = m_-$, where $m_\pm$ stands for the sum of the multiplicities of the eigenvalues of $L$ that are bigger than $\frac{1}{\lambda_i \pm \delta}$. Note that $m_+ = m_- \pm m_i$ (with $m_i$ as in the statement of the theorem). Combining this with (7.21)–(7.23), we obtain

$$d_{\mathrm{LS}}(\varphi_r, V_i(\delta + r), 0) = (-1)^{1 + m_- \pm m_i} + (-1)^{m_-} = (-1)^{m_-}\left(1 - (-1)^{m_i}\right).$$

This proves Claim 4.

Let $I \subset \{1,\dots,p\}$ be the subset of indices $i$ such that $\frac{1}{\lambda_i}$ has an odd multiplicity as an eigenvalue of $L$. Combining (7.15), Claim 2, (7.19), and Claim 4, we obtain

$$0 = \sum_{i \in I} 2\varepsilon_i,$$

with $\varepsilon_i \in \{-1,1\}$. This clearly imposes that the cardinal of $I$ is even, which means that the set $C$ satisfies condition (b) of the statement. This completes the proof of the theorem.                                                                                           $\square$

We conclude the topological approach with an example. We present an application of Theorems 7.9 and 7.10 to the study of the Dirichlet problem

$$\begin{cases} u''(t) = \lambda u(t) + f(t,\lambda,u(t),u'(t)) \text{ for all } t \in [0,1], \\ u(0) = u(1) = 0, \end{cases} \tag{7.24}$$

where $\lambda \in \mathbb{R}$ and $f : [0,1] \times \mathbb{R}^3 \to \mathbb{R}$ is a continuous function such that

$$\lim_{(s,\xi)\to(0,0)} \frac{f(t,\lambda,s,\xi)}{\sqrt{s^2+\xi^2}} = 0 \text{ uniformly for all } t \in [0,1], \text{ all } \lambda \in \mathbb{R}. \tag{7.25}$$

We deal with the Banach space

$$X = C_0^2([0,1]) := \{u \in C^2([0,1]) : u(0) = u(1) = 0\}$$

endowed with the norm $\|u\| = \|u\|_\infty + \|u'\|_\infty + \|u''\|_\infty$. Specifically, we study the structure of the set of nontrivial solutions for problem (7.24):

$$\mathscr{S} = \{(\lambda,u) \in \mathbb{R} \times C_0^2([0,1]) : (\lambda,u) \text{ is a solution of (7.24)}, u \neq 0\}.$$

Some preliminary constructions are needed.

- First, for every $h \in C([0,1])$ there is a unique $S(h) \in C_0^2([0,1])$ such that $S(h)'' = h$ characterized by

$$S(h)(t) = \int_0^t \int_0^s h(\tau)\,d\tau\,ds - t\int_0^1 \int_0^s h(\tau)\,d\tau\,ds \text{ for all } t \in [0,1].$$

The map $S : C([0,1]) \to C_0^2([0,1])$ thus obtained is clearly continuous and linear.
- Moreover, we recall that the Arzelà–Ascoli theorem implies that the embedding $i : C_0^2([0,1]) \hookrightarrow C([0,1])$ is compact (e.g., Brezis [52, p. 111]). Thus, the composition

$$L := S \circ i : C_0^2([0,1]) \to C_0^2([0,1])$$

is a compact linear operator. It is readily seen that the eigenvalues of $L$ are the numbers $\mu_k = -\frac{1}{(k\pi)^2}$ (for integers $k \geq 1$), each eigenvalue is simple, and an eigenfunction corresponding to $\mu_k$ is $\hat{u}_k(t) = \sin(k\pi t)$.

- Furthermore, we define a map $\tilde{f} : \mathbb{R} \times C_0^2([0,1]) \to C_0^2([0,1])$ by

$$\tilde{f}(\lambda, u) = S(f(\cdot, \lambda, u(\cdot), u'(\cdot))).$$

This map is obtained as the composition $\tilde{f} = S \circ N_f \circ j$, where $j : \mathbb{R} \times C_0^2([0,1]) \hookrightarrow \mathbb{R} \times C^1([0,1])$ is a compact embedding (by the Arzelà–Ascoli theorem) and $N_f : \mathbb{R} \times C^1([0,1]) \to C([0,1])$ given by $N_f(\lambda, u) = f(\cdot, \lambda, u(\cdot), u'(\cdot))$ is clearly continuous. All together, we deduce that $\tilde{f}$ is a compact operator. Moreover, because of (7.25), we have

$$\lim_{u \to 0} \frac{\tilde{f}(\lambda, u)}{\|u\|} = 0 \quad \text{uniformly for all } \lambda \in \mathbb{R}. \tag{7.26}$$

We are now able to interpret problem (7.24) in terms of a bifurcation equation of the same form as in the statements of Theorems 7.9 and 7.10: clearly, $u \in C_0^2([0,1])$ is a solution of (7.24) if and only if

$$\varphi(\lambda, u) := u - \lambda L(u) - \tilde{f}(\lambda, u) = 0. \tag{7.27}$$

Thus, the solution set becomes

$$\mathscr{S} = \{(\lambda, u) \in \mathbb{R} \times C_0^2([0,1]) : \varphi(\lambda, u) = 0, \ u \neq 0\}.$$

In this way, bifurcation theory related to (7.27) can be applied to the study of the solution set of the Dirichlet problem (7.24). We obtain the following result.

**Proposition 7.12.**

(a) *For every integer* $k \geq 1$, $-(k\pi)^2$ *is a bifurcation point for (7.27), so* $(-(k\pi)^2, 0) \in \overline{\mathscr{S}}$. *These are the only points of* $\overline{\mathscr{S}}$ *of the form* $(\lambda, 0)$.

   *By* $\mathscr{S}_k$ *we denote the connected component of* $\overline{\mathscr{S}}$ *containing* $(-(k\pi)^2, 0)$. *Then:*

(b) *If* $(\lambda, u) \in \mathscr{S}_k$ *with* $u \neq 0$, *then* $u$ *has exactly* $k+1$ *zeros in* $[0,1]$, *all simple. In particular,* $\mathscr{S}_k \cap \mathscr{S}_\ell = \emptyset$ *whenever* $k \neq \ell$.

(c) *For every* $k \geq 1$, *the component* $\mathscr{S}_k$ *is unbounded.*

*Proof.* Part (a) is an easy consequence of Proposition 7.7, Theorem 7.9, and the spectral properties of $L$ described earlier. Part (c) is a consequence of part (b) and of Theorem 7.10. Thus, it remains to establish part (b). Let

$$\mathscr{D}_k = \{u \in C_0^2([0,1]) : u \text{ has exactly } k+1 \text{ zeros, all simple}\}.$$

It is straightforward to check that $\mathscr{D}_k$ is open in $C_0^2([0,1])$. Thus, $\mathscr{S}_k \cap (\mathbb{R} \times \mathscr{D}_k)$ is open in $\mathscr{S}_k$. Note that part (b) will be obtained once we check

$$\mathscr{G}_k := \mathscr{S}_k \cap ((\mathbb{R} \times \mathscr{D}_k) \cup \{(-(k\pi)^2, 0)\}) \text{ is open and closed in } \mathscr{S}_k. \tag{7.28}$$

Indeed, the connectedness of $\mathscr{S}_k$ then implies that $\mathscr{S}_k \setminus \{(-(k\pi)^2, 0)\} \subset \mathbb{R} \times \mathscr{D}_k$, which completes the proof of (b). Thus, it remains to check (7.28).

For the first part of (7.28), since $\mathscr{D}_k$ is open, it suffices to check that $\mathscr{G}_k$ contains a neighborhood of $(-(k\pi)^2, 0)$ in $\mathscr{S}_k$. Arguing by contradiction, assume that we can find a sequence $\{(\lambda_n, u_n)\}_{n \geq 1} \subset \mathscr{S}_k \setminus \mathscr{G}_k$ such that $\lambda_n \to -(k\pi)^2$ and $u_n \to 0$ in $C_0^2([0,1])$ as $n \to \infty$. The sequence $v_n = \frac{u_n}{\|u_n\|}$ $(n \geq 1)$ is bounded in $C_0^2([0,1])$. Using the compactness of the embedding $i: C_0^2([0,1]) \hookrightarrow C([0,1])$ and the compactness of $L = S \circ i$, up to considering a subsequence, we may assume that

$$v_n \to v \text{ in } C([0,1]) \text{ and } L(v_n) \to S(v) \text{ in } C^2([0,1]) \text{ as } n \to \infty. \tag{7.29}$$

For all $n \geq 1$ we have $\varphi(\lambda_n, u_n) = 0$; thus,

$$v_n = \lambda_n L(v_n) + \frac{\tilde{f}(\lambda_n, u_n)}{\|u_n\|}. \tag{7.30}$$

On the one hand, in view of (7.26), (7.29), and (7.30), we get

$$v_n \to -(k\pi)^2 S(v) \text{ in } C^2([0,1]) \text{ as } n \to \infty,$$

which guarantees that $(k\pi)^2 \|S(v)\| = 1$ (since $\|v_n\| = 1$ for all $n \geq 1$), hence $v \neq 0$. On the other hand, we see that $v = -(k\pi)^2 S(v)$. Thus, $v$ is a nonzero scalar multiple of $\hat{u}_k(t) = \sin(k\pi t)$, whence $v \in \mathscr{D}_k$. Since $\mathscr{D}_k$ is open, we have that $u_n \in \mathscr{D}_k$ for $n \geq 1$ large, a contradiction. The first part of (7.28) ensues.

It remains to establish the second part of (7.28). Arguing by contradiction, assume that we can find $(\lambda, u) \in \overline{\mathscr{G}_k} \setminus \mathscr{G}_k$. By construction of $\mathscr{G}_k$, this implies that $u \in \overline{\mathscr{D}_k} \setminus \mathscr{D}_k$. The first step is to show that

$$\text{there is } t_0 \in [0,1] \text{ such that } u(t_0) = u'(t_0) = 0. \tag{7.31}$$

Since the sets $\{\mathscr{D}_\ell\}_{\ell \geq 1}$ are open in $C_0^2([0,1])$ and pairwise disjoint, the fact that $u \in \partial \mathscr{D}_k$ ensures that $u \notin \mathscr{D}_\ell$ for all $\ell \geq 1$. Thus, either $u$ has a zero of multiplicity $\geq 2$ in $[0,1]$ (and we are done) or $u$ has an infinite sequence $\{t_i\}_{i \geq 1}$ of simple zeros in $[0,1]$. In the latter case, we may assume that $t_i \to t_0 \in [0,1]$ as $i \to \infty$ and, clearly, $t_0$ is a zero of multiplicity $\geq 2$. This shows (7.31).

The next step is to show that $u = 0$. Since $(\lambda, u) \in \mathscr{S}_k \subset \overline{\mathscr{S}}$, we have $\varphi(\lambda, u) = 0$, hence $u'' = \lambda u + f(t, \lambda, u, u')$. This can be rewritten as

$$u''(t) = \lambda u(t) + a(t)u(t) + b(t)u'(t) \text{ for all } t \in [0,1],$$

where $a, b : [0,1] \to \mathbb{R}$ are given by $a(t) = q(t)u(t)$ and $b(t) = q(t)u'(t)$, where

$$q(t) = \begin{cases} \dfrac{f(t,\lambda,u(t),u'(t))}{u(t)^2 + u'(t)^2} & \text{if } u(t)^2 + u'(t)^2 > 0, \\ 0 & \text{if } u(t)^2 + u'(t)^2 = 0, \end{cases}$$

for all $t \in [0,1]$. It follows from (7.25) that $q$, $a$, and $b$ are continuous. Hence, $u$ is the solution of a second-order linear differential equation with continuous coefficients. In view of (7.31), we infer that $u = 0$.

Now, part (a) of the statement implies that $\lambda = -(\ell\pi)^2$ for some $\ell \geq 1$. Thus, $(\lambda, u) = (\lambda, 0) \in \mathscr{S}_k \cap \mathscr{S}_\ell$, whence $\mathscr{S}_\ell = \mathscr{S}_k$. We have $(\lambda, 0) \in \mathscr{G}_\ell \cap \mathscr{G}_k$. The sets $\mathscr{G}_\ell$ and $\mathscr{G}_k$ are open in $\mathscr{S}_\ell = \mathscr{S}_k$, so they cannot be disjoint, which imposes that $\ell = k$ and $(\lambda, u) = (-(k\pi)^2, 0)$. However, we know that $(-(k\pi)^2, 0) \in \mathscr{G}_k$, whereas the fact that $(\lambda, u) \notin \mathscr{G}_k$ is assumed. This is a contradiction. We have checked the second part of (7.28). The proof of the proposition is complete. $\square$

**Approach Through Implicit Function Theorem**

The previous results illustrate that the spectrum of the linearization $\varphi_x'(\lambda, 0)$ plays a crucial role when looking for sufficient conditions for bifurcation. From Proposition 7.4, we know that in order for $\lambda^*$ to be a bifurcation point for (7.1), $\varphi_x'(\lambda^*, 0)$ cannot be an isomorphism. In Theorems 7.9 and 7.10, we deal with a map $(\lambda, x) \mapsto \varphi(\lambda, x)$ whose linearization $\varphi_x'(\lambda, 0) = \mathrm{id} - \lambda L$ is a compact perturbation of the identity. A more general situation to address is when $\varphi_x'(\lambda, 0)$ is a Fredholm operator.

**Definition 7.13.** Let $X, Y$ be Banach spaces and $A \in \mathscr{L}(X, Y)$. We say that $A$ is a *Fredholm operator* if its kernel $\ker A$ has finite dimension and its image $\mathrm{im}\, A$ has finite codimension (i.e., $\mathrm{codim}\, \mathrm{im}\, A := \dim Y / \mathrm{im}\, A < +\infty$). Then the number

$$i(A) := \dim \ker A - \mathrm{codim}\, \mathrm{im}\, A \in \mathbb{Z}$$

is called the *index* of $A$. We denote by $\Phi(X, Y)$ the set of Fredholm operators $A : X \to Y$, and we abbreviate $\Phi(X) = \Phi(X, X)$.

The next proposition, for which we refer the reader to Abramovich and Aliprantis [1, Sect. 4.4], summarizes the basic properties of Fredholm operators. In particular, part (d) shows that they are more general than compact perturbations of the identity.

**Proposition 7.14.** *Let $X, Y$ be Banach spaces. Then:*

(a) *If $A \in \Phi(X, Y)$, then $\mathrm{im}\, A$ is closed.*

(b) *If $A \in \Phi(X, Y)$, then there are closed subspaces $Z \subset X$ and $E \subset Y$ such that*

$$X = \ker A \oplus Z \quad \text{and} \quad Y = \mathrm{im}\, A \oplus E.$$

*Moreover, the restriction $A : Z \to \mathrm{im}\, A$ is an isomorphism.*

(c) *If $A \in \Phi(X,Y)$, then $A^* : Y^* \to X^*$ is a Fredholm operator of index $-i(A)$. Moreover, if $B \in \Phi(Y,Z)$, then the composition $B \circ A : X \to Z$ is a Fredholm operator of index $i(A) + i(B)$.*

(d) *$A \in \mathscr{L}(X,Y)$ is a Fredholm operator if and only if there is $B \in \mathscr{L}(Y,X)$ such that $\mathrm{id}_Y - A \circ B$ and $\mathrm{id}_X - B \circ A$ are compact.*

In what follows, given $\varphi \in C^1(\mathbb{R} \times X, Y)$ such that $\varphi(\lambda, 0) = 0$ for all $\lambda \in \mathbb{R}$, we look for sufficient conditions to guarantee that $\lambda^* \in \mathbb{R}$ is a bifurcation point for (7.1) under the assumption that $\varphi'_x(\lambda^*, 0)$ is a Fredholm operator. Actually, we will focus mainly on the particular situation where

$$\dim \ker \varphi'_x(\lambda^*, 0) = \mathrm{codim\,im}\, \varphi'_x(\lambda^*, 0) = 1 \qquad (7.32)$$

[i.e., $\varphi'_x(\lambda^*, 0)$ is a Fredholm operator of index zero with a one-dimensional kernel]. This situation is called a *bifurcation from a simple eigenvalue*.

*Example 7.15.* If $\varphi(\lambda, x) = \lambda x - f(x)$, with $f \in C^1(X,X)$, $f(0) = 0$, $f'(0) \in \mathscr{L}_c(X)$, and $\lambda^* \in \mathbb{R}$ is a simple eigenvalue of $f'(0)$, then (7.32) is satisfied.

We will prove the following theorem for bifurcation from a simple eigenvalue. Recall that $\varphi'_\lambda(\lambda^*, x_0) \in \mathscr{L}(\mathbb{R}, Y)$ denotes the differential at $\lambda^*$ of the map $\varphi(\cdot, x_0)$. In the same way, for $\varphi$ of class $C^2$, by $\varphi''_{\lambda, x}(\lambda^*, x_0)$ we denote the differential at $x_0$ of the map $\varphi'_\lambda(\lambda^*, \cdot)$. A priori $\varphi''_{\lambda, x}(\lambda^*, x_0) \in \mathscr{L}(X, \mathscr{L}(\mathbb{R}, Y))$; thus, it can be seen as a bilinear map $\mathbb{R} \times X \to Y$.

**Theorem 7.16.** *Let $X, Y$ be Banach spaces, $\varphi \in C^2(\mathbb{R} \times X, Y)$ such that $\varphi(\lambda, 0) = 0$ for all $\lambda \in \mathbb{R}$, and $\lambda^* \in \mathbb{R}$. We assume that $A := \varphi'_x(\lambda^*, 0)$ satisfies*

$$\ker A = \mathbb{R}\hat{v} \ (\text{with } \hat{v} \in X \setminus \{0\}) \ \text{ and } \ \mathrm{codim\,im}\, A = 1.$$

*If $\varphi''_{\lambda, x}(\lambda^*, 0)(1, \hat{v}) \notin \mathrm{im}\, A$, then $\lambda^*$ is a bifurcation point for (7.1). More precisely, for any topological complement $Z \subset X$ of $\ker A$, we can find $\varepsilon > 0$ and a $C^1$-map $(-\varepsilon, \varepsilon) \to \mathbb{R} \times X$, $s \mapsto (\lambda(s), x(s))$ with $\lambda(0) = \lambda^*$, $x(0) = 0$, $x(s) \in s\hat{v} + Z$, and $\varphi(\lambda(s), x(s)) = 0$ for all $s \in (-\varepsilon, \varepsilon)$.*

The proof of Theorem 7.16 is given below. The method used is the so-called *Lyapunov–Schmidt reduction method*, which we describe next. For the moment we deal with a more general situation than in the statement of the theorem: we assume $\varphi \in C^k(\mathbb{R} \times X, Y)$ for some $k \geq 1$ and $\lambda^* \in \mathbb{R}$ such that $A := \varphi'_x(\lambda^*, 0) \in \mathscr{L}(X,Y)$ is a general Fredholm operator with $\ker A \neq 0$. By Proposition 7.14(b), we have decompositions

$$\begin{cases} X = \ker A \oplus Z & \text{with } 0 < \dim \ker A < +\infty, \\ Y = \mathrm{im}\, A \oplus E & \text{with } \dim E < +\infty. \end{cases}$$

Let

$$P : Y \to E \text{ and } Q : Y \to \operatorname{im} A$$

be the linear projections with respect to the decomposition $Y = \operatorname{im} A \oplus E$. Evidently, $P = \operatorname{id} - Q$. Thus, $(\lambda, x)$ is a solution of (7.1) if and only if

$$P\varphi(\lambda, x) = 0 \text{ and } Q\varphi(\lambda, x) = 0. \tag{7.33}$$

The second equation in (7.33) is known as the *auxiliary equation*. It is treated in the following lemma.

**Lemma 7.17.** *Let* $X, Y, \varphi, \lambda^*, A, Z, E$ *be as above. There exist* $T_0$ *neighborhood of* $\lambda^*$ *in* $\mathbb{R}$, $V_0$ *neighborhood of* $0$ *in* $\ker A$, $Z_0$ *neighborhood of* $0$ *in* $Z$, *and a map* $h \in C^k(T_0 \times V_0, Z_0)$ *such that for all* $(\lambda, v, z) \in T_0 \times V_0 \times Z_0$ *we have*

$$Q\varphi(\lambda, v + z) = 0 \iff z = h(\lambda, v).$$

*Moreover, we have* $h(\lambda, 0) = 0$ *for all* $\lambda \in T_0$ *and* $h'_v(\lambda^*, 0) = 0$.

*Proof.* Consider $\psi \in C^k(\mathbb{R} \times \ker A \times Z, \operatorname{im} A)$ defined by

$$\psi(\lambda, v, z) = Q\varphi(\lambda, v + z) \text{ for all } (\lambda, v, z) \in \mathbb{R} \times \ker A \times Z.$$

We see that $\psi'_z(\lambda^*, 0, 0) \in \mathscr{L}(Z, \operatorname{im} A)$ is given by

$$\psi'_z(\lambda^*, 0, 0)(z) = Q\varphi'_x(\lambda^*, 0)(z) = QA(z) = A(z) \text{ for all } z \in Z,$$

which yields $\psi'_z(\lambda^*, 0, 0) = A|_Z$. Hence $\psi'_z(\lambda^*, 0, 0)$ is invertible. By the implicit function theorem (Theorem 7.3), there exist $T_0 \times V_0$ neighborhood of $(\lambda^*, 0)$ in $\mathbb{R} \times \ker A$, $Z_0$ neighborhood of $0$ in $Z$, and a map $h \in C^k(T_0 \times V_0, Z_0)$ such that for all $(\lambda, v, z) \in T_0 \times V_0 \times Z_0$ we have

$$(Q\varphi(\lambda, v + z) =) \ \psi(\lambda, v, z) = 0 \iff z = h(\lambda, v).$$

Note that $\psi(\lambda, 0, 0) = Q\varphi(\lambda, 0) = Q(0) = 0$, so in particular $h(\lambda, 0) = 0$ for all $\lambda \in T_0$. Moreover,

$$h'(\lambda^*, 0) = -\psi'_z(\lambda^*, 0, 0)^{-1} \circ \psi'_{(\lambda, v)}(\lambda^*, 0, 0).$$

Hence, for $v \in \ker A$ we have

$$h'_v(\lambda^*, 0)(v) = h'(\lambda^*, 0)(0, v) = -\psi'_z(\lambda^*, 0, 0)^{-1}(\psi'_{(\lambda, v)}(\lambda^*, 0, 0)(0, v))$$

$$= -\psi'_z(\lambda^*, 0, 0)^{-1}(\psi'(\lambda^*, 0, 0)(0, v, 0))$$

$$= -\psi'_z(\lambda^*, 0, 0)^{-1}(Q\varphi'(\lambda^*, 0)(0, v))$$

$$= -\psi'_z(\lambda^*, 0, 0)^{-1}(QA(v)) = 0.$$

The proof is complete.                                                                          □

Lemma 7.17 implies that a pair of the form $(\lambda, x)$, with $\lambda \in T_0$ and $x = v + z \in V_0 + Z_0$, satisfies (7.33) if and only if $z = h(\lambda, v)$ and

$$P\varphi(\lambda, v + h(\lambda, v)) = 0. \tag{7.34}$$

Equation (7.34) is known as the *bifurcation equation*. To deal with it, we consider $\xi \in C^k(T_0 \times V_0, E)$ given by $\xi(\lambda, v) = P\varphi(\lambda, v + h(\lambda, v))$ for all $(\lambda, v) \in T_0 \times V_0$. The main principle of the Lyapunov–Schmidt reduction method is stated in the next proposition.

**Proposition 7.18.** *Let* $X, Y, \varphi, \lambda^*, h, \xi$ *be as above. Set*

$$S_\xi := \{(\lambda, v) \in T_0 \times V_0 : \xi(\lambda, v) = 0, \ v \neq 0\}.$$

(a) *If* $(\lambda, v) \in S_\xi$, *then* $(\lambda, v + h(\lambda, v)) \in S_\varphi$.
(b) *If* $\lambda^*$ *is a bifurcation point for (7.34), that is,* $(\lambda^*, 0) \in \overline{S_\xi}$, *then* $\lambda^*$ *is a bifurcation point for (7.1).*

*Proof.*

(a) Let $(\lambda, v) \in S_\xi$, and set $x = v + h(\lambda, v)$. We have $x \neq 0$ because $v \neq 0$. Moreover, $P\varphi(\lambda, x) = 0$ (by assumption) and $Q\varphi(\lambda, x) = 0$ (by Lemma 7.17), whence $\varphi(\lambda, x) = 0$, i.e., $(\lambda, x) \in S_\varphi$.
(b) Assume that $(\lambda^*, 0) \in \overline{S_\xi}$. Thus, there is a sequence $\{(\lambda_n, v_n)\}_{n \geq 1} \subset S_\xi$ such that $\lambda_n \to \lambda^*$ and $v_n \to 0$ as $n \to \infty$. The continuity of $h$ implies that $x_n := v_n + h(\lambda_n, x_n) \to h(\lambda^*, 0) = 0$ (Lemma 7.17). The sequence $\{(\lambda_n, x_n)\}_{n \geq 1}$ lies in $S_\varphi$ [by (a)] and converges to $(\lambda^*, 0)$, hence $\lambda^*$ is a bifurcation point for (7.1).
                                                                                               □

According to this proposition, the search for a sufficient condition for $\lambda^*$ to be a bifurcation point for (7.1) is reduced to the study of the bifurcation equation (7.34), involving $\xi(\lambda, v) = P\varphi(\lambda, v + h(\lambda, v))$. This new bifurcation equation is easier to deal with because the map $\xi : T_0 \times V_0 \subset \mathbb{R} \times \ker A \to E$ is defined between two finite-dimensional spaces. The simplest situation for studying the new equation is that considered in Theorem 7.16, where

$$\dim \ker A = 1 \quad \text{and} \quad \dim E = \operatorname{codim} \operatorname{im} A = 1.$$

We then focus on this situation as we are now in a position to prove Theorem 7.16.

*Proof (of Theorem 7.16).* The theorem involves the assumptions that $\varphi : \mathbb{R} \times X \to Y$ is of class $C^2$, $\ker A = \mathbb{R}\hat{v}$ (with $\hat{v} \in X \setminus \{0\}$), and $\operatorname{codim} \operatorname{im} A = 1$. The last fact yields $y^* \in Y^* \setminus \{0\}$ such that

$$\operatorname{im} A = \{y \in Y : \langle y^*, y \rangle = 0\}, \tag{7.35}$$

where $\langle \cdot, \cdot \rangle$ denote the duality brackets for the pair $(Y^*, Y)$. It is also assumed that

$$\varphi''_{\lambda,x}(\lambda^*, 0)(1, \hat{v}) \notin \mathrm{im} A. \tag{7.36}$$

Let $I_0 \subset \mathbb{R}$ be an open interval containing 0 such that $\{s\hat{v} : s \in I_0\} \subset V_0$. We consider the map $\zeta \in C^2(T_0 \times I_0, \mathbb{R})$ given by

$$\zeta(\lambda, s) = \langle y^*, \varphi(\lambda, s\hat{v} + h(\lambda, s\hat{v})) \rangle \quad \text{for all } (\lambda, s) \in T_0 \times I_0.$$

In view of Proposition 7.18, the theorem will be proved once we show the existence of $\varepsilon > 0$, with $(-\varepsilon, \varepsilon) \subset I_0$ and a map $\lambda : (-\varepsilon, \varepsilon) \to T_0$ of class $C^1$ such that

$$\lambda(0) = \lambda^* \quad \text{and} \quad \zeta(\lambda(s), s) = 0 \quad \text{for all } s \in (-\varepsilon, \varepsilon). \tag{7.37}$$

As a preliminary step, we point out some properties of the map $\zeta$. First, since $h(\lambda, 0) = 0$ (Lemma 7.17), we have $\zeta(\lambda, 0) = 0$ and $\zeta'_\lambda(\lambda, 0) = 0$ for all $\lambda \in T_0$. We evaluate the partial derivative of $\zeta$ with respect to $s$. From the chain rule we have

$$\zeta'_s(\lambda, s) = \langle y^*, \varphi'_x(\lambda, s\hat{v} + h(\lambda, s\hat{v}))(\hat{v} + h'_v(\lambda, s\hat{v})(\hat{v})) \rangle \tag{7.38}$$

for all $(\lambda, s) \in T_0 \times I_0$. Thus,

$$\zeta'_s(\lambda, 0) = \langle y^*, \varphi'_x(\lambda, 0)(\hat{v} + h'_v(\lambda, 0)(\hat{v})) \rangle \quad \text{for all } \lambda \in T_0. \tag{7.39}$$

In particular, by (7.35), we have

$$\zeta'_s(\lambda^*, 0) = \langle y^*, A(\hat{v} + h'_v(\lambda^*, 0)(\hat{v})) \rangle = 0. \tag{7.40}$$

From (7.39), the chain rule, the fact that $h'_v(\lambda^*, 0) = 0$ (Lemma 7.17), (7.35), and (7.36), we have

$$\zeta''_{\lambda,s}(\lambda^*, 0) = \langle y^*, \varphi''_{\lambda,x}(\lambda^*, 0)(1, \hat{v} + h'_v(\lambda^*, 0)(\hat{v})) + A(\hat{v} + h''_{\lambda,v}(\lambda^*, 0)(1, \hat{v})) \rangle$$

$$= \langle y^*, \varphi''_{\lambda,x}(\lambda^*, 0)(1, \hat{v}) \rangle \neq 0. \tag{7.41}$$

We are now ready to construct a map $s \mapsto \lambda(s)$ satisfying (7.37). Define $\eta : T_0 \times I_0 \to \mathbb{R}$ by

$$\eta(\lambda, s) = \begin{cases} \dfrac{\zeta(\lambda, s)}{s} & \text{if } s \neq 0 \\ \zeta'_s(\lambda, 0) & \text{if } s = 0, \end{cases} \quad \text{for all } (\lambda, s) \in T_0 \times I_0.$$

Evidently, $\eta \in C^1(T_0 \times I_0, \mathbb{R})$. By (7.40), we have $\eta(\lambda^*, 0) = 0$. Moreover, from (7.41) it follows that

$$\eta'_\lambda(\lambda^*, 0) = \zeta''_{\lambda,s}(\lambda^*, 0) \neq 0.$$

We can therefore apply the implicit function theorem (Theorem 7.3) to the map $\eta$, which yields a neighborhood $(-\varepsilon, \varepsilon)$ of 0 in $I_0$, a neighborhood $T_0^*$ of $\lambda^*$ in $T_0$, and a map $\lambda \in C^1((-\varepsilon, \varepsilon), T_0^*)$ such that, for $(\lambda, s) \in T_0^* \times (-\varepsilon, \varepsilon)$, we have

$$\eta(\lambda, s) = 0 \iff \lambda = \lambda(s).$$

In particular, we get $\lambda(0) = \lambda^*$ and, for all $s \in (-\varepsilon, \varepsilon) \setminus \{0\}$,

$$\zeta(\lambda(s), s) = s\eta(\lambda(s), s) = 0.$$

Thus, the map $s \mapsto \lambda(s)$ satisfies (7.37). The proof is now complete.                   □

*Remark 7.19.* In addition to the curve $\Gamma_1 = \{(\lambda(s), x(s)) : s \in (-\varepsilon, \varepsilon)\}$ provided by Theorem 7.16, another curve of solutions to (7.1) passing through $(\lambda^*, 0)$ is the trivial curve $\Gamma_0 = \{(\lambda, 0) : \lambda \in \mathbb{R}\}$. Thus, for every neighborhood $U \subset \mathbb{R} \times X$ of $(\lambda^*, 0)$, we have

$$\{(\lambda, x) \in \mathbb{R} \times X : (\lambda, x) \text{ solution of } (7.1)\} \cap U \supset (\Gamma_0 \cup \Gamma_1) \cap U. \qquad (7.42)$$

Actually, the conclusion of Theorem 7.16 can be strengthened: when the neighborhood $U$ is chosen small enough, equality holds in (7.42). This follows from a careful application of the implicit function theorem and suitable estimates of the form $x(s) = s\hat{v} + o(s)$ (see Crandall and Rabinowitz [93, Theorem 1.7]). This local uniqueness of the nontrivial curve of solutions passing through $(\lambda^*, 0)$ is specific to the situation of bifurcation from a simple eigenvalue.

Applying Theorem 7.16 to Example 7.15, we obtain the following corollary.

**Corollary 7.20.** *Assume that $f \in C^2(X, X)$, $f(0) = 0$, $f'(0) \in \mathscr{L}_c(X)$. If $\lambda^* \in \mathbb{R}$ is a simple eigenvalue of $f'(0)$, then $\lambda^*$ is a bifurcation point for the equation $\lambda u - f(u) = 0$.*

We conclude this section with an example of the application of Theorem 7.16. We study the bifurcation points for the periodic problem

$$\begin{cases} u''(t) = \lambda u(t) + f(t, \lambda, u(t), u'(t)) & \text{for all } t \in [0, 2\pi], \\ u(0) = u(2\pi), \ u'(0) = u'(2\pi), \end{cases} \qquad (7.43)$$

where $\lambda \in \mathbb{R}$ and $f : \mathbb{R}^4 \to \mathbb{R}$ is a continuous function such that $t \mapsto f(t, \lambda, s, \xi)$ is $2\pi$-periodic for all $(\lambda, s, \xi) \in \mathbb{R}^3$, $(\lambda, s, \xi) \mapsto f(t, \lambda, s, \xi)$ is of class $C^2$ for all $t \in \mathbb{R}$, and

$$f(t,\lambda,0,0) = f_s'(t,\lambda,0,0) = f_\xi'(t,\lambda,0,0) = 0 \text{ for all } (t,\lambda) \in \mathbb{R}^2. \qquad (7.44)$$

We deal with the Banach spaces

$$X = C_{\text{per}}^2(\mathbb{R}) = \{u \in C^2(\mathbb{R}) : u \text{ is } 2\pi\text{-periodic}\},$$

$$Y = C_{\text{per}}(\mathbb{R}) = \{u \in C(\mathbb{R}) : u \text{ is } 2\pi\text{-periodic}\},$$

endowed with the norms $\|u\|_X = \|u\|_\infty + \|u'\|_\infty + \|u''\|_\infty$ and $\|u\|_Y = \|u\|_\infty$, respectively. Let $\varphi : \mathbb{R} \times X \to Y$ be the map defined by

$$\varphi(\lambda,u)(t) = u''(t) - \lambda u(t) - f(t,\lambda,u(t),u'(t)).$$

A solution of (7.43) is then an element $u \in X$ such that $\varphi(\lambda,u) = 0$. Evidently, $\varphi(\lambda,0) = 0$ for all $\lambda \in \mathbb{R}$, hence $u = 0$ is a (trivial) solution of (7.43). Let

$$\mathscr{S} = \{(\lambda,u) \in \mathbb{R} \times X : \varphi(\lambda,u) = 0,\ u \neq 0\}$$

be the set of nontrivial solutions to (7.43). We say that $\lambda^* \in \mathbb{R}$ is a bifurcation point for problem (7.43) if $\lambda^*$ is a bifurcation point for the equation $\varphi(\lambda,u) = 0$ in the sense of Definition 7.1, that is, if $(\lambda^*,0) \in \overline{\mathscr{S}}$.

To apply Theorem 7.16, we need to consider the linearization $A_\lambda := \varphi_u'(\lambda,0)$ at a point $(\lambda,0) \in \mathbb{R} \times X$. In view of (7.44), for every $\lambda \in \mathbb{R}$ we have

$$A_\lambda(u) = u'' - \lambda u \text{ for all } u \in X.$$

Clearly, the kernel of $A_\lambda$ is nontrivial if and only if $\lambda = -k^2$ for some integer $k \geq 0$. If $k \geq 1$, then $\dim \ker A_{-k^2} = 2$, which does not fit into the setting of Theorem 7.16. If $k = 0$, then the kernel of $A := A_0$ is the subspace of constant functions $\mathbb{R} \subset X$, whereas

$$\operatorname{im} A = \left\{ u \in Y : \int_0^{2\pi} u(t)\,dt = 0 \right\}.$$

Thus, $\dim \ker A = \operatorname{codim} \operatorname{im} A = 1$. In this situation, Theorem 7.16 can be applied. We obtain the following proposition.

**Proposition 7.21.** $\lambda^* = 0$ *is a bifurcation point for problem (7.43). More precisely, for every decomposition $C_{\text{per}}^2(\mathbb{R}) = \mathbb{R} \oplus Z$:*

(a) *There exist $\varepsilon > 0$ and a $C^1$-map $(-\varepsilon,\varepsilon) \to C_{\text{per}}^2(\mathbb{R})$, $s \mapsto (\lambda_s, s + u_s)$, with $(\lambda_0, u_0) = (0,0)$, such that $u_s \in Z$ for all $s \in (-\varepsilon,\varepsilon)$ and $s + u_s$ is a nontrivial solution of (7.43) with respect to $\lambda = \lambda_s$ whenever $s \neq 0$.*

(b) *If $U \subset \mathbb{R} \times C_{\text{per}}^2(\mathbb{R})$ is a neighborhood of $(0,0)$ sufficiently small, then every pair $(\lambda,u) \in U$, with $u$ a nontrivial solution of (7.43), is of the form $(\lambda_s, s + u_s)$ for some $s \in (-\varepsilon,\varepsilon)$, $s \neq 0$.*

*Proof.* We write $\ker A = \mathbb{R}\hat{v}$, where $\hat{v} \in X$ is the constant function $\hat{v} \equiv 1$. For every $\lambda \in \mathbb{R}$ we have $\varphi''_{\lambda,u}(\lambda,0)(1,u) = u'' - u$. Thus,

$$\varphi''_{\lambda,u}(0,0)(1,\hat{v}) = -\hat{v} \notin \operatorname{im} A.$$

Therefore, all the assumptions of Theorem 7.16 are satisfied. Part (a) of the proposition ensues, whereas part (b) is a consequence of Remark 7.19.     □

*Remark 7.22.* By playing on the choice of the topological complement of $\mathbb{R}$ in $C^2_{\text{per}}(\mathbb{R})$ in Proposition 7.21, we can emphasize various properties of the branch of nontrivial solutions to (7.43) that passes through $(0,0)$. Here are two examples:

(i) Taking $Z = \{u \in C^2_{\text{per}}(\mathbb{R}) : u(0) = 0\}$, we obtain that $(0,0)$ belongs to a branch $\{(\lambda_s, v_s) : s \in (-\varepsilon, \varepsilon)\}$ of solutions to (7.43) satisfying the initial condition $v_s(0) = s$.

(ii) Taking $Z = \{u \in C^2_{\text{per}}(\mathbb{R}) : \int_0^{2\pi} u(t)\,dt = 0\}$, we obtain that $(0,0)$ belongs to a branch $\{(\mu_s, w_s) : s \in (-\varepsilon, \varepsilon)\}$ of solutions to (7.43) such that $\int_0^{2\pi} w_s(t)\,dt = s$. Proposition 7.21(b) shows that, near $(0,0)$, the branches obtained in (i) and (ii) coincide up to reindexing.

## 7.2 Remarks

**Section 7.1:** Bifurcation theory has its roots in the work of Poincaré on the equilibrium forms for a rotating ideal fluid. The strong relation between bifurcation points and eigenvalues of linear compact operators emphasized in Theorem 7.9 is due to Krasnosel'skiĭ [202, Chap. 4]. Theorem 7.10 is the well-known global bifurcation theorem of Rabinowitz [334]. Proposition 7.12 can be found in Rabinowitz [334, Sect. 2], among other examples of application of bifurcation theory to the study of differential equations and integral equations. Theorem 7.16 is due to Crandall and Rabinowitz [93]. The method used in the proof is a procedure known as the Lyapunov–Schmidt reduction, based on the implicit function theorem, and it goes back to the classical works of Lyapunov [227] and Schmidt [356].

    More about bifurcation theory can be found in the books by Ambrosetti and Prodi [16], Deimling [108], Kielhöfer [195], Krasnosel'skiĭ [202], Krawcewicz and Wu [205], Nirenberg [309], and Zeidler [387].

# Chapter 8
# Regularity Theorems and Maximum Principles

**Abstract** This chapter provides a comprehensive presentation of regularity theorems and maximum principles that are essential for the subsequent study of nonlinear elliptic boundary value problems. In addition to the presentation of fundamental results, the chapter offers, to a large extent, a novel approach with clarification of tedious arguments and simplification of proofs. The first section of this chapter treats two major topics related to weak solutions of nonlinear elliptic problems: boundedness and regularity. The second section has as its objective to report on maximum and comparison principles. It comprises two parts: local results and strong maximum principles. Comments and related references are given in a remarks section.

## 8.1 Regularity of Solutions

In this section, we prove regularity results for the weak solutions of certain nonlinear elliptic problems, which include as a particular case problems driven by the $p$-Laplace differential operator.

Usually, the regularity of a weak solution of a nonlinear elliptic problem is obtained by arguing in two steps: first, one shows that the weak solution is bounded, and second, relying on this boundedness property, one establishes the regularity of the weak solution up to the boundary of its domain of definition. Accordingly, this section is organized in two parts: in the first part, we provide a criterion that guarantees the boundedness of weak solutions, and in the second part, we present results that establish the regularity of bounded weak solutions.

### Boundedness of Weak Solutions of Nonlinear Elliptic Problems

Let $\Omega$ be a bounded domain in $\mathbb{R}^N$ ($N \geq 1$) with a Lipschitz boundary $\partial\Omega$, and let $p \in (1, +\infty)$. We consider a general operator $a : \overline{\Omega} \times \mathbb{R}^N \to \mathbb{R}^N$ satisfying the following hypotheses:

D. Motreanu et al., *Topological and Variational Methods with Applications to Nonlinear Boundary Value Problems*, DOI 10.1007/978-1-4614-9323-5_8,
© Springer Science+Business Media, LLC 2014

H$(a)_1$ (i) $a : \overline{\Omega} \times \mathbb{R}^N \to \mathbb{R}^N$ is continuous;

(ii) There is a constant $c_1 > 0$ such that

$$|a(x,\xi)| \leq c_1(1+|\xi|^{p-1}) \text{ for all } x \in \overline{\Omega}, \text{ all } \xi \in \mathbb{R}^N;$$

(iii) There are constants $c_0 > 0$ and $R, \sigma \geq 0$ such that

$$(a(x,\xi),\xi)_{\mathbb{R}^N} \geq c_0(R+|\xi|)^{p-\sigma}|\xi|^\sigma \text{ for all } x \in \overline{\Omega}, \text{ all } \xi \in \mathbb{R}^N.$$

Hypothesis H$(a)_1$ (ii) implies that we have $a(\cdot, \nabla u(\cdot)) \in L^{p'}(\Omega, \mathbb{R}^N)$ whenever $u \in W^{1,p}(\Omega)$, where $p' = \frac{p}{p-1}$. In particular, the divergence $\operatorname{div} a(x, \nabla u)$ (in the distributional sense) is well defined.

*Example 8.1.* Many interesting operators fit the setting of hypotheses H$(a)_1$:

(a) $a(x,\xi) = |\xi|^{p-2}\xi$, so that $\operatorname{div} a(x, \nabla u)$ is the $p$-Laplacian in this case.

If $a_1$ satisfies H$(a)_1$ and $a_2$ satisfies H$(a)_1$ (i), (ii), and $(a_2(x,\xi),\xi)_{\mathbb{R}^N} \geq 0$, then the sum $a_1 + a_2$ also satisfies H$(a)_1$. Thus, we can derive other examples from (a):

(b) $a(x,\xi) = |\xi|^{p-2}\xi + \ln(1+|\xi|^{p-2})\xi;$

(c) $a(x,\xi) = \begin{cases} |\xi|^{p-2}\xi + |\xi|^{q-2}\xi & \text{if } |\xi| \leq 1 \\ |\xi|^{p-2}\xi + \frac{q-2}{\tau-2}|\xi|^{\tau-2}\xi - \frac{q-\tau}{\tau-2}\xi & \text{if } |\xi| > 1, \end{cases}$ with $1 < \tau \leq p \leq q$ and $\tau \neq 2;$

(d) $a(x,\xi) = |\xi|^{p-2}\xi + c\frac{|\xi|^{p-2}\xi}{1+|\xi|^p}$, with $c > 0$, in this case $\operatorname{div} a(x, \nabla u)$ is the sum of the $p$-Laplacian and a generalized mean curvature operator.

In each example (a)–(d), the map $a$ satisfies H$(a)_1$ (iii) with $R = 0$. Finally, we may note that if $a$ satisfies H$(a)_1$ and $\theta \in C(\overline{\Omega}, (0,+\infty))$, then the product $(x,\xi) \mapsto \theta(x)a(x,\xi)$ also satisfies H$(a)_1$. Thus, further examples can be derived from (a)–(d).

In addition to the operator $a(x,\xi)$, we consider a function $f : \Omega \times \mathbb{R} \to \mathbb{R}$ satisfying the following hypotheses. Recall that $p^*$ denotes the critical exponent of $p$, i.e., $p^* = \frac{Np}{N-p}$ if $N > p$ and $p^* = +\infty$ if $N \leq p$ [Remark 1.50(b), (c)].

H$(f)_1$    $f : \Omega \times \mathbb{R} \to \mathbb{R}$ is a Carathéodory function, i.e., $f(\cdot, s)$ is measurable for all $s \in \mathbb{R}$ and $f(x,\cdot)$ is continuous for a.a. $x \in \Omega$. Moreover, there are constants $c > 0$ and $r \in [p, p^*)$ such that

$$|f(x,s)| \leq c(1+|s|^{r-1}) \text{ for a.a. } x \in \Omega, \text{ all } s \in \mathbb{R}.$$

By Theorems 1.49 and 2.76, H$(f)_1$ guarantees that the Nemytskii operator $N_f :$ $W^{1,p}(\Omega) \to L^{r'}(\Omega)$, $u \mapsto f(\cdot, u(\cdot))$ is well defined and continuous. In particular, $N_f(u) \in (W^{1,p}(\Omega))^*$ whenever $u \in W^{1,p}(\Omega)$. Now, given the operator $a$ and the function $f$ as above, we consider the following nonlinear elliptic problems under Dirichlet boundary conditions:

$$\begin{cases} -\operatorname{div} a(x, \nabla u(x)) = f(x, u(x)) & \text{in } \Omega, \\ u = 0 & \text{on } \partial\Omega, \end{cases} \tag{8.1}$$

and under Neumann boundary conditions:

$$\begin{cases} -\operatorname{div} a(x, \nabla u(x)) = f(x, u(x)) & \text{in } \Omega, \\ \frac{\partial u}{\partial n_a} = 0 & \text{on } \partial\Omega. \end{cases} \tag{8.2}$$

In (8.2), we denote $\frac{\partial u}{\partial n_a} = \gamma_n(a(\cdot, \nabla u(\cdot)))$, where $\gamma_n$ is the generalized normal derivative operator (Theorem 1.38).

**Definition 8.2.**

(a) A *weak solution* to problem (8.1) is a function $u \in W_0^{1,p}(\Omega)$ such that

$$\int_\Omega (a(x, \nabla u), \nabla v)_{\mathbb{R}^N}\, dx = \int_\Omega f(x, u(x))\, v(x)\, dx \ \text{ for all } v \in W_0^{1,p}(\Omega).$$

(b) A *weak solution* to problem (8.2) is a function $u \in W^{1,p}(\Omega)$ such that

$$\int_\Omega (a(x, \nabla u), \nabla v)_{\mathbb{R}^N}\, dx = \int_\Omega f(x, u(x))\, v(x)\, dx \ \text{ for all } v \in W^{1,p}(\Omega).$$

*Remark 8.3.* Recall that $\Omega \subset \mathbb{R}^N$ is a bounded domain with Lipschitz boundary.

(a) The condition that $u = 0$ on $\partial\Omega$ in problem (8.1) is translated in Definition 8.2(a) into the fact that a weak solution of (8.1) is expected to belong to the space $W_0^{1,p}(\Omega)$, so its trace $\gamma(u)$ is zero.

(b) Let $u$ be a weak solution of (8.2) in the sense of Definition 8.2(b), and assume that $u$ satisfies the further property that $f(\cdot, u(\cdot)) \in L^{p'}(\Omega)$. Since the equality $-\operatorname{div} a(x, \nabla u) = f(x, u)$ holds a fortiori in distributions, we get $-\operatorname{div} a(\cdot, \nabla u(\cdot)) \in L^{p'}(\Omega)$. This implies that $a(\cdot, u(\cdot)) \in V^{p'}(\Omega, \operatorname{div})$, and so $\frac{\partial u}{\partial n_a}$ is well defined (Theorem 1.38). Using nonsmooth Green's identity (Theorem 1.38), from Definition 8.2(b) we infer that $\frac{\partial u}{\partial n_a} = 0$. In this way, the weak solution $u$ satisfies the boundary condition in problem (8.2). Actually, it will follow from Corollary 8.7 that [under the hypotheses H($a$)$_1$ and H($f$)$_1$] a weak solution $u$ of (8.2) is always such that $f(\cdot, u(\cdot)) \in L^{p'}(\Omega)$, so it always satisfies the boundary condition $\frac{\partial u}{\partial n_a} = 0$.

(c) By an argument based on the density of regular functions in $W_0^{1,p}(\Omega)$ and $W^{1,p}(\Omega)$ (Definition 1.8 and Theorem 1.19), we obtain equivalent versions of Definition 8.2(a) and (b) if we take the test functions $v$ in $C_c^\infty(\Omega)$ and $C^\infty(\overline{\Omega})$, respectively.

Now we can state a general result from which we will derive criteria of boundedness for weak solutions of problems (8.1) and (8.2).

**Theorem 8.4.** *Assume that* $H(a)_1$ *and* $H(f)_1$ *hold. Let* $u \in W^{1,p}(\Omega)$ *be a function such that the inequality*

$$\int_\Omega (a(x,\nabla u),\nabla v)_{\mathbb{R}^N}\, dx \leq \int_\Omega f(x,u(x))\, v(x)\, dx$$

*holds whenever* $v$ *is of the form* $v = \min\{u^+,\lambda\}^\alpha$ *or* $v = -\min\{u^-,\lambda\}^\alpha$ *for* $\lambda > 0$ *and* $\alpha \geq 1$. *Let* $\theta \in (r,p^*]$ *if* $p < N$ *and* $\theta \in (r,+\infty)$ *if* $p \geq N$. *Then* $u \in L^\infty(\Omega)$ *and*

$$\|u\|_\infty \leq M(1+\|u\|_\theta)^{\frac{\theta-p}{\theta-r}},$$

*where* $M > 0$ *depends only on* $c_0$, $c$, $R$, $\sigma$, $\Omega$, $p$, $\theta$, *and* $N$.

*Proof.* By Theorem 1.49, there is a constant $M_0 = M_0(\Omega,p,\theta,N) > 0$ such that

$$\|v\|_\theta \leq M_0(\|v\|_p + \|\nabla v\|_p) \quad \text{for all } v \in W^{1,p}(\Omega). \tag{8.3}$$

We denote $v = u^+$.

*Claim 1*: There exists $M_1 = M_1(c_0,c,R,\sigma,\Omega,p,\theta,N) > 0$ such that for all $\ell \in [0,+\infty)$ we have

$$1 + \int_\Omega v^{\theta(\ell+1)}\, dx \leq M_1(\ell+1)^\theta \left(1+\int_\Omega v^{p\ell+r}\, dx\right)^{\frac{\theta}{p}}$$

(where the integrals may be infinite a priori).

We set

$$M_2 = M_2(R,\sigma,p) := \inf_{t\in[1,+\infty)} \left(\frac{t}{R+t}\right)^{\sigma-p} > 0, \tag{8.4}$$

with $R,\sigma \geq 0$ from $H(a)_1$ (iii). Fix $\ell \in [0,+\infty)$ and $\lambda \in (0,+\infty)$, and denote $v_\lambda = \min\{v,\lambda\}$. The assumption yields

$$\int_\Omega (a(x,\nabla u),\nabla(v_\lambda^{p\ell+1}))_{\mathbb{R}^N}\, dx \leq \int_\Omega f(x,u)v_\lambda^{p\ell+1}\, dx. \tag{8.5}$$

On the one hand, by $H(f)_1$ and the fact that $v_\lambda = 0$ a.e. on $\{x \in \Omega : u(x) \leq 0\}$, we have

$$\int_\Omega f(x,u)v_\lambda^{p\ell+1}\, dx = \int_\Omega f(x,v)v_\lambda^{p\ell+1}\, dx \leq \int_\Omega c(1+v^{r-1})v_\lambda^{p\ell+1}\, dx$$

$$\leq M_3 \left(1+\int_\Omega v^{p\ell+r}\, dx\right) \tag{8.6}$$

for some $M_3 = M_3(c, |\Omega|_N) > 0$. On the other hand, using that $v_\lambda$ is constant (and so $\nabla v_\lambda$ vanishes) on the set $\{x \in \Omega : u(x) \neq v_\lambda(x)\}$ and invoking H($a$)$_2$ (iii), we obtain

$$\int_\Omega (a(x, \nabla u), \nabla(v_\lambda^{p\ell+1}))_{\mathbb{R}^N}\, dx = (p\ell + 1) \int_\Omega (a(x, \nabla u), \nabla v_\lambda)_{\mathbb{R}^N}\, v_\lambda^{p\ell}\, dx$$

$$= (p\ell + 1) \int_\Omega (a(x, \nabla v_\lambda), \nabla v_\lambda)_{\mathbb{R}^N}\, v_\lambda^{p\ell}\, dx$$

$$\geq c_0 \int_\Omega (R + |\nabla v_\lambda|)^{p-\sigma} |\nabla v_\lambda|^\sigma v_\lambda^{p\ell}\, dx. \qquad (8.7)$$

The next computation combines (8.4)–(8.7):

$$\frac{1}{(\ell+1)^p} \int_\Omega |\nabla(v_\lambda^{\ell+1})|^p\, dx = \int_\Omega |\nabla v_\lambda|^p v_\lambda^{p\ell}\, dx$$

$$\leq \int_{\{|\nabla v_\lambda| < 1\}} v_\lambda^{p\ell}\, dx + \frac{1}{M_2} \int_{\{|\nabla v_\lambda| \geq 1\}} v_\lambda^{p\ell}(R + |\nabla v_\lambda|)^{p-\sigma} |\nabla v_\lambda|^\sigma\, dx$$

$$\leq M_4 \left( 1 + \int_\Omega v^{p\ell+r}\, dx \right), \qquad (8.8)$$

where $M_4 = \max\{1, |\Omega|_N\} + \frac{M_3}{c_0 M_2}$. Since $r \geq p$, for $M_5 = \max\{1, |\Omega|_N\}$ we have

$$\int_\Omega |v_\lambda^{\ell+1}|^p\, dx = \int_\Omega v_\lambda^{p\ell+p}\, dx \leq \int_\Omega (1 + v_\lambda^{p\ell+r})\, dx \leq M_5 \left( 1 + \int_\Omega v^{p\ell+r}\, dx \right).$$

Combining this inequality with (8.3) and (8.8) yields

$$\int_\Omega v_\lambda^{\theta(\ell+1)}\, dx \leq M_0^\theta (\|\nabla(v_\lambda^{\ell+1})\|_p + \|v_\lambda^{\ell+1}\|_p)^\theta \leq M_6(\ell+1)^\theta \left( 1 + \int_\Omega v^{p\ell+r}\, dx \right)^{\frac{\theta}{p}},$$

where $M_6 = M_0^\theta \left( M_4^{\frac{1}{p}} + M_5^{\frac{1}{p}} \right)^\theta$, whence

$$1 + \int_\Omega v_\lambda^{\theta(\ell+1)}\, dx \leq (1 + M_6)(\ell+1)^\theta \left( 1 + \int_\Omega v^{p\ell+r}\, dx \right)^{\frac{\theta}{p}}.$$

Claim 1 follows by passing to the limit as $\lambda \to +\infty$.

Since $p \leq r < \theta$, we have $\alpha := \frac{r-p}{\theta-p}\theta \in [0, \theta)$. Let $\{q_i\}_{i \geq 0} \subset [\theta, +\infty)$ be the sequence defined by

$$q_i = \alpha + \left( \frac{\theta}{p} \right)^i (\theta - \alpha). \qquad (8.9)$$

Thus, $\{q_i\}_{i \geq 0}$ is increasing and $q_i \to +\infty$ as $i \to \infty$. Moreover, we have the inductive relation

$$q_{i+1} = \frac{\theta}{p}(q_i - r + p) \quad \text{for all } i \geq 0. \tag{8.10}$$

For $i \in \mathbb{N}_0$, set $\ell_i = \frac{q_i - r}{p} \in (0, +\infty)$. Thus, $q_i = p\ell_i + r$ and $q_{i+1} = \theta(\ell_i + 1)$ [see (8.10)]. Applying Claim 1 with $\ell = \ell_i$, we obtain the relation

$$1 + \int_\Omega v^{q_{i+1}} \, dx \leq \frac{M_1}{p^\theta}(q_i - r + p)^\theta \left(1 + \int_\Omega v^{q_i} \, dx\right)^{\frac{\theta}{p}} \quad \text{for all } i \geq 0. \tag{8.11}$$

On the basis of (8.11), we will prove the following claim.

*Claim 2:* There is $M_7 = M_7(c_0, c, R, \sigma, \Omega, p, \theta, N) > 0$ such that

$$J_i := \left(1 + \int_\Omega v^{q_i} \, dx\right)^{\frac{1}{q_i}} \leq M_7(1 + \|u\|_\theta)^{\frac{\theta - p}{\theta - r}} \quad \text{for all } i \geq 0.$$

First, since $q_0 = \theta$, we know that $J_0 < +\infty$ [see (8.3)]. Then, from (8.11) we obtain that $\{J_i\}_{i \geq 0} \subset (0, +\infty)$. Let $S_i = q_i \ln J_i$ and $M_8 = \left(\frac{M_1}{p^\theta}\right)^{\frac{1}{\theta}}$. By (8.11), we have

$$S_{i+1} \leq \theta \ln M_8 + \theta \ln(q_i - r + p) + \frac{\theta}{p} S_i \leq \theta \ln M_8 + \theta \ln q_i + \frac{\theta}{p} S_i \quad \text{for all } i \geq 0$$

(since $q_i - r + p \leq q_i$). Clearly, $q_i \leq \left(\frac{\theta}{p}\right)^i \theta$ [see (8.9)], whence

$$S_{i+1} \leq (i+1)M_9 + \frac{\theta}{p} S_i \quad \text{for all } i \geq 0 \tag{8.12}$$

for some $M_9 = M_9(c_0, c, R, \sigma, \Omega, p, \theta, N) > 0$. By easy induction and straightforward computations, from (8.12) we derive

$$S_i \leq M_9 \sum_{j=0}^{i-1}(i-j)\left(\frac{\theta}{p}\right)^j + \left(\frac{\theta}{p}\right)^i S_0 = M_9 \frac{\left(\frac{\theta}{p}\right)^{i+1} - (i+1)\frac{\theta}{p} + i}{\left(\frac{\theta}{p} - 1\right)^2} + \left(\frac{\theta}{p}\right)^i S_0$$

$$\leq \left(\frac{\theta}{p}\right)^i \left(M_9 \frac{\theta}{p}\left(\frac{\theta}{p} - 1\right)^{-2} + \ln(1 + \|u\|_\theta^\theta)\right)$$

$$\leq \left(\frac{\theta}{p}\right)^i \ln\left(M_{10}(1 + \|u\|_\theta)^\theta\right) \quad \text{for all } i \geq 0,$$

for some $M_{10} = M_{10}(c_0, c, R, \sigma, \Omega, p, \theta, N) > 0$. Combining with (8.9) we get

$$\ln J_i = \frac{1}{q_i} S_i \leq \frac{\left(\frac{\theta}{p}\right)^i \ln\left(M_{10}(1 + \|u\|_\theta)^\theta\right)}{\alpha + \left(\frac{\theta}{p}\right)^i (\theta - \alpha)} \leq \frac{\ln\left(M_{10}(1 + \|u\|_\theta)^\theta\right)}{\theta - \alpha},$$

hence

$$J_i \leq M_7 \left(1 + \|u\|_\theta\right)^{\frac{\theta-p}{\theta-r}} \quad \text{for all } i \geq 0,$$

with $M_7 = M_{10}^{\frac{\theta-p}{\theta(\theta-r)}}$. This proves Claim 2.

*Claim 3:* $v \in L^\infty(\Omega)$ and $\|v\|_\infty \leq \hat{M}_u := M_7 \left(1 + \|u\|_\theta\right)^{\frac{\theta-p}{\theta-r}}$.
Let $w \in L^1(\Omega)$. For $i \geq 0$ we define $w_i \in L^{q_i'}(\Omega)$ by

$$w_i(x) = \begin{cases} |w(x)| & \text{if } |w(x)| < 1, \\ |w(x)|^{\frac{1}{q_i'}} & \text{if } |w(x)| \geq 1. \end{cases}$$

Since the sequence $\{q_i\}_{i\geq 0} \subset (1, +\infty)$ is increasing and $q_i \to +\infty$ as $i \to \infty$, the conjugate sequence $\{q_i'\}_{i\geq 0} \subset (1, +\infty)$ is decreasing and $q_i' \to 1$ as $i \to \infty$. Hence, the sequence of functions $\{w_i\}_{i\geq 0}$ is nondecreasing and, for all $x \in \Omega$, we have $w_i(x) \to |w(x)|$ as $i \to \infty$. For all $i \geq 0$, we see that

$$\int_\Omega v w_i \, dx \leq \|v\|_{q_i} \|w_i\|_{q_i'} \leq \hat{M}_u \left(\int_{\{|w|<1\}} |w|^{q_i'} \, dx + \int_{\{|w|\geq 1\}} |w| \, dx\right)^{\frac{1}{q_i'}}$$

$$\leq \hat{M}_u \left(\int_\Omega |w| \, dx\right)^{\frac{1}{q_i'}} \tag{8.13}$$

(since $q_i' > 1$). Passing to the limit as $i \to \infty$ in (8.13) by invoking the Beppo Levi monotone convergence theorem, we obtain

$$\int_\Omega |vw| \, dx \leq \hat{M}_u \|w\|_1 \quad \text{for all } w \in L^1(\Omega).$$

This implies that the map $\psi : w \mapsto \int_\Omega vw \, dx$ belongs to $(L^1(\Omega))^*$ and satisfies $\|\psi\| \leq \hat{M}_u$. Therefore, $v \in L^\infty(\Omega)$ and $\|v\|_\infty \leq \hat{M}_u$. This proves Claim 3.

By Claim 3, we obtain $\|u^+\|_\infty \leq M_7 \left(1 + \|u\|_\theta\right)^{\frac{\theta-p}{\theta-r}}$. Applying the first part of the proof to the maps $\tilde{a}(x, \xi) := -a(x, -\xi)$, $\tilde{f}(x, s) := -f(x, -s)$, and to the function $\tilde{u} := -u$ instead of $u$, we also derive $u^- \in L^\infty(\Omega)$ and $\|u^-\|_\infty \leq M_7 \left(1 + \|u\|_\theta\right)^{\frac{\theta-p}{\theta-r}}$. The proof of the theorem is now complete. □

*Remark 8.5.* In the literature, the method used in the proof of Theorem 8.4 is called the *Moser iteration technique*.

**Corollary 8.6.** *Assume that* $H(a)_1$ *and* $H(f)_1$ *hold. If* $u \in W_0^{1,p}(\Omega)$ *is a weak solution of (8.1), then* $u \in L^\infty(\Omega)$, *and for any* $\theta \in (r, +\infty)$, $\theta \leq p^*$ *we have*

$$\|u\|_\infty \leq M \left(1 + \|u\|_\theta\right)^{\frac{\theta-p}{\theta-r}},$$

*with M as in Theorem 8.4.*

*Proof.* Since $u \in W_0^{1,p}(\Omega)$, we have $\min\{u^+, \lambda\}^\alpha \in W_0^{1,p}(\Omega)$ and $\min\{u^-, \lambda\}^\alpha \in W_0^{1,p}(\Omega)$ for all $\lambda > 0$, all $\alpha \geq 1$. Hence Theorem 8.4 can be applied to $u$. $\quad\square$

**Corollary 8.7.** *Assume that* $H(a)_1$ *and* $H(f)_1$ *hold. If* $u \in W^{1,p}(\Omega)$ *is a weak solution of (8.2), then* $u \in L^\infty(\Omega)$, *and for any* $\theta \in (r, +\infty)$, $\theta \leq p^*$, *we have*

$$\|u\|_\infty \leq M \left(1 + \|u\|_\theta\right)^{\frac{\theta-p}{\theta-r}},$$

*with M as in Theorem 8.4. Moreover, u satisfies the boundary condition* $\frac{\partial u}{\partial n_a} = 0$.

*Proof.* From Theorem 8.4 we know that $u \in L^\infty(\Omega)$, with $\|u\|_\infty \leq M \left(1 + \|u\|_\theta\right)^{\frac{\theta-p}{\theta-r}}$. Then $H(f)_1$ yields $f(\cdot, u(\cdot)) \in L^\infty(\Omega) \subset L^{p'}(\Omega)$. As noted in Remark 8.3(b), this fact ensures that $\frac{\partial u}{\partial n_a} = 0$. $\quad\square$

## Regularity of Weak Solutions of Nonlinear Elliptic Problems

Next we present criteria that guarantee regularity up to the boundary of a weak solution of a boundary value problem, provided that we know that this weak solution is bounded. Here we consider $\Omega \subset \mathbb{R}^N$ ($N \geq 1$) a bounded domain with a $C^2$-boundary and $p \in (1, +\infty)$. We will state the relevant regularity results for an elliptic equation in divergence form involving a general operator $a$ and a Carathéodory function $f$. The hypotheses on $a$ and $f$ are as follows.

$H(a)_2$ (i) $a : \overline{\Omega} \times \mathbb{R} \times \mathbb{R}^N \to \mathbb{R}^N$ is a continuous map whose restriction to $\overline{\Omega} \times \mathbb{R} \times (\mathbb{R}^N \setminus \{0\})$ is of class $C^1$, and $a(x, s, 0) = 0$ for all $(x, s) \in \overline{\Omega} \times \mathbb{R}$;

　　　(ii) There are a constant $R \geq 0$ and a nonincreasing map $\mu_1 : [0, +\infty) \to (0, +\infty)$ such that, for all $x \in \overline{\Omega}$, all $s \in \mathbb{R}$, and all $\xi, \eta \in \mathbb{R}^N$, $\xi \neq 0$, we have

$$(a'_\xi(x, s, \xi)\eta, \eta)_{\mathbb{R}^N} \geq \mu_1(|s|)(R + |\xi|)^{p-2}|\eta|^2,$$

　　　　where $a'_\xi(x, s, \xi)$ stands for the differential of the map $a(x, s, \cdot)$ evaluated at $\xi$;

(iii) There is a nondecreasing map $\mu_2 : [0,+\infty) \to (0,+\infty)$ such that, for all $x \in \overline{\Omega}$, all $s \in \mathbb{R}$, and all $\xi \in \mathbb{R}^N \setminus \{0\}$, we have

$$\|a'_\xi(x,s,\xi)\| \leq \mu_2(|s|)(R+|\xi|)^{p-2},$$

where $R \geq 0$ is the same constant as in (ii);
(iv) There are constants $\alpha, \beta \in (0,1)$ such that, for all $x,y \in \overline{\Omega}$, all $s,t \in \mathbb{R}$, and all $\xi \in \mathbb{R}^N$, we have

$$|a(x,s,\xi) - a(y,t,\xi)| \leq \mu_2(\max\{|s|,|t|\})(|x-y|^\alpha + |s-t|^\beta)(1+|\xi|)^{p-2}|\xi|,$$

where $\mu_2 : [0,+\infty) \to (0,+\infty)$ is the same as in (iii).

H$(f)_2$  $f{:}\Omega \times \mathbb{R} \times \mathbb{R}^N \to \mathbb{R}$ is a Carathéodory function, i.e., $f(\cdot,s,\xi)$ is measurable for all $(s,\xi) \in \mathbb{R} \times \mathbb{R}^N$ and $f(x,\cdot,\cdot)$ is continuous for a.a. $x \in \Omega$. Moreover,

$$|f(x,s,\xi)| \leq \mu_2(|s|)(1+|\xi|^p) \text{ for a.a. } x \in \Omega, \text{ all } (s,\xi) \in \mathbb{R} \times \mathbb{R}^N,$$

where $\mu_2 : [0,+\infty) \to (0,+\infty)$ is the same as in H$(a)_2$ (iii).

*Remark 8.8.*  Hypothesis H$(f)_2$ implies that we have $f(\cdot,u(\cdot),\nabla u(\cdot)) \in L^1(\Omega)$ whenever $u \in W^{1,p}(\Omega) \cap L^\infty(\Omega)$. Similarly, it easily follows from H$(a)_2$ (iii) that we have $a(\cdot,u(\cdot),\nabla u(\cdot)) \in L^{p'}(\Omega,\mathbb{R}^N)$ whenever $u \in W^{1,p}(\Omega) \cap L^\infty(\Omega)$.

*Example 8.9.*  A typical example of operator $a$ satisfying hypotheses H$(a)_2$ is

$$a(x,s,\xi) = \theta(x,s)(R+|\xi|)^{p-2}\xi \text{ for all } (x,s,\xi) \in \overline{\Omega} \times \mathbb{R} \times \mathbb{R}^N,$$

where $R \geq 0$ is a constant and $\theta \in C^1(\overline{\Omega} \times \mathbb{R}, \mathbb{R})$ is a bounded function such that $\inf_{\overline{\Omega} \times \mathbb{R}} \theta > 0$. In particular, if $R = 0$ and $\theta \equiv 1$, then the resulting differential operator $\operatorname{div} a(x,u,\nabla u)$ is the $p$-Laplacian.

We can now state the following regularity up to the boundary result due to Lieberman [228]. For $\lambda \in (0,1)$, we denote by $(C^{1,\lambda}(\overline{\Omega}), \|\cdot\|_{C^{1,\lambda}(\overline{\Omega})})$ the space of $C^1$-functions $u: \overline{\Omega} \to \mathbb{R}$ whose differential is Hölder continuous with exponent $\lambda$.

**Theorem 8.10.**  *Assume that* H$(a)_2$ *and* H$(f)_2$ *hold. Let* $u \in W^{1,p}_0(\Omega) \cap L^\infty(\Omega)$ *[resp.* $u \in W^{1,p}(\Omega) \cap L^\infty(\Omega)$*] be a function such that the equality*

$$\int_\Omega (a(x,u,\nabla u),\nabla v)_{\mathbb{R}^N} dx = \int_\Omega f(x,u,\nabla u)\, v(x)\, dx$$

*holds for all* $v \in W^{1,p}_0(\Omega) \cap L^\infty(\Omega)$ *[resp. all* $v \in W^{1,p}(\Omega) \cap L^\infty(\Omega)$*]. Fix* $m \geq \|u\|_\infty$*. Then there are constants* $\lambda \in (0,1)$ *and* $M > 0$ *depending only on* $m$, $\mu_1(m)$, $\mu_2(m)$, $R$, $\alpha$, $\beta$, $\Omega$, $p$, *and* $N$ *such that*

$$u \in C^{1,\lambda}(\overline{\Omega}) \quad \text{and} \quad \|u\|_{C^{1,\lambda}(\overline{\Omega})} \leq M.$$

*Remark 8.11.*

(a) Lieberman's result is actually more general because it addresses boundary value problems with nonhomogeneous Dirichlet or Neumann boundary conditions. Also, it says more precisely that the constant $\lambda$ is independent of $\Omega$, whereas in the case where the domain $\Omega$ is convex, the constant $M$ depends on diam $\Omega$.

(b) If $\Omega$ has only a Lipschitz boundary, then one can get only a local regularity result, i.e., $u \in C_{\mathrm{loc}}^{1,\lambda}(\Omega)$, where $\lambda \in (0,1)$. This means that for every $\Omega_0 \subset \Omega$ with $\overline{\Omega_0} \subset \Omega$ we have $u \in C^{1,\lambda}(\overline{\Omega_0})$.

Combining this result with Theorem 8.4, we obtain the following corollary.

**Corollary 8.12.** *Let* $a : \overline{\Omega} \times \mathbb{R}^N \to \mathbb{R}^N$ *and* $f : \Omega \times \mathbb{R} \to \mathbb{R}$ *satisfy hypotheses* $\mathrm{H}(a)_1$, $\mathrm{H}(a)_2$, *and* $\mathrm{H}(f)_1$. *Let* $u \in W_0^{1,p}(\Omega)$ *[resp.* $u \in W^{1,p}(\Omega)$*] such that*

$$\int_\Omega (a(x, \nabla u), \nabla v)_{\mathbb{R}^N}\, dx = \int_\Omega f(x, u) v\, dx$$

*for all* $v \in W_0^{1,p}(\Omega)$ *[resp. all* $v \in W^{1,p}(\Omega)$*]. Then* $u \in C^{1,\lambda}(\overline{\Omega})$ *for some* $\lambda \in (0,1)$.

*Proof.* By Theorem 8.4, we know that $u \in L^\infty(\Omega)$. Hence, $\mathrm{H}(f)_1$ yields $f(\cdot, u(\cdot)) \in L^\infty(\Omega)$. Now, applying Theorem 8.10 to $u$ and the maps $\hat{a}(x, s, \xi) = a(x, \xi)$ and $\hat{f}(x, s, \xi) = f(x, u(x))$, we infer that $u \in C^{1,\lambda}(\overline{\Omega})$, with $\lambda \in (0,1)$.  □

**Corollary 8.13.** *Let* $f : \Omega \times \mathbb{R} \to \mathbb{R}$ *satisfy* $\mathrm{H}(f)_1$. *If* $u \in W_0^{1,p}(\Omega)$ *[resp.* $u \in W^{1,p}(\Omega)$*] satisfies*

$$\int_\Omega |\nabla u|^{p-2}(\nabla u, \nabla v)_{\mathbb{R}^N}\, dx = \int_\Omega f(x, u) v\, dx$$

*for all* $v \in W_0^{1,p}(\Omega)$ *[resp.* $v \in W^{1,p}(\Omega)$*], then* $u \in C^{1,\lambda}(\overline{\Omega})$ *for some* $\lambda \in (0,1)$.

## 8.2  Maximum Principles and Comparison Results

The maximum principle type results and the related comparison theorems are basic tools in the study of second-order elliptic partial differential equations. Their origin can be traced back to the maximum principle for harmonic functions, already known to Gauss (1839). The next major breakthrough was achieved by Hopf (1927) for classical solutions of elliptic differential inequalities. In more recent decades starting with the pioneering works of Serrin, Vázquez, and Díaz, there have been extensions to nonlinear elliptic differential inequalities. The aim of this section is to survey some of the nonlinear results that we will need later in the study of elliptic equations.

**Local Maximum Principles and Local Comparison Principles**

We consider $\Omega \subset \mathbb{R}^N$ $(N \geq 1)$ a domain. In the first part of this section, we present local maximum principles and local comparison principles for functions defined on $\Omega$, so we do not need any boundedness assumption on $\Omega$ or regularity assumption on its boundary $\partial \Omega$.

From the classical theory of harmonic functions we know that if $u \in C^2(\Omega)$ is a nonnegative, harmonic function and $\Omega_0$ is a subdomain of $\Omega$ such that $\overline{\Omega_0} \subset \Omega$, then there exists a constant $c = c(N, \Omega_0, \Omega) > 0$ such that

$$\sup_{\Omega_0} u \leq c \inf_{\Omega_0} u.$$

The result is known in the literature as *Harnack's inequality*. This result was extended by Trudinger [378] to nonnegative solutions of nonlinear elliptic inequalities of the form

$$-\operatorname{div} a(x, u(x), \nabla u(x)) + f(x, u(x), \nabla u(x)) \geq 0 \text{ in } \Omega, \tag{8.14}$$

comprising a general operator $a$ and a function $f$, the assumptions on which are as follows. Here, we fix $p \in (1, +\infty)$.

H$(a)_1$ (i) $a : \Omega \times \mathbb{R} \times \mathbb{R}^N \to \mathbb{R}^N$ is a Carathéodory function [i.e., $a(\cdot, s, \xi)$ is measurable for all $(s, \xi) \in \mathbb{R} \times \mathbb{R}^N$ and $a(x, \cdot, \cdot)$ is continuous for a.a. $x \in \Omega$];
   (ii) There are constants $c_1 > 0$ and $c_2 \geq 0$ such that

$$(a(x, s, \xi), \xi)_{\mathbb{R}^N} \geq c_1 |\xi|^p - c_2 |s|^p \text{ for a.a. } x \in \Omega, \text{ all } s \in \mathbb{R}, \text{ all } \xi \in \mathbb{R}^N;$$

   (iii) There are constants $c_3, c_4 \geq 0$ such that

$$|a(x, s, \xi)| \leq c_3 |\xi|^{p-1} + c_4 |s|^{p-1} \text{ for a.a. } x \in \Omega, \text{ all } s \in \mathbb{R}, \text{ all } \xi \in \mathbb{R}^N.$$

H$(f)_1$ $f : \Omega \times \mathbb{R} \times \mathbb{R}^N \to \mathbb{R}$ is a Carathéodory function satisfying the growth condition

$$|f(x, s, \xi)| \leq c_5 |\xi|^p + c_6 |s|^{p-1} \text{ for a.a. } x \in \Omega, \text{ all } s \in \mathbb{R}, \text{ all } \xi \in \mathbb{R}^N,$$

for some constants $c_5, c_6 \geq 0$.

*Example 8.14.* Again, the typical example for the hypotheses H$(a)_1$ is the operator $a(x, s, \xi) = |\xi|^{p-2}\xi$, for which $\operatorname{div} a(x, u, \nabla u) = \Delta_p u$ is the $p$-Laplacian.

The following result is due to Trudinger [378, Theorem 1.2].

**Theorem 8.15.** *Assume that* H$(a)_1$ *and* H$(f)_1$ *hold. Let* $u \in W^{1,p}_{\text{loc}}(\Omega) \cap C(\Omega)$, *with* $u \geq 0$ *in* $\Omega$, *such that (8.14) holds in the distributional sense, that is,*

$$\int_\Omega (a(x,u,\nabla u),\nabla v)_{\mathbb{R}^N}\,dx + \int_\Omega f(x,u,\nabla u)v\,dx \geq 0 \quad \text{for all } v \in C_c^\infty(\Omega),\ v \geq 0.$$

*Then, for every $x_0 \in \Omega$ and $\rho > 0$ with $\overline{B_{3\rho}(x_0)} \subset \Omega$, there is $M > 0$ depending only on $N, p, \|u|_{B_{3\rho}(x_0)}\|_\infty, \rho, c_1, \ldots, c_6$, such that the inequality*

$$\left(\int_{B_\rho(x_0)} u^q\,dx\right)^{\frac{1}{q}} \leq M\rho^{\frac{N}{q}} \inf_{B_\rho(x_0)} u$$

*holds for all $q \in (0, \frac{N(p-1)}{N-p})$ in the case $p < N$ [resp. all $q \in (0,+\infty]$ in the case $p \geq N$].*

*Remark 8.16.*

(a) In fact, Trudinger's result holds under the weaker assumption that $u \in W_{\text{loc}}^{1,p}(\Omega) \cap L_{\text{loc}}^\infty(\Omega)$ (in this case the infimum on the right-hand side stands for the essential infimum).

(b) Trudinger [378, Theorem 1.1] obtained a refined version of Theorem 8.15 under the stronger assumption that equality holds (instead of inequality) in (8.14), that is, $u \in W_{\text{loc}}^{1,p}(\Omega) \cap C(\Omega)$, $u \geq 0$, is such that

$$\int_\Omega (a(x,u,\nabla u),\nabla v)_{\mathbb{R}^N}\,dx + \int_\Omega f(x,u,\nabla u)v\,dx = 0 \quad \text{for all } v \in C_c^\infty(\Omega).$$

Then, for every $x_0 \in \Omega$, every $\rho \in (0,1)$ such that $\overline{B_{3\rho}(x_0)} \subset \Omega$, we have

$$\sup_{B_\rho(x_0)} u \leq M \inf_{B_\rho(x_0)} u,$$

with $M > 0$ as in Theorem 8.15.

This Harnack-type inequality leads to the following maximum principle.

**Corollary 8.17.** *Assume that* $H(a)_1$ *and* $H(f)_1$ *hold. Let $u \in W_{\text{loc}}^{1,p}(\Omega) \cap C(\Omega)$, with $u \geq 0$, satisfy (8.14) in the distributional sense. Then, we have either $u \equiv 0$ in $\Omega$ or $u(x) > 0$ for all $x \in \Omega$.*

*Proof.* It suffices to show that the set $Z := \{x \in \Omega : u(x) = 0\}$ is open and closed in $\Omega$. Since $u$ is continuous, $Z$ is clearly closed. Let $x_0 \in Z$. Choose $\rho > 0$ small so that $\overline{B_{3\rho}(x_0)} \subset \Omega$. Evidently, $\inf_{B_\rho(x_0)} u = 0$ and so, from Theorem 8.15, it follows that $\int_{B_\rho(x_0)} u^q\,dx = 0$ for some $q > 0$, hence $u \equiv 0$ on $B_\rho(x_0)$, whence $B_\rho(x_0) \subset Z$, so $Z$ is open. The proof is complete.                                                                        $\square$

A result related to Theorem 8.15 is the following Harnack-type comparison inequality due to Damascelli [99, Theorem 1.3]. It involves an operator $a$ satisfying the following (stronger) assumptions.

H($a$)$_2$ (i) $a : \overline{\Omega} \times \mathbb{R}^N \to \mathbb{R}^N$ is a continuous map whose restriction to $\overline{\Omega} \times (\mathbb{R}^N \setminus \{0\})$
is of class $C^1$, and $a(x,0) = 0$ for all $x \in \overline{\Omega}$;

(ii) There is a constant $c_1 > 0$ such that

$$(a'_\xi(x,\xi)\eta, \eta)_{\mathbb{R}^N} \geq c_1 |\xi|^{p-2} |\eta|^2 \text{ for all } x \in \overline{\Omega}, \text{ all } \xi, \eta \in \mathbb{R}^N, \xi \neq 0;$$

(iii) There is a constant $c_3 > 0$ such that

$$\|a'_\xi(x,\xi)\| \leq c_3 |\xi|^{p-2} \text{ for all } x \in \overline{\Omega}, \text{ all } \xi \in \mathbb{R}^N \setminus \{0\}.$$

*Remark 8.18.* Note that H($a$)$_2$ (ii) implies that the operator $a$ is strictly monotone.
More precisely, we establish the formula

$$(a(x,\xi) - a(x,\eta), \xi - \eta)_{\mathbb{R}^N} \geq \tilde{c}_1 |\xi - \eta|^2 (|\xi| + |\eta|)^{p-2} \tag{8.15}$$

for a.a. $x \in \overline{\Omega}$, all $\xi, \eta \in \mathbb{R}^N$, for some $\tilde{c}_1 > 0$. We may assume that $\eta \neq \xi$ and
$|\eta| \geq |\xi|$ (up to exchanging the roles of $\eta$ and $\xi$). The function $t \mapsto (a(x, \eta + t(\xi - \eta)), \xi - \eta)_{\mathbb{R}^N}$ is continuous on $[0,1]$. Its derivative exists a.e. in $[0,1]$ and is positive
(by H($a$)$_2$ (ii)), hence integrable. Thus, using H($a$)$_2$ (ii), we can write

$$(a(x,\xi) - a(x,\eta), \xi - \eta)_{\mathbb{R}^N} = \int_0^1 (a'_\xi(x, \eta + t(\xi - \eta))(\xi - \eta), \xi - \eta)_{\mathbb{R}^N} \, dt$$

$$\geq c_1 |\xi - \eta|^2 \int_0^{\frac{1}{4}} |\eta + t(\xi - \eta)|^{p-2} \, dt. \tag{8.16}$$

For every $t \in [0, \frac{1}{4}]$, using that $|\eta| \geq |\xi|$, we see that

$$\frac{1}{4}(|\eta| + |\xi|) \leq |\eta| - \frac{1}{4}(|\xi| + |\eta|) \leq |\eta + t(\xi - \eta)| \leq |\eta| + |\xi|. \tag{8.17}$$

Relation (8.15) easily follows by combining (8.16) and (8.17).

**Theorem 8.19.** *Assume that* H($a$)$_2$ *holds and* $N \geq 2$. *Let* $u, v \in W_{\text{loc}}^{1,\infty}(\Omega)$ *in the case*
$p \neq 2$ *[resp.* $u, v \in W_{\text{loc}}^{1,2}(\Omega) \cap L_{\text{loc}}^\infty(\Omega)$ *in the case* $p = 2$] *satisfy* $u \leq v$ *a.e. in* $\Omega$ *and*

$$\int_\Omega (a(x, \nabla u), \nabla w)_{\mathbb{R}^N} \, dx + \lambda \int_\Omega uw \, dx \leq \int_\Omega (a(x, \nabla v), \nabla w)_{\mathbb{R}^N} \, dx + \lambda \int_\Omega vw \, dx \tag{8.18}$$

*for all* $w \in C_c^\infty(\Omega)$, $w \geq 0$, *and for a fixed* $\lambda \in \mathbb{R}$. *Let* $x_0 \in \Omega$ *and* $\rho > 0$ *such that*
$\overline{B_{5\rho}(x_0)} \subset \Omega$. *In the case* $p \neq 2$, *we assume moreover that*

$$m_1 := \underset{B_{5\rho}(x_0)}{\text{ess inf}} (|\nabla u(\cdot)| + |\nabla v(\cdot)|) > 0.$$

*Then there is* $M > 0$, *which depends only on* $N, p, \lambda, c_1, c_3, \rho$, *and in the case* $p \neq 2$
*also on* $m_1$ *and* $m_2 := \underset{B_{5\rho}(x_0)}{\text{ess sup}} (|\nabla u(\cdot)| + |\nabla v(\cdot)|)$ *such that*

$$\left(\int_{B_{2\rho}(x_0)} (v-u)^q \, dx\right)^{\frac{1}{q}} \leq M\rho^{\frac{N}{q}} \operatorname*{ess\,inf}_{B_\rho(x_0)}(v-u) \ \ for \ all \ q \in (0, \tfrac{N}{N-2}).$$

Theorem 8.19 yields the following comparison result.

**Corollary 8.20.** *Assume that* $H(a)_2$ *holds and* $N \geq 2$. *Let* $u, v \in C^1(\Omega)$, $u \leq v$, *satisfy (8.18) for some* $\lambda \in \mathbb{R}$. *Set*

$$\begin{cases} D = \{x \in \Omega : |\nabla u(x)| + |\nabla v(x)| = 0\} & if \ p \neq 2, \\ D = \emptyset & if \ p = 2. \end{cases}$$

*For every connected component* $C \subset \Omega \setminus D$ *we have either* $u \equiv v$ *in* $C$ *or* $u(x) < v(x)$ *for all* $x \in C$.

*Proof.* Again, it suffices to check that $Z := \{x \in C : u(x) = v(x)\}$ is open and closed in $C$. Evidently, it is closed in $C$. Now, let $x_0 \in Z$. Thus, $u(x_0) = v(x_0)$ and, in the case $p \neq 2$, $|\nabla u(x_0)| + |\nabla v(x_0)| > 0$. By continuity, we can find $\rho > 0$ such that

$$\overline{B_{5\rho}(x_0)} \subset \Omega \quad \text{and, for } p \neq 2, \quad \inf_{x \in B_{5\rho}(x_0)} (|\nabla u(x)| + |\nabla v(x)|) > 0.$$

Since $0 = v(x_0) - u(x_0) = \inf_{B_\rho(x_0)} (v-u)$, invoking Theorem 8.19, we have

$$\int_{B_{2\rho}(x_0)} (v-u) \, dx = 0, \ \text{i.e., } u = v \ \text{in } B_{2\rho}(x_0).$$

Hence $B_{2\rho}(x_0) \subset Z$, and so $Z$ is also open in $C$.                                   □

We now state a tangency principle from which we will derive further comparison principles. In addition to the operator $a$ satisfying $H(a)_2$, we consider a function $f$ satisfying the following assumptions.

$H(f)_2$ (i) $f : \Omega \times \mathbb{R} \to \mathbb{R}$ is Carathéodory;
    (ii) There exist $c \geq 0$ and $r \in (1, p^*)$ such that

$$|f(x,s)| \leq c(1 + |s|^{r-1}) \ \text{for a.a. } x \in \Omega, \text{ all } s \in \mathbb{R};$$

    (iii) For a.a. $x \in \Omega$ the function $s \mapsto f(x,s)$ is nonincreasing.

**Theorem 8.21.** *Let* $\Omega \subset \mathbb{R}^N$ $(N \geq 1)$ *be a bounded domain with Lipschitz boundary. Assume that* $H(a)_2$ *and* $H(f)_2$ *hold. Let* $u, v \in W^{1,p}(\Omega) \cap C(\overline{\Omega})$ *be such that the inequality*

$$-\operatorname{div} a(x, \nabla u) - f(x,u) \leq -\operatorname{div} a(x, \nabla v) - f(x,v) \tag{8.19}$$

*holds in the distributional sense. Assume that $v \geq u + \mu$ on $\partial\Omega$ for some $\mu \in \mathbb{R}$ and that*

$$f(x, u(x)) \leq f(x, u(x) + \mu) \quad \text{for a.a. } x \in \Omega. \tag{8.20}$$

*Then we have $v \geq u + \mu$ in $\Omega$.*

*Proof.* Let $w = \max\{0, u + \mu - v\} \in W^{1,p}(\Omega) \cap C(\overline{\Omega})$. The assumption that $(v - u)|_{\partial\Omega} \geq \mu$ implies that $w|_{\partial\Omega} = 0$. Since $\partial\Omega$ is Lipschitz, this guarantees that $w \in W_0^{1,p}(\Omega)$ (Theorem 1.33). The fact that $w \geq 0$ and the density of $C_c^\infty(\Omega)$ in $W_0^{1,p}(\Omega)$ imply that we can act on (8.19) with $w$. Using also (8.20) and H$(f)_2$ (iii) and letting $S := \{x \in \Omega : u(x) + \mu > v(x)\}$, we obtain

$$\int_\Omega (a(x, \nabla u) - a(x, \nabla v), \nabla w)_{\mathbb{R}^N} \, dx \leq \int_\Omega (f(x, u) - f(x, v)) w \, dx$$

$$\leq \int_S (f(x, u + \mu) - f(x, v))(u + \mu - v) \, dx \leq 0. \tag{8.21}$$

Note that $\nabla w = \nabla(u - v)$ a.e. in $S$ and $\nabla w = 0$ a.e. in $\Omega \setminus S$. Combining this observation with (8.21) and Remark 8.18, we obtain

$$(a(x, \nabla u) - a(x, \nabla v), \nabla(u - v))_{\mathbb{R}^N} = 0 \quad \text{for a.a. } x \in S$$

and

$$\nabla w = 0 \quad \text{a.e. in } S.$$

Hence $\nabla w = 0$ a.e. in $\Omega$. Since $w$ is continuous in $\overline{\Omega}$ and vanishes on $\partial\Omega$, we obtain that $w \equiv 0$, whence $v(x) \geq u(x) + \mu$ for all $x \in \Omega$. $\square$

*Remark 8.22.* The hypothesis (8.20) of the theorem is guaranteed, for instance, if $\mu \leq 0$ [see H$(f)_2$ (iii)] or if $f(x, s) = f(x)$ does not depend on $s$.

In the case where $f \equiv 0$, Theorem 8.21 yields the following comparison principles. Here, $\Omega \subset \mathbb{R}^N$ ($N \geq 1$) is a domain.

**Corollary 8.23.** *Assume that H$(a)_2$ holds. Let $u, v \in W_{loc}^{1,p}(\Omega) \cap C(\Omega)$, with $u \leq v$ in $\Omega$, such that the inequality*

$$-\operatorname{div} a(x, u, \nabla u) \leq -\operatorname{div} a(x, v, \nabla v) \tag{8.22}$$

*holds in $\Omega$ in the distributional sense. Assume that the set $Z := \{x \in \Omega : u(x) = v(x)\}$ is compact or discrete in $\Omega$. Then $Z = \emptyset$, i.e., we have $u(x) < v(x)$ for all $x \in \Omega$.*

*Proof.* Arguing by contradiction, we assume that $Z$ contains at least one element $x_0$. In the case where $Z$ is compact, we let $\Omega_0 \subset \overline{\Omega_0} \subset \Omega$ be a bounded subdomain with Lipschitz boundary $\partial\Omega_0$ such that $\Omega_0 \supset Z$. In the case where $Z$ is discrete, we choose the subdomain $\Omega_0$ such that $\Omega_0 \cap Z = \{x_0\}$. In both cases, the boundary

$\partial\Omega_0$ does not intersect the set $Z$, hence we have $(v-u)|_{\partial\Omega_0} \geq \mu$ for some $\mu > 0$. Applying Theorem 8.21, we obtain that $v(x_0) \geq u(x_0)+\mu > u(x_0)$, a contradiction. The proof is now complete. □

**Corollary 8.24.** *Assume that* H$(a)_2$ *holds. Let* $u,v \in C^1(\Omega)$ *with* $u \leq v$ *in* $\Omega$, *satisfy* (8.22). *Assume that* $D := \{x \in \Omega : \nabla u(x) = \nabla v(x)\}$ *is compact or discrete. Then* $u(x) < v(x)$ *for all* $x \in \Omega$.

*Proof.* The set $\Omega \setminus D$ is open, and we have $|\nabla u(x) - \nabla v(x)| > 0$ for all $x \in \Omega \setminus D$. Any connected component $C \subset \Omega \setminus D$ is open. Clearly, $u \not\equiv v$ in $C$ (otherwise, we would have $C \subset D$, a contradiction). Then, clearly (if $N = 1$) or by Corollary 8.20 with $\lambda = 0$ (if $N \geq 2$), we have $u(x) < v(x)$ for all $x \in \Omega \setminus D$. Hence $Z := \{x \in \Omega : u(x) = v(x)\} \subset D$. Thus, $Z$ is compact or discrete. Now the conclusion follows by applying Corollary 8.23. □

We also state the following particular case of Corollary 8.23.

**Corollary 8.25.** *Let* $u,v \in W^{1,p}_{loc}(\Omega) \cap C(\Omega)$, *with* $u \leq v$ *in* $\Omega$, *such that we have* $-\Delta_p u \leq -\Delta_p v$ *in* $\Omega$. *If the set* $\{x \in \Omega : u(x) = v(x)\}$ *is compact or discrete, then it is empty, i.e., we have* $u(x) < v(x)$ *for all* $x \in \Omega$.

Finally, we mention the following strong comparison principle, which involves the $p$-Laplacian and a nontrivial function $f$ and whose proof can be found in Roselli and Sciunzi [347].

**Theorem 8.26.** *Let* $\Omega \subset \mathbb{R}^N$ $(N \geq 2)$ *be a bounded domain with* $C^2$-*boundary* $\partial\Omega$, *and let* $\Omega_0 \subset \Omega$ *be a subdomain. Assume that* $\frac{2(N+1)}{N+2} < p \leq 2$ *or* $p \geq 2$. *Let* $f : [0,+\infty) \to \mathbb{R}$ *be a continuous function that is locally Lipschitz on* $(0,+\infty)$. *Let* $u,v \in C^{1,\alpha}(\overline{\Omega})$ $(0 < \alpha < 1)$, *with* $u \geq 0$, $v \geq 0$, *be such that the inequality*

$$-\Delta_p u - f(u) \leq -\Delta_p v - f(v) \ \ in \ \Omega$$

*holds in the distributional sense, and there is* $w \in \{u,v\}$, *which is a solution of the problem*

$$\begin{cases} -\Delta_p w = f(w) & in \ \Omega, \\ w = 0 & on \ \partial\Omega, \end{cases}$$

*and such that* $f(w(\cdot))$ *is either positive or negative on* $\Omega_0$. *Assume that* $u \leq v$, $u \neq v$ *in* $\Omega_0$. *Then we have* $u(x) < v(x)$ *for all* $x \in \Omega$.

## Strong Maximum Principles

The last part of this section is devoted to a nonlinear version of the classical Hopf maximum principle and its consequences. Since the results concern the boundary of $\Omega$, it requires a regularity assumption, namely, in what follows, $\Omega$ denotes a bounded domain of $\mathbb{R}^N$ $(N \geq 1)$ with $C^2$-boundary $\partial\Omega$. This assumption guarantees that every boundary point $z_1 \in \partial\Omega$ satisfies the *interior sphere condition* (in fact, the assumption that $\partial\Omega$ is $C^1$ would be sufficient for this property), that is, whenever $r > 0$ is small enough, there is $z_0 \in \Omega$ such that $z_1$ is the unique point of $\partial\Omega \cap \partial B_r(z_0)$. Then the outward unit normal vector at $z_1$ is

$$n(z_1) = \frac{1}{r}(z_1 - z_0).$$

The version of the strong maximum principle that we prove here involves a general operator $-\operatorname{div} a(x, \nabla u)$ satisfying the following assumptions [which coincide with H$(a)_2$ apart from the new hypothesis (iv)].

H$(a)_3$  (i) $a : \overline{\Omega} \times \mathbb{R}^N \to \mathbb{R}^N$ is a continuous map whose restriction to $\overline{\Omega} \times (\mathbb{R}^N \setminus \{0\})$ is of class $C^1$, and $a(x, 0) = 0$ for all $x \in \overline{\Omega}$;

(ii) There is a constant $c_1 > 0$ such that

$$(a'_{\xi}(x, \xi)\eta, \eta)_{\mathbb{R}^N} \geq c_1 |\xi|^{p-2} |\eta|^2 \text{ for all } x \in \overline{\Omega}, \text{ all } \xi, \eta \in \mathbb{R}^N, \xi \neq 0;$$

(iii) There is a constant $c_3 \geq 0$ such that

$$\|a'_{\xi}(x, \xi)\| \leq c_3 |\xi|^{p-2} \text{ for all } x \in \overline{\Omega} \text{ and all } \xi \in \mathbb{R}^N \setminus \{0\};$$

(iv) There are constants $c_7 > 0$ and $\delta \in (0, 1]$ such that

$$\|a'_x(x, \xi)\| \leq c_7 |\xi|^{p-1} \text{ for all } x \in \overline{\Omega} \text{ and all } \xi \in \mathbb{R}^N, 0 < |\xi| < \delta.$$

**Theorem 8.27.** *Let $a : \overline{\Omega} \times \mathbb{R}^N \to \mathbb{R}^N$ satisfy H$(a)_3$. Let $z_1 \in \partial\Omega$. Assume that $u \in C^1(\Omega \cup \{z_1\}) \setminus \{0\}$, $u \geq 0$, satisfies the inequality*

$$\operatorname{div} a(x, \nabla u(x)) \leq c u(x)^{p-1} \text{ in } \Omega \tag{8.23}$$

*in the distributional sense, where $c > 0$ is a constant. Then $u > 0$ in $\Omega$. Moreover, if $u(z_1) = 0$, then we have*

$$\frac{\partial u}{\partial n}(z_1) := (\nabla u(z_1), n(z_1))_{\mathbb{R}^N} < 0,$$

*where $n(\cdot)$ denotes, as previously, the outward unit normal on $\partial\Omega$.*

*Proof.* The hypotheses H$(a)_3$ are clearly stronger than H$(a)_1$; thus, the fact that $u > 0$ in $\Omega$ is implied by Corollary 8.17. By the interior sphere condition at $z_1$, we find $r > 0$ and $z_0 \in \Omega$ such that $\partial\Omega \cap \partial B_{2r}(z_0) = \{z_1\}$. There is no loss of generality in assuming that $z_0 = 0$. Let

$$\Omega_0 = \{x \in \Omega : r < |x| < 2r\}.$$

We fix $k \in (0, +\infty)$ such that

$$k \geq \frac{2N}{rc_1}(c_3 + rc_7) \quad \text{and} \quad k \geq \left(\frac{2c}{c_1}\right)^{\frac{1}{p}}, \tag{8.24}$$

and then $R \in (0, +\infty)$ such that

$$R \leq \min_{\partial B_r(0)} u \quad \text{and} \quad kR \frac{e^{kr}}{e^{kr} - 1} < \delta. \tag{8.25}$$

Finally, we define $\xi : [0, r] \to [0, +\infty)$ by letting $\xi(t) = R\frac{e^{kt}-1}{e^{kr}-1}$, and we consider $v \in C^\infty(\overline{\Omega_0})$ given by

$$v(x) = \xi(2r - |x|) \quad \text{for all } x \in \overline{\Omega_0}.$$

We first note that $v \equiv 0$ on $\partial B_{2r}(0)$ and $v \equiv R$ on $\partial B_r(0)$, whence

$$v \leq u \quad \text{on } \partial \Omega_0 \tag{8.26}$$

[see (8.25)]. Fix $x \in \overline{\Omega_0}$, and set $t = 2r - |x|$. For $i, j \in \{1, \dots, N\}$ we have

$$\frac{\partial v}{\partial x_j}(x) = -\xi'(t)\frac{x_j}{|x|}, \qquad |\nabla v(x)| = \xi'(t),$$

and

$$\frac{\partial^2 v}{\partial x_i \partial x_j}(x) = \frac{k|x| + 1}{|x|^3}\xi'(t)x_i x_j - \frac{1}{|x|}\xi'(t)\delta_{i,j}. \tag{8.27}$$

By $a_i(\cdot, \cdot)$ we denote the $i$th component of the map $a(\cdot, \cdot)$. Let $e_i = (\delta_{i,j})_{j=1}^N \in \mathbb{R}^N$. We have

$$\operatorname{div} a(x, \nabla v(x)) = \sum_{i=1}^N \frac{\partial}{\partial x_i}(a_i(x, \nabla v))$$

$$= \sum_{i=1}^N \frac{\partial a_i}{\partial x_i}(x, \nabla v) + \sum_{i,j=1}^N \frac{\partial a_i}{\partial \xi_j}(x, \nabla v)\frac{\partial^2 v}{\partial x_i \partial x_j}(x)$$

$$= \sum_{i=1}^N \left((a_x'(x, \nabla v)e_i, e_i)_{\mathbb{R}^N} - \frac{\xi'(t)}{|x|}(a_\xi'(x, \nabla v)e_i, e_i)_{\mathbb{R}^N}\right)$$

$$+ \frac{k|x| + 1}{|x|^3}\xi'(t)(a_\xi'(x, \nabla v)x, x)_{\mathbb{R}^N}$$

$$\geq \left(-Nc_7 - \frac{Nc_3}{r} + kc_1\right)\xi'(t)^{p-1} \geq \frac{kc_1}{2}\xi'(t)^{p-1},$$

where we use (8.27), H(a)$_3$ (ii), (iii), (iv) [since we have $0 < |\nabla v(x)| = \xi'(t) < \delta$ by (8.25)], and (8.24). On the other hand, we see that

$$\frac{kc_1}{2}\xi'(t)^{p-1} = \frac{kc_1}{2}\left(kR\frac{e^{kt}}{e^{kr}-1}\right)^{p-1} \geq \frac{k^p c_1}{2c}c\left(R\frac{e^{kt}-1}{e^{kr}-1}\right)^{p-1} \geq cv(x)^{p-1}$$

[see (8.24)]. All together, we obtain

$$\operatorname{div} a(x, \nabla v) \geq cv(x)^{p-1} \quad \text{for all } x \in \overline{\Omega_0}. \tag{8.28}$$

In view of (8.23), (8.26), and (8.28), we may apply Theorem 8.21, which yields

$$v(x) \leq u(x) \quad \text{for all } x \in \overline{\Omega_0}.$$

Because $z_0 = 0$, we have $n(z_1) = \frac{z_1}{|z_1|}$. Using that $u(z_1) = v(z_1) = 0$ and the definition of $v$, we obtain

$$\frac{\partial u}{\partial n}(z_1) = -\lim_{t\downarrow 0}\frac{u((1-t)z_1)}{t|z_1|} \leq -\lim_{t\downarrow 0}\frac{v((1-t)z_1)}{t|z_1|} = -\frac{Rk}{e^{kr}-1} < 0.$$

The proof of the theorem is complete. $\qquad\square$

*Remark 8.28.* In Theorem 8.27, it is needed that the operator $a$ must be continuously differentiable. However, there is a version of the strong maximum principle, due to Finn and Gilbarg [139, pp. 31–35], which is valid for elliptic operators whose coefficients are only required to be Hölder continuous. This result can be stated as follows. Let $A = (a_{i,j})_{1 \leq i,j \leq N}$, where $a_{i,j} : \overline{\Omega} \to \mathbb{R}$ (for $i, j \in \{1, \ldots, N\}$) are Hölder continuous maps, and assume that there is a constant $c_1 > 0$ such that

$$(A(x)\xi, \xi)_{\mathbb{R}^N} \geq c_1|\xi|^2 \quad \text{for all } x \in \overline{\Omega}, \text{ all } \xi \in \mathbb{R}^N.$$

Let $u \in C^1(\overline{\Omega})$, with $u > 0$ in $\Omega$, be such that the inequality

$$\operatorname{div}(A(x)\nabla u) \leq 0$$

holds in the distributional sense. Then, for $z_1 \in \partial\Omega$ such that $u(z_1) = 0$, we have

$$\frac{\partial u}{\partial n}(z_1) < 0.$$

We conclude this section by proving a useful comparison result involving the $p$-Laplacian $\Delta_p u = \operatorname{div}(|\nabla u|^{p-2}\nabla u)$. First we introduce the Banach space

$$C_0^1(\overline{\Omega}) = \{u \in C^1(\overline{\Omega}) : u|_{\partial\Omega} = 0\}.$$

This is an ordered Banach space for the usual pointwise order. The positive cone is

$$C_0^1(\overline{\Omega})_+ = \{u \in C_0^1(\overline{\Omega}) : u(x) \geq 0 \text{ for all } x \in \overline{\Omega}\}.$$

This cone has a nonempty interior given by

$$\text{int}(C_0^1(\overline{\Omega})_+) = \left\{u \in C_0^1(\overline{\Omega})_+ : u(x) > 0 \text{ for all } x \in \Omega, \frac{\partial u}{\partial n}(x) < 0 \text{ for all } x \in \partial\Omega\right\}.$$

Here, as earlier, $n(\cdot)$ denotes the outward unit normal on $\partial\Omega$.

**Proposition 8.29.** *Let* $f, g \in L^\infty(\Omega)$, *with* $f > g \geq 0$ *a.e. in* $\Omega$, *and let* $u, v \in W_0^{1,p}(\Omega)$ *be such that the equalities*

$$-\Delta_p u(x) = f(x) \quad \text{and} \quad -\Delta_p v(x) = g(x) \quad \text{in } \Omega \tag{8.29}$$

*hold in the distributional sense [or, equivalently, in* $(W_0^{1,p}(\Omega))^*$*]. Then,*

$$u - v \in \text{int}(C_0^1(\overline{\Omega})_+).$$

*Proof.* First, invoking Corollary 8.13, we obtain that $u, v \in C^{1,\lambda}(\overline{\Omega})$ for some $\lambda \in (0, 1)$. It also follows from Theorem 8.21 that we have $u \geq v \geq 0$ in $\Omega$. If $v = 0$, then the result follows from Theorem 8.27. Thus, in what follows, we may assume that $v \neq 0$. Applying Theorem 8.27 to the function $v$, we obtain

$$u \geq v > 0 \text{ in } \Omega, \quad \frac{\partial u}{\partial n} \leq \frac{\partial v}{\partial n} < 0 \text{ on } \partial\Omega. \tag{8.30}$$

Let $w = u - v$. Hence $w \in C_0^1(\overline{\Omega})_+$. We need to show that $w \in \text{int}(C_0^1(\overline{\Omega})_+)$. As a first step, let us prove that

$$\frac{\partial w}{\partial n}(z) < 0 \text{ for all } z \in \partial\Omega. \tag{8.31}$$

So we fix $z_1 \in \partial\Omega$. For $t \in [0, 1]$ let $\zeta_t = v + t(u - v)$. Relation (8.30) implies that

$$\frac{\partial \zeta_t}{\partial n}(z_1) < 0 \text{ for all } t \in [0, 1],$$

hence there is a constant $\delta > 0$ such that

$$|\nabla\zeta_t(z_1)| \geq 2\delta \text{ for all } t \in [0, 1].$$

Fix a ball $B = B_r(z_0) \subset \Omega$ such that $\partial B \cap \partial\Omega = \{z_1\}$ with $r > 0$ small so that

$$|\nabla\zeta_t(x)| \geq \delta \text{ for all } x \in \overline{B}, \text{ all } t \in [0, 1].$$

From (8.29) we can see that $w$ satisfies the following inequality in the distributional sense:

$$\operatorname{div}(A(x)\nabla w) = g - f \leq 0 \text{ in } B, \qquad (8.32)$$

where $A(x) = (a_{i,j}(x))_{1 \leq i,j \leq N}$ is the matrix operator defined by

$$a_{i,j}(x) = \int_0^1 \hat{a}_{i,j}(t,x)\,dt$$

for all $x \in \bar{B}$, all $i,j \in \{1,\dots,N\}$, with

$$\hat{a}_{i,j}(t,x) = |\nabla \zeta_t(x)|^{p-4}\left(\delta_{i,j}|\nabla\zeta_t(x)|^2 + (p-2)\frac{\partial \zeta_t}{\partial x_i}(x)\frac{\partial \zeta_t}{\partial x_j}(x)\right).$$

Since $u,v \in C^{1,\lambda}(\bar{\Omega})$, the coefficients $a_{i,j}$ are Hölder continuous on $\bar{B}$. Moreover, one can see that the only eigenvalues of $\hat{A}(t,x) := (\hat{a}_{i,j}(t,x))_{1 \leq i,j \leq N}$ are $|\nabla\zeta_t(x)|^{p-2}$ (with multiplicity $N-1$) and $(p-1)|\nabla\zeta_t(x)|^{p-2}$ (with multiplicity one). It follows that there is a constant $c > 0$ such that

$$(A(x)\xi,\xi)_{\mathbb{R}^N} = \int_0^1 (\hat{A}(t,x)\xi,\xi)_{\mathbb{R}^N}\,dt \geq c|\xi|^2 \text{ for all } x \in \bar{B}, \text{ all } \xi \in \mathbb{R}^N.$$

We have checked that the operator $(x,\xi) \mapsto A(x)\xi$ satisfies hypotheses H(a)$_1$ (for $p = 2$). In view of (8.32), we can apply Corollary 8.17, which implies that we have either $w \equiv 0$ or $w > 0$ in $B$. Since $f > g$ a.e. in $B$, we know that $w \not\equiv 0$ in $B$, so $w > 0$ in $B$. Now, invoking the version of the strong maximum principle of Finn–Gilbarg (Remark 8.28), we obtain that $\frac{\partial w}{\partial n}(z_1) < 0$. Thus, (8.31) holds.

To get that $w \in \operatorname{int}(C_0^1(\bar{\Omega})_+)$, it remains to check that $w > 0$ in $\Omega$. According to Corollary 8.24, for this it is sufficient to note that the set $D = \{x \in \Omega : \nabla u(x) = \nabla v(x)\}$ is compact. Arguing by contradiction, assume that there is a sequence $\{z_n\}_{n \geq 1} \subset D$ and a point $z_0 \in \partial\Omega$ such that $z_n \to z_0$ as $n \to \infty$. But then we have

$$\frac{\partial w}{\partial n}(z_0) = (\nabla w(z_0), n(z_0))_{\mathbb{R}^N} = \lim_{n\to\infty}(\nabla u(z_n) - \nabla v(z_n), n(z_0))_{\mathbb{R}^N} = 0,$$

which contradicts (8.31). The proof is then complete. $\qquad\square$

*Remark 8.30.*

(a) This result was first stated by Guedda and Véron [167] under the assumption that $f \geq g$ are such that the set $\{x \in \Omega : f(x) = g(x)\}$ has an empty interior. However, this assumption does not seem sufficient to imply the result [take, for instance, $\Omega = (0,1)$, $f = 1 + \chi_{\mathbb{Q}}$ – where $\chi_{\mathbb{Q}}$ is the characteristic function of the set of rational numbers – $g = 1$, $u \in W_0^{1,p}((0,1))$, satisfying $-\Delta_p u = 1$, and $v = u$]. Here we have replaced this assumption with the fact that $f > g$ a.e. in $\Omega$.

Moreover, we emphasize that the property of $u, v$ belonging to $C^{1,\lambda}(\overline{\Omega})$ is crucial in the proof because it implies that the operator $A(x)$ has Hölder continuous coefficients, which enables us to invoke the strong maximum principle for $w$.

(b) A version of this result, where it is only assumed that $f, g$ differ on a set of positive measure but where in compensation the boundary of $\Omega$ is required to be connected, was proved by Cuesta and Takáč [97, Appendix].

## 8.3  Remarks

**Section 8.1:** Regularity results for general nonlinear elliptic equations can be found in DiBenedetto [118] and Tolksdorf [376] (local regularity results) and Lieberman [228] (regularity results up to the boundary). These papers build upon the earlier fundamental work of Ladyzhenskaya and Ural'tseva [215]. The method of proof of Theorem 8.4 is the well-known Moser iteration technique (see Moser [269]).

**Section 8.2:** Other forms of the nonlinear Harnack-type inequality (Theorem 8.15) can be found in Damascelli [99], Pucci and Serrin [332, Chap. 7], and Serrin [359]. A detailed discussion of the nonlinear maximum principle can be found in the books of Gilbarg and Trudinger [158] and Pucci and Serrin [332], where one can also find various weak and strong comparison principles. In this direction, we also mention the works of Cuesta and Takáč [96], Damascelli [99], Damascelli and Sciunzi [100], Lucia and Prashanth [238], Pigola et al. [329], and Roselli and Sciunzi [347]. The version of the strong maximum principle that we provide (Theorem 8.27) deals with a nonhomogeneous differential operator $\operatorname{div} a(x, \nabla u)$, and it extends the well-known result of Vázquez [379]. Other versions of the strong maximum principle for nonhomogeneous differential operators can be found in the works of Miyajima et al. [261], Montenegro [262], and Zhang [391].

# Chapter 9
# Spectrum of Differential Operators

**Abstract** This chapter provides a self-contained account of the spectral properties of the following fundamental differential operators: Laplacian, $p$-Laplacian, and $p$-Laplacian plus an indefinite potential, with any $1 < p < +\infty$. The first section of the chapter examines the spectrum of the Laplacian separately under Dirichlet and Neumann boundary conditions, taking advantage of the essential feature that this refers to a linear operator. The second section addresses the spectrum of the $p$-Laplacian, again considering separately the Dirichlet and Neumann boundary conditions. Here the methods are completely different with respect to the Laplacian because the $p$-Laplacian is a nonlinear operator for $p \neq 2$, making use of topological tools such as the Lyusternik–Schnirelmann principle. The third section extends this study to the more general class of nonlinear operators expressed as the sum of $p$-Laplacian and certain indefinite potential. Powerful related techniques are developed, for instance, the antimaximum principle, which is presented in a novel form. The fourth section addresses the Fučík spectrum, which incorporates the ordinary spectrum. The last section contains comments and information on relevant literature.

## 9.1 Spectrum of the Laplacian

Given a domain $\Omega \subset \mathbb{R}^N$, the *Laplacian operator* $u \mapsto \Delta u = \mathrm{div}(\nabla u)$ is a map from the Sobolev space $H^1(\Omega)$ into, a priori, the space of distributions on $\Omega$. Actually the range of $\Delta$ is contained in the dual of $H^1(\Omega)$. The aim of this section is to develop the spectral properties of the Laplacian under Dirichlet and Neumann boundary conditions. For this analysis, we rely on the spectral properties of self-adjoint compact linear operators in Hilbert spaces. This method requires that the space of functions [here $H^1(\Omega)$] be a Hilbert space, in particular we will not be able to use the same method in Sect. 9.2 for studying the spectral properties of the $p$-Laplacian. First, we present the Dirichlet case with all the details. Then, we will indicate how similar results can be obtained in the Neumann case.

D. Motreanu et al., *Topological and Variational Methods with Applications to Nonlinear Boundary Value Problems*, DOI 10.1007/978-1-4614-9323-5_9,
© Springer Science+Business Media, LLC 2014

**Spectrum of Laplacian Under Dirichlet Boundary Conditions**

Let $\Omega \subset \mathbb{R}^N$ ($N \geq 1$) be a bounded domain with Lipschitz boundary. We start by studying the following eigenvalue problem, where $\lambda \in \mathbb{R}$ is a parameter:

$$\begin{cases} -\Delta u(x) = \lambda u(x) & \text{in } \Omega, \\ u = 0 & \text{on } \partial\Omega. \end{cases} \tag{9.1}$$

**Definition 9.1.** We say that $\lambda \in \mathbb{R}$ is an *eigenvalue* of the negative Laplacian under Dirichlet boundary conditions (of $-\Delta^D$, for short) if problem (9.1) admits a nontrivial weak solution, i.e., there exists $u \in H_0^1(\Omega)$, $u \neq 0$, such that

$$\int_\Omega (\nabla u, \nabla v)_{\mathbb{R}^N}\, dx = \lambda \int_\Omega uv\, dx \quad \text{for all } v \in H_0^1(\Omega).$$

The function $u$ is called an *eigenfunction* of $-\Delta^D$ corresponding to the eigenvalue $\lambda$. We denote by $E(\lambda) \subset H_0^1(\Omega)$ the space of weak solutions of problem (9.1).

*Remark 9.2.* (a) Note that every eigenvalue of $-\Delta^D$ is positive. Indeed, if $\lambda \in \mathbb{R}$ is an eigenvalue with corresponding eigenfunction $u$, then we have in particular $\lambda = \frac{\|\nabla u\|_2^2}{\|u\|_2^2}$. Thus, $\lambda \geq \frac{1}{c^2}$, where $c = c(\Omega, 2) > 0$ is the constant in Poincaré's inequality (Theorem 1.41).

(b) If $\Omega$ has a $C^2$-boundary $\partial\Omega$, then we know from Corollary 8.13 that every eigenfunction $u$ of $-\Delta^D$ belongs to $C^1(\overline{\Omega})$. Actually, even in the case where $\partial\Omega$ is Lipschitz, we have $u \in L^\infty(\Omega)$ (Corollary 8.6), and the interior regularity theory implies $u \in C^\infty(\Omega)$ (see Gilbarg and Trudinger [158]).

The development of the spectral properties of the negative Dirichlet Laplacian will be based on Theorem 2.23, the spectral theorem for compact self-adjoint operators. The basic tool in this direction is an *essential inverse* of $-\Delta|_{H_0^1(\Omega)}$ introduced in the following proposition.

**Proposition 9.3.** (a) *Given $h \in L^2(\Omega)$, the boundary value problem*

$$\begin{cases} -\Delta u(x) = h(x) & \text{in } \Omega, \\ u = 0 & \text{on } \partial\Omega \end{cases} \tag{9.2}$$

*admits a unique solution $u \in H_0^1(\Omega)$; we set $S(h) = u$.*

(b) *The operator $S : L^2(\Omega) \to L^2(\Omega)$ thus obtained is a self-adjoint, positive definite, compact, linear map.*

(c) *$\lambda \in (0, +\infty)$ is an eigenvalue of $S$ if and only if $\frac{1}{\lambda}$ is an eigenvalue of $-\Delta^D$.*

*Proof.* (a) The operator $A \in \mathscr{L}(H_0^1(\Omega), H^{-1}(\Omega))$ defined by

$$\langle A(u), y \rangle = \int_\Omega (\nabla u, \nabla y)_{\mathbb{R}^N}\, dx \quad \text{for all } u, y \in H_0^1(\Omega)$$

is strictly monotone, continuous, coercive, and, hence, surjective (Theorem 2.55). Thus, problem (9.2) has a solution. The strict monotonicity of $A$ guarantees the uniqueness of this solution.

(b) The linearity of $S$ is clear. The other properties of $S$ can be checked as follows.

*Continuity:* From the definition of $S$ we have

$$\int_\Omega (\nabla S(h), \nabla v)_{\mathbb{R}^N} \, dx = \int_\Omega hv \, dx \text{ for all } v \in H_0^1(\Omega).$$

Choosing $v = S(h) \in H_0^1(\Omega)$ and invoking the Cauchy–Schwarz inequality, we obtain

$$\|\nabla S(h)\|_2^2 = \int_\Omega hS(h) \, dx \le \|h\|_2 \|S(h)\|_2. \tag{9.3}$$

From Poincaré's inequality (Theorem 1.41) we can find $c > 0$ independent of $h$ such that $\|S(h)\|_2 \le c\|\nabla S(h)\|_2$. Using this in (9.3), we obtain

$$\|S(h)\|_2 \le c\|\nabla S(h)\|_2 \le c^2 \|h\|_2,$$

which establishes the continuity of $S : L^2(\Omega) \to L^2(\Omega)$ as well as the continuity of $S$ seen as a map from $L^2(\Omega)$ into $H_0^1(\Omega)$. We will need this sharper conclusion in establishing the compactness of $S$, which we do next.

*Compactness:* The compactness of $S$ follows from the fact that $S$ is obtained as the composition of the bounded linear operator $L^2(\Omega) \to H_0^1(\Omega)$, $h \mapsto S(h)$, with the compact embedding operator $H_0^1(\Omega) \hookrightarrow L^2(\Omega)$ (Theorem 1.49).

*Self-adjointness:* Let $h, f \in L^2(\Omega)$. We have

$$\int_\Omega (\nabla S(h), \nabla v)_{\mathbb{R}^N} \, dx = \int_\Omega hv \, dx \text{ and } \int_\Omega (\nabla S(f), \nabla v)_{\mathbb{R}^N} \, dx = \int_\Omega fv \, dx$$

for all $v \in H_0^1(\Omega)$. In the first equality we choose $v = S(f) \in H_0^1(\Omega)$ and in the second $v = S(h) \in H_0^1(\Omega)$. We obtain

$$\int_\Omega (\nabla S(h), \nabla S(f))_{\mathbb{R}^N} \, dx = (h, S(f))_{L^2(\Omega)}$$

and $\int_\Omega (\nabla S(f), \nabla S(h))_{\mathbb{R}^N} \, dx = (f, S(h))_{L^2(\Omega)},$

which implies that $(h, S(f))_{L^2(\Omega)} = (S(h), f)_{L^2(\Omega)}$, i.e., $S$ is self-adjoint.

*Positive definiteness:* First, note that $(S(h), h)_{L^2(\Omega)} = \|\nabla S(h)\|_{L^2(\Omega)}^2 \ge 0$ for all $h \in L^2(\Omega)$ (see (9.3)). Now, if $(S(h), h)_{L^2(\Omega)} = 0$, then $\nabla S(h) = 0$, so $\int_\Omega hv \, dx = 0$ for all $v \in H_0^1(\Omega)$, and the density of $H_0^1(\Omega)$ in $L^2(\Omega)$ yields $h = 0$.

(c) If $\lambda$ is an eigenvalue of $S$, i.e., there is $h \in L^2(\Omega)$, $h \neq 0$, such that $S(h) = \lambda h$, then $h = \frac{1}{\lambda}S(h) \in H_0^1(\Omega)$, and we obtain $-\Delta h = \frac{1}{\lambda}h$, so $\frac{1}{\lambda}$ is an eigenvalue of $-\Delta^D$. Conversely, if there is $u \in H_0^1(\Omega)$, $u \neq 0$, such that $-\Delta u = \frac{1}{\lambda}u$, then this yields $S(u) = \lambda u$, i.e., $\lambda$ is an eigenvalue of $S$. The proof is now complete.  $\square$

Recall that $H_0^1(\Omega)$ is a Hilbert space for the inner product

$$(u,v)_{H_0^1(\Omega)} = \int_\Omega (\nabla u, \nabla v)_{\mathbb{R}^N} \, dx \ \text{ for all } u,v \in H_0^1(\Omega).$$

We derive the following description of the spectrum of negative Dirichlet Laplacian.

**Theorem 9.4.** (a) *The eigenvalues of $-\Delta^D$ consist of a sequence $\{\lambda_n\}_{n\geq 1}$ such that*

$$0 < \lambda_1 \leq \lambda_2 \leq \cdots \leq \lambda_n \leq \dots, \ \text{ with } \lim_{n\to\infty} \lambda_n = +\infty,$$

*where each eigenvalue is repeated according to its multiplicity (which is finite).*
(b) *The Hilbert space $L^2(\Omega)$ admits an orthonormal basis $\{\hat{u}_n\}_{n\geq 1}$, where each $\hat{u}_n$ belongs to $H_0^1(\Omega)$ and is an eigenfunction of $-\Delta^D$ corresponding to $\lambda_n$.*
(c) *Moreover, $\{\frac{1}{\sqrt{\lambda_n}}\hat{u}_n\}_{n\geq 1}$ is an orthonormal basis of $(H_0^1(\Omega), (\cdot,\cdot)_{H_0^1(\Omega)})$.*

*Proof.* Since the operator $S : L^2(\Omega) \to L^2(\Omega)$ is compact, self-adjoint, and positive definite [by Proposition 9.3(b)], Theorems 2.19 and 2.23 imply that the eigenvalues of $S$ consist of a sequence

$$\mu_1 \geq \mu_2 \geq \dots \geq \mu_n \geq \dots > 0, \ \text{ with } \lim_{n\to\infty} \mu_n = 0$$

[each eigenvalue in the sequence repeated according to its (finite) multiplicity] and that there is an orthonormal basis $\{\hat{u}_n\}_{n\geq 1}$ of $L^2(\Omega)$ consisting of eigenfunctions of $S$ corresponding to the eigenvalues $\{\mu_n\}_{n\geq 1}$. By Proposition 9.3(c), the numbers $\lambda_n = \frac{1}{\mu_n}$, for $n \geq 1$, are exactly the eigenvalues of $-\Delta^D$ and $\hat{u}_n = \frac{1}{\mu_n}S(\hat{u}_n) \in H_0^1(\Omega)$ is an eigenfunction corresponding to the eigenvalue $\lambda_n$. This proves parts (a) and (b) of the theorem. It remains to check (c). The orthonormality of the family $\{\frac{1}{\sqrt{\lambda_n}}\hat{u}_n\}_{n\geq 1}$ in $H_0^1(\Omega)$ can be checked as follows: for $m,n \geq 1$, applying the relation $-\Delta\hat{u}_n = \lambda_n\hat{u}_n$ to the test function $\hat{u}_m$, we obtain

$$\left(\frac{1}{\sqrt{\lambda_n}}\hat{u}_n, \frac{1}{\sqrt{\lambda_m}}\hat{u}_m\right)_{H_0^1} = \frac{1}{\sqrt{\lambda_n\lambda_m}}\int_\Omega (\nabla\hat{u}_n, \nabla\hat{u}_m)_{\mathbb{R}^N}\, dx$$

$$= \frac{1}{\sqrt{\lambda_n\lambda_m}}\int_\Omega \lambda_n\hat{u}_n(x)\hat{u}_m(x)\, dx = \frac{\sqrt{\lambda_n}}{\sqrt{\lambda_m}}(\hat{u}_n, \hat{u}_m)_{L^2(\Omega)} = \delta_{m,n}.$$

Finally, we show that $\overline{\operatorname{span}}\{\frac{1}{\sqrt{\lambda_n}}\hat{u}_n\}_{n\geq 1} = H_0^1(\Omega)$. To this end, take $h \in H_0^1(\Omega)$ such that $(h, \hat{u}_n)_{H_0^1(\Omega)} = 0$ for all $n \geq 1$ and let us show that $h = 0$. We have

$$0 = (h, \hat{u}_n)_{H_0^1(\Omega)} = \int_\Omega (\nabla h, \nabla\hat{u}_n)_{\mathbb{R}^N}\, dx = \lambda_n \int_\Omega h\hat{u}_n\, dx.$$

Since $\lambda_n \neq 0$, we get $\int_\Omega h\hat{u}_n\, dx = 0$ for all $n \geq 1$, and so $h = 0$, as $\{\hat{u}_n\}_{n\geq 1}$ is an orthonormal basis of $L^2(\Omega)$. □

Recall that if $\lambda > 0$ is an eigenvalue of $-\Delta^D$ with eigenfunction $\hat{u}$, then

$$\lambda = \frac{\|\nabla\hat{u}\|_2^2}{\|\hat{u}\|_2^2}. \tag{9.4}$$

The quotient on the right-hand side plays an important role in the derivation of variational expressions for the eigenvalues of $-\Delta^D$. This motivates the next definition.

**Definition 9.5.** The *Rayleigh quotient* is the map $R : H^1(\Omega) \setminus \{0\} \to [0, +\infty)$ defined by

$$R(u) = \frac{\|\nabla u\|_2^2}{\|u\|_2^2} \quad \text{for all } u \in H^1(\Omega),\ u \neq 0.$$

We start with the variational characterization of the first eigenvalue $\lambda_1$.

**Proposition 9.6.** *The first eigenvalue of* $-\Delta^D$ *is characterized by*

$$\lambda_1 = \inf\{R(u) : u \in H_0^1(\Omega),\ u \neq 0\}$$

*and the infimum is attained exactly on the set* $E(\lambda_1) \setminus \{0\}$.

*Proof.* From (9.4) we see that $\lambda_1 \geq \inf\{R(u) : u \in H_0^1(\Omega),\ u \neq 0\}$. Therefore, it remains to show that the opposite inequality also holds. Recall that $\{\hat{u}_n\}_{n\geq 1}$ is an orthonormal basis of $L^2(\Omega)$ and $\{\frac{1}{\sqrt{\lambda_n}}\hat{u}_n\}_{n\geq 1}$ is an orthonormal basis of $H_0^1(\Omega)$ (Theorem 9.4). Then, for any $u \in H_0^1(\Omega)$, using the Bessel–Parseval identity (e.g., Brezis [52, p. 141]), we have

$$\|\nabla u\|_2^2 = \sum_{n\geq 1}\left(u, \frac{1}{\sqrt{\lambda_n}}\hat{u}_n\right)_{H_0^1(\Omega)}^2 \quad \text{and} \quad \|u\|_2^2 = \sum_{n\geq 1}(u, \hat{u}_n)_{L^2(\Omega)}^2. \tag{9.5}$$

Using (9.5) and the equality $-\Delta\hat{u}_n = \lambda_n\hat{u}_n$, we obtain

$$\|\nabla u\|_2^2 = \sum_{n\geq 1}\frac{1}{\lambda_n}\left(\int_\Omega (\nabla\hat{u}_n, \nabla u)_{\mathbb{R}^N}\, dx\right)^2 = \sum_{n\geq 1}\frac{1}{\lambda_n}\left(\int_\Omega \lambda_n\hat{u}_n(x)u(x)\, dx\right)^2$$

$$= \sum_{n\geq 1}\lambda_n(u, \hat{u}_n)_{L^2(\Omega)}^2 \geq \lambda_1\sum_{n\geq 1}(u, \hat{u}_n)_{L^2(\Omega)}^2 = \lambda_1\|u\|_2^2, \tag{9.6}$$

whence $R(u) \geq \lambda_1$ for all $u \in H_0^1(\Omega)$, $u \neq 0$. Therefore, we conclude that $\lambda_1 = \inf\{R(u) : u \in H_0^1(\Omega),\ u \neq 0\}$. From (9.4) it follows that every eigenfunction $\hat{u}$ corresponding to $\lambda_1$ realizes this infimum. Conversely, suppose that $R(u) = \lambda_1$. Then

$$\lambda_1 = R(u) = \frac{\sum\limits_{n\geq 1} \lambda_n(u,\hat{u}_n)^2_{L^2(\Omega)}}{\sum\limits_{n\geq 1} (u,\hat{u}_n)^2_{L^2(\Omega)}}$$

[see (9.6)], which yields $\sum\limits_{n\geq 1}(\lambda_n - \lambda_1)(u,\hat{u}_n)^2_{L^2(\Omega)} = 0$. Hence $u \perp \hat{u}_n$ in $L^2(\Omega)$ for all $n$ such that $\lambda_n > \lambda_1$. In this way, $u \in E(\lambda_1)$.                                       □

*Remark 9.7.* Here is an alternative proof of Proposition 9.6, based on the Lagrange multiplier rule. We consider the maps $\varphi, \psi \in C^1(H^1_0(\Omega), \mathbb{R})$ defined by

$$\varphi(u) = \|\nabla u\|^2_2 \quad \text{and} \quad \psi(u) = \|u\|^2_2 \quad \text{for all } u \in H^1_0(\Omega).$$

By Lyusternik's theorem (Theorem 5.72), the subset $M = \{u \in H^1_0(\Omega) : \psi(u) = 1\}$ is a $C^1$-submanifold of $H^1_0(\Omega)$ of codimension 1. Given $u \in M$, the following equivalences hold:

$u$ is an eigenfunction of $-\Delta^D$ corresponding to $\lambda$,

$\Leftrightarrow \varphi'(u) = \lambda \psi'(u)$,

$\Leftrightarrow T_u M = \ker \psi'(u) \subset \ker \varphi'(u)$ and $\varphi(u) = \lambda$,

$\Leftrightarrow u$ is a constrained critical point of $\varphi$ on $M$ with critical value $\lambda$     (9.7)

(Definition 5.73). Thus, the $L^2$-normalized eigenfunctions of $-\Delta^D$ coincide with the constrained critical points of $\varphi$ on $M$ and the eigenvalues of $-\Delta^D$ coincide with the critical values of $\varphi|_M$. Because $M$ is weakly closed in $H^1_0(\Omega)$ [by the compact embedding $H^1_0(\Omega) \hookrightarrow L^2(\Omega)$] and the functional $\varphi$ is coercive and weakly l.s.c., the minimization problem

$$\varphi(u) = \inf_M \varphi, \quad u \in M \tag{9.8}$$

has at least one solution $u$. Then $u$ is a constrained critical point of $\varphi$ on $M$, so that $\inf_M \varphi = \varphi(u)$ is an eigenvalue of $-\Delta^D$, necessarily the minimal one [see (9.7)]. Thus,

$$\lambda_1 = \inf_M \varphi = \inf_{H^1_0(\Omega)\setminus\{0\}} R.$$

Moreover, if $v \in H^1_0(\Omega) \setminus \{0\}$ satisfies $R(v) = \lambda_1$, then $\frac{1}{\|v\|_2}v$ is a solution of (9.8), which implies that $v$ is an eigenfunction of $-\Delta^D$ corresponding to $\lambda_1$ [see (9.7)]. This completes the proof of Proposition 9.6.

   This method is a particular case of the so-called Lyusternik–Schnirelmann theory, which will be used in Sect. 9.2 to extend the spectral analysis to the case of the $p$-Laplacian, where, due to the nonlinearity of the operator, we cannot invoke a spectral theorem analogous to Theorem 2.23.

Continuing with the examination of the first eigenvalue $\lambda_1 > 0$, we obtain the following proposition.

**Proposition 9.8.** *The first eigenvalue $\lambda_1$ of $-\Delta^D$ is simple [i.e., $\dim E(\lambda_1) = 1$] and any corresponding eigenfunction $\hat{u}$ does not vanish in $\Omega$ (in particular it is of constant sign).*

*Proof.* We first show that every $\hat{u} \in E(\lambda_1) \setminus \{0\}$ is either positive or negative. We know that $\hat{u} \in C^\infty(\Omega)$ [see Remark 9.2(b)] and $\hat{u}^+, \hat{u}^- \in H_0^1(\Omega)$ (Proposition 1.29). We have

$$\|\nabla \hat{u}^+\|_2^2 = \int_\Omega (\nabla \hat{u}, \nabla \hat{u}^+)_{\mathbb{R}^N} \, dx = \lambda_1 \int_\Omega \hat{u} \hat{u}^+ \, dx = \lambda_1 \|\hat{u}^+\|_2^2 \qquad (9.9)$$

$$\text{and} \quad \|\nabla \hat{u}^-\|_2^2 = \int_\Omega (\nabla \hat{u}, -\nabla \hat{u}^-)_{\mathbb{R}^N} \, dx = \lambda_1 \int_\Omega \hat{u}(-\hat{u}^-) \, dx = \lambda_1 \|\hat{u}^-\|_2^2. \quad (9.10)$$

Suppose that $\hat{u}$ is nodal (i.e., sign changing). Then $\hat{u}^+ \neq 0$, $\hat{u}^- \neq 0$, and, from (9.9), (9.10), we have $R(\hat{u}^+) = R(\hat{u}^-) = \lambda_1$. By virtue of Proposition 9.6, this implies that $\hat{u}^+, \hat{u}^- \in E(\lambda_1)$, and so $-\Delta \hat{u}^+ = \lambda_1 \hat{u}^+$, $-\Delta \hat{u}^- = \lambda_1 \hat{u}^-$. Then, from Corollary 8.17, we have $\hat{u}^+(x) > 0$, $\hat{u}^-(x) > 0$ for all $x \in \Omega$, which is impossible. This shows that $\hat{u}$ has a constant sign. Moreover, invoking again Corollary 8.17, we obtain that $\hat{u}$ does not vanish in $\Omega$, so it is either positive or negative.

Finally, we show that $\lambda_1$ is simple. Arguing by contradiction, suppose that $\dim E(\lambda_1) > 1$. Then the first two functions $\hat{u}_1, \hat{u}_2$ in the orthonormal basis of Theorem 9.4 correspond to the same eigenvalue $\lambda_1 = \lambda_2$. Since $\hat{u}_1, \hat{u}_2$ have constant signs, we have $\int_\Omega \hat{u}_1 \hat{u}_2 \, dx \neq 0$, but this relation contradicts the fact that $\hat{u}_1, \hat{u}_2$ are orthogonal in $L^2(\Omega)$. The proof is now complete. $\qquad \square$

Now we look for variational characterizations of the higher eigenvalues. Let $\{\hat{u}_n\}_{n \geq 1} \subset H_0^1(\Omega)$ be the orthonormal basis of $L^2(\Omega)$ consisting of eigenfunctions of $-\Delta^D$, which is provided by Theorem 9.4. We set

$$H_m = \text{span}\{\hat{u}_k\}_{k=1}^m \subset H_0^1(\Omega).$$

The following result characterizes the eigenvalues $\{\lambda_m\}_{m \geq 2}$.

**Proposition 9.9.** *For all $m \geq 2$, we have*

$$\lambda_m = R(\hat{u}_m) = \max\{R(v) : v \in H_m, v \neq 0\} = \min\{R(v) : v \in H_{m-1}^\perp, v \neq 0\},$$

*where $H_{m-1}^\perp$ denotes the orthogonal complement of $H_{m-1}$ in $H_0^1(\Omega)$.*

*Proof.* From (9.4) we know that

$$\inf\{R(v) : v \in H_{m-1}^\perp, v \neq 0\} \leq R(\hat{u}_m) = \lambda_m \leq \sup\{R(v) : v \in H_m, v \neq 0\}.$$

Recall that any element $v \in H_0^1(\Omega) \setminus \{0\}$ can be written as

$$v = \sum_{k \geq 1} (v, \hat{u}_k)_{L^2(\Omega)} \hat{u}_k,$$

and, by the Bessel–Parseval identity, we have

$$R(v) = \frac{\sum\limits_{k \geq 1} \lambda_k (v, \hat{u}_k)^2_{L^2(\Omega)}}{\sum\limits_{k \geq 1} (v, \hat{u}_k)^2_{L^2(\Omega)}} \tag{9.11}$$

[see (9.6)]. If $v \in H_m \setminus \{0\}$, then $(v, \hat{u}_k)_{L^2(\Omega)} = 0$ for all $k \geq m+1$, so that (9.11) yields $R(v) \leq \lambda_m$, whence $\sup\{R(v) : v \in H_m, \ v \neq 0\} \leq \lambda_m$. On the other hand, for $v \in H_{m-1}^\perp \setminus \{0\}$ we have $(v, \hat{u}_k)_{L^2(\Omega)} = 0$ for all $k \in \{1, \ldots, m-1\}$, and so $R(v) \geq \lambda_m$ [by (9.11)]; this yields $\inf\{R(v) : v \in H_{m-1}^\perp, \ v \neq 0\} \geq \lambda_m$. The proof is now complete. $\qquad \square$

*Remark 9.10.* (a) It can be easily noted from the foregoing proof that, in the statement of Proposition 9.9, the maximum is attained exactly on the set $H_m \cap E(\lambda_m) \setminus \{0\}$ and the minimum is attained exactly on the set $H_{m-1}^\perp \cap E(\lambda_m) \setminus \{0\}$.
(b) If $\{\hat{\lambda}_k\}_{k \geq 1}$ denotes the *increasing* sequence formed by the eigenvalues of $-\Delta^D$ (i.e., counted without multiplicity), then Proposition 9.9 implies the following characterization:

$$\hat{\lambda}_k = \max\{R(v) : v \in \hat{E}_k, \ v \neq 0\} = \min\{R(v) : v \in \hat{E}_{k-1}^\perp, \ v \neq 0\} \ \text{for all } k \geq 1,$$

where $\hat{E}_k = E(\hat{\lambda}_1) \oplus \ldots \oplus E(\hat{\lambda}_k)$, and so $\hat{E}_{k-1}^\perp = \bigoplus\limits_{\ell \geq k} E(\hat{\lambda}_\ell)$. Moreover, these maximum and minimum are attained exactly on $E(\hat{\lambda}_k) \setminus \{0\}$.

The drawback in the preceding variational characterization is that it is recursive, i.e., we need to know $H_{m-1}$ in order to obtain $\lambda_m$. By contrast, the next theorem provides a direct minimax characterization of the eigenvalues. The result is known as the *Courant–Fischer theorem*.

**Theorem 9.11.** *For all $m \geq 2$ we have*

$$\lambda_m = \min_{Y \in \mathscr{S}_m} \max_{v \in Y \setminus \{0\}} R(v) = \max_{Y \in \mathscr{S}_{m-1}} \min_{v \in Y^\perp \setminus \{0\}} R(v),$$

*where $\mathscr{S}_m$ is the family of all $m$-dimensional vector subspaces of $H_0^1(\Omega)$ and $Y^\perp$ denotes the orthogonal complement of $Y$ in $H_0^1(\Omega)$.*

*Proof.* On the one hand, using Proposition 9.9, we have

$$\inf_{Y \in \mathscr{S}_m} \max_{v \in Y \setminus \{0\}} R(v) \leq \lambda_m \leq \sup_{Y \in \mathscr{S}_{m-1}} \min_{v \in Y^\perp \setminus \{0\}} R(v).$$

On the other hand, every $Y_1 \in \mathscr{S}_m$ satisfies $Y_1 \cap H_{m-1}^\perp \neq 0$ and every $Y_2 \in \mathscr{S}_{m-1}$ satisfies $Y_2^\perp \cap H_m \neq 0$, so we can find nonzero elements $v_1 \in Y_1 \cap H_{m-1}^\perp$ and $v_2 \in Y_2^\perp \cap H_m$. From Proposition 9.9 we have

$$\inf\{R(v) : v \in Y_2^\perp \setminus \{0\}\} \leq R(v_2) \leq \lambda_m \leq R(v_1) \leq \sup\{R(v) : v \in Y_1 \setminus \{0\}\},$$

whence

$$\sup_{Y \in \mathscr{S}_{m-1}} \min_{v \in Y^\perp \setminus \{0\}} R(v) \leq \lambda_m \leq \inf_{Y \in \mathscr{S}_m} \max_{v \in Y \setminus \{0\}} R(v).$$

The theorem ensues. □

We derive from Theorem 9.11 the following monotonicity of the eigenvalues with respect to the domain.

**Proposition 9.12.** *Let* $\Omega' \subset \mathbb{R}^N$ *be another bounded domain with Lipschitz boundary. Let* $\{\lambda_m(\Omega)\}_{m \geq 1}$ *[resp.* $\{\lambda_m(\Omega')\}_{m \geq 1}$*] denote the nondecreasing sequence of eigenvalues of* $(-\Delta, H_0^1(\Omega))$ *[resp. of* $(-\Delta, H_0^1(\Omega'))$*]. If* $\Omega \subset \Omega'$*, then* $\lambda_m(\Omega') \leq \lambda_m(\Omega)$ *for all* $m \geq 1$.

*Proof.* Recall that, for $u \in H_0^1(\Omega)$, its extension by zero,

$$\tilde{u}(x) = \begin{cases} u(x) & \text{if } x \in \Omega, \\ 0 & \text{if } x \in \Omega' \setminus \Omega, \end{cases}$$

belongs to $H_0^1(\Omega')$ (Proposition 1.10). This yields a linear continuous embedding $H_0^1(\Omega) \subset H_0^1(\Omega')$. Moreover, for $u \in H_0^1(\Omega) \setminus \{0\}$ we clearly have $R_\Omega(u) = R_{\Omega'}(\tilde{u})$, where $R_\Omega$ (resp. $R_{\Omega'}$) stands for the Rayleigh quotient with respect to $\Omega$ (resp. $\Omega'$) (Definition 9.5). The proposition follows from these observations and the minimax characterization of $\lambda_m(\Omega)$ and $\lambda_m(\Omega')$ provided by Theorem 9.11. □

*Remark 9.13.* This monotonicity property is not true for the Neumann eigenvalue problem, even for the first positive eigenvalue and planar regular domains $\Omega, \Omega'$ (see Ni and Wang [308]).

We have already seen that the nontrivial elements of $E(\lambda_1)$ (the eigenspace corresponding to $\lambda_1 > 0$) have a constant sign in $\Omega$ and $E(\lambda_1) = \mathbb{R}\hat{u}_1$ (Proposition 9.8). In contrast, we have the following proposition.

**Proposition 9.14.** *Any eigenfunction* $\hat{u}$ *of* $-\Delta^D$ *corresponding to a higher eigenvalue* $\lambda_m$ *(for* $m \geq 2$*) is nodal (sign changing).*

*Proof.* We know that $\hat{u}$ is orthogonal to $\hat{u}_1$ in $L^2(\Omega)$, i.e., we have $\int_\Omega \hat{u}_1 \hat{u} \, dx = 0$. Since $\hat{u}_1$ is either positive or negative in $\Omega$, this implies that $\hat{u}$ is nodal. □

*Remark 9.15.* (a) If $m \geq 2$ and $\hat{u}$ is an eigenfunction corresponding to $\lambda_m$, then the connected components of the open sets $\Omega_+ = \{x \in \Omega : \hat{u}(x) > 0\}$ and $\Omega_- =$

$\{x \in \Omega : \hat{u}(x) < 0\}$ are called the *nodal domains* of $\hat{u}$ (Definition 1.60). Thus, Proposition 9.14 implies that $\hat{u}$ admits at least two nodal domains. In fact, the *Courant nodal domain theorem* (see Courant and Hilbert [92, Sect. VI.2]) states that $\hat{u}$ has at most $m$ nodal domains.

(b) If $\tilde{\Omega}_m$ is one of those nodal domains, then $\lambda_1(\tilde{\Omega}_m) = \lambda_m$. Indeed, we clearly have $\hat{u}|_{\partial \tilde{\Omega}_m} = 0$, hence $\hat{u}|_{\tilde{\Omega}_m} \in H_0^1(\tilde{\Omega}_m)$, and so $\hat{u}|_{\tilde{\Omega}_m}$ is an eigenfunction of $(-\Delta, H_0^1(\tilde{\Omega}_m))$ corresponding to the eigenvalue $\lambda_m$. Since $\hat{u}$ has constant sign on $\tilde{\Omega}_m$, Proposition 9.14 implies that $\lambda_m = \lambda_1(\tilde{\Omega}_m)$. Note that this equality is no longer true for the Neumann case.

The explicit computation of the spectrum of $-\Delta^D$ is only known for very special types of domains. One such case is the scalar one, i.e., $N = 1$ with $\Omega = (0,1)$. It is easy to check the following proposition.

**Proposition 9.16.** *If $\Omega = (0,1)$, then the eigenvalues of $-\Delta^D$ are $\lambda_n = (n\pi)^2$, $n \geq 1$, and an orthonormal basis $\{\hat{u}_n\}_{n\geq 1}$ of $L^2((0,1))$ consisting of eigenfunctions of $-\Delta^D$ is given by $\hat{u}_n(t) = \sqrt{2}\sin(n\pi t)$. In this case, every eigenvalue $\lambda_n$ is simple.*

### Spectrum of Laplacian Under Neumann Boundary Conditions

Again, $\Omega \subset \mathbb{R}^N$ denotes a bounded domain with Lipschitz boundary. We now consider the following eigenvalue problem, under Neumann boundary conditions:

$$\begin{cases} -\Delta u(x) = \lambda u(x) & \text{in } \Omega, \\ \frac{\partial u}{\partial n} = 0 & \text{on } \partial\Omega. \end{cases} \tag{9.12}$$

**Definition 9.17.** We say that $\lambda \in \mathbb{R}$ is an *eigenvalue* of the negative Laplacian under Neumann boundary conditions (of $-\Delta^N$, for short) if problem (9.12) admits a nontrivial weak solution, i.e., there exists $u \in H^1(\Omega)$, $u \neq 0$, such that

$$\int_\Omega (\nabla u, \nabla v)_{\mathbb{R}^N}\, dx = \lambda \int_\Omega uv\, dx \text{ for all } v \in H^1(\Omega).$$

In particular, $\frac{\partial u}{\partial n} := \gamma_n(\nabla u) = 0$ (Remark 8.3(b)). The function $u$ is called an *eigenfunction* of $-\Delta^N$ corresponding to $\lambda$.

*Remark 9.18.* (a) Clearly, $\lambda = 0$ is an eigenvalue of $-\Delta^N$, and the corresponding eigenspace is the space of constant functions $\mathbb{R} \subset H^1(\Omega)$.

(b) All eigenvalues $\lambda$ of $-\Delta^N$ are nonnegative. Indeed, let $u \in H^1(\Omega)$ be an eigenfunction corresponding to $\lambda$. Then $\lambda = \frac{\|\nabla u\|_2^2}{\|u\|_2^2} \geq 0$.

(c) If $\Omega$ has a $C^2$-boundary, then we know from Corollary 8.13 that $u \in C^1(\overline{\Omega})$. In the case where $\partial\Omega$ is Lipschitz, we have $u \in L^\infty(\Omega)$ (Corollary 8.7) and the interior regularity theory guarantees that $u \in C^\infty(\Omega)$ (see Gilbarg and Trudinger [158]).

The spectral analysis in the Neumann case is similar to the analysis in the Dirichlet case. There is, however, a slight difference in the sense that we construct an essential inverse of $-\Delta + \mathrm{id}$ (instead of $-\Delta$) on $\{u \in H^1(\Omega) : \frac{\partial u}{\partial n} = 0\}$ as shown by the following proposition.

**Proposition 9.19.** (a) *Given* $h \in L^2(\Omega)$, *the boundary value problem*

$$\begin{cases} -\Delta u(x) + u(x) = h(x) & \text{in } \Omega, \\ \frac{\partial u}{\partial n} = 0 & \text{on } \partial\Omega \end{cases} \tag{9.13}$$

*admits a unique solution* $u \in H^1(\Omega)$; *we set* $T(h) = u$.
(b) *The operator* $T : L^2(\Omega) \to L^2(\Omega)$ *thus obtained is a self-adjoint, positive definite, compact, linear map.*
(c) *The eigenvalues of* $T$ *are contained in* $(0,1]$. *Moreover,* $\lambda \in (0,1]$ *is an eigenvalue of* $T$ *if and only if* $\frac{1}{\lambda} - 1$ *is an eigenvalue of* $-\Delta^{\mathrm{N}}$.

*Proof.* (a) Since $(h, \cdot)_{L^2(\Omega)}$ is a continuous linear form on $H^1(\Omega)$ [due to the continuous embedding $H^1(\Omega) \hookrightarrow L^2(\Omega)$], by the Riesz representation theorem (e.g., Brezis [52, p. 135]), there is $u \in H^1(\Omega)$ unique such that $(u,v)_{H^1(\Omega)} = (h,v)_{L^2(\Omega)}$ for all $v \in H^1(\Omega)$. This means that $u$ is the unique solution of (9.13).
(b) Is obtained as in the proof of Proposition 9.3.
(c) If $h \in L^2(\Omega)$ is an eigenvector of $T$ for the eigenvalue $\lambda$, then $h \in H^1(\Omega), h \neq 0$, and we have

$$\lambda \int_\Omega (\nabla h, \nabla v)_{\mathbb{R}^N}\, dx = (1 - \lambda) \int_\Omega hv\, dx \quad \text{for all } v \in H^1(\Omega).$$

Thus, $\lambda = \frac{\|h\|_2^2}{\|h\|_2^2 + \|\nabla h\|_2^2}$ (this yields $\lambda \in (0,1]$), and $\frac{1}{\lambda} - 1$ is an eigenvalue of $-\Delta^{\mathrm{N}}$ with corresponding eigenfunction $h$. Conversely, one easily checks that, if $\mu \in [0, +\infty)$ is an eigenvalue of $-\Delta^{\mathrm{N}}$, then $\frac{1}{\mu+1}$ is an eigenvalue of $T$.  □

Combining the previous proposition with Theorem 2.23, we obtain an analog of Theorem 9.4, as follows.

**Theorem 9.20.** (a) *The eigenvalues of* $-\Delta^{\mathrm{N}}$ *consist of a sequence* $\{\lambda_n\}_{n \geq 0}$ *such that*

$$0 = \lambda_0 < \lambda_1 \leq \lambda_2 \leq \cdots \leq \lambda_n \leq \ldots \quad \text{with } \lim_{n \to \infty} \lambda_n = +\infty,$$

*repeated according to their (finite) multiplicity.*
(b) *The Hilbert space* $L^2(\Omega)$ *admits an orthonormal basis* $\{\hat{u}_n\}_{n \geq 0}$, *where each* $\hat{u}_n$ *belongs to* $H^1(\Omega)$ *and is an eigenfunction of* $-\Delta^{\mathrm{N}}$ *corresponding to* $\lambda_n$.
(c) *Moreover,* $\{\frac{1}{\sqrt{1+\lambda_n}}\,\hat{u}_n\}_{n \geq 0}$ *form an orthonormal basis of* $H^1(\Omega)$.

Relying on this theorem, our study of the Dirichlet eigenvalue problem (9.1) can now be transposed to the situation of the Neumann eigenvalue problem (9.12). This is summarized in the next statement.

**Proposition 9.21.** *Let $\{\lambda_n\}_{n\geq 0}$ be the sequence of eigenvalues of $-\Delta^N$, and let $\{\hat{u}_n\}_{n\geq 0}$ be the corresponding basis of eigenfunctions provided by Theorem 9.20.*

(a) *The first eigenvalue $\lambda_0 = 0$ is simple, with the eigenspace the space of constant functions $\mathbb{R} \subset H^1(\Omega)$. Any eigenfunction of $-\Delta^N$ corresponding to a higher eigenvalue $\lambda_n > 0$ must be nodal.*

(b) *For $n \geq 1$ we have*

$$\lambda_n = \max_{v \in H_n \setminus \{0\}} R(v) = \min_{v \in H_{n-1}^\perp \setminus \{0\}} R(v) = \min_{Y \in \mathscr{S}_{n+1}} \max_{v \in Y \setminus \{0\}} R(v) = \max_{Y \in \mathscr{S}_n} \min_{v \in Y^\perp \setminus \{0\}} R(v),$$

*where $H_n = \mathrm{span}\{\hat{u}_k\}_{k=0}^n \subset H^1(\Omega)$, $\mathscr{S}_n$ denotes the family of $n$-dimensional vector subspaces of $H^1(\Omega)$, and $R(v) = \frac{\|\nabla v\|_2^2}{\|v\|_2^2}$ is the Rayleigh quotient.*

(c) *If $\Omega = (0,1) \subset \mathbb{R}$, then the eigenvalues of $-\Delta^N$ are $\lambda_n = (n\pi)^2$, for $n \geq 0$, and the corresponding $L^2$-normalized eigenfunctions are $\hat{u}_n(t) = \sqrt{2}\cos(n\pi t)$.*

To distinguish between the Dirichlet and Neumann cases, we will sometimes denote by $\{\lambda_n^D\}_{n\geq 1}$ the eigenvalues of $-\Delta^D$ [problem (9.1)] and by $\{\lambda_n^N\}_{n\geq 1}$ the eigenvalues of $-\Delta^N$ [problem (9.12)]. Both sequences of eigenvalues are related. Indeed, the inclusion $H_0^1(\Omega) \subset H^1(\Omega)$ and the variational characterizations stated in Theorem 9.11 and Proposition 9.21(b) yield the following proposition.

**Proposition 9.22.** $\lambda_{n-1}^N \leq \lambda_n^D$ *for all $n \geq 1$.*

We conclude our review of the linear eigenvalue theory by mentioning an important property of eigenfunctions (see Garofalo and Lin [148]).

**Proposition 9.23.** *The eigenfunctions of $-\Delta^D$ and $-\Delta^N$ satisfy the unique continuation property: if $\hat{u}$ is an eigenfunction (in particular $\hat{u} \neq 0$), then its vanishing set $\{x \in \Omega : \hat{u}(x) = 0\}$ is a Lebesgue-null set.*

Counterexamples due to Martio [250] suggest that it is unlikely that this property can extend to the case of eigenfunctions of the $p$-Laplacian for $p \neq 2$. See, however, Ling [230] for a weaker statement.

## 9.2  Spectrum of $p$-Laplacian

In this section, $\Omega \subset \mathbb{R}^N$ ($N \geq 1$) denotes a bounded domain with a $C^2$-boundary, we fix $p \in (1, +\infty)$, and we consider the $p$-Laplacian operator

$$\Delta_p u = \mathrm{div}(|\nabla u|^{p-2}\nabla u) \ \text{ for } u \in W^{1,p}(\Omega),$$

defined from the Sobolev space $W^{1,p}(\Omega)$ into, a priori, the space of distributions on $\Omega$, but whose image lies in fact in the dual space $(W^{1,p}(\Omega))^*$. The aim of this section is to develop the spectral analysis of the operator $-\Delta_p$ under Dirichlet and Neumann boundary conditions inspired by the one presented in Sect. 9.1 for the Laplacian. Actually, we deal here with more general eigenvalue problems involving a weight function on the right-hand side. Thus, we fix $\xi \in L^\infty(\Omega)$ such that $\xi \geq 0$, $\xi \neq 0$, and we consider the following eigenvalue problems, under Dirichlet boundary conditions,

$$\begin{cases} -\Delta_p u(x) = \lambda \xi(x)|u(x)|^{p-2}u(x) & \text{in } \Omega, \\ u = 0 & \text{on } \partial\Omega, \end{cases} \tag{9.14}$$

and under Neumann boundary conditions,

$$\begin{cases} -\Delta_p u(x) = \lambda \xi(x)|u(x)|^{p-2}u(x) & \text{in } \Omega, \\ \frac{\partial u}{\partial n_p} = 0 & \text{on } \partial\Omega. \end{cases} \tag{9.15}$$

Recall the notion of weak solution for problems (9.14) and (9.15) (Definition 8.2). The basic definition for this section is as follows.

**Definition 9.24.** (a) We say that $\lambda \in \mathbb{R}$ is an *eigenvalue of the negative $p$-Laplacian under Dirichlet boundary conditions* (of $-\Delta_p^D$, for short) *in the domain $\Omega$, with respect to the weight $\xi$*, if problem (9.14) admits a nontrivial weak solution $u \in W_0^{1,p}(\Omega)$. Then $u$ is called an *eigenfunction* of $-\Delta_p^D$ corresponding to $\lambda$.

(b) Similarly, an *eigenvalue of the negative $p$-Laplacian under Neumann boundary conditions* (of $-\Delta_p^N$, for short) *in $\Omega$ with respect to $\xi$* is a real $\lambda$ such that problem (9.15) admits a nontrivial weak solution $u \in W^{1,p}(\Omega)$. Such a $u$ is called an *eigenfunction* of $-\Delta_p^N$ corresponding to $\lambda$.

*Remark 9.25.* (a) If $\lambda$ is an eigenvalue of $-\Delta_p^D$ with eigenfunction $u$, then

$$\|\nabla u\|_p^p = \lambda \int_\Omega \xi |u|^p \, dx.$$

As $u \in W_0^{1,p}(\Omega) \setminus \{0\}$, we have $\nabla u \neq 0$, hence $\lambda > 0$.

(b) $\lambda = 0$ is an eigenvalue of $-\Delta_p^N$ with respect to $\xi$, and the corresponding eigenfunctions are the constant functions. All other eigenvalues of $-\Delta_p^N$ are positive.

(c) If $u$ is an eigenfunction of $-\Delta_p^D$ or $-\Delta_p^N$, then $u \in C^{1,\alpha}(\overline{\Omega})$ for some $\alpha \in (0,1)$ (Corollary 8.13).

(d) Unlike in the case of the Laplacian, for a given $\lambda$, the set of solutions of problems (9.14) and (9.15) is not in general a linear space.

The techniques used in Sect. 9.1, which rely extensively on the linearity of the eigenvalue problems, cannot be transposed to the case of the $p$-Laplacian. The analysis of the eigenvalue problems (9.14) and (9.15) is based on the Lyusternik–Schnirelmann theory for critical points, which in turn has its starting point in the Courant–Fischer minimax characterizations of the eigenvalues in the linear theory. In the linear case, as we note in Remark 9.7, the eigenvalues of $-\Delta^D$ are the critical values of the quadratic functional $u \mapsto \|\nabla u\|_2^2$ on the submanifold $M = \{u \in H_0^1(\Omega) : \|u\|_2 = 1\}$. The Lyusternik–Schnirelmann theory aims to extend the theory to a general smooth functional. First we state the general principle, and then we will apply it to the case of the $p$-Laplacian.

## Lyusternik–Schnirelmann Principle

We consider the setting of an infinite-dimensional reflexive Banach space $X$ and maps $\varphi, \psi \in C^1(X, \mathbb{R})$ subject to the following assumptions. By $\langle \cdot, \cdot \rangle$ we denote the duality brackets for the pair $(X^*, X)$.

H($\varphi, \psi$)  (i)  $\varphi, \psi \in C^1(X, \mathbb{R})$ are even maps, $\varphi(0) = \psi(0) = 0$, and the level set
$M := \{u \in X : \psi(u) = 1\}$ is bounded;

(ii)  $\varphi'$ is completely continuous [i.e., $u_n \overset{w}{\to} u$ in $X$ implies $\varphi'(u_n) \to \varphi'(u)$ in $X^*$], and for $u \in \overline{\mathrm{conv}}\, M$ we have

$$\langle \varphi'(u), u \rangle = 0 \iff \varphi(u) = 0;$$

(iii)  $\psi'$ is bounded, and

if $u_n \overset{w}{\to} u$ in $X$, $\psi'(u_n) \overset{w}{\to} v$ in $X^*$, $\langle \psi'(u_n), u_n \rangle \to \langle v, u \rangle$, then $u_n \to u$ in $X$;

(iv)  For $u \in X \setminus \{0\}$ we have $\langle \psi'(u), u \rangle > 0$ and $\lim\limits_{t \to +\infty} \psi(tu) = +\infty$; moreover, $\inf\limits_{u \in M} \langle \psi'(u), u \rangle > 0$.

In particular, by H($\varphi, \psi$) (iv), Lyusternik's theorem (Theorem 5.72) implies that the level set $M = \{u \in X : \psi(u) = 1\}$ is a $C^1$-Banach submanifold of $X$ of codimension 1. We now consider the following nonlinear eigenvalue problem:

$$\mu \psi'(u) = \varphi'(u), \quad (\mu, u) \in \mathbb{R} \times M. \tag{9.16}$$

If a couple $(\mu, u)$ solves (9.16), then we call $\mu$ (resp. $u$) an eigenvalue (resp. a solution) of (9.16).

*Remark 9.26.* (a) As in Remark 9.7, $u$ is a constrained critical point of $\varphi$ on $M$ if and only if $u$ is a solution of (9.16) for some eigenvalue $\mu \in \mathbb{R}$. Unlike in Remark 9.7, without further assumption, $\mu$ does not necessarily coincide with the value $\varphi(u)$. This property holds, however, when $\varphi$ and $\psi$ are positively

$p$-homogeneous for some $p > 1$, i.e., $\varphi(tu) = t^p \varphi(u)$ and $\psi(tu) = t^p \psi(u)$ for all $t > 0$. Indeed, differentiating these relations at $t = 1$, we have $\varphi(u) = \frac{1}{p}\langle \varphi'(u), u \rangle$ and $\psi(u) = \frac{1}{p}\langle \psi'(u), u \rangle$. Thus, when $(\mu, u)$ solves (9.16), we obtain

$$\varphi(u) = \frac{1}{p}\langle \varphi'(u), u \rangle = \frac{\mu}{p}\langle \psi'(u), u \rangle = \mu \psi(u) = \mu.$$

(b) In problem (9.16), we have intentionally reversed the equality with respect to (9.14) and (9.15) by writing the eigenvalue $\mu$ on the left-hand side. The reason is that the following abstract result will not be applied directly to the eigenvalue problems (9.14) and (9.15) but to their inverse problems, exactly as in the linear case where we first apply an abstract result to describe the spectrum $\{\mu_n\}_{n \geq 1}$ of the essential inverse of the negative Dirichlet Laplacian $-\Delta^D$ (Proposition 9.3) and then deduce the spectrum $\{\frac{1}{\mu_n}\}_{n \geq 1}$ of $-\Delta^D$ itself (Theorem 9.4). Actually Remark 9.7 does not reflect faithfully the reasoning that we will pursue here since the roles of $\varphi$ and $\psi$ in what follows will be reversed with respect to Remark 9.7.

Recall the notion of Krasnosel'skiĭ genus $A \mapsto \operatorname{gen} A$ introduced in Definition 5.61. For $n \in \mathbb{N}$, let

$$\mathscr{K}_n = \{K \subset M : K \text{ is symmetric, compact, with } \varphi|_K > 0 \text{ and } \operatorname{gen} K \geq n\}. \quad (9.17)$$

The set $\mathscr{K}_n$ plays here the same role as the set $\mathscr{S}_n$ in the Courant–Fischer theorem (Theorem 9.11). In particular, we define max-min values

$$c_n = \begin{cases} \sup\limits_{K \in \mathscr{K}_n} \min\limits_{u \in K} \varphi(u) & \text{if } \mathscr{K}_n \neq \emptyset, \\ 0 & \text{if } \mathscr{K}_n = \emptyset. \end{cases}$$

Thus, $\{c_n\}_{n \geq 1}$ form a nonincreasing sequence

$$+\infty \geq c_1 \geq c_2 \geq \ldots \geq c_n \geq \ldots \geq 0.$$

The Lyusternik–Schnirelmann principle can be stated as follows.

**Theorem 9.27.** *Assume that* H$(\varphi, \psi)$ *hold. Then:*

(a) $c_1 < +\infty$ *and* $c_n \to 0$ *as* $n \to \infty$;
(b) *If* $c := c_n > 0$, *then we can find an element* $u \in M$ *that is a solution of (9.16) for an eigenvalue* $\mu \neq 0$ *and such that* $\varphi(u) = c$;
(c) *More generally, if* $c := c_n = c_{n+k} > 0$ *for some* $k \geq 0$, *then the set of solutions* $u \in M$ *of (9.16) such that* $\varphi(u) = c$ *has genus* $\geq k + 1$;
(d) *If* $c_n > 0$ *for all* $n \geq 1$, *then there is a sequence* $\{(\mu_n, u_n)\}_{n \geq 1}$ *of solutions of (9.16) with* $\varphi(u_n) = c_n$, $\mu_n \neq 0$ *for all* $n \geq 1$, *and* $\mu_n \to 0$ *as* $n \to \infty$;
(e) *We strengthen* H$(\varphi, \psi)$ *(ii) by assuming that, for* $u \in \overline{\operatorname{conv}} M$, *we have*

$$\langle \varphi'(u), u \rangle = 0 \iff \varphi(u) = 0 \iff u = 0.$$

*Then, $c_n > 0$ for all $n \geq 1$, and there is a sequence $\{(\mu_n, u_n)\}_{n \geq 1}$ of solutions of (9.16) such that $\varphi(u_n) = c_n$, $\mu_n \neq 0$, $\mu_n \to 0$, and $u_n \overset{w}{\to} 0$ in $X$.*

*Remark 9.28.* A proof can be found in Zeidler [387, pp. 326–328]. Also note the similarity between Theorem 9.27(b), (c) and Theorem 5.60.

## Spectrum of $p$-Laplacian Under Neumann Boundary Conditions

We now apply Theorem 9.27 to describe the spectral properties of the negative $p$-Laplacian under Dirichlet and Neumann boundary conditions. This time, we start with the Neumann case for which we provide full details, then we will sketch the analysis in the Dirichlet case. We deal with the reflexive Banach space $X = W^{1,p}(\Omega)$ and the maps $\varphi, \psi : W^{1,p}(\Omega) \to \mathbb{R}$ defined by

$$\varphi(u) = \frac{1}{p} \int_\Omega \xi(x)|u(x)|^p \, dx \quad \text{and} \quad \psi(u) = \frac{1}{p} \left( \int_\Omega |\nabla u|^p \, dx + \int_\Omega \xi(x)|u(x)|^p \, dx \right)$$

for all $u \in W^{1,p}(\Omega)$. We know that $\varphi, \psi \in C^1(W^{1,p}(\Omega), \mathbb{R})$, and for $u, v \in W^{1,p}(\Omega)$ we have

$$\langle \varphi'(u), v \rangle = \int_\Omega \xi |u|^{p-2} uv \, dx, \tag{9.18}$$

$$\langle \psi'(u), v \rangle = \int_\Omega |\nabla u|^{p-2} (\nabla u, \nabla v)_{\mathbb{R}^N} \, dx + \int_\Omega \xi |u|^{p-2} uv \, dx. \tag{9.19}$$

The following lemma will be necessary to check the hypotheses H($\varphi, \psi$).

**Lemma 9.29.** *The map $u \mapsto \|u\|_\xi := \left( \int_\Omega |\nabla u|^p \, dx + \int_\Omega \xi |u|^p \, dx \right)^{\frac{1}{p}}$ is a norm on $W^{1,p}(\Omega)$, equivalent to the Sobolev norm $\| \cdot \|$.*

*Proof.* Since we have $\xi \in L^\infty(\Omega)$, $\xi \geq 0$, and $\xi \neq 0$, it is clear that $\| \cdot \|_\xi$ is a norm, and there is a constant $M_1 > 0$ such that $\|u\|_\xi \leq M_1 \|u\|$ for all $u \in W^{1,p}(\Omega)$. Note that

$$\{x \in \Omega : \xi(x) > 0\} = \bigcup_{n \in \mathbb{N}} \left\{ x \in \Omega : \xi(x) \geq \frac{1}{n} \right\}.$$

Since the set $\{x \in \Omega : \xi(x) > 0\}$ has positive Lebesgue measure, we can find $n_0 \geq 1$ such that $D := \{x \in \Omega : \xi(x) \geq \frac{1}{n_0}\}$ has positive Lebesgue measure. By Theorem 1.44, there is $M_2 > 0$ such that

$$\|u - u_D\|_p \leq M_2 \|\nabla u\|_p \quad \text{for all } u \in W^{1,p}(\Omega), \tag{9.20}$$

with $u_D = \frac{1}{|D|_N}\int_D u\,dx$. Note that

$$|u_D| \le \frac{n_0}{|D|_N}\int_D \xi|u|\,dx \le \frac{n_0}{|D|_N}\Big(\int_\Omega \xi|u|^p\,dx\Big)^{\frac{1}{p}}\Big(\int_\Omega \xi\,dx\Big)^{\frac{p-1}{p}}. \tag{9.21}$$

Using (9.20), (9.21), we find a constant $M_3 > 0$ such that

$$\|u\|_p \le \|u - u_D\|_p + |\Omega|_N^{\frac{1}{p}}|u_D| \le M_3\|u\|_\xi \quad \text{for all } u \in W^{1,p}(\Omega).$$

This clearly yields $M_4 > 0$, with $\|u\| \le M_4\|u\|_\xi$ for all $u \in W^{1,p}(\Omega)$. The proof is now complete. $\square$

Then we check that the Lyusternik–Schnirelmann principle can be applied here.

**Proposition 9.30.** *The maps $\varphi, \psi$ satisfy hypotheses* H($\varphi, \psi$).

*Proof.* (i) Clearly, the maps $\varphi, \psi$ are even and, by Lemma 9.29, the set

$$M := \{u \in W^{1,p}(\Omega) : \psi(u) = 1\} = \{u \in W^{1,p}(\Omega) : \|u\|_\xi^p = p\} \tag{9.22}$$

is bounded in $W^{1,p}(\Omega)$.

(ii) The second part of H($\varphi, \psi$) (ii) is immediately obtained by noting that

$$\langle \varphi'(u), u \rangle = \int_\Omega \xi|u|^p\,dx = p\varphi(u) \quad \text{for all } u \in W^{1,p}(\Omega).$$

Thus, it remains to check that $\varphi'$ is completely continuous. Suppose that $u_n \xrightarrow{w} u$ in $W^{1,p}(\Omega)$. It suffices to check that from any relabeled subsequence of $\{\varphi'(u_n)\}_{n\ge 1}$ we can extract a subsequence converging to $\varphi'(u)$. Clearly, by (9.18),

$$\|\varphi'(u_n) - \varphi'(u)\| \le \|\xi\|_\infty \||u_n|^{p-2}u_n - |u|^{p-2}u\|_{p'} \tag{9.23}$$

(with $p' = \frac{p}{p-1}$). Since $u_n \xrightarrow{w} u$ in $W^{1,p}(\Omega)$, up to extracting a subsequence, we may assume that $u_n \to u$ in $L^p(\Omega)$ (Theorem 1.49). By the elementary properties of $L^p$-spaces, up to a subsequence, we may assume that $u_n \to u$ a.e. in $\Omega$ and there is $h \in L^p(\Omega)$ such that $|u_n| \le h$ a.e. in $\Omega$ (e.g., Brezis [52, p. 94]). By Lebesgue's dominated convergence theorem, it follows that $\||u_n|^{p-2}u_n - |u|^{p-2}u\|_{p'} \to 0$ as $n \to \infty$, whence $\varphi'(u_n) \to \varphi'(u)$ in $W^{1,p}(\Omega)^*$ [see (9.23)]. Thus, we obtain that $\varphi'$ is completely continuous.

(iii) By (9.19), we have

$$\|\psi'(u)\| \le \|\nabla u\|_p^{p-1} + \|\xi\|_\infty\|u\|_p^{p-1},$$

hence $\psi'$ is bounded. Also, by (9.19), the map $\psi'$ is monotone and continuous and, hence, generalized pseudomonotone (Corollary 2.42 and Propositions 2.60 and 2.67). To show the second part of H($\varphi, \psi$) (iii), assume that $u_n \xrightarrow{w} u$ in $X$, $\psi'(u_n) \xrightarrow{w} v$ in $X^*$, and $\langle \psi'(u_n), u_n \rangle \to \langle v, u \rangle$. On the one hand, arguing as in part (ii) above on the basis of Theorem 1.49 and Lebesgue's dominated convergence theorem, we obtain that $\|u_n\|_p \to \|u\|_p$ and $\int_\Omega \xi |u_n|^p \, dx \to \int_\Omega \xi |u|^p \, dx$. On the other hand, the generalized pseudomonotonicity of $\psi'$ implies that $\langle \psi'(u_n), u_n \rangle \to \langle \psi'(u), u \rangle$. Combining these two observations, we infer that $\|u_n\| \to \|u\|$ as $n \to \infty$. Since the space $(W^{1,p}(\Omega), \|\cdot\|)$ enjoys the Kadec–Klee property (Remark 2.47(a), (c)), we then conclude that $u_n \to u$ in $W^{1,p}(\Omega)$ as $n \to \infty$.

(iv) For all $u \in W^{1,p}(\Omega) \setminus \{0\}$ we have $\langle \psi'(u), u \rangle = \|u\|_\xi^p > 0$, $\langle \psi'(u), u \rangle = p$ if $u \in M$ [see (9.22)], and $\psi(tu) = \frac{t}{p} \|u\|_\xi^p \to +\infty$ as $t \to +\infty$.  $\square$

In our situation we consider $M = \{u \in W^{1,p}(\Omega) : \|u\|_\xi^p = p\}$, where $\|\cdot\|_\xi$ is the norm defined in Lemma 9.29. Recall the set $\mathscr{K}_n$ defined in (9.17),

$$\mathscr{K}_n = \{K \subset M : K \text{ is compact, symmetric, with } \varphi|_K > 0 \text{ and gen } K \geq n\}.$$

We can prove the following result on the basis of Theorem 9.27.

**Theorem 9.31.** *The operator $-\Delta_p^N$ admits a sequence of eigenvalues $\{\lambda_n(\xi)\}_{n \geq 0}$ with respect to the weight $\xi$ such that*

$$0 = \lambda_0(\xi) \leq \lambda_1(\xi) \leq \ldots \leq \lambda_n(\xi) \leq \ldots, \qquad \lim_{n \to \infty} \lambda_n(\xi) = +\infty,$$

*characterized by*

$$\frac{1}{\lambda_n(\xi) + 1} = \sup_{K \in \mathscr{K}_{n+1}} \min_{u \in K} \varphi(u) \text{ for all } n \geq 0.$$

*Proof.* We first prove the following claim.

*Claim 1:* $\mathscr{K}_n \neq \emptyset$ and $c_n := \sup_{K \in \mathscr{K}_n} \min_{u \in K} \varphi(u) > 0$ for all $n \geq 1$.

Since the set $\{x \in \Omega : \xi(x) > 0\}$ has positive measure, we can find functions $u_1, \ldots, u_n \in C_c^\infty(\Omega)$ whose supports are pairwise disjoint and such that

$$\int_\Omega \xi(x) |u_k|^p \, dx > 0 \text{ for all } k \in \{1, \ldots, n\}.$$

Then the subspace $V := \text{span} \{u_k\}_{k=1}^n \subset W^{1,p}(\Omega)$ has dimension $n$. By construction, we have $\varphi|_{V \setminus \{0\}} > 0$. Invoking Remark 5.62(a), we obtain that the set $K := \{u \in V : \|u\|_\xi^p = p\}$ belongs to $\mathscr{K}_n$. Moreover, $c_n \geq \min_{u \in K} \varphi(u) > 0$. This proves Claim 1.

By virtue of Proposition 9.30 and Claim 1, we can apply Theorem 9.27(d), which implies that problem (9.16) admits a sequence of solutions $\{(\mu_n, u_n)\}_{n \geq 1}$ such that $\mu_n \neq 0$, $\mu_n \to 0$ as $n \to \infty$, and

$$\varphi(u_n) = c_n = \sup_{K \in \mathscr{K}_n} \min_{u \in K} \varphi(u) > 0. \tag{9.24}$$

Actually, since here both maps $\varphi$ and $\psi$ are positively $p$-homogeneous, we have $\varphi(u_n) = \mu_n$ for all $n \geq 1$ [Remark 9.26(a)]. In particular, by (9.24), $\mu_n > 0$ for all $n \geq 1$ and the sequence $\{\mu_n\}_{n \geq 1}$ is nonincreasing. Note also that

$$\mu_1 = \sup_{K \in \mathscr{K}_1} \min_{u \in K} \varphi(u) = 1,$$

where the supremum is realized by the compact set $K = \{v, -v\} \subset M$, with $v$ being the constant function $v \equiv \left(\frac{p}{\|\xi\|_1}\right)^{\frac{1}{p}}$. Now, in view of our choices of $\varphi$ and $\psi$, it is clear that $\mu \neq 0$ is an eigenvalue of (9.16) if and only if $\frac{1}{\mu} - 1$ is an eigenvalue of $-\Delta_p^N$ with respect to $\xi$. Letting $\lambda_n(\xi) = \frac{1}{\mu_{n+1}} - 1$ for all $n \geq 0$, we obtain a sequence of eigenvalues of $-\Delta_p^N$ satisfying all the claimed properties. $\qquad\square$

*Remark 9.32.* (a) In the case where the weight function satisfies $\xi > 0$ a.e. in $\Omega$, the foregoing proof can be simplified by invoking Theorem 9.27(e) (instead of Theorem 9.27(d)), in which case Claim 1 is unnecessary.

(b) The eigenvalues of $-\Delta_p^N$ provided by Theorem 9.31 are usually called the (LS)-*eigenvalues*. They are not a priori all the eigenvalues of $-\Delta_p^N$.

Next we look for additional properties of the eigenvalues of $-\Delta_p^N$ (with respect to the weight $\xi$). We already know that $\lambda_0(\xi) = 0$ is the minimal eigenvalue (Remark 9.25), with the nonzero constant functions as a corresponding set of eigenfunctions [so of course, all the eigenfunctions corresponding to $\lambda_0(\xi)$ have constant sign]. We also know that any eigenfunction of $-\Delta_p^N$ corresponding to any eigenvalue $\lambda$ belongs to $C^{1,\alpha}(\overline{\Omega})$ for some $\alpha \in (0, 1)$ (Remark 9.25).

**Proposition 9.33.** *Let $\lambda > 0$ be an eigenvalue of $-\Delta_p^N$ with respect to $\xi$. Then every eigenfunction $u$ corresponding to $\lambda$ is nodal (i.e., sign changing).*

*Proof.* As mentioned previously, $u \in C^1(\overline{\Omega})$. Arguing by contradiction, assume that $u$ has a constant sign, say, $u \geq 0$ on $\Omega$. Then, by Corollary 8.17, we have $u > 0$ in $\Omega$. On the other hand, acting on (9.15) with the test function $v \equiv 1$, we obtain $\lambda \int_\Omega \xi u^{p-1} dx = 0$. This is impossible. Hence $u$ is nodal. $\qquad\square$

We denote by $\sigma_p^N(\xi) \subset [0, +\infty)$ the set of all the eigenvalues of $-\Delta_p^N$ with respect to the weight $\xi$.

**Proposition 9.34.** $\lambda_0(\xi) = 0$ *is isolated in* $\sigma_p^N(\xi)$.

*Proof.* Arguing by contradiction, suppose that we can find a sequence $\{\lambda_n\}_{n\geq 1} \subset \sigma_p^N(\xi)$ such that $\lambda_n > 0$ for all $n \geq 1$ and $\lambda_n \to 0$ as $n \to \infty$. Let $\{u_n\}_{n\geq 1} \subset W^{1,p}(\Omega)$ be a sequence of corresponding eigenfunctions with $\|u_n\| = 1$ for all $n \geq 1$. In particular, since $\{u_n\}_{n\geq 1}$ is bounded in $W^{1,p}(\Omega)$, by virtue of Theorems 8.4 and 8.10, we can find $\alpha \in (0,1)$ and $M > 0$ such that $u_n \in C^{1,\alpha}(\overline{\Omega})$ and $\|u_n\|_{C^{1,\alpha}(\overline{\Omega})} \leq M$ for all $n \geq 1$. Since the embedding $C^{1,\alpha}(\overline{\Omega}) \hookrightarrow C^1(\overline{\Omega})$ is compact (by the Arzelà–Ascoli theorem; see, e.g., Brezis [52, p. 111]), we may assume that there is $u \in C^1(\overline{\Omega})$ such that $u_n \to u$ in $C^1(\overline{\Omega})$. By the fact that $u_n$ is an eigenfunction corresponding to $\lambda_n$, we have

$$\|\nabla u_n\|_p^p = \lambda_n \int_\Omega \xi |u_n|^p \, dx \ \text{ for all } n \geq 1.$$

Passing to the limit as $n \to \infty$, it follows that $\|\nabla u\|_p = 0$, and so $u \equiv c \in \mathbb{R}$. Moreover, $\|u\| = \lim_{n\to\infty} \|u_n\| = 1$. Hence $c \neq 0$. We can find $n_0 \geq 1$ such that $\|u_{n_0} - c\|_\infty < |c|$, thus $u_{n_0}$ has a constant sign. This contradicts Proposition 9.33. $\qquad\square$

**Proposition 9.35.** *The set $\sigma_p^N(\xi)$ is closed in $[0,+\infty)$.*

*Proof.* Let $\{\lambda_n\}_{n\geq 1} \subset \sigma_p^N(\xi)$ be such that $\lambda_n \to \lambda > 0$ as $n \to \infty$. Let $\{u_n\}_{n\geq 1} \subset W^{1,p}(\Omega)$, $\|u_n\| = 1$ for all $n \geq 1$, be corresponding eigenfunctions. We have

$$A(u_n) = \lambda_n \xi |u_n|^{p-2} u_n \ \text{ for all } n \geq 1, \tag{9.25}$$

with $A : W^{1,p}(\Omega) \to W^{1,p}(\Omega)^*$ defined by (2.28), and we may assume that

$$u_n \xrightarrow{\text{w}} u \text{ in } W^{1,p}(\Omega) \text{ and } u_n \to u \text{ in } L^p(\Omega) \text{ as } n \to \infty. \tag{9.26}$$

Acting on (9.25) with $u_n - u$, passing to the limit as $n \to \infty$, and using (9.26), we obtain $\lim_{n\to\infty} \langle A(u_n), u_n - u \rangle = 0$, which implies that $u_n \to u$ in $W^{1,p}(\Omega)$ (Proposition 2.72), and so $\|u\| = 1$. Passing to the limit as $n \to \infty$ in (9.25), we have $A(u) = \lambda \xi |u|^{p-2} u$. Since $u \neq 0$, we deduce that $\lambda \in \sigma_p^N(\xi)$, which proves the closedness of $\sigma_p^N(\xi)$. $\qquad\square$

Our next purpose is to obtain a result analogous to the Courant nodal domain theorem mentioned in Remark 9.15.

**Definition 9.36.** Let $\lambda \in \sigma_p^N(\xi)$, $\lambda \neq 0$. Let $u$ be an eigenfunction of $-\Delta_p^N$ corresponding to $\lambda$; thus, $u \in C^1(\overline{\Omega})$ and $u$ is nodal. Let $N(u)$ denote the number of nodal domains of $u$, i.e., of connected components of $\{x \in \Omega : u(x) \neq 0\}$ (Definition 1.60). We set

$$N(\lambda) = \sup\{N(u) : u \text{ is an eigenfunction corresponding to } \lambda\}.$$

**Proposition 9.37.** *For every $\lambda \in \sigma_p^N(\xi)$, $\lambda \neq 0$, we have $2 \leq N(\lambda) < +\infty$.*

*Proof.* Since every eigenfunction corresponding to $\lambda$ is nodal (Proposition 9.33), it is clear that $N(\lambda) \geq 2$. Now, let $u$ be an eigenfunction corresponding to $\lambda$, and let $\Omega_0 \subset \Omega$ be a nodal domain of $u$. Let $u_0 = \chi_{\Omega_0} u \in W^{1,p}(\Omega)$ (Proposition 1.61). Acting on (9.15) with the test function $v = u_0$, we obtain

$$\|\nabla u_0\|_p^p = \lambda \int_{\Omega_0} \xi |u|^p \, dx,$$

which through Hölder's inequality implies that

$$\|u_0\|^p \leq (\lambda \|\xi\|_\infty + 1) \int_{\Omega_0} |u_0|^p \, dx \leq (\lambda \|\xi\|_\infty + 1) |\Omega_0|_N^{\frac{\theta-p}{\theta}} \|u_0\|_\theta^p,$$

where $\theta = \frac{Np}{N-p}$ if $p < N$ and $\theta = p+1$ if $p \geq N$. Using Theorem 1.49, we can find $c_0 > 0$ such that

$$\|u_0\|^p \leq c_0 (\lambda \|\xi\|_\infty + 1) |\Omega_0|_N^{\frac{\theta-p}{\theta}} \|u_0\|^p,$$

and thus

$$\left( c_0 (\lambda \|\xi\|_\infty + 1) \right)^{-\frac{\theta}{\theta-p}} \leq |\Omega_0|_N.$$

This forces $N(u) \leq |\Omega|_N \left( c_0 (\lambda \|\xi\|_\infty + 1) \right)^{\frac{\theta}{\theta-p}}$. Hence $N(\lambda) < +\infty$. □

The next result compares a general eigenvalue of $-\Delta_p^N$ with an (LS)-eigenvalue.

**Proposition 9.38.** *For every* $\lambda \in \sigma_p^N(\xi)$, $\lambda \neq 0$, *we have* $\lambda_{N(\lambda)-1}(\xi) \leq \lambda$.

*Proof.* Let $u$ be an eigenfunction corresponding to $\lambda$ such that $N(u) = N(\lambda)$, and let $\Omega_1, \ldots, \Omega_{N(\lambda)}$ be the nodal domains of $u$. We define

$$u_i = \chi_{\Omega_i} u \in W^{1,p}(\Omega), \quad i = 1, \ldots, N(\lambda)$$

(Proposition 1.61). In particular, acting on (9.15) with the test function $u_i$, we obtain

$$\int_\Omega |\nabla u_i|^p \, dx = \lambda \int_\Omega \xi |u_i|^p \, dx \quad \text{for all } i \in \{1, \ldots, N(\lambda)\}. \tag{9.27}$$

Let $V_\lambda = \text{span}\{u_i\}_{i=1}^{N(\lambda)}$. Since the elements $u_i$ are linearly independent, we have $\dim V_\lambda = N(\lambda)$. Let

$$C(\lambda) = \{v \in V_\lambda : \|v\|_\xi^p = p\}.$$

The set $C(\lambda)$ is a sphere in the space $V_\lambda$ (by Lemma 9.29), so for its genus we have $\text{gen}\, C(\lambda) = N(\lambda)$ (by Remark 5.62), and thus $C(\lambda) \in \mathcal{K}_{N(\lambda)}$. Hence, by Theorem 9.31,

$$\frac{1}{1+\lambda_{N(\lambda)-1}(\xi)} \geq \min_{v \in C(\lambda)} \frac{1}{p} \int_\Omega \xi |v|^p \, dx. \tag{9.28}$$

On the other hand, for $v = \sum_{i=1}^{N(\lambda)} \beta_i u_i \in C(\lambda)$, using (9.27), we see that

$$\frac{1}{p} \int_\Omega \xi |v|^p \, dx = \frac{1}{p} \sum_{i=1}^{N(\lambda)} |\beta_i|^p \int_\Omega \xi |u_i|^p \, dx = \frac{1}{p(1+\lambda)} \sum_{i=1}^{N(\lambda)} |\beta_i|^p \|u_i\|_\xi^p$$

$$= \frac{1}{p(1+\lambda)} \|v\|_\xi^p = \frac{1}{1+\lambda}. \tag{9.29}$$

Combining (9.28) and (9.29), we finally deduce $\lambda_{N(\lambda)-1}(\xi) \leq \lambda$.                    □

**Corollary 9.39.** *Let $m, n \in \mathbb{N}$ such that $\lambda_n(\xi) < \lambda_m(\xi)$. Then $N(\lambda_n(\xi)) \leq m$.*

*Proof.* Apply Proposition 9.38 with $\lambda = \lambda_n(\xi)$.                    □

From Propositions 9.34 and 9.35 we see that

$$\lambda_1^* := \inf\{\lambda \in \sigma_p^N(\xi) : \lambda > 0\}$$

still belongs to $\sigma_p^N(\xi)$ and is positive. The next result shows that $\lambda_1^*$ actually coincides with the second (LS)-eigenvalue of $-\Delta_p^N$.

**Proposition 9.40.** $\lambda_1^* = \lambda_1(\xi)$.

*Proof.* The inequality $\lambda_1^* \leq \lambda_1(\xi)$ is due to the definition of $\lambda_1^*$. By Proposition 9.37, we have $N(\lambda_1^*) \geq 2$, whence, by Proposition 9.38, we obtain

$$\lambda_1^* \geq \lambda_{N(\lambda_1^*)-1}(\xi) \geq \lambda_1(\xi).$$

The proof is complete.                    □

We focus on the second eigenvalue $\lambda_1(\xi)$. It admits a variational characterization provided by the Lyusternik–Schnirelmann theory. The next two propositions present alternative variational characterizations of $\lambda_1(\xi)$.

**Proposition 9.41.** *Let*

$$C_1(p) = \left\{ u \in W^{1,p}(\Omega) : \int_\Omega \xi |u|^p \, dx = 1, \int_\Omega \xi |u|^{p-2} u \, dx = 0 \right\}.$$

(a) *We always have $\lambda_1(\xi) \geq \inf\{\|\nabla u\|_p^p : u \in C_1(p)\}$.*
(b) *Moreover, if $p \geq 2$, then $\lambda_1(\xi) = \inf\{\|\nabla u\|_p^p : u \in C_1(p)\}$.*

*Proof.* Set

$$\hat{\lambda}_1 = \inf\{\|\nabla u\|_p^p : u \in C_1(p)\}.$$

(a) Let $u \in W^{1,p}(\Omega)$ be an eigenfunction corresponding to $\lambda_1(\xi)$. Take $\hat{u} = \alpha u$, with $\alpha > 0$ chosen such that $\|\nabla \hat{u}\|_p^p = \lambda_1(\xi)$. We have

$$\int_\Omega |\nabla \hat{u}|^{p-2} (\nabla \hat{u}, \nabla v)_{\mathbb{R}^N} \, dx = \lambda_1(\xi) \int_\Omega \xi |\hat{u}|^{p-2} \hat{u} v \, dx \quad \text{for all } v \in W^{1,p}(\Omega).$$

In particular, choosing $v = 1$ (resp. $v = \hat{u}$) in this relation, we see that

$$\int_\Omega \xi |\hat{u}|^{p-2} \hat{u} \, dx = 0 \quad \text{and} \quad \int_\Omega \xi |\hat{u}|^p \, dx = \frac{\|\nabla \hat{u}\|_p^p}{\lambda_1(\xi)} = 1,$$

hence $\hat{u} \in C_1(p)$. Therefore,

$$\hat{\lambda}_1 \leq \|\nabla \hat{u}\|_p^p = \lambda_1(\xi).$$

(b) Since the map $u \mapsto \|u\|_\xi$ is coercive and weakly l.s.c. on $W^{1,p}(\Omega)$ (Lemma 9.29), whereas the set $C_1(p) \subset W^{1,p}(\Omega)$ is sequentially weakly closed, there is $u \in C_1(p)$ such that $\|u\|_\xi = \inf_{v \in C_1(p)} \|v\|_\xi$. Thus,

$$\|\nabla u\|_p^p = \hat{\lambda}_1. \tag{9.30}$$

Since $p \geq 2$, the Lagrange multiplier rule implies that there exist $a, b \in \mathbb{R}$ such that

$$p \int_\Omega |\nabla u|^{p-2} (\nabla u, \nabla v)_{\mathbb{R}^N} \, dx = ap \int_\Omega \xi |u|^{p-2} u v \, dx + b(p-1) \int_\Omega \xi |u|^{p-2} v \, dx$$

for all $v \in W^{1,p}(\Omega)$. Choosing the test function $v = b$, since $u \in C_1(p)$, we have

$$b^2 (p-1) \int_\Omega \xi |u|^{p-2} \, dx = 0, \quad \text{i.e., } b = 0.$$

In this way, we obtain

$$\int_\Omega |\nabla u|^{p-2} (\nabla u, \nabla v)_{\mathbb{R}^N} \, dx = a \int_\Omega \xi |u|^{p-2} u v \, dx \quad \text{for all } v \in W^{1,p}(\Omega). \tag{9.31}$$

Hence $u$ is an eigenfunction of $-\Delta_p^N$ corresponding to the eigenvalue $a$. Since $u \in C_1(p)$, we have $\int_\Omega \xi |u|^{p-2} u \, dx = 0$, which implies that the function $u$ is not constant, so its eigenvalue $a$ is positive. Thus, Proposition 9.40 yields

$$a \geq \lambda_1(\xi).$$

Finally, choosing $v = u$ in (9.31) and invoking (9.30) and the fact that $u \in C_1(p)$, we get

$$\hat{\lambda}_1 = \|\nabla u\|_p^p = a \int_\Omega \xi |u|^p \, dx = a.$$

Therefore, $\hat{\lambda}_1 \geq \lambda_1(\xi)$. The proof is now complete.  $\square$

The following variational characterization of $\lambda_1(\xi)$ is valid for all $p \in (1, +\infty)$.

**Theorem 9.42.** *We consider*

$$S = \{u \in W^{1,p}(\Omega) : \int_\Omega \xi |u|^p \, dx = 1\}, \quad \hat{u}_0 \equiv \left( \int_\Omega \xi \, dx \right)^{-\frac{1}{p}} \in S,$$

$$and \quad \Gamma = \{\gamma \in C([-1,1], S) : \gamma(-1) = -\hat{u}_0, \; \gamma(1) = \hat{u}_0\}.$$

*The second eigenvalue of $-\Delta_p^N$ with respect to $\xi$ is characterized by*

$$\lambda_1(\xi) = \min_{\gamma \in \Gamma} \max_{-1 \leq t \leq 1} \|\nabla \gamma(t)\|_p^p.$$

*Proof.* We start with a preliminary claim.

*Claim 1:* Let $X$ be a Banach space and $\gamma \in C([-1,1], X)$ such that $\gamma(-1) = -\gamma(1)$. The set

$$C_\gamma := \{\gamma(t) : t \in [-1,1]\} \cup \{-\gamma(t) : t \in [-1,1]\}$$

has genus $\geq 2$.

In view of Definition 5.59(b) and Remark 5.62(a), it suffices to construct an odd map $h \in C(S^1, C_\gamma)$, where $S^1 = \{(\cos\theta, \sin\theta) : \theta \in [0, 2\pi]\} \subset \mathbb{R}^2$. This is done by letting

$$h(\cos\theta, \sin\theta) = \begin{cases} \gamma(\cos(\theta + \pi)) & \text{if } \theta \in [0, \pi], \\ -\gamma(\cos\theta) & \text{if } \theta \in [\pi, 2\pi]. \end{cases}$$

Claim 1 is proved.

Now we can prove the following claim.

*Claim 2:* $\lambda_1(\xi) \leq \inf_{\gamma \in \Gamma} \max_{-1 \leq t \leq 1} \|\nabla \gamma(t)\|_p^p.$

Let $\gamma \in \Gamma$. Let $\alpha \in C([-1,1], (0, +\infty))$ be such that

$$\alpha(t) = \left( \frac{1}{p} + \frac{1}{p} \|\nabla \gamma(t)\|_p^p \right)^{-\frac{1}{p}} \quad \text{for all } t \in [-1,1],$$

and let $\hat{\gamma} \in C([-1,1], W^{1,p}(\Omega))$ be given by $\hat{\gamma} = \alpha\gamma$. Then the image of $\hat{\gamma}$ lies in $M = \{u \in W^{1,p}(\Omega) : \|u\|_\xi^p = p\}$. In particular, the set $C_{\hat{\gamma}}$ from Claim 1 is compact, symmetric, and contained in $M$, and gen $C_{\hat{\gamma}} \geq 2$. By Theorem 9.31, this implies

$$\frac{1}{1+\lambda_1(\xi)} \geq \min_{u \in C_{\hat{\gamma}}} \frac{1}{p} \int_\Omega \xi |u|^p \, dx$$

$$= \min_{-1 \leq t \leq 1} \frac{1}{p} \int_\Omega \xi |\alpha(t)\gamma(t)|^p \, dx = \min_{-1 \leq t \leq 1} \frac{1}{1+\|\nabla\gamma(t)\|_p^p},$$

whence

$$\lambda_1(\xi) \leq \max_{-1 \leq t \leq 1} \|\nabla\gamma(t)\|_p^p \quad \text{for all } \gamma \in \Gamma.$$

This proves Claim 2.

In view of Claim 2, to complete the proof of the theorem, it suffices to construct $\gamma_* \in \Gamma$ such that

$$\max_{-1 \leq t \leq 1} \|\nabla\gamma_*(t)\|_p^p = \lambda_1(\xi). \tag{9.32}$$

To this end, let $\hat{u}_1 \in C^1(\overline{\Omega})$ be an eigenfunction corresponding to $\lambda_1(\xi)$. Let $\sigma : \mathbb{R} \to S$ be defined by

$$\sigma(r) = \frac{\hat{u}_1 + r}{\left(\int_\Omega \xi |\hat{u}_1 + r|^p \, dx\right)^{\frac{1}{p}}} \quad \text{for all } r \in \mathbb{R}.$$

Note that

$$\sigma(r) \to \pm\left(\int_\Omega \xi \, dx\right)^{-\frac{1}{p}} = \pm\hat{u}_0 \quad \text{in } C^1(\overline{\Omega}) \text{ [so in } W^{1,p}(\Omega)] \text{ as } r \to \pm\infty.$$

Hence, the map $\gamma_* : [-1,1] \to S$ given by

$$\gamma_*(t) = \begin{cases} \sigma(\frac{t}{1-t^2}) & \text{if } t \in (-1,1), \\ \pm\hat{u}_0 & \text{if } t = \pm 1 \end{cases}$$

is well defined and continuous and so belongs to $\Gamma$. Let us show that $\gamma_*$ satisfies (9.32). We have

$$\|\nabla\sigma(r)\|_p^p = \frac{\|\nabla\hat{u}_1\|_p^p}{\int_\Omega \xi |\hat{u}_1 + r|^p \, dx},$$

hence

$$\frac{d}{dr}\|\nabla\sigma(r)\|_p^p = \frac{-p\|\nabla\hat{u}_1\|_p^p}{(\int_\Omega \xi |\hat{u}_1 + r|^p \, dx)^2} \int_\Omega \xi |\hat{u}_1 + r|^{p-2}(\hat{u}_1 + r) \, dx. \tag{9.33}$$

From (9.33), since $\int_\Omega \xi |\hat{u}_1|^{p-2}\hat{u}_1\,dx = 0$ (by the choice of $\hat{u}_1$) and $s \mapsto |s|^{p-2}s$ is increasing on $\mathbb{R}$, we see that the function $r \mapsto \|\nabla \sigma(r)\|_p^p$ attains its maximum at $r = 0$, i.e.,

$$\|\nabla \sigma(r)\|_p^p \le \|\nabla \sigma(0)\|_p^p \text{ for all } r \in \mathbb{R}. \tag{9.34}$$

By (9.34) and the choice of $\hat{u}_1$, we finally obtain

$$\max_{-1 \le t \le 1} \|\nabla \gamma_*(t)\|_p^p = \|\nabla \sigma(0)\|_p^p = \frac{\|\nabla \hat{u}_1\|_p^p}{\int_\Omega \xi |\hat{u}_1|^p\,dx} = \lambda_1(\xi).$$

This completes the proof. $\qquad\qquad\qquad\qquad\qquad\qquad\qquad\qquad\qquad\qquad\qquad\square$

We conclude our analysis of the spectral properties of the negative Neumann $p$-Laplacian with the following result, which establishes a monotonicity property of the second eigenvalue $\lambda_1(\xi)$ with respect to the weight function $\xi$.

**Proposition 9.43.** *If* $\xi, \xi' \in L^\infty(\Omega)_+$, $\xi \ne 0$, *and* $\xi(x) < \xi'(x)$ *for a.a.* $x \in \Omega$, *then* $\lambda_1(\xi') < \lambda_1(\xi)$.

*Proof.* Let $\hat{u}_1 \in C^1(\overline{\Omega})$ be an eigenfunction of $-\Delta^N$ with respect to $\xi$, corresponding to $\lambda_1(\xi)$. From Proposition 9.33 we know that $\hat{u}_1$ is nodal. Thus, $\hat{u}_1^+$ and $\hat{u}_1^-$ are nonzero and linearly independent. So $V := \operatorname{span}\{\hat{u}_1^+, \hat{u}_1^-\}$ is a two-dimensional subspace of $W^{1,p}(\Omega)$. The choice of $\hat{u}_1$ as eigenfunction implies that, for $v = \beta_1 \hat{u}_1^+ + \beta_2 \hat{u}_1^- \in V$ (where $\beta_1, \beta_2 \in \mathbb{R}$), we have

$$\int_\Omega |\nabla v|^p\,dx = |\beta_1|^p \int_\Omega |\nabla \hat{u}_1^+|^p\,dx + |\beta_2|^p \int_\Omega |\nabla \hat{u}_1^-|^p\,dx$$

$$= \lambda_1(\xi)\left(|\beta_1|^p \int_\Omega \xi |\hat{u}_1^+|^p\,dx + |\beta_2|^p \int_\Omega \xi |\hat{u}_1^-|^p\,dx\right). \tag{9.35}$$

Since $v \mapsto \|v\|_{\xi'}$ is a norm on $V$ (Lemma 9.29), the set

$$C(V, \xi') := \{v \in V : \|v\|_{\xi'}^p = p\}$$

has genus $\operatorname{gen} C(V, \xi') = 2$ (Remark 5.62). The expression of $\lambda_1(\xi')$ in Theorem 9.31 yields

$$\frac{1}{\lambda_1(\xi') + 1} \ge \min_{v \in C(V, \xi')} \frac{1}{p} \int_\Omega \xi' |v|^p\,dx. \tag{9.36}$$

Let $v = \beta_1 \hat{u}_1^+ + \beta_2 \hat{u}_1^- \in C(V, \xi')$, achieving its minimum in (9.36). The assumption that $0 \le \xi < \xi'$ implies that

$$\int_\Omega \xi |v|^p\,dx < \int_\Omega \xi' |v|^p\,dx.$$

Moreover, since $v \in C(V, \xi')$, we have $\|v\|_{\xi'}^p = p$. Combining this with (9.35) and using that $t \mapsto \frac{t}{t+a}$ is increasing on $[0, +\infty)$ for $a = \int_\Omega |\nabla v|^p \, dx > 0$, we obtain

$$\frac{1}{\lambda_1(\xi') + 1} \geq \frac{1}{p} \int_\Omega \xi' |v|^p \, dx = \frac{\int_\Omega \xi' |v|^p \, dx}{\int_\Omega |\nabla v|^p \, dx + \int_\Omega \xi' |v|^p \, dx}$$
$$> \frac{\int_\Omega \xi |v|^p \, dx}{\int_\Omega |\nabla v|^p \, dx + \int_\Omega \xi |v|^p \, dx} = \frac{1}{\lambda_1(\xi) + 1}.$$

This implies the desired conclusion.                                                              □

### Spectrum of $p$-Laplacian Under Dirichlet Boundary Conditions

We now explain how spectral analysis works in the case of the negative Dirichlet $p$-Laplacian $-\Delta_p^D$, i.e., for problem (9.14). This time, we consider the Banach space $X = W_0^{1,p}(\Omega)$ and the maps $\varphi, \psi \in C^1(W_0^{1,p}(\Omega), \mathbb{R})$ defined by

$$\varphi(u) = \frac{1}{p} \int_\Omega \xi |u|^p \, dx \quad \text{and} \quad \psi(u) = \frac{1}{p} \int_\Omega |\nabla u|^p \, dx \quad \text{for all } u \in W_0^{1,p}(\Omega).$$

The next proposition is the analog of Proposition 9.30 and can be proved in the same way.

**Proposition 9.44.** *The maps $\varphi, \psi$ defined above satisfy hypotheses $H(\varphi, \psi)$.*

This proposition allows us to apply Theorem 9.27. We obtain the following theorem.

**Theorem 9.45.** *Let $M = \{u \in W_0^{1,p}(\Omega) : \|\nabla u\|_p^p = p\}$ and*

$$\mathcal{K}_n = \{K \subset M : K \text{ is compact, symmetric, with } \varphi|_K > 0 \text{ and } \mathrm{gen}\, K \geq n\}.$$

*The operator $-\Delta_p^D$ admits a sequence of eigenvalues $\{\lambda_n(\xi)\}_{n \geq 1}$ with respect to the weight $\xi$ such that*

$$0 < \lambda_1(\xi) \leq \lambda_2(\xi) \leq \ldots \leq \lambda_n(\xi) \leq \ldots, \quad \lim_{n \to \infty} \lambda_n(\xi) = +\infty,$$

*characterized by*

$$\frac{1}{\lambda_n(\xi)} = \sup_{K \in \mathcal{K}_n} \min_{u \in K} \frac{1}{p} \int_\Omega \xi |u|^p \, dx \quad \text{for all } n \geq 1.$$

*Remark 9.46.* (a) The sequence of eigenvalues produced by Theorem 9.45 are called the (LS)-eigenvalues of $-\Delta_p^D$ with respect to $\xi$.

(b) It easily follows from the formula given in the theorem that, for all $n \geq 1$, the map $\xi \mapsto \lambda_n(\xi)$ is continuous on $L^\infty(\Omega)_+ \setminus \{0\}$.

The proof of this result is similar to the proof of Theorem 9.31, its analog in the Neumann case. It is even simpler since we deal directly with the map $u \mapsto \|\nabla u\|_p^p$ instead of dealing with the norm $u \mapsto \|u\|_\xi^p$. In contrast, whereas the properties of the first Neumann eigenvalue are immediate, the proof of the next proposition is more involved. By $\sigma_p^D(\xi)$ we denote the set of eigenvalues of $-\Delta_p^D$ with respect to the weight $\xi$.

**Proposition 9.47.** (a) *The (LS)-eigenvalue $\lambda_1(\xi)$ of $-\Delta_p^D$ can be characterized by*

$$\lambda_1(\xi) = \min \sigma_p^D(\xi) = \lambda_1^* := \inf \{\|\nabla u\|_p^p : u \in C\}, \qquad (9.37)$$

*where*

$$C = \{u \in W_0^{1,p}(\Omega) : \int_\Omega \xi |u|^p \, dx = 1\}.$$

*Moreover, every $u \in C$ realizing the infimum in (9.37) is an eigenfunction of $-\Delta_p^D$ corresponding to $\lambda_1(\xi)$.*

(b) *Every eigenfunction $u$ corresponding to $\lambda_1(\xi)$ has a constant sign [more precisely, we have $u \in \text{int}(C_0^1(\overline{\Omega})_+)$ or $u \in -\text{int}(C_0^1(\overline{\Omega})_+)$]. In contrast, all eigenfunctions corresponding to another eigenvalue $\lambda > \lambda_1(\xi)$ are nodal (sign changing).*

(c) *$\lambda_1(\xi)$ is simple [i.e., any two eigenfunctions $u,v$ corresponding to $\lambda_1(\xi)$ are scalar multiple one of another] and isolated in $\sigma_p^D(\xi)$. The set $\sigma_p^D(\xi)$ is closed.*

(d) *For $\xi, \xi' \in L^\infty(\Omega)$ such that $0 \leq \xi \leq \xi'$ a.e. in $\Omega$, $\xi \neq 0$ and $\xi \neq \xi'$, we have $\lambda_1(\xi) > \lambda_1(\xi')$.*

*Proof.* (a) Every $\lambda \in \sigma_p^D(\xi)$ admits an eigenfunction $u \in C$; thus, we have $\lambda_1^* \leq \|\nabla u\|_p^p = \lambda$, whence $\lambda_1^* \leq \inf \sigma_p^D(\xi) \leq \lambda_1(\xi)$. On the other hand, for every $u \in C$, the set $K := \{-p^{\frac{1}{p}} \frac{u}{\|\nabla u\|_p}, p^{\frac{1}{p}} \frac{u}{\|\nabla u\|_p}\}$ belongs to $\mathscr{K}_1$. Thus, from Theorem 9.45 we have

$$\lambda_1(\xi) \leq \frac{1}{\int_\Omega \xi \left|\frac{u}{\|\nabla u\|_p}\right|^p dx} = \|\nabla u\|_p^p \quad \text{for all } u \in C.$$

Therefore, $\lambda_1(\xi) = \lambda_1^* = \min \sigma_p^D(\xi)$. If $u \in C$ realizes the infimum in (9.37), then the Lagrange multiplier rule implies that $u$ is an eigenfunction of $-\Delta_p^D$ corresponding to some $\lambda \in \sigma_p^D(\xi)$, and, moreover, $\lambda_1(\xi) = \|\nabla u\|_p^p = \lambda$. This completes the proof of (a).

Parts (b) and (c) of the proposition will be shown in Sect. 9.3 in a more general context.

(d) By (a) and (b), there is $u \in \operatorname{int}(C_0^1(\overline{\Omega})_+)$ such that $\int_\Omega \xi u^p \, dx = 1$ and $\|\nabla u\|_p^p = \lambda_1(\xi)$. Set $v = \left( \int_\Omega \xi' u^p \, dx \right)^{-\frac{1}{p}} u$. Invoking (a) again, we have

$$\lambda_1(\xi') \leq \|\nabla v\|_p^p = \frac{\|\nabla u\|_p^p}{\int_\Omega \xi' u^p \, dx} < \frac{\|\nabla u\|_p^p}{\int_\Omega \xi u^p \, dx} = \lambda_1(\xi).$$

The proof is now complete. □

The rest of the spectral analysis of $-\Delta_p^D$ can be performed as in the Neumann case. Given $\lambda \in \sigma_p^D(\xi) \setminus \{\lambda_1(\xi)\}$, we denote by $N(\lambda)$ the supremum of the number of nodal domains of the eigenfunctions corresponding to $\lambda$. Transposing the reasoning in the proofs of Propositions 9.37 and 9.38, we can prove the following proposition.

**Proposition 9.48.** (a) *For every $\lambda \in \sigma_p^D(\xi) \setminus \{\lambda_1(\xi)\}$ we have*

$$2 \leq N(\lambda) < +\infty \quad \text{and} \quad \lambda_{N(\lambda)}(\xi) \leq \lambda.$$

(b) *In particular, for $m, n \geq 2$ such that $\lambda_n(\xi) < \lambda_m(\xi)$ we have $N(\lambda_n(\xi)) < m$.*

We now focus on the second (LS)-eigenvalue of $-\Delta_p^D$. First, like its counterpart in the Neumann case, it can be characterized as the second smallest eigenvalue of $-\Delta_p^D$ (Proposition 9.40).

**Proposition 9.49.** *We have $\lambda_2(\xi) = \inf\{\lambda \in \sigma_p^D(\xi) : \lambda > \lambda_1(\xi)\}$.*

As in the Neumann case, we dispose of alternative characterizations of $\lambda_2(\xi)$. The proof of the next theorem can be modeled on the proof of Theorem 9.42.

**Theorem 9.50.** *Let $S = \{u \in W_0^{1,p}(\Omega) : \int_\Omega \xi |u|^p \, dx = 1\}$, and let $\hat{u}_1 \in \operatorname{int}(C_0^1(\overline{\Omega})_+)$ be an eigenfunction for $\lambda_1(\xi)$ normalized so that $\hat{u}_1 \in S$. Let*

$$\Gamma = \{\gamma \in C([-1,1], S) : \gamma(-1) = -\hat{u}_1, \ \gamma(1) = \hat{u}_1\}.$$

*The second eigenvalue of $-\Delta_p^D$ with respect to $\xi$ is characterized by*

$$\lambda_2(\xi) = \min_{\gamma \in \Gamma} \max_{-1 \leq t \leq 1} \|\nabla \gamma(t)\|_p^p.$$

Finally, much as we did in Proposition 9.43, we can prove the following proposition.

**Proposition 9.51.** *If $\xi, \xi' \in L^\infty(\Omega)_+$, $\xi \neq 0$, and $\xi(x) < \xi'(x)$ for a.a. $x \in \Omega$, then $\lambda_2(\xi) > \lambda_2(\xi')$.*

## (LS)-eigenvalues in Semilinear Case

In the semilinear case ($p = 2$), the Neumann (LS)-eigenvalues $\{\lambda_n^N(\xi)\}_{n\geq0}$ and the Dirichlet (LS)-eigenvalues $\{\lambda_n^D(\xi)\}_{n\geq1}$ are actually all the eigenvalues of the negative Neumann Laplacian $-\Delta^N$ and of the negative Dirichlet Laplacian $-\Delta^D$ with respect to the weight $\xi \in L^\infty(\Omega)_+ \setminus \{0\}$. Moreover, in addition to the results of this section, most of the properties of the eigenvalues of $-\Delta$ stated in Sect. 9.1 (for the weight $\xi \equiv 1$) are still valid when the weight $\xi$ is general. Specifically, an alternative to the characterization of the eigenvalues provided by Theorems 9.31 and 9.45 is the Courant–Fischer characterization (Courant and Hilbert [92]), as follows.

**Proposition 9.52.** *For all $n \geq 1$, denoting by $\mathscr{S}_n^D$ and $\mathscr{S}_n^N$ the sets of $n$-dimensional subspaces of $H_0^1(\Omega)$ and $H^1(\Omega)$, respectively, we have*

$$\lambda_n^D(\xi) = \min_{Y \in \mathscr{S}_n^D} M_\xi(Y) = \max_{Y \in \mathscr{S}_{n-1}^D} m_\xi(Y^\perp),$$

$$\lambda_{n-1}^N(\xi) = \min_{Y \in \mathscr{S}_n^N} M_\xi(Y) = \max_{Y \in \mathscr{S}_{n-1}^N} m_\xi(Y^\perp),$$

*where the orthogonal $Y^\perp$ is with respect to $H_0^1(\Omega)$ in the first relation, with respect to $H^1(\Omega)$ in the second one, and where $m_\xi(Y)$ and $M_\xi(Y)$ stand respectively for the minimum and the maximum of $\{\|\nabla v\|_2^2 : v \in Y, \int_\Omega \xi v^2 \, dx = 1\}$.*

Moreover, it remains true that all eigenfunctions $\hat{u} \in C^1(\overline{\Omega}) \setminus \{0\}$ of $-\Delta^N$ or $-\Delta^D$ with respect to $\xi$ satisfy the unique continuation property, that is, $\hat{u}(x) \neq 0$ for a.a. $x \in \Omega$ (Garofalo and Lin [148]). This fact and Proposition 9.52 yield the following monotonicity property of eigenvalues (stronger than Propositions 9.43 and 9.51).

**Proposition 9.53.** *If $\xi, \xi' \in L^\infty(\Omega)_+ \setminus \{0\}$ satisfy $\xi \leq \xi'$ a.e. in $\Omega$ with strict inequality on a set of positive measure, then we have $\lambda_n^N(\xi) > \lambda_n^N(\xi')$ and $\lambda_n^D(\xi) > \lambda_n^D(\xi')$ for all $n \geq 1$.*

## Spectral Properties of Scalar $p$-Laplacian Operators

Finally, let us briefly recall the spectral properties of the negative scalar $p$-Laplacian (i.e., $N = 1$) for the weight $\xi \equiv 1$. This is the only case so far where the eigenvalues can be explicitly determined. We consider the domain $\Omega = (0, b)$, with $b > 0$. In the present case, we have $p > N = 1$; hence the Rellich–Kondrachov theorem (Theorem 1.49) yields a compact embedding

$$W^{1,p}((0,b)) \hookrightarrow C^{0, \frac{p-1}{p}}([0,b]).$$

We have Banach subspaces $W_0^{1,p}((0,b)) \subset W_{\text{per}}^{1,p}((0,b)) \subset W^{1,p}((0,b))$ given by

$$W_0^{1,p}((0,b)) = \{u \in W^{1,p}((0,b)) : u(0) = u(b) = 0\},$$

$$W_{\text{per}}^{1,p}((0,b)) = \{u \in W^{1,p}((0,b)) : u(0) = u(b)\}.$$

In the scalar case, the Dirichlet and Neumann eigenvalue problems are written as follows:

$$\begin{cases} -(|u'(t)|^{p-2}u'(t))' = \lambda |u(t)|^{p-2}u(t) & \text{in } (0,b), \\ u(0) = u(b) = 0, \end{cases} \tag{9.38}$$

$$\begin{cases} -(|u'(t)|^{p-2}u'(t))' = \lambda |u(t)|^{p-2}u(t) & \text{in } (0,b), \\ u'(0) = u'(b) = 0. \end{cases} \tag{9.39}$$

In addition, we consider the following periodic eigenvalue problem:

$$\begin{cases} -(|u'(t)|^{p-2}u'(t))' = \lambda |u(t)|^{p-2}u(t) & \text{in } (0,b), \\ u(0) = u(b), \ u'(0) = u'(b). \end{cases} \tag{9.40}$$

**Definition 9.54.** We say that $\lambda \in \mathbb{R}$ is an *eigenvalue* of the negative Dirichlet (resp. Neumann, resp. periodic) $p$-Laplacian in the domain $(0,b)$ if problem (9.38) [resp. (9.39), resp. (9.40)] has a nontrivial weak solution, i.e., there is $u \neq 0$ in $W_0^{1,p}((0,b))$ [resp. $W^{1,p}((0,b))$, resp. $W_{\text{per}}^{1,p}((0,b))$] such that

$$\int_0^b |u'(t)|^{p-2}u'(t)v'(t)\,dt = \lambda \int_0^b |u(t)|^{p-2}u(t)v(t)\,dt$$

for every test function $v$ in $W_0^{1,p}((0,b))$ [resp. $W^{1,p}((0,b))$, resp. $W_{\text{per}}^{1,p}((0,b))$]. Such a $u$ is called an *eigenfunction* corresponding to $\lambda$.

*Remark 9.55.* As in the Neumann case, $\lambda = 0$ is an eigenvalue of the negative periodic $p$-Laplacian (of $-\Delta_p^P$, for short) and the corresponding eigenfunctions are the nonzero constants. All other eigenvalues of $-\Delta_p^P$ are positive.

The following theorem provides a full description of the eigenvalues of $-\Delta_p^D$, $-\Delta_p^N$, $-\Delta_p^P$ in the scalar case. The description involves the constant

$$\pi_p := \frac{2\pi(p-1)^{\frac{1}{p}}}{p\sin\frac{\pi}{p}} = \int_0^{(p-1)^{\frac{1}{p}}} \frac{2\,ds}{(1 - \frac{s^p}{p-1})^{\frac{1}{p}}}$$

and the $2\pi_p$-periodic, odd $C^1$-function $\sin_p : \mathbb{R} \to \mathbb{R}$ defined by the relation

$$\int_0^{\sin_p t} \frac{ds}{(1 - \frac{s^p}{p-1})^{\frac{1}{p}}} = t \quad \text{for } t \in [0, \tfrac{\pi_p}{2}]$$

and then extended to $\mathbb{R}$ in a similar way as for $\sin$. Note that $\pi_2 = \pi$ and $\sin_2 = \sin$. In the case where $p = 2$, we recover Propositions 9.16 and 9.21(c).

**Theorem 9.56.** *In the domain $(0,b)$, the sets of eigenvalues of the operators $-\Delta_p^D$, $-\Delta_p^N$, $-\Delta_p^P$ consist of sequences $\{\lambda_k^D\}_{k \geq 1}$, $\{\lambda_k^N\}_{k \geq 0}$, $\{\lambda_k^P\}_{k \geq 0}$ such that*

$$\lambda_k^D = \left(\frac{k\pi_p}{b}\right)^p \quad \text{for all } k \geq 1,$$

$$\lambda_k^N = \left(\frac{k\pi_p}{b}\right)^p \quad \text{and} \quad \lambda_k^P = \left(\frac{2k\pi_p}{b}\right)^p \quad \text{for all } k \geq 0.$$

*The corresponding eigenfunctions are described as follows.*

(a) *For $k \geq 1$ all eigenfunctions of $-\Delta_p^D$ corresponding to $\lambda_k^D$ are scalar multiples of $\hat{u}_k^D$ defined by*

$$\hat{u}_k^D(t) = \sin_p\left(\frac{k\pi_p}{b} t\right) \quad \text{for all } t \in [0,b].$$

(b) *For $k \geq 0$ all eigenfunctions of $-\Delta_p^N$ corresponding to $\lambda_k^N$ are scalar multiples of $\hat{u}_k^N$ defined by*

$$\hat{u}_k^N(t) = \sin_p\left(\frac{k\pi_p}{b} t - \frac{\pi_p}{2}\right) \quad \text{for all } t \in [0,b].$$

(c) *The eigenfunctions of $-\Delta_p^P$ corresponding to $\lambda_0^P = 0$ are nonzero constant functions. For $k \geq 1$ the eigenfunctions of $-\Delta_p^P$ corresponding to $\lambda_k^P$ are the nonzero scalar multiples of the functions $\{\hat{u}_{k,\mu}^P\}_{\mu \in \mathbb{R}}$ given by*

$$\hat{u}_{k,\mu}^P(t) = \sin_p\left(\frac{2k\pi_p}{b}(t + \mu)\right) \quad \text{for all } t \in [0,b], \text{ for } \mu \in \mathbb{R}.$$

*Remark 9.57.* The proof of this theorem can be found in Gasiński and Papageorgiou [151, Sect. 6.3]. Note that each eigenvalue is simple in the Dirichlet and the Neumann cases, but not in the periodic case.

## 9.3  Spectrum of $p$-Laplacian Plus an Indefinite Potential

Let $\Omega \subset \mathbb{R}^N$ ($N \geq 1$) be a bounded domain with a $C^2$-boundary $\partial\Omega$. In this section we deal with the following generalized eigenvalue problem involving, in addition to the nonnegative weight function $\xi \in L^\infty(\Omega)_+$, $\xi \neq 0$, an indefinite weight function

$\beta \in L^q(\Omega)$, with $q \in (N, +\infty]$, on the left-hand side:

$$\begin{cases} -\Delta_p u(x) + \beta(x)|u(x)|^{p-2}u(x) = \lambda\xi(x)|u(x)|^{p-2}u(x) & \text{in } \Omega, \\ u = 0 & \text{on } \partial\Omega. \end{cases} \quad (9.41)$$

We focus on the Dirichlet boundary conditions. A similar analysis can be conducted for the analogous Neumann eigenvalue problem (Mugnai and Papageorgiou [305]).

*Remark 9.58.* The assumption that $q > N$ ensures that we can find $r < p^*$ such that $\beta|u|^{p-2}u \in L^r(\Omega) \subset W^{-1,p'}(\Omega)$ whenever $u \in W_0^{1,p}(\Omega)$ (recall that $p^*$ stands for the Sobolev critical exponent; see Remark 1.50(b), (c)). Hence the operator on the left-hand side of (9.41) is well defined from $W_0^{1,p}(\Omega)$ into $W^{-1,p'}(\Omega)$.

**Definition 9.59.** An *eigenvalue* for problem (9.41) is a real number $\lambda$ such that problem (9.41) has a nontrivial weak solution $u \in W_0^{1,p}(\Omega)$, i.e.,

$$\int_\Omega |\nabla u|^{p-2}(\nabla u, \nabla v)_{\mathbb{R}^N}\, dx + \int_\Omega \beta(x)|u|^{p-2}uv\, dx = \int_\Omega \lambda\xi(x)|u|^{p-2}uv\, dx$$

for all $v \in W_0^{1,p}(\Omega)$. Such a $u$ is called an eigenfunction corresponding to $\lambda$.

*Remark 9.60.* (a) Due to the presence of the indefinite potential $\beta(x)|u|^{p-2}u$ in the equation, we cannot guarantee that the eigenvalues of (9.41) are nonnegative.
(b) If $\beta \in L^\infty(\Omega)$, then all eigenfunctions $u$ of (9.41) belong to $C^{1,\alpha}(\overline{\Omega})$ for some $\alpha \in (0,1)$ (Corollary 8.13).

In the sequel, we will need the following generalized version of the well-known Picone's identity. This generalization was proved by Allegretto and Huang [8].

**Proposition 9.61.** *Let* $u, v : \Omega \to \mathbb{R}$ *be differentiable functions with* $v(x) > 0$ *and* $u(x) \geq 0$ *for all* $x \in \Omega$. *Let*

$$L(u,v) = |\nabla u|^p + (p-1)\left(\frac{u}{v}\right)^p|\nabla v|^p - p\left(\frac{u}{v}\right)^{p-1}(|\nabla v|^{p-2}\nabla v, \nabla u)_{\mathbb{R}^N},$$

$$R(u,v) = |\nabla u|^p - \left(|\nabla v|^{p-2}\nabla v, \nabla\left(\frac{u^p}{v^{p-1}}\right)\right)_{\mathbb{R}^N}.$$

*Then we have*

$$L(u,v)(x) = R(u,v)(x) \geq 0 \text{ for a.a. } x \in \Omega.$$

*Moreover, the equality* $L(u,v) = 0$ *holds a.e. in* $\Omega$ *if and only if* $u = \mu v$ *for some* $\mu \in [0, +\infty)$.

We show the existence of a smallest eigenvalue for problem (9.41).

**Proposition 9.62.** *Recall that $\beta \in L^q(\Omega)$ with $q \in (N, +\infty]$. We assume that the weight functions $\beta, \xi$ involved in (9.41) satisfy*

$$\beta \geq 0 \text{ a.e. in } \{x \in \Omega : \xi(x) = 0\}.$$

*Then problem (9.41) has a smallest eigenvalue $\lambda_1(\beta, \xi) \in \mathbb{R}$ characterized by*

$$\lambda_1(\beta, \xi) = \inf \left\{ \|\nabla u\|_p^p + \int_\Omega \beta(x)|u|^p \, dx : u \in C \right\}, \qquad (9.42)$$

*where*

$$C = \left\{ u \in W_0^{1,p}(\Omega) : \int_\Omega \xi(x)|u|^p \, dx = 1 \right\}.$$

*Moreover, every $u \in C$ realizing the infimum in (9.42) is an eigenfunction corresponding to $\lambda_1(\beta, \xi)$.*

*Proof.* Lyusternik's theorem (Theorem 5.72) implies that $C$ is a $C^1$-Banach submanifold of $W_0^{1,p}(\Omega)$ of codimension 1. Let $\varphi \in C^1(W_0^{1,p}(\Omega), \mathbb{R})$ be defined by

$$\varphi(u) = \|\nabla u\|_p^p + \int_\Omega \beta(x)|u|^p \, dx \text{ for all } u \in W_0^{1,p}(\Omega).$$

The proposition is obtained by combining Claims 1–4 below.

*Claim 1:* $\hat{\lambda}_1 := \inf_{u \in C} \varphi(u) > -\infty$.

In the case where $p < N$, the assumption that $q > N$ yields $pq' < \frac{Np}{N-p}$. Hence, in any case, there is a compact embedding $W_0^{1,p}(\Omega) \overset{c}{\hookrightarrow} L^{pq'}(\Omega)$ (Theorem 1.49). Let $D_\xi = \{x \in \Omega : \xi(x) > 0\}$ and fix $\varepsilon \in (0, 1)$. We claim that there is $c(\varepsilon) > 0$ with

$$\left| \int_{D_\xi} \beta |u|^p \, dx \right| \leq \varepsilon \|\nabla u\|_p^p + c(\varepsilon) \int_\Omega \xi |u|^p \, dx \text{ for all } u \in W_0^{1,p}(\Omega). \qquad (9.43)$$

Arguing by contradiction, suppose that (9.43) does not hold. Hence we can find a sequence $\{u_n\}_{n \geq 1} \subset W_0^{1,p}(\Omega)$ such that, for all $n \geq 1$,

$$\left| \int_{D_\xi} \beta |u_n|^p \, dx \right| > \varepsilon \|\nabla u_n\|_p^p + n \int_\Omega \xi |u_n|^p \, dx. \qquad (9.44)$$

Let $v_n = \frac{u_n}{\|\nabla u_n\|_p}$. Thus, $\|\nabla v_n\|_p = 1$ for all $n > 1$. Since the embedding $W_0^{1,p}(\Omega) \hookrightarrow L^{pq'}(\Omega)$ is compact, up to extracting a subsequence, we may assume that $v_n \to v$ in $L^{pq'}(\Omega)$ as $n \to \infty$. Passing to the limit in (9.44), we obtain

$$\left| \int_{D_\xi} \beta |v|^p \, dx \right| \geq \varepsilon \quad \text{and} \quad \int_\Omega \xi |v|^p \, dx = 0,$$

a contradiction. This proves (9.43).

Using (9.43), for all $u \in C$, we see that

$$\varphi(u) = \|\nabla u\|_p^p + \int_\Omega \beta |u|^p \, dx \geq \|\nabla u\|_p^p + \int_{D_\xi} \beta |u|^p \, dx$$

$$\geq (1 - \varepsilon) \|\nabla u\|_p^p - c(\varepsilon) \int_\Omega \xi |u|^p \, dx > -c(\varepsilon). \qquad (9.45)$$

Therefore, $\hat{\lambda}_1 \geq -c(\varepsilon) > -\infty$. This proves Claim 1.

*Claim 2:* $\hat{\lambda}_1 \leq \lambda$ for every eigenvalue $\lambda$ of (9.41).

Let $u \in W_0^{1,p}(\Omega)$, $u \neq 0$, be an eigenfunction corresponding to $\lambda$. Thus,

$$\|\nabla u\|_p^p + \int_\Omega \beta(x) |u|^p \, dx = \lambda \int_\Omega \xi(x) |u|^p \, dx. \qquad (9.46)$$

We claim that

$$\int_\Omega \xi(x) |u|^p \, dx \neq 0. \qquad (9.47)$$

Note that, once we know that (9.47) holds, we let $v = (\int_\Omega \xi(x) |u|^p \, dx)^{-\frac{1}{p}} u$, which belongs to $C$ and satisfies $\varphi(v) = \lambda$, whence we obtain $\lambda \geq \hat{\lambda}_1$. Thus it suffices to show (9.47) in order to complete the proof of Claim 2. Arguing by contradiction, assume that $\int_\Omega \xi |u|^p \, dx = 0$. This implies that $\xi = 0$ a.e. in $\{x \in \Omega : u(x) \neq 0\}$, so that the assumption on $\beta, \xi$ yields $\int_\Omega \beta |u|^p \, dx \geq 0$. Hence (9.46) gives $\|\nabla u\|_p = 0$, whence $u = 0$, a contradiction. This proves Claim 2.

*Claim 3:* There exists $u \in C$ such that $\hat{\lambda}_1 = \varphi(u)$.

Let $\{u_n\}_{n \geq 1} \subset C$ be such that $\varphi(u_n) \to \hat{\lambda}_1$ as $n \to \infty$. From (9.45) we see that $\{u_n\}_{n \geq 1}$ is bounded in $W_0^{1,p}(\Omega)$. Thus, we may assume that

$$u_n \xrightarrow{w} u \text{ in } W_0^{1,p}(\Omega) \text{ and } u_n \to u \text{ in } L^{pq'}(\Omega) \text{ as } n \to \infty \qquad (9.48)$$

for some $u \in C$. From (9.48) it follows that

$$\|\nabla u\|_p^p \leq \liminf_{n \to \infty} \|\nabla u_n\|_p^p \text{ and } \int_\Omega \beta |u_n|^p \, dx \to \int_\Omega \beta |u|^p \, dx \text{ as } n \to \infty,$$

and thus $\varphi(u) \leq \hat{\lambda}_1$. We infer that $\varphi(u) = \hat{\lambda}_1$, and this proves Claim 3.

*Claim 4:* Every $u \in C$ such that $\varphi(u) = \hat{\lambda}_1$ is a weak solution of (9.41) for $\lambda = \hat{\lambda}_1$.

The Lagrange multiplier rule yields $\lambda \in \mathbb{R}$ such that the equality

$$|\nabla u|^{p-2} \nabla u + \beta(x)|u|^{p-2} u = \lambda \xi(x)|u|^{p-2} u$$

holds in $W^{-1,p'}(\Omega)$. Acting with the test function $u$ and taking into account that $u \in C$, we derive

$$\varphi(u) = \|\nabla u\|_p^p + \int_\Omega \beta(x)|u|^p \, dx = \lambda \int_\Omega \xi(x)|u|^p \, dx = \lambda.$$

Thus, $\lambda = \hat{\lambda}_1$. This proves Claim 4.                                                    $\square$

If the weight function $\beta$ is in $L^\infty(\Omega)$, then we can improve the conclusion of the previous proposition.

**Proposition 9.63.** *Assume that $\beta \in L^\infty(\Omega)$ and $\beta \geq 0$ a.e. in $\{x \in \Omega : \xi(x) = 0\}$.*

(a) *Every eigenfunction $u$ of (9.41) corresponding to the smallest eigenvalue $\lambda_1(\beta, \xi)$ has a constant sign, more precisely, $u \in \mathrm{int}\,(C_0^1(\overline{\Omega})_+)$ or $u \in -\mathrm{int}\,(C_0^1(\overline{\Omega})_+)$.*
(b) *The eigenvalue $\lambda_1(\beta, \xi)$ is simple, i.e., any two eigenfunctions $u_1, u_2$ corresponding to $\lambda_1(\beta, \xi)$ coincide up to scalar multiplication.*

*Proof.* We abbreviate $\lambda_1 = \lambda_1(\beta, \xi)$.

(a) Since $\beta \in L^\infty(\Omega)$, we have $u \in C_0^1(\overline{\Omega})$ (Remark 9.60(b)). Since $u \neq 0$, at least one of $u^+, u^-$ is nonzero. Say that $u^+ \neq 0$. Acting on the equality $-\Delta_p u + \beta|u|^{p-2} u = \lambda_1 \xi |u|^{p-2} u$ with the test function $u^+$, we obtain

$$\|\nabla u^+\|_p^p + \int_\Omega \beta(x)(u^+)^p \, dx = \lambda_1 \int_\Omega \xi(x)(u^+)^p \, dx.$$

The fact that $\beta \geq 0$ on $\{x \in \Omega : \xi(x) = 0\}$ ensures that $\int_\Omega \xi(x)(u^+)^p \, dx > 0$. Hence $v := (\int_\Omega \xi(x)(u^+)^p \, dx)^{-\frac{1}{p}} u^+ \in C$ realizes the infimum in (9.42). Then Proposition 9.62 implies that $u^+$ is a solution of (9.41) corresponding to the eigenvalue $\lambda_1$. Remark 9.60(b) and the strong maximum principle (Theorem 8.27) yield $u^+ \in \mathrm{int}\,(C_0^1(\overline{\Omega})_+)$. Therefore, $u = u^+ \in \mathrm{int}\,(C_0^1(\overline{\Omega})_+)$.

(b) By part (a) of the proposition, we may assume that $u_1, u_2 \in \mathrm{int}\,(C_0^1(\overline{\Omega})_+)$. Let $\Omega_0 \subset \Omega$ be a subdomain such that $\overline{\Omega_0} \subset \Omega$. Fix $\varepsilon > 0$. By Proposition 9.61, we have

$$0 \leq \int_{\Omega_0} L(u_1, u_2 + \varepsilon) \, dx \leq \int_\Omega L(u_1, u_2 + \varepsilon) \, dx = \int_\Omega R(u_1, u_2 + \varepsilon) \, dx$$

$$= \int_\Omega \left[ |\nabla u_1|^p - \left( |\nabla u_2|^{p-2} \nabla u_2, \nabla \left( \frac{u_1^p}{(u_2 + \varepsilon)^{p-1}} \right) \right)_{\mathbb{R}^N} \right] dx$$

$$= \int_\Omega \left( |\nabla u_1|^p + (\beta(x) - \lambda_1 \xi(x)) u_2^{p-1} \frac{u_1^p}{(u_2 + \varepsilon)^{p-1}} \right) dx$$

$$= \|\nabla u_1\|_p^p + \int_\Omega (\beta(x) - \lambda_1 \xi(x)) u_1^p \left( \frac{u_2}{u_2 + \varepsilon} \right)^{p-1} dx.$$

Passing to the limit as $\varepsilon \to 0$ by means of the Lebesgue dominated convergence theorem (using that $\inf_{\Omega_0} u_2 > 0$), we obtain

$$0 \le \int_{\Omega_0} L(u_1, u_2)\, dx \le \|\nabla u_1\|_p^p + \int_\Omega \beta(x) u_1^p\, dx - \lambda_1 \int_\Omega \xi(x) u_1^p\, dx = 0.$$

Hence $L(u_1, u_2) = 0$ in $\Omega$, which, by Proposition 9.61, implies that $u_1, u_2$ are scalar multiples one of another. We conclude that $\lambda_1$ is simple. □

Our next objective is to show that any eigenfunction of (9.41) corresponding to an eigenvalue other than $\lambda_1(\beta, \xi)$ must be nodal. This will be a consequence of the following more general statement. Henceforth, we continue assuming that

$$\beta \in L^\infty(\Omega) \quad \text{and} \quad \beta \ge 0 \text{ a.e. in } \{x \in \Omega : \xi(x) = 0\}. \tag{9.49}$$

**Proposition 9.64.** *Given $h \in L^\infty(\Omega)_+$, $h \ne 0$, the Dirichlet problem*

$$\begin{cases} -\Delta_p u + \beta(x)|u|^{p-2} u = \lambda_1(\beta, \xi) \xi(x)|u|^{p-2} u + h(x) & \text{in } \Omega, \\ u = 0 & \text{on } \partial\Omega \end{cases} \tag{9.50}$$

*has no weak solution.*

*Proof.* We abbreviate $\lambda_1 = \lambda_1(\beta, \xi)$. Arguing by contradiction, assume that there is a $u \in W_0^{1,p}(\Omega)$ solution of (9.50). We know that $u \in C_0^1(\overline{\Omega})$ (Corollary 8.13).

*Claim 1: $u \ge 0$ in $\Omega$.*

We argue by contradiction and assume that $u^- \ne 0$. Acting on (9.50) with the test function $-u^-$, we get

$$\|\nabla u^-\|_p^p + \int_\Omega \beta(x)|u^-|^p\, dx = \lambda_1 \int_\Omega \xi(x)|u^-|^p\, dx - \int_\Omega h(x) u^-\, dx. \tag{9.51}$$

Note that $\int_\Omega \xi(x)|u^-|^p\, dx > 0$. Indeed, otherwise, due to the assumption that $\beta \ge 0$ on $\{x \in \Omega : \xi(x) = 0\}$, we would have $\|\nabla u^-\|_p^p \le 0$, a contradiction. Hence we can consider $v = (\int_\Omega \xi|u^-|^p\, dx)^{-\frac{1}{p}} u^-$. We thus have

$$\int_\Omega \xi |v|^p\, dx = 1 \quad \text{and} \quad \|\nabla v\|_p^p + \int_\Omega \beta(x)|v|^p\, dx \le \lambda_1;$$

hence Proposition 9.62 implies that $v$ is an eigenfunction corresponding to $\lambda_1$. Then Proposition 9.63(a) yields $v \in \text{int}\,(C_0^1(\overline{\Omega})_+)$. But then (9.51) implies that

$$\|\nabla v\|_p^p + \int_\Omega \beta(x)|v|^p\,dx < \lambda_1,$$

a contradiction of (9.42). This proves Claim 1.

Claim 1 and the strong maximum principle (Theorem 8.27) imply that $u \in \text{int}\,(C_0^1(\overline{\Omega})_+)$. Let $u_1 \in \text{int}\,(C_0^1(\overline{\Omega})_+)$ be an eigenfunction corresponding to $\lambda_1$. Let $\Omega_0 \subset \Omega$ be a subdomain with $\overline{\Omega_0} \subset \Omega$ and $h \not\equiv 0$ in $\Omega_0$. Fix $\varepsilon > 0$. Applying Proposition 9.61 to the functions $u_1, u + \varepsilon$, we have

$$0 \le \int_\Omega R(u_1, u+\varepsilon)\,dx = \int_\Omega \left[ |\nabla u_1|^p - \left(|\nabla u|^{p-2}\nabla u, \nabla\left(\frac{u_1^p}{(u+\varepsilon)^{p-1}}\right)\right)_{\mathbb{R}^N} \right] dx$$

$$= \int_\Omega \left( |\nabla u_1|^p + \left((\beta(x) - \lambda_1\xi(x))u^{p-1} - h(x)\right)\frac{u_1^p}{(u+\varepsilon)^{p-1}} \right) dx$$

$$\le \|\nabla u_1\|_p^p + \int_\Omega (\beta(x) - \lambda_1\xi(x))u_1^p\left(\frac{u}{u+\varepsilon}\right)^{p-1} dx - \int_{\Omega_0} h(x)\frac{u_1^p}{(u+\varepsilon)^{p-1}}\,dx.$$

Passing to the limit as $\varepsilon \to 0$ (by means of the Lebesgue dominated convergence theorem) and using the fact that $u_1$ is an eigenfunction corresponding to $\lambda_1$, we obtain

$$0 \le \|\nabla u_1\|_p^p + \int_\Omega \beta(x)u_1^p\,dx - \lambda_1\int_\Omega \xi(x)u_1^p\,dx - \int_{\Omega_0} h(x)\frac{u_1^p}{u^{p-1}}\,dx < 0,$$

a contradiction. The proof of the proposition is now complete.                                        □

**Corollary 9.65.** *If $\lambda \in \mathbb{R}$ is an eigenvalue for (9.41) different from $\lambda_1(\beta,\xi)$, then all eigenfunctions corresponding to $\lambda$ are nodal.*

*Proof.* Since $\lambda_1 := \lambda_1(\beta,\xi)$ is the smallest eigenvalue of (9.41), we have $\lambda > \lambda_1$. Arguing by contradiction, assume that there is an eigenfunction $u$ corresponding to $\lambda$ such that $u \ge 0$ in $\Omega$. Note that $u \in \text{int}\,(C_0^1(\overline{\Omega})_+)$ (Corollary 8.13 and Theorem 8.27). Then $u$ is a weak solution of problem (9.50) for the choice $h := (\lambda - \lambda_1)\xi u^{p-1} \in L^\infty(\Omega)_+ \setminus \{0\}$. This contradicts Proposition 9.64. The proof is now complete.                                        □

We denote by $\sigma_p^D(\beta,\xi)$ the set of eigenvalues of problem (9.41). We still assume (9.49).

**Proposition 9.66.** (a) *The set $\sigma_p^D(\beta,\xi)$ is closed.*
(b) *The smallest eigenvalue $\lambda_1(\beta,\xi)$ is isolated in $\sigma_p^D(\beta,\xi)$.*

*Proof.* (a) Let $\{\lambda_n\}_{n\ge 1} \subset \sigma_p^D(\beta,\xi)$ be such that $\lambda_n \to \lambda \in \mathbb{R}$ as $n \to \infty$. Let $\{u_n\}_{n\ge 1} \subset W_0^{1,p}(\Omega) \setminus \{0\}$ be a corresponding sequence of $L^p$-normalized eigenfunctions (i.e., $\|u_n\|_p = 1$ for all $n \geq 1$). This implies that

$$\|\nabla u_n\|_p^p \leq (\|\beta\|_\infty + |\lambda_n| \|\xi\|_\infty) \|u_n\|_p^p \leq \|\beta\|_\infty + \Big(\sup_{n \geq 1} |\lambda_n|\Big) \|\xi\|_\infty < +\infty$$

for all $n \geq 1$. Thus, we may assume that

$$u_n \overset{w}{\to} u \text{ in } W_0^{1,p}(\Omega) \text{ and } u_n \to u \text{ in } L^p(\Omega).$$

In particular, we have $\|u\|_p = 1$, so $u \neq 0$, and the equality $-\Delta_p u + \beta |u|^{p-2} u = \lambda \xi |u|^{p-2} u$ holds in $W^{-1,p'}(\Omega)$ by passing to the limit. Hence $u$ is an eigenfunction of (9.41) corresponding to $\lambda$; thus, $\lambda \in \sigma_p^D(\beta, \xi)$.

(b) Arguing by contradiction, assume that there is a sequence $\{\lambda_n\}_{n \geq 1} \subset \sigma_p^D(\beta, \xi) \setminus \{\lambda_1(\beta, \xi)\}$ such that $\lambda_n \to \lambda_1(\beta, \xi)$ as $n \to \infty$. As in (a), up to considering a subsequence, we can find a corresponding sequence of eigenfunctions $\{u_n\}_{n \geq 1} \subset W_0^{1,p}(\Omega) \setminus \{0\}$ and an eigenfunction $u \in W_0^{1,p}(\Omega)$ corresponding to $\lambda_1(\beta, \xi)$ such that $u_n \to u$ in $L^p(\Omega)$. By Proposition 9.63(a), $u$ belongs to $\pm \operatorname{int}(C_0^1(\overline{\Omega})_+)$, say, $u \in \operatorname{int}(C_0^1(\overline{\Omega})_+)$.

*Claim 1:* There is $M_1 > 0$ such that $|\{x \in \Omega : u_n(x) < 0\}|_N \geq M_1$ for all $n \geq 1$.

Note that $u_n \in C_0^1(\overline{\Omega})$ (Remark 9.60(b)). For each $n \geq 1$, fix a connected component $D_n$ of $\{x \in \Omega : u_n(x) < 0\}$. Since $u_n$ is nodal (Corollary 9.65), $D_n$ is nonempty and open. Thus, $\chi_{D_n} u_n \in W_0^{1,p}(\Omega) \setminus \{0\}$ (Proposition 1.61). Using $\chi_{D_n} u_n$ as a test function, we have

$$\int_{D_n} |\nabla u_n|^p \, dx = \int_{D_n} (\lambda_n \xi(x) - \beta(x)) |u_n|^p \, dx \leq (|\lambda_n| \|\xi\|_\infty + \|\beta\|_\infty) \int_{D_n} |u_n|^p \, dx$$

for all $n \geq 1$. Let $\theta \in (p, p^*)$. Invoking the continuity of the embedding $W_0^{1,p}(\Omega) \hookrightarrow L^\theta(\Omega)$ (Remark 1.50(c)), Hölder's inequality, and the boundedness of the sequence $\{\lambda_n\}_{n \geq 1}$, we find constants $M_2, M_3 > 0$ such that

$$\|\chi_{D_n} u_n\|_\theta^p \leq M_2 \|\nabla(\chi_{D_n} u_n)\|_p^p = M_2 \int_{D_n} |\nabla u_n|^p \, dx$$

$$\leq M_2 (\|\beta\|_\infty + |\lambda_n| \|\xi\|_\infty) \int_{D_n} |u_n|^p \, dx$$

$$\leq M_2 \Big(\|\beta\|_\infty + \big(\sup_{n \geq 1} |\lambda_n|\big) \|\xi\|_\infty\Big) |D_n|_N^{\frac{\theta-p}{\theta}} \Big(\int_{D_n} |u_n|^\theta \, dx\Big)^{\frac{p}{\theta}}$$

$$\leq M_3 |D_n|_N^{\frac{\theta-p}{\theta}} \|\chi_{D_n} u_n\|_\theta^p \text{ for all } n \geq 1.$$

Therefore,

$$|\{x \in \Omega : u_n(x) < 0\}|_N \geq |D_n|_N \geq M_3^{\frac{\theta}{p-\theta}} \text{ for all } n \geq 1.$$

This proves Claim 1, with $M_1 = M_3^{\frac{\theta}{p-\theta}}$.

Let

$$\Omega_n^- = \{x \in \Omega : u_n(x) < 0\} \text{ for all } n \geq 1 \text{ and } \Omega^- = \{x \in \Omega : u(x) \leq 0\}.$$

Up to extracting a subsequence, we may assume that $u_n(x) \to u(x)$ as $n \to \infty$ for a.a. $x \in \Omega$. This easily implies

$$\limsup_{n \to \infty} \chi_{\Omega_n^-} \leq \chi_{\Omega^-} \text{ a.e. in } \Omega.$$

Invoking Fatou's lemma and Claim 1, we deduce

$$|\Omega^-|_N \geq \int_\Omega \limsup_{n \to \infty} \chi_{\Omega_n^-} \, dx \geq \limsup_{n \to \infty} \int_\Omega \chi_{\Omega_n^-} \, dx = \limsup_{n \to \infty} |\Omega_n^-|_N \geq M_1 > 0.$$

However, $\Omega^- = \emptyset$ [because $u \in \text{int}\,(C_0^1(\overline{\Omega})_+)$], a contradiction. The proof of the proposition is now complete.　　　　　　　　　　　　　　　　　　　□

Proposition 9.66 allows us to consider the second eigenvalue for problem (9.41):

$$\lambda_2(\beta, \xi) := \inf\{\lambda \in \sigma_p^D(\beta, \xi) : \lambda > \lambda_1(\beta, \xi)\}.$$

Then this second eigenvalue can be studied like we did in Sect. 9.2. In particular, it can be proved that it admits a variational characterization similar to that provided in Theorem 9.50. Such a characterization is also valid in the Neumann case (Mugnai and Papageorgiou [305]).

We mention that the eigenvalue problem (9.41) can be addressed in the more general situation where both weights $\beta, \xi$ are indefinite and may be unbounded. We refer to Cuesta and Ramos Quoirin [95]. However, the study in this situation is more involved. For simplicity of exposition, we restricted ourselves to the case where only $\beta$ is indefinite.

**Antimaximum Principles**

We end this section with two *antimaximum principles*. The first one is related to the eigenvalue problem under Dirichlet boundary conditions studied in Sect. 9.2:

$$\begin{cases} -\Delta_p u = \lambda \xi(x)|u|^{p-2}u & \text{in } \Omega, \\ u = 0 & \text{on } \partial\Omega. \end{cases} \tag{9.52}$$

Here $\xi \in L^\infty(\Omega)_+ \setminus \{0\}$. Let $\lambda_1(\xi) > 0$ be the first eigenvalue for problem (9.52).

**Theorem 9.67.** *Given $\xi, h \in L^\infty(\Omega)_+ \setminus \{0\}$, there is a number $\delta > 0$ such that, if $\zeta \in L^\infty(\Omega)_+ \setminus \{0\}$ and $\lambda \in \mathbb{R}$ satisfy $\|\zeta - \xi\|_\infty < \delta$ and $\lambda_1(\zeta) < \lambda < \lambda_1(\zeta) + \delta$, then any weak solution of the Dirichlet problem*

$$\begin{cases} -\Delta_p u = \lambda\zeta(x)|u|^{p-2}u + h(x) & \text{in } \Omega, \\ u = 0 & \text{on } \partial\Omega \end{cases}$$

*belongs to $-\mathrm{int}(C_0^1(\overline{\Omega})_+)$.*

*Proof.* Arguing by contradiction, assume that there exist sequences $\{\zeta_n\}_{n\geq 1} \subset L^\infty(\Omega)_+$ with $\zeta_n \to \xi$ uniformly on $\Omega$, $\{\lambda_n\}_{n\geq 1} \subset \mathbb{R}$, with $\lambda_1(\zeta_n) < \lambda_n < \lambda_1(\zeta_n) + \frac{1}{n}$ [hence $\lambda_n \to \lambda_1(\xi)$; see Remark 9.46(b)], and $\{u_n\}_{n\geq 1} \subset W_0^{1,p}(\Omega)$ such that

$$\begin{cases} -\Delta_p u_n = \lambda_n\zeta_n(x)|u_n|^{p-2}u_n + h(x) & \text{in } \Omega, \\ u_n = 0 & \text{on } \partial\Omega \end{cases} \tag{9.53}$$

and $u_n \notin -\mathrm{int}(C_0^1(\overline{\Omega})_+)$. Note that $u_n \in L^\infty(\Omega)$ (Corollary 8.6). If $\{u_n\}_{n\geq 1}$ were bounded in $L^\infty(\Omega)$, then, due to Theorem 8.10, $\{u_n\}_{n\geq 1}$ would be bounded in $C^{1,\alpha}(\overline{\Omega})$ for some $\alpha \in (0,1)$, so along a subsequence, $u_n \to u$ in $C^1(\overline{\Omega})$, with $u \in C^1(\overline{\Omega})$ solving

$$\begin{cases} -\Delta_p u = \lambda_1(\xi)\xi(x)|u|^{p-2}u + h(x) & \text{in } \Omega, \\ u = 0 & \text{on } \partial\Omega, \end{cases}$$

contradicting Proposition 9.64. Thus, along a relabeled subsequence, we have that $\|u_n\|_\infty \to +\infty$ as $n \to \infty$. Let $v_n = \frac{u_n}{\|u_n\|_\infty}$. By (9.53), we have that

$$\begin{cases} -\Delta_p v_n = \lambda_n\zeta_n(x)|v_n|^{p-2}v_n + \dfrac{h(x)}{\|u_n\|_\infty^{p-1}} & \text{in } \Omega, \\ v_n = 0 & \text{on } \partial\Omega. \end{cases} \tag{9.54}$$

The sequence $\{v_n\}_{n\geq 1}$ is bounded in $C^{1,\alpha}(\overline{\Omega})$ for some $\alpha \in (0,1)$ (by Theorem 8.10), hence up to considering a subsequence we have $v_n \to v$ in $C^1(\overline{\Omega})$ as $n \to \infty$ for some $v \in C^1(\overline{\Omega})$. Passing to the limit in (9.54), we obtain

$$\begin{cases} -\Delta_p v = \lambda_1(\xi)\xi(x)|v|^{p-2}v & \text{in } \Omega, \\ v = 0 & \text{on } \partial\Omega. \end{cases} \tag{9.55}$$

Moreover, $\|v\|_\infty = \lim\limits_{n\to\infty} \|v_n\|_\infty = 1$, hence $v \neq 0$. By Proposition 9.47(b), it follows that either $v \in \mathrm{int}(C_0^1(\overline{\Omega})_+)$ or $v \in -\mathrm{int}(C_0^1(\overline{\Omega})_+)$. The case where $v \in \mathrm{int}(C_0^1(\overline{\Omega})_+)$ cannot occur because otherwise we would have $v_n \in C_0^1(\overline{\Omega})_+$ for $n$ large enough, but then (9.54) would contradict Proposition 9.64 applied to $\beta \equiv 0$ and $\hat{h} \in L^\infty(\Omega)_+ \setminus \{0\}$ given by

$$\hat{h}(x) = (\lambda_n - \lambda_1(\zeta_n))v_n(x)^{p-1} + \frac{h(x)}{\|u_n\|_\infty^{p-1}}.$$

The case $v \in -\mathrm{int}(C_0^1(\overline{\Omega})_+)$ is also impossible because, as we have $v_n \to v$ in $C^1(\overline{\Omega})$, it would imply that $v_n \in -\mathrm{int}(C_0^1(\overline{\Omega})_+)$ for $n$ large enough, which contradicts the assumption that $u_n \notin -\mathrm{int}(C_0^1(\overline{\Omega})_+)$. In all the cases, we reach a contradiction. The proof of the theorem is now complete.  □

We now provide a version of this antimaximum principle in the Neumann case.

**Theorem 9.68.** *Given $\xi, h \in L^\infty(\Omega)_+ \setminus \{0\}$, there is a number $\delta > 0$ such that if $\zeta \in L^\infty(\Omega)_+ \setminus \{0\}$ and $\lambda \in \mathbb{R}$ satisfy $\|\zeta - \xi\|_\infty < \delta$ and $0 < \lambda < \delta$, then any weak solution $u$ of the Neumann problem*

$$\begin{cases} -\Delta_p u = \lambda \zeta(x)|u|^{p-2}u + h(x) & \text{in } \Omega, \\ \frac{\partial u}{\partial n_p} = 0 & \text{on } \partial\Omega \end{cases}$$

*satisfies $u \in C^1(\overline{\Omega})$ and $u < 0$ in $\overline{\Omega}$.*

*Proof.* Recall that the first eigenvalue of the negative Neumann $p$-Laplacian with respect to any weight $\zeta$ is the trivial one $\lambda_1(\zeta) = 0$, and the corresponding eigenfunctions are the nonzero constant functions. Taking this into account, the proof of Theorem 9.67 can be paraphrased, except that the following claim is invoked in place of Proposition 9.64.

*Claim 1:* For every $\hat{h} \in L^\infty(\Omega)_+ \setminus \{0\}$ the Neumann problem

$$-\Delta_p u = \hat{h}(x) \text{ in } \Omega, \quad \frac{\partial u}{\partial n_p} = 0 \text{ on } \partial\Omega$$

has no weak solution.

The claim follows at once by noting that if the problem admits a weak solution, then acting with the test function $v \equiv 1$ yields $\int_\Omega \hat{h}(x)\,dx = 0$, a contradiction.  □

*Remark 9.69.* Similar antimaximum principles were established by Godoy et al. [160] in the situation where the weight $\xi$ is not assumed to have a constant sign but where one takes $\zeta \equiv \xi$.

## 9.4  Fučík Spectrum

The Fučík spectrum is useful in problems involving a so-called jumping nonlinearity. More precisely, consider the problem

$$\begin{cases} -\Delta u(x) = f(x, u(x)) \text{ in } \Omega, \\ u = 0 & \text{on } \partial\Omega, \end{cases} \tag{9.56}$$

where $\Omega \subset \mathbb{R}^N$ is a bounded domain with a $C^2$-boundary $\partial\Omega$ and $f$ is a Carathéodory function. We say that (9.56) is *asymptotically resonant at infinity* if

$$\frac{f(x,s)}{s} \to \lambda_n \text{ as } |s| \to +\infty, \tag{9.57}$$

where $\lambda_n > 0$ ($n \geq 1$) is one of the eigenvalues of the negative Dirichlet Laplacian $-\Delta^D$ (Sect. 9.1). Resonant problems are more difficult to deal with because, for $|u(x)|$ large, problem (9.56) approximates the linear eigenvalue problem

$$\begin{cases} -\Delta u(x) = \lambda_n u(x) \text{ in } \Omega, \\ u = 0 \qquad\qquad\quad \text{on } \partial\Omega, \end{cases} \tag{9.58}$$

and (9.58) has inherent instabilities. If (9.56) has an *asymmetric nonlinearity* (*jumping nonlinearity*), that is, if instead of (9.57) we have

$$\frac{f(x,s)}{s} \to a \text{ as } s \to -\infty \text{ and } \frac{f(x,s)}{s} \to b \text{ as } s \to +\infty, \tag{9.59}$$

then we encounter the same difficulties whenever $a, b$ are such that the problem

$$\begin{cases} -\Delta u(x) = b u^+(x) - a u^-(x) \text{ in } \Omega, \\ u = 0 \qquad\qquad\qquad\qquad \text{on } \partial\Omega \end{cases} \tag{9.60}$$

has a nontrivial solution. In fact, the eigenvalue problem (9.60) is inherently more difficult to deal with compared to (9.58) since its right-hand side is not linear.

This section is a short survey of the known facts on the eigenvalue problem (9.60) and on its extension to the $p$-Laplacian case. We consider $\Omega \subset \mathbb{R}^N$ ($N \geq 1$) a bounded domain with a $C^2$-boundary. First, we focus on the negative Dirichlet Laplacian operator $-\Delta^D$. The basic definition is as follows.

**Definition 9.70.** The set

$$\Sigma_2 = \{(a,b) \in \mathbb{R}^2 : \text{problem (9.60) has a nontrivial weak solution } u \in H_0^1(\Omega)\}$$

is called the *Fučík spectrum* of $-\Delta^D$.

We first point out the elementary properties of $\Sigma_2$. In the next statement, $\{\lambda_n\}_{n\geq 1}$ stands for the nondecreasing sequence of eigenvalues of $-\Delta^D$ in $\Omega$ (Sect. 9.1). This result easily follows from Definition 9.70 and Proposition 9.8.

**Proposition 9.71.** (a) *$\Sigma_2$ is closed in $\mathbb{R}^2$ and symmetric by the diagonal, i.e.,*
   *$(a,b) \in \Sigma_2$ if and only if $(b,a) \in \Sigma_2$.*
(b) *$(\lambda,\lambda) \in \Sigma_2$ if and only if $\lambda = \lambda_n$ for some $n \geq 1$.*
(c) *The lines $\mathbb{R} \times \{\lambda_1\}$ and $\{\lambda_1\} \times \mathbb{R}$ are contained in $\Sigma_2$.*

While the spectrum of $-\Delta^D$ is a sequence of points, the frame of the Fučík spectrum $\Sigma_2$ consists of a family of curves. In particular, the lines $(\{\lambda_1\} \times \mathbb{R}) \cup (\mathbb{R} \times \{\lambda_1\})$ can be considered as the *first curve* in $\Sigma_2$. Like $\lambda_1$ in the spectrum of $-\Delta^D$, a noticeable feature of this curve is to be isolated in $\Sigma_2$ (Dancer [101]).

**Proposition 9.72.** $\Sigma_2$ *is the union of* $(\{\lambda_1\} \times \mathbb{R}) \cup (\mathbb{R} \times \{\lambda_1\})$ *and a closed set contained in* $(\lambda_1, +\infty) \times (\lambda_1, +\infty)$.

The basic approach to the Fučík spectrum is to extend Proposition 9.71(c) by constructing, for each eigenvalue $\lambda_n$ of $-\Delta^D$, a pair of curves (possibly coinciding) contained in $\Sigma_2$ and passing through the point $(\lambda_n, \lambda_n)$. This construction is outlined in the next theorem. Instead of dealing with the nondecreasing sequence $\{\lambda_n\}_{n\geq 1}$, where the eigenvalues are repeated according to their multiplicities, we consider

$$(0 <)\hat{\lambda}_1 < \hat{\lambda}_2 < \ldots < \hat{\lambda}_k < \ldots$$

the increasing sequence of eigenvalues of $-\Delta^D$ written without repetition.

**Theorem 9.73.** *For all $k \geq 2$ there are functions*

$$\mu_k, \nu_k : (\hat{\lambda}_{k-1}, \hat{\lambda}_{k+1}) \to (\hat{\lambda}_{k-1}, \hat{\lambda}_{k+1})$$

*with the following properties:*

(a) $\mu_k$ *and* $\nu_k$ *are continuous and decreasing, and* $\mu_k(\hat{\lambda}_k) = \nu_k(\hat{\lambda}_k) = \hat{\lambda}_k$.
(b) *The curves* $C_{k,1} := \{(a, \mu_k(a)): a \in (\hat{\lambda}_{k-1}, \hat{\lambda}_{k+1})\}$ *and* $C_{k,2} := \{(a, \nu_k(a)): a \in (\hat{\lambda}_{k-1}, \hat{\lambda}_{k+1})\}$ *are contained in* $\Sigma_2$.
(c) *A point* $(a, b) \in S_k := (\hat{\lambda}_{k-1}, \hat{\lambda}_{k+1}) \times (\hat{\lambda}_{k-1}, \hat{\lambda}_{k+1})$ *belongs to* $\Sigma_2$ *only if it lies between the curves* $C_{k,1}, C_{k,2}$, *i.e.,* $\min\{\mu_k(a), \nu_k(a)\} \leq b \leq \max\{\mu_k(a), \nu_k(a)\}$.
(d) *If the eigenvalue* $\hat{\lambda}_k$ *is simple, then* $\Sigma_2 \cap S_k = C_{k,1} \cup C_{k,2}$.

The construction of the functions $\mu_k, \nu_k$ is explained in Gallouët and Kavian [146] (for $\hat{\lambda}_k$ simple) and Schechter [351] (in the general case). It relies on critical point theory, variational principles, and topological arguments. Theorem 9.73 shows that the curves $C_{k,1}, C_{k,2}$ delimit two regions of the square $S_k$:

- The set of points of $S_k$ that are either above or below both curves, usually referred to as a *region of type (I)*; it lies outside of $\Sigma_2$;
- The part of $S_k$ between the two curves, usually called a *region of type (II)*; it can be empty because the curves $C_{k,1}, C_{k,2}$ may coincide.

Points in a region of type (II) have unknown status: when $\hat{\lambda}_k$ is simple, none of them belongs to $\Sigma_2$, but when $\hat{\lambda}_k$ is a multiple eigenvalue of $-\Delta^D$, there can be other curves in $\Sigma_2$ that emanate from the point $(\hat{\lambda}_k, \hat{\lambda}_k)$ (Margulies and Margulies [248]) and that evidently pass in a region of type (II). So, thus far there is no complete description of $\Sigma_2$. Note also that the construction outlined in Theorem 9.73 is local and restricted to the square $S_k = (\hat{\lambda}_{k-1}, \hat{\lambda}_{k+1}) \times (\hat{\lambda}_{k-1}, \hat{\lambda}_{k+1})$.

In the scalar case (i.e., $N = 1$), we have a complete description of $\Sigma_2$ due to Fučík [143]. Since in this case all the eigenvalues of $-\Delta^D$ are simple, at most two curves emanate from $(\hat{\lambda}_k, \hat{\lambda}_k)$ for all $k$. In addition, these curves can be extended to the infinity. The result is as follows.

**Theorem 9.74.** *Assume that* $\Omega = (0,1) \subset \mathbb{R}$. *Thus, the eigenvalues of* $-\Delta^D$ *are* $\lambda_k = (k\pi)^2$ *for all* $k \geq 1$. *The Fučík spectrum is the union* $\Sigma_2 = \bigcup_{k \geq 1} C_k$, *where*

$$C_1 = (\{\lambda_1\} \times \mathbb{R}) \cup (\mathbb{R} \times \{\lambda_1\}),$$

$$C_{2k} = \left\{ (a,b) \in \mathbb{R}^2 : a > 0,\ b > 0,\ \text{and}\ \frac{1}{\sqrt{a}} + \frac{1}{\sqrt{b}} = \frac{1}{k\pi} \right\}\ \text{for}\ k \geq 1,$$

$$C_{2k+1} = C_{2k+1,1} \cup C_{2k+1,2}\ \text{for}\ k \geq 1,$$

*with*

$$C_{2k+1,1} = \left\{ (a,b) \in \mathbb{R}^2 : a > 0,\ b > 0,\ \text{and}\ \frac{k+1}{k\sqrt{a}} + \frac{1}{\sqrt{b}} = \frac{1}{k\pi} \right\},$$

$$C_{2k+1,2} = \left\{ (a,b) \in \mathbb{R}^2 : a > 0,\ b > 0,\ \text{and}\ \frac{1}{\sqrt{a}} + \frac{k+1}{k\sqrt{b}} = \frac{1}{k\pi} \right\}.$$

In the remainder of this section, we deal with the following extension of the eigenvalue problem (9.60):

$$\begin{cases} -\Delta_p u(x) = bu^+(x)^{p-1} - au^-(x)^{p-1} & \text{in}\ \Omega, \\ u = 0 & \text{on}\ \partial\Omega, \end{cases} \tag{9.61}$$

where $p \in (1, +\infty)$.

**Definition 9.75.** The set

$$\Sigma_p = \{(a,b) \in \mathbb{R}^2 : \text{problem (9.61) has a nontrivial weak solution}\ u \in W_0^{1,p}(\Omega)\}$$

is called the *Fučík spectrum* of the negative Dirichlet $p$-Laplacian $-\Delta_p^D$.

The elementary properties of $\Sigma_p$ are the same as in the case $p = 2$, as follows.

**Proposition 9.76.** (a) $\Sigma_p$ *is a closed subset of* $\mathbb{R}^2$, *symmetric by the diagonal.*
(b) *We have that* $(\lambda, \lambda)$ *belongs to* $\Sigma_p$ *if and only if* $\lambda$ *is an eigenvalue of* $-\Delta_p^D$.
(c) $(\{\lambda_1\} \times \mathbb{R}) \cup (\mathbb{R} \times \{\lambda_1\}) \subset \Sigma_p$, *where* $\lambda_1 > 0$ *is the first eigenvalue of* $-\Delta_p^D$.

*Remark 9.77.* (a) In the case where $a = \lambda_1$ or $b = \lambda_1$, every nontrivial solution $u$ of problem (9.61) is an eigenfunction of $-\Delta_p^D$ corresponding to $\lambda_1$; in particular, it belongs to $\pm \text{int}\,(C_0^1(\overline{\Omega})_+)$. Indeed, say $b = \lambda_1$. Then we have that either $u^+ = 0$, in which case $u = -u^-$ is an eigenfunction of $-\Delta_p^D$ of constant sign, hence corresponding to $\lambda_1$ (by Proposition 9.47(b)), or $u^+ \neq 0$, in which case Proposition 9.47(a) implies that $u^+$ is an eigenfunction of $-\Delta_p^D$ for $\lambda_1$, hence $u^+ \in \text{int}\,(C_0^1(\overline{\Omega})_+)$, and so $u = u^+$.
(b) In contrast, every nontrivial solution $u$ of (9.61) corresponding to a pair $(a,b)$ with $a \neq \lambda_1$ and $b \neq \lambda_1$ must change sign (Proposition 9.47(b)).

(c) The set $(\{\lambda_1\} \times \mathbb{R}) \cup (\mathbb{R} \times \{\lambda_1\})$ can be interpreted as the *first curve* in $\Sigma_p$ in the sense that, if we have $a \neq \lambda_1$, $b \neq \lambda_1$, and $\min\{a,b\} < \lambda_1$, then $(a,b) \notin \Sigma_p$. Indeed, let $a,b$ be so (with, e.g., $a < \lambda_1$), and assume that there is a nontrivial solution $u$ of (9.61). Then $u^+, u^- \neq 0$ [by (b)] and $\|\nabla u^-\|_p^p = a\|u^-\|_p^p$, which contradicts the variational characterization of $\lambda_1$ in Proposition 9.47(a).

(d) As in the case where $p = 2$, it can be shown that the set $(\{\lambda_1\} \times \mathbb{R}) \cup (\mathbb{R} \times \{\lambda_1\})$ is isolated in $\Sigma_p$.

Let $\lambda_2 \in (\lambda_1, +\infty)$ be the second eigenvalue of $-\Delta_p^D$ (Proposition 9.49). We now present the construction due to Cuesta et al. [98] of a curve $\mathscr{C} \subset \Sigma_p$ passing through the point $(\lambda_2, \lambda_2)$ (the *first nontrivial curve* of $\Sigma_p$). This construction extends an earlier result by de Figueiredo and Gossez [106] in the case where $p = 2$. Given $s \in [0, +\infty)$, let $\varphi_s \in C^1(W_0^{1,p}(\Omega), \mathbb{R})$ be the functional defined by

$$\varphi_s(u) = \|\nabla u\|_p^p - s\|u^+\|_p^p.$$

Let $S = \{u \in W_0^{1,p}(\Omega) : \|u\|_p = 1\}$ and let $\hat{u}_1 \in S \cap \mathrm{int}(C_0^1(\overline{\Omega})_+)$ be an eigenfunction of $-\Delta_p^D$ corresponding to $\lambda_1$. We set

$$\Gamma = \{\gamma \in C([-1,1],S) : \gamma(-1) = -\hat{u}_1, \ \gamma(1) = \hat{u}_1\}$$

and define

$$c(s) = \inf_{\gamma \in \Gamma} \max_{-1 \leq t \leq 1} \varphi_s(\gamma(t)).$$

Cuesta et al. [98] proved the following theorem.

**Theorem 9.78.** (a) *The map* $c : [0, +\infty) \to (\lambda_1, +\infty)$ *is Lipschitz continuous and nonincreasing. The set*

$$\mathscr{C} = \{(s + c(s), c(s)) : s \in [0, +\infty)\} \cup \{(c(s), s + c(s)) : s \in [0, +\infty)\}$$

*is a continuous, decreasing curve contained in* $\Sigma_p$.

(b) *For every* $s \in [0, +\infty)$, *the point* $(s + c(s), c(s))$ *[resp.* $(c(s), s + c(s))$*] is the first nontrivial point of* $\Sigma_p$ *[i.e., not in* $(\{\lambda_1\} \times \mathbb{R}) \cup (\mathbb{R} \times \{\lambda_1\})$*] on the parallel to the diagonal through* $(s,0)$ *[resp.* $(0,s)$*]. In particular,*

$$\lambda_2 = c(0) = \inf_{\gamma \in \Gamma} \max_{-1 \leq t \leq 1} \|\nabla \gamma(t)\|_p^p.$$

(c) $\lim_{s \to +\infty} c(s) = \lambda_1$. *Hence* $\mathscr{C}$ *is asymptotic to the lines* $(\{\lambda_1\} \times \mathbb{R}) \cup (\mathbb{R} \times \{\lambda_1\})$.

*Remark 9.79.* Theorem 9.78(b) recovers the variational characterization of $\lambda_2$ known from Theorem 9.50.

## 9.5 Remarks

**Section 9.1:** The spectrum of the Dirichlet Laplacian is discussed in Brezis [52], Evans [130], Gasiński and Papageorgiou [151], and Jost [186]. In the paper by de Figueiredo [104], we find a spectral analysis of more general second-order elliptic differential operators. Finally, we mention the classic volume of Courant and Hilbert [92], where we find the fundamental minimax characterizations of the eigenvalues (the Courant–Fischer expressions; see Theorem 9.11) and the nodal domain theorem (Remark 9.15).

**Section 9.2:** For the Lyusternik–Schnirelmann theory, consult Zeidler [386] (see also Gasiński and Papageorgiou [151] and Zeidler [387]). The systematic study of the spectrum of $-\Delta_p^D$ started with Anane [19] (see also Anane and Tsouli [20]), whose results were improved upon by Lindqvist [229]. The Lyusternik–Schnirelmann minimax scheme provides a whole sequence of eigenvalues of $-\Delta_p^N$ and $-\Delta_p^D$. In this way we have a first variational characterization of the second eigenvalue $\lambda_1 > 0$ of $-\Delta_p^N$ (recall that $\lambda_0 = 0$ is the first eigenvalue), while Theorem 9.42 (due to Aizicovici et al. [4] in the case $\xi \equiv 1$) provides an alternative minimax characterization of $\lambda_1$. The analogous result for $-\Delta_p^D$ can be found in Cuesta et al. [98] (see also Theorem 9.50), but where the proof involves different arguments. Apart from the (LS)-eigenvalues, we mention that an alternative construction of a sequence of eigenvalues of the negative $p$-Laplacian can be found in Perera [327], where the Yang index (Yang [384]) is used instead of the Krasnosel'skiĭ genus. In Lê [220], we find a discussion of the spectral properties of $-\Delta_p$ under various boundary conditions (Dirichlet, Neumann, Robin, Steklov, and no-flux). For a spectral analysis of the scalar $p$-Laplacian (Theorem 9.56), we refer readers to Drábek and Manásevich [121], Gasiński and Papageorgiou [151], Papageorgiou and Kyritsi [318], and Rynne [349].

**Section 9.3:** A spectral analysis of the $p$-Laplacian plus an indefinite potential can be found in Cuesta and Ramos Quoirin [95], Del Pezzo and Fernández Bonder [109, 137] (in the Dirichlet case), and Mugnai and Papageorgiou [305] (in the Neumann case). For analogous studies for the periodic scalar $p$-Laplacian, consult the papers of Binding and Rynne [45, 46], Rynne [349], and Zhang [390]. Theorems 9.67 and 9.68 are formulations of the antimaximum principle for the Dirichlet and Neumann $p$-Laplacian operators with weight. Theorem 9.67 is due to Motreanu et al. [289]. It is based on an earlier result of Godoy et al. [160].

**Section 9.4:** The Fučík spectrum $\Sigma_2$ (Definition 9.70) was first introduced by Fučík [143] in order to deal with problems that are asymptotically linear but exhibit an asymmetric behavior at $\pm\infty$ (jumping nonlinearity). Fučík [143] gave a complete description of $\Sigma_2$ when $N = 1$ (ordinary differential equations). For the case where $N \geq 2$ (partial differential equations), Dancer [101] showed that the curve formed by the two lines $\{\lambda_1\} \times \mathbb{R}$ and $\mathbb{R} \times \{\lambda_1\}$ is isolated in $\Sigma_2$ (Proposition 9.72). Gallouët and Kavian [146] and Các [63] proved that from each pair $(\lambda_n, \lambda_n) \in \Sigma_2$ emanates a curve in $\Sigma_2$. A variational characterization of the first nontrivial curve (emanating from $(\lambda_2, \lambda_2)$) is given in de Figueiredo and Gossez [106]. The spectrum in the radial case was investigated by Arias and Campos [24], and further properties of parts of the Fučík spectrum can be found in the works of Schechter [351]. For the Dirichlet $p$-Laplacian, the first nontrivial curve of the Fučík spectrum $\Sigma_p$ was studied by Cuesta et al. [98]. Their work was extended to more general eigenvalue problems with weights by Arias et al. [25]. For the Neumann $p$-Laplacian, we refer to the works of Arias et al. [25], Miyajima et al. [261], and Motreanu and Tanaka [280]. See also Motreanu and Winkert [282, 283] for a study on the Fučík spectrum under various boundary conditions.

# Chapter 10
# Ordinary Differential Equations

**Abstract** This chapter examines the existence and multiplicity of periodic solutions for nonlinear ordinary differential equations. The first section of the chapter investigates a nonlinear periodic problem involving the scalar $p$-Laplacian for $1 < p < +\infty$ in the principal part and a smooth potential. The results cover cases of resonance at any eigenvalue of the principal part. They are obtained through variational methods and Morse theory. The second section presents results on the existence of multiple solutions for a second-order periodic system in the form of a differential inclusion. The multivalued term is expressed as a generalized gradient of a locally Lipschitz function. The approach is based on nonsmooth critical point theory. Comments and relevant references are given in a remarks section.

## 10.1 Nonlinear Periodic Problems

In this section, we deal with the following nonlinear periodic problem driven by the scalar $p$-Laplacian $(1 < p < +\infty)$

$$\begin{cases} -(|u'(t)|^{p-2}u'(t))' = f(t,u(t)) & \text{in } (0,b), \\ u(0) = u(b), \ u'(0) = u'(b), \end{cases} \tag{10.1}$$

where $b > 0$ and $f : (0,b) \times \mathbb{R} \to \mathbb{R}$ is a Carathéodory function. The purpose of the section is to provide existence and multiplicity results for problem (10.1) in different situations depending on the asymptotic behavior of the quotient $\frac{f(t,s)}{|s|^{p-2}s}$ as $|s| \to +\infty$ with respect to the eigenvalues of the negative periodic scalar $p$-Laplacian.

Let us present the background in a more precise way. The solution space for problem (10.1) is the periodic Sobolev space

$$W_{per}^{1,p}((0,b)) := \{u \in W^{1,p}((0,b)) : u(0) = u(b)\} \tag{10.2}$$

D. Motreanu et al., *Topological and Variational Methods with Applications to Nonlinear Boundary Value Problems*, DOI 10.1007/978-1-4614-9323-5_10,
© Springer Science+Business Media, LLC 2014

endowed with the Sobolev norm

$$\|u\| := (\|u'\|_p^p + \|u\|_p^p)^{\frac{1}{p}} \quad \text{for all } u \in W_{\text{per}}^{1,p}((0,b)).$$

Recall that $W^{1,p}((0,b))$ is embedded compactly in $C([0,b])$ (Theorem 1.55(b)) and so, in (10.2), the evaluations at $t = 0$ and $t = b$ make sense. The periodic $p$-Laplacian is the operator

$$\Delta_p u := \Delta_p^P u = (|u'|^{p-2} u')' \quad \text{for all } u \in W_{\text{per}}^{1,p}((0,b)).$$

We know that the eigenvalues of $-\Delta_p^P$ consist of an increasing sequence $\{\lambda_m\}_{m \geq 0}$ tending to $+\infty$, which can be determined explicitly (Theorem 9.56). The first eigenvalue is $\lambda_0 = 0$, and the corresponding eigenfunctions are the nonzero constants.

Now we focus on problem (10.1).

**Definition 10.1.** We say that a function $u \in W_{\text{per}}^{1,p}((0,b))$ is a *solution* of problem (10.1) if $f(\cdot, u(\cdot)) \in L^1((0,b))$ and

$$\langle A(u), v \rangle = \int_0^b f(t, u(t)) v(t) \, dt \quad \text{for all } v \in W_{\text{per}}^{1,p}((0,b)), \qquad (10.3)$$

where $A : W_{\text{per}}^{1,p}((0,b)) \to W_{\text{per}}^{1,p}((0,b))^*$ is the nonlinear map defined by

$$\langle A(u), v \rangle = \int_0^b |u'(t)|^{p-2} u'(t) v'(t) \, dt \quad \text{for all } u, v \in W_{\text{per}}^{1,p}((0,b)). \qquad (10.4)$$

*Remark 10.2.* We note that if $u$ is a solution of (10.1) in the sense of Definition 10.1, then we have $u \in C^1([0,b])$ and $u$ satisfies the initial conditions

$$u(0) = u(b) \quad \text{and} \quad u'(0) = u'(b).$$

Indeed, from the embedding $W_{\text{per}}^{1,p}((0,b)) \subset C([0,b])$ we already have $u \in C([0,b])$ and $u(0) = u(b)$. By (10.3), the equality $-(|u'|^{p-2} u')' = f(\cdot, u(\cdot))$ holds in distributions, hence $(|u'|^{p-2} u')' \in L^1((0,b))$. This yields $|u'|^{p-2} u' \in W^{1,1}((0,b)) \subset C([0,b])$, therefore $u' \in C([0,b])$, and so $u \in C^1([0,b])$ [since $u \in W^{1,1}((0,b))$]. Finally, acting on (10.3) with the test function $v = 1$, we find

$$0 = \int_0^b f(t, u(t)) \, dt = -\int_0^b (|u'|^{p-2} u')' \, dt = -|u'(b)|^{p-2} u'(b) + |u'(0)|^{p-2} u'(0),$$

whence $u'(0) = u'(b)$.

We will prove the existence of nontrivial solutions for problem (10.1) in two different situations, corresponding to two sets of hypotheses $H(f)_1$ and $H(f)_2$ on the nonlinearity $f(t,s)$ of (10.1).

- The first situation corresponds to the case where the nonlinearity $f(t,s)$ is $p$-linear at infinity, i.e., we assume that asymptotically at $+\infty$ the quotient $\frac{f(t,s)}{s^{p-1}}$ is within an interval $[c_1, c_2]$, with $0 < c_1 < c_2 < +\infty$ [see hypothesis H$(f)_1$ (iii) below]. Our assumption allows resonance at infinity with respect to any nonzero eigenvalue $\lambda_m$ $(m \geq 1)$.
- In the second situation we will assume resonance at infinity with respect to the first eigenvalue $\lambda_0 = 0$ [see hypotheses H$(f)_2$ (iii)].

In the first situation, through variational methods, we study the existence of constant sign solutions. In fact, for simplicity, we concentrate on the existence of positive solutions, and for this reason all our hypotheses H$(f)_1$ concern the positive half-line $[0, +\infty)$. By requiring analogous hypotheses on the negative half-line, one can similarly obtain existence results for negative solutions. In the second situation, we prove a multiplicity result through variational methods and Morse theory.

Our first set of hypotheses on $f$ is as follows. Denote $F(t,s) = \int_0^s f(t,\tau)d\tau$.

H$(f)_1$ (i) $f : (0,b) \times \mathbb{R} \to \mathbb{R}$ is a Carathéodory function [i.e., $f(\cdot,s)$ is measurable for all $s \in \mathbb{R}$ and $f(t,\cdot)$ is continuous for a.a. $t \in (0,b)$], with $f(t,0) = 0$ for a.a. $t \in (0,b)$;

(ii) For every $\rho > 0$ there exists $a_\rho \in L^{p'}((0,b))$ such that

$$|f(t,s)| \leq a_\rho(t) \text{ for a.a. } t \in (0,b), \text{ all } s \in [0,\rho];$$

(iii) There exist constants $c_1, c_2 > 0$ such that

$$c_1 \leq \liminf_{s \to +\infty} \frac{f(t,s)}{s^{p-1}} \leq \limsup_{s \to +\infty} \frac{f(t,s)}{s^{p-1}} \leq c_2 \text{ uniformly for a.a. } t \in (0,b);$$

(iv) There exists $\theta \in L^\infty((0,b))$ with $\theta \geq 0$ a.e. in $(0,b)$, $\theta \neq 0$, such that

$$\limsup_{s \downarrow 0} \frac{F(t,s)}{s^p} \leq -\theta(t) \text{ uniformly for a.a. } t \in (0,b).$$

*Example 10.3.* The following function $f$ satisfies hypotheses H$(f)_1$ (for the sake of simplicity, we drop the $t$-dependence):

$$f(s) = \begin{cases} \mu|s|^{r-2}s - |s|^{\tau-2}s & \text{if } |s| \leq 1, \\ \mu|s|^{p-2}s - |s|^{q-2}s & \text{if } |s| > 1, \end{cases}$$

with $\mu \in (0, +\infty)$, $1 < q < p < r < +\infty$ and $\tau \in (1, p]$.

*Remark 10.4.* If $u \in W^{1,p}_{\text{per}}((0,b))$, with $u \geq 0$, then H$(f)_1$ (ii) ensures that $f(\cdot, u(\cdot)) \in L^1((0,b))$. Indeed, due to the embedding $W^{1,p}_{\text{per}}((0,b)) \subset C([0,b])$, we find a constant $\rho > 0$ such that $0 \leq u(t) \leq \rho$ for all $t \in [0,b]$, whence

$|f(t, u(t))| \leq a_\rho(t)$ for a.a. $t \in (0, b)$, with $a_\rho \in L^{p'}((0, b)) \subset L^1((0, b))$ [see H$(f)_1$ (ii)]. This observation implies that, in order to have that $u \geq 0$ is a solution of (10.1), it is enough to check (10.3).

Our existence result is as follows.

**Theorem 10.5.** *If hypotheses* H$(f)_1$ *hold, then problem (10.1) has at least one solution* $u_0 \in C^1([0, b])$ *such that* $u_0 \geq 0$ *on* $[0, b]$, $u_0 \neq 0$.

*Proof.* We prove the theorem through variational methods. First we note that, by H$(f)_1$ (ii), (iii), we can find constants $\rho > 0$ and $c_3, c_4 > 0$ such that

$$c_3 s^{p-1} - a_\rho(t) \leq f(t, s) \leq c_4 s^{p-1} + a_\rho(t) \quad \text{for a.a. } t \in (0, b), \text{ all } s \geq 0. \quad (10.5)$$

We consider the functional $\psi_+ : W^{1,p}_{per}((0, b)) \to \mathbb{R}$ defined by

$$\psi_+(u) = \frac{1}{p} \|u'\|_p^p + \frac{1}{p} \|u^-\|_p^p - \int_0^b F(t, u^+(t)) \, dt \quad \text{for all } u \in W^{1,p}_{per}((0, b)),$$

where $u^+ = \max\{0, u\}$ and $u^- = \max\{0, -u\}$. In view of H$(f)_1$ (i), (10.5), and using Proposition 2.78, we see that $\psi_+ \in C^1(W^{1,p}_{per}((0, b)), \mathbb{R})$, and for all $u, v \in W^{1,p}_{per}((0, b))$ we have

$$\langle \psi'_+(u), v \rangle = \langle A(u), v \rangle - \int_0^b (u^-)^{p-1} v \, dt - \int_0^b f(t, u^+) v \, dt, \quad (10.6)$$

with $A$ given in (10.4). The proof of the theorem is divided into several steps. The first one relates the nonnegative solutions of problem (10.1) to the critical points of the functional $\psi_+$.

*Step 1:* If $u \in W^{1,p}_{per}((0, b))$ is a critical point of $\psi_+$, then we have $u \in C^1([0, b])$, $u \geq 0$, and $u$ is a solution of (10.1).

Acting with the test function $v = -u^- \in W^{1,p}_{per}((0, b))$ on the equality $\psi'_+(u) = 0$, we obtain $\|(u^-)'\|_p^p + \|u^-\|_p^p = 0$ [see (10.6)]. Hence $u^- = 0$. We have shown that $u \geq 0$ on $[0, b]$. Then the equality $\psi'_+(u) = 0$ yields

$$\langle A(u), v \rangle = \int_0^b f(t, u) v \, dt \quad \text{for all } v \in W^{1,p}_{per}((0, b)),$$

so $u$ is a solution of (10.1). By Remark 10.2, we have $u \in C^1([0, b])$.

*Step 2:* The functional $\psi_+ : W^{1,p}_{per}((0, b)) \to \mathbb{R}$ satisfies the (C)-condition.

Let $\{u_n\}_{n \geq 1} \subset W^{1,p}_{per}((0, b))$ be a sequence such that

$$|\psi_+(u_n)| \leq M_1 \quad \text{for all } n \geq 1 \quad (10.7)$$

for some $M_1 > 0$, and

$$(1 + \|u_n\|)\psi'_+(u_n) \to 0 \text{ in } W^{1,p}_{\text{per}}((0,b))^* \text{ as } n \to \infty. \tag{10.8}$$

From (10.8) we have

$$\left| \langle A(u_n), h \rangle - \int_0^b (u_n^-)^{p-1} h \, dt - \int_0^b f(t, u_n^+) h \, dt \right| \leq \frac{\varepsilon_n \|h\|}{1 + \|u_n\|} \tag{10.9}$$

for all $h \in W^{1,p}_{\text{per}}((0,b))$ and $n \geq 1$, with $\varepsilon_n \downarrow 0$. Choosing $h = -u_n^- \in W^{1,p}_{\text{per}}((0,b))$ in (10.9), we obtain

$$\|(u_n^-)'\|_p^p + \|u_n^-\|_p^p \leq \varepsilon_n \text{ for all } n \geq 1,$$

which implies that

$$u_n^- \to 0 \text{ in } W^{1,p}_{\text{per}}((0,b)) \text{ as } n \to \infty. \tag{10.10}$$

We claim that

$$\{u_n^+\}_{n \geq 1} \text{ is bounded in } W^{1,p}_{\text{per}}((0,b)). \tag{10.11}$$

Arguing by contradiction, suppose that (10.11) is not true. Then we may assume that $\|u_n^+\| \to +\infty$ as $n \to \infty$. We set $y_n = \frac{u_n^+}{\|u_n^+\|}$, $n \geq 1$. Then $\|y_n\| = 1$ for all $n \geq 1$, and so we may assume that

$$y_n \xrightarrow{w} y \text{ in } W^{1,p}_{\text{per}}((0,b)) \text{ and } y_n \to y \text{ in } C([0,b]) \text{ as } n \to \infty. \tag{10.12}$$

From (10.9) and (10.10) we have

$$\left| \langle A(y_n), h \rangle - \int_0^b \frac{(u_n^-)^{p-1}}{\|u_n^+\|^{p-1}} h \, dt - \int_0^b \frac{f(t, u_n^+)}{\|u_n^+\|^{p-1}} h \, dt \right| \leq \varepsilon'_n \|h\| \tag{10.13}$$

for all $h \in W^{1,p}_{\text{per}}((0,b))$ and $n \geq 1$, with $\varepsilon'_n \downarrow 0$. In (10.13), we choose $h = y_n - y \in W^{1,p}_{\text{per}}((0,b))$ and we pass to the limit as $n \to \infty$ using (10.5) and (10.12). This yields $\lim_{n \to \infty} \langle A(y_n), y_n - y \rangle = 0$, which implies that

$$y_n \to y \text{ in } W^{1,p}_{\text{per}}((0,b)), \text{ and so } \|y\| = 1, \ y \geq 0 \tag{10.14}$$

(Proposition 2.72). Relations (10.5) and (10.12) imply that $\left\{ \frac{f(\cdot, u_n^+(\cdot))}{\|u_n^+\|^{p-1}} \right\}_{n \geq 1}$ is bounded in $L^{p'}((0,b))$. Since $p' > 1$, we may assume that

$$\frac{f(\cdot, u_n^+(\cdot))}{\|u_n^+\|^{p-1}} \xrightarrow{w} g \text{ in } L^{p'}((0,b)) \tag{10.15}$$

for some $g \in L^{p'}((0,b))$. By (10.5), we have

$$c_3 \, y_n(t)^{p-1} - \frac{a_\rho(t)}{\|u_n^+\|^{p-1}} \leq \frac{f(t, u_n^+(t))}{\|u_n^+\|^{p-1}} \leq c_4 \, y_n(t)^{p-1} + \frac{a_\rho(t)}{\|u_n^+\|^{p-1}}$$

for a.a. $t \in (0,b)$, all $n \geq 1$. Invoking Mazur's theorem (e.g., Brezis [52, p. 61]) and using (10.14) and (10.15), we deduce $c_3 y^{p-1} \leq g \leq c_4 y^{p-1}$ a.e. in $(0,b)$. Hence there exists $\xi \in L^\infty((0,b))$, with $c_3 \leq \xi \leq c_4$ a.e. in $(0,b)$ such that

$$g(t) = \xi(t) y(t)^{p-1} \quad \text{for a.a. } t \in (0,b). \tag{10.16}$$

Then, returning to (10.13), passing to the limit as $n \to \infty$, and using (10.14), (10.15), and (10.16), we obtain $\langle A(y), h \rangle = \int_0^b \xi y^{p-1} h \, dt$ for all $h \in W_{\mathrm{per}}^{1,p}((0,b))$. Choosing the test function $h = 1$, we derive $y = 0$, a contradiction of the fact that $\|y\| = 1$. This establishes (10.11).

Combining (10.10) and (10.11), it follows that $\{u_n\}_{n \geq 1}$ is bounded in $W_{\mathrm{per}}^{1,p}((0,b))$. Thus, we may assume that

$$u_n \xrightarrow{w} u \text{ in } W_{\mathrm{per}}^{1,p}((0,b)) \text{ and } u_n \to u \text{ in } C([0,b]) \text{ as } n \to \infty. \tag{10.17}$$

In (10.9) we choose $h = u_n - u \in W_{\mathrm{per}}^{1,p}((0,b))$, pass to the limit as $n \to \infty$, and use (10.17). We obtain $\lim_{n \to \infty} \langle A(u_n), u_n - u \rangle = 0$, which implies that $u_n \to u$ in $W_{\mathrm{per}}^{1,p}((0,b))$ (Proposition 2.72). This proves that $\psi_+$ satisfies the (C)-condition.

*Step 3:* 0 is a strict local minimizer of $\psi_+$.

First, we note that, by Lemma 9.29, there exists a constant $c_0 > 0$ such that

$$\frac{1}{p} \|u'\|_p^p + \int_0^b \theta(t) |u(t)|^p \, dt \geq c_0 \|u\|^p \quad \text{for all } u \in W_{\mathrm{per}}^{1,p}((0,b)), \tag{10.18}$$

where $\theta$ is the function from H$(f)_1$ (iv). Fix $\delta \in (0, c_0)$. By virtue of hypotheses H$(f)_1$ (ii)–(iv), we can find $\xi \in L^1((0,b))$, $\xi \geq 0$, such that

$$F(t,s) \leq (-\theta(t) + \delta) s^p + \xi(t) s^\eta \quad \text{for a.a. } t \in (0,b), \text{ all } s \geq 0, \tag{10.19}$$

with $\eta > p$. Then for $u \in W_{\mathrm{per}}^{1,p}((0,b))$ we have

$$\psi_+(u) = \frac{1}{p} \|u'\|_p^p + \frac{1}{p} \|u^-\|_p^p - \int_0^b F(t, u^+(t)) \, dt \geq \frac{1}{p} \|u'\|_p^p - \int_0^b F(t, u^+(t)) \, dt$$

$$\geq \frac{1}{p} \|u'\|_p^p + \int_0^b \theta |u|^p \, dt - \delta \|u\|_p^p - \|u\|_\infty^\eta \|\xi\|_1$$

$$\geq (c_0 - \delta) \|u\|^p - c_5 \|u\|^\eta \tag{10.20}$$

for some constant $c_5 > 0$ [see (10.18) and (10.19)]. Since $c_0 > \delta$ and $\eta > p$, from (10.20) we infer that if $\rho > 0$ is small, then

$$\psi_+(u) > 0 = \psi_+(0) \quad \text{for all } u \in W_{\text{per}}^{1,p}((0,b)), \ 0 < \|u\| \le \rho,$$

so 0 is a strict local minimizer of $\psi_+$.

*Step 4:* The functional $\psi_+$ has at least one nontrivial critical point $u_0 \in W_{\text{per}}^{1,p}((0,b))$.
   By $H(f)_1$ (iii), we have

$$\liminf_{s \to +\infty} \frac{pF(t,s)}{s^p} \ge c_1 > 0 \quad \text{uniformly for a.a. } t \in (0,b).$$

Thus, we can find $M \in (0,+\infty)$ such that $F(t,M) \ge 0$ for a.a. $t \in (0,b)$. In this way, the constant function $v_0 := M$ satisfies $\psi_+(v_0) = -\int_0^b F(t,M)\,dt \le 0 = \psi_+(0)$. Invoking Steps 2 and 3 and Proposition 5.42, we find $u_0 \in W_{\text{per}}^{1,p}((0,b))$, with $u_0 \ne 0$, which is a critical point of $\psi_+$.

   Finally, the theorem follows by combining Steps 1 and 4. $\qquad\square$

*Remark 10.6.* Assume that the nonlinearity $f$ satisfies the additional hypothesis: for every $R > 0$, there exists a constant $c_R > 0$ such that

$$f(t,s) \ge -c_R\, s^{p-1} \quad \text{for a.a. } t \in (0,b), \text{ all } s \in [0,R].$$

Then we can reinforce the conclusion of Theorem 10.5 by observing that every solution $u_0 \in C^1([0,b])$ of (10.1) such that $u_0 \ge 0$, $u_0 \ne 0$, satisfies

$$u_0(t) > 0 \quad \text{for all } t \in [0,b]. \tag{10.21}$$

Indeed, letting $R = \|u_0\|_\infty$ and taking $c_R > 0$ as above, we obtain the inequality $(|u_0'(t)|^{p-2}u_0'(t))' \le \xi_r u_0(t)^{p-1}$ a.e. in $(0,b)$. Invoking Theorem 8.27, we infer that $u_0 > 0$ in $(0,b)$. Moreover, since $u'(0) = u'(b)$ (Remark 10.2), we have necessarily $u_0(0) = u_0(b) \ne 0$ (by Theorem 8.27), whence (10.21).

   Now we establish a multiplicity result for problem (10.1) in the situation of resonance with respect to the first eigenvalue $\lambda_0 = 0$, i.e., when $f(t,\cdot)$ is strictly $(p-1)$-sublinear at $\pm\infty$. We consider the following hypotheses on the nonlinearity $f$ and on its primitive $F(t,s) = \int_0^s f(t,\tau)\,d\tau$.

$H(f)_2$  (i) $f : (0,b) \times \mathbb{R} \to \mathbb{R}$ is a Carathéodory function such that $f(t,0) = 0$ for a.a. $t \in (0,b)$;
   (ii) There exist $r \in (1,+\infty)$, $a \in L^{r'}((0,b))$, and $c_1 > 0$ such that

$$|f(t,s)| \le a(t) + c_1|s|^{r-1} \quad \text{for a.a. } t \in (0,b), \text{ all } s \in \mathbb{R};$$

(iii) $\lim\limits_{s \to \pm \infty} \dfrac{F(t,s)}{|s|^p} = 0$ and $\lim\limits_{s \to \pm \infty} F(t,s) = -\infty$ uniformly for a.a. $t \in (0,b)$;

(iv) There exist $\delta > 0$ and $\mu \in (1,p)$ such that, for all $s \in [-\delta,\delta] \setminus \{0\}$,

$$\mu F(t,s) - f(t,s)s > 0 \quad \text{for a.a. } t \in (0,b) \quad \text{and} \quad \operatorname*{essinf}_{t \in (0,b)} F(t,s) > 0;$$

(v) For every $R > 0$, there exists $c_R > 0$ such that

$$f(t,s)s \ge -c_R|s|^p \quad \text{for a.a. } t \in (0,b), \text{ all } s \in [-R,R].$$

*Remark 10.7.* (a) The first formula in hypothesis $H(f)_2$ (iii) incorporates the situation of resonance at infinity of the nonlinearity with respect to the eigenvalue $\lambda_0 = 0$, that is,

$$\lim\limits_{s \to \pm \infty} \dfrac{f(t,s)}{|s|^{p-2}s} = 0 \quad \text{uniformly for a.a. } t \in (0,b).$$

(b) Hypothesis $H(f)_2$ (iv) assumes for $f(t,\cdot)$ a different behavior near $s = 0$ than hypothesis $H(f)_1$ (iv). This is illustrated by the next example.

*Example 10.8.* The following function $f : \mathbb{R} \to \mathbb{R}$ satisfies hypotheses $H(f)_2$ (for the sake of simplicity, we drop the $t$-dependence):

$$f(s) = \begin{cases} |s|^{\tau-2}s & \text{if } |s| \le 1, \\ 2|s|^{q-2}s - |s|^{\ell-2}s & \text{if } |s| > 1, \end{cases}$$

with $1 < \tau < p$ and $1 < q < \ell < p$.

The announced multiplicity result is the following three solutions theorem.

**Theorem 10.9.** *If hypotheses* $H(f)_2$ *hold, then problem (10.1) has at least three nontrivial solutions* $u_0, v_0, y_0 \in C^1([0,b])$, *with* $v_0 < 0 < u_0$ *on* $[0,b]$.

*Proof.* We first introduce the functionals $\varphi, \psi_+, \psi_- : W^{1,p}_{\mathrm{per}}((0,b)) \to \mathbb{R}$ by letting

$$\varphi(u) = \frac{1}{p}\|u'\|_p^p - \int_0^b F(t,u(t))\,dt,$$

$$\psi_+(u) = \frac{1}{p}\|u'\|_p^p + \frac{1}{p}\|u^-\|_p^p - \int_0^b F(t,u^+(t))\,dt,$$

$$\psi_-(u) = \frac{1}{p}\|u'\|_p^p + \frac{1}{p}\|u^+\|_p^p - \int_0^b F(t,u^-(t))\,dt$$

for all $u \in W^{1,p}_{\mathrm{per}}((0,b))$. Using hypotheses $H(f)_2$ (i), (ii), we see that $\varphi, \psi_+, \psi_- \in C^1(W^{1,p}_{\mathrm{per}}((0,b))$. The proof of the theorem is divided into several steps.

*Step 1:* If $u \in W^{1,p}_{per}((0,b))$ is a critical point of $\varphi$, $\psi_+$, or $\psi_-$, then we have $u \in C^1([0,b])$, and $u$ is a solution of (10.1). Moreover, if $u$ is a nontrivial critical point of $\psi_+$ (resp. of $\psi_-$), then $u > 0$ on $[0,b]$ (resp. $u < 0$ on $[0,b]$).

If $u$ is a critical point of $\varphi$, then we have $\varphi'(u) = A(u) - f(\cdot, u(\cdot)) = 0$ in $W^{1,p}_{per}((0,b))^*$, which implies that $u$ is a solution of (10.1). By Remark 10.2, we have $u \in C^1([0,b])$. Next assume that $u$ is a nontrivial critical point of $\psi_+$. Reasoning as in Step 1 of the proof of Theorem 10.5, we get that $u \geq 0$ on $[0,b]$, $u \in C^1([0,b])$, and $u$ is a solution of (10.1). The fact that $u > 0$ on $[0,b]$ is deduced by applying the strong maximum principle (Theorem 8.27) on the basis of $H(f)_2$ (v), as in Remark 10.6. The reasoning is similar if $u$ is a nontrivial critical point of $\psi_-$.

*Step 2:* The functionals $\varphi, \psi_+, \psi_- : W^{1,p}_{per}((0,b)) \to \mathbb{R}$ are coercive, sequentially weakly l.s.c., and satisfy the (PS)-condition.

We start by showing that $\psi_+$ is coercive. Arguing by contradiction, suppose that $\psi_+$ is not coercive. Then we can find $\{u_n\}_{n \geq 1} \subset W^{1,p}_{per}((0,b))$ and $M_1 > 0$ such that

$$\|u_n\| \to +\infty \quad \text{and} \quad \psi_+(u_n) \leq M_1 \quad \text{for all } n \geq 1. \tag{10.22}$$

Hence

$$\frac{1}{p}\|u_n'\|_p^p + \frac{1}{p}\|u_n^-\|_p^p - \int_0^b F(t, u_n^+(t))\, dt \leq M_1 \quad \text{for all } n \geq 1. \tag{10.23}$$

Relation (10.23) yields $\|u_n^+\| \to +\infty$ [otherwise, $H(f)_2$ (ii) and (10.23) imply that $\{u_n^-\}_{n \geq 1}$ (and so $\{u_n\}_{n \geq 1}$) is bounded in $W^{1,p}_{per}((0,b))$, a contradiction]. Let $y_n = \frac{u_n^+}{\|u_n^+\|}$, $n \geq 1$. Then $\|y_n\| = 1$ for all $n \geq 1$, and so we may assume that

$$y_n \xrightarrow{w} y \text{ in } W^{1,p}_{per}((0,b)) \quad \text{and} \quad y_n \to y \text{ in } C([0,b]) \quad \text{as } n \to \infty. \tag{10.24}$$

From (10.23) we have

$$\frac{1}{p}\|y_n'\|_p^p - \int_0^b \frac{F(t, u_n^+)}{\|u_n^+\|^p}\, dt < \frac{M_1}{\|u_n^+\|^p} \quad \text{for all } n \geq 1. \tag{10.25}$$

Note that we have that

$$\lim_{n \to \infty} \int_0^b \frac{F(t, u_n^+(t))}{\|u_n^+\|^p}\, dt = 0. \tag{10.26}$$

Indeed, invoking $H(f)_2$ (ii), (iii), we can find $c_\varepsilon > 0$ such that

$$|F(t,s)| \leq \frac{\varepsilon}{2}|s|^p + c_\varepsilon \quad \text{for a.a. } t \in (0,b), \text{ all } s \in \mathbb{R},$$

which yields, for $n \geq 1$ large enough (recall that $\|u_n^+\| \to +\infty$ as $n \to \infty$),

$$\int_0^b \frac{|F(t, u_n^+)|}{\|u_n^+\|^p} \, dt \leq \frac{\varepsilon}{2} + \frac{bc_\varepsilon}{\|u_n^+\|^p} \leq \varepsilon.$$

This proves (10.26). Thus, if in (10.25) we pass to the lim sup as $n \to \infty$ and use (10.24) and (10.26), then

$$\limsup_{n \to \infty} \|y_n'\|_p^p \leq 0 \leq \|y'\|_p^p. \tag{10.27}$$

On the other hand, from (10.24) we have $\|y'\|_p^p \leq \liminf_{n \to \infty} \|y_n'\|_p^p$. Combining this with (10.27), we obtain that $y_n' \to y' = 0$ in $L^p((0, b))$. Hence,

$$y_n \to y \text{ in } W_{\mathrm{per}}^{1,p}((0, b)) \text{ [see (10.24)], and so } \|y\| = 1, \ y \geq 0. \tag{10.28}$$

Since $y' = 0$, we have in fact $y \equiv (\frac{1}{b})^{\frac{1}{p}}$. Then $u_n^+(t) \to +\infty$ for all $t \in [0, b]$, and so, by virtue of hypothesis $H(f)_2$ (ii), (iii) and Fatou's lemma, we have

$$-\int_0^b F(t, u_n^+) \, dt \to +\infty \quad \text{as } n \to \infty. \tag{10.29}$$

For all $n \geq 1$, by (10.23), we have

$$M_1 \geq \frac{1}{p} \|u_n'\|_p^p + \frac{1}{p} \|u_n^-\|_p^p - \int_0^b F(t, u_n^+(t)) \, dt \geq -\int_0^b F(t, u_n^+) \, dt. \tag{10.30}$$

Comparing (10.29) and (10.30), we reach a contradiction. We have shown that $\psi_+$ is coercive.

The coercivity of $\psi_-$ can be shown similarly by reversing the roles of $u_n^+$ and $u_n^-$. The coercivity of $\varphi$ is obtained through analogous reasoning, in fact simpler because we deal directly with the sequence $y_n = \frac{u_n}{\|u_n\|}$.

The fact that $\varphi$, $\psi_+$, and $\psi_-$ are sequentially weakly l.s.c. easily follows from $H(f)_2$ (ii), the compactness of the embedding $W_{\mathrm{per}}^{1,p}((0, b)) \hookrightarrow C([0, b])$, and the weak lower semicontinuity of the $L^p$-norm.

Finally, we show that $\varphi$ satisfies the (PS)-condition (the proof is similar for $\psi_+$ and $\psi_-$). Let $\{u_n\}_{n \geq 1} \subset W_{\mathrm{per}}^{1,p}((0, b))$ be such that

$$\{\varphi(u_n)\}_{n \geq 1} \text{ is bounded and } \varphi'(u_n) \to 0 \text{ in } W_{\mathrm{per}}^{1,p}((0, b))^* \text{ as } n \to \infty. \tag{10.31}$$

Since $\varphi$ is coercive, the fact that $\{\varphi(u_n)\}_{n \geq 1}$ is bounded implies that $\{u_n\}_{n \geq 1}$ is bounded in $W_{\mathrm{per}}^{1,p}((0, b))$. Hence, up to considering a subsequence, we may assume that

$$u_n \xrightarrow{w} u \text{ in } W_{\mathrm{per}}^{1,p}((0, b)) \text{ and } u_n \to u \text{ in } C([0, b]) \text{ as } n \to \infty. \tag{10.32}$$

By (10.31), we have

$$\langle A(u_n), u_n - u \rangle - \int_0^b f(t, u_n(t))(u_n(t) - u(t)) \, dt \to 0 \quad \text{as } n \to \infty. \qquad (10.33)$$

The second convergence in (10.32) ensures that $\int_0^b f(t, u_n)(u_n - u) \, dt \to 0$ as $n \to \infty$, so that (10.33) results in

$$\lim_{n \to \infty} \langle A(u_n), u_n - u \rangle = 0.$$

Since $A$ is an $(S)_+$-map (Proposition 2.72), we deduce that $u_n \to u$ in $W_{\text{per}}^{1,p}((0,b))$ as $n \to \infty$. In this way, $\varphi$ satisfies the (PS)-condition. This completes Step 2.

*Step 3:* There exist $u_0, v_0 \in C^1([0,b])$ global minimizers (hence critical points) of $\psi_+, \psi_-$, respectively, with $v_0 < 0 < u_0$ in $[0,b]$. Moreover, $u_0, v_0$ are local minimizers of $\varphi$.

By Step 2, $\psi_+$ is coercive and sequentially weakly l.s.c. This implies that $m_+ := \inf\{\psi_+(u) : u \in W_{\text{per}}^{1,p}((0,b))\} > -\infty$ and that there exists $u_0 \in W_{\text{per}}^{1,p}((0,b))$ such that $m_+ = \psi_+(u_0)$. On the other hand, for the constant $\delta > 0$ provided by H$(f)_2$ (iv) we have

$$\psi_+(u_0) \le \psi_+(\delta) = -\int_0^b F(t, \delta) \, dt < 0 = \psi_+(0),$$

hence $u_0 \ne 0$. Thus, $u_0$ is a nontrivial critical point of $\psi_+$, and so we have $u_0 \in C^1([0,b])$ and $u_0 > 0$ on $[0,b]$ (Step 1). The existence of $v_0$ can be shown similarly.

By virtue of the compact embedding $W_{\text{per}}^{1,p}((0,b)) \hookrightarrow C([0,b])$, there is a constant $c > 0$ such that

$$\|u\|_\infty \le c\|u\| \quad \text{for all } u \in W_{\text{per}}^{1,p}((0,b)).$$

Let $m_0 = \inf_{[0,b]} u_0 > 0$ and $V = \{u \in W_{\text{per}}^{1,p}((0,b)) : \|u - u_0\| < \frac{m_0}{c}\}$. Every $u \in V$ satisfies $\|u - u_0\|_\infty < m_0$, hence $u > 0$ on $[0,b]$. In view of the definition of $\psi_+$, this implies that

$$\varphi(u) = \psi_+(u) \ge \psi_+(u_0) = \varphi(u_0) \quad \text{for all } u \in V.$$

Hence $u_0$ is a local minimizer of $\varphi$. Similarly, $v_0$ is a local minimizer of $\varphi$. This concludes Step 3.

For our purposes, we may assume that $\varphi$ has a finite number of critical points (otherwise we are done). In particular, each critical point of $\varphi$ is isolated, and so its critical groups are well defined. The assumption that $f(t,0) = 0$ for a.a. $t \in (0,b)$ implies that 0 is a critical point of $\varphi$. Our next step is the computation of the critical groups of $\varphi$ at 0.

*Step 4:*    $C_k(\varphi,0) = 0$ for all $k \in \mathbb{N}_0$.

Recall that we have $C_k(\varphi,0) = H_k(B_\rho(0) \cap \varphi^0, B_\rho(0) \cap \varphi^0 \setminus \{0\})$ for $\rho > 0$ small enough, where $B_\rho(0) = \{u \in W^{1,p}_{\text{per}}((0,b)) : \|u\| < \rho\}$ (Definition 6.43). In view of Proposition 6.21, Step 4 will be complete once we show that, for $\rho > 0$ small,

$$H_k(B_\rho(0) \cap \varphi^0, *) = H_k(B_\rho(0) \cap \varphi^0 \setminus \{0\}, *) = 0 \quad \text{for all } k \in \mathbb{N}_0, \qquad (10.34)$$

where $H_k(A, *)$ stands for the reduced homology groups. To do this, we will establish the following three properties for all $u \in W^{1,p}_{\text{per}}((0,b)) \setminus \{0\}$:

$$\text{there is } \lambda^* = \lambda^*(u) \text{ such that } \varphi(\lambda u) < 0 \text{ for all } \lambda \in (0,\lambda^*), \qquad (10.35)$$

and there is $\rho > 0$ such that for all $u \in B_\rho(0) \setminus \{0\}$:

$$\text{if } \varphi(u) = 0, \text{ then we have } \left.\frac{d}{d\lambda}\varphi(\lambda u)\right|_{\lambda=1} > 0, \qquad (10.36)$$

$$\text{if } \varphi(u) \leq 0, \text{ then we have } \varphi(\lambda u) \leq 0 \text{ for all } \lambda \in [0,1]. \qquad (10.37)$$

Assume for the moment that we have established relations (10.35)–(10.37). Fix $\rho > 0$ satisfying (10.36) and (10.37). Relation (10.37) implies that the map $h_0 : [0,1] \times B_\rho(0) \cap \varphi^0 \to B_\rho(0) \cap \varphi^0$ defined by $h_0(\lambda,u) = \lambda u$ is a well-defined homotopy between $h_0(0,\cdot) = 0$ and $h_0(1,\cdot) = \text{id}_{B_\rho(0)\cap\varphi^0}$. This shows that $B_\rho(0) \cap \varphi^0$ is contractible. By Proposition 6.24, we get

$$H_k(B_\rho(0) \cap \varphi^0, *) = 0 \quad \text{for all } k \in \mathbb{N}_0.$$

Given $u \in B_\rho(0) \setminus \{0\}$ such that $\varphi(u) \geq 0$, we claim that there is $\lambda(u) \in (0,1]$ (necessarily unique) such that $\varphi(\lambda(u)u) = 0$ and

$$\varphi(\lambda u) < 0 \text{ if } \lambda \in (0,\lambda(u)) \text{ and } \varphi(\lambda u) > 0 \text{ if } \lambda \in (\lambda(u),1]. \qquad (10.38)$$

Indeed, set $\lambda(u) = \sup\{\lambda \in (0,1] : \varphi(\lambda u) \leq 0\}$. By (10.35), we have $\lambda(u) \in (0,1]$. By construction, we have $\varphi(\lambda(u)u) = 0$ and $\varphi(\lambda u) > 0$ for $\lambda \in (\lambda(u),1]$, whereas (10.37) implies that $\varphi(\lambda u) \leq 0$ for $\lambda \in (0,\lambda(u))$. If there is $\hat{\lambda} \in (0,\lambda(u))$ such that $\varphi(\hat{\lambda}u) = 0$, then, using (10.37), we see that

$$\left.\frac{d}{d\lambda}\varphi(\lambda\hat{\lambda}u)\right|_{\lambda=1} = \lim_{\lambda\downarrow 1}\frac{\varphi(\lambda\hat{\lambda}u) - \varphi(\hat{\lambda}u)}{\lambda - 1} \leq 0,$$

which contradicts (10.36). We have proven (10.38).

We further set $\lambda(u) = 1$ if $u \in B_\rho(0) \setminus \{0\}$ is such that $\varphi(u) \leq 0$. The map $\lambda : B_\rho(0) \setminus \{0\} \to (0,1]$ thus obtained is well defined.

We claim that the map $u \mapsto \lambda(u)$ is continuous on $B_\rho(0) \setminus \{0\}$. It is sufficient to check the continuity of $\lambda$ on the closed subsets $\{u \in B_\rho(0) \setminus \{0\} : \varphi(u) \leq 0\}$ and $\{u \in B_\rho(0) \setminus \{0\} : \varphi(u) \geq 0\}$. The continuity on the first subset is immediate, so it remains to check the continuity on the second subset. Let $\{u_n\}_{n \geq 1} \subset B_\rho(0) \setminus \{0\}$ be such that $\varphi(u_n) \geq 0$ for all $n \geq 1$ and $\lim_{n \to \infty} u_n = u \in B_\rho(0) \setminus \{0\}$. Up to taking a subsequence, we may assume that $\lambda(u_n) \to \overline{\lambda} \in [0,1]$. Assume by contradiction that $\overline{\lambda} < \lambda(u)$; hence, fixing $\hat{\lambda} \in (\overline{\lambda}, \lambda(u))$, for every $n \geq 1$ large enough, we have $\lambda(u_n) < \hat{\lambda}$, and so (10.38) implies $\varphi(\hat{\lambda} u_n) > 0$. In this way, $\varphi(\hat{\lambda} u) = \lim_{n \to \infty} \varphi(\hat{\lambda} u_n) \geq 0$, which contradicts (10.38). This yields $\overline{\lambda} \geq \lambda(u)$, and similarly we can prove that $\overline{\lambda} \leq \lambda(u)$, so $\overline{\lambda} = \lambda(u)$. This proves our claim.

By the continuity of $u \mapsto \lambda(u)$, the map $\zeta : B_\rho(0) \setminus \{0\} \to B_\rho(0) \cap \varphi^0 \setminus \{0\}$ defined by $\zeta(u) = \lambda(u) u$ is a well-defined retraction. Because $W_{per}^{1,p}((0,b))$ is infinite dimensional, $B_\rho(0) \setminus \{0\}$ is contractible (Benyamini and Sternfeld [42]), hence $H_k(B_\rho(0) \setminus \{0\}, *) = 0$ for $k \in \mathbb{N}_0$. By Proposition 6.16, we deduce that

$$H_k(B_\rho(0) \cap \varphi^0 \setminus \{0\}, *) = 0 \quad \text{for all } k \in \mathbb{N}_0.$$

We obtain (10.34).

Thus, to complete the proof of Step 4, it remains to establish relations (10.35)–(10.37).

By H$(f)_2$ (iv), we know that $\frac{f(t,s)}{F(t,s)} < \frac{\mu}{s}$ [resp. $\frac{f(t,s)}{F(t,s)} > \frac{\mu}{s}$] for a.a. $t \in (0,b)$, all $s \in (0,\delta)$ [resp. $s \in (-\delta, 0)$]. By integrating on $[s, \delta]$ (resp. $[-\delta, s]$) and using that $m_\delta := \operatorname*{essinf}_{t \in (0,b)} F(t, \pm\delta) > 0$ in $[-\delta, \delta]$ [H$(f)_2$ (iv)], we obtain

$$F(t,s) \geq \frac{m_\delta}{\delta^\mu} |s|^\mu \quad \text{for a.a. } t \in (0,b), \text{ all } s \in [-\delta, \delta]. \tag{10.39}$$

Let $u \in W_{per}^{1,p}((0,b)) \setminus \{0\}$. Then $u \in C([0,b])$; hence there is $\lambda_0 > 0$ such that, for every $\lambda \in (0, \lambda_0)$, we have $\|\lambda u\|_\infty \leq \delta$. Using (10.39), we obtain

$$\varphi(\lambda u) = \frac{\lambda^p}{p} \|u'\|_p^p - \int_0^b F(t, \lambda u) \, dt \leq \frac{\lambda^p}{p} \|u'\|_p^p - \frac{m_\delta \lambda^\mu}{\delta^\mu} \|u\|_\mu^\mu. \tag{10.40}$$

Since $1 < \mu < p$, we find $\lambda^* \in (0, \lambda_0)$ such that, for all $\lambda \in (0, \lambda^*)$, we have $\varphi(\lambda u) < 0$. This establishes (10.35).

In view of the compact embedding $W_{per}^{1,p}((0,b)) \hookrightarrow C([0,b])$, we can find $\rho > 0$ such that $\|u\|_\infty \leq \delta$ whenever $u \in B_\rho(0)$. Let $u \in B_\rho(0) \setminus \{0\}$ such that $\varphi(u) = 0$. Then, by the chain rule, we have

$$\frac{d}{d\lambda} \varphi(\lambda u) \Big|_{\lambda=1} = \langle \varphi'(u), u \rangle - \mu\varphi(u)$$

$$= \left(1 - \frac{\mu}{p}\right) \|u'\|_p^p + \int_0^b (\mu F(t, u(t)) - f(t, u(t))u(t)) \, dt > 0,$$

where we also used that $\mu \in (1,p)$ and that $\mu F(t,u(t)) - f(t,u(t))u(t) > 0$ whenever $u(t) \neq 0$ [H($f$)$_2$ (iv)]. This proves (10.36).

It remains to check (10.37). Arguing indirectly, suppose that (10.37) does not hold. Hence we can find $\hat{\lambda} \in (0,1)$ such that $\varphi(\hat{\lambda}u) > 0$. Since $\varphi(u) \leq 0$ and $\varphi$ is continuous, we can define

$$\lambda_1 = \min\{\lambda \in [\hat{\lambda},1] : \varphi(\lambda u) = 0\} > \hat{\lambda} > 0.$$

Then we have

$$\varphi(\lambda u) > 0 \text{ for all } \lambda \in [\hat{\lambda},\lambda_1). \tag{10.41}$$

Let $v = \lambda_1 u$. Then $0 < \|v\| \leq \|u\| \leq \rho$ and $\varphi(v) = 0$. By virtue of (10.36), we have

$$\frac{d}{d\lambda}\varphi(\lambda v)\Big|_{\lambda=1} > 0. \tag{10.42}$$

On the other hand, from (10.41) we have $\varphi(\lambda_1 u) = 0 < \varphi(\lambda u)$ for all $\lambda \in [\hat{\lambda},\lambda_1)$, hence

$$\frac{d}{d\lambda}\varphi(\lambda v)\Big|_{\lambda=1} = \lambda_1 \frac{d}{d\lambda}\varphi(\lambda u)\Big|_{\lambda=\lambda_1} = \lambda_1 \lim_{\lambda\uparrow\lambda_1} \frac{\varphi(\lambda u)}{\lambda - \lambda_1} \leq 0. \tag{10.43}$$

Comparing (10.42) and (10.43), we reach a contradiction. This proves (10.37), and this completes Step 4.

*Step 5:* $\varphi$ admits a fourth critical point $y_0 \in C^1([0,b]) \setminus \{u_0, v_0, 0\}$.

Arguing by contradiction, we assume that $u_0$, $v_0$, and 0 are the only critical points of $\varphi$. Since $u_0$ and $v_0$ are local minimizers of $\varphi$ (Step 3), we have

$$C_k(\varphi, u_0) = C_k(\varphi, v_0) = \delta_{k,0}\mathbb{F} \text{ for all } k \in \mathbb{N}_0 \tag{10.44}$$

(Example 6.45(a)). Moreover, since $\varphi$ is bounded below, by Proposition 6.64(a), we have

$$C_k(\varphi, \infty) = \delta_{k,0}\mathbb{F} \text{ for all } k \in \mathbb{N}_0. \tag{10.45}$$

Combining Step 4 with (10.44), (10.45), and Theorem 6.62(b) (for $t = -1$), we obtain

$$(-1)^0 + (-1)^0 = (-1)^0,$$

a contradiction. This completes Step 5.

The theorem is obtained by comparing Steps 1, 3, and 5.                    $\square$

*Remark 10.10.* A different reasoning can be pursued in Step 5 of the proof of Theorem 10.9. Step 3, together with the assumption made in Step 5 that $\varphi$ has only a finite number of critical points, implies that $u_0, v_0$ are strict local minimizers of $\varphi$. Due to Proposition 5.42, we can apply Theorem 6.99, which yields $y_0 \in W_{\mathrm{per}}^{1,p}((0,b)) \setminus \{u_0, v_0\}$ critical point of $\varphi$ of mountain pass type. In particular, $C_1(\varphi, y_0) \neq 0$ (Proposition 6.100). This, in conjunction with Step 4, ensures that $y_0 \neq 0$.

## 10.2  Nonsmooth Periodic Systems

In this section, we study the following second-order periodic system:

$$\begin{cases} -u''(t) - Au(t) \in \partial F(t, u(t)) & \text{in } (0,b), \\ u(0) = u(b), \; u'(0) = u'(b), \end{cases} \tag{10.46}$$

where $b > 0$, $A$ is a diagonal $N \times N$ matrix ($N \geq 1$) with real coefficients, and $F : (0,b) \times \mathbb{R}^N \to \mathbb{R}$ is a Carathéodory function that is locally Lipschitz with respect to the second variable. The main feature of the problem is that the function $\xi \mapsto F(t, \xi)$ is not assumed to be differentiable. For this reason, the right-hand part of (10.46) involves the generalized subdifferential $\partial F(t, \cdot)$ of $F(t, \cdot)$ (Sect. 3.2). In what follows, we identify the duality brackets for the pair $((\mathbb{R}^N)^*, \mathbb{R}^N)$ with the scalar product $(\cdot, \cdot)_{\mathbb{R}^N}$, in particular $\partial F(t, \xi) \subset \mathbb{R}^N$.

To deal with problem (10.46), we consider the solution space

$$W_{\mathrm{per}}^{1,2}((0,b), \mathbb{R}^N) = \{ u \in W^{1,2}((0,b))^N : u(0) = u(b) \},$$

which is a Hilbert space endowed with the norm defined by

$$\|u\|^2 = \|u\|_2^2 + \|u'\|_2^2 \text{ for all } u \in W_{\mathrm{per}}^{1,2}((0,b), \mathbb{R}^N),$$

where $u' = (u_1', \ldots, u_N')$ whenever $u = (u_1, \ldots, u_N) \in W_{\mathrm{per}}^{1,2}((0,b), \mathbb{R}^N)$. The space $W^{1,2}((0,b))$ is compactly embedded in $C([0,b])$ [Theorem 1.49(c)], so that the pointwise evaluations at $t = 0$ and $t = b$ in the definition of $W_{\mathrm{per}}^{1,2}((0,b), \mathbb{R}^N)$ make sense.

**Definition 10.11.** We say that $u \in W_{\mathrm{per}}^{1,2}((0,b), \mathbb{R}^N)$ is a *solution* of problem (10.46) if there exists $w \in L^1((0,b), \mathbb{R}^N)$, with $w(t) \in \partial F(t, u(t))$ for a.a. $t \in (0,b)$, such that

$$\int_0^b (u'(t), v'(t))_{\mathbb{R}^N} \, dt - \int_0^b (Au(t), v(t))_{\mathbb{R}^N} \, dt = \int_0^b (w(t), v(t))_{\mathbb{R}^N} \, dt$$

for all $v \in W_{\mathrm{per}}^{1,2}((0,b), \mathbb{R}^N)$.

*Remark 10.12.* If $u = (u_1, \ldots, u_N) \in W^{1,2}_{\text{per}}((0,b), \mathbb{R}^N)$ is a solution of problem (10.46), then $u \in C^1([0,b], \mathbb{R}^N)$, and the conditions

$$u(0) = u(b) \text{ and } u'(0) = u'(b)$$

are satisfied. Indeed, the condition $u(0) = u(b)$ follows from the definition of the space $W^{1,2}_{\text{per}}((0,b), \mathbb{R}^N)$. We have $u'' = -Au - w \in L^1((0,b), \mathbb{R}^N)$, with $w \in L^1((0,b), \mathbb{R}^N)$ as in Definition 10.11, which implies that $u' \in W^{1,1}((0,b))^N \subset C([0,b])^N$, whence $u \in C^1([0,b], \mathbb{R}^N)$. Finally, acting on the equality $u'' = -Au - w$ with the test functions $(\delta_{i,j})_{j=1}^N \in W^{1,2}_{\text{per}}((0,b), \mathbb{R}^N)$ for each $i \in \{1, \ldots, N\}$, we see that

$$0 = -\int_0^b (Au(t) + w(t)) \, dt = \int_0^b u''(t) \, dt = u'(b) - u'(0).$$

The purpose of this section is to prove existence results and a multiplicity result for the solutions of problem (10.46). The methods used are variational. Because the potential $F(t, \cdot)$ is not assumed to be smooth, we cannot associate a $C^1$-functional to the problem, hence we cannot rely on smooth critical point theory. Our approach will be based on the calculus with generalized gradients for locally Lipschitz functions presented in Sect. 3.2 and on minimax principles from nonsmooth critical point theory (Sect. 5.5).

Let us focus for a while on the linear differential operator $L(u) := -u'' - Au$ involved in the left-hand side of problem (10.46). Recall that $A$ is assumed to be a diagonal matrix

$$A = \begin{pmatrix} a_1 & & (0) \\ & \ddots & \\ (0) & & a_N \end{pmatrix},$$

with $a_1, \ldots, a_N \in \mathbb{R}$. An eigenvalue of $L$ is a number $\lambda \in \mathbb{R}$ such that the problem

$$\begin{cases} -u''(t) - Au(t) = \lambda u(t) & \text{in } (0,b), \\ u(0) = u(b), \ u'(0) = u'(b) \end{cases}$$

has a nontrivial solution $u \in W^{1,2}_{\text{per}}((0,b), \mathbb{R}^N)$ (called an eigenfunction). The next result describes the spectrum of the operator $L$. It is a straightforward consequence of the properties of the spectrum of the negative periodic scalar Laplacian (Theorem 9.56, with $p = 2$).

**Proposition 10.13.** *The eigenvalues of the operator $L(u) = -u'' - Au$ are exactly the numbers*

$$\lambda_{k,i} := \left(\frac{2k\pi}{b}\right)^2 - a_i \text{ for } (k,i) \in \mathbb{Z} \times \{1, \ldots, N\}.$$

*An eigenfunction corresponding to* $\lambda_{k,i}$ *is* $\hat{u}_{k,i} = \left((\hat{u}_{k,i})_j\right)_{j=1}^N \in C^\infty([0,b],\mathbb{R}^N)$ *given by* $(\hat{u}_{k,i})_j = 0$ *for* $j \neq i$ *and*

$$(\hat{u}_{k,i})_i(t) = \sin\left(\frac{2k\pi}{b}t\right) \text{ if } k < 0 \text{ and } (\hat{u}_{k,i})_i(t) = \cos\left(\frac{2k\pi}{b}t\right) \text{ if } k \geq 0.$$

*Moreover, the functions* $\{\hat{u}_{k,i} : (k,i) \in \mathbb{Z} \times \{1,\ldots,N\}\}$ *form an orthogonal basis of* $L^2((0,b),\mathbb{R}^N)$ *and* $W^{1,2}_{\text{per}}((0,b),\mathbb{R}^N)$.

From the proposition we obtain an orthogonal direct sum decomposition

$$W^{1,2}_{\text{per}}((0,b),\mathbb{R}^N) = H_- \oplus H_0 \oplus H_+, \tag{10.47}$$

where

$$H_- = \text{span}\{\hat{u}_{k,i} : \lambda_{k,i} < 0\}, \quad H_0 = \ker L = \text{span}\{\hat{u}_{k,i} : \lambda_{k,i} = 0\}$$

$$\text{and } H_+ = \overline{\text{span}}\{\hat{u}_{k,i} : \lambda_{k,i} > 0\}.$$

Note that $H_-$ and $H_0$ are finite dimensional and their dimensions can be explicitly determined in terms of the relative position of the numbers $a_1,\ldots,a_N$ with respect to $\left(\frac{2k\pi}{b}\right)^2$, $k \in \mathbb{Z}$. For instance, we have $H_- = 0$ if and only if $A$ is negative semidefinite (i.e., $a_i \leq 0$ for all $i$) and we have $H_- = H_0 = 0$ if and only if $A$ is negative definite (i.e., $a_i < 0$ for all $i$). The dimensions of $H_-$ and $H_0$ play an important role in the following existence and multiplicity results.

*Remark 10.14.* The preceding analysis of the spectrum of $L(u) = -u'' - Au$ remains valid if $A$ is a symmetric $N \times N$ matrix (instead of diagonal) with eigenvalues $a_1,\ldots,a_N$. In this case, the eigenvalues are still the numbers $\lambda_{k,i} = \left(\frac{2k\pi}{b}\right)^2 - a_i$, and one can determine a corresponding orthogonal basis by means of a change of basis. In particular, the dimensions of $H_-$ and $H_0$ are the same as in the diagonal case. In what follows, we do not use the fact that $A$ is diagonal, so the following results are still valid in the case where $A$ is only supposed to be symmetric.

We conclude the preliminary part with a technical lemma. The statement is general, although we will use it only in the case where $\lambda = 0$ (in which case $H(\lambda) = H_0 \oplus H_+$) and in the case where $\lambda = \hat{\lambda}_+ > 0$ is the smallest positive eigenvalue of $L(u) = -u'' - Au$ [in which case $H(\lambda) = H_+$].

**Lemma 10.15.** *Let* $\lambda$ *be any eigenvalue of* $L(u) = -u'' - Au$, *and let* $\sigma \in L^\infty((0,b))$ *such that* $\sigma(t) \leq \lambda$ *for a.a.* $t \in (0,b)$ *with strict inequality on a set of positive measure. Then there exists a constant* $c_1 > 0$ *such that*

$$\psi_\sigma(u) := \|u'\|_2^2 - \int_0^b (Au(t),u(t))_{\mathbb{R}^N}\, dt - \int_0^b \sigma(t)|u(t)|^2\, dt \geq c_1\|u\|^2$$

*for all* $u \in H(\lambda) := \overline{\text{span}}\{\hat{u}_{k,i} : \lambda_{k,i} \geq \lambda\} \subset W^{1,2}_{\text{per}}((0,b),\mathbb{R}^N)$.

*Proof.* Since $\sigma \leq \lambda$ a.e. in $(0,b)$, we have $\psi_\sigma(u) \geq 0$ for all $u \in H(\lambda)$. Arguing by contradiction, assume that there is a sequence $\{u_n\}_{n\geq 1} \subset W^{1,2}_{\mathrm{per}}((0,b),\mathbb{R}^N)$ such that

$$\psi_\sigma(u_n) \to 0 \text{ as } n \to \infty \text{ and } \|u_n\| = 1 \text{ for all } n \geq 1.$$

We may assume that

$$u_n \xrightarrow{\mathrm{w}} u \text{ in } W^{1,2}_{\mathrm{per}}((0,b),\mathbb{R}^N) \text{ and } u_n \to u \text{ in } C([0,b],\mathbb{R}^N).$$

Note that $u \neq 0$, because otherwise the facts that $\psi_\sigma(u_n) \to 0$ and $u_n \to u$ in $C([0,b],\mathbb{R}^N)$ yield $\|u'_n\|_2 \to 0$ (and so $\|u_n\| \to 0$), which is impossible since $\|u_n\| = 1$ for all $n \geq 1$. Since $\psi_\sigma(u_n) \to 0$ as $n \to \infty$, using that $\|\cdot\|_2$ is weakly l.s.c., we obtain

$$\psi(u) := \|u'\|^2_2 - \int_0^b (Au(t),u(t))_{\mathbb{R}^N}\,dt \leq \int_0^b \sigma(t)|u(t)|^2\,dt \leq \lambda\|u\|^2_2. \quad (10.48)$$

The subspaces $H(\lambda)_0 := \mathrm{span}\{\hat{u}_{k,i} : \lambda_{k,i} = \lambda\}$ (the eigenspace of $L$ corresponding to $\lambda$, which is finite dimensional) and $H(\lambda)_+ := \mathrm{span}\{\hat{u}_{k,i} : \lambda_{k,i} > \lambda\}$ are orthogonal with respect to the scalar product of $L^2((0,b),\mathbb{R}^N)$, and they form a decomposition $H(\lambda) = H(\lambda)_0 \oplus H(\lambda)_+$ (Proposition 10.13). Thus, writing $u = v + \hat{u}$ with $v \in H(\lambda)_0$ and $\hat{u} \in H(\lambda)_+$, we obtain $\psi(u) = \psi(v) + \psi(\hat{u})$ and $\|u\|^2_2 = \|v\|^2_2 + \|\hat{u}\|^2_2$. From Proposition 10.13 we know that every $z \in H(\lambda)_0 \setminus \{0\}$ satisfies $z(t) \neq 0$ for a.a. $t \in (0,b)$. Hence

$$\psi(v) = \lambda\|v\|^2_2 \geq \int_0^b \sigma(t)|v(t)|^2\,dt,$$

with equality if and only if $v \equiv 0$. On the other hand, by the definition of $H(\lambda)_+$, we have $\psi(\hat{u}) \geq \min\{\lambda_{k,i} : \lambda_{k,i} > \lambda\}\|\hat{u}\|^2_2 \geq \lambda\|\hat{u}\|^2_2$, with equality between the last two terms if and only if $\hat{u} \equiv 0$. Comparing these observations with (10.48), we necessarily have $v \equiv 0$ and $\hat{u} \equiv 0$, so $u \equiv 0$. This is a contradiction since we noted earlier that $u \neq 0$. The lemma is proved.                                  □

For the first existence result we need the following hypotheses on the nonsmooth potential $F$. By $\partial F(t,\xi)$ we denote the generalized subdifferential of $F(t,\cdot)$ at $\xi \in \mathbb{R}^N$ (Definition 3.24), and by $F^0(t,\xi;\eta)$ we denote the generalized directional derivative of $F(t,\cdot)$ at $\xi$ in the direction $\eta$ (Definition 3.22).

$\mathrm{H}(F)_1$  (i)  $F : (0,b) \times \mathbb{R}^N \to \mathbb{R}$ is a function such that for all $\xi \in \mathbb{R}^N, t \mapsto F(t,\xi)$ is measurable, for a.a. $t \in (0,b), \xi \mapsto F(t,\xi)$ is locally Lipschitz, and $F(\cdot,0) \in L^1((0,b))_+$;

  (ii)  There exist $\alpha \in L^1((0,b))_+$ and $1 \leq p < +\infty$ such that

$$|w| \leq \alpha(t)(1+|\xi|^{p-1}) \text{ for a.a. } t \in (0,b), \text{ all } \xi \in \mathbb{R}^N, \text{ all } w \in \partial F(t,\xi);$$

(iii) There exist $\mu > 2$, $M > 0$, and $c_0 > 0$ such that

$$c_0 \leq \mu F(t,\xi) \leq -F^0(t,\xi;-\xi) \quad \text{for a.a. } t \in (0,b), \text{ all } |\xi| \geq M;$$

(iv) $\limsup\limits_{\xi \to 0} \dfrac{F(t,\xi)}{|\xi|^2} \leq 0$ uniformly for a.a. $t \in (0,b)$;

(v) $F(t,\xi) \geq \dfrac{\lambda_-}{2}|\xi|^2$ for a.a. $t \in (0,b)$, all $\xi \in \mathbb{R}^N$, where $\lambda_- \leq 0$ denotes the biggest nonpositive eigenvalue of $L(u) = -u'' - Au$.

*Example 10.16.* The following function satisfies hypotheses $H(F)_1$:

$$F(t,\xi) = \begin{cases} \dfrac{\lambda}{2}|\xi|^2 & \text{if } |\xi| \leq 1 \\ \dfrac{\alpha(t)}{p}|\xi|^p + \dfrac{\lambda}{2} - \dfrac{\alpha(t)}{p} & \text{if } |\xi| > 1, \end{cases}$$

with $\alpha \in L^1((0,b))$, $\alpha(t) \geq \alpha_0$ for a.a. $t \in (0,b)$, for constants $\alpha_0 > 0$ and $2 < p < +\infty$. Assumptions $H(F)_1$ are verified taking $\mu \in (2,p)$ in $H(F)_1$ (iii).

The next lemma shows that hypotheses $H(F)_1$ imply that the potential $F(t,\cdot)$ is strictly superquadratic.

**Lemma 10.17.** *If hypotheses $H(F)_1$ hold, then there exist $\alpha_1, \alpha_2 \in L^1((0,b))_+$, with $\frac{c_0}{M^\mu} \leq \alpha_1(t)$ for a.a. $t \in (0,b)$ such that*

$$F(t,\xi) \geq \alpha_1(t)|\xi|^\mu - \alpha_2(t) \quad \text{for a.a. } t \in (0,b), \text{ all } \xi \in \mathbb{R}^N,$$

*where $M, c_0 > 0$ and $\mu > 2$ are as in $H(F)_1$ (iii).*

*Proof.* Let $T_0$ be the Lebesgue-null subset of $[0,b]$ outside of which hypotheses $H(F)_1$ (i)–(v) hold for all $t$. Let $t \in [0,b] \setminus T_0$ and $\xi \in \mathbb{R}^N$.

First, we assume that $|\xi| \leq M$. Proposition 3.31 yields $\theta_t \in (0,1)$ and $\eta_t^* \in \partial F(t,\theta_t \xi)$ such that

$$F(t,\xi) = F(t,0) + (\eta_t^*, \xi)_{\mathbb{R}^N}.$$

Using $H(F)_1$ (ii), for $|\xi| \leq M$ we get

$$|F(t,\xi)| \leq |F(t,0)| + |\eta_t^*||\xi| \leq |F(t,0)| + \alpha(t)(1 + M^{p-1})M =: \tilde{\alpha}_2(t), \quad (10.49)$$

with $\tilde{\alpha}_2 \in L^1((0,b))_+$ [by $H(F)_1$ (i)].

Now assume that $|\xi| \geq M$. Set $\beta(t,r) = F(t,r\xi)$, $r \geq 1$. Clearly, $\beta(t,\cdot)$ is locally Lipschitz. Moreover, from the chain rule (Proposition 3.34) we have that

$$-r\partial\beta(t,r) \subset (\partial F(t,r\xi), -r\xi)_{\mathbb{R}^N}. \quad (10.50)$$

Recall that $\beta(t, \cdot)$ is differentiable a.e. on $\mathbb{R}$, and at every point of differentiability $r > 1$ we have $\frac{d}{dr}\beta(t, r) \in \partial\beta(t, r)$. So from (10.50) and hypothesis $H(F)_1$ (iii) we have

$$r\frac{d}{dr}\beta(t, r) \geq -F^0(t, r\xi; -r\xi) \geq \mu F(t, r\xi) = \mu\beta(t, r) > 0 \text{ for a.a. } r > 1,$$

which implies that

$$\frac{\mu}{r} \leq \frac{\frac{d}{dr}\beta(t, r)}{\beta(t, r)} \text{ for a.a. } r > 1.$$

Integrating from 1 to $r > 1$, we obtain

$$\ln r^\mu \leq \ln\frac{\beta(t, r)}{\beta(t, 1)},$$

and thus we have $r^\mu\beta(t, 1) \leq \beta(t, r)$. So we have shown that for all $t \in [0, b] \setminus T_0$, all $\xi \in \mathbb{R}^N$ with $|\xi| \geq M$, all $r \geq 1$, we have

$$r^\mu F(t, \xi) \leq F(t, r\xi). \tag{10.51}$$

In view of (10.51), for $|\xi| \geq M$ we have

$$F(t, \xi) = F\left(t, \frac{|\xi|}{M}\frac{M\xi}{|\xi|}\right) \geq \frac{|\xi|^\mu}{M^\mu}F\left(t, \frac{M\xi}{|\xi|}\right)$$

$$\geq \frac{|\xi|^\mu}{M^\mu}\min\{F(t, \eta) : |\eta| = M\} = \alpha_1(t)|\xi|^\mu, \tag{10.52}$$

with $\alpha_1 \in L^1((0, b))_+$ [by (10.49)], $\alpha_1(t) \geq \frac{c_0}{M^\mu}$ for a.a. $t \in (0, b)$. Finally, let $\alpha_2(t) = \tilde{\alpha}_2(t) + \alpha_1(t)M^\mu$. The lemma ensues by combining (10.49) and (10.52).   □

We are ready for the first existence theorem concerning problem (10.46).

**Theorem 10.18.** *If hypotheses* $H(F)_1$ *hold, then problem (10.46) has a nontrivial solution* $u \in C^1([0, b], \mathbb{R}^N)$.

*Proof.* Consider the locally Lipschitz functional $\varphi : W^{1,2}_{per}((0, b), \mathbb{R}^N) \to \mathbb{R}$ for problem (10.46) defined by

$$\varphi(u) = \frac{1}{2}\|u'\|_2^2 - \frac{1}{2}\int_0^b (Au(t), u(t))_{\mathbb{R}^N}\, dt - \int_0^b F(t, u(t))\, dt$$

for all $u \in W^{1,2}_{per}((0, b), \mathbb{R}^N)$. Recall from Definition 3.38 that $u \in W^{1,2}_{per}((0, b), \mathbb{R}^N)$ is called a critical point of $\varphi$ if $0 \in \partial\varphi(u)$, where $\partial\varphi(u)$ denotes the generalized subdifferential of $\varphi$ at $u$. Let $V \in \mathcal{L}(W^{1,2}_{per}((0, b), \mathbb{R}^N), W^{1,2}_{per}((0, b), \mathbb{R}^N)^*)$ be defined by

$$\langle V(u), v \rangle = \int_0^b (u'(t), v'(t))_{\mathbb{R}^N} \, dt \quad \text{for all } u, v \in W^{1,2}_{\text{per}}((0,b), \mathbb{R}^N).$$

*Step 1:* For every $u \in W^{1,2}_{\text{per}}((0,b), \mathbb{R}^N)$ we have

$$\partial \varphi(u) \subset \{V(u) - Au - w : \ w \in L^1((0,b), \mathbb{R}^N), \ w(t) \in \partial F(t, u(t)) \text{ for a.a. } t \in (0,b)\};$$

in particular, if $u$ is a critical point of $\varphi$, then $u$ is a solution of problem (10.46), and we have $u \in C^1([0,b], \mathbb{R}^N)$.

The inclusion follows from Propositions 3.27(a), 3.45, and 3.49, while the second part of the statement in Step 1 follows from Definition 10.11 and Remark 10.12.

*Step 2:* $\varphi$ satisfies the (PS)-condition (Definition 5.80).

To this end, let $\{u_n\}_{n \geq 1} \subset W^{1,2}_{\text{per}}((0,b), \mathbb{R}^N)$ and $\{u_n^*\} \subset W^{1,2}_{\text{per}}((0,b), \mathbb{R}^N)^*$ be sequences such that

$$|\varphi(u_n)| \leq M_1, \ u_n^* \in \partial \varphi(u_n) \text{ for all } n \geq 1, \text{ and } \|u_n^*\| \to 0 \text{ as } n \to \infty$$

for some $M_1 > 0$. The fact that $u_n^* \in \partial \varphi(u_n)$ implies that

$$u_n^* = V(u_n) - Au_n - w_n \quad \text{for all } n \geq 1, \tag{10.53}$$

where $w_n \in L^1((0,b), \mathbb{R}^N)$ is such that $w_n(t) \in \partial F(t, u_n(t))$ for a.a. $t \in (0,b)$ (Step 1). Evidently, $V$ is monotone and continuous, so it is maximal monotone (Corollary 2.42). Let $\eta \in (2, \mu)$. From the choice of the sequences $\{u_n\}_{n \geq 1}$ and $\{u_n^*\}_{n \geq 1}$, for all $n \geq 1$, we have

$$\frac{\eta}{2} \|u_n'\|^2 - \frac{\eta}{2} \int_0^b (Au_n(t), u_n(t))_{\mathbb{R}^N} \, dt - \int_0^b \eta F(t, u_n(t)) \, dt \leq \eta M_1 \tag{10.54}$$

and

$$-\|u_n'\|_2^2 + \int_0^b (Au_n(t), u_n(t))_{\mathbb{R}^N} \, dt + \int_0^b (w_n(t), u_n(t))_{\mathbb{R}^N} \, dt \leq \varepsilon_n \|u_n\|,$$

with $\varepsilon_n = \|u_n^*\| \to 0$ as $n \to \infty$, which implies that

$$-\|u_n'\|_2^2 + \int_0^b (Au_n(t), u_n(t))_{\mathbb{R}^N} \, dt - \int_0^b F^0(t, u_n(t); -u_n(t)) \, dt \leq \varepsilon_n \|u_n\| \tag{10.55}$$

[Proposition 3.26(b)]. Adding (10.54) and (10.55), we obtain

$$
\left(\frac{\eta}{2}-1\right)\|u_n'\|_2^2 - \left(\frac{\eta}{2}-1\right)\int_0^b (Au_n(t),u_n(t))_{\mathbb{R}^N}\,dt
$$

$$
-\int_0^b \left(\eta F(t,u_n(t)) + F^0(t,u_n(t);-u_n(t))\right)dt
$$

$$
\leq \varepsilon_n\|u_n\| + \eta M_1. \tag{10.56}
$$

Recall that $u_n = v_n + \hat{u}_n$, with $v_n \in H_- \oplus H_0$ and $\hat{u}_n \in H_+$ [see (10.47)]. Exploiting the orthogonality of the decomposition in (10.47), from (10.56) we obtain

$$
\left(\frac{\eta}{2}-1\right)\left(\|\hat{u}_n'\|_2^2 - \int_0^b (A\hat{u}_n(t),\hat{u}_n(t))_{\mathbb{R}^N}\,dt\right)
$$

$$
+\left(\frac{\eta}{2}-1\right)\left(\|v_n'\|_2^2 - \int_0^b (Av_n(t),v_n(t))_{\mathbb{R}^N}\,dt\right)
$$

$$
-\int_0^b \left(\mu F(t,u_n(t)) + F^0(t,u_n(t);-u_n(t))\right)dt + (\mu-\eta)\int_0^b F(t,u_n(t))\,dt
$$

$$
\leq \varepsilon_n\|u_n\| + \eta M_1. \tag{10.57}
$$

From Lemma 10.15 (applied for $H(\lambda) = H_+$ and $\sigma \equiv 0$) we know that

$$
\|\hat{u}_n'\|_2^2 - \int_0^b (A\hat{u}_n(t),\hat{u}_n(t))_{\mathbb{R}^N}\,dt \geq c_1\|\hat{u}_n\|^2 \quad \text{for all } n \geq 1, \tag{10.58}
$$

with $c_1 > 0$. Also, we have

$$
\|v_n'\|_2^2 - \int_0^b (Av_n(t),v_n(t))_{\mathbb{R}^N}\,dt \geq \lambda_1\|v_n\|_2^2 \geq -|\lambda_1|\,\|v_n\|_2^2, \tag{10.59}
$$

where $\lambda_1$ is the minimal eigenvalue of $L(u) = -u'' - Au$. Moreover, from hypotheses $H(F)_1$ (ii) and (iii) and the mean value theorem for locally Lipschitz functions (Proposition 3.31) we have

$$
-\int_0^b \left(\mu F(t,u_n(t)) + F^0(t,u_n(t);-u_n(t))\right)dt
$$

$$
= -\int_{\{|u_n(t)|\geq M\}} \left(\mu F(t,u_n(t)) + F^0(t,u_n(t);-u_n(t))\right)dt
$$

$$
-\int_{\{|u_n(t)|<M\}} \left(\mu F(t,u_n(t)) + F^0(t,u_n(t);-u_n(t))\right)dt
$$

$$
\geq -c_2 \quad \text{for all } n \geq 1 \tag{10.60}
$$

for some $c_2 > 0$. Finally Lemma 10.17 implies that

$$(\mu - \eta) \int_0^b F(t, u_n(t)) \, dt \geq \frac{c_0(\mu - \eta)}{M^\mu} \|u_n\|_\mu^\mu - (\mu - \eta) \|\alpha_2\|_1 \qquad (10.61)$$

for all $n \geq 1$. Returning to (10.57) and using (10.58)–(10.61), we obtain

$$\left(\frac{\eta}{2} - 1\right)\left(c_1 \|\hat{u}_n\|^2 - |\lambda_1| \|v_n\|_2^2\right) + c_3 \|u_n\|_\mu^\mu \leq M_2 + \varepsilon_n \|u_n\| \quad \text{for all } n \geq 1$$

for some $c_3, M_2 > 0$. Note that there is $M_3 > 0$ such that $\varepsilon_n \leq M_3$ for all $n \geq 1$. Note also that $\|u_n\| \leq \|\hat{u}_n\| + \|v_n\|$, and since $\mu > 2$, there is a constant $c_4 > 0$ such that $\|u_n\|_\mu \geq c_4 \|u_n\|_2 \geq c_4 \|v_n\|_2$ for all $n \geq 1$. Moreover, the fact that $H_- \oplus H_0$ is finite dimensional yields $c_5 > 0$ such that $\|v_n\| \leq c_5 \|v_n\|_2$ for all $n \geq 1$. All together, we obtain

$$\left(\frac{\eta}{2} - 1\right) c_1 \|\hat{u}_n\|^2 + c_3 c_4^\mu \|v_n\|_2^\mu \leq M_2 + M_3 \|\hat{u}_n\| + M_3 c_5 \|v_n\|_2 + \left(\frac{\eta}{2} - 1\right) |\lambda_1| \|v_n\|_2^2$$

for all $n \geq 1$. Since $\eta > 2$ and $\mu > 2$, we infer that $\{\hat{u}_n\}_{n \geq 1}$ is bounded in $W_{per}^{1,2}((0,b), \mathbb{R}^N)$ and that $\{v_n\}_{n \geq 1}$ is bounded in $L^2((0,b), \mathbb{R}^N)$ and, therefore, in $W_{per}^{1,2}((0,b), \mathbb{R}^N)$ (due to the finite dimensionality of $H_- \oplus H_0$). In this way, $\{u_n\}_{n \geq 1}$ is bounded in $W_{per}^{1,2}((0,b), \mathbb{R}^N)$. By passing to a suitable subsequence if necessary, we may assume that

$$u_n \xrightarrow{w} u \text{ in } W_{per}^{1,2}((0,b), \mathbb{R}^N) \text{ and } u_n \to u \text{ in } C([0,b], \mathbb{R}^N). \qquad (10.62)$$

Recall that

$$\left| \langle V(u_n), u_n - u \rangle - \int_0^b (Au_n, u_n - u)_{\mathbb{R}^N} \, dt - \int_0^b (w_n, u_n - u)_{\mathbb{R}^N} \, dt \right| \leq \varepsilon_n \|u_n - u\|$$

[see (10.53)]. The second part in (10.62) and hypothesis H(F)$_1$ (ii) imply that

$$\int_0^b (Au_n, u_n - u)_{\mathbb{R}^N} \, dt \to 0 \text{ and } \int_0^b (w_n, u_n - u)_{\mathbb{R}^N} \, dt \to 0 \text{ as } n \to \infty.$$

It follows that $\langle V(u_n), u_n - u \rangle \to 0$; therefore, $u_n \to u$ in $W_{per}^{1,2}((0,b), \mathbb{R}^N)$ as $n \to \infty$ (Proposition 2.72). This completes Step 2.

*Step 3:* There exist $\rho > 0$ and $\beta > 0$ such that $\varphi(u) \geq \beta$ for all $u \in H_+$, $\|u\| = \rho$.

Because of hypothesis H(F)$_1$ (iv), given $\varepsilon > 0$, we can find $\delta = \delta(\varepsilon)$ such that

$$F(t, \xi) \leq \frac{\varepsilon}{2} |\xi|^2 \text{ for a.a. } t \in (0,b), \text{ all } |\xi| \leq \delta. \qquad (10.63)$$

On the other hand by the mean value theorem for locally Lipschitz functions (Proposition 3.31) and H($F$)$_1$(i), (ii), we have

$$|F(t,\xi)| \leq |F(t,0)| + \alpha(t)(1+|\xi|^{p-1})|\xi| \leq \alpha_3(t)|\xi|^{2+p}$$

for a.a. $t \in (0,b)$, all $\xi \in \mathbb{R}^N$, with $|\xi| \geq \delta$, for some $\alpha_3 \in L^1((0,b))_+$. Combining this with (10.63), it follows that

$$F(t,\xi) \leq \frac{\varepsilon}{2}|\xi|^2 + \alpha_3(t)|\xi|^{2+p} \text{ for a.a. } t \in (0,b), \text{ all } \xi \in \mathbb{R}^N. \tag{10.64}$$

Then, in view of Lemma 10.15 [applied to $H(\lambda) = H_+$ and $\sigma \equiv 0$] and (10.64), and invoking the continuity of the embedding $W_{per}^{1,2}((0,b),\mathbb{R}^N) \hookrightarrow C([0,b],\mathbb{R}^N)$, we find constants $c_1, c_6 > 0$ such that, for all $u \in H_+$, we have

$$\varphi(u) \geq \frac{c_1}{2}\|u\|^2 - \frac{\varepsilon}{2}\|u\|_2^2 - \int_0^b \alpha_3(t)|u(t)|^{2+p}\,dt$$

$$\geq \frac{c_1}{2}\|u\|^2 - \frac{\varepsilon}{2}\|u\|^2 - c_6\|\alpha_3\|_1\|u\|^{2+p}.$$

Therefore, choosing $\varepsilon > 0$ sufficiently small, we obtain $\varphi(u) \geq c_7\|u\|^2 - c_8\|u\|^{2+p}$ for all $u \in H_+$, for some constants $c_7, c_8 > 0$. Because $2+p > 2$, we can find $\rho > 0$ such that

$$\beta := \inf\{\varphi(u) : u \in H_+, \|u\| = \rho\} > 0.$$

This completes Step 3.

*Step 4:* We have $\varphi(v) \leq 0$ for all $v \in H_- \oplus H_0$.

For $v \in H_- \oplus H_0$ we have by H($F$)$_1$ (v) that

$$\varphi(v) \leq \frac{1}{2}\|v'\|_2^2 - \frac{1}{2}\int_0^b (Av(t),v(t))_{\mathbb{R}^N}\,dt - \frac{\lambda_-}{2}\|v\|_2^2 \leq 0,$$

which establishes Step 4.

*Step 5:* The functional $\varphi$ admits a critical point $u \in W_{per}^{1,2}((0,b),\mathbb{R}^N)$ with $u \neq 0$.

Let $\hat{\lambda}_+ > 0$ be the smallest positive eigenvalue of $L(u) = -u'' - Au$, and let $\hat{u}_+ \in C^1([0,b],\mathbb{R}^N)$ be an eigenfunction corresponding to $\hat{\lambda}_+$ and satisfying $\|\hat{u}_+\| = \rho$ with $\rho > 0$ as in Step 3. Let $u = v + s\hat{u}_+$ with $v \in H_- \oplus H_0$ and $s > 0$. Exploiting the orthogonality of the component spaces, the facts that $v \in H_- \oplus H_0$ and $\mu > 2$, and using Lemma 10.17, we obtain

$$\varphi(u) \leq \frac{1}{2}\left(\|v'\|_2^2 - \int_0^b (Av,v)_{\mathbb{R}^N}\,dt\right)$$

$$+ \frac{s^2}{2}\left(\|\hat{u}'_+\|_2^2 - \int_0^b (A\hat{u}_+,\hat{u}_+)_{\mathbb{R}^N}\,dt\right) - \int_0^b F(t,u)\,dt$$

$$\leq \frac{s^2\hat{\lambda}_+}{2}\|\hat{u}_+\|_2^2 - \frac{c_0}{M^\mu}\|u\|_\mu^\mu + \|\alpha_2\|_1$$

$$\leq \frac{\hat{\lambda}_+}{2}\|u\|_2^2 - c_9\|u\|_2^\mu + \|\alpha_2\|_1$$

for some $c_9 > 0$ independent of $u$. Since $\mu > 2$, we conclude that $\varphi(u) \to -\infty$ as $\|u\| \to +\infty$ (since $H_- \oplus H_0$ is of finite dimension). Therefore, we can find $R > \rho$ large enough so that

$$\varphi(u) < 0 \quad \text{whenever } u = v + s\hat{u}_+ \text{ with } v \in H_- \oplus H_0, \, s > 0, \text{ and } \|u\| = R. \quad (10.65)$$

We consider the half-ball

$$E = \{u = v + s\hat{u}_+ : v \in H_- \oplus H_0, \|u\| \leq R, \, s \geq 0\}.$$

Then

$$E_0 := \partial E = \{u = v + s\hat{u}_+ : v \in H_- \oplus H_0, (\|u\| = R, \, s \geq 0) \text{ or } (\|u\| \leq R, \, s = 0)\}.$$

Moreover, let $D = \{u \in H_+ : \|u\| = \rho\}$. From Example 5.38 (d) we know that $\{E_0, E, D\}$ are linking in $W^{1,2}_{\text{per}}((0,b),\mathbb{R}^N)$. By Steps 3 and 4 and (10.65), we have

$$\varphi(0) \leq \sup_{E_0} \varphi \leq 0 < \inf_D \varphi.$$

By virtue of this inequality and of Step 2, we can apply Theorem 5.83 and obtain $u \in W^{1,2}_{\text{per}}((0,b),\mathbb{R}^N)$ such that $0 \in \partial\varphi(u)$ (i.e., $u$ is a critical point of $\varphi$) and $\varphi(u) \geq \inf_D \varphi > \varphi(0)$ (so $u \neq 0$).

The theorem follows by comparing Steps 1 and 5. $\qquad\qquad\qquad\square$

We can weaken the hypotheses on the nonsmooth potential $F$ if in compensation we assume that $H_- = H_0 = 0$. In this case, the linear differential operator $L(u) = -u'' - Au$ is maximal monotone and coercive and the matrix $A$ is negative definite. Now the hypotheses on the nonsmooth potential $F$ are as follows:

H$(F)_2$ (i) $F : (0,b) \times \mathbb{R}^N \to \mathbb{R}$ is a function such that for all $\xi \in \mathbb{R}^N, t \mapsto F(t,\xi)$ is measurable, for a.a. $t \in (0,b), \xi \mapsto F(t,\xi)$ is locally Lipschitz, and $F(t,0) = 0$ for a.a. $t \in (0,b)$;

(ii) There exist $\alpha \in L^1((0,b))_+$ and $1 \leq p < +\infty$ such that

$$|w| \leq \alpha(t)(1 + |\xi|^{p-1}) \quad \text{for a.a. } t \in (0,b), \text{ all } \xi \in \mathbb{R}^N, \text{ all } w \in \partial F(t,\xi);$$

(iii) There exists $\mu > 2$ such that

$$\mu F(t,\xi) \leq -F^0(t,\xi;-\xi) \quad \text{for a.a. } t \in (0,b), \text{ all } \xi \in \mathbb{R}^N;$$

(iv) There exists $\zeta \in L^\infty((0,b))_+$ such that $\zeta(t) \leq \hat{\lambda}_+$ for a.a. $t \in (0,b)$ with strict inequality on a set of positive measure and

$$\limsup_{\xi \to 0} \frac{2F(t,\xi)}{|\xi|^2} \leq \zeta(t) \quad \text{uniformly for a.a. } t \in (0,b),$$

where $\hat{\lambda}_+ > 0$ is the smallest positive eigenvalue of $L(u) = -u'' - Au$;

(v) There exists $\xi_0 \in \mathbb{R}^N \setminus \{0\}$ such that $\int_0^b F(t,\xi_0)\,dt > 0$.

*Remark 10.19.* Note that hypotheses $H(F)_2$ do not contain conditions on the asymptotic behavior of $F$ at $\pm\infty$. In particular, they allow the potential $F$ to be subquadratic or superquadratic.

*Example 10.20.* The following function satisfies hypotheses $H(F)_2$ (for simplicity we drop the $t$-dependence):

$$F(\xi) = \begin{cases} -\frac{1}{r}|\xi|^r & \text{if } |\xi| \leq 1, \\ \frac{1}{\mu}|\xi|^\mu - \frac{1}{\mu} - \frac{1}{r} & \text{if } |\xi| > 1, \end{cases}$$

with $r < 2 < \mu$.

**Theorem 10.21.** *If hypotheses $H(F)_2$ hold and $H_- = H_0 = 0$, then problem (10.46) has a nontrivial solution $u \in C^1([0,b],\mathbb{R}^N)$.*

*Proof.* Consider the locally Lipschitz functional $\varphi : W^{1,2}_{per}((0,b),\mathbb{R}^N) \to \mathbb{R}$ given by

$$\varphi(u) = \frac{1}{2}\|u'\|_2^2 - \frac{1}{2}\int_0^b (Au(t),u(t))_{\mathbb{R}^N}\,dt - \int_0^b F(t,u(t))\,dt$$

for all $u \in W^{1,2}_{per}((0,b),\mathbb{R}^N)$.

*Claim 1.* $\varphi$ satisfies the (PS)-condition.

Let $\{u_n\}_{n\geq 1} \subset W^{1,2}_{per}((0,b),\mathbb{R}^N)$ and $\{u_n^*\}_{n\geq 1} \subset W^{1,2}_{per}((0,b),\mathbb{R}^N)^*$ be such that

$$|\varphi(u_n)| \leq M_1, \quad u_n^* \in \partial\varphi(u_n) \text{ for all } n \geq 1, \text{ and } \|u_n^*\| \to 0 \text{ as } n \to \infty$$

for some $M_1 > 0$. Reasoning as in Step 2 of the proof of Theorem 10.18 (with $\mu$ in place of $\eta$), we obtain

$$\left(\frac{\mu}{2}-1\right)\left(\|u_n'\|_2^2 - \int_0^b (Au_n(t),u_n(t))_{\mathbb{R}^N}\,dt\right)$$

$$+ \int_0^b \left( -F^0(t, u_n(t); -u_n(t)) - \mu F(t, u_n(t)) \right) dt \le \varepsilon_n \|u_n\| + \mu M_1$$

for all $n \ge 1$, with $\varepsilon_n = \|u_n^*\|$ [see (10.56)]. Invoking Lemma 10.15 [for $H(\lambda) = H_+ = W_{\mathrm{per}}^{1,2}((0,b), \mathbb{R}^N)$ and $\sigma \equiv 0$] and H($F$)$_2$ (iii), we find $c_1 > 0$ and $M_2 > 0$ such that

$$\left( \frac{\mu}{2} - 1 \right) c_1 \|u_n\|^2 \le M_2 \|u_n\| + \mu M_1 \quad \text{for all } n \ge 1,$$

which clearly implies that the sequence $\{u_n\}_{n \ge 1}$ is bounded in $W_{\mathrm{per}}^{1,2}((0,b), \mathbb{R}^N)$. From this, as in Step 2 of the proof of Theorem 10.18, we see that one can extract from $\{u_n\}_{n \ge 1}$ a strongly convergent subsequence. We conclude that $\varphi$ satisfies the (PS)-condition.

*Claim 2.* There exist $\rho, \beta > 0$ such that $\varphi(u) \ge \beta$ for all $u \in W_{\mathrm{per}}^{1,2}((0,b), \mathbb{R}^N)$, with $\|u\| = \rho$.

Arguing as in Step 3 of the proof of Theorem 10.18, from hypotheses H($F$)$_2$ (i), (ii), and (iv) we see that, given $\varepsilon > 0$, we can find $\alpha_\varepsilon \in L^1((0,b))_+$ such that

$$F(t, \xi) \le \frac{1}{2} (\zeta(t) + \varepsilon) |\xi|^2 + \alpha_\varepsilon(t) |\xi|^{2+p} \quad \text{for a.a. } t \in (0,b), \text{ all } \xi \in \mathbb{R}^N. \quad (10.66)$$

Then, using (10.66), Lemma 10.15 (for $\lambda = \hat{\lambda}_+$ and $\sigma = \zeta$), the assumption that $H_- = H_0 = 0$, and the continuity of the embedding $W_{\mathrm{per}}^{1,2}((0,b), \mathbb{R}^N) \hookrightarrow C([0,b], \mathbb{R}^N)$, we have

$$\varphi(u) \ge \frac{1}{2} \|u'\|_2^2 - \frac{1}{2} \int_0^b (Au(t), u(t))_{\mathbb{R}^N} \, dt - \frac{1}{2} \int_0^b \zeta(t) |u(t)|^2 \, dt$$

$$- \frac{\varepsilon}{2} \|u\|_2^2 - \int_0^b \alpha_\varepsilon(t) |u(t)|^{2+p} \, dt$$

$$\ge \frac{c_1}{2} \|u\|^2 - \frac{\varepsilon}{2} \|u\|^2 - c_2 \|\alpha_\varepsilon\|_1 \|u\|^{2+p} \quad (10.67)$$

for all $u \in W_{\mathrm{per}}^{1,2}((0,b), \mathbb{R}^N)$, for some $c_1, c_2 > 0$ independent of $\varepsilon$. Using (10.67) and choosing $\varepsilon > 0$ such that $\varepsilon < \frac{c_1}{2}$, we obtain

$$\varphi(u) \ge \frac{c_1}{4} \|u\|^2 - \|\alpha_\varepsilon\|_1 \|u\|^{2+p} \quad \text{for all } u \in W_{\mathrm{per}}^{1,2}((0,b), \mathbb{R}^N).$$

Because $2 + p > 2$, we can find $\rho > 0$ small such that

$$\inf \{ \varphi(u) : u \in W_{\mathrm{per}}^{1,2}((0,b), \mathbb{R}^N), \|u\| = \rho \} > 0.$$

This proves Claim 2.

*Claim 3.* For a.a. $t \in (0,b)$, all $\xi \in \mathbb{R}^N$, all $r \geq 1$, we have $r^\mu F(t,\xi) \leq F(t,r\xi)$.

The function $r \mapsto \frac{1}{r^\mu}$ is of class $C^1$ on $(0,+\infty)$; hence, for a.a. $t \in (0,b)$ the function $r \mapsto \frac{1}{r^\mu} F(t,r\xi)$ is locally Lipschitz on $(0,+\infty)$ and

$$\partial_r \left( \frac{1}{r^\mu} F(t,r\xi) \right) \subset -\frac{\mu}{r^{\mu+1}} F(t,r\xi) + \frac{1}{r^\mu} (\partial F(t,r\xi),\xi)_{\mathbb{R}^N}$$

(Proposition 3.34 and Corollary 3.35). In the preceding inclusion by $\partial_r$ we denote the subdifferential with respect to $r$. Using the mean value theorem for locally Lipschitz functions (Proposition 3.31), we can find $s \in (1,r)$ (depending in general on $t$) with $r > 1$ such that

$$\frac{1}{r^\mu} F(t,r\xi) - F(t,\xi) = \frac{r-1}{s^{\mu+1}} (-\mu F(t,s\xi) + (\eta^*,s\xi)_{\mathbb{R}^N}),$$

where $\eta^* \in \partial F(t,s\xi)$. Because of hypothesis H$(F)_2$ (iii), we have that

$$-\mu F(t,s\xi) + (\eta^*,s\xi)_{\mathbb{R}^N} \geq -\mu F(t,s\xi) - F^0(t,s\xi;-s\xi)) \geq 0$$

for a.a. $t \in (0,b)$, which implies that $F(t,r\xi) \geq r^\mu F(t,\xi)$ for a.a. $t \in (0,b)$, all $\xi \in \mathbb{R}^N$, all $r \geq 1$. This proves Claim 3.

Let $\xi_0 \in \mathbb{R}^N \setminus \{0\}$ be as in H$(F)_2$ (v). Using the definition of $\varphi$ and Claim 3, for $r \geq 1$ we have

$$\varphi(r\xi_0) \leq -\frac{r^2}{2} \int_0^b (A\xi_0,\xi_0)_{\mathbb{R}^N} \, dt - r^\mu \int_0^b F(t,\xi_0) \, dt. \qquad (10.68)$$

By H$(F)_2$ (v) and (10.68), and since $\mu > 2$, it follows that

$$\varphi(r\xi_0) \to -\infty \text{ as } r \to +\infty.$$

So for $r \geq 1$ large, we will have

$$\varphi(r\xi_0) \leq \varphi(0) = 0 < \beta \leq \inf\{\varphi(u) : u \in W^{1,2}_{per}((0,b),\mathbb{R}^N), \|u\| = \rho\}$$

(Claim 2). This and Claim 1 permit the use of Theorem 5.83 applied to the linking sets from Example 5.38 (a). So we obtain $u \in W^{1,2}_{per}((0,b),\mathbb{R}^N)$ such that

$$\varphi(0) = 0 < \beta \leq \varphi(u) \text{ and } 0 \in \partial\varphi(u).$$

From the inequality we see that $u \neq 0$, while from the inclusion, as in Step 1 of the proof of Theorem 10.18, we get that $u \in C^1([0,b],\mathbb{R}^N)$ is a solution of (10.46).  $\square$

In the case of Theorem 10.21 the kernel of the linear differential operator $L(u) = -u'' - Au$ is trivial. This convenient situation allowed us to incorporate into our framework both subquadratic and superquadratic systems. In the next multiplicity theorem, we still require that $H_- = 0$, but now $H_0 \neq 0$ [i.e., the linear differential operator $L(u) = -u'' - Au$ has a nontrivial kernel]. Now our hypotheses on $F$ incorporate into our setting quadratic or superquadratic systems.

The hypotheses on the nonsmooth potential $F$ are as follows:

$H(F)_3$  (i) $F : (0,b) \times \mathbb{R}^N \to \mathbb{R}$ is a function such that for all $\xi \in \mathbb{R}^N$, $t \mapsto F(t,\xi)$ is measurable, for a.a. $t \in (0,b)$, $\xi \mapsto F(t,\xi)$ is locally Lipschitz, and $F(t,0) = 0$ for a.a. $t \in (0,b)$;

(ii) There exist $\alpha \in L^1((0,b))_+$ and $1 \leq p < +\infty$ such that

$$|w| \leq \alpha(t)(1 + |\xi|^{p-1}) \quad \text{for a.a. } t \in (0,b), \text{ all } \xi \in \mathbb{R}^N, \text{ all } w \in \partial F(t,\xi);$$

(iii) There exists $\theta \in L^\infty((0,b))$ such that $\theta(t) \leq 0$ for a.a. $t \in (0,b)$ with strict inequality on a set of positive measure and

$$\limsup_{|\xi| \to +\infty} \frac{F(t,\xi)}{|\xi|^2} \leq \theta(t) \quad \text{uniformly for a.a. } t \in (0,b);$$

(iv) There exists $\zeta \in L^\infty((0,b))_+$ such that $\zeta(t) \leq \hat{\lambda}_+$ for a.a. $t \in (0,b)$ with strict inequality on a set of positive measure, where $\hat{\lambda}_+ > 0$ is the first positive eigenvalue of $L(u) = -u'' - Au$, and

$$\limsup_{|\xi| \to 0} \frac{2F(t,\xi)}{|\xi|^2} \leq \zeta(t) \quad \text{uniformly for a.a. } t \in (0,b);$$

(v) There exists $\delta > 0$ such that $F(t,\xi) \geq 0$ for a.a. $t \in (0,b)$, all $|\xi| \leq \delta$.

*Example 10.22.* The following locally Lipschitz function satisfies $H(F)_3$:

$$F(t,\xi) = \begin{cases} \frac{\zeta(t)}{r}|\xi|^r & \text{if } |\xi| \leq 1, \\ \theta(t)|\xi|^2 - \theta(t) + \frac{\zeta(t)}{r} & \text{if } |\xi| > 1, \end{cases}$$

with $2 \leq r < +\infty$ and where $\theta$ and $\zeta$ are as in hypotheses $H(F)_3$ (iii) and (iv), respectively.

**Theorem 10.23.** *If hypotheses $H(F)_3$ hold, $H_- = 0$, and $H_0 \neq 0$, then problem (10.46) has at least two nontrivial solutions $u_1, u_2 \in C^1([0,b], \mathbb{R}^N)$.*

*Proof.* Consider the locally Lipschitz functional $\varphi : W^{1,2}_{per}((0,b), \mathbb{R}^N) \to \mathbb{R}$ given by

$$\varphi(u) = \frac{1}{2}\|u'\|_2^2 - \frac{1}{2}\int_0^b (Au(t), u(t))_{\mathbb{R}^N}\, dt - \int_0^b F(t, u(t))\, dt$$

for all $u \in W^{1,2}_{per}((0,b), \mathbb{R}^N)$.

*Claim 1.* $\varphi$ is coercive and bounded below and satisfies the (PS)-condition.

By virtue of hypotheses H$(F)_3$ (ii) and (iii) and Proposition 3.31, given $\varepsilon > 0$, we can find $\alpha_\varepsilon \in L^1((0,b))_+$ such that

$$F(t,\xi) \leq (\theta(t) + \varepsilon)|\xi|^2 + \alpha_\varepsilon(t) \quad \text{for a.a. } t \in (0,b), \text{ all } \xi \in \mathbb{R}^N. \tag{10.69}$$

Applying Lemma 10.15 (for $\lambda = 0$ and $\sigma = 2\theta$), and using also that $H_- = 0$, we find a constant $c_1 > 0$ such that

$$\varphi(u) \geq \frac{1}{2}\|u'\|_2^2 - \frac{1}{2}\int_0^b (Au(t),u(t))_{\mathbb{R}^N}\, dt - \int_0^b \theta(t)|u(t)|^2\, dt - \varepsilon\|u\|_2^2 - \|\alpha_\varepsilon\|_1$$

$$\geq \frac{c_1}{2}\|u\|^2 - \varepsilon\|u\|^2 - \|\alpha_\varepsilon\|_1 \quad \text{for all } u \in W_{\text{per}}^{1,2}((0,b),\mathbb{R}^N).$$

Choosing $\varepsilon > 0$ such that $\varepsilon < \frac{c_1}{2}$, we obtain that $\varphi$ is coercive and bounded below. Arguing as at the end of Step 2 of the proof of Theorem 10.18, we deduce that $\varphi$ satisfies the (PS)-condition. This proves Claim 1.

*Claim 2.* There is $\rho_1 > 0$ such that $\varphi(u) \geq 0$ for all $u \in H_+$, $\|u\| \leq \rho_1$.

Using hypotheses H$(F)_3$ (ii) and (iv) and Proposition 3.31, for all $\varepsilon > 0$ there is $\beta_\varepsilon \in L^1((0,b))_+$ such that

$$F(t,\xi) \leq \frac{1}{2}(\zeta(t) + \varepsilon)|\xi|^2 + \beta_\varepsilon(t)|\xi|^{2+p} \quad \text{for a.a. } t \in (0,b), \text{ all } \xi \in \mathbb{R}^N. \tag{10.70}$$

In view of (10.70), applying Lemma 10.15 (for $\lambda = \hat{\lambda}_+$ and $\sigma = \zeta$), and using the continuity of the embedding $W^{1,2}((0,b),\mathbb{R}^N) \hookrightarrow C([0,b],\mathbb{R}^N)$, we find constants $c_1, c_2 > 0$ (independent of $\varepsilon$) such that

$$\varphi(u) \geq \frac{1}{2}\|u'\|^2 - \frac{1}{2}\int_0^b (Au(t),u(t))_{\mathbb{R}^N}\, dt - \frac{1}{2}\int_0^b \zeta(t)|u(t)|^2\, dt$$

$$- \frac{\varepsilon}{2}\|u\|_2^2 - \|\beta_\varepsilon\|_1\|u\|_\infty^{2+p}$$

$$\geq \frac{c_1 - \varepsilon}{2}\|u\|^2 - c_2\|\beta_\varepsilon\|_1\|u\|^{2+p} \quad \text{for all } u \in H_+. \tag{10.71}$$

The conclusion of Claim 2 follows by choosing $\varepsilon < c_2$ and using that $2 + p > 2$.

*Claim 3.* There is $\rho_2 > 0$ such that $\varphi(u) \leq 0$ for all $u \in H_0$, $\|u\| \leq \rho_2$.

The continuity of the embedding $W^{1,2}((0,b),\mathbb{R}^N) \hookrightarrow C([0,b],\mathbb{R}^N)$ implies that we can find $\rho_2 > 0$ such that $\|u\|_\infty \leq \delta$ whenever $u \in W_{\text{per}}^{1,2}((0,b),\mathbb{R}^N)$ satisfies $\|u\| \leq \rho_2$, where $\delta > 0$ is as in H$(F)_3$ (v). Using hypothesis H$(F)_3$ (v), for all $u \in H_0$ such that $\|u\| \leq \rho_2$ we have

$$\varphi(u) = \frac{1}{2}\|u'\|_2^2 - \frac{1}{2}\int_0^b (Au(t),u(t))_{\mathbb{R}^N}\, dt - \int_0^b F(t,u(t))\, dt$$

$$= -\int_0^b F(t,u(t))\, dt \leq 0.$$

This proves Claim 3.

Note that Claims 2 and 3 yield in particular $\varphi(0) = 0$. Now let

$$\eta_\varphi = \inf \{ \varphi(v) : v \in W^{1,2}_{\text{per}}((0,b), \mathbb{R}^N) \}.$$

Evidently, $\eta_\varphi \leq \varphi(0) = 0$. If $\eta_\varphi = 0$, then, since $H_0 \neq 0$, Claim 3 provides infinitely many global minimizers (hence critical points) of $\varphi$. If $\eta_\varphi < 0$, then on the basis of Claims 1–3, we can apply Theorem 5.85 and obtain that $\varphi$ admits at least two nontrivial critical points. In both cases, arguing as in Step 1 of the proof of Theorem 10.18, we deduce that problem (10.46) has at least two nontrivial solutions belonging to $C^1([0,b], \mathbb{R}^N)$. The proof is complete. $\square$

## 10.3 Remarks

**Section 10.1:** Theorem 10.5 is a basic existence result for a periodic problem involving the scalar $p$-Laplacian. The method used [verification of the (C)-condition, truncation techniques, and application of an appropriate minimax principle (here the mountain pass theorem)] is the prototype of the variational method that we extensively use in Chaps. 10–12. Multiplicity results require in general finer hypotheses (mostly related to the behavior of the nonlinearity $f$ with respect to the spectrum of the differential operator) and to combine different techniques. In Theorem 10.9, we provide a multiplicity result under hypotheses allowing resonance at infinity of $f$ with respect to the first eigenvalue $\lambda_0 = 0$ of the negative scalar periodic $p$-Laplacian, and where we combine the variational method with Morse theory. We mention that the first version of the computation of critical groups in Step 4 of the proof of Theorem 10.9 can be found in the semilinear work of Moroz [264], and subsequent extensions can be found in Jiu and Su [185], Motreanu et al. [297], and Motreanu [270].

Section 10.1 is inspired by the work of Motreanu et al. [296], where a further multiplicity result is provided for problem (10.1) under hypotheses allowing resonance with respect to a higher eigenvalue of the negative periodic scalar $p$-Laplacian. Doubly resonant situations (i.e., hypotheses allowing resonance with respect to two consecutive eigenvalues) were investigated by Fabry and Fonda [132], Papageorgiou and Staicu [322] for semilinear periodic problems (i.e., $p = 2$), and by Kyritsi and Papageorgiou [211] for nonlinear periodic problems with nonsmooth potential. In Fabry and Fonda [132] and Papageorgiou and Staicu [322], the authors employ certain Landesman–Lazer type conditions, inspired by the seminal work on resonant equations, due to Landesman and Lazer [216]. In Kyritsi and Papageorgiou [211], the double resonance is compensated by a nonuniform nonresonance condition on the potential $F(t, \cdot)$, in the spirit of Gossez and Omari [163, 164] (see also Motreanu et al. [296]). The approach of Fabry and Fonda [132] is degree theoretic, while Papageorgiou and Staicu [322] combine variational methods with degree theory.

The approach of Kyritsi and Papageorgiou [211] is purely variational based on the nonsmooth critical point theory (Sect. 5.5).

Other existence and multiplicity results for solutions of equations driven by the periodic scalar $p$-Laplacian can be found in Aizicovici et al. [5, 6], Ben-Naoum and De Coster [41], del Pino et al. [111], Gasiński [149], Gasiński and Papageorgiou [152], Kyritsi and Papageorgiou [212], Njoku and Zanolin [310], Papageorgiou and Papalini [320], Wang [380], Yang [385], and Zhang and Liu [393].

**Section 10.2:** When $A = 0$, problem (10.46) has been studied extensively and various existence results have been proved under the assumption that the potential function $F(t, \cdot)$ is smooth (i.e., a $C^1$-function). We refer to the works of Mawhin and Willem [253], Tang [370], and Tang and Wu [371]. The case where $A = k^2 \omega^2 I$, with $k \in \mathbb{N}$, $\omega = \frac{2N}{b}$, and $I$ the $N \times N$ identity matrix, was considered by Mawhin and Willem [253, p. 61] under the assumption that the right-hand-side nonlinearity has the form $\nabla F(t, \cdot)$, where $F(t, \cdot)$ is a convex function of class $C^1$, and their approach uses the dual action principle. In Mawhin and Willem [253, p. 88], one finds a general problem with $A$ a symmetric matrix and the right-hand-side nonlinearity $\nabla F(t, \cdot)$ satisfying

$$|F(t, \xi)| \le g(t) \quad \text{and} \quad |\nabla F(t, \xi)| \le g(t)$$

for a.a. $t \in (0, b)$, all $\xi \in \mathbb{R}^N$, with $g \in L^1((0, b))_+$. The potential function $F(t, \cdot)$ is still of class $C^1$ but no longer assumed convex. Tang and Wu [372] extended the work of Mawhin and Willem to systems with subquadratic smooth potential, that is, they assumed that $F(t, \cdot) \in C^1(\mathbb{R}^N)$, and for a.a. $t \in (0, b)$, all $\xi \in \mathbb{R}^N$, we have

$$|\nabla F(t, \xi)| \le g(t) + f(t)|\xi|^\alpha$$

with $g, f \in L^1((0, b))_+$ and $0 \le \alpha < 1$.

This section is based on the work of Motreanu et al. [286]. It complements the aforementioned work of Tang and Wu [372] by considering systems where the potential function is quadratic or superquadratic. In addition, the potential function is in general nonsmooth. Moreover, we also present a multiplicity result in Theorem 10.23. Actually, the study in Motreanu et al. [286], where the matrix $A(t)$ depends on $t \in (0, b)$, is more general. Two related works are by Motreanu et al. [290, 300], where the method of proof is a nonsmooth version of the so-called reduction method, developed for smooth boundary value problems by Amann [10], Castro and Lazer [74], and Thews [375].

Additional existence and multiplicity results for second-order periodic systems can be found in the works by Barletta and Papageorgiou [32], Cordaro [87], Faraci [135], Hu and Papageorgiou [177], and Papageorgiou and Papalini [321]. We also refer readers to Manásevich and Mawhin [239], Mawhin [252], and Rabinowitz [337].

# Chapter 11
# Nonlinear Elliptic Equations with Dirichlet Boundary Conditions

**Abstract** This chapter studies nonlinear Dirichlet boundary value problems through various methods such as degree theory, variational methods, lower and upper solutions, Morse theory, and nonlinear operators techniques. The combined application of these methods enables us to handle, under suitable hypotheses, a large variety of cases: sublinear, asymptotically linear, superlinear, coercive, noncoercive, parametric, resonant, and near resonant. In many situations we are able to provide multiple solutions with additional information about their properties, for instance, constant-sign (i.e., positive or negative) solutions and nodal (sign-changing) solutions. The first section of the chapter is devoted to the study of nonlinear elliptic problems through degree theory. The second section focuses on the variational approach, specifically for investigating coercive problems and $(p-1)$-superlinear parametric problems. The third section makes use of Morse theory in studying $(p-1)$-linear noncoercive equations and $p$-Laplace equations with concave terms. The fourth section deals with general elliptic inclusion problems treated via nonlinear, possibly multivalued, operators. The last section highlights related remarks and bibliographical comments.

## 11.1 Nonlinear Dirichlet Problems Using Degree Theory

Let $\Omega \subset \mathbb{R}^N$ be a bounded domain with a $C^2$-boundary $\partial\Omega$. In this section, we are interested in the existence of multiple smooth solutions of constant sign for the following nonlinear Dirichlet problem:

$$\begin{cases} -\Delta_p u = f(x,u) & \text{in } \Omega, \\ u = 0 & \text{on } \partial\Omega, \end{cases} \tag{11.1}$$

driven by the $p$-Laplacian $\Delta_p u = \operatorname{div}(|\nabla u|^{p-2}\nabla u)$, with $p \in (1,+\infty)$ (where $|\cdot|$ stands for the Euclidean norm of $\mathbb{R}^N$), and involving a Carathéodory function $f$ :

D. Motreanu et al., *Topological and Variational Methods with Applications to Nonlinear Boundary Value Problems*, DOI 10.1007/978-1-4614-9323-5_11,
© Springer Science+Business Media, LLC 2014

$\Omega \times \mathbb{R} \to \mathbb{R}$. In our study of problem (11.1), we consider as solution space the Sobolev space $W_0^{1,p}(\Omega)$. This is a Banach space for the norm $u \mapsto \|\nabla u\|_p$, which, by virtue of Poincaré's inequality (Theorem 1.41), is equivalent to the usual Sobolev norm $u \mapsto (\|\nabla u\|_p^p + \|u\|_p^p)^{\frac{1}{p}}$.

**Definition 11.1.** An element $u \in W_0^{1,p}(\Omega)$ is called a *weak solution* of problem (11.1) if $f(\cdot, u(\cdot)) \in L^{r'}(\Omega)$ for some $r \in (1, p^*)$ and we have

$$\int_\Omega |\nabla u|^{p-2}(\nabla u, \nabla v)_{\mathbb{R}^N} \, dx = \int_\Omega f(x, u(x))v(x) \, dx$$

for all $v \in W_0^{1,p}(\Omega)$.

*Remark 11.2.* Recall that $p^* \in (1, +\infty]$ stands for the Sobolev critical exponent (Remark 1.50). In Definition 11.1, the extra condition involving $f(\cdot, u(\cdot))$ is needed (in the absence of growth conditions on $f$) to ensure that the integral on the right-hand side of the formula is well defined.

  In this section, we focus on the existence of smooth, positive solutions of problem (11.1) belonging to the interior of the positive cone $C_0^1(\overline{\Omega})_+ = \{u \in C_0^1(\overline{\Omega}) : u \geq 0 \text{ in } \Omega\}$ of the Banach space $C_0^1(\overline{\Omega}) = \{u \in C^1(\overline{\Omega}) : u|_{\partial\Omega} = 0\}$ (Sect. 8.2). Our approach here is degree theoretic and makes use of the degree map for operators of the monotone type developed in Sect. 4.3.

  We prove multiplicity results for both the coercive and noncoercive cases, i.e., the hypotheses on the nonlinearity $f$ are such that the energy functional of the problem is coercive (resp. noncoercive). The occurrence of these two situations depends on the asymptotic behavior at $+\infty$ of the nonlinearity $f$ with respect to the first eigenvalue $\lambda_1 > 0$ of the negative Dirichlet $p$-Laplacian. For the multiplicity result in the coercive case, we rely on Corollary 4.46, and the proposed argument is in fact an alternative to the mountain pass theorem (more precisely, Proposition 5.42 could also be invoked). In the noncoercive case, the advantage of the use of degree theory (instead of critical point theory) is that it does not require checking the compactness conditions of the functional [(PS)- or (C)-conditions]. In both situations, we use the degree for showing the existence of a second nontrivial solution, after having already established the existence of a first nontrivial solution of the problem through minimization, truncation techniques, and the so-called lower and upper solutions method. This section is organized as follows. First, we provide preliminary results and present the truncation techniques that will also be used in the following sections of this chapter. Second, we apply these techniques, together with degree theory of $(S)_+$-maps, to obtain multiplicity results.

## Preliminaries

Here we state three preliminary results for later use in this section and in other sections of the chapter. The first preliminary result is the following lemma.

**Lemma 11.3.** *Let $\lambda_1 > 0$ be the first eigenvalue of the negative Dirichlet $p$-Laplacian, and let $\zeta \in L^\infty(\Omega)_+$ be such that $\zeta(x) \leq \lambda_1$ for a.a. $x \in \Omega$, with strict inequality on a set of positive measure. Then there is a constant $c_1 > 0$ such that*

$$\psi_\zeta(u) := \|\nabla u\|_p^p - \int_\Omega \zeta(x)|u(x)|^p \, dx \geq c_1 \|\nabla u\|_p^p$$

*for all $u \in W_0^{1,p}(\Omega)$.*

*Proof.* From the variational characterization of $\lambda_1$ in Proposition 9.47 (with $\xi \equiv 1$) we have that $\psi_\zeta \geq 0$. Arguing by contradiction, suppose that the lemma is not true. Then we can find a sequence $\{u_n\}_{n\geq 1} \subset W_0^{1,p}(\Omega)$ such that

$$\|\nabla u_n\|_p = 1 \text{ for all } n \geq 1 \text{ and } \psi_\zeta(u_n) \to 0 \text{ as } n \to \infty.$$

By passing to a relabeled subsequence if necessary, we may assume that

$$u_n \xrightarrow{w} u \text{ in } W_0^{1,p}(\Omega), \ u_n \to u \text{ in } L^p(\Omega), \ u_n(x) \to u(x) \text{ a.e. in } \Omega,$$

and $|u_n(x)| \leq k(x)$ a.e. in $\Omega$, for all $n \geq 1$, with some $k \in L^p(\Omega)_+$. Since

$$\|\nabla u\|_p^p \leq \liminf_{n\to\infty} \|\nabla u_n\|_p^p \text{ and } \int_\Omega \zeta(x)|u_n(x)|^p \, dx \to \int_\Omega \zeta(x)|u(x)|^p \, dx,$$

from the convergence $\psi_\zeta(u_n) \to 0$ we obtain

$$\|\nabla u\|_p^p \leq \int_\Omega \zeta(x)|u(x)|^p \, dx \leq \lambda_1 \|u\|_p^p. \tag{11.2}$$

From Proposition 9.47(a) and (11.2) we infer that

$$\|\nabla u\|_p^p = \lambda_1 \|u\|_p^p, \text{ and so } u = t\hat{u}_1 \text{ with } t \in \mathbb{R}, \tag{11.3}$$

where $\hat{u}_1$ denotes the $L^p$-normalized positive eigenfunction corresponding to $\lambda_1$. If $u = 0$, from the fact that $\psi_\zeta(u_n) \to 0$ and since $\int_\Omega \zeta(x)|u_n(x)|^p \, dx \to 0$, it follows that $\|\nabla u_n\|_p \to 0$, which is a contradiction of the fact that $\|\nabla u_n\|_p = 1$ for all $n \geq 1$. Thus, $u = t\hat{u}_1$, with $t \neq 0$. Then, from the first inequality in (11.2) and since $\zeta < \lambda_1$ on a set of positive measure and $\hat{u}_1(x) > 0$ for all $x \in \Omega$, we deduce $\|\nabla u\|_p^p < \lambda_1 \|u\|_p^p$, which contradicts (11.3). □

In our approach to problem (11.1), we use the variational method. This means that we associate to the problem a $C^1$-functional whose critical points coincide with the (weak) solutions of (11.1). Assume that the nonlinearity $f$ satisfies the following growth condition.

H($f$)$_0$ $f : \Omega \times \mathbb{R} \to \mathbb{R}$ is a Carathéodory function and there exist $c > 0$ and $r \in (p, p^*)$ such that

$$|f(x,s)| \leq c(1+|s|^{r-1}) \text{ for a.a. } x \in \Omega, \text{ all } s \in \mathbb{R}.$$

We denote $F(x,s) = \int_0^s f(x,t)\,dt$. Under hypothesis $\mathrm{H}(f)_0$, from Proposition 2.78 we know that the functional $\varphi : W_0^{1,p}(\Omega) \to \mathbb{R}$ given by

$$\varphi(u) = \frac{1}{p}\|\nabla u\|_p^p - \int_\Omega F(x,u(x))\,dx \text{ for all } u \in W_0^{1,p}(\Omega)$$

is well defined and of class $C^1$, and for all $u,v \in W_0^{1,p}(\Omega)$ we have

$$\langle \varphi'(u),v \rangle = \int_\Omega |\nabla u|^{p-2}(\nabla u, \nabla v)_{\mathbb{R}^N}\,dx - \int_\Omega f(x,u)v(x)\,dx.$$

Thus, the critical points of $\varphi$ are exactly the solutions of problem (11.1). Solutions obtained as local minimizers of $\varphi$ are of particular interest because their existence can be shown via the direct method (minimization of $\varphi$ or of a suitable truncated functional) and because one can rely on them for showing the existence of further critical points (using the mountain pass theorem or degree or Morse theory, for instance). However, it is usually easier to show that $u$ is a local minimizer of $\varphi$ with respect to the topology of $C_0^1(\overline{\Omega})$ than with respect to the topology of $W_0^{1,p}(\Omega)$. In this situation, the next result is helpful.

**Proposition 11.4.** *Assume that* $\mathrm{H}(f)_0$ *holds, and let* $\varphi$ *be as above. If* $u_0 \in W_0^{1,p}(\Omega)$ *is a local* $C_0^1(\overline{\Omega})$*-minimizer of* $\varphi$, *i.e., there exists* $\rho_0 > 0$ *such that*

$$\varphi(u_0) \leq \varphi(u_0+h) \text{ for all } h \in C_0^1(\overline{\Omega}) \text{ with } \|h\|_{C^1(\overline{\Omega})} \leq \rho_0,$$

*then* $u_0$ *is also a local* $W_0^{1,p}(\Omega)$*-minimizer of* $\varphi$, *i.e., there exists* $\rho_1 > 0$ *such that*

$$\varphi(u_0) \leq \varphi(u_0+h) \text{ for all } h \in W_0^{1,p}(\Omega) \text{ with } \|\nabla h\|_p \leq \rho_1.$$

This result is based on a classic result of Brezis and Nirenberg [54]. Its proof is given in wider generality in Sect. 12.2.

Finally, we describe truncation techniques and a *lower and upper solutions principle*, which are useful in the study of problem (11.1). This method is independent of the variational method in the sense that it does not require that $f$ satisfy the growth condition in $\mathrm{H}(f)_0$ so that the functional associated to the problem is not necessarily well defined. This functional is actually replaced by a suitable truncated functional. The basic definition is as follows.

**Definition 11.5.** Let $u \in W^{1,p}(\Omega)$. We say that $u$ is an *upper* (resp. *lower*) *solution* of problem (11.1) if $u|_{\partial\Omega} \geq 0$ (resp. $u|_{\partial\Omega} \leq 0$), $f(\cdot,u(\cdot)) \in L^r(\Omega)$ for some $r \in (1,p^*)$, and

$$\int_\Omega |\nabla u|^{p-2}(\nabla u, \nabla v)_{\mathbb{R}^N}\,dx - \int_\Omega f(x,u(x))v(x)\,dx \text{ is } \geq 0 \text{ (resp. } \leq 0)$$

for all $v \in W_0^{1,p}(\Omega)$ with $v \geq 0$ a.e. in $\Omega$.

*Remark 11.6.*

(a) In Definition 11.5, the notation $u|_{\partial\Omega}$ stands for the trace $\gamma(u)$ (Theorem 1.33). If $u \in C(\overline{\Omega})$, then this actually coincides with the restriction of $u$ to $\partial\Omega$.

(b) Evidently, $u$ is a solution of (11.1) if and only if it is both a lower and an upper solution of (11.1).

Let $\underline{u} : \Omega \to \mathbb{R} \cup \{-\infty\}$ and $\overline{u} : \Omega \to \mathbb{R} \cup \{+\infty\}$ be measurable functions such that $\underline{u}(x) \leq \overline{u}(x)$ for a.a. $x \in \Omega$. Let $[\underline{u}, \overline{u}]$ be the order interval defined by

$$[\underline{u}, \overline{u}] := \{u \in W_0^{1,p}(\Omega) : \underline{u}(x) \leq u(x) \leq \overline{u}(x) \text{ for a.a. } x \in \Omega\}.$$

We define a truncated function $f_{[\underline{u},\overline{u}]} : \Omega \times \mathbb{R} \to \mathbb{R}$ by letting, for a.a. $x \in \Omega$, all $s \in \mathbb{R}$,

$$f_{[\underline{u},\overline{u}]}(x,s) = \begin{cases} f(x,\underline{u}(x)) & \text{if } s \leq \underline{u}(x), \\ f(x,s) & \text{if } \underline{u}(x) < s < \overline{u}(x), \\ f(x,\overline{u}(x)) & \text{if } s \geq \overline{u}(x). \end{cases} \tag{11.4}$$

Clearly, $f_{[\underline{u},\overline{u}]}$ is a Carathéodory function. We denote by $F_{[\underline{u},\overline{u}]}(x,s) = \int_0^s f_{[\underline{u},\overline{u}]}(x,t)\,dt$ its primitive. Finally, when $f_{[\underline{u},\overline{u}]}$ satisfies $H(f)_0$, we define the truncated functional $\varphi_{[\underline{u},\overline{u}]} \in C^1(W_0^{1,p}(\Omega),\mathbb{R})$ by

$$\varphi_{[\underline{u},\overline{u}]}(u) = \frac{1}{p}\|\nabla u\|_p^p - \int_\Omega F_{[\underline{u},\overline{u}]}(x,u(x))\,dx \text{ for all } u \in W_0^{1,p}(\Omega). \tag{11.5}$$

*Remark 11.7.*

(a) If $f$ satisfies hypothesis $H(f)_0$, then so does $f_{[\underline{u},\overline{u}]}$.

(b) In the case where $\underline{u}, \overline{u} \in L^\infty(\Omega)$, a sufficient condition for $f_{[\underline{u},\overline{u}]}$ to satisfy hypothesis $H(f)_0$ is that $f$ is bounded on bounded subsets of $\Omega \times \mathbb{R}$, i.e., for every $\rho > 0$ there is $M_\rho > 0$ such that $f(x,s) \leq M_\rho$ for a.a. $x \in \Omega$, all $s \in [-\rho,\rho]$.

(c) If $f_{[\underline{u},\overline{u}]}$ satisfies $H(f)_0$, then we have in particular

$$|f_{[\underline{u},\overline{u}]}(x,s)| \leq c(1 + (|\underline{u}(x)| + |\overline{u}(x)|)^{r-1}) \text{ for a.a. } x \in \Omega, \text{ all } s \in \mathbb{R},$$

with $c > 0$ and $r \in (p,p^*)$.

The lower and upper solutions method is based on the following principle.

**Proposition 11.8.**

(a) *Let $\underline{u}$ be either a lower solution of (11.1) or $-\infty$, and let $\bar{u}$ be either an upper solution of (11.1) or $+\infty$. Assume that $\underline{u} \leq \bar{u}$ a.e. in $\Omega$ and that $f_{[\underline{u},\bar{u}]}$ satisfies hypothesis $\mathrm{H}(f)_0$. If $u \in W_0^{1,p}(\Omega)$ is a critical point of $\varphi_{[\underline{u},\bar{u}]}$, then we have $u \in [\underline{u},\bar{u}] \cap C_0^1(\overline{\Omega})$ and $u$ is a solution of (11.1).*

(b) *Let $\underline{u}, \bar{u} \in W^{1,p}(\Omega)$ be respectively a lower and an upper solution of problem (11.1) such that $\underline{u} \leq \bar{u}$ a.e. in $\Omega$ and assume that $f_{[\underline{u},\bar{u}]}$ satisfies $\mathrm{H}(f)_0$. Then, the functional $\varphi_{[\underline{u},\bar{u}]}$ is coercive, sequentially weakly l.s.c., and it satisfies the (PS)-condition. In particular, there is a $u \in [\underline{u},\bar{u}] \cap C_0^1(\overline{\Omega})$ solution of (11.1) that is obtained as a global minimizer of $\varphi_{[\underline{u},\bar{u}]}$.*

*Proof.* (a) The fact that $u$ is a critical point of $\varphi_{[\underline{u},\bar{u}]}$ is equivalent to saying that $u$ solves the problem

$$\begin{cases} -\Delta_p u = f_{[\underline{u},\bar{u}]}(x, u(x)) & \text{in } \Omega, \\ u = 0 & \text{on } \partial\Omega. \end{cases} \tag{11.6}$$

Then, regularity theory (Corollary 8.13) implies that $u \in C_0^1(\overline{\Omega})$.

Let us check that $u \in [\underline{u}, \bar{u}]$. In the case where $\underline{u} \equiv -\infty$, we clearly have $u \geq \underline{u}$ in $\Omega$. Thus, let us assume that $\underline{u}$ is a lower solution of (11.1). This property implies in particular that $u - \underline{u} \geq 0$ on $\partial\Omega$, hence $(u - \underline{u})^- \in W_0^{1,p}(\Omega)$ (Remark 1.35). Acting on (11.6) with the test function $(u - \underline{u})^-$, we obtain on the one hand

$$\int_{\{u<\underline{u}\}} (|\nabla u|^{p-2}\nabla u, \nabla(u-\underline{u}))_{\mathbb{R}^N}\, dx = \int_{\{u<\underline{u}\}} f(x,\underline{u}(x))(u(x)-\underline{u}(x))\, dx,$$

whereas using the fact that $\underline{u}$ is a lower solution, we get on the other hand

$$\int_{\{u<\underline{u}\}} (|\nabla \underline{u}|^{p-2}\nabla \underline{u}, \nabla(u-\underline{u}))_{\mathbb{R}^N}\, dx \geq \int_{\{u<\underline{u}\}} f(x,\underline{u}(x))(u(x)-\underline{u}(x))\, dx.$$

Comparing both relations, we have

$$\int_{\{u<\underline{u}\}} (|\nabla u|^{p-2}\nabla u - |\nabla \underline{u}|^{p-2}\nabla \underline{u}, \nabla u - \nabla \underline{u})_{\mathbb{R}^N}\, dx \leq 0.$$

Now, invoking Remark 8.18, we conclude that the set $\{x \in \Omega : u(x) < \underline{u}(x)\}$ has zero measure. Thus $\underline{u} \leq u$ a.e. in $\Omega$. Similarly, we can see that $u \leq \bar{u}$ a.e. in $\Omega$. Therefore, $u \in [\underline{u}, \bar{u}]$.

Finally, we note that this relation yields $f_{[\underline{u},\bar{u}]}(x, u(x)) = f(x, u(x))$ for a.a. $x \in \Omega$. Thus, from (11.6) we obtain that $u$ is a solution of (11.1).

(b) Using the constants $c > 0$, $r \in (p, p^*)$ of Remark 11.7 (c) and the continuity of the embedding $W_0^{1,p}(\Omega) \hookrightarrow L^r(\Omega)$, we have

$$\varphi_{[\underline{u},\overline{u}]}(u) \geq \frac{1}{p} \|\nabla u\|_p^p - \int_\Omega c(1 + (|\underline{u}(x) + \overline{u}(x)|)^{r-1})|u(x)| \, dx$$

$$\geq \frac{1}{p} \|\nabla u\|_p^p - c \|1 + |\underline{u}| + |\overline{u}| \|_r^{r-1} \|u\|_r$$

$$\geq \frac{1}{p} \|\nabla u\|_p^p - c_1 \|\nabla u\|_p \quad \text{for all } u \in W_0^{1,p}(\Omega)$$

for some constant $c_1 > 0$. This relation clearly implies that $\varphi_{[\underline{u},\overline{u}]}$ is coercive.

Next, the weak lower semicontinuity of the $L^p$-norm, together with the assumption that $f_{[\underline{u},\overline{u}]}$ satisfies $H(f)_0$, easily ensures that $\varphi_{[\underline{u},\overline{u}]}$ is sequentially weakly l.s.c.

Finally, as was already seen at the end of Step 2 of the proof of Theorem 10.9, the fact that $\varphi_{[\underline{u},\overline{u}]}$ is coercive together with the property of $u \mapsto -\Delta_p u$ of being an $(S)_+$-map (Proposition 2.72) allows us to obtain that $\varphi_{[\underline{u},\overline{u}]}$ satisfies the (PS)-condition. □

## Multiplicity Results

We start by treating an auxiliary problem that will play a role in our first multiplicity result.

**Proposition 11.9.** *Let $q \in (p, p^*)$, and let $c_q > 0$ be the norm of the bounded linear embedding $(W_0^{1,p}(\Omega), \|\nabla \cdot \|_p) \hookrightarrow L^q(\Omega)$. The problem*

$$\begin{cases} -\Delta_p u = |u|^{q-2}u & \text{in } \Omega, \\ u = 0 & \text{on } \partial\Omega \end{cases} \tag{11.7}$$

*admits a solution $\underline{u}_q \in \text{int}(C_0^1(\overline{\Omega})_+)$ realizing the equality $\|\underline{u}_q\|_q = c_q \|\nabla \underline{u}_q\|_p$. Moreover, we have $\|\nabla \underline{u}_q\|_p^p = \|\underline{u}_q\|_q^q = (c_q)^{\frac{pq}{p-q}}$.*

*Proof.* By the compactness of the embedding $W_0^{1,p}(\Omega) \hookrightarrow L^q(\Omega)$, the set

$$C = \{v \in W_0^{1,p}(\Omega) : \|v\|_q^q = (c_q)^{\frac{pq}{p-q}}\}$$

is sequentially weakly closed in $W_0^{1,p}(\Omega)$; moreover, the definition of $c_q$ yields

$$\|\nabla v\|_p^p \geq \frac{1}{(c_q)^p} \|v\|_q^p = (c_q)^{\frac{pq}{p-q}} \quad \text{for all } v \in C. \tag{11.8}$$

By the definition of $c_q$, we can also find a sequence $\{\tilde{u}_n\}_{n \geq 1} \subset W_0^{1,p}(\Omega)$ such that

$$\lim_{n \to \infty} \frac{\|\nabla \tilde{u}_n\|_p}{\|\tilde{u}_n\|_q} = \frac{1}{c_q}.$$

Up to replacing $\tilde{u}_n$ by $|\tilde{u}_n|$, we may assume that $\tilde{u}_n \geq 0$ a.e. in $\Omega$ for all $n \geq 1$ (Proposition 1.29). Setting $u_n = \frac{(c_q)^{\frac{p}{p-q}}}{\|\tilde{u}_n\|_q} \tilde{u}_n$, we have $u_n \in C$ for all $n \geq 1$ and

$$\lim_{n \to \infty} \|\nabla u_n\|_p^p = (c_q)^{\frac{pq}{p-q}}. \tag{11.9}$$

Since $\{u_n\}_{n \geq 1}$ is bounded in $W_0^{1,p}(\Omega)$, up to considering a subsequence, we find $u \in C$ with $u \geq 0$ a.e. in $\Omega$ such that $u_n \xrightarrow{w} u$ in $W_0^{1,p}(\Omega)$ as $n \to \infty$. Then, using (11.8) and (11.9) and the weak lower semicontinuity of the $L^p$-norm, we have

$$(c_q)^{\frac{pq}{p-q}} \leq \|\nabla u\|_p^p \leq \lim_{n \to \infty} \|\nabla u_n\|_p^p = (c_q)^{\frac{pq}{p-q}}.$$

Thus, since $u \in C$,

$$\|\nabla u\|_p^p = \|u\|_q^q = (c_q)^{\frac{pq}{p-q}} \tag{11.10}$$

and [by (11.8)]

$$\|\nabla u\|_p^p = \inf\{\|\nabla v\|_p^p : v \in C\}.$$

By the Lagrange multiplier rule, there exists $\mu \in \mathbb{R}$ such that

$$-\Delta_p u = \mu u^{q-1} \text{ in } W^{-1,p'}(\Omega).$$

In particular, $\|\nabla u\|_p^p = \mu \|u\|_q^q$. Comparing with (11.10), we obtain $\mu = 1$. Therefore, $u$ is a nontrivial solution of problem (11.7). By Corollary 8.13, we have $u \in C_0^1(\overline{\Omega})$, and, since $u \geq 0$ in $\Omega$, $u \neq 0$, the strong maximum principle (Theorem 8.27) implies that $u \in \text{int}(C_0^1(\overline{\Omega})_+)$. Therefore, $\underline{u}_q := u$ satisfies the required properties. $\qquad\square$

In what follows, we fix $\underline{u}_q \in \text{int}(C_0^1(\overline{\Omega})_+)$, satisfying the conditions of Proposition 11.9, and we let $M_q = \|\underline{u}_q\|_\infty$.

*Remark 11.10.* We know from Proposition 11.9 that $\|\nabla \underline{u}_q\|_p = c_q^{\frac{q}{p-q}}$. Thus, by Theorem 8.4, there is $\tilde{M} > 0$ depending only on $p$, $q$, $N$, and $\Omega$ such that $M_q \leq \tilde{M}$, and $\tilde{M}$ can be estimated through the procedure described in the proof of Theorem 8.4.

Here and later in this section, by $\lambda_1 > 0$ we denote the first eigenvalue of the negative Dirichlet $p$-Laplacian (Sect. 9.2). For our first multiplicity result, we impose the following conditions on the nonlinearity $f$ in problem (11.1):

$H(f)_1$  (i) $f : \Omega \times \mathbb{R} \to \mathbb{R}$ is a Carathéodory function such that $f(x,0) = 0$ a.e. in $\Omega$, and there are $c > 0$ and $r \in (p, p^*)$ such that

$$0 \le f(x,s) \le c(1 + s^{r-1}) \quad \text{for a.a. } x \in \Omega, \text{ all } s \ge 0;$$

(ii) There exists $\vartheta \in L^\infty(\Omega)_+$ satisfying $\vartheta(x) \le \lambda_1$ a.e. in $\Omega$ with strict inequality on a set of positive measure such that

$$\limsup_{s \to +\infty} \frac{pF(x,s)}{s^p} \le \vartheta(x) \quad \text{uniformly for a.a } x \in \Omega;$$

(iii) There exists $q \in (p, p^*)$ such that

$$f(x,s) > \min\{s, M_q\}^{q-1} \quad \text{for a.a. } x \in \Omega, \text{ all } s > 0;$$

(iv) There exists $\eta \in L^\infty(\Omega)_+$ satisfying $\eta(x) \le \lambda_1$ a.e. in $\Omega$ with strict inequality on a set of positive measure such that

$$\limsup_{s \downarrow 0} \frac{pF(x,s)}{s^p} \le \eta(x) \quad \text{uniformly for a.a. } x \in \Omega.$$

*Remark 11.11.* Hypotheses $H(f)_1$ (ii) and (iv) are nonuniform, nonresonance conditions at $+\infty$ and at $0^+$. Hypothesis $H(f)_1$ (ii) implies that the truncated functional $\varphi_{[0,+\infty]}$ associated to the problem is coercive.

*Example 11.12.* Let numbers $q$, $a$, and $b$, with $p < q < p^*$, $b > a > M_q$, and let functions $\vartheta, c_0 \in L^\infty(\Omega)_+$ be such that $\vartheta(x) \le \lambda_1$ a.e. in $\Omega$ with strict inequality on a set of positive measure, and $1 < c_0(x) \le \vartheta(x)a^{p-q}$ a.e. in $\Omega$. The map $f : \Omega \times \mathbb{R} \to \mathbb{R}$ defined by

$$f(x,s) = \begin{cases} 0 & \text{if } s \le 0, \\ c_0(x)s^{q-1} & \text{if } 0 < s \le a, \\ \frac{1}{b-a}\left(\vartheta(x)(s-a)s^{p-1} + c_0(x)(b-s)s^{q-1}\right) & \text{if } a < s < b, \\ \vartheta(x)s^{p-1} & \text{if } s \ge b \end{cases}$$

satisfies $H(f)_1$.

**Theorem 11.13.** *Assume that $H(f)_1$ holds. Then problem (11.1) admits at least two nontrivial positive solutions $u_1, u_2 \in \text{int}(C_0^1(\overline{\Omega})_+)$.*

*Proof.* Let $\psi = \varphi_{[0,+\infty]}$, i.e.,

$$\psi(u) = \frac{1}{p}\|\nabla u\|_p^p - \int_\Omega F(x,u^+)\,dx \quad \text{for all } u \in W_0^{1,p}(\Omega),$$

where $F(x,s) = \int_0^s f(x,t)\,dt$. The proof is divided into five steps.

*Step 1*: Every nontrivial critical point $u$ of $\psi$ is a solution of (11.1) and belongs to $\mathrm{int}(C_0^1(\overline{\Omega})_+)$.

It follows from Proposition 11.8(a) that $u$ is a solution of (11.1) belonging to $C_0^1(\overline{\Omega})_+$. Moreover, from H$(f)_1$ (i) we have

$$-\Delta_p u = f(x,u(x)) \geq 0 \quad \text{for a.a. } x \in \Omega;$$

hence the strong maximum principle (Theorem 8.27) yields $u \in \mathrm{int}(C_0^1(\overline{\Omega})_+)$.

*Step 2*: $\psi$ is coercive.

Given $\varepsilon > 0$, hypotheses H$(f)_1$ (i), (ii) imply that we can find $c_\varepsilon > 0$ such that

$$F(x,s) \leq \frac{\vartheta(x)+\varepsilon}{p}s^p + c_\varepsilon \quad \text{for a.a. } x \in \Omega, \text{ all } s \geq 0.$$

Using Lemma 11.3 (for $\zeta = \vartheta$) and the variational characterization of $\lambda_1$ in Proposition 9.47 (for $\xi = 1$), we have

$$\psi(u) \geq \frac{1}{p}\|\nabla u\|_p^p - \frac{\varepsilon}{p}\|u\|_p^p - \frac{1}{p}\int_\Omega \vartheta(x)|u|^p\,dx - c_\varepsilon|\Omega|_N$$

$$\geq \frac{c_1}{p}\|\nabla u\|_p^p - \frac{\varepsilon}{\lambda_1 p}\|\nabla u\|_p^p - c_\varepsilon|\Omega|_N \quad \text{for all } u \in W_0^{1,p}(\Omega)$$

for some $c_1 > 0$. Choosing $\varepsilon > 0$ such that $\varepsilon < \lambda_1 c_1$, we obtain that $\psi(u) \to +\infty$ as $\|\nabla u\|_p \to +\infty$. Thus, $\psi$ is coercive.

*Step 3*: 0 is a local minimizer of $\psi$.

Let $\varepsilon > 0$. In view of hypothesis H$(f)_1$ (iv), there exists $\delta = \delta(\varepsilon) > 0$ such that

$$F(x,s) \leq \frac{\eta(x)+\varepsilon}{p}s^p \quad \text{for a.a. } x \in \Omega, \text{ all } s \in [0,\delta].$$

Thus, applying Lemma 11.3 (for $\zeta = \eta$) and using Proposition 9.47, we can find a constant $c_1 > 0$ such that, for every $u \in C_0^1(\overline{\Omega})$ such that $\|u\|_\infty \leq \delta$, we have

$$\psi(u) \geq \frac{1}{p}\|\nabla u\|_p^p - \frac{\varepsilon}{p}\|u\|_p^p - \frac{1}{p}\int_\Omega \eta(x)|u|^p\,dx \geq \left(\frac{c_1}{p} - \frac{\varepsilon}{\lambda_1 p}\right)\|\nabla u\|_p^p.$$

Choosing $\varepsilon > 0$ such that $\varepsilon < \lambda_1 c_1$, we obtain that 0 is a local $C_0^1(\overline{\Omega})$-minimizer of $\psi$. By Proposition 11.4, it follows that 0 is a local minimizer of $\psi$ for the topology of $W_0^{1,p}(\Omega)$. This establishes Step 3.

*Step 4:* $\psi$ admits a local minimizer $u_1 \in W_0^{1,p}(\Omega) \setminus \{0\}$.

Using $H(f)_1$ (iii), we see that the function $\underline{u} := \underline{u}_q \in \text{int}(C_0^1(\overline{\Omega})_+)$ from Proposition 11.9 satisfies

$$\langle -\Delta_p \underline{u}, v \rangle = \int_\Omega \underline{u}(x)^{q-1} v(x)\, dx < \int_\Omega f(x, \underline{u}(x)) v(x)\, dx$$

for all $v \in W_0^{1,p}(\Omega)$, with $v \geq 0$ a.e. in $\Omega$. Hence, $\underline{u}$ is a lower solution of problem (11.1). Clearly, the truncated Carathéodory function $f_{[\underline{u},+\infty]}$ satisfies $H(f)_1$ (ii) and the growth condition in $H(f)_1$ (i). Thus, arguing as in Step 2, we can see that the corresponding truncated functional $\varphi_{[\underline{u},+\infty]}$ is coercive, bounded below, and sequentially weakly l.s.c. Hence there is a $u_1 \in W_0^{1,p}(\Omega)$ global minimizer of $\varphi_{[\underline{u},+\infty]}$. Proposition 11.8 (a) implies that $u_1 \in C_0^1(\overline{\Omega})$, $0 < \underline{u}(x) \leq u_1(x)$ for all $x \in \Omega$ [so $u_1 \in \text{int}(C_0^1(\overline{\Omega})_+)$], and $u_1$ is a solution of (11.1). In particular, the following equalities hold in $W^{-1,p'}(\Omega)$:

$$-\Delta_p u_1 = f(x, u_1(x)) \quad \text{and} \quad -\Delta_p \underline{u} = \underline{u}(x)^{q-1}. \qquad (11.11)$$

On the other hand, in view of $H(f)_1$ (iii), for a.a. $x \in \Omega$ we have

$$f(x, u_1(x)) > \min\{u_1(x), M_q\}^{q-1} \geq \underline{u}(x)^{q-1}. \qquad (11.12)$$

On the basis of (11.11) and (11.12), we can apply Proposition 8.29, which yields

$$u_1 - \underline{u} \in \text{int}(C_0^1(\overline{\Omega})_+).$$

Therefore, $V := \{u \in C_0^1(\overline{\Omega}) : u \geq \underline{u} \text{ in } \Omega\}$ is a neighborhood of $u_1$ in $C_0^1(\overline{\Omega})$. Clearly,

$$\psi(u) = \varphi_{[\underline{u},+\infty]}(u) - \int_\Omega \left(F(x, \underline{u}(x)) - \underline{u}(x) f(x, \underline{u}(x))\right) dx \quad \text{for all } u \in V.$$

Thus, $u_1$ is a local $C_0^1(\overline{\Omega})$-minimizer of $\psi$. Proposition 11.4 implies that $u_1$ is a local minimizer of $\psi$ for the topology of $W_0^{1,p}(\Omega)$. This concludes Step 4.

*Step 5:* $\psi$ admits a third critical point $u_2 \in W_0^{1,p}(\Omega) \setminus \{0, u_1\}$.

The differential $\psi' : W_0^{1,p}(\Omega) \to W^{-1,p'}(\Omega)$, which is given by

$$\langle \psi'(u), v \rangle = \int_\Omega |\nabla u|^{p-2} (\nabla u, \nabla v)_{\mathbb{R}^N}\, dx - \int_\Omega f(x, u^+) v\, dx \quad \text{for } u, v \in W_0^{1,p}(\Omega),$$

is the sum of two terms, respectively, a continuous $(S)_+$-map (Proposition 2.72) and a completely continuous map [by $H(f)_1$ (i)]. By Proposition 2.70 (d), we deduce that $\psi'$ is an $(S)_+$-map. Therefore, we can apply the degree theory for demicontinuous $(S)_+$-maps presented in Sect. 4.3 to the study of the critical points of $\psi$.

Arguing by contradiction, assume that $0, u_1$ are the only critical points of $\psi$. First, because $\psi$ is coercive (Step 2), from Corollary 4.46 we know that there is $R_0 > 0$ such that for all $R \geq R_0$ we have

$$d_{(S)_+}(\psi', B_R(0), 0) = 1, \tag{11.13}$$

where $B_R(0)$ stands for the open ball in $W_0^{1,p}(\Omega)$. Choosing $R \geq R_0$ large enough, we may assume that $0, u_0 \in B_R(0)$. Second, since $0, u_0$ are local minimizers of $\psi$ (Steps 3 and 4), by Corollary 4.49, we can find $\rho_0 > 0$ such that for all $\rho \in (0, \rho_0]$ we have

$$d_{(S)_+}(\psi', B_\rho(0), 0) = d_{(S)_+}(\psi', B_\rho(u_0), 0) = 1.$$

We choose $\rho \in (0, \rho_0]$ sufficiently small so that both $B_\rho(0)$ and $B_\rho(u_0)$ are contained in $B_R(0)$ and are pairwise disjoint. Combining the excision and domain additivity properties of the degree [Theorem 4.42(b), (d)], we get

$$d_{(S)_+}(\psi', B_R(0), 0) = d_{(S)_+}(\psi', B_\rho(0), 0) + d_{(S)_+}(\psi', B_\rho(u_0), 0) = 2,$$

a contradiction of (11.13). In this way, we obtain that $\psi$ admits at least one more critical point $u_2 \in W_0^{1,p}(\Omega)$. This completes Step 5.

The theorem is now obtained by combining Steps 1, 4, and 5. $\qquad\square$

Next, we focus on the noncoercive case. Now the hypotheses on $f$ in problem (11.1) are as follows:

$H(f)_2$ (i) $f : \Omega \times \mathbb{R} \to \mathbb{R}$ is a Carathéodory function with $f(x, 0) = 0$ for a.a. $x \in \Omega$, and there exists $c > 0$ such that

$$|f(x, s)| \leq c(1 + s^{p-1}) \quad \text{for a.a. } x \in \Omega, \text{ all } s \geq 0;$$

(ii) There exist constants $c_+, M_+ > 0$ such that

$$\limsup_{s \uparrow c_+} \frac{f(x, s)}{(c_+ - s)^{p-1}} \leq M_+ \quad \text{uniformly for a.a. } x \in \Omega;$$

(iii) There exist $\eta_1, \eta_2 \in L^\infty(\Omega)$, with $\eta_1(x) \geq \lambda_1$ a.e. in $\Omega$, $\eta_1 \neq \lambda_1$, such that

$$\eta_1(x) \leq \liminf_{s \to +\infty} \frac{f(x, s)}{s^{p-1}} \leq \limsup_{s \to +\infty} \frac{f(x, s)}{s^{p-1}} \leq \eta_2(x),$$

$$\eta_1(x) \leq \liminf_{s \downarrow 0} \frac{f(x, s)}{s^{p-1}} \leq \limsup_{s \downarrow 0} \frac{f(x, s)}{s^{p-1}} \leq \eta_2(x)$$

uniformly for a.a. $x \in \Omega$.

*Example 11.14.* Let $\eta \in L^\infty(\Omega)$ such that $\eta(x) \geq \lambda_1$ a.e. in $\Omega$ with strict inequality on a set of positive measure. Let $f : \Omega \times \mathbb{R} \to \mathbb{R}$ be defined by

$$f(x,s) = \begin{cases} 0 & \text{if } s \leq 0, \\ \eta(x)(1-s)^{p-1}(1 - e^{-s^{p-1}}) & \text{if } 0 < s < 1, \\ \eta(x)(s^{p-1} - 1) & \text{if } s \geq 1. \end{cases}$$

It is straightforward to check that hypotheses $H(f)_2$ are satisfied with $c_+ = 1$ and $\eta_1 = \eta_2 = \eta$.

**Theorem 11.15.** *Assume that* $H(f)_2$ *holds. Then problem (11.1) admits at least two nontrivial positive solutions* $u_1, u_2 \in \mathrm{int}(C_0^1(\overline{\Omega})_+)$.

*Proof.* As in the proof of Theorem 11.13, we consider the functional $\psi = \varphi_{[0,+\infty]}$. It follows from Proposition 11.8(a) that every critical point $u$ of $\psi$ is a solution of (11.1) and belongs to $C_0^1(\overline{\Omega})_+$, and from $H(f)_2$ (i) and (iii) we find a constant $\tilde{c} > 0$ such that

$$-\Delta_p u = f(x, u(x)) \geq -\tilde{c} u(x)^{p-1} \quad \text{for a.a. } x \in \Omega;$$

thus, by Theorem 8.27, every nontrivial solution $u$ of (11.1) belongs to $\mathrm{int}(C_0^1(\overline{\Omega})_+)$. Thus, it suffices to show that the functional $\psi$ admits at least two nontrivial critical points. The proof is divided into several steps.

*Step 1:* There is $u_1 \in W_0^{1,p}(\Omega)$, which is a local minimizer of $\psi$.

It follows from $H(f)_2$ (ii) that $f(x, c_+) \leq 0$ for a.a. $x \in \Omega$, hence $\overline{u} := c_+$ is an upper solution of (11.1). By Proposition 11.8(b), the functional $\varphi_{[0,c_+]} : W_0^{1,p}(\Omega) \to \mathbb{R}$ admits a global minimizer $u_1 \in C_0^1(\overline{\Omega})$, which satisfies $0 \leq u_1(x) \leq c_+$ for all $x \in \overline{\Omega}$.

We claim that

$$\max_{x \in \overline{\Omega}} u_1(x) < c_+. \tag{11.14}$$

Arguing by contradiction, assume that there is $x_0 \in \overline{\Omega}$ such that $u_1(x_0) = c_+$. Since $(c_+ - u_1)|_{\partial\Omega} = c_+ > 0$, we have $x_0 \in \Omega$. By $H(f)_2$ (ii), there are constants $M_1 > 0$ and $\delta \in (0, c_+)$ such that

$$f_{[0,c_+]}(x,s) = f(x,s) \leq M_1(c_+ - s)^{p-1} \quad \text{for a.a. } x \in \Omega, \text{ all } s \in (\delta, c_+]. \tag{11.15}$$

Let $\Omega_0$ be the connected component of the set $\{x \in \Omega : u_1(x) > \delta\}$ containing $x_0$. Thus, $\delta < u_1(x) \leq c_+$ for all $x \in \Omega_0$ and $u_1|_{\partial\Omega_0} \equiv \delta$, hence $u_1|_{\Omega_0} \not\equiv c_+$. For every $v \in C_c^\infty(\Omega_0)$ such that $v \geq 0$ in $\Omega_0$, using the fact that $u_1$ is in particular a critical point of $\varphi_{[0,c_+]}$, and invoking relation (11.15), we have

$$\int_{\Omega_0} |\nabla(c_+ - u_1)|^{p-2} (\nabla(c_+ - u_1), \nabla v)_{\mathbb{R}^N} \, dx$$

$$= -\int_{\Omega_0} |\nabla u_1|^{p-2} (\nabla u_1, \nabla v)_{\mathbb{R}^N} \, dx$$

$$= -\int_{\Omega_0} f_{[0,c_+]}(x, u_1(x)) v(x) \, dx$$

$$\geq -\int_{\Omega_0} M_1 (c_+ - u_1(x))^{p-1} v(x) \, dx. \tag{11.16}$$

On the basis of (11.16), we can apply Corollary 8.17, which yields $c_+ - u_1(x) > 0$ for all $x \in \Omega_0$, so $u_1(x_0) < c_+$, a contradiction. This establishes (11.14).

From (11.14) we obtain that $V := \{u \in C_0^1(\overline{\Omega}) : u(x) < c_+ \text{ for all } x \in \Omega\}$ is an open neighborhood of $u_1$ in $C_0^1(\overline{\Omega})$. Since we have $(\varphi_{[0,c_+]})|_V = \psi|_V$, we deduce that $u_1$ is a local $C_0^1(\overline{\Omega})$-minimizer of $\psi$. Applying Proposition 11.4, we conclude that $u_1$ is a local minimizer of $\psi$ for the topology of $W_0^{1,p}(\Omega)$. This ends Step 1.

The differential $\psi'$ is a continuous $(S)_+$-map, so the degree theory for $(S)_+$-maps can be applied to the study of the critical points of $\psi$. This is done in the following steps.

*Step 2:* There is $R_0 > 0$ such that, for $R \geq R_0$, we have $d_{(S)_+}(\psi', B_R(0), 0) = 0$.

Let $N_+, K_+ : W_0^{1,p}(\Omega) \to L^{p'}(\Omega) \subset W^{-1,p'}(\Omega)$ be defined by

$$N_+(u)(x) = f(x, u^+(x)) \quad \text{and} \quad K_+(u)(x) = \frac{\eta_1(x) + \eta_2(x)}{2} u^+(x)^{p-1}$$

for a.a. $x \in \Omega$, where $\eta_1, \eta_2 \in L^\infty(\Omega)_+$ are as in H($f$)$_2$ (iii). Then we consider the homotopy $h_1 : [0,1] \times W_0^{1,p}(\Omega) \to W^{-1,p'}(\Omega)$ given by

$$h_1(t, u) = -\Delta_p u - t N_+(u) - (1-t) K_+(u) \quad \text{for all } t \in [0,1], \text{ all } u \in W_0^{1,p}(\Omega).$$

We know that $u \mapsto -\Delta_p u$ is an $(S)_+$-map (Proposition 2.72) and $N_+$ and $K_+$ are completely continuous, hence $h_1$ is a homotopy of class $(S)_+$ [Propositions 2.70(d) and 4.41].

*Claim 1:* There is $R_0 > 0$ such that

$$h_1(t, u) \neq 0 \quad \text{for all } t \in [0,1], \text{ all } u \in W_0^{1,p}(\Omega) \text{ with } \|\nabla u\|_p \geq R_0.$$

Arguing by contradiction, assume that there are sequences $\{t_n\}_{n \geq 1} \subset [0,1]$ and $\{u_n\}_{n \geq 1} \subset W_0^{1,p}(\Omega)$ satisfying

$$\lim_{n \to \infty} t_n = t \in [0,1], \quad \lim_{n \to \infty} \|\nabla u_n\|_p = +\infty,$$

and

$$-\Delta_p u_n = t_n N_+(u_n) + (1-t_n)K_+(u_n) \quad \text{for all } n \geq 1. \tag{11.17}$$

Acting on (11.17) with the test function $-u_n^-$, we easily check that $u_n \geq 0$ a.e. in $\Omega$ for all $n \geq 1$. Let $y_n = \frac{u_n}{\|\nabla u_n\|_p}$ for $n \geq 1$. Then $\|\nabla y_n\|_p = 1$, and so we may assume that

$$y_n \overset{w}{\to} y \text{ in } W_0^{1,p}(\Omega), \ y_n \to y \text{ in } L^p(\Omega), \ y_n(x) \to y(x) \text{ a.e. in } \Omega, \tag{11.18}$$

and $y_n(x) \leq k(x)$ for a.a. $x \in \Omega$, all $n \geq 1$, for some $k \in L^p(\Omega)$, with $k, y \geq 0$ a.e. in $\Omega$. From (11.17) we have

$$-\Delta_p y_n = t_n \frac{N_+(u_n)}{\|\nabla u_n\|_p^{p-1}} + (1-t_n)K_+(y_n) \quad \text{for all } n \geq 1. \tag{11.19}$$

Acting on (11.19) with $y_n - y$ and passing to the limit as $n \to \infty$ through (11.18), from the fact that $-\Delta_p$ is an $(S)_+$-map we obtain that

$$y_n \to y \text{ in } W_0^{1,p}(\Omega) \text{ as } n \to \infty, \tag{11.20}$$

and thus $\|\nabla y\|_p = 1$, so $y \neq 0$.

Hypothesis H$(f)_2$ (i) implies that $\left\{ \frac{N_+(u_n)}{\|\nabla u_n\|_p^{p-1}} \right\}_{n \geq 1}$ is bounded in $L^{p'}(\Omega)$. Hence we may assume that

$$\frac{N_+(u_n)}{\|\nabla u_n\|_p^{p-1}} \overset{w}{\to} h \text{ in } L^{p'}(\Omega) \text{ as } n \to \infty, \tag{11.21}$$

with $h \in L^{p'}(\Omega)$. For a while we fix $\varepsilon > 0$. Then, for every $n \geq 1$, let

$$C_{\varepsilon,n} = \left\{ x \in \Omega : u_n(x) > 0, \ \eta_1(x) - \varepsilon \leq \frac{f(x, u_n(x))}{u_n(x)^{p-1}} \leq \eta_2(x) + \varepsilon \right\}.$$

Since $u_n(x) \to +\infty$ a.e. in the set $\{x \in \Omega : y(x) > 0\}$, from hypothesis H$(f)_2$ (iii) we have

$$\chi_{C_{\varepsilon,n}}(x) \to 1 \text{ for a.a. } x \in \{y > 0\} \text{ as } n \to \infty,$$

where $\chi_{C_{\varepsilon,n}}$ denotes the characteristic function of the set $C_{\varepsilon,n}$. Thus,

$$\chi_{C_{\varepsilon,n}} \frac{N_+(u_n)}{\|\nabla u_n\|_p^{p-1}} \overset{w}{\to} h \text{ in } L^{p'}(\{y > 0\}) \text{ as } n \to \infty.$$

On the other hand, from the definition of $C_{\varepsilon,n}$ we have

$$\chi_{C_{\varepsilon,n}}(x)(\eta_1(x)-\varepsilon)y_n(x)^{p-1} \leq \chi_{C_{\varepsilon,n}}(x)\frac{N_+(u_n)(x)}{\|\nabla u_n\|_p^{p-1}} \leq \chi_{C_{\varepsilon,n}}(x)(\eta_2(x)+\varepsilon)y_n(x)^{p-1}$$

for a.a. $x \in \Omega$, all $n \geq 1$. Invoking Mazur's theorem (e.g., Brezis [52, p. 61]), we obtain

$$(\eta_1(x)-\varepsilon)y(x)^{p-1} \leq h(x) \leq (\eta_2(x)+\varepsilon)y(x)^{p-1} \quad \text{for a.a. } x \in \{y>0\}.$$

In addition, because $\varepsilon > 0$ is arbitrary, we get

$$\eta_1(x)y(x)^{p-1} \leq h(x) \leq \eta_2(x)y(x)^{p-1} \quad \text{for a.a. } x \in \{y>0\}.$$

Furthermore, by H$(f)_2$ (i), (iii), there are constants $c_1, c_2 \in \mathbb{R}$ such that

$$c_1 y_n(x)^{p-1} \leq \frac{N_+(u_n)(x)}{\|\nabla u_n\|_p^p} \leq c_2 y_n(x)^{p-1} \quad \text{for a.a. } x \in \Omega, \text{ all } n \geq 1,$$

whence $h = 0$ a.e. in $\{y=0\}$. We conclude that there is $\xi \in L^\infty(\Omega)$ such that $\eta_1 \leq \xi \leq \eta_2$ a.e. in $\Omega$ and

$$h(x) = \xi(x)y(x)^{p-1} \quad \text{for a.a. } x \in \Omega.$$

Now, passing to the limit as $n \to \infty$ in (11.19), on the basis of (11.21) and (11.20), we see that

$$-\Delta_p y = \hat{\xi}y^{p-1} \text{ in } W^{-1,p'}(\Omega), \tag{11.22}$$

where $\hat{\xi} = t\xi + (1-t)\frac{\eta_1+\eta_2}{2}$. Thus, $y$ is a solution of the problem

$$\begin{cases} -\Delta_p y = \hat{\xi}y^{p-1} \text{ in } \Omega, \\ y = 0 \quad\quad\quad\quad \text{ on } \partial\Omega. \end{cases} \tag{11.23}$$

In other words, $y$ is an eigenfunction of the negative Dirichlet $p$-Laplacian $-\Delta_p^D$ with respect to the weight $\hat{\xi}$ corresponding to the eigenvalue 1. We have $\hat{\xi}(x) \geq \eta_1(x) \geq \lambda_1$ for a.a. $x \in \Omega$, where the second inequality is strict on a set of positive measure [see H$(f)_2$ (iii)]. By the monotonicity of the first eigenvalue of $-\Delta_p^D$ with respect to the weight [Proposition 9.47(d)], we have

$$\hat{\lambda}_1(\hat{\xi}) \leq \hat{\lambda}_1(\eta_1) < \hat{\lambda}_1(\lambda_1) = 1,$$

where $\hat{\lambda}_1(\eta) > 0$ stands for the first eigenvalue of $-\Delta_p^D$ with respect to a weight $\eta$. Thus, $\hat{\lambda}_1(\hat{\xi}) \neq 1$. From Proposition 9.47(b) we infer that $y$ must be nodal. Since we noted that $y \geq 0$ a.e. in $\Omega$, this is a contradiction. Claim 1 is proved.

Because of Claim 1, we can invoke the homotopy invariance property of the degree map $d_{(S)_+}$ [Theorem 4.42(c)], which ensures that

$$d_{(S)_+}(-\Delta_p - N_+, B_R(0), 0) = d_{(S)_+}(-\Delta_p - K_+, B_R(0), 0) \quad \text{for all } R \geq R_0. \quad (11.24)$$

Fix $R \geq R_0$. Fix an element $w \in W_0^{1,p}(\Omega) \cap L^\infty(\Omega)_+$, $w \neq 0$. We consider the homotopy $h_2 : [0,1] \times W_0^{1,p}(\Omega) \to W^{-1,p'}(\Omega)$ of class $(S)_+$ defined by

$$h_2(t, u) = -\Delta_p u - K_+(u) - tw \quad \text{for all } (t, u) \in [0,1] \times W_0^{1,p}(\Omega).$$

By Claim 1, we already know that $h_2(0, u) \neq 0$ for all $u \in \partial B_R(0)$.

*Claim 2:* We have $h_2(t, u) \neq 0$ for all $t \in (0,1]$, all $u \in W_0^{1,p}(\Omega)$.

Arguing by contradiction, suppose that we can find $t \in (0,1]$ and $u \in W_0^{1,p}(\Omega)$ such that

$$-\Delta_p u = K_+(u) + tw \quad \text{in } W^{-1,p'}(\Omega).$$

Acting with the test function $-u^-$, we easily infer that $u \geq 0$ a.e. in $\Omega$. Moreover, Corollary 8.13 implies that $u \in C_0^1(\overline{\Omega})$. Thus $u$ is a nontrivial solution of the problem

$$\begin{cases} -\Delta_p u = \lambda_1 u^{p-1} + \tilde{h} & \text{in } \Omega, \\ u = 0 & \text{on } \partial\Omega, \end{cases}$$

with $\tilde{h} = (\frac{\eta_1 + \eta_2}{2} - \lambda_1) u^{p-1} + tw \in L^\infty(\Omega)_+ \setminus \{0\}$, which contradicts Proposition 9.64. We obtain Claim 2.

Using Claim 2 and Theorem 4.42(c), (e), we get

$$d_{(S)_+}(-\Delta_p - K_+, B_R(0), 0) = d_{(S)_+}(-\Delta_p - K_+ - w, B_R(0), 0) = 0. \quad (11.25)$$

Comparing (11.24) and (11.25), we infer that $d_{(S)_+}(\psi', B_R(0), 0) = 0$. This concludes Step 2.

*Step 3:* There is $\rho_0 > 0$ such that, for $\rho \in (0, \rho_0]$, we have $d_{(S)_+}(\psi', B_\rho(0), 0) = 0$.

Consider the maps $N_+, K_+$ and the homotopy $h_1$ of class $(S)_+$ given in Step 2.

*Claim 3:* There exists $\rho_0 > 0$ such that

$$h(t, u) \neq 0 \quad \text{for all } t \in [0,1], \text{ all } u \in W_0^{1,p}(\Omega) \text{ with } 0 < \|\nabla u\|_p \leq \rho_0.$$

Arguing by contradiction, suppose we can find sequences $\{t_n\}_{n \geq 1} \subset [0,1]$ and $\{u_n\}_{n \geq 1} \subset W_0^{1,p}(\Omega) \setminus \{0\}$ such that $t_n \to t \in [0,1]$, $\|\nabla u_n\|_p \to 0$, and

$$-\Delta_p u_n = t_n N_+(u_n) + (1 - t_n) K_+(u_n) \quad \text{for all } n \geq 1. \tag{11.26}$$

Acting with the test function $-u_n^-$, we note that $u_n \geq 0$ a.e. in $\Omega$. Let $y_n = \frac{u_n}{\|\nabla u_n\|_p}$. We may assume that

$$y_n \xrightarrow{w} y \text{ in } W_0^{1,p}(\Omega) \text{ and } y_n \to y \text{ in } L^p(\Omega) \text{ as } n \to \infty.$$

As in Step 2, we can see that $y_n \to y$ in $W_0^{1,p}(\Omega)$, and thus $\|\nabla y\|_p = 1$, so $y \neq 0$. Because of H$(f)_1$, we can find $g \in L^{p'}(\Omega), g \geq 0$ a.e. in $\Omega$, such that

$$\frac{N_+(u_n)}{\|\nabla u_n\|_p^{p-1}} \xrightarrow{w} g \text{ in } L^{p'}(\Omega).$$

Again arguing as in Step 2, we find $\eta \in L^\infty(\Omega)$, with $\eta_1 \leq \eta \leq \eta_2$ a.e. in $\Omega$, such that

$$g(x) = \eta(x) y(x)^{p-1} \quad \text{for a.a. } x \in \Omega.$$

Dividing (11.26) by $\|\nabla u_n\|_p^{p-1}$ and then passing to the limit as $n \to \infty$ in (11.26), we see that

$$-\Delta_p y = \hat\eta y^{p-1} \text{ in } W^{-1,p'}(\Omega),$$

where $\hat\eta = t\eta + (1-t)\frac{\eta_1 + \eta_2}{2} \in L^\infty(\Omega)_+ \setminus \{0\}$. In other words, $y$ is an eigenfunction of $-\Delta_p$ with respect to $\hat\eta$, corresponding to the eigenvalue 1. From the monotonicity of the first eigenvalue of $-\Delta_p^D$ with respect to the weight [Proposition 9.47(d)], we have

$$\hat\lambda_1(\hat\eta) \leq \hat\lambda_1(\eta_1) < \hat\lambda_1(\lambda_1) = 1$$

[see H$(f)_2$ (iii)]. Thus, in view of Proposition 9.47(b), $y$ must change sign, a contradiction. This proves Claim 3.

Fix $\rho \in (0, \rho_0]$, with $\rho_0 > 0$ provided by Claim 3. The homotopy invariance of the degree map ensures that

$$d_{(S)_+}(-\Delta_p - N_+, B_\rho(0), 0) = d_{(S)_+}(-\Delta_p - K_+, B_\rho(0), 0). \tag{11.27}$$

Consider $w \in W_0^{1,p}(\Omega) \cap L^\infty(\Omega)_+ \setminus \{0\}$ and the homotopy $h_2$ of class $(S)_+$ involved in Step 2. Arguing exactly as in Claim 2, we see that

$$h_2(t, u) \neq 0 \quad \text{for all } t \in (0, 1], \text{ all } u \in W_0^{1,p}(\Omega).$$

Using the homotopy invariance and the solution property of the degree [Theorem 4.42(c), (e)], we obtain

$$d_{(S)_+}(-\Delta_p - K_+, B_\rho(0), 0) = d_{(S)_+}(-\Delta_p - K_+ - w, B_\rho(0), 0) = 0.$$

Therefore, $d_{(S)_+}(\psi', B_\rho(0), 0) = 0$ [see (11.27)]. This concludes Step 3.

Note that we may assume that $\psi$ admits only a finite number of critical points (otherwise we are done).

*Step 4:* We have $u_1 \neq 0$, and the functional $\psi$ admits a third critical point $u_2 \in W_0^{1,p}(\Omega) \setminus \{0, u_1\}$.

Since $u_1$ is a local minimizer and an isolated critical point of $\psi$, by Corollary 4.49, we find $\rho_1 > 0$ such that, for all $\rho \in (0, \rho_1)$, we have

$$d_{(S)_+}(\psi', B_\rho(u_1), 0) = 1. \tag{11.28}$$

Comparing with Step 3, we deduce that $u_1 \neq 0$. Arguing by contradiction, we assume that $0, u_1$ are the only critical points of $\psi$. Let $R_0 > 0$ be as in Step 2, and fix $R > R_0$ such that $0, u_1 \in B_R(0)$. Moreover, we fix $\rho \in (0, \min\{\rho_0, \rho_1\})$ (with $\rho_0 > 0$ as in Step 3) sufficiently small so that $B_\rho(0)$ and $B_\rho(u_1)$ are contained in $B_R(0)$ and are pairwise disjoint. From Steps 2 and 3 we have

$$d_{(S)_+}(\psi', B_R(0), 0) = d_{(S)_+}(\psi', B_\rho(0), 0) = 0. \tag{11.29}$$

On the other hand, applying the domain additivity and excision properties of the degree [Theorem 4.42(b), (d)], we know that

$$d_{(S)_+}(\psi', B_R(0), 0) = d_{(S)_+}(\psi', B_\rho(0), 0) + d_{(S)_+}(\psi', B_\rho(u_1), 0).$$

In view of relations (11.28) and (11.29), we are led to a contradiction. This completes Step 4 and the proof of the theorem. □

*Remark 11.16.* Results guaranteeing the existence of *negative* solutions of problem (11.1) can be similarly obtained, through the same proofs, by requiring the counterparts of $H(f)_1$ and $H(f)_2$ on the negative half-line.

## 11.2 Nonlinear Dirichlet Problems Using Variational Methods

In this section, we let $\Omega \subset \mathbb{R}^N$ be a bounded domain with a $C^2$-boundary $\partial\Omega$ and we fix $p \in (1, +\infty)$. We study the following nonlinear elliptic problem:

$$\begin{cases} -\Delta_p u = f(x, u) & \text{in } \Omega, \\ u = 0 & \text{on } \partial\Omega, \end{cases} \tag{11.30}$$

where $u \mapsto -\Delta_p u$ stands for the negative $p$-Laplacian operator and $f : \Omega \times \mathbb{R} \to \mathbb{R}$ is a Carathéodory function.

Whereas Sect. 11.1 focuses on the existence of positive solutions of problem (11.30), here our goal is to prove multiplicity results for (11.30) involving both constant-sign and nodal (sign-changing) solutions. The results provided in this section cover both problems with coercive and indefinite energy functionals. In fact, we do not assume in general that the nonlinearity $f$ satisfies a subcritical growth condition, so that the variational method is not applicable. Instead, we use suitable truncation techniques combined with the lower and upper solutions method described in the beginning of Sect. 11.1. We are then able to apply the variational method to the functionals obtained by truncation.

The section is organized as follows. First, we present preliminary results related to the method of upper and lower solutions followed by abstract results on the existence of constant-sign solutions and of nodal solutions (the results are abstract in the sense that they rely on the assumption that the problem admits a negative lower and a positive upper solutions). In the last part, we will prove the existence of constant-sign lower and upper solutions for problem (11.30) (in both the coercive and indefinite cases), and then, in view of the preliminary abstract results, we will obtain multiple solutions for coercive problems and for $(p-1)$-superlinear problems.

Throughout this section, we use the following notation:

- $\lambda_2 > \lambda_1 > 0$ are the first two eigenvalues of the negative Dirichlet $p$-Laplacian $-\Delta_p^D$ (Sect. 9.2);
- More generally, $\hat{\lambda}_k(\xi)$ ($k \in \{1,2\}$) denotes the $k$th eigenvalue of $-\Delta_p^D$ with respect to a weight $\xi \in L^\infty(\Omega)_+ \setminus \{0\}$ (Sect. 9.2);
- $\hat{u}_1 \in \text{int}(C_0^1(\overline{\Omega})_+)$ is the $L^p$-normalized positive eigenfunction of $-\Delta_p^D$ corresponding to $\lambda_1$ [Proposition 9.47(b), (c)];
- $[\underline{u}, \overline{u}]$ denotes the set $\{u \in W_0^{1,p}(\Omega) : \underline{u}(x) \leq u(x) \leq \overline{u}(x) \text{ for a.a. } x \in \Omega\}$, where $\underline{u}$ (resp. $\overline{u}$) may be $-\infty$ (resp. $+\infty$);
- $f_{[\underline{u},\overline{u}]}$ is the truncated Carathéodory function defined in (11.4);
- $F_{[\underline{u},\overline{u}]}(x,s) = \int_0^s f_{[\underline{u},\overline{u}]}(x,t)\, dt$;
- $\varphi_{[\underline{u},\overline{u}]}$ is the corresponding functional defined in (11.5).

## Abstract Results on Constant-Sign Solutions, Extremal Solutions, Nodal Solutions

Our basic assumptions on the nonlinearity $f$ in (11.30) are as follows [in the rest of the section, we will consider other sets of hypotheses that will always strengthen $\text{H}(f)_1^+$ or its negative counterpart $\text{H}(f)_1^-$]:

$\text{H}(f)_1^+$ (i) $f : \Omega \times \mathbb{R} \to \mathbb{R}$ is a Carathéodory function with $f(x,0) = 0$ a.e. in $\Omega$, and there are constants $c > 0$ and $r \in [1, +\infty)$ such that

$$|f(x,s)| \leq c(1 + |s|^{r-1}) \quad \text{for a.a. } x \in \Omega, \text{ all } s \in \mathbb{R};$$

(ii) There exists $\eta \in L^{\infty}(\Omega)$ such that $\eta(x) \geq \lambda_1$ a.e. in $\Omega$, $\eta \neq \lambda_1$ and

$$\liminf_{s \downarrow 0} \frac{f(x,s)}{s^{p-1}} \geq \eta(x) \quad \text{uniformly for a.a. } x \in \Omega.$$

In a symmetric way, we state:

$H(f)_1^-$ (i) $f : \Omega \times \mathbb{R} \to \mathbb{R}$ is a Carathéodory function with $f(x,0) = 0$ a.e. in $\Omega$, and there are constants $c > 0$ and $r \in [1, +\infty)$ such that

$$|f(x,s)| \leq c(1 + |s|^{r-1}) \quad \text{for a.a. } x \in \Omega, \text{ all } s \in \mathbb{R};$$

(ii) There exists $\eta \in L^{\infty}(\Omega)$ such that $\eta(x) \geq \lambda_1$ a.e. in $\Omega$, $\eta \neq \lambda_1$ and

$$\liminf_{s \uparrow 0} \frac{f(x,s)}{|s|^{p-2}s} \geq \eta(x) \quad \text{uniformly for a.a. } x \in \Omega.$$

Hypothesis $H(f)_1^{\pm}$ (i) involves a polynomial growth condition on the nonlinearity $f$ with arbitrary exponent (not necessarily subcritical). This assumption is not sufficient for guaranteeing that the energy functional associated to problem (11.30) is well defined on the Sobolev space $W_0^{1,p}(\Omega)$. To avoid this difficulty, in what follows, we will use truncation techniques based on the lower and upper solutions method (Definition 11.5).

Moreover, condition $H(f)_1^{\pm}$ (i) does not allow us to apply Corollary 8.6, so it is not ensured that all the solutions of (11.30) are bounded. However, we have the following regularity property of the bounded solutions of (11.30).

**Proposition 11.17.** *Assume that* $H(f)_1^{\pm}$ (i) *holds, and let* $u \in W_0^{1,p}(\Omega)$ *be a solution of (11.30) such that* $u \in L^{\infty}(\Omega)$. *Then* $u \in C_0^1(\overline{\Omega})$. *Moreover:*

(a) *If* $H(f)_1^+$ *holds and* $u \geq 0$ *in* $\Omega$, $u \neq 0$, *then* $u \in \operatorname{int}(C_0^1(\overline{\Omega})_+)$.
(b) *If* $H(f)_1^-$ *holds and* $u \leq 0$ *in* $\Omega$, $u \neq 0$, *then* $-u \in \operatorname{int}(C_0^1(\overline{\Omega})_+)$.

*Proof.* The fact that $u \in C_0^1(\overline{\Omega})$ is implied by Corollary 8.13. In the case where $H(f)_1^+$ holds and $u \geq 0$, $u \neq 0$, from $H(f)_1^+$ (i), (ii) we find a constant $c_0 > 0$ such that

$$\Delta_p u = -f(\cdot, u) \leq c_0 u^{p-1} \quad \text{in } W^{-1,p'}(\Omega).$$

Then, by the strong maximum principle (Theorem 8.27), we conclude that $u \in \operatorname{int}(C_0^1(\overline{\Omega})_+)$. This proves (a). Part (b) can be checked similarly. $\qquad \square$

We deduce from Proposition 11.8 a first abstract result of the existence of constant-sign solutions of (11.30).

**Proposition 11.18.**

(a) *Under* $H(f)_1^+$, *given an upper solution* $\overline{u} \in \operatorname{int}(C_0^1(\overline{\Omega})_+)$ *and a lower solution* $\underline{u} \in W_0^{1,p}(\Omega)$ *of problem (11.30), with* $\overline{u} \geq \underline{u} \geq 0$ *a.e. in* $\Omega$, *there exists a solution*

$u_0 \in \text{int}\,(C_0^1(\overline{\Omega})_+)$ of (11.30) satisfying $u_0 \in [\underline{u}, \overline{u}]$ and obtained as a global minimizer of the functional $\varphi_{[\underline{u},\overline{u}]}$ [see (11.5)].

(b) Under $H(f)_1^-$, given a lower solution $\underline{v} \in -\text{int}\,(C_0^1(\overline{\Omega})_+)$ and an upper solution $\overline{v} \in W_0^{1,p}(\Omega)$ of (11.30), with $\underline{v} \le \overline{v} \le 0$ a.e. in $\Omega$, there exists a solution $v_0 \in -\text{int}\,(C_0^1(\overline{\Omega})_+)$ of (11.30) satisfying $v_0 \in [\underline{v}, \overline{v}]$ and obtained as a global minimizer of the functional $\varphi_{[\underline{v},\overline{v}]}$.

*Proof.* We only prove assertion (a) because (b) holds by the same argument. Since $\underline{u}, \overline{u} \in L^\infty(\Omega)$ and $f$ is bounded on bounded sets of $\Omega \times [0, +\infty)$ [by $H(f)_1^+$ (i)], the truncated function $f_{[\underline{u},\overline{u}]}$ satisfies a subcritical growth condition [Remark 11.7(b)]. This allows us to apply Proposition 11.8(b), which shows that there is a $u_0 \in [\underline{u}, \overline{u}] \cap C_0^1(\overline{\Omega})$ solution of (11.30) obtained as a global minimizer of the functional $\varphi_{[\underline{u},\overline{u}]}$.

Let us justify that $u_0 \ne 0$. Clearly, it suffices to check this when $\underline{u} = 0$. Letting $\eta \in L^\infty(\Omega)_+$ be as in hypothesis $H(f)_1^+$ (ii), we have that

$$\gamma := \lambda_1 - \int_\Omega \eta(x)\hat{u}_1(x)^p\,dx = \int_\Omega (\lambda_1 - \eta(x))\hat{u}_1(x)^p\,dx < 0.$$

From $H(f)_1^+$ (ii) we know that, for each $\varepsilon \in (0, -\gamma)$, there is $\delta = \delta(\varepsilon) > 0$ such that

$$\frac{1}{p}(\eta(x) - \varepsilon)s^p \le \int_0^s f(x,t)\,dt \quad \text{for a.a. } x \in \Omega, \text{ all } s \in [0, \delta).$$

Since $\overline{u} \in \text{int}\,(C_0^1(\overline{\Omega})_+)$, we can find $t \in (0, \frac{\delta}{\|\hat{u}_1\|_\infty})$ such that $t\hat{u}_1 \in (0, \overline{u}]$. Then the definition of $\varphi_{[0,\overline{u}]}$ in (11.5) yields

$$\varphi_{[0,\overline{u}]}(t\hat{u}_1) \le \frac{\lambda_1 t^p}{p} - \frac{t^p}{p}\int_\Omega (\eta(x) - \varepsilon)\hat{u}_1(x)^p\,dx \le \frac{t^p}{p}(\gamma + \varepsilon) < 0 = \varphi_{[0,\overline{u}]}(0).$$

Because $u_0$ is a global minimizer of $\varphi_{[0,\overline{u}]}$, we deduce that $u_0 \ne 0$.

Finally, as $u_0 \ge 0$ in $\Omega$, $u_0 \ne 0$, from Proposition 11.17(a) we obtain that $u_0 \in \text{int}\,(C_0^1(\overline{\Omega})_+)$. The proof is then complete. $\qquad\square$

Next, we need the following useful property of lower and upper solutions.

**Lemma 11.19.** *We assume* $H(f)_1^\pm$ (i).

(a) *If* $\overline{u}_1, \overline{u}_2 \in W^{1,p}(\Omega)$ *are upper solutions of problem (11.30), then so is* $\overline{u} :=$ $\min\{\overline{u}_1, \overline{u}_2\}$.

(b) *If* $\underline{v}_1, \underline{v}_2 \in W^{1,p}(\Omega)$ *are lower solutions for problem (11.30), then so is* $\underline{v} :=$ $\max\{\underline{v}_1, \underline{v}_2\}$.

*Proof.* We only prove part (a) as the proof of part (b) is similar. Given $\varepsilon > 0$, we define $\hat{\tau}_\varepsilon : \mathbb{R} \to \mathbb{R}$ by

$$\hat{\tau}_\varepsilon(s) = \begin{cases} -\varepsilon & \text{if } s \le -\varepsilon, \\ s & \text{if } -\varepsilon < s < \varepsilon, \\ \varepsilon & \text{if } s \ge \varepsilon. \end{cases}$$

It follows from Remark 1.30 that for every $u \in W^{1,p}(\Omega)$ we have $\hat{\tau}_\varepsilon(u(\cdot)) \in W^{1,p}(\Omega)$ and

$$\nabla \hat{\tau}_\varepsilon(u) = \begin{cases} 0 & \text{a.e. in } \{x \in \Omega : |u(x)| \geq \varepsilon\}, \\ \nabla u & \text{a.e. in } \{x \in \Omega : |u(x)| < \varepsilon\}. \end{cases} \tag{11.31}$$

Let $\psi \in C_c^\infty(\Omega)$, with $\psi \geq 0$ in $\Omega$. Since $\bar{u}_1, \bar{u}_2$ are upper solutions of (11.30), we have

$$\int_\Omega f(x, \bar{u}_1) \hat{\tau}_\varepsilon((\bar{u}_1 - \bar{u}_2)^-) \psi \, dx \leq \langle -\Delta_p \bar{u}_1, \hat{\tau}_\varepsilon((\bar{u}_1 - \bar{u}_2)^-) \psi \rangle, \tag{11.32}$$

$$\int_\Omega f(x, \bar{u}_2)(\varepsilon - \hat{\tau}_\varepsilon((\bar{u}_1 - \bar{u}_2)^-)) \psi \, dx \leq \langle -\Delta_p \bar{u}_2, (\varepsilon - \hat{\tau}_\varepsilon((\bar{u}_1 - \bar{u}_2)^-)) \psi \rangle. \tag{11.33}$$

Moreover, in view of (11.31), we have

$$\langle -\Delta_p \bar{u}_1, \hat{\tau}_\varepsilon((\bar{u}_1 - \bar{u}_2)^-) \psi \rangle + \langle -\Delta_p \bar{u}_2, (\varepsilon - \hat{\tau}_\varepsilon((\bar{u}_1 - \bar{u}_2)^-)) \psi \rangle$$

$$\leq \int_\Omega |\nabla \bar{u}_1|^{p-2} (\nabla \bar{u}_1, \nabla \psi)_{\mathbb{R}^N} \, \hat{\tau}_\varepsilon((\bar{u}_1 - \bar{u}_2)^-) \, dx$$

$$+ \int_\Omega |\nabla \bar{u}_2|^{p-2} (\nabla \bar{u}_2, \nabla \psi)_{\mathbb{R}^N} (\varepsilon - \hat{\tau}_\varepsilon((\bar{u}_1 - \bar{u}_2)^-)) \, dx. \tag{11.34}$$

Adding (11.32) and (11.33) and using (11.34), we obtain

$$\int_\Omega f(x, \bar{u}_1) \frac{1}{\varepsilon} \hat{\tau}_\varepsilon((\bar{u}_1 - \bar{u}_2)^-) \psi \, dx + \int_\Omega f(x, \bar{u}_2) \left(1 - \frac{1}{\varepsilon} \hat{\tau}_\varepsilon((\bar{u}_1 - \bar{u}_2)^-) \right) \psi \, dx$$

$$\leq \int_\Omega |\nabla \bar{u}_1|^{p-2} (\nabla \bar{u}_1, \nabla \psi)_{\mathbb{R}^N} \frac{1}{\varepsilon} \hat{\tau}_\varepsilon((\bar{u}_1 - \bar{u}_2)^-) \, dx$$

$$+ \int_\Omega |\nabla \bar{u}_2|^{p-2} (\nabla \bar{u}_2, \nabla \psi)_{\mathbb{R}^N} \left(1 - \frac{1}{\varepsilon} \hat{\tau}_\varepsilon((\bar{u}_1 - \bar{u}_2)^-) \right) dx. \tag{11.35}$$

Note that

$$\frac{1}{\varepsilon} \hat{\tau}_\varepsilon((\bar{u}_1 - \bar{u}_2)^-(x)) \to \chi_{\{\bar{u}_1 < \bar{u}_2\}}(x) \text{ a.e. in } \Omega \text{ as } \varepsilon \downarrow 0.$$

Hence, passing to the limit as $\varepsilon \downarrow 0$ in (11.35), we get

$$\int_\Omega f(x, \bar{u}) \psi \, dx \leq \int_\Omega |\nabla \bar{u}|^{p-2} (\nabla \bar{u}, \nabla \psi)_{\mathbb{R}^N} \, dx.$$

Since $C_c^\infty(\Omega)$ is dense in $W_0^{1,p}(\Omega)$, the same inequality holds for all $\psi \in W_0^{1,p}(\Omega)$ such that $\psi \geq 0$ a.e. in $\Omega$. The proof is complete. $\qquad \square$

In the next result, we show the existence of extremal constant-sign solutions.

**Proposition 11.20.**

(a) *If hypotheses* H$(f)_1^+$ *hold, then for each upper solution* $\bar{u} \in$ int $(C_0^1(\overline{\Omega})_+)$ *and each lower solution* $\underline{u} \in W_0^{1,p}(\Omega)$ *of (11.30), with* $\bar{u} \geq \underline{u} \geq 0$ *a.e. in* $\Omega$, $\underline{u} \neq 0$, *problem (11.30) admits a smallest solution* $u_*$ *in the order interval* $[\underline{u}, \bar{u}]$. *In addition,* $u_* \in$ int $(C_0^1(\overline{\Omega})_+)$.
(b) *If hypotheses* H$(f)_1^-$ *hold, then for each lower solution* $\underline{v} \in -$int $(C_0^1(\overline{\Omega})_+)$ *and each upper solution* $\bar{v} \in W_0^{1,p}(\Omega)$ *of (11.30), with* $\underline{v} \leq \bar{v} \leq 0$ *a.e. in* $\Omega$, $\bar{v} \neq 0$, *problem (11.30) admits a biggest solution* $v^*$ *in* $[\underline{v}, \bar{v}]$. *In addition,* $v^* \in -$int $(C_0^1(\overline{\Omega})_+)$.

*Proof.* We only prove part (a) because part (b) can be obtained similarly. Given $\bar{u}$ and $\underline{u}$ as in the statement, we let

$$\mathscr{S} = \{u \in [\underline{u}, \bar{u}] : u \text{ is a solution of (11.30)}\}.$$

From Proposition 11.17(a), since $\underline{u} \neq 0$, we know that $\mathscr{S} \subset$ int$(C_0^1(\overline{\Omega})_+)$. Moreover, Proposition 11.18 ensures that $\mathscr{S}$ is nonempty. To show the proposition, we need to check that $\mathscr{S}$ has a smallest element.

*Claim 1:* For every $u_1, u_2 \in \mathscr{S}$ there exists $u \in \mathscr{S}$ such that $u \leq u_1$ and $u \leq u_2$.

Since $u_1, u_2$ are solutions of (11.30), they are a fortiori upper solutions. Then, by virtue of Lemma 11.19(a), $\hat{u} := \min\{u_1, u_2\} \in W_0^{1,p}(\Omega)$ is also an upper solution of (11.30). Proposition 11.18(a), applied with the pair $\{\underline{u}, \hat{u}\}$ of lower and upper solutions, yields a solution $u$ of (11.30) such that $\underline{u} \leq u \leq \hat{u} = \min\{u_1, u_2\}$. This proves Claim 1.

*Claim 2:* There is $\alpha \in (0, 1)$ such that the set $\mathscr{S}$ is a bounded subset of $C^{1,\alpha}(\overline{\Omega})$.

By definition, for each $u \in \mathscr{S}$ we have $\|u\|_\infty \leq \|\bar{u}\|_\infty$. Then the claim is implied by Theorem 8.10.

Let $\{x_k\}_{k \geq 1}$ be a dense subset of $\Omega$. For each $k \geq 1$ we let $m_k = \inf_{u \in \mathscr{S}} u(x_k) \geq 0$.

*Claim 3:* For all $n \geq 1$ there is $u_n \in \mathscr{S}$ such that

$$m_k \leq u_n(x_k) \leq m_k + \frac{1}{n} \text{ for all } k \in \{1, \ldots, n\}.$$

By the definition of $m_k$, we find $u_{n,1}, \ldots, u_{n,n} \in \mathscr{S}$, with $u_{n,k}(x_k) \leq m_k + \frac{1}{n}$ for all $k \in \{1, \ldots, n\}$. By Claim 1, we can find $u_n \in \mathscr{S}$ such that $u_n \leq u_{n,k}$ for all $k \in \{1, \ldots, n\}$. This function $u_n$ satisfies Claim 3.

From Claim 3 we obtain a sequence $\{u_n\}_{n \geq 1} \subset \mathscr{S}$. By Claim 2, this sequence is bounded in $C^{1,\alpha}(\overline{\Omega})$, so up to considering a subsequence we may assume that $u_n \to u_0$ in $C^1(\overline{\Omega})$ as $n \to \infty$, for some $u_0 \in C^1(\overline{\Omega})$. It is clear that $u_0 \in \mathscr{S}$. Moreover, passing to the limit as $n \to \infty$ in the relation in Claim 3, we have $u_0(x_k) = m_k$ for all $k \geq 1$. Hence, $u_0(x_k) \leq u(x_k)$ for all $k \geq 1$, all $u \in \mathscr{S}$. Since $\{x_k\}_{k \geq 1}$ is dense in $\Omega$, we deduce that $u_0 \leq u$ for all $u \in \mathscr{S}$. Therefore, $u_0$ is the smallest element of $\mathscr{S}$. This completes the proof. $\qquad\square$

The next result produces positive lower solutions and negative upper solutions.

**Proposition 11.21.**

(a) *Under* $\mathrm{H}(f)_1^+$, *for each function* $\overline{u} \in \operatorname{int}(C_0^1(\overline{\Omega})_+)$ *there exists a lower solution* $\underline{u} \in \operatorname{int}(C_0^1(\overline{\Omega})_+)$ *of* (11.30) *satisfying* $\overline{u} - \underline{u} \in \operatorname{int}(C_0^1(\overline{\Omega})_+)$. *Moreover, for every* $\varepsilon \in (0,1)$, $\varepsilon\underline{u}$ *is a lower solution of* (11.30).
(b) *Under* $\mathrm{H}(f)_1^-$, *for each function* $\underline{v} \in -\operatorname{int}(C_0^1(\overline{\Omega})_+)$ *there exists an upper solution* $\overline{v} \in -\operatorname{int}(C_0^1(\overline{\Omega})_+)$ *of* (11.30) *satisfying* $\overline{v} - \underline{v} \in \operatorname{int}(C_0^1(\overline{\Omega})_+)$. *Moreover, for every* $\varepsilon \in (0,1)$, $\varepsilon\overline{v}$ *is an upper solution of* (11.30).

*Proof.* Let $V := \{u \in W_0^{1,p}(\Omega) : \int_\Omega \hat{u}_1^{p-1} u \, dx = 0\}$. We have the direct sum decomposition $W_0^{1,p}(\Omega) = \mathbb{R}\hat{u}_1 \oplus V$. We claim that

$$\lambda_V := \inf\left\{ \frac{\|\nabla u\|_p^p}{\|u\|_p^p} : u \in V, \, u \neq 0 \right\} > \lambda_1. \tag{11.36}$$

Indeed, arguing by contradiction, assume that there is a sequence $\{u_n\}_{n \geq 1} \subset V$ such that $\|u_n\|_p = 1$ and $\|\nabla u_n\|_p \to \lambda_1$ as $n \to \infty$. Then $\{u_n\}_{n \geq 1}$ is bounded in $W_0^{1,p}(\Omega)$, so we may assume that $u_n \xrightarrow{w} u$ in $W_0^{1,p}(\Omega)$ and $u_n \to u$ in $L^p(\Omega)$ as $n \to \infty$, for some $u \in W_0^{1,p}(\Omega)$. Hence, $u \in V$, $\|u\|_p = 1$, and $\|\nabla u\|_p \leq \lambda_1$. By Proposition 9.47(a), (c), we have that $u = \pm\hat{u}_1$, which contradicts the fact that $u \in V$. This proves (11.36).

Let $\delta > 0$ be the constant given by Theorem 9.67 applied to $h := \hat{u}_1^{p-1}$ and $\xi := \lambda_1$. For $\zeta \in L^\infty(\Omega)_+ \setminus \{0\}$, recall that $\hat{\lambda}_1(\zeta) > 0$ denotes the first eigenvalue of $-\Delta_p^D$ with respect to $\zeta$. Since the map $\zeta \mapsto \hat{\lambda}_1(\zeta)$ is continuous [by Remark 9.46(b)], we find $\varepsilon > 0$ such that for all $\zeta \in L^\infty(\Omega)$, with $\|\zeta - \lambda_1\|_\infty \leq \varepsilon$ a.e. in $\Omega$, we have $|\hat{\lambda}_1(\zeta) - 1| < \delta$. We may assume that $0 < \varepsilon < \min\{\lambda_V - \lambda_1, \lambda_2 - \lambda_1, \delta\}$. We define the weight

$$\zeta := \min\{\eta, \lambda_1 + \varepsilon\} \in L^\infty(\Omega)_+, \tag{11.37}$$

with $\eta \in L^\infty(\Omega)_+$ as in $\mathrm{H}(f)_1^+$ (ii). Thus, $\lambda_1 \leq \zeta < \lambda_2$ a.e. in $\Omega$, $\zeta \neq \lambda_1$, so, by the monotonicity property of $\hat{\lambda}_1(\cdot)$ and $\hat{\lambda}_2(\cdot)$ [Propositions 9.47(c) and 9.51], we get

$$1 - \delta < \hat{\lambda}_1(\zeta) < \hat{\lambda}_1(\lambda_1) = 1 = \hat{\lambda}_2(\lambda_2) < \hat{\lambda}_2(\zeta). \tag{11.38}$$

We consider the auxiliary boundary value problem

$$\begin{cases} -\Delta_p u = \zeta(x)|u|^{p-2}u - \hat{u}_1(x)^{p-1} & \text{in } \Omega, \\ u = 0 & \text{on } \partial\Omega. \end{cases} \tag{11.39}$$

The functional $\varphi_0 : W_0^{1,p}(\Omega) \to \mathbb{R}$ defined by

$$\varphi_0(u) = \frac{1}{p} \|\nabla u\|_p^p - \frac{1}{p} \int_\Omega \zeta |u|^p \, dx + \int_\Omega \hat{u}_1^{p-1} u \, dx \quad \text{for all } u \in W_0^{1,p}(\Omega)$$

is of class $C^1$, and its critical points are the solutions of (11.39).

*Claim 1:* $\varphi_0$ satisfies the (PS)-condition.

Let $\{u_n\}_{n \geq 1} \subset W_0^{1,p}(\Omega)$ be a sequence such that $\{\varphi_0(u_n)\}_{n \geq 1}$ is bounded and $\varphi_0'(u_n) \to 0$ in $W^{-1,p'}(\Omega)$. First, we show that $\{u_n\}_{n \geq 1}$ is bounded in $W_0^{1,p}(\Omega)$. Arguing by contradiction, we assume that along a subsequence $\|\nabla u_n\|_p \to +\infty$ as $n \to \infty$ and set $y_n = \frac{u_n}{\|\nabla u_n\|_p}$ for $n \geq 1$. We may suppose that $y_n \overset{w}{\to} y$ in $W_0^{1,p}(\Omega)$ and $y_n \to y$ in $L^p(\Omega)$, for some $y \in W_0^{1,p}(\Omega)$. From the fact that $\varphi_0'(u_n) \to 0$ it easily follows that $\langle -\Delta_p y_n, y_n - y \rangle \to 0$ as $n \to \infty$. According to Proposition 2.72, we deduce that $y_n \to y$ in $W_0^{1,p}(\Omega)$, and so $\|\nabla y\|_p = 1$, and

$$-\Delta_p y = \zeta |y|^{p-2} y \quad \text{in } W^{-1,p'}(\Omega). \tag{11.40}$$

By (11.38), we infer that $y = 0$, which is a contradiction. So $\{u_n\}_{n \geq 1} \subset W_0^{1,p}(\Omega)$ is bounded, and along a relabeled subsequence we have $u_n \overset{w}{\to} u$ in $W_0^{1,p}(\Omega)$ and $u_n \to u$ in $L^p(\Omega)$, for some $u \in W_0^{1,p}(\Omega)$. As before, we deduce that $u_n \to u$ in $W_0^{1,p}(\Omega)$.

*Claim 2:* $\varphi_0|_V \geq 0$.

Since $0 < \varepsilon < \lambda_V - \lambda_1$, we have that $\zeta(x) < \lambda_V$ a.e. in $\Omega$ [see (11.37)]. Then Claim 2 follows from the definitions of $\varphi_0$, $V$, and $\lambda_V$.

*Claim 3:* For $t > 0$ large we have $\varphi_0(\pm t \hat{u}_1) < 0$.

Using that $\|\hat{u}_1\|_p = 1$, for $t > 0$ we see that

$$\varphi_0(\pm t \hat{u}_1) = \frac{t^p}{p} \beta \pm t, \quad \text{where } \beta := \int_\Omega (\lambda_1 - \zeta(x)) \hat{u}_1(x)^p \, dx.$$

Since $\zeta \geq \lambda_1$ a.e. in $\Omega$, $\zeta \neq \lambda_1$ [see (11.37)], we have $\beta < 0$. This yields Claim 3.

*Claim 4:* The auxiliary problem (11.39) has a solution $\hat{\underline{u}} \in \text{int}(C_0^1(\overline{\Omega})_+)$.

Claims 1–3 allow us to apply the saddle point theorem (Theorem 5.41), which provides $\hat{\underline{u}} \in W_0^{1,p}(\Omega)$ such that $\varphi_0'(\hat{\underline{u}}) = 0$; thus, $\hat{\underline{u}}$ is a solution of problem (11.39), hence $\hat{\underline{u}} \neq 0$. Finally, since $\|\zeta - \lambda_1\|_\infty < \delta$ [by (11.37) and because $0 < \varepsilon < \delta$] and $\hat{\lambda}_1(\zeta) < 1 < \hat{\lambda}_1(\zeta) + \delta$ [see (11.38)], we can apply Theorem 9.67 to the function $u = -\hat{\underline{u}}$, which yields $\hat{\underline{u}} \in \text{int}(C_0^1(\overline{\Omega})_+)$. This establishes Claim 4.

Since $\hat{u}_1 \in \text{int}(C_0^1(\overline{\Omega})_+)$ and $\hat{\underline{u}} \in C_0^1(\overline{\Omega})$, we can find $t > 0$ such that

$$\hat{u}_1 - t\hat{\underline{u}} \in \text{int}(C_0^1(\overline{\Omega})_+). \tag{11.41}$$

By (11.37) and hypothesis H$(f)_1^+$ (ii), we can find $\tilde{\delta} = \tilde{\delta}(t) > 0$ such that

$$(\zeta(x) - t^{p-1})s^{p-1} \leq f(x,s) \quad \text{for a.a. } x \in \Omega, \text{ all } s \in [0, \tilde{\delta}]. \tag{11.42}$$

Finally, since $\overline{u}, \hat{\underline{u}} \in \text{int}(C_0^1(\overline{\Omega})_+)$, there is $\rho > 0$ satisfying

$$\overline{u} - \rho\hat{\underline{u}} \in \text{int}(C_0^1(\overline{\Omega})_+) \quad \text{and} \quad 0 \leq \rho\hat{\underline{u}}(x) \leq \tilde{\delta} \text{ for all } x \in \overline{\Omega}. \tag{11.43}$$

We set $\underline{u} := \rho\hat{\underline{u}}$. By Claim 4, we know that $\underline{u} \in \text{int}(C_0^1(\overline{\Omega})_+)$, whereas (11.43) yields $\overline{u} - \underline{u} \in \text{int}(C_0^1(\overline{\Omega})_+)$. Using (11.41)–(11.43), we obtain

$$-\Delta_p \underline{u} = \zeta \underline{u}^{p-1} - \rho^{p-1}\hat{u}_1^{p-1} < (\zeta - t^{p-1})\underline{u}^{p-1} \leq f(\cdot, \underline{u}(\cdot)) \quad \text{a.e. in } \Omega. \tag{11.44}$$

This implies that $\underline{u}$ is a lower solution of problem (11.30) (Definition 11.5). Clearly, $\varepsilon\underline{u}$ is also a lower solution of (11.30) for all $\varepsilon \in (0,1)$. This proves part (a) of the proposition. The proof of part (b) follows the same scheme. □

Our next purpose is to produce extremal constant-sign solutions for problem (11.30). To do this, we rely on strengthened versions of hypotheses H$(f)_1^\pm$, which require that the nonlinearity $f(x, \cdot)$ is $(p-1)$-linear near the origin.

H$(f)_2^+$ (i) $f: \Omega \times \mathbb{R} \to \mathbb{R}$ is a Carathéodory function with $f(x,0) = 0$ a.e. in $\Omega$, and there are constants $c > 0$ and $r \in [1, +\infty)$ such that

$$|f(x,s)| \leq c(1 + |s|^{r-1}) \quad \text{for a.a. } x \in \Omega, \text{ all } s \in \mathbb{R};$$

(ii) there exist $\eta, \hat{\eta} \in L^\infty(\Omega)$ such that $\eta(x) \geq \lambda_1$ a.e. in $\Omega$, $\eta \neq \lambda_1$, and

$$\eta(x) \leq \liminf_{s \downarrow 0} \frac{f(x,s)}{s^{p-1}} \leq \limsup_{s \downarrow 0} \frac{f(x,s)}{s^{p-1}} \leq \hat{\eta}(x) \quad \text{uniformly for a.a. } x \in \Omega.$$

Symmetrically, we consider:

H$(f)_2^-$ (i) $f: \Omega \times \mathbb{R} \to \mathbb{R}$ is a Carathéodory function with $f(x,0) = 0$ a.e. in $\Omega$, and there are constants $c > 0$ and $r \in [1, +\infty)$ such that

$$|f(x,s)| \leq c(1 + |s|^{r-1}) \quad \text{for a.a. } x \in \Omega, \text{ all } s \in \mathbb{R};$$

(ii) There exist $\eta, \hat{\eta} \in L^\infty(\Omega)$ such that $\eta(x) \geq \lambda_1$ a.e. in $\Omega$, $\eta \neq \lambda_1$, and

$$\eta(x) \leq \liminf_{s \uparrow 0} \frac{f(x,s)}{|s|^{p-2}s} \leq \limsup_{s \uparrow 0} \frac{f(x,s)}{|s|^{p-2}s} \leq \hat{\eta}(x) \quad \text{uniformly for a.a. } x \in \Omega.$$

We know from Proposition 11.17 that, under H$(f)_1^\pm$, every bounded, nontrivial, nonnegative (resp. nonpositive) solution $u$ of (11.30) belongs to $\pm\text{int}(C_0^1(\overline{\Omega})_+)$.

330 Nonlinear Elliptic Equations with Dirichlet Boundary Conditions

In the next statement, under $H(f)_2^\pm$, we show that $\pm\mathrm{int}\,(C_0^1(\overline{\Omega})_+)$ actually contains a smallest (resp. biggest) element with the property of being a solution of (11.30).

**Proposition 11.22.**

(a) *Under* $H(f)_2^+$, *for each upper solution* $\overline{u} \in \mathrm{int}\,(C_0^1(\overline{\Omega})_+)$, *problem (11.30) has a smallest positive solution* $u_+$ *in* $[0, \overline{u}]$, *which in addition satisfies* $u_+ \in \mathrm{int}\,(C_0^1(\overline{\Omega})_+)$.
(b) *Under* $H(f)_2^-$, *for each lower solution* $\underline{v} \in -\mathrm{int}\,(C_0^1(\overline{\Omega})_+)$, *problem (11.30) has a biggest negative solution* $v_-$ *in* $[\underline{v}, 0]$, *which in addition satisfies* $v_- \in -\mathrm{int}\,(C_0^1(\overline{\Omega})_+)$.

*Proof.* We only prove part (a), as part (b) can be obtained similarly. Let $\underline{u} \in \mathrm{int}\,(C_0^1(\overline{\Omega})_+)$ be the lower solution of problem (11.30) obtained in Proposition 11.21(a) applied to $\overline{u}$. We fix a sequence $\{\varepsilon_n\}_{n\geq 1} \subset (0,1)$ converging to 0, and for $n \geq 1$ we set $\underline{u}_n = \varepsilon_n \underline{u}$, which is also a lower solution of (11.30) by virtue of Proposition 11.21(a). By Proposition 11.20(a), we can find $u_n^*$, which is a smallest solution of (11.30) in the order interval $[\underline{u}_n, \overline{u}]$. From the relation $-\Delta_p u_n^* = f(\cdot, u_n^*(\cdot))$, hypothesis $H(f)_2^+$ (i), and the fact that $0 \leq u_n^* \leq \overline{u}$, we obtain that the sequence $\{u_n^*\}_{n\geq 1}$ is bounded in $W_0^{1,p}(\Omega)$, so we may assume that $u_n^* \xrightarrow{w} u_+$ in $W_0^{1,p}(\Omega)$ and $u_n^* \to u_+$ in $L^p(\Omega)$ as $n \to \infty$, for some $u_+ \in W_0^{1,p}(\Omega)$. As in Claim 1 of the proof of Proposition 11.21, we have

$$u_n^* \to u_+ \text{ in } W_0^{1,p}(\Omega) \text{ as } n \to \infty. \tag{11.45}$$

From (11.45) it follows that $u_+$ is a solution of (11.30). Moreover, up to considering a subsequence, we may assume that we have $u_n^*(x) \to u_+(x)$ for a.a. $x \in \Omega$. This implies that $u_+ \in [0, \overline{u}]$.

*Claim 1:* $u_+ \neq 0$.

Arguing by contradiction, assume that $u_+ = 0$. For $n \geq 1$ we set $y_n^* = \frac{u_n^*}{\|\nabla u_n^*\|_p}$. We may suppose that $y_n^* \xrightarrow{w} y$ in $W_0^{1,p}(\Omega)$, $y_n^* \to y$ in $L^p(\Omega)$ as $n \to \infty$, for some $y \in W_0^{1,p}(\Omega)$. Denoting $h_n := \frac{f(\cdot, u_n^*(\cdot))}{\|\nabla u_n^*\|_p^{p-1}}$, we have

$$-\Delta_p y_n^* = h_n \text{ in } W^{-1,p'}(\Omega) \text{ for all } n \geq 1. \tag{11.46}$$

Hypothesis $H(f)_2^+$ implies that there exists $c_0 > 0$ such that $|f(x,s)| \leq c_0 s^{p-1}$ for a.a. $x \in \Omega$, all $s \in [0, \|\overline{u}\|_\infty]$. Thus, $\{h_n\}_{n\geq 1}$ is bounded in $L^{p'}(\Omega)$. Therefore, acting on (11.46) with the test function $y_n^* - y \in W_0^{1,p}(\Omega)$, we obtain $\lim_{n\to\infty} \langle -\Delta_p y_n^*, y_n^* - y \rangle = 0$, and so $y_n^* \to y$ in $W_0^{1,p}(\Omega)$ (by Proposition 2.72) and $\|\nabla y\|_p = 1$. Since $y_n^*(x) \to y(x)$ for a.a. $x \in \Omega$ (at least along a subsequence), we have $y \geq 0$ a.e. in $\Omega$, $y \neq 0$.

Since $\{h_n\}_{n\geq 1}$ is bounded in $L^{p'}(\Omega)$, we may assume that $h_n \xrightarrow{w} h$ in $L^{p'}(\Omega)$, for some $h \in L^{p'}(\Omega)$. Arguing on the basis of $H(f)_2^+$ (ii) as in Claim 3 of the proof of Theorem 11.15, we can see that

$$\eta(x)y(x)^{p-1} \le h(x) \le \hat{\eta}(x)y(x)^{p-1} \quad \text{for a.a. } x \in \Omega;$$

therefore, $h(x) = \kappa(x)y(x)^{p-1}$ a.e. in $\Omega$, with $\kappa \in L^\infty(\Omega)$ such that $\eta \le \kappa \le \hat{\eta}$ a.e. in $\Omega$. Passing to the limit as $n \to \infty$ in (11.46), we obtain that $y$ solves the problem

$$\begin{cases} -\Delta_p y = \kappa y^{p-1} & \text{in } \Omega, \\ y = 0 & \text{on } \partial\Omega. \end{cases}$$

Since $y \ne 0$, we deduce that 1 is an eigenvalue of $-\Delta_p^D$ with respect to the weight $\kappa$ and, since $y$ has a constant sign, we know that $1 = \hat{\lambda}_1(\kappa)$ [see Proposition 9.47(b)]. However, by $H(f)_2^+$ (ii), we have $\kappa \ge \lambda_1$ a.e. in $\Omega$ with strict inequality on a set of positive measure; hence, by virtue of the monotonicity property of $\hat{\lambda}_1(\cdot)$ [see Proposition 9.47 (d)], we must have $\hat{\lambda}_1(\kappa) < \hat{\lambda}_1(\lambda_1) = 1$, a contradiction. This proves Claim 1.

*Claim 2:* For every nontrivial solution $u$ of (11.30) belonging to $[0, \bar{u}]$ we have $u_+ \le u$ in $\Omega$.

From Proposition 11.17(a) we know that $u \in \text{int}(C_0^1(\overline{\Omega})_+)$. Using that the sequence $\{\varepsilon_n\}_{n \ge 1}$ converges to 0, for $n$ large enough we have $\underline{u}_n = \varepsilon_n \underline{u} \le u \le \bar{u}$ in $\Omega$. Since $u_n^*$ is the smallest solution of (11.30) in $[\underline{u}_n, \bar{u}]$, we derive $u_n^* \le u$ in $\Omega$, whence $u_+ \le u$ in $\Omega$, which proves Claim 2.

The proposition is obtained by combining Claims 1 and 2. □

To obtain an intermediate nontrivial solution (possibly nodal) between any negative and positive solutions of (11.30), we strengthen hypotheses $H(f)_1^\pm$ (and $H(f)_2^\pm$):

$H(f)_3$ (i) $f : \Omega \times \mathbb{R} \to \mathbb{R}$ is a Carathéodory function, with $f(x,0) = 0$ a.e. in $\Omega$, and there are constants $c > 0$ and $r \in [1, +\infty)$ such that

$$|f(x,s)| \le c(1 + |s|^{r-1}) \quad \text{for a.a. } x \in \Omega, \text{ all } s \in \mathbb{R};$$

(ii) One of the following conditions is satisfied: either

(ii.a) There exists a constant $\mu_0 > \lambda_2$ such that

$$\mu_0 < \liminf_{s \to 0} \frac{f(x,s)}{|s|^{p-2}s} \quad \text{uniformly for a.a. } x \in \Omega;$$

or (stronger)

(ii.b) There are $\mu_0 > \lambda_2$ and $\hat{\eta} \in L^\infty(\Omega)$ such that

$$\mu_0 < \liminf_{s \to 0} \frac{f(x,s)}{|s|^{p-2}s} \le \limsup_{s \to 0} \frac{f(x,s)}{|s|^{p-2}s} \le \hat{\eta}(x) \quad \text{uniformly for a.a. } x \in \Omega.$$

**Theorem 11.23.**

(a) *Under* $H(f)_3$, *for each upper solution* $\bar{u} \in \operatorname{int}(C_0^1(\overline{\Omega})_+)$ *and each lower solution* $\underline{v} \in -\operatorname{int}(C_0^1(\overline{\Omega})_+)$, *problem (11.30) has at least three distinct, nontrivial solutions* $u_0 \in \operatorname{int}(C_0^1(\overline{\Omega})_+)$, $v_0 \in -\operatorname{int}(C_0^1(\overline{\Omega})_+)$, *and* $y_0 \in C_0^1(\overline{\Omega})$, *satisfying*

$$\underline{v} \leq v_0 \leq y_0 \leq u_0 \leq \bar{u} \text{ in } \Omega.$$

(b) *If, in addition, hypothesis* $H(f)_3$ *(ii.b) holds, then the solution* $y_0$ *can be chosen to be nodal.*

*Proof.* The existence of opposite constant-sign solutions $u_0 \in \operatorname{int}(C_0^1(\overline{\Omega})_+)$ and $v_0 \in -\operatorname{int}(C_0^1(\overline{\Omega})_+)$ of problem (11.30) follows from Proposition 11.18 by choosing $\underline{u} = \bar{v} = 0$. To produce the third nontrivial solution, we define the functions $u_+ \in \operatorname{int}(C_0^1(\overline{\Omega})_+)$ and $v_- \in -\operatorname{int}(C_0^1(\overline{\Omega})_+)$ as follows: in case (a) set $u_+ := u_0$, $v_- := v_0$, while for (b) let $u_+$ and $v_-$ be the minimal positive solution and maximal negative solution of (11.30), respectively, obtained in Proposition 11.22.

We consider the $C^1$-functionals $\varphi_{[0,u_+]}$, $\varphi_{[v_-,0]}$, and $\varphi_{[v_-,u_+]}$, obtained by truncation with respect to the pairs $\{0,u_+\}$, $\{v_-,0\}$, and $\{v_-,u_+\}$, respectively [see (11.5)].

By hypothesis $H(f)_3$ (ii) we find $\mu \in (\lambda_2,\mu_0)$ and $\delta > 0$ such that

$$\frac{f(x,s)}{|s|^{p-2}s} > \mu \text{ for a.a. } x \in \Omega, \text{ all } s \in [-\delta,\delta], s \neq 0. \tag{11.47}$$

For $\varepsilon > 0$ with $\varepsilon \hat{u}_1(x) \leq \min\{\delta, u_+(x)\}$ in $\Omega$, by (11.47), we see that

$$\max\{\varphi_{[v_-,0]}(-\varepsilon\hat{u}_1), \varphi_{[0,u_+]}(\varepsilon\hat{u}_1)\} < \frac{\varepsilon^p}{p}\int_\Omega (\lambda_1 - \mu)\hat{u}_1(x)^p \, dx < 0. \tag{11.48}$$

Note that, in case (b), the minimality of $u_+$ implies that $0, u_+$ are the only critical points of $\varphi_{[0,u_+]}$ (Proposition 11.8), and, similarly, $0, v_-$ are the only critical points of $\varphi_{[v_-,0]}$. In case (a), we may also suppose that $0, u_+$ are the only critical points of $\varphi_{[0,u_+]}$ and that $0, v_-$ are the only critical points of $\varphi_{[v_-,0]}$ [because otherwise we deduce that there is a third nontrivial solution of problem (11.30) belonging to either $[0, u_+]$ or $[v_-, 0]$, and we are done]. From Proposition 11.18 and (11.48) we deduce that

$$u_+ \text{ is the unique global minimizer of } \varphi_{[0,u_+]} \tag{11.49}$$

and

$$v_- \text{ is the unique global minimizer of } \varphi_{[v_-,0]}. \tag{11.50}$$

Note that the restrictions of the functionals $\varphi_{[0,u_+]}$ and $\varphi_{[v_-,u_+]}$ to $C_0^1(\overline{\Omega})_+$ coincide, so (11.49) implies that $u_+$ is a local $C_0^1(\overline{\Omega})$-minimizer of $\varphi_{[v_-,u_+]}$. By

Proposition 11.4, we deduce that $u_+$ is a local minimizer of $\varphi_{[v_-,u_+]}$ for the topology of $W_0^{1,p}(\Omega)$. Similarly, we can see that $v_-$ is a local minimizer of $\varphi_{[v_-,u_+]}$.

Note that we may assume that $v_-, u_+$ are isolated critical points of $\varphi_{[v_-,u_+]}$ because otherwise we find a sequence $\{u_n\}_{n \geq 1} \subset W_0^{1,p}(\Omega)$ of distinct solutions of (11.30) belonging to the order interval $[v_-, u_+]$, so in case (a) we deduce the existence of a third nontrivial solution $y_0 \in [v_-, u_+]$, whereas in case (b) the extremality of $v_-$ and $u_+$ implies that $y_0$ is nodal.

From Proposition 11.8 we know that $\varphi_{[v_-,u_+]}$ has a global minimizer $z_0 \in [v_-, u_+]$, and we have $\varphi_{[v_-,u_+]}(z_0) < 0$ [see (11.48)], hence $z_0 \neq 0$. If $z_0 \neq u_+$ and $z_0 \neq v_-$, then $z_0$ is the third desired solution of (11.30) [nodal in case (b)].

It remains to study the case where $z_0 = u_+$ or $z_0 = v_-$. Say $z_0 = u_+$ (the other case can be analogously treated). Since $u_+, v_-$ are strict local minimizers of $\varphi_{[v_-,u_+]}$, we can apply Proposition 5.42, which yields a critical point $y_0 \in W_0^{1,p}(\Omega)$ of $\varphi_{[v_-,u_+]}$ [hence a solution of problem (11.30) belonging to $C_0^1(\overline{\Omega}) \cap [v_-, u_+]$; see Propositions 11.8(a) and 11.17] satisfying

$$\varphi_{[v_-,u_+]}(u_+) \leq \varphi_{[v_-,u_+]}(v_-) < \varphi_{[v_-,u_+]}(y_0) = \inf_{\gamma \in \Gamma} \max_{t \in [-1,1]} \varphi_{[v_-,u_+]}(\gamma(t)), \quad (11.51)$$

where $\Gamma = \{\gamma \in C([-1,1], W_0^{1,p}(\Omega)) : \gamma(-1) = v_-, \ \gamma(1) = u_+\}$.

Clearly, (11.51) implies that $y_0$ is distinct from $v_-, u_+$. If we know that $y_0 \neq 0$, then $y_0$ is the desired third nontrivial solution of problem (11.30) [nodal in case (b) in view of the extremality of $v_-, u_+$]. Hence, to complete the proof of the theorem, it remains to check that $y_0 \neq 0$. To do this, we show that

$$\varphi_{[v_-,u_+]}(y_0) < 0. \quad (11.52)$$

To this end [by (11.51)], it is sufficient to construct a path $\overline{\gamma}_0 \in \Gamma$ such that

$$\varphi_{[v_-,u_+]}(\overline{\gamma}_0(t)) < 0 \quad \text{for all } t \in [-1,1]. \quad (11.53)$$

The rest of the proof is devoted to this purpose.

Denote $S = \{u \in W_0^{1,p}(\Omega) : \|u\|_p = 1\}$ endowed with the $W_0^{1,p}(\Omega)$-topology and $S_C = S \cap C_0^1(\overline{\Omega})$ equipped with the $C_0^1(\overline{\Omega})$-topology. Evidently, $S_C$ is dense in $S$ in the $W_0^{1,p}(\Omega)$-topology. Setting $\Gamma_0 = \{\gamma \in C([-1,1], S) : \gamma(-1) = -\hat{u}_1, \ \gamma(1) = \hat{u}_1\}$ and $\Gamma_{0,C} = \{\gamma \in C([-1,1], S_C) : \gamma(-1) = -\hat{u}_1, \ \gamma(1) = \hat{u}_1\}$, we have that $\Gamma_{0,C}$ is dense in $\Gamma_0$. Recall from Theorem 9.50 the following variational characterization of $\lambda_2 > 0$:

$$\lambda_2 = \inf_{\gamma \in \Gamma_0} \max_{u \in \gamma([-1,1])} \|\nabla u\|_p^p.$$

Since $\mu > \lambda_2$ [see (11.47)], we can find $\hat{\gamma}_0 \in \Gamma_{0,C}$ such that

$$\max\{\|\nabla u\|_p^p : u \in \hat{\gamma}_0([-1,1])\} < \mu. \quad (11.54)$$

*Claim 1:* There is $\varepsilon > 0$ such that $\|\varepsilon u\|_\infty \le \delta$ and $\varepsilon u \in [v_-, u_+]$ for all $u \in \hat{\gamma}_0([-1,1])$.

The set $\hat{\gamma}_0([-1,1])$ is compact, hence it is bounded in $C_0^1(\overline{\Omega})$, and so in $L^\infty(\Omega)$. Thus, we can find $\varepsilon_1 > 0$ satisfying the first property of Claim 1. To show the second property, note that for each $u \in \hat{\gamma}_0([-1,1])$ we can find a constant $\varepsilon_u > 0$ such that $-v_- - \varepsilon_u u$ and $u_+ - \varepsilon_u u$ belong to $\mathrm{int}\,(C_0^1(\overline{\Omega})_+)$. Thus, we also find a neighborhood $V_u \subset C_0^1(\overline{\Omega})$ such that $-v_- - \varepsilon_u v, u_+ - \varepsilon_u v \in \mathrm{int}\,(C_0^1(\overline{\Omega})_+)$ for all $v \in V_u$. Since $\hat{\gamma}_0([-1,1])$ is compact, it is covered by a finite number $V_{u_1}, \dots, V_{u_\ell}$ of such neighborhoods. It follows that the number $\varepsilon_2 := \min\{\varepsilon_{u_1}, \dots, \varepsilon_{u_\ell}\}$ satisfies the second property of Claim 1. Thus, $\varepsilon := \min\{\varepsilon_1, \varepsilon_2\}$ satisfies Claim 1.

Fix $\varepsilon > 0$ as in Claim 1. Then, from (11.47) and (11.54), and since $\hat{\gamma}_0([-1,1]) \subset S$, we obtain

$$\varphi_{[v_-, u_+]}(\varepsilon u) \le \frac{\varepsilon^p}{p} \|\nabla u\|_p^p - \frac{\varepsilon^p}{p} \mu \|u\|_p^p < 0 \quad \text{for all } u \in \hat{\gamma}_0([-1,1]).$$

So the path $\gamma_0 := \varepsilon \hat{\gamma}_0$ joining $-\varepsilon \hat{u}_1$ and $\varepsilon \hat{u}_1$ verifies

$$\varphi_{[v_-, u_+]}(u) < 0 \quad \text{for all } u \in \gamma_0([-1,1]). \tag{11.55}$$

Next we construct a path $\gamma_+$ joining $\varepsilon \hat{u}_1$ with $u_+$ along which $\varphi_{[v_-, u_+]}$ is negative. To do this, we may assume that $u_+ \ne \varepsilon \hat{u}_1$ (otherwise the path $\gamma_+ \equiv u_+$ satisfies our requirements). We rely on the second deformation lemma (Theorem 5.34), which we will apply to the functional $\varphi_{[0,u_+]}$. Let $a = \varphi_{[0,u_+]}(u_+)$ and $b = \varphi_{[0,u_+]}(\varepsilon \hat{u}_1)$. Thus, $a < b < 0$, and $u_+$ is the only critical point of $\varphi_{[0,u_+]}$ with critical value $a$ [by (11.48) and (11.49)]; moreover, $(a,b]$ contains no critical value of $\varphi_{[0,u_+]}$ (since $0, u_+$ are the only critical points of $\varphi_{[0,u_+]}$). These properties, together with the fact that $\varphi_{[0,u_+]}$ satisfies the (PS)-condition [Proposition 11.8(b)], allow us to apply Theorem 5.34, which provides a continuous mapping $h : [0,1] \times \varphi_{[0,u_+]}^b \to \varphi_{[0,u_+]}^b$ (with $\varphi_{[0,u_+]}^b = \{u \in W_0^{1,p}(\Omega) : \varphi_{[0,u_+]}(u) \le b\}$) such that, for all $u \in \varphi_{[0,u_+]}^b$, we have

$$h(0,u) = u, \quad h(1,u) = u_+, \quad \text{and } \varphi_{[0,u_+]}(h(t,u)) \le \varphi_{[0,u_+]}(u) \text{ for all } t \in [0,1]$$

[recall that $\varphi_{[0,u_+]}^a = \{u_+\}$; cf. (11.49)]. Then we consider the path $\gamma_+ : [0,1] \to W_0^{1,p}(\Omega)$ defined by

$$\gamma_+(t) = h(t, \varepsilon \hat{u}_1)^+ \quad \text{for all } t \in [0,1].$$

Clearly, $\gamma_+$ is continuous, and we have $\gamma_+(0) = \varepsilon \hat{u}_1$ and $\gamma_+(1) = u_+$. We claim that

$$\varphi_{[v_-, u_+]}(u) < 0 \quad \text{for all } u \in \gamma_+([0,1]). \tag{11.56}$$

Indeed, let $u \in \gamma_+([0,1])$, so we have that $u = h(t, \varepsilon \hat{u}_1)^+$ for some $t \in [0,1]$. Since $F_{[0,u_+]}(-h(t, \varepsilon \hat{u}_1)^-) = 0$ [see (11.4)], the definition of $\varphi_{[0,u_+]}$ [see (11.5)] yields

$\varphi_{[0,u_+]}(u) \leq \varphi_{[0,u_+]}(h(t, \varepsilon \hat{u}_1))$, whence

$$\varphi_{[v_-,u_+]}(u) = \varphi_{[0,u_+]}(u) \leq \varphi_{[0,u_+]}(h(t, \varepsilon \hat{u}_1)) \leq \varphi_{[0,u_+]}(\varepsilon \hat{u}_1) < 0,$$

where the last inequality follows from (11.55). Therefore, we have checked (11.56). Similarly, we construct a path $\gamma_- : [0,1] \to W_0^{1,p}(\Omega)$ satisfying

$$\gamma_-(0) = -\varepsilon \hat{u}_1, \quad \gamma_-(1) = v_-, \quad \text{and} \quad \varphi_{[v_-,u_+]}(u) < 0 \quad \text{for all } u \in \gamma_-([0,1]). \quad (11.57)$$

Concatenating the paths $\gamma_-, \gamma_0, \gamma_+$, we obtain a path $\overline{\gamma}_0 \in \Gamma$ that satisfies (11.53) [see (11.55), (11.56), (11.57)]. This implies (11.52). The proof of the theorem is now complete. $\qquad \square$

To obtain explicit existence and multiplicity results for problem (11.30), we will apply the previous results, and to do this we need to construct a positive upper solution and a negative lower solution of problem (11.30). We focus on two particular situations: coercive problems and $(p-1)$-superlinear parametric problems.

## Multiple Solutions for Coercive Problems

Here we deal with problem (11.30) in the case where the corresponding energy functional is coercive. In this respect, we formulate the following conditions.

$H(f)_4^+$ (i) $f : \Omega \times \mathbb{R} \to \mathbb{R}$ is a Carathéodory function with $f(x,0) = 0$ a.e. in $\Omega$, and there are constants $c > 0$ and $r \in [1, +\infty)$ such that

$$|f(x,s)| \leq c(1 + |s|^{r-1}) \quad \text{for a.a. } x \in \Omega, \text{ all } s \in \mathbb{R}.$$

(ii) There exists $\eta \in L^\infty(\Omega)$ such that $\eta(x) \geq \lambda_1$ a.e. in $\Omega$, $\eta \neq \lambda_1$ and

$$\liminf_{s \downarrow 0} \frac{f(x,s)}{s^{p-1}} \geq \eta(x) \quad \text{uniformly for a.a. } x \in \Omega.$$

(iii) There exists $\vartheta \in L^\infty(\Omega)$ such that $\vartheta(x) \leq \lambda_1$ a.e. in $\Omega$, $\vartheta \neq \lambda_1$ and

$$\limsup_{s \to +\infty} \frac{f(x,s)}{s^{p-1}} \leq \vartheta(x) \quad \text{uniformly for a.a. } x \in \Omega.$$

Symmetrically,

$H(f)_4^-$ (i) $f : \Omega \times \mathbb{R} \to \mathbb{R}$ is a Carathéodory function with $f(x,0) = 0$ a.e. in $\Omega$, and there are constants $c > 0$ and $r \in [1, +\infty)$ such that

$$|f(x,s)| \leq c(1 + |s|^{r-1}) \quad \text{for a.a. } x \in \Omega, \text{ all } s \in \mathbb{R}.$$

(ii) There exists $\eta \in L^\infty(\Omega)$ such that $\eta(x) \geq \lambda_1$ a.e. in $\Omega$, $\eta \neq \lambda_1$ and

$$\liminf_{s\uparrow 0} \frac{f(x,s)}{|s|^{p-2}s} \geq \eta(x) \text{ uniformly for a.a. } x \in \Omega.$$

(iii) There exists $\vartheta \in L^\infty(\Omega)$ such that $\vartheta(x) \leq \lambda_1$ a.e. in $\Omega$, $\vartheta \neq \lambda_1$ and

$$\limsup_{s\to-\infty} \frac{f(x,s)}{|s|^{p-2}s} \leq \vartheta(x) \text{ uniformly for a.a. } x \in \Omega.$$

*Remark 11.24.* Note that hypothesis $H(f)_4^+$ (iii) is of the same type as hypothesis $H(f)_1$ (ii) of Sect. 11.1, and both imply that the energy functional associated to the problem is coercive. Here, hypothesis $H(f)_4^+$ (ii), indicating the behavior of the nonlinearity $f$ near 0 (nonuniform nonresonance from the right with respect to the first eigenvalue $\lambda_1$), complements hypothesis $H(f)_1$ (iv) of Sect. 11.1 (nonuniform nonresonance from the left with respect to the first eigenvalue $\lambda_1$).

Under these hypotheses, we provide the following existence result for constant-sign solutions.

**Proposition 11.25.**

(a) *Assume that hypothesis $H(f)_4^+$ holds. Then problem (11.30) has a solution $u_+ \in \text{int}(C_0^1(\overline{\Omega})_+)$. If, in addition, $H(f)_2^+$ (ii) holds, then problem (11.30) has a smallest positive solution $u_+ \in \text{int}(C_0^1(\overline{\Omega})_+)$.*
(b) *Assume that hypothesis $H(f)_4^-$ holds. Then problem (11.30) has a solution $v_- \in -\text{int}(C_0^1(\overline{\Omega})_+)$. If, in addition, $H(f)_2^-$ (ii) holds, then problem (11.30) has a biggest negative solution $v_- \in -\text{int}(C_0^1(\overline{\Omega})_+)$.*

*Proof.* We only prove (a) because the proof of (b) follows the same pattern.

*Claim 1:* There exists an upper solution $\overline{u} \in \text{int}(C_0^1(\overline{\Omega})_+)$ of problem (11.30).

Applying Lemma 11.3 to the function $\vartheta$ of $H(f)_4^+$ (iii), we find a constant $c_1 > 0$ such that

$$\|\nabla u\|_p^p - \int_\Omega \vartheta(x)|u(x)|^p \, dx \geq c_1 \|\nabla u\|_p^p \text{ for all } u \in W_0^{1,p}(\Omega). \tag{11.58}$$

By virtue of $H(f)_4^+$ (i), (iii), given $\varepsilon \in (0, c_1\lambda_1)$, we can find $c_\varepsilon > 0$ such that

$$f(x,s) < (\vartheta(x)+\varepsilon)s^{p-1} + c_\varepsilon(x) \text{ for a.a. } x \in \Omega, \text{ all } s \geq 0. \tag{11.59}$$

Let $K_\varepsilon : L^p(\Omega) \to L^{p'}(\Omega)$ be the nonlinear operator defined by $K_\varepsilon(u)(\cdot) = (\vartheta(\cdot) + \varepsilon)|u(\cdot)|^{p-2}u(\cdot)$. Clearly, $K_\varepsilon$ is bounded and continuous, and the restriction $K_\varepsilon|_{W_0^{1,p}(\Omega)}$ is compact and, thus, pseudomonotone (Remark 2.59). Taking into account that $-\Delta_p : W_0^{1,p}(\Omega) \to W^{-1,p'}(\Omega)$ is maximal monotone (Corollary 2.42), and thus pseudomonotone (Proposition 2.60), the operator $-\Delta_p - K_\varepsilon : W_0^{1,p}(\Omega) \to W^{-1,p'}(\Omega)$ is pseudomonotone (Proposition 2.61). Moreover, using (11.58) and Proposition 9.47(a), we have

$$\frac{1}{\|\nabla u\|_p}\langle -\Delta_p u - K_\varepsilon(u), u\rangle \geq \left(c_1 - \frac{\varepsilon}{\lambda_1}\right)\|\nabla u\|_p^{p-1} \text{ for all } u \in W_0^{1,p}(\Omega).$$

Since $\varepsilon < c_1\lambda_1$, we infer that the operator $-\Delta_p - K_\varepsilon$ is strongly coercive (Definition 2.50), so $-\Delta_p - K_\varepsilon$ is surjective (Theorem 2.63). Thus, there exists $\bar{u} \in W_0^{1,p}(\Omega)$ satisfying

$$\begin{cases} -\Delta_p\bar{u} = (\vartheta(x) + \varepsilon)|\bar{u}|^{p-2}\bar{u} + c_\varepsilon & \text{in } \Omega, \\ \bar{u} = 0 & \text{on } \partial\Omega. \end{cases} \tag{11.60}$$

Acting on (11.60) with the test function $-\bar{u}^- \in W_0^{1,p}(\Omega)$, by (11.58) and Proposition 9.47(a), we obtain

$$c_1\|\nabla\bar{u}^-\|_p^p \leq \|\nabla\bar{u}^-\|_p^p - \int_\Omega \vartheta|\bar{u}^-|^p\,dx \leq \varepsilon\|\bar{u}^-\|_p^p \leq \frac{\varepsilon}{\lambda_1}\|\nabla\bar{u}^-\|_p^p.$$

Again, since $\varepsilon < c_1\lambda_1$, we deduce that $\bar{u} \geq 0$. From (11.60) and since $c_\varepsilon > 0$, we have $\bar{u} \neq 0$. Nonlinear regularity theory (Corollary 8.13) and the strong maximum principle (Theorem 8.27) yield $\bar{u} \in \text{int}(C_0^1(\overline{\Omega})_+)$. Finally, the fact that $\bar{u}$ is an upper solution of problem (11.30) can be seen by combining (11.59) and (11.60). This establishes Claim 1.

Now we can apply Proposition 11.18(a) to the upper solution $\bar{u}$ obtained in Claim 1 and by taking $\underline{u} = 0$, which yields a solution $u_+ \in \text{int}(C_0^1(\overline{\Omega})_+)$ of problem (11.30). If $H(f)_2^+$ (ii) holds, then, in view of Proposition 11.22(a), we can choose $u_+$ as the smallest positive solution of (11.30).                                           $\square$

Let us now state:

$H(f)_5$ (i) $f : \Omega \times \mathbb{R} \to \mathbb{R}$ is a Carathéodory function with $f(x,0) = 0$ a.e. in $\Omega$, and there are constants $c > 0$ and $r \in [1, +\infty)$ such that

$$|f(x,s)| \leq c(1 + |s|^{r-1}) \text{ for a.a. } x \in \Omega, \text{ all } s \in \mathbb{R};$$

(ii)  One of the following conditions is satisfied: either

(ii.a)  There exists a constant $\mu_0 > \lambda_2$ such that

$$\mu_0 < \liminf_{s \to 0} \frac{f(x,s)}{|s|^{p-2}s} \text{ uniformly for a.a. } x \in \Omega$$

or

(ii.b)  There are $\mu_0 > \lambda_2$ and $\hat{\eta} \in L^\infty(\Omega)$ such that

$$\mu_0 < \liminf_{s \to 0} \frac{f(x,s)}{|s|^{p-2}s} \leq \limsup_{s \to 0} \frac{f(x,s)}{|s|^{p-2}s} \leq \hat{\eta}(x) \quad \text{uniformly for a.a. } x \in \Omega.$$

(iii)  There exists $\vartheta \in L^\infty(\Omega)$ such that $\vartheta(x) \leq \lambda_1$ a.e. in $\Omega$, $\vartheta \neq \lambda_1$, and

$$\limsup_{s \to \pm\infty} \frac{f(x,s)}{|s|^{p-2}s} \leq \vartheta(x) \quad \text{uniformly for a.a. } x \in \Omega.$$

**Theorem 11.26.**

(a) *Under* $H(f)_5$, *problem (11.30) has at least three distinct, nontrivial solutions* $u_0 \in \mathrm{int}\,(C_0^1(\overline{\Omega})_+)$, $v_0 \in -\mathrm{int}\,(C_0^1(\overline{\Omega})_+)$, *and* $y_0 \in C_0^1(\overline{\Omega})$ *satisfying* $v_0 \leq y_0 \leq u_0$ *in* $\Omega$.

(b) *If, in addition, hypothesis* $H(f)_5$ (ii.b) *holds, then the solution* $y_0$ *can be chosen to be nodal.*

*Proof.* The solutions $u_+$ and $v_-$ obtained in Proposition 11.25 are in particular upper and lower solutions of (11.30). Then the result is obtained by applying Theorem 11.23 to the functions $\overline{u} = u_+$ and $\underline{v} = v_-$. $\qquad\Box$

*Example 11.27.* The following nonlinearity satisfies $H(f)_5$ [including $H(f)_5$ (ii.b)]:

$$f(x,s) = \begin{cases} \vartheta(x)|s|^{p-2}s - \eta(x) & \text{if } s < -1, \\ \vartheta(x)|s|^{r-2}s + \eta(x)|s|^{p-2}s & \text{if } -1 \leq s < 0, \\ \eta(x)s^{p-1} & \text{if } 0 \leq s \leq 1, \\ \mu(x)s^{q-1} + \eta(x) - \mu(x) & \text{if } s > 1, \end{cases}$$

where $1 < q < p < r$ and $\vartheta, \eta, \mu \in L^\infty(\Omega)$ are such that $\vartheta(x) \leq \lambda_1$ a.e. in $\Omega$, $\vartheta \neq \lambda_1$, and $\eta(x) > \mu_0 > \lambda_2$ a.e. in $\Omega$. Thus, Theorem 11.26 shows that problem (11.30) admits at least three nontrivial solutions: one positive, one negative, and one nodal.

**Multiple Solutions for ($p$-1)-Superlinear Problems**

Another important class of problems that fit into the preliminary results of this section consists of certain parametric problems:

$$\begin{cases} -\Delta_p u = f(x, u(x), \lambda) & \text{in } \Omega, \\ u = 0 & \text{on } \partial\Omega, \end{cases} \tag{11.61}$$

where $\lambda$ is a parameter belonging to the interval $\Lambda := (0, \overline{\lambda})$, with $\overline{\lambda} > 0$. We consider the following hypotheses on the nonlinearity $f(x, s, \lambda)$:

H$(f)_6^+$ (i) $f : \Omega \times \mathbb{R} \times \Lambda \to \mathbb{R}$ is such that $f(\cdot, \cdot, \lambda)$ is a Carathéodory function, $f(x, 0, \lambda) = 0$ a.e. in $\Omega$, for all $\lambda \in \Lambda$; moreover, there are numbers $a(\lambda) > 0$, with $a(\lambda) \to 0$ as $\lambda \downarrow 0$, and $c > 0$, $r > p$ (independent of $\lambda$), such that

$$|f(x, s, \lambda)| \leq a(\lambda) + c|s|^{r-1} \text{ for a.a. } x \in \Omega, \text{ all } s \in \mathbb{R}, \text{ all } \lambda \in \Lambda;$$

(ii) For every $\lambda \in \Lambda$ there exists $\eta_\lambda \in L^\infty(\Omega)$ such that $\eta_\lambda \geq \lambda_1$ a.e. in $\Omega$, $\eta_\lambda \neq \lambda_1$, and

$$\liminf_{s \downarrow 0} \frac{f(x, s, \lambda)}{s^{p-1}} \geq \eta_\lambda(x) \text{ uniformly for a.a. } x \in \Omega.$$

Symmetrically, we formulate the following conditions:

H$(f)_6^-$ (i) $f : \Omega \times \mathbb{R} \times \Lambda \to \mathbb{R}$ is such that $f(\cdot, \cdot, \lambda)$ is a Carathéodory function, $f(x, 0, \lambda) = 0$ a.e. in $\Omega$, for all $\lambda \in \Lambda$; moreover, there are numbers $a(\lambda) > 0$ with $a(\lambda) \to 0$ as $\lambda \downarrow 0$, and $c > 0$, $r > p$ (independent of $\lambda$), such that

$$|f(x, s, \lambda)| \leq a(\lambda) + c|s|^{r-1} \text{ for a.a. } x \in \Omega, \text{ all } s \in \mathbb{R}, \text{ all } \lambda \in \Lambda;$$

(ii) For every $\lambda \in \Lambda$ there exists $\eta_\lambda \in L^\infty(\Omega)$ such that $\eta_\lambda \geq \lambda_1$ a.e. in $\Omega$, $\eta_\lambda \neq \lambda_1$, and

$$\liminf_{s \uparrow 0} \frac{f(x, s, \lambda)}{|s|^{p-2}s} \geq \eta_\lambda(x) \text{ uniformly for a.a. } x \in \Omega.$$

First we are concerned with constant-sign solutions.

**Proposition 11.28.**

(a) *Under* H$(f)_6^+$, *for all* $b > 0$, *there exists* $\lambda^* \in \Lambda$ *such that for* $\lambda \in (0, \lambda^*)$, *problem (11.61) has a solution* $u_0 \in \text{int}(C_0^1(\overline{\Omega})_+)$, *with* $\|u_0\|_\infty < b$.
(b) *Under* H$(f)_6^-$, *for all* $b > 0$, *there exists* $\lambda^* \in \Lambda$ *such that for* $\lambda \in (0, \lambda^*)$, *problem (11.61) has a solution* $v_0 \in -\text{int}(C_0^1(\overline{\Omega})_+)$, *with* $\|v_0\|_\infty < b$.

*Proof.* We only prove part (a) because the proof of part (b) is similar.

*Claim 1:* There exists $e \in \text{int}(C_0^1(\overline{\Omega})_+)$ such that $-\Delta_p e = 1$ in $W^{-1,p'}(\Omega)$.

The operator $-\Delta_p : W_0^{1,p}(\Omega) \to W^{-1,p'}(\Omega)$ is maximal monotone (Corollary 2.42), coercive, and so surjective (Theorem 2.55). Hence, there is $e \in W_0^{1,p}(\Omega)$, $e \neq 0$, with $-\Delta_p e = 1$ in $W^{-1,p'}(\Omega)$. This relation yields $\|\nabla e^-\|_p^p = \int_\Omega (-e^-) \, dx \leq 0$; thus, $e \geq 0$ in $\Omega$. Finally, Corollary 8.13 and Theorem 8.27 imply that $e \in \text{int}(C_0^1(\overline{\Omega})_+)$.

Fix $b > 0$.

*Claim 2:* There is $\lambda^* \in \Lambda$ such that, for all $\lambda \in (0,\lambda^*)$, there is $t_\lambda \in (0, \frac{b}{\|e\|_\infty})$, with

$$a(\lambda) + c(t_\lambda \|e\|_\infty)^{r-1} < t_\lambda^{p-1},$$

where $a(\lambda), c > 0$ and $r > p$ are as in H$(f)_6^+$ (i).

Arguing by contradiction, assume that we can find a sequence $\{\lambda_n\}_{n\geq1} \subset \Lambda$ such that $\lambda_n \to 0$ as $n \to \infty$ and

$$a(\lambda_n) + c(t\|e\|_\infty)^{r-1} \geq t^{p-1} \quad \text{for all } t \in (0, \tfrac{b}{\|e\|_\infty}), \text{ all } n \geq 1.$$

Letting $n \to \infty$ in this relation [using that $a(\lambda_n) \to 0$ as $n \to \infty$, by H$(f)_6^+$ (i)], we obtain that $c\|e\|_\infty^{r-1} t^{r-p} \geq 1$ for all $t \in (0, \frac{b}{\|e\|_\infty})$. Since $r - p > 0$, this is impossible. This proves Claim 2.

*Claim 3:* For every $\lambda \in (0,\lambda^*)$, problem (11.61) has an upper solution $\bar{u}_\lambda \in$ int $(C_0^1(\overline{\Omega})_+)$, with $\|\bar{u}_\lambda\|_\infty < b$.

Fix $\lambda \in (0,\lambda^*)$, and let $t_\lambda \in (0, \frac{b}{\|e\|_\infty})$ be the number provided by Claim 2. We set $\bar{u}_\lambda = t_\lambda e$. Then $\bar{u}_\lambda \in$ int $(C_0^1(\overline{\Omega})_+)$, $\|\bar{u}_\lambda\|_\infty < b$, and we have $-\Delta_p \bar{u}_\lambda = t_\lambda^{p-1}$ in $W^{-1,p'}(\Omega)$. By Claim 2 and hypothesis H$(f)_6^+$ (i), we see that

$$-\Delta_p \bar{u}_\lambda > a(\lambda) + c\|\bar{u}_\lambda\|_\infty^{r-1} \geq f(x,s,\lambda) \quad \text{for a.a. } x \in \Omega, \text{ all } s \in [0,\bar{u}_\lambda(x)]. \quad (11.62)$$

This implies that $\bar{u}_\lambda$ is an upper solution of problem (11.61), so Claim 3 is proven.

Part (a) of the statement is then obtained by applying Proposition 11.18 (a) with $\bar{u} = \bar{u}_\lambda$ from Claim 3 and $\underline{u} = 0$. $\qquad\square$

Now we strengthen our assumptions on $f$ to be $(p-1)$-superlinear at infinity [see hypotheses (iii) below]. In Theorem 11.30, we show that these hypotheses yield the existence of additional constant-sign solutions for problem (11.61).

H$(f)_7^+$ (i) $f : \Omega \times \mathbb{R} \times \Lambda \to \mathbb{R}$ is such that $f(\cdot,\cdot,\lambda)$ is a Carathéodory function, $f(x,0,\lambda) = 0$ a.e. in $\Omega$, for all $\lambda \in \Lambda$; moreover, there are numbers $a(\lambda) > 0$ with $a(\lambda) \to 0$ as $\lambda \downarrow 0$, and $c > 0$, $r \in (p,p^*)$ (independent of $\lambda$), such that

$$|f(x,s,\lambda)| \leq a(\lambda) + c|s|^{r-1} \quad \text{for a.a. } x \in \Omega, \text{ all } s \in \mathbb{R}, \text{ all } \lambda \in \Lambda;$$

(ii) For every $\lambda \in \Lambda$ there exists $\eta_\lambda \in L^\infty(\Omega)$ such that $\eta_\lambda \geq \lambda_1$ a.e. in $\Omega$, $\eta_\lambda \neq \lambda_1$, and

$$\liminf_{s\downarrow0} \frac{f(x,s,\lambda)}{s^{p-1}} \geq \eta_\lambda(x) \quad \text{uniformly for a.a. } x \in \Omega;$$

(iii) For every $\lambda \in \Lambda$ there exist $M_\lambda > 0$ and $\mu_\lambda > p$ such that

$$0 < \mu_\lambda F(x,s,\lambda) \leq f(x,s,\lambda)s \text{ for a.a. } x \in \Omega, \text{ all } s \geq M_\lambda,$$

where $F(x,s,\lambda) = \int_0^s f(x,\tau,\lambda)\,d\tau$;

(iv) There exists $\rho > 0$ such that $f(x,s,\lambda) > 0$ for a.a. $x \in \Omega$, all $s \in (0,\rho)$, and all $\lambda \in \Lambda$.

Symmetrically, we state:

$H(f)_7^-$ (i) $f : \Omega \times \mathbb{R} \times \Lambda \to \mathbb{R}$ is such that $f(\cdot,\cdot,\lambda)$ is a Carathéodory function, $f(x,0,\lambda) = 0$ a.e. in $\Omega$, for all $\lambda \in \Lambda$; moreover, there are numbers $a(\lambda) > 0$ with $a(\lambda) \to 0$ as $\lambda \downarrow 0$, and $c > 0$, $r \in (p,p^*)$ (independent of $\lambda$), such that

$$|f(x,s,\lambda)| \leq a(\lambda) + c|s|^{r-1} \text{ for a.a. } x \in \Omega, \text{ all } s \in \mathbb{R}, \text{ all } \lambda \in \Lambda;$$

(ii) For every $\lambda \in \Lambda$ there exists $\eta_\lambda \in L^\infty(\Omega)$ such that $\eta_\lambda \geq \lambda_1$ a.e. in $\Omega$, $\eta_\lambda \neq \lambda_1$, and

$$\liminf_{s \uparrow 0} \frac{f(x,s,\lambda)}{|s|^{p-2}s} \geq \eta_\lambda(x) \text{ uniformly for a.a. } x \in \Omega;$$

(iii) For every $\lambda \in \Lambda$ there exist $M_\lambda > 0$ and $\mu_\lambda > p$ such that

$$0 < \mu_\lambda F(x,s,\lambda) \leq f(x,s,\lambda)s \text{ for a.a. } x \in \Omega, \text{ all } s \leq -M_\lambda;$$

(iv) There exists $\rho > 0$ such that $f(x,s,\lambda) < 0$ for a.a. $x \in \Omega$, all $s \in (-\rho,0)$, and all $\lambda \in \Lambda$.

*Remark 11.29.*

(a) Hypothesis $H(f)_7^\pm$ (iii) is a nonuniform version of the so-called Ambrosetti–Rabinowitz condition, which is needed to guarantee that the energy functional associated to the problem satisfies the (PS)-condition. It forces the nonlinearity $f(x,s)$ to be $(p-1)$-superlinear at infinity.

(b) Note that the sign condition [hypothesis $H(f)_7^\pm$ (iv)] is uniform with respect to $\lambda$ (i.e., satisfied in a neighborhood of 0 that is independent of $\lambda$). A nonuniform sign condition [i.e., satisfied by $f(x,s,\lambda)$ for a fixed $\lambda$] is already implied by hypothesis $H(f)_7^\pm$ (ii).

**Theorem 11.30.**

(a) *Under $H(f)_7^+$, for all $b > 0$ there exists $\lambda^* \in \Lambda$ such that for $\lambda \in (0,\lambda^*)$ problem (11.61) has at least two distinct solutions $u_0, \hat{u} \in \text{int}(C_0^1(\overline{\Omega})_+)$, with $u_0 \leq \hat{u}$ in $\Omega$ and $\|u_0\|_\infty < b$.*

(b) *Under $H(f)_7^-$, for all $b > 0$ there exists $\lambda^* \in \Lambda$ such that for $\lambda \in (0,\lambda^*)$ problem (11.61) has at least two distinct solutions $v_0, \hat{v} \in -\text{int}(C_0^1(\overline{\Omega})_+)$, with $\hat{v} \leq v_0$ in $\Omega$ and $\|v_0\|_\infty < b$.*

*Proof.* Again, we only prove part (a) of the statement since the proof of part (b) is similar. Note that, up to dealing with $\min\{b,\rho\}$ instead of $b$, we may assume that $b \leq \rho$, where $\rho$ is as in $H(f)_7^+$ (iv).

First, we apply Proposition 11.28(a) to $b$, and this yields $\lambda^* \in \Lambda$ such that, for every $\lambda \in (0,\lambda^*)$, we find a $u_0 \in \mathrm{int}\,(C_0^1(\overline{\Omega})_+)$ solution of (11.61), with $\|u_0\|_\infty < b$. Moreover, let $\overline{u} := \overline{u}_\lambda$ be the upper solution of (11.61) constructed in Claim 3 of the proof of Proposition 11.28. In particular, it satisfies relation (11.62), and, according to the proof of Proposition 11.28, we have $u_0 \in [0,\overline{u}]$. In the rest of the proof, we fix $\lambda \in (0,\lambda^*)$ and abbreviate $f(x,s) = f(x,s,\lambda)$. We will apply to $f$ the truncation techniques developed in Sect. 11.1.

In particular, the fact that $u_0 \leq \overline{u}$ in $\Omega$ allows us to consider the truncation $f_{[u_0,\overline{u}]}$ and the corresponding energy functional $\varphi_{[u_0,\overline{u}]}$ [see (11.4) and (11.5)]. Applying Proposition 11.18 to $\overline{u}$ and $\underline{u} = u_0$, we know that there is $\tilde{u} \in C_0^1(\overline{\Omega}) \cap [u_0,\overline{u}]$ realizing the infimum of $\varphi_{[u_0,\overline{u}]}$. Actually, we may assume that $u_0 = \tilde{u}$ [otherwise $\tilde{u}$ is the second positive solution of (11.61) that we were looking for (Proposition 11.8(a))]. Therefore, we may assume that

$$u_0 \text{ is a global minimizer of } \varphi_{[u_0,\overline{u}]}. \tag{11.63}$$

*Claim 1:* $\overline{u} - u_0 \in \mathrm{int}\,(C_0^1(\overline{\Omega})_+)$.

Using that $u_0$ is a solution of (11.61), the fact that $\|u_0\|_\infty < b \leq \rho$, and $H(f)_7^+$ (iv), we have that

$$-\Delta_p u_0 = f(x,u_0(x),\lambda) \text{ in } W^{-1,p'}(\Omega), \text{ with } f(x,u_0,\lambda) \geq 0 \text{ a.e. in } \Omega. \tag{11.64}$$

On the other hand, in Claim 3 of the proof of Proposition 11.28, the upper solution $\overline{u} = \overline{u}_\lambda$ is constructed so that $-\Delta_p \overline{u} = t_\lambda^{p-1}$ in $W^{-1,p'}(\Omega)$, for some $t_\lambda \in (0, \frac{b}{\|e\|_\infty})$, and by (11.62) and the fact that $0 \leq u_0 \leq \overline{u}$ in $\Omega$, we have

$$-\Delta_p \overline{u} = t_\lambda^{p-1} > f(x,u_0(x),\lambda) \text{ a.e. in } \Omega. \tag{11.65}$$

On the basis of (11.64) and (11.65), we can invoke Proposition 8.29. Claim 1 ensues.

Now we consider the truncation

$$f_{[u_0,+\infty]}(x,s) = \begin{cases} f(x,u_0(x),\lambda) & \text{if } s \leq u_0(x), \\ f(x,s,\lambda) & \text{if } s > u_0(x) \end{cases} \tag{11.66}$$

for a.a. $x \in \Omega$, all $s \in \mathbb{R}$, the primitive $F_{[u_0,+\infty]}(x,s) = \int_0^s f_{[u_0,+\infty]}(x,t)\,dt$, and the corresponding $C^1$-functional $\varphi_{[u_0,+\infty]} : W_0^{1,p}(\Omega) \to \mathbb{R}$ given by

$$\varphi_{[u_0,+\infty]}(u) = \frac{1}{p}\|\nabla u\|_p^p - \int_\Omega F_{[u_0,+\infty]}(x,u(x))\,dx \text{ for all } u \in W_0^{1,p}(\Omega),$$

which is well defined due to the growth condition in $H(f)_7^+$ (i) [where $r \in (p, p^*)$]. To complete the proof of the theorem, it suffices to show that the functional $\varphi_{[u_0, +\infty]}$ admits a critical point $\hat{u} \in W_0^{1,p}(\Omega)$, with $\hat{u} \neq u_0$ [Propositions 11.8(a) and 11.17(a)]. Note that the functionals $\varphi_{[u_0, +\infty]}$ and $\varphi_{[u_0, \bar{u}]}$ coincide on the set

$$V := \{u \in C_0^1(\overline{\Omega}) : \bar{u} - u \in \text{int}(C_0^1(\overline{\Omega}))\}.$$

The latter is an open subset of $C_0^1(\overline{\Omega})$, and, by (11.63), we have that $u_0$ is a minimizer of $\varphi_{[u_0, \bar{u}]}$ on $V$. Thus, $u_0$ is a local $C_0^1(\overline{\Omega})$-minimizer of $\varphi_{[u_0, +\infty]}$. Therefore, applying Proposition 11.4, we obtain that $u_0$ is a local minimizer of $\varphi_{[u_0, +\infty]}$ with respect to the topology of $W_0^{1,p}(\Omega)$. In the case where $u_0$ is not a strict local minimizer of $\varphi_{[u_0, +\infty]}$, we deduce the existence of further critical points of $\varphi_{[u_0, +\infty]}$, and then we are done. In this way, we may assume that

$$u_0 \text{ is a strict local minimizer of } \varphi_{[u_0, +\infty]}. \tag{11.67}$$

The next two claims point out additional properties of the functional $\varphi_{[u_0, +\infty]}$.

*Claim 2:* The functional $\varphi_{[u_0, +\infty]}$ satisfies the (PS)-condition.

Let $\{u_n\}_{n \geq 1} \subset W_0^{1,p}(\Omega)$ be a sequence such that $\{\varphi_{[u_0, +\infty]}(u_n)\}_{n \geq 1}$ is bounded and $\varphi'_{[u_0, +\infty]}(u_n) \to 0$ in $W^{-1,p'}(\Omega)$ as $n \to \infty$. The former property implies that

$$\frac{1}{p}\|\nabla u_n\|_p^p - \int_\Omega F_{[u_0, +\infty]}(x, u_n)\, dx \leq M_1 \text{ for all } n \geq 1, \tag{11.68}$$

for some $M_1 > 0$, and the latter property yields

$$\langle -\Delta_p u_n, v \rangle - \int_\Omega f_{[u_0, +\infty]}(x, u_n) v\, dx \leq \varepsilon_n \|\nabla v\|_p \tag{11.69}$$

for all $v \in W_0^{1,p}(\Omega)$, all $n \geq 1$, with $\varepsilon_n \to 0$ as $n \to \infty$. We take $v = -u_n^- \in W_0^{1,p}(\Omega)$ in (11.69). Note that $f_{[u_0, +\infty]}(x, -u_n^-)u_n^- = 0$ a.e. in $\{x \in \Omega : u_n(x) \geq 0\}$, whereas $f_{[u_0, +\infty]}(x, -u_n^-)u_n^- = f(x, u_0, \lambda)u_n^- \geq 0$ a.e. in $\{x \in \Omega : u_n(x) < 0\}$ [by (11.66), $H(f)_7^+$ (iv), and the fact that $\|u_0\|_\infty < b \leq \rho$]. Thus, from (11.69) we infer that

$$\|\nabla u_n^-\|_p^p \leq \|\nabla u_n^-\|_p^p + \int_\Omega f_{[u_0, +\infty]}(x, u_n)u_n^-\, dx \leq \varepsilon_n \|\nabla u_n^-\|_p \text{ for all } n \geq 1.$$

Since $p > 1$, it follows that $\{u_n^-\}_{n \geq 1}$ is bounded in $W_0^{1,p}(\Omega)$. Let $\mu_\lambda > p$ and $M_\lambda > 0$ be as in $H(f)_7^+$ (iii). Taking $v = u_n^+$ in (11.69), combining with (11.68), and using (11.66) and $H(f)_7^+$ (i), we obtain

$$\left(\frac{\mu_\lambda}{p} - 1\right)\|\nabla u_n^+\|_p^p + \int_{\{u_n \geq M_0\}} (f(x, u_n, \lambda)u_n - \mu_\lambda F(x, u_n, \lambda))\, dx$$

$$\leq M_2(1 + \|\nabla u_n^+\|_p)$$

for all $n \geq 1$, with some $M_2 > 0$, where $M_0 := \max\{M_\lambda, \|u_0\|_\infty\}$. From $H(f)_7^+$ (iii) we obtain that $\{u_n^+\}_{n\geq 1}$ is bounded in $W_0^{1,p}(\Omega)$. Therefore, $\{u_n\}_{n\geq 1}$ is bounded in $W_0^{1,p}(\Omega)$, so along a relabeled subsequence we have $u_n \xrightarrow{w} u$ in $W_0^{1,p}(\Omega)$, $u_n \to u$ in $L^r(\Omega)$, for some $u \in W_0^{1,p}(\Omega)$. Taking $v = u_n - u$ in (11.69), it follows that $\langle -\Delta_p u_n, u_n - u\rangle \to 0$ as $n \to \infty$. Then, invoking Proposition 2.72, we infer that $u_n \to u$ in $W_0^{1,p}(\Omega)$. Therefore, we have shown that $\varphi_{[u_0,+\infty]}$ satisfies the (PS)-condition.

*Claim 3:* $\lim\limits_{t \to +\infty} \varphi_{[u_0,+\infty]}(t\hat{u}_1) = -\infty$.

Note that hypotheses $H(f)_7^+$ (i), (iii) imply that $F(x,s,\lambda) \geq c_1 s^{\mu_\lambda} - c_2$ for a.a. $x \in \Omega$ and all $s \geq 0$, with $c_1, c_2 > 0$, whence

$$F_{[u_0,+\infty]}(x,s) \geq c_1 s^{\mu_\lambda} - \tilde{c}_2 \text{ for a.a. } x \in \Omega, \text{ all } s \geq 0,$$

for some $\tilde{c}_2 > 0$ [see (11.66)]. We infer that

$$\varphi_{[u_0,+\infty]}(t\hat{u}_1) \leq \frac{t^p}{p}\|\nabla\hat{u}_1\|_p^p - c_1 t^{\mu_\lambda}\|\hat{u}_1\|_{\mu_\lambda}^{\mu_\lambda} + \tilde{c}_2|\Omega|_N \to -\infty \text{ as } t \to +\infty. \quad (11.70)$$

This proves Claim 3.

Combining (11.67) with Claims 2 and 3, we can apply Proposition 5.42, which yields a critical point $\hat{u} \neq u_0$ of the functional $\varphi_{[u_0,+\infty]}$, and so a second positive solution of (11.61). The proof of the theorem is complete. $\qquad\square$

More insight into our multiplicity study can be achieved under further conditions on $f(x,s,\lambda)$, as follows.

$H(f)_8$   (i)   $f : \Omega \times \mathbb{R} \times \Lambda \to \mathbb{R}$ is such that $f(\cdot,\cdot,\lambda)$ is a Carathéodory function, $f(x,0,\lambda) = 0$ a.e. in $\Omega$, for all $\lambda \in \Lambda$; moreover, there are numbers $a(\lambda) > 0$ with $a(\lambda) \to 0$ as $\lambda \downarrow 0$, and $c > 0$, $r > p$ (independent of $\lambda$), such that

$$|f(x,s,\lambda)| \leq a(\lambda) + c|s|^{r-1} \text{ for a.a. } x \in \Omega, \text{ all } s \in \mathbb{R}, \text{ all } \lambda \in \Lambda;$$

     (ii)   One of the following conditions holds: either

        (ii.a)   For all $\lambda \in \Lambda$ there exists $\theta_\lambda > \lambda_2$ for which we have

$$\theta_\lambda < \liminf_{s \to 0} \frac{f(x,s,\lambda)}{|s|^{p-2}s} \text{ uniformly for a.a. } x \in \Omega$$

     or

        (ii.b)   For all $\lambda \in \Lambda$ there exist $\theta_\lambda > \lambda_2$ and $\hat{\eta}_\lambda \in L^\infty(\Omega)$ such that

$$\theta_\lambda < \liminf_{s \to 0} \frac{f(x,s,\lambda)}{|s|^{p-2}s} \leq \limsup_{s \to 0} \frac{f(x,s,\lambda)}{|s|^{p-2}s} \leq \hat{\eta}_\lambda(x)$$

        uniformly for a.a. $x \in \Omega$;

(iii)  For every $\lambda \in \Lambda$, there exist $M_\lambda > 0$ and $\mu_\lambda > p$ such that

$$0 < \mu_\lambda F(x,s,\lambda) \leq f(x,s,\lambda)s \ \text{ for a.a. } x \in \Omega, \text{ all } s \in \mathbb{R} \text{ with } |s| \geq M_\lambda \ ;$$

(iv)  There exists $\rho > 0$ such that $f(x,s,\lambda)s > 0$ for a.a. $x \in \Omega$, all $s \in [-\rho,\rho]$, $s \neq 0$, and all $\lambda \in \Lambda$.

**Theorem 11.31.**

(a) *Assume that* $\mathrm{H}(f)_8$ *(i), (ii) hold. Then, for all* $b > 0$ *there exists* $\lambda^* \in \Lambda$ *such that, for* $\lambda \in (0,\lambda^*)$, *problem (11.61) has at least three distinct, nontrivial solutions:* $u_0 \in \mathrm{int}\,(C_0^1(\overline{\Omega})_+)$, $v_0 \in -\mathrm{int}\,(C_0^1(\overline{\Omega})_+)$, *and* $y_0 \in C_0^1(\overline{\Omega})$ *with*

$$-b < v_0 \leq y_0 \leq u_0 < b \ \text{ in } \overline{\Omega}.$$

*If, in addition,* $\mathrm{H}(f)_8$ *(ii.b) holds, then* $y_0$ *can be chosen to be nodal.*

(b) *Assume that* $\mathrm{H}(f)_8$ *holds, with* $r < p^*$ *in* $\mathrm{H}(f)_8$ *(i). Then, for all* $b > 0$ *there exists* $\lambda^* \in \Lambda$ *such that, for* $\lambda \in (0,\lambda^*)$, *problem (11.61) has at least five distinct, nontrivial solutions:* $u_0, \hat{u} \in \mathrm{int}\,(C_0^1(\overline{\Omega})_+)$, $v_0, \hat{v} \in -\mathrm{int}\,(C_0^1(\overline{\Omega})_+)$, *and* $y_0 \in C_0^1(\overline{\Omega})$, *with*

$$\hat{v} \leq v_0 \leq y_0 \leq u_0 \leq \hat{u} \ \text{ in } \overline{\Omega}, \quad \|u_0\|_\infty < b, \ \text{ and } \ \|v_0\|_\infty < b.$$

*If, in addition,* $\mathrm{H}(f)_8$ *(ii.b) holds, then* $y_0$ *can be chosen to be nodal.*

*Proof.* (a) Consider $\lambda^*$ given by Proposition 11.28(a). Fix $\lambda \in (0,\lambda^*)$. Then Proposition 11.28 shows that problem (11.61) admits at least two solutions, $u_+ \in \mathrm{int}\,(C_0^1(\overline{\Omega})_+)$ and $v_- \in -\mathrm{int}\,(C_0^1(\overline{\Omega})_+)$, such that $\|u_+\|_\infty \leq b$ and $\|v_-\|_\infty \leq b$. Moreover, in the case where $\mathrm{H}(f)_8$ (ii.b) is satisfied, $u_+$ and $v_-$ can be chosen to be the smallest positive solution and the biggest negative solution of (11.61), respectively (Proposition 11.20). Then the three solutions $u_0, v_0, y_0$ satisfying the conclusion of the statement are obtained by applying Theorem 11.23, with $\overline{u} = u_+$ and $\underline{v} = v_-$.

(b) We proceed as for (a) by applying in addition Theorem 11.30 to obtain the two additional constant-sign solutions $\hat{u} \in \mathrm{int}\,(C_0^1(\overline{\Omega})_+)$, $\hat{v} \in -\mathrm{int}\,(C_0^1(\overline{\Omega})_+)$.  $\square$

*Example 11.32.* Let $p < r < p^*$. A typical nonlinearity fulfilling $\mathrm{H}(f)_8$ is of the form

$$f(x,s,\lambda) = |s|^{r-2}s + \lambda g(x,s), \tag{11.71}$$

where $g : \Omega \times \mathbb{R} \to \mathbb{R}$ is a Carathéodory function, with $g(x,0) = 0$ a.e. in $\Omega$, which satisfies the following conditions:

(i) There exist $\hat{c}_0 > 0$ and $1 \leq q < p$ such that

$$|g(x,s)| \leq \hat{c}_0(1+|s|^{q-1}) \text{ for a.a. } x \in \Omega, \text{ all } s \in \mathbb{R};$$

(ii) $\displaystyle\liminf_{s\to 0} \frac{g(x,s)}{|s|^{p-2}s} = +\infty$ uniformly for a.a. $x \in \Omega$;

(iii) There exist $M_0 > 0$, $\mu \in (p,r)$, $c_1, c_2 > 0$, and $r_0 \in [0,r)$ such that

$$-c_1|s|^r \leq \mu G(x,s) \leq g(x,s)s + c_2|s|^{r_0} \text{ for a.a. } x \in \Omega, \text{ all } |s| \geq M_0;$$

(iv) There exists $\rho > 0$ such that $g(x,s)s \geq 0$ for a.a. $x \in \Omega$, all $s \in [-\rho,\rho]$.

Under these conditions, it can be seen that $f$ given in (11.71) satisfies H$(f)_8$ for $\lambda \in \Lambda := (0, \frac{\mu}{rc_1})$. Thus, Theorem 11.31(b) yields five nontrivial solutions for problem (11.61): two positive, two negative, and an intermediate one. A particular case of $g$ satisfying (i)–(iv) given previously is $g(x,s) = |s|^{q-2}s$ with $q \in (1,p)$, so

$$f(x,s,\lambda) = |s|^{r-2}s + \lambda|s|^{q-2}s.$$

This nonlinearity is usually referred to as a concave–convex nonlinearity. Indeed, when $p = 2$, it is the sum of a concave and of a convex term. This kind of nonlinearity [for $p \in (1,+\infty)$] is also considered in Sect. 11.3.

Finally, we can obtain an additional nodal solution by strengthening the assumptions H$(f)_8$:

H$(f)_9$ (i) $f : \overline{\Omega} \times \mathbb{R} \times \Lambda \to \mathbb{R}$ is such that $f(\cdot,\cdot,\lambda)$ is a continuous function, $f(x,0,\lambda) = 0$ in $\Omega$, for all $\lambda \in \Lambda$; moreover, there are numbers $a(\lambda) > 0$ with $a(\lambda) \to 0$ as $\lambda \downarrow 0$, and $c > 0$, $r \in (p,p^*)$ (independent of $\lambda$), such that

$$|f(x,s,\lambda)| \leq a(\lambda) + c|s|^{r-1} \text{ for all } x \in \Omega, \text{ all } s \in \mathbb{R}, \text{ all } \lambda \in \Lambda;$$

(ii) For all $\lambda \in \Lambda$ there exist $\theta_\lambda > \lambda_2$ and $\hat{\eta}_\lambda \in L^\infty(\Omega)$ such that

$$\theta_\lambda < \liminf_{s\to 0} \frac{f(x,s,\lambda)}{|s|^{p-2}s} \leq \limsup_{s\to 0} \frac{f(x,s,\lambda)}{|s|^{p-2}s} \leq \hat{\eta}_\lambda(x)$$

uniformly for all $x \in \Omega$;

(iii) For all $\lambda \in \Lambda$ there exist $M_\lambda > 0$ and $\mu_\lambda > p$ such that

$$0 < \mu_\lambda F(x,s,\lambda) \leq f(x,s,\lambda)s \text{ for all } x \in \Omega, \text{ all } s \in \mathbb{R} \text{ with } |s| \geq M_\lambda;$$

(iv) There exist $\rho_- < 0 < \rho_+$ such that for all $\lambda \in \Lambda$ we have

$$f(x,\rho_-,\lambda) = 0 = f(x,\rho_+,\lambda) \text{ for all } x \in \Omega,$$

$$f(x,s,\lambda)s > 0 \text{ for all } x \in \Omega, \text{ all } s \in (\rho_-,\rho_+), s \neq 0.$$

*Remark 11.33.* Under hypotheses that are implied by $H(f)_9$, it is established in Bartsch et al. [38, Theorem 1.1] that problem (11.61) admits a nodal solution $w_0 \in C_0^1(\overline{\Omega})$ satisfying $\max_{\overline{\Omega}} w_0 \geq \rho_+$ and $\min_{\overline{\Omega}} w_0 \leq \rho_-$. We use this observation in the proof of the next result.

**Theorem 11.34.** *Assume that $H(f)_9$ holds. Then there exists $\lambda^* \in \Lambda$ such that for all $\lambda \in (0, \lambda^*)$ problem (11.61) has at least six distinct, nontrivial solutions: $u_0, \hat{u} \in \mathrm{int}(C_0^1(\overline{\Omega})_+)$, $v_0, \hat{v} \in -\mathrm{int}(C_0^1(\overline{\Omega})_+)$, and $y_0, w_0 \in C_0^1(\overline{\Omega})$ both nodal.*

*Proof.* Applying Theorem 11.31(b) with $b := \min\{\rho_+, |\rho_-|\}$, we find $\lambda^* \in \Lambda$ such that for $\lambda \in (0, \lambda^*)$ problem (11.61) admits five solutions $u_0, \hat{u} \in \mathrm{int}(C_0^1(\overline{\Omega})_+)$, $v_0, \hat{v} \in -\mathrm{int}(C_0^1(\overline{\Omega})_+)$, $y_0 \in C_0^1(\overline{\Omega})$ nodal, and, moreover, $\|y_0\|_\infty < b$. An additional nodal solution $w_0 \in C_0^1(\overline{\Omega})$ such that $\|w_0\|_\infty \geq \max\{\rho_+, |\rho_-|\}$ is obtained from Bartsch et al. [38, Theorem 1.1] (Remark 11.33). Finally, the fact that $\|y_0\|_\infty < b \leq \|w_0\|_\infty$ justifies that $y_0 \neq w_0$. The proof of the theorem is complete. $\square$

*Example 11.35.* We consider the following nonlinearity for $\lambda \in (0, +\infty)$:

$$
f(x,s,\lambda) = \begin{cases} |s|^{r-2}s + 1 & \text{if } s \leq -1, \\ -\theta(x,s) \min\{\lambda, |s|^{p-1}\} & \text{if } -1 < s \leq 0, \\ \theta(x,s) \min\{\lambda, s^{p-1}\} & \text{if } 0 < s \leq 1, \\ s^{r-1} - 1 & \text{if } s > 1, \end{cases}
$$

where $r \in (p, p^*)$, and $\theta : \overline{\Omega} \times [-1, 1] \to \mathbb{R}$ is a continuous function satisfying $\theta(x,0) > \lambda_2$, $\theta(x,-1) = \theta(x,1) = 0$ for all $x \in \overline{\Omega}$, and $\theta(x,s) > 0$ for all $x \in \Omega$, all $s \in (-1,1)$; for example, we can take $\theta(x,s) = (e^{|x|} + \lambda_2)(1 - |s|)$. Then the function $f(x,s,\lambda)$ satisfies $H(f)_9$, with $\rho_- = -1$, $\rho_+ = 1$. Therefore, Theorem 11.34 implies that, for the preceding nonlinearity $f$ and $\lambda > 0$ small, problem (11.61) admits at least six nontrivial solutions: two positive, two negative, and two nodal.

## 11.3 Nonlinear Dirichlet Problems Using Morse Theory

The setting of this section is the same as in Sects. 11.1 and 11.2, namely $\Omega \subset \mathbb{R}^N$ is a bounded domain with $C^2$-boundary $\partial\Omega$, we fix $p \in (1, +\infty)$, and we consider the following nonlinear elliptic Dirichlet problem:

$$
\begin{cases} -\Delta_p u(x) = f(x, u(x)) & \text{in } \Omega, \\ u = 0 & \text{on } \partial\Omega, \end{cases} \tag{11.72}
$$

driven by the $p$-Laplacian $\Delta_p : W_0^{1,p}(\Omega) \to W^{-1,p'}(\Omega)$ and involving a Carathéodory function $f : \Omega \times \mathbb{R} \to \mathbb{R}$.

In Sects. 11.1 and 11.2, problem (11.72) was treated in the coercive case (i.e., when the energy functional associated to the problem is coercive). In this case, the existence of a nontrivial solution of the problem is usually ensured via the direct method, and the real challenge is to prove the existence of a further nontrivial solution (Theorems 11.13 and 11.26). In Sect. 11.1, we provide a second multiplicity result in the case where the nonlinearity $f$ is $(p-1)$-linear at 0, at $+\infty$, and $(p-1)$-sublinear at some $c_+ \in (0,+\infty)$ (Theorem 11.15). In Sect. 11.2, we also study a parametric version of problem (11.72) [see problem (11.61)] and we establish multiplicity results when the parameter $\lambda$ is small enough. For the main multiplicity results (Theorems 11.30 and 11.34), we use (a slightly more general version of) the so-called Ambrosetti–Rabinowitz condition [see H$(f)_8$ (iii) in Sect. 11.2], which forces the nonlinearity $f$ to be strictly $(p-1)$-superlinear at $\pm\infty$.

Here we complete the results of the previous sections in two directions:

- In the first part of this section, we give an existence result for a nonparametric, noncoercive Dirichlet problem involving a nonlinearity that is asymptotically $(p-1)$-linear at $\pm\infty$, including the situation of resonance with respect to the first eigenvalue of $-\Delta_p^D$.
- In the second part we give a multiplicity result for a parametric problem of a special form (where the parameter appears as the coefficient of a "concave" term), involving a nonlinearity that is asymptotically $(p-1)$-superlinear at $\pm\infty$ but without assuming the Ambrosetti–Rabinowitz condition. The multiplicity result will guarantee the existence of four constant-sign solutions and a nodal solution [although the nonlinearity $f$ will not satisfy H$(f)_8$ (ii.b) in Sect. 11.2].

For this purpose, we will use the tools of Morse theory presented in Chap. 6.

## Existence Result for (p-1)-Linear, Noncoercive Elliptic Equations

In the first part of this section, we study problem (11.72) in the noncoercive case and when the nonlinearity is $(p-1)$-linear at $\pm\infty$. More precisely, we require that asymptotically at $\pm\infty$, the ratio $\frac{f(x,s)}{|s|^{p-2}s}$ must lie in the interval $[\lambda_1, \lambda_2)$, where $0 < \lambda_1 < \lambda_2$ denote the first two eigenvalues of the negative Dirichlet $p$-Laplacian. Since the limit can be equal to $\lambda_1$, our setting incorporates resonant problems, where we encounter the lack of compactness of the energy functionals. Here, the challenge is to seek conditions for the existence of at least one nontrivial solution of the problem.

We state the precise hypotheses on $f$ and its primitive $F(x,s) = \int_0^s f(x,t)\,dt$:

H$(f)_1$   (i)  $f : \Omega \times \mathbb{R} \to \mathbb{R}$ is a Carathéodory function with $f(x,0) = 0$ a.e. in $\Omega$, and there is $c > 0$ such that

$$|f(x,s)| \le c(1 + |s|^{p-1}) \text{ for a.a. } x \in \Omega, \text{ all } s \in \mathbb{R};$$

(ii)  There exist $\delta > 0$ and $\tau \in (1,p)$ such that for all $s \in [-\delta,\delta] \setminus \{0\}$

$$\tau F(x,s) - f(x,s)s \ge 0 \text{ for a.a. } x \in \Omega \text{ and } \operatorname*{ess\,inf}_{x\in\Omega} F(x,s) > 0;$$

(iii) There exists $\theta_0 < \lambda_2$ such that

$$\lambda_1 \leq \liminf_{s \to \pm\infty} \frac{pF(x,s)}{|s|^p} \leq \limsup_{s \to \pm\infty} \frac{pF(x,s)}{|s|^p} \leq \theta_0 \quad \text{uniformly for a.a. } x \in \Omega;$$

(iv) There exist $\beta_0 > 0$ and $\mu \in [1,p]$ such that

$$\beta_0 \leq \liminf_{s \to \pm\infty} \frac{pF(x,s) - f(x,s)s}{|s|^\mu} \quad \text{uniformly for a.a. } x \in \Omega.$$

*Remark 11.36.*

(a) Hypothesis $H(f)_1$ (ii) is fulfilled in particular if the nonlinearity takes the form $f(x,s) = \lambda|s|^{q-2}s + g(x,s)$ with $\lambda > 0$, $q \in (1,p)$, and $\lim_{s \to 0} \frac{g(x,s)}{|s|^{p-1}} = 0$.

(b) Hypothesis $H(f)_1$ (iii) incorporates into our setting problems that at $\pm\infty$ are resonant with respect to the first eigenvalue $\lambda_1 > 0$, that is, $f$ satisfies the condition

$$\lim_{s \to \pm\infty} \frac{f(x,s)}{|s|^{p-2}s} = \lambda_1 \quad \text{uniformly for a.a. } x \in \Omega.$$

*Example 11.37.* The following function $f$ satisfies $H(f)_1$ (where for the sake of simplicity we drop the $x$-dependence):

$$f(s) = \begin{cases} |s|^{q-2}s - |s|^{p-2}s + (\theta+\eta)|s|^{r-2}s & \text{if } |s| \leq 1, \\ \theta|s|^{p-2}s + \eta|s|^{\mu-2}s & \text{if } |s| > 1, \end{cases}$$

with $q, \mu \in (1,p)$, $r \in (p, +\infty)$, $\lambda_1 \leq \theta < \lambda_2$, and $\eta > 0$.

Under hypothesis $H(f)_1$ (i), we can define the energy functional $\varphi : W_0^{1,p}(\Omega) \to \mathbb{R}$ for problem (11.72), given by

$$\varphi(u) = \frac{1}{p}\|\nabla u\|_p^p - \int_\Omega F(x,u(x))\,dx \quad \text{for all } u \in W_0^{1,p}(\Omega).$$

Evidently, $\varphi \in C^1(W_0^{1,p}(\Omega), \mathbb{R})$ and

$$\varphi'(u) = -\Delta_p u - N_f(u) \quad \text{for all } u \in W_0^{1,p}(\Omega), \tag{11.73}$$

where $N_f(u)(\cdot) = f(\cdot, u(\cdot)) \in L^{p'}(\Omega)$ for all $u \in W_0^{1,p}(\Omega)$ [see hypothesis $H(f)_1$ (i)].

**Proposition 11.38.** *If hypotheses* $H(f)_1$ *(i), (iv) hold, then the functional* $\varphi$ *satisfies the* (C)-*condition.*

*Proof.* Let $\{u_n\}_{n\geq 1} \subset W_0^{1,p}(\Omega)$ be a sequence such that

$$|\varphi(u_n)| \leq M_1 \text{ for all } n \geq 1, \tag{11.74}$$

for some $M_1 > 0$, and

$$(1 + \|\nabla u_n\|_p)\varphi'(u_n) \to 0 \text{ in } W^{-1,p'}(\Omega) \text{ as } n \to \infty. \tag{11.75}$$

We claim that

$$\{u_n\}_{n\geq 1} \text{ is bounded in } W_0^{1,p}(\Omega). \tag{11.76}$$

From (11.75) and (11.73), for all $n \geq 1$, all $h \in W_0^{1,p}(\Omega)$, we have

$$\left| \langle -\Delta_p u_n, h \rangle - \int_\Omega f(x, u_n) h \, dx \right| \leq \frac{\varepsilon_n \|\nabla h\|_p}{1 + \|\nabla u_n\|_p}, \tag{11.77}$$

with $\varepsilon_n \to 0$. Choosing $h = u_n$ in (11.77), we obtain

$$\left| \|\nabla u_n\|_p^p - \int_\Omega f(x, u_n) u_n \, dx \right| \leq \varepsilon_n \text{ for all } n \geq 1. \tag{11.78}$$

On the other hand, (11.74) yields

$$-\|\nabla u_n\|_p^p + \int_\Omega pF(x, u_n) \, dx \leq pM_1 \text{ for all } n \geq 1. \tag{11.79}$$

Adding (11.78) and (11.79), we get

$$\int_\Omega (pF(x, u_n) - f(x, u_n) u_n) \, dx \leq M_2 \text{ for all } n \geq 1, \tag{11.80}$$

for some $M_2 > 0$. By H$(f)_1$ (iv), we can find $\beta_1 \in (0, \beta_0)$ and $M_3 = M_3(\beta_1) > 0$ with

$$0 < \beta_1 |s|^\mu \leq pF(x, s) - f(x, s)s \text{ for a.a. } x \in \Omega, \text{ all } |s| \geq M_3. \tag{11.81}$$

In view of hypothesis H$(f)_1$ (i), from (11.81) it follows that

$$\beta_1 |s|^\mu - M_4 \leq pF(x, s) - f(x, s)s \text{ for a.a. } x \in \Omega, \text{ all } s \in \mathbb{R}, \tag{11.82}$$

with $M_4 > 0$. Using (11.82) in (11.80), we infer that

$$\{u_n\}_{n\geq 1} \text{ is bounded in } L^\mu(\Omega). \tag{11.83}$$

Since $\mu \leq p < p^*$ [by H$(f)_1$ (iv)], fixing $r \in (p, p^*)$, there is $t \in [0, 1)$ such that $\frac{1}{p} = \frac{1-t}{\mu} + \frac{t}{r}$. From the interpolation inequality (e.g., Brezis [52, p. 93]) we have

$$\|u_n\|_p \leq \|u_n\|_\mu^{1-t} \|u_n\|_r^t \text{ for all } n \geq 1.$$

On the basis of Theorem 1.49 and (11.83), this ensures that

$$\|u_n\|_p^p \leq M_5 \|\nabla u_n\|_p^{tp} \text{ for all } n \geq 1, \tag{11.84}$$

with $M_5 > 0$. Returning to (11.78) and using (11.84) and hypothesis H$(f)_1$ (i), we derive that

$$\|\nabla u_n\|_p^p \leq c_1 (1 + \|\nabla u_n\|_p^{tp}) \text{ for all } n \geq 1,$$

with $c_1 > 0$. Since $t \in [0, 1)$, this proves (11.76).

Because of (11.76), along a relabeled subsequence, we have

$$u_n \overset{w}{\to} u \text{ in } W_0^{1,p}(\Omega) \text{ and } u_n \to u \text{ in } L^p(\Omega) \text{ as } n \to \infty. \tag{11.85}$$

In (11.77) we choose $h = u_n - u$ and pass to the limit as $n \to \infty$ through (11.85). Then we obtain $\lim_{n \to \infty} \langle -\Delta_p u_n, u_n - u \rangle = 0$. By Proposition 2.72, it follows that $u_n \to u$ in $W_0^{1,p}(\Omega)$, which completes the proof. $\qquad \square$

As in Sects. 11.1 and 11.2, we denote by $\hat{u}_1 \in \text{int}(C_0^1(\overline{\Omega})_+)$ the $L^p$-normalized positive eigenfunction of $-\Delta_p^D$ corresponding to $\lambda_1$ (Proposition 9.47).

**Proposition 11.39.** *If hypotheses* H$(f)_1$ (i), (iii), (iv) *hold, then* $\varphi|_{\mathbb{R}\hat{u}_1}$ *is anticoercive, i.e., if* $|t| \to +\infty$, *then* $\varphi(t\hat{u}_1) \to -\infty$.

*Proof.* Clearly, in hypothesis H$(f)_1$ (iv), without any loss of generality, we may assume that $\mu < p$. By (11.81) we see that

$$\frac{d}{ds}\left(\frac{F(x,s)}{s^p}\right) = \frac{f(x,s)s - pF(x,s)}{s^{p+1}} \leq -\beta_1 s^{\mu-p-1} \text{ for a.a. } x \in \Omega, \text{ all } s \geq M_3.$$

Integrating, we have

$$\frac{F(x,s)}{s^p} - \frac{F(x,t)}{t^p} \leq \frac{\beta_1}{p-\mu}\left(\frac{1}{s^{p-\mu}} - \frac{1}{t^{p-\mu}}\right) \text{ for a.a. } x \in \Omega, \text{ all } s \geq t \geq M_3.$$

Letting $s \to +\infty$, since $\mu < p$ and using hypothesis H$(f)_1$ (iii), we obtain

$$\frac{\lambda_1}{p} t^p - F(x,t) \leq -\frac{\beta_1}{p-\mu} t^\mu \text{ for a.a. } x \in \Omega, \text{ all } t \geq M_3.$$

Combining with $H(f)_1$ (i), it follows that

$$\varphi(t\hat{u}_1) = \int_\Omega \left( \frac{\lambda_1}{p} t^p \hat{u}_1^p - F(x, t\hat{u}_1) \right) dx \leq -\frac{t^\mu \beta_1}{p - \mu} \int_{\{t\hat{u}_1 \geq M_3\}} \hat{u}_1^\mu \, dx + M_6$$

for all $t > 0$, for some $M_6 > 0$. From this, we see that $\varphi(t\hat{u}_1) \to -\infty$ as $t \to +\infty$. In a similar fashion, we show $\varphi(t\hat{u}_1) \to -\infty$ as $t \to -\infty$. $\qquad\square$

We introduce the set

$$D = \{u \in W_0^{1,p}(\Omega) : \|\nabla u\|_p^p = \lambda_2 \|u\|_p^p\}.$$

**Proposition 11.40.** *If hypotheses* $H(f)_1$ (i), (iii) *hold, then* $\varphi|_D$ *is coercive and bounded below.*

*Proof.* Hypotheses $H(f)_1$ (i), (iii) imply that we can find $\theta_1 \in (\theta_0, \lambda_2)$ and $c_2 > 0$ such that

$$F(x, s) \leq \frac{\theta_1}{p} |s|^p + c_2 \text{ for a.a. } x \in \Omega, \text{ all } s \in \mathbb{R}.$$

Then we obtain

$$\varphi(u) \geq \frac{1}{p} \left( 1 - \frac{\theta_1}{\lambda_2} \right) \|\nabla u\|_p^p - c_2 |\Omega|_N \text{ for all } u \in D.$$

Since $\theta_1 < \lambda_2$, we conclude that $\varphi|_D$ is coercive and bounded below. $\qquad\square$

By Propositions 11.39 and 11.40, we can find $\hat{t} > 0$ such that

$$\varphi(\pm\hat{t}\hat{u}_1) < \inf_D \varphi =: \eta_D. \tag{11.86}$$

Set $E_0 = \{\pm\hat{t}\hat{u}_1\}$ and $E = \{-s\hat{t}\hat{u}_1 + (1-s)\hat{t}\hat{u}_1 : s \in [0,1]\}$.

**Proposition 11.41.** *If hypotheses* $H(f)_1$ (i), (iii), (iv) *hold, then* $\{E_0, E\}$ *and* $D$ *are homologically linking in dimension one (Definition 6.77).*

*Proof.* First, we note that $-\hat{t}\hat{u}_1$ and $\hat{t}\hat{u}_1$ belong to distinct connected components of $W_0^{1,p}(\Omega) \setminus D$. To see this, let $\gamma \in C([-1,1], W_0^{1,p}(\Omega))$ such that $\gamma(-1) = -\hat{t}\hat{u}_1$ and $\gamma(1) = \hat{t}\hat{u}_1$, and let us check that $\gamma([-1,1]) \cap D \neq \emptyset$. If this is not the case, then we have $\gamma(t) \neq 0$ and

$$\frac{\|\nabla\gamma(t)\|_p^p}{\|\gamma(t)\|_p^p} < \lambda_2 \text{ for all } t \in [-1,1],$$

a contradiction of Theorem 9.50. So we have $\hat{t}\hat{u}_1 \in C_1$, $-\hat{t}\hat{u}_1 \in C_2$, where $C_1, C_2$ are distinct connected components of $Y := W_0^{1,p}(\Omega) \setminus D$.

Let $j : E_0 \to Y$ denote the inclusion map, and let $* = \hat{t}\hat{u}_1$. Moreover, let $R : Y \to E_0$ be the map such that $R(u) = \hat{t}\hat{u}_1$ for all $u \in C_1$ and $R(u) = -\hat{t}\hat{u}_1$ for all $u \in Y \setminus C_1$. Thus, $R$ is well defined and continuous, and $R \circ j = \mathrm{id}_{E_0}$. This implies that $j_* : H_0(E_0, *) \to H_0(Y, *)$ is injective (Definition 6.9) and so nontrivial because $H_0(E_0, *) = \mathbb{F} \neq 0$ (Example 6.42). According to Remark 6.78(b), this implies that $\{E_0, E\}$ homologically links $D$ in dimension one. $\qquad\square$

Next we compute the critical groups of $\varphi$ at the origin.

**Proposition 11.42.** *If hypotheses* $H(f)_1$ *(i), (ii) hold and* $0$ *is an isolated critical point of* $\varphi$, *then* $C_k(\varphi, 0) = 0$ *for all* $k \geq 0$.

*Proof.* Hypothesis $H(f)_1$ (ii) implies that we can find $\delta_1 \in (0, \delta)$ and $c_3 > 0$ such that

$$F(x, s) \geq c_3 |s|^\tau \text{ for a.a. } x \in \Omega, \text{ all } |s| \leq \delta_1. \tag{11.87}$$

Combining (11.87) with $H(f)_1$ (i), it follows that

$$F(x, s) \geq c_3 |s|^\tau - c_4 |s|^r \text{ for a.a. } x \in \Omega, \text{ all } s \in \mathbb{R},$$

for $r \in (p, p^*)$ and some $c_4 > 0$. Then we obtain

$$\varphi(tu) \leq \frac{t^p}{p} \|\nabla u\|_p^p - c_3 t^\tau \|u\|_\tau^\tau + c_4 t^r \|u\|_r^r \text{ for all } u \in W_0^{1,p}(\Omega), \text{ all } t > 0. \tag{11.88}$$

Since $\tau < p < r$, relation (11.88) implies that for each $u \in W_0^{1,p}(\Omega) \setminus \{0\}$ there is $t^*(u) > 0$ small satisfying

$$\varphi(tu) < 0 \text{ for all } t \in (0, t^*(u)). \tag{11.89}$$

From $H(f)_1$ (i), (ii) we find $c_5 > 0$ such that

$$\tau F(x, s) - f(x, s)s \geq -c_5 |s|^r \text{ for a.a. } x \in \Omega, \text{ all } s \in \mathbb{R}.$$

Then for all $u \in W_0^{1,p}(\Omega) \setminus \{0\}$ such that $\varphi(u) = 0$ we have

$$\frac{d}{dt} \varphi(tu) \Big|_{t=1} = \left(1 - \frac{\tau}{p}\right) \|\nabla u\|_p^p + \int_\Omega (\tau F(x, u) - f(x, u)u) \, dx$$

$$\geq \left(1 - \frac{\tau}{p}\right) \|\nabla u\|_p^p - c_6 \|\nabla u\|_p^r, \tag{11.90}$$

with $c_6 > 0$. Because $r > p > \tau$, by (11.90) there exists $\rho > 0$ small such that, for all $u \in W_0^{1,p}(\Omega) \setminus \{0\}$, we have

$$\frac{d}{dt}\varphi(tu)\Big|_{t=1} > 0 \quad \text{whenever } \|\nabla u\|_p < \rho \text{ and } \varphi(u) = 0. \tag{11.91}$$

Arguing as at the end of Step 4 of the proof of Theorem 10.9, we deduce for (11.91) that, for all $u \in W_0^{1,p}(\Omega) \setminus \{0\}$, with $\|\nabla u\|_p \leq \rho$ and $\varphi(u) \leq 0$, we have

$$\varphi(tu) \leq 0 \quad \text{for all } t \in [0,1]. \tag{11.92}$$

Finally, arguing on the basis of (11.89), (11.91), and (11.92) [which are the analogs of (10.35), (10.36), and (10.37)], as in Step 4 of the proof of Theorem 10.9, we obtain that $C_k(\varphi,0) = 0$ for all $k \geq 0$, which completes the proof of the proposition. $\square$

*Remark 11.43.* A careful reading of the previous proof shows that Proposition 11.42 remains true if, instead of H$(f)_1$ (i), we assume a general subcritical growth condition [i.e., where the exponent $p$ is replaced by some $r \in (p, p^*)$].

Now we present our existence theorem for problem (11.72).

**Theorem 11.44.** *Under hypotheses* H$(f)_1$, *problem (11.72) has a nontrivial solution $u_0 \in C_0^1(\overline{\Omega})$.*

*Proof.* In view of (11.73), the solutions of problem (11.72) coincide with the critical points of the functional $\varphi$. Moreover, from Corollary 8.13 we know that every solution belongs to $C_0^1(\overline{\Omega})$. Thus, it suffices to show that $\varphi$ admits at least one nontrivial critical point. We may suppose that $\varphi$ has a finite number of critical points (otherwise, we are done). Relation (11.86) and Propositions 11.38 and 11.41 permit the use of Proposition 6.80 (b), which yields a $u_0 \in W_0^{1,p}(\Omega)$ critical point of $\varphi$ with

$$C_1(\varphi, u_0) \neq 0. \tag{11.93}$$

Comparing (11.93) with Proposition 11.42, we conclude that $u_0 \neq 0$. The proof of the theorem is now complete. $\square$

Next we study a semilinear version of problem (11.72):

$$\begin{cases} -\Delta u(x) = f(x, u(x)) & \text{in } \Omega, \\ u = 0 & \text{on } \partial\Omega. \end{cases} \tag{11.94}$$

The hypotheses on $f$ are similar to H$(f)_1$ except H$(f)_2$ (iii), where we assume that the quotient $\frac{2F(x,s)}{|s|^2}$ lies asymptotically at $\pm\infty$ between two consecutive eigenvalues $\lambda_m < \lambda_{m+1}$ ($m \geq 1$) of the negative Dirichlet Laplacian:

H$(f)_2$ (i) $f : \Omega \times \mathbb{R} \to \mathbb{R}$ is a Carathéodory function with $f(x,0) = 0$ a.e. in $\Omega$, and there is $c > 0$ such that

$$|f(x,s)| \leq c(1 + |s|) \quad \text{for a.a. } x \in \Omega, \text{ all } s \in \mathbb{R};$$

(ii) There exist $\delta > 0$ and $\tau \in (1,2)$ such that for all $s \in [-\delta, \delta] \setminus \{0\}$

$$\tau F(x,s) - f(x,s)s \geq 0 \quad \text{for a.a. } x \in \Omega \quad \text{and} \quad \operatorname*{ess\,inf}_{x \in \Omega} F(x,s) > 0;$$

(iii) There exists $\theta_0 < \lambda_{m+1}$ ($m \geq 1$) such that

$$\lambda_m \leq \liminf_{s \to \pm\infty} \frac{2F(x,s)}{|s|^2} \leq \limsup_{s \to \pm\infty} \frac{2F(x,s)}{|s|^2} \leq \theta_0 \quad \text{uniformly for a.a. } x \in \Omega;$$

(iv) There exist $\beta_0 > 0$ and $\mu \in [1,2)$ such that

$$\beta_0 \leq \liminf_{s \to \pm\infty} \frac{2F(x,s) - f(x,s)s}{|s|^\mu} \quad \text{uniformly for a.a. } x \in \Omega.$$

*Example 11.45.* Taking $p = 2$ and choosing $\theta \in [\lambda_m, \lambda_{m+1})$ in Example 11.37 [instead of $\theta \in [\lambda_1, \lambda_2)$], we obtain an example of function $f$ fulfilling H$(f)_2$.

**Theorem 11.46.** *Assume that* H$(f)_2$ *holds. Then problem (11.94) admits at least one nontrivial solution* $u_0 \in C_0^1(\overline{\Omega})$.

*Proof.* As in the proof of Theorem 11.44, the solutions of (11.94) coincide with the critical points of the $C^1$-functional $\varphi : H_0^1(\Omega) \to \mathbb{R}$ given by

$$\varphi(u) = \frac{1}{2} \|\nabla u\|_2^2 - \int_\Omega F(x,u)\, dx \quad \text{for all } u \in H_0^1(\Omega),$$

and all of them belong to $C_0^1(\overline{\Omega})$. Thus, it suffices to show that $\varphi$ admits at least one nontrivial critical point. To do this, we may assume that $\varphi$ has only a finite number of critical points. We know from Proposition 11.38 that $\varphi$ satisfies the (C)-condition, and, by Proposition 11.42, we have

$$C_k(\varphi, 0) = 0 \quad \text{for all } k \geq 0. \tag{11.95}$$

As in Theorem 9.4, by $\{\lambda_n\}_{n \geq 1}$ we denote the nondecreasing sequence of eigenvalues of $-\Delta^D$ repeated according to their multiplicities, and by $\{\hat{u}_n\}_{n \geq 1}$ we denote an orthogonal basis of $H_0^1(\Omega)$ made of corresponding eigenfunctions. We consider the decomposition $H_0^1(\Omega) = H_m \oplus H_m^\perp$, where

$$H_m = \operatorname{span}\{\hat{u}_n : 1 \leq n \leq m\} \quad \text{and} \quad H_m^\perp = \overline{\operatorname{span}}\{\hat{u}_n : n \geq m+1\}.$$

*Claim 1:* $\varphi|_{H_m}$ is anticoercive, i.e., $\varphi(u) \to -\infty$ as $\|\nabla u\|_2 \to +\infty$, $u \in H_m$.

Arguing as in the proof of Proposition 11.39, we find $\beta_1, M_1 > 0$ such that

$$\frac{\lambda_m}{2} |t|^2 - F(x,t) \leq -\frac{\beta_1}{2-\mu} |t|^\mu \quad \text{for a.a. } x \in \Omega, \text{ whenever } |t| \geq M_1.$$

Combining with H($f$)$_2$ (i), it follows that

$$\frac{\lambda_m}{2}|t|^2 - F(x,t) \le \beta_2 - \frac{\beta_1}{2-\mu}|t|^\mu \quad \text{for a.a. } x \in \Omega, \text{ all } t \in \mathbb{R},$$

for some $\beta_2 > 0$. By Proposition 9.9, each $u \in H_m$ satisfies $\|\nabla u\|_2^2 \le \lambda_m \|u\|_2^2$, whence

$$\varphi(u) \le \int_\Omega \left(\frac{\lambda_m}{2}|u|^2 - F(x,u)\right) dx \le -\frac{\beta_1}{2-\mu}\|u\|_\mu^\mu + \beta_2|\Omega|_N \quad \text{for all } u \in H_m.$$

This yields $\varphi(u) \to -\infty$ as $\|u\|_\mu \to +\infty$, $u \in H_m$. Since all the norms are equivalent in $H_m$ (which is finite dimensional), we conclude that Claim 1 holds true.

*Claim 2:* $\varphi|_{H_m^\perp}$ is bounded below.

By H($f$)$_2$ (i), (iii), we can find $\theta_1 \in (\lambda_m, \lambda_{m+1})$ and $\theta_2 > 0$ such that

$$F(x,s) \le \theta_2 + \frac{\theta_1}{2}|s|^2 \quad \text{for a.a. } x \in \Omega, \text{ all } s \in \mathbb{R}.$$

Since each $u \in H_m^\perp$ satisfies $\|\nabla u\|_2^2 \ge \lambda_{m+1}\|u\|_2^2$ (Proposition 9.9), we get

$$\varphi(u) \ge \frac{\lambda_{m+1} - \theta_1}{2}\|u\|_2^2 - \theta_2|\Omega|_N \quad \text{for all } u \in H_m^\perp.$$

Knowing that $\lambda_{m+1} > \theta_1$, this yields Claim 2.

Claims 1 and 2 allow us to apply Proposition 6.63, which yields $C_m(\varphi, \infty) \ne 0$. Then Theorem 6.62 (a) implies that $\varphi$ admits a critical point $u_0 \in H_0^1(\Omega)$ such that $C_m(\varphi, u_0) \ne 0$. Comparing this with (11.95), we deduce that $u_0$ is nontrivial. The proof of the theorem is complete.                                                    $\square$

## Multiplicity Result for a $p$-Laplace Equation with Concave Term

Now, for $p \in (1, +\infty)$, we consider the following Dirichlet problem:

$$\begin{cases} -\Delta_p u(x) = \beta(x)|u(x)|^{q-2}u(x) + g(x, u(x)) & \text{in } \Omega, \\ u = 0 & \text{on } \partial\Omega, \end{cases} \tag{11.96}$$

where $\beta \in L^\infty(\Omega) \setminus \{0\}$, $\beta(x) \ge 0$ a.e. in $\Omega$, $q \in (1, p)$, and $g : \Omega \times \mathbb{R} \to \mathbb{R}$ is a Carathéodory function. The term $\beta(x)|u(x)|^{q-2}u$ is called a concave term. We see the function $\beta$ as a parameter since we study the problem when $\beta$ varies, that is, when $\|\beta\|_\infty$ becomes small. Specifically, in the last part of our study we will suppose that $\beta \equiv \lambda \in (0, \overline{\lambda})$ is constant. The Carathéodory function $g$ will be assumed to be $(p-1)$-superlinear near $\pm\infty$. Thus, in problem (11.96), we have the combined effects of a concave term and a convex nonlinearity.

First, we look for constant-sign solutions to problem (11.96) under the following hypotheses. We denote $G(x,s) = \int_0^s g(x,t)\,dt$.

$H(g)_1^+$   (i)   $g : \Omega \times \mathbb{R} \to \mathbb{R}$ is a Carathéodory function with $g(x,0) = 0$ a.e. in $\Omega$, and there are $c > 0$ and $r \in (p,p^*)$ such that

$$|g(x,s)| \le c(1+|s|^{r-1}) \quad \text{for a.a. } x \in \Omega, \text{ all } s \in \mathbb{R};$$

   (ii)   There exist $\vartheta, \hat{\vartheta} \in L^\infty(\Omega)_+$ such that $\vartheta(x) \le \lambda_1$ a.e. in $\Omega$, $\vartheta \ne \lambda_1$, and

$$-\hat{\vartheta}(x) \le \liminf_{s\downarrow 0} \frac{g(x,s)}{s^{p-1}} \le \limsup_{s\downarrow 0} \frac{g(x,s)}{s^{p-1}} \le \vartheta(x)$$

uniformly for a.a. $x \in \Omega$;

  (iii)   The following asymptotic conditions at $\pm\infty$ are satisfied:

   (iii.a)   $\displaystyle\lim_{s\to+\infty} \frac{G(x,s)}{s^p} = +\infty$ uniformly for a.a. $x \in \Omega$;

   (iii.b)   There exist $\tau \in ((r-p)\max\{\frac{N}{p},1\}, p^*)$, $\tau > q$, and $\gamma_0 > 0$ such that

$$\liminf_{s\to+\infty} \frac{g(x,s)s - pG(x,s)}{s^\tau} \ge \gamma_0 \quad \text{uniformly for a.a. } x \in \Omega.$$

*Remark 11.47.* Hypothesis $H(g)_1^+$ (iii.a) implies that, for a.a. $x \in \Omega$, $G(x,\cdot)$ is $p$-superlinear near $+\infty$. Note that, in contrast to Sect. 11.2, we do not require the Ambrosetti–Rabinowitz condition, which is common in such cases. Hypothesis $H(g)_1^+$ (ii) expresses that near zero, $g(x,\cdot)$ satisfies a nonuniform nonresonance condition at the first eigenvalue $\lambda_1$ of the negative Dirichlet $p$-Laplacian.

*Example 11.48.* The functions $g_1(s) = |s|^{r-2}s$ for all $s \in \mathbb{R}$, with $p < r < p^*$, and $g_2(s) = |s|^{p-2}s\ln(1+|s|^p)$ for all $s \in \mathbb{R}$ satisfy $H(g)_1^+$. Note that $g_1$ satisfies the Ambrosetti–Rabinowitz condition, but $g_2$ does not.

**Theorem 11.49.** *Assume that $H(g)_1^+$ holds. Then, there is $\lambda^* > 0$ such that, whenever $\|\beta\|_\infty < \lambda^*$, problem (11.96) has two distinct solutions $u_0, \hat{u} \in \text{int}(C_0^1(\overline{\Omega})_+)$.*

*Proof.* Let $f(x,s) = \beta(x)|s|^{q-2}s + g(x,s)$ for a.a. $x \in \Omega$, all $s \subset \mathbb{R}$. We consider the truncation $f_{[0,+\infty]}(x,s) = \beta(x)(s^+)^{p-1} + g(x,s^+)$ and the corresponding functional

$$\varphi_{[0,+\infty]}(u) = \frac{1}{p}\|\nabla u\|_p^p - \frac{1}{q}\int_\Omega \beta(x)(u^+)^q\,dx - \int_\Omega G(x,u^+)\,dx$$

for all $u \in W_0^{1,p}(\Omega)$ [see (11.4), (11.5)].

*Step 1:* Every nontrivial critical point of $\varphi_{[0,+\infty]}$ is a solution of (11.96) belonging to $\text{int}(C_0^1(\overline{\Omega})_+)$.

By Proposition 11.8(a), a critical point $u \in W_0^{1,p}(\Omega) \setminus \{0\}$ of $\varphi_{[0,+\infty]}$ is a solution of (11.96) belonging to $C_0^1(\overline{\Omega})_+$. Moreover, by H$(g)_1^+$ (i), (ii), we have $-\Delta_p u \geq -\tilde{c} u^{p-1}$ in $W^{-1,p'}(\Omega)$, for some $\tilde{c} > 0$. Then Theorem 8.27 yields $u \in \text{int}(C_0^1(\overline{\Omega})_+)$.

*Step 2:* $\varphi_{[0,+\infty]}$ satisfies the (C)-condition.

Let $\{u_n\}_{n \geq 1} \subset W_0^{1,p}(\Omega)$ be a sequence such that

$$|\varphi_{[0,+\infty]}(u_n)| \leq M_1 \quad \text{for all } n \geq 1, \tag{11.97}$$

with some $M_1 > 0$, and

$$(1 + \|\nabla u_n\|_p)\varphi'_{[0,+\infty]}(u_n) \to 0 \quad \text{in } W^{-1,p'}(\Omega) \text{ as } n \to \infty. \tag{11.98}$$

From (11.98) we have

$$\left| \langle -\Delta_p u_n, h \rangle - \int_\Omega \beta(x)(u_n^+)^{q-1} h \, dx - \int_\Omega g(x, u_n^+) h \, dx \right| \leq \frac{\varepsilon_n \|\nabla h\|_p}{1 + \|\nabla u_n\|_p} \tag{11.99}$$

for all $h \in W_0^{1,p}(\Omega)$, all $n \geq 1$, with $\varepsilon_n \to 0$. Choosing $h = -u_n^- \in W_0^{1,p}(\Omega)$ in (11.99), we obtain $\|\nabla u_n^-\|_p^p \leq \varepsilon_n$ for all $n \geq 1$, from which we infer that

$$u_n^- \to 0 \quad \text{in } W_0^{1,p}(\Omega) \text{ as } n \to \infty. \tag{11.100}$$

Next, we want to show that

$$\{u_n^+\}_{n \geq 1} \text{ is bounded in } W_0^{1,p}(\Omega). \tag{11.101}$$

Choosing $h = u_n^+ \in W_0^{1,p}(\Omega)$ in (11.99), we have

$$-\|\nabla u_n^+\|_p^p + \int_\Omega \beta(x)(u_n^+)^q \, dx + \int_\Omega g(x, u_n^+) u_n^+ \, dx \leq \varepsilon_n. \tag{11.102}$$

On the other hand, from (11.97) it follows that

$$\|\nabla u_n^+\|_p^p - \frac{p}{q} \int_\Omega \beta(x)(u_n^+)^q \, dx - \int_\Omega p G(x, u_n^+) \, dx \leq p M_1 \quad \text{for all } n \geq 1. \tag{11.103}$$

Adding (11.102) and (11.103), we obtain

$$\int_\Omega (g(x, u_n^+) u_n^+ - p G(x, u_n^+)) \, dx \leq M_2 + \|\beta\|_\infty \left( \frac{p}{q} - 1 \right) \|u_n^+\|_q^q \tag{11.104}$$

for all $n \geq 1$, for some $M_2 > 0$. By hypotheses H$(g)_1^+$ (i), (iii.b), we can find constants $\gamma_1 \in (0, \gamma_0)$ and $M_3 > 0$ such that

$$\gamma_1 s^\tau - M_3 \le g(x,s)s - pG(x,s) \quad \text{for a.a. } x \in \Omega, \text{ all } s \ge 0. \tag{11.105}$$

Using (11.104), (11.105), and the fact that $\tau > q$, we find $M_4 > 0$ such that

$$\gamma_1 \|u_n^+\|_\tau^\tau \le M_4(1 + \|u_n^+\|_q^q) \quad \text{for all } n \ge 1. \tag{11.106}$$

From (11.106) and since $\tau > q$, it follows that

$$\{u_n^+\}_{n \ge 1} \text{ is bounded in } L^\tau(\Omega). \tag{11.107}$$

Choosing $h = u_n^+ \in W_0^{1,p}(\Omega)$ in (11.99) and using H$(g)_1^+$ (i) also shows that

$$\|\nabla u_n^+\|_p^p \le \varepsilon_n + M_5(1 + \|u_n^+\|_q^q + \|u_n^+\|_r^r) \quad \text{for all } n \ge 1, \tag{11.108}$$

for some $M_5 > 0$. If $\tau \ge r$, then (11.101) follows from (11.107), (11.108), the continuity of the inclusion $W_0^{1,p}(\Omega) \hookrightarrow L^q(\Omega)$, and the fact that $q < p$. Thus, we may suppose that $\tau < r$. The assumption that $\tau \in ((r-p)\max\{\frac{N}{p},1\}, p^*)$ implies that we can always find $\ell \in (r, p^*)$ such that $\ell > \frac{p\tau}{p+\tau-r}$. Since $\tau < r < \ell$, we can find $t \in (0,1)$ such that

$$\frac{1}{r} = \frac{1-t}{\tau} + \frac{t}{\ell}. \tag{11.109}$$

Invoking the interpolation inequality (e.g., Brezis [52, p. 93]), we have $\|u_n^+\|_r \le \|u_n^+\|_\tau^{1-t}\|u_n^+\|_\ell^t$ for all $n \ge 1$. Because of (11.107) and recalling that $W_0^{1,p}(\Omega) \hookrightarrow L^\ell(\Omega)$ (Theorem 1.49), there is $M_6 > 0$ such that

$$\|u_n^+\|_r^r \le M_6 \|\nabla u_n^+\|_p^{tr} \quad \text{for all } n \ge 1. \tag{11.110}$$

The fact that $\ell > \frac{p\tau}{p+\tau-r}$ ensures that the number $t \in (0,1)$ from (11.109) satisfies $tr < p$. This fact, combined with (11.110), the continuity of the inclusion $W_0^{1,p}(\Omega) \hookrightarrow L^q(\Omega)$, and the fact that $q < p$, allows us to conclude from (11.108) that (11.101) holds true.

From (11.100) and (11.101) it follows that $\{u_n\}_{n \ge 1}$ is bounded in $W_0^{1,p}(\Omega)$. Arguing as at the end of the proof of Proposition 11.38, we deduce that $\{u_n\}_{n \ge 1}$ has a convergent subsequence in $W_0^{1,p}(\Omega)$. This completes Step 2.

*Step 3:* There exists $\lambda^* > 0$ such that for $\|\beta\|_\infty < \lambda^*$ we find $\rho = \rho(\|\beta\|_\infty) > 0$ with

$$\hat\eta_\rho := \inf\{\varphi_{[0,+\infty]}(u) : \|\nabla u\|_p = \rho\} > 0.$$

By hypotheses H$(g)_1^+$ (i), (ii), given $\varepsilon > 0$, we can find $c_\varepsilon > 0$ such that

$$G(x,s) \le \frac{1}{p}(\vartheta(x) + \varepsilon)s^p + c_\varepsilon s^r \quad \text{for a.a. } x \in \Omega, \text{ all } s \ge 0. \tag{11.111}$$

Then, using (11.111), Lemma 11.3, and Proposition 9.47(a), we have

$$\varphi_{[0,+\infty]}(u) \geq \frac{1}{p}\left(c_1 - \frac{\varepsilon}{\lambda_1}\right)\|\nabla u\|_p^p - \|\beta\|_\infty c_2 \|\nabla u\|_p^q - c_\varepsilon c_3 \|\nabla u\|_p^r$$

for all $u \in W_0^{1,p}(\Omega)$, with $c_1, c_2, c_3 > 0$. Choosing $\varepsilon \in (0, c_1\lambda_1)$, we obtain

$$\varphi_{[0,+\infty]}(u) \geq \left(c_4 - \|\beta\|_\infty c_2 \|\nabla u\|_p^{q-p} - c_5 \|\nabla u\|_p^{r-p}\right)\|\nabla u\|_p^p \qquad (11.112)$$

for all $u \in W_0^{1,p}(\Omega)$, with constants $c_4, c_5 > 0$ (depending on the choice of $\varepsilon$). Consider the function $\sigma : (0, +\infty) \to \mathbb{R}$ defined by

$$\sigma(t) = \|\beta\|_\infty c_2 t^{q-p} + c_5 t^{r-p} \quad \text{for all } t > 0. \qquad (11.113)$$

There is a unique $t_0 > 0$ such that $\sigma(t_0) = \inf_{(0,+\infty)} \sigma$, namely,

$$t_0 = \left(\frac{\|\beta\|_\infty c_2(p-q)}{c_5(r-p)}\right)^{\frac{1}{r-q}}.$$

Then, estimating $\sigma(t_0)$ [from (11.113)], we can see that there is $\lambda^* > 0$ such that $\sigma(t_0) < c_4$ whenever $\|\beta\|_\infty < \lambda^*$. From (11.112) it follows that $\inf\{\varphi_{[0,+\infty]}(u) : \|\nabla u\|_p = \rho\} > 0$ for $\rho = \rho(\|\beta\|_\infty) := t_0$. This completes Step 3.

*Step 4:* For every $u \in C_0^1(\overline{\Omega})_+ \setminus \{0\}$ we have $\varphi_{[0,+\infty]}(tu) \to -\infty$ as $t \to +\infty$.

By hypotheses $H(g)_1^+$ (i), (iii.a), given $M > 0$, we find $M_7 = M_7(M) > 0$ such that

$$G(x,s) \geq Ms^p - M_7 \quad \text{for a.a. } x \in \Omega, \text{ all } s \geq 0.$$

Thus

$$\varphi_{[0,+\infty]}(tu) \leq \frac{t^p}{p}\|\nabla u\|_p^p - Mt^p\|u\|_p^p + M_7|\Omega|_N \quad \text{for all } t \geq 0.$$

Since $M > 0$ is arbitrary, we can choose it such that $M\|u\|_p^p > \frac{1}{p}\|\nabla u\|_p^p$. The conclusion of Step 4 follows.

*Step 5:* $\varphi_{[0,+\infty]}$ admits a critical point $u_0 \in W_0^{1,p}(\Omega) \setminus \{0\}$ with $\varphi_{[0,+\infty]}(u_0) > 0$.

Steps 2–4 permit the application of the mountain pass theorem (Theorem 5.40), which yields a $u_0 \in W_0^{1,p}(\Omega)$ critical point of $\varphi_{[0,+\infty]}$ such that

$$\varphi_{[0,+\infty]}(u_0) \geq \hat{\eta}_\rho > 0 = \varphi_{[0,+\infty]}(0).$$

This completes Step 5.

*Step 6:* $\varphi_{[0,+\infty]}$ admits a local minimizer $\hat{u} \in W_0^{1,p}(\Omega) \setminus \{0\}$ with $\varphi_{[0,+\infty]}(\hat{u}) < 0$.

Let $\rho, \hat{\eta}_\rho > 0$ be as in Step 3. We consider the ball $B_\rho(0) = \{u \in W_0^{1,p}(\Omega) : \|\nabla u\|_p < \rho\}$. In view of H$(g)_1^+$ (i), we know that $\inf_{\overline{B}_\rho(0)} \varphi_{[0,+\infty]} \in (-\infty, 0]$. Thus, $\eta_0 := \hat{\eta}_\rho - \inf_{\overline{B}_\rho(0)} \varphi_{[0,+\infty]} > 0$. Let $\varepsilon \in (0, \eta_0)$. By the Ekeland variational principle (Corollary 5.9), there exists $v_\varepsilon \in \overline{B}_\rho(0)$ such that

$$\varphi_{[0,+\infty]}(v_\varepsilon) \leq \inf_{\overline{B}_\rho(0)} \varphi_{[0,+\infty]} + \varepsilon \tag{11.114}$$

and

$$\varphi_{[0,+\infty]}(v_\varepsilon) \leq \varphi_{[0,+\infty]}(y) + \varepsilon \|\nabla(y - v_\varepsilon)\|_p \text{ for all } y \in \overline{B}_\rho(0). \tag{11.115}$$

Since $\varepsilon < \eta_0$, from (11.114) we have $\varphi_{[0,+\infty]}(v_\varepsilon) < \hat{\eta}_\rho$, hence $v_\varepsilon \in B_\rho(0)$. Thus, for any $h \in W_0^{1,p}(\Omega)$ we have $v_\varepsilon + th \in B_\rho(0)$ whenever $t > 0$ is sufficiently small. Taking $y = v_\varepsilon + th$ in (11.115), dividing by $t$, and then letting $t \to 0$, we obtain $-\varepsilon \|\nabla h\|_p \leq \langle \varphi'_{[0,+\infty]}(v_\varepsilon), h \rangle$. This establishes that

$$\|\varphi'_{[0,+\infty]}(v_\varepsilon)\| \leq \varepsilon. \tag{11.116}$$

Consider a sequence $\varepsilon_n \to 0$ and denote $u_n = v_{\varepsilon_n}$. Then, from (11.116) we have $\varphi'_{[0,+\infty]}(u_n) \to 0$ in $W^{-1,p'}(\Omega)$ and $(1 + \|\nabla u_n\|_p)\varphi'_{[0,+\infty]}(u_n) \to 0$ in $W^{-1,p'}(\Omega)$ as $n \to \infty$ [recall that $u_n \in B_\rho(0)$ for all $n \geq 1$]. Step 2 implies that we may assume that $u_n \to \hat{u}$ in $W_0^{1,p}(\Omega)$ as $n \to \infty$ for some $\hat{u} \in \overline{B}_\rho(0)$. From (11.114) we have

$$\varphi_{[0,+\infty]}(\hat{u}) = \inf_{\overline{B}_\rho(0)} \varphi_{[0,+\infty]} \leq 0. \tag{11.117}$$

Since $\inf_{\partial B_\rho(0)} \varphi_{[0,+\infty]} = \hat{\eta}_\rho > 0$, we necessarily have $\hat{u} \in B_\rho(0)$, and thus $\hat{u}$ is a local minimizer of $\varphi_{[0,+\infty]}$. We claim that

$$\inf_{\overline{B}_\rho(0)} \varphi_{[0,+\infty]} < 0. \tag{11.118}$$

By virtue of hypothesis H$(g)_1^+$ (ii), we can find $c_6 > 0$ and $\hat{\delta} > 0$ such that

$$G(x, s) \geq -c_6 s^p \text{ for a.a. } x \in \Omega, \text{ all } s \in [0, \hat{\delta}]. \tag{11.119}$$

Let $v \in \text{int}(C_0^1(\overline{\Omega})_+)$ with $\|v\|_\infty \leq \hat{\delta}$. Due to (11.119), for $t \in (0, 1)$ we have

$$\varphi_{[0,+\infty]}(tv) \leq \frac{t^p}{p} \|\nabla v\|_p^p - \frac{t^q}{q} \int_\Omega \beta(x) v^q \, dx + t^p c_6 \|v\|_p^p.$$

Since $q < p$, for $t \in (0,1)$ small, we have $\varphi_{[0,+\infty]}(tv) < 0$ and $tv \in B_\rho(0)$. This yields (11.118). Finally, comparing (11.117) and (11.118), we obtain that $\hat{u}$ fulfills the requirements of Step 6.

The theorem is now obtained by combining Steps 1, 5, and 6.    □

Next we look for nodal (sign-changing) solutions. We will produce such a solution for a restricted version of problem (11.96) in which $\beta(\cdot)$ is constant:

$$\begin{cases} -\Delta_p u(x) = \lambda |u(x)|^{q-2} u(x) + g(x, u(x)) & \text{in } \Omega, \\ u = 0 & \text{on } \partial\Omega, \end{cases} \tag{11.120}$$

with $\lambda > 0$ and $q \in (1, p)$. The first step is to establish the existence of a smallest positive solution for problem (11.120). To this end, we strengthen the hypotheses on $g(x, \cdot)$ at the origin:

$H(g)_2^+$  (i)  $g : \Omega \times \mathbb{R} \to \mathbb{R}$ is a Carathéodory function with $g(x, 0) = 0$ a.e. in $\Omega$ and there exist $c > 0$ and $r \in [1, +\infty)$ such that

$$|g(x,s)| \le c(1 + |s|^{r-1}) \quad \text{for a.a. } x \in \Omega, \text{ all } s \in \mathbb{R};$$

(ii)  $\lim_{s \downarrow 0} \dfrac{g(x,s)}{s^{p-1}} = 0$ uniformly for a.a. $x \in \Omega$;

(iii)  There exists $\delta_0 > 0$ such that $g(x,s) \ge 0$ for a.a. $x \in \Omega$, all $s \in [0, \delta_0]$.

**Proposition 11.50.** *Assume that* $H(g)_2^+$ *holds. Then there is* $\lambda^* > 0$ *such that for* $\lambda \in (0, \lambda^*)$ *problem (11.120) has a smallest positive solution* $u_+ \in \text{int}(C_0^1(\overline{\Omega})_+)$. *Furthermore, it satisfies* $\|u_+\|_\infty < \delta_0$.

*Proof.* Let $e \in \text{int}(C_0^1(\overline{\Omega})_+)$ be the unique solution of the equation $-\Delta_p e = 1$ in $W^{-1,p'}(\Omega)$ (see Claim 1 of proof of Proposition 11.28). We fix $\varepsilon \in (0, \frac{1}{\|e\|_\infty^{p-1}})$. By $H(g)_2^+$ (ii), there is $\delta_\varepsilon \in (0, \delta_0)$ [see $H(g)_2^+$ (iii)] such that

$$0 \le g(x,s) \le \varepsilon s^{p-1} \quad \text{for a.a. } x \in \Omega, \text{ all } s \in [0, \delta_\varepsilon]. \tag{11.121}$$

Set $\lambda^* = \delta_\varepsilon^{p-q}(\|e\|_\infty^{1-p} - \varepsilon) > 0$ and fix $\lambda \in (0, \lambda^*)$. It is straightforward to check that the number $\eta_\lambda := \left(\lambda \|e\|_\infty^{q-1}(1 - \varepsilon \|e\|_\infty^{p-1})^{-1}\right)^{\frac{1}{p-q}}$ satisfies

$$0 < \eta_\lambda \|e\|_\infty < \delta_\varepsilon \quad \text{and} \quad \lambda(\eta_\lambda \|e\|_\infty)^{q-1} + \varepsilon(\eta_\lambda \|e\|_\infty)^{p-1} = \eta_\lambda^{p-1}. \tag{11.122}$$

Let $\overline{u} = \eta_\lambda e \in \text{int}(C_0^1(\overline{\Omega})_+)$. Then, by (11.121) and (11.122), we see that

$$-\Delta_p \overline{u} = \eta_\lambda^{p-1} = \lambda(\eta_\lambda \|e\|_\infty)^{q-1} + \varepsilon(\eta_\lambda \|e\|_\infty)^{p-1} \ge \lambda \overline{u}^{q-1} + g(x, \overline{u}) \text{ in } W^{-1,p'}(\Omega),$$

hence $\overline{u}$ is an upper solution of problem (11.120) (Definition 11.5). Moreover, we have $\|\overline{u}\|_\infty < \delta_\varepsilon < \delta_0$.

Note that the function $f(x,s) = \lambda |s|^{q-2}s + g(x,s)$ satisfies hypothesis $H(f)_1^+$ of Sect. 11.2. Thus, we can apply Proposition 11.21(a), which yields $\underline{u} \in \text{int}(C_0^1(\overline{\Omega})_+)$, satisfying $\underline{u} \leq \overline{u}$ in $\Omega$ and such that $\tilde{\varepsilon}\underline{u}$ is a lower solution of problem (11.120) whenever $\tilde{\varepsilon} \in (0,1]$ (Definition 11.5). Then we fix a sequence $\{\tilde{\varepsilon}_n\}_{n \geq 1} \subset (0,1]$, with $\tilde{\varepsilon}_n \to 0$ as $n \to \infty$, and we let $\underline{u}_n = \tilde{\varepsilon}_n \hat{u}_1$. From Proposition 11.20(a) we know that problem (11.120) has a smallest solution $u_n^*$ in the order interval $[\underline{u}_n, \overline{u}]$, which in addition belongs to $\text{int}(C_0^1(\overline{\Omega})_+)$. Thus,

$$-\Delta_p u_n^* = \lambda (u_n^*)^{q-1} + g(x, u_n^*) \text{ in } W^{-1,p'}(\Omega) \text{ for all } n \geq 1. \tag{11.123}$$

From (11.123), the fact that $0 \leq u_n^* \leq \overline{u} < \delta_\varepsilon$ in $\Omega$, and (11.121), we see that $\{u_n^*\}_{n \geq 1}$ is bounded in $W_0^{1,p}(\Omega)$, and thus there is $u_+ \in W_0^{1,p}(\Omega)$ such that

$$u_n^* \xrightarrow{w} u_+ \text{ in } W_0^{1,p}(\Omega) \text{ and } u_n^* \to u_+ \text{ in } L^p(\Omega) \text{ as } n \to \infty \tag{11.124}$$

along a relabeled subsequence. On (11.123) we act with $u_n^* - u_+ \in W_0^{1,p}(\Omega)$ and let $n \to \infty$. By (11.121) and (11.124), we obtain $\lim_{n \to \infty} \langle -\Delta_p u_n^*, u_n^* - u_+ \rangle = 0$. Invoking Proposition 2.72, it follows that

$$u_n^* \to u_+ \text{ in } W_0^{1,p}(\Omega) \text{ as } n \to \infty. \tag{11.125}$$

Passing to the limit in (11.123) and using (11.125), we obtain that $u_+$ is a solution of (11.120).

We show that $u_+ \neq 0$. To this end, take $\tilde{u} = \lambda^{\frac{1}{p-q}} \underline{u}_q \in \text{int}(C_0^1(\overline{\Omega})_+)$, with $\underline{u}_q$ given by Proposition 11.9, so $\tilde{u}$ satisfies

$$-\Delta_p \tilde{u}(x) = \lambda \tilde{u}(x)^{q-1} \text{ in } W^{-1,p'}(\Omega).$$

Since $u_n^* \in \text{int}(C_0^1(\overline{\Omega})_+)$, we know that there is $t > 0$ such that $t\tilde{u} \leq u_n^*$ in $\Omega$. Let $t_n = \max\{t > 0 : t\tilde{u} \leq u_n^* \text{ in } \Omega\}$ for all $n \geq 1$. We claim that $t_n \geq 1$ for all $n \geq 1$. Suppose that there is $n \geq 1$ with $t_n < 1$. By (11.123), the fact that $-\Delta_p(t_n\tilde{u}) = \lambda(t_n\tilde{u})^{q-1}$ in $W^{-1,p'}(\Omega)$, and the relation

$$\lambda u_n^*(x)^{q-1} + g(x, u_n^*(x)) \geq \lambda(t_n\tilde{u}(x))^{q-1} > \lambda t_n^{p-1}\tilde{u}(x)^{q-1} \text{ a.e. in } \Omega$$

[using $H(g)_2^+$ (iii) and $0 \leq u_n^* \leq \overline{u} < \delta_0$ in $\Omega$], we can apply Proposition 8.29, which yields $u_n^* - t_n\tilde{u} \in \text{int}(C_0^1(\overline{\Omega})_+)$. This contradicts the maximality of $t_n$. Thus, we conclude that $t_n \geq 1$ for all $n \geq 1$. Therefore, we have $u_n^* \geq \tilde{u}$ in $\Omega$ for all $n \geq 1$. Letting $n \to \infty$, we obtain $u_+ \geq \tilde{u}$ in $\Omega$, and so $u_+ \neq 0$.

Since $0 \leq u_+ \leq \overline{u} < \delta_0$ in $\Omega$, we have $-\Delta_p u_+ \geq 0$ in $W^{-1,p'}(\Omega)$. Hence, by Corollary 8.13 and the strong maximum principle (Theorem 8.27), we deduce that $u_+ \in \text{int}(C_0^1(\overline{\Omega})_+)$.

Finally, we claim that $u_+$ is the smallest positive solution of (11.120). To justify this, let $u \in W_0^{1,p}(\Omega)$ be a nontrivial solution of (11.120) such that $u \geq 0$ a.e. in $\Omega$. Thus, $u \in C_0^1(\overline{\Omega})$ (Corollary 8.13) and, due to H$(g)_2^+$ (i), (ii), we have $-\Delta_p u \geq -\tilde{c} u^{p-1}$ in $W^{-1,p'}(\Omega)$, whence $\hat{u} \in \mathrm{int}(C_0^1(\overline{\Omega})_+)$ (by Theorem 8.27). Note that $\overline{u}_0 := \min\{u, \overline{u}\}$ is an upper solution of (11.120) (Lemma 11.19). Using that $u, \overline{u} \in \mathrm{int}(C_0^1(\overline{\Omega})_+)$, for $n \geq 1$ large we have $\underline{u}_n = \tilde{\varepsilon}_n \underline{u} \leq \overline{u}_0$ in $\Omega$. By Proposition 11.8(b), there exists a solution $\tilde{u}_n$ of (11.120) in the order interval $[\underline{u}_n, \overline{u}_0]$. Since $u_n^*$ is the smallest solution of (11.120) in $[\underline{u}_n, \overline{u}]$, it follows that $u_n^* \leq \tilde{u}_n \leq \overline{u}_0 \leq u$ in $\Omega$, which yields $u_+ \leq u$ in $\Omega$. This proves the minimality of $u_+$.                               $\square$

Now we collect all the hypotheses on $g$ considered in this section together with their counterparts on the negative half-line:

H$(g)_3$ (i) $g : \Omega \times \mathbb{R} \to \mathbb{R}$ is a Carathéodory function with $g(x,0) = 0$ a.e. in $\Omega$, and there exist $c > 0$ and $r \in (p, p^*)$ such that

$$|g(x,s)| \leq c(1 + |s|^{r-1}) \quad \text{for a.a. } x \in \Omega, \text{ all } s \in \mathbb{R};$$

  (ii) $\displaystyle\lim_{s \to 0} \frac{g(x,s)}{|s|^{p-1}} = 0$ uniformly for a.a. $x \in \Omega$;

  (iii) There exist $\tau \in ((r-p)\max\{\frac{N}{p}, 1\}, p^*)$, $\tau > q$, and $\gamma_0 > 0$ such that

$$\lim_{s \to \pm\infty} \frac{G(x,s)}{|s|^p} = +\infty \quad \text{and} \quad \liminf_{s \to \pm\infty} \frac{g(x,s)s - pG(x,s)}{|s|^\tau} \geq \gamma_0$$

   uniformly for a.a. $x \in \Omega$;
  (iv) There exists $\delta_0 > 0$ such that $g(x,s)s \geq 0$ for a.a. $x \in \Omega$, all $s \in [-\delta_0, \delta_0]$.

*Example 11.51.* The functions $g_1$ and $g_2$ in Example 11.48 also satisfy H$(g)_3$.

*Remark 11.52.* Note that the form of the nonlinearity on which we focus here [that is, obtained as the sum of the concave term $\lambda |s|^{q-2}s$ and a convex term $g(x,s)$] is the counterpoint of the nonlinearity studied in Example 11.32 [sum of the convex term $|s|^{r-2}s$ and a concave perturbation $\lambda g(x,s)$]. In this respect, the next result complements Theorem 11.31.

**Theorem 11.53.** *Assume that* H$(g)_3$ *holds. Then there is $\lambda^* > 0$ such that for $\lambda \in (0, \lambda^*)$ problem (11.120) has at least five distinct, nontrivial solutions: $u_0, \hat{u} \in \mathrm{int}(C_0^1(\overline{\Omega})_+)$, $v_0, \hat{v} \in -\mathrm{int}(C_0^1(\overline{\Omega})_+)$, and $y_0 \in C_0^1(\overline{\Omega})$ nodal.*

*Proof.* Theorem 11.49 and Proposition 11.50 yield $\lambda^* > 0$ such that, given $\lambda \in (0, \lambda^*)$, problem (11.120) admits two distinct positive solutions $u_0, \hat{u} \in \mathrm{int}(C_0^1(\overline{\Omega})_+)$ as well as a smallest positive solution $u_+ \in \mathrm{int}(C_0^1(\overline{\Omega})_+)$ with $\|u_+\|_\infty < \delta_0$ (possibly equal to $u_0$ or $\hat{u}$). Since the hypotheses are symmetric with respect to the origin, the same reasoning as in Theorem 11.49 and Proposition 11.50 shows that, up to choosing $\lambda^* > 0$ smaller, we can also find $v_0, \hat{v} \in -\mathrm{int}(C_0^1(\overline{\Omega})_+)$ distinct solutions of (11.120) as well as a biggest negative solution $v_- \in -\mathrm{int}(C_0^1(\overline{\Omega})_+)$, with $\|v_-\|_\infty <$

$\delta_0$. It remains to show that we can find a solution $y_0 \in C_0^1(\overline{\Omega})$ of (11.120) in the order interval $[v_-, u_+]$ distinct from $0, v_-, u_+$: then the extremality property of $v_-, u_+$ will ensure that $y_0$ must be nodal.

Recall that we denote $f(x,s) = \lambda |s|^{q-2}s + g(x,s)$. We consider the Carathéodory function $f_{[v_-,u_+]}$ obtained by truncation:

$$f_{[v_-,u_+]}(x,s) = \begin{cases} \lambda |v_-(x)|^{q-2}v_-(x) + g(x,v_-(x)) & \text{if } s < v_-(x), \\ \lambda |s|^{q-2}s + g(x,s) & \text{if } v_-(x) \leq s \leq u_+(x), \\ \lambda u_+(x)^{q-1} + g(x,u_+(x)) & \text{if } s > u_+(x), \end{cases}$$

(11.126)

and the corresponding $C^1$-functional $\varphi_{[v_-,u_+]}$ [see (11.5)]. According to Proposition 11.8(a), it suffices to show that $\varphi_{[v_-,u_+]}$ admits a critical point distinct from $0, v_-, u_+$. We may assume that $\varphi_{[v_-,u_+]}$ has only a finite number of critical points (otherwise we are done).

*Claim 1:* $v_-$ and $u_+$ are strict local minimizers of $\varphi_{[v_-,u_+]}$.

We only argue for $u_+$ (the proof in the case of $v_-$ is similar). Consider the truncation $\varphi_{[0,u_+]}$ [see (11.5)]. From Proposition 11.8(b) we know that $\varphi_{[0,u_+]}$ admits a global minimizer $v \in C_0^1(\overline{\Omega}) \cap [0, u_+]$. Arguing as at the end of Step 6 in the proof of Theorem 11.49, we can see that $\varphi_{[0,u_+]}(tu_+) < 0$ for $t \in (0,1)$ small, which guarantees that $v \neq 0$. By the minimality of $u_+$ and Proposition 11.8(a), we get that $u_+ = v$ is the unique global minimizer of $\varphi_{[0,u_+]}$. The functionals $\varphi_{[0,u_+]}$ and $\varphi_{[v_-,u_+]}$ coincide on $C_0^1(\overline{\Omega})_+$, so $u_+$ is also a local $C^1(\overline{\Omega})$-minimizer of $\varphi_{[v_-,u_+]}$ and so, in view of Proposition 11.4, a local minimizer of $\varphi_{[v_-,u_+]}$ with respect to the topology of $W_0^{1,p}(\Omega)$. In fact, $u_+$ is a strict local minimizer because $\varphi_{[v_-,u_+]}$ is assumed to have only a finite number of critical points. This proves Claim 1.

*Claim 2:* There is $y_0 \in W_0^{1,p}(\Omega)$ critical point of $\varphi_{[v_-,u_+]}$ distinct from $v_-, u_+$ such that $C_1(\varphi_{[v_-,u_+]}, y_0) \neq 0$.

Say that $\varphi_{[v_-,u_+]}(v_-) \leq \varphi_{[v_-,u_+]}(u_+)$ (the analysis is similar in the other situation). Arguing as in the proof of Proposition 5.42, we can find $\rho > 0$ small such that

$$\varphi_{[v_-,u_+]}(u_+) < \inf\{\varphi_{[v_-,u_+]}(u) : u \in W_0^{1,p}(\Omega), \|\nabla(u - u_+)\|_p = \rho\}.$$

Then Claim 2 follows by applying Corollary 6.81.

*Claim 3:* $C_k(\varphi_{[v_-,u_+]}, 0) = 0$ for all $k \geq 0$.

It is straightforward to check that the truncated function $f_{[v_-,u_+]}$ given in (11.126) satisfies conditions H$(f)_1$ (i), (ii) stated at the beginning of the section [relying on H$(g)_3$ (ii), (iv) and taking any $\tau \in (q,p)$ for H$(f)_1$ (ii)]. Therefore, Claim 3 is obtained by invoking Proposition 11.42.

Comparing Claims 2 and 3, we obtain that $y_0$ is a critical point of $\varphi_{[v_-,u_+]}$ distinct from $v_-, u_+, 0$. The proof of the theorem is now complete. $\qquad\square$

*Remark 11.54.* The nodal solution $y_0$ in Theorem 11.53 satisfies the a priori estimate $\|y_0\|_\infty < \delta_0$, with the constant $\delta_0 > 0$ in hypothesis H$(g)_3$ (iv). In Theorem 11.53, we can choose $v_0$ to be the biggest negative solution and $u_0$ the smallest positive solution, and thus we can order the solutions as $\hat{v} \leq v_0 \leq y_0 \leq u_0 \leq \hat{u}$.

## 11.4 Nonlinear Dirichlet Problems Using Nonlinear Operator Theory

Let $\Omega \subset \mathbb{R}^N$ be a bounded domain with a $C^1$-boundary $\partial\Omega$. In this section, we study a very general boundary value problem that takes the form of the following nonlinear elliptic differential inclusion:

$$\begin{cases} -\operatorname{div} A(x,u(x),\nabla u(x)) + \beta(u(x)) + G(x,u(x),\nabla u(x)) \ni e(x) & \text{in } \Omega, \\ u|_{\partial\Omega} = 0. \end{cases} \tag{11.127}$$

Throughout this section we fix a number $p \in [2,+\infty)$. Then, as solution space for problem (11.127), we consider the Sobolev space $W_0^{1,p}(\Omega)$ equipped with the norm $u \mapsto \|\nabla u\|_p$. As before, we let $p' = \frac{p}{p-1}$. Thus $p' \in (1,2]$, and so $L^2(\Omega) \subset L^{p'}(\Omega)$.

Let us present in more detail the different data involved in (11.127). The maps $A : \Omega \times \mathbb{R} \times \mathbb{R}^N \to 2^{\mathbb{R}^N} \setminus \{\emptyset\}$ and $G : \Omega \times \mathbb{R} \times \mathbb{R}^N \to 2^{\mathbb{R}} \setminus \{\emptyset\}$ are multifunctions, whereas $\beta : \mathbb{R} \to 2^{\mathbb{R}}$ is a maximal monotone map. Finally, $e \in L^2(\Omega)$. The precise assumptions on the multifunctions $A, G, \beta$ (stated below) will ensure that they give rise to well-defined Nemytskii operators $N_A, N_G, N_\beta : W_0^{1,p}(\Omega) \to 2^{L^{p'}(\Omega)}$ defined by

$$N_A(u) = \{v \in L^{p'}(\Omega,\mathbb{R}^N) : v(x) \in A(x,u(x),\nabla u(x)) \text{ for a.a. } x \in \Omega\},$$

$$N_G(u) = \{v \in L^{p'}(\Omega) : v(x) \in G(x,u(x),\nabla u(x)) \text{ for a.a. } x \in \Omega\},$$

$$N_\beta(u) = \{v \in L^{p'}(\Omega) : v(x) \in \beta(u(x)) \text{ for a.a. } x \in \Omega\}.$$

The term $\operatorname{div} A(x,u(x),\nabla u(x))$ in (11.127) must be understood as

$$\operatorname{div} A(x,u(x),\nabla u(x)) = \{\operatorname{div} v : v \in N_A(u)\}.$$

Theorem 1.31 implies that for every $v \in N_A(u)$ we have $\operatorname{div} v \in W^{-1,p'}(\Omega)$. Then we consider the following notion of solution for problem (11.127).

**Definition 11.55.** An element $u \in W_0^{1,p}(\Omega)$ is called a *weak solution* of problem (11.127) if there exist $v \in N_A(u)$, $w \in N_G(u)$, and $y \in N_\beta(u)$ such that the equality

$$e = -\operatorname{div} v + w + y$$

holds in $W^{-1,p'}(\Omega)$.

The statement of the problem is very general. It incorporates the following situations:

- If $\beta \equiv 0$ while $A$ and $G$ are single-valued and such that $A(x,s,\xi) = |\xi|^{p-2}\xi$ and $G(x,s,\xi) = f(x,s)$ where $f : \Omega \times \mathbb{R} \to \mathbb{R}$ is a Carathéodory function, then we recover the type of problems studied in Sects. 11.1–11.3. In fact, the hypotheses on the multifunction $A$ [see hypotheses H(A) below] are such that they incorporate in the expression $\mathrm{div}\, A(\cdot, u(\cdot), \nabla u(\cdot))$ several interesting generalizations of the $p$-Laplace differential operator.
- In general, $\beta$ will be the convex subdifferential of a lower semicontinuous convex function $\vartheta : \mathbb{R} \to \mathbb{R} \cup \{+\infty\}$ (Sect. 3.1); in particular, problem (11.127) includes variational constraints [Remark 5.78(a)].
- The multifunction $G$ can be, for instance, of the form $G(x,s,\xi) = \partial F(x,s)$, where $F(x,s)$ is a function measurable in $x \in \Omega$ and locally Lipschitz in $s \in \mathbb{R}$, and $\partial F(x,\cdot)$ stands for the generalized subdifferential of $F(x,\cdot)$ (Sect. 3.2). Thus, the statement of problem (11.127) also includes certain hemivariational and variational–hemivariational inequalities [Remark 5.78(c), (d)].

Accordingly, our treatment of problem (11.127) will be by means of general tools of operator theory presented in Sect. 2.2.

Let us present some conventions and notation used in this section. We denote by $2^Y$ or $\mathscr{P}(Y)$ the set of all subsets of $Y$. Moreover, when $Y$ is a normed space, we let

$$\mathscr{P}_{\mathrm{wkc}}(Y) = \{C \in \mathscr{P}(Y) : C \text{ is nonempty, weakly compact, and convex}\}.$$

Recall that the graph of a multimap $H : X \to 2^Y$ is given by

$$\mathrm{Gr}\, H = \{(x,y) \in X \times Y : y \in H(x)\}.$$

When $X, Y$ are measure spaces, we say that the multimap $H$ is *graph measurable* if $\mathrm{Gr}\, H$ is a measurable subset of $X \times Y$. We will need the following proposition (see, for instance, Hu and Papageorgiou [175, p. 175]).

**Proposition 11.56.** *Let $H : \Omega \to 2^{\mathbb{R}^M} \setminus \{\emptyset\}$ (with $M \geq 1$) be a graph-measurable multifunction. Let $r \in (1, +\infty)$, and assume that there is $h_0 \in L^r(\Omega)_+$ such that*

$$\inf\{|z| : z \in H(x)\} \leq h_0(x) \text{ for a.a. } x \in \Omega.$$

*Then there is $h \in L^r(\Omega, \mathbb{R}^M)$ such that $h(x) \in H(x)$ for a.a. $x \in \Omega$.*

A function $h$ as in Proposition 11.56 is called a *selection* of $H$. This proposition, which will be useful in particular to guarantee that the Nemytskii operators $N_A$, $N_G$, $N_\beta$ have nonempty values, is in fact a consequence of a much more general result called the *Yankov–von Neumann–Aumann selection theorem*. See Hu and Papageorgiou [175, p. 158] for more details.

Now we state our hypotheses on the multifunctions involved in problem (11.127). We start with the hypotheses on $A$.

**H($A$)**   (i)   $A : \Omega \times \mathbb{R} \times \mathbb{R}^N \to 2^{\mathbb{R}^N}$ is a multifunction with nonempty, compact, convex values, with measurable graph, and such that

         (i.a)   For a.a. $x \in \Omega$ the graph of $(s, \xi) \mapsto A(x, s, \xi)$ is closed;

         (i.b)   For a.a. $x \in \Omega$, all $s \in \mathbb{R}$, the mapping $\xi \mapsto A(x, s, \xi)$ is strictly monotone and $0 \in A(x, s, 0)$;

         (i.c)   For a.a. $x \in \Omega$, all $\xi \in \mathbb{R}^N$, the mapping $s \mapsto A(x, s, \xi)$ is l.s.c.;

     (ii)   There exists $c_0 > 0$ such that

$$|z| \le c_0 (1 + |s|^{p-1} + |\xi|^{p-1})$$

         for a.a. $x \in \Omega$, all $(s, \xi) \in \mathbb{R} \times \mathbb{R}^N$, all $z \in A(x, s, \xi)$;

     (iii)   There exist $c_1, c_2 > 0$ such that

$$(z, \xi)_{\mathbb{R}^N} \ge c_1 |\xi|^p - c_2$$

         for a.a. $x \in \Omega$, all $(s, \xi) \in \mathbb{R} \times \mathbb{R}^N$, all $z \in A(x, s, \xi)$.

*Example 11.57.* The preceding hypotheses are general and incorporate a broad family of nonlinear differential operators. In what follows, we give examples of maps satisfying hypotheses H($A$). For simplicity, we take single-valued maps. In these examples, we suppose that $p \in [2, +\infty)$. The following maps satisfy hypotheses H($A$):

$A_1(x, s, \xi) = |\xi|^{p-2} \xi$   (then the resulting operator div$A_1$ is the $p$-Laplacian),

$A_2(x, s, \xi) = |\xi|^{p-2} \xi + |\xi|^{r-2} \xi, r \in (1, p)$ (then div$A_2$ is the $(p, r)$-Laplacian),

$A_3(x, s, \xi) = (1 + |\xi|^2)^{\frac{p-2}{2}} \xi$,

$$A_4(x, s, \xi) = \begin{cases} |\xi|^{p-2} \xi & \text{if } |\xi| \le 1 \\ \frac{1}{2}(|\xi|^{p-2} \xi + |\xi|^{r-2} \xi) & \text{if } |\xi| > 1 \end{cases} \quad \text{with } 1 \le r < p,$$

$$A_5(x, s, \xi) = \begin{cases} |\xi|^{p-2} \xi + \left( \frac{\ln(1+|\xi|)}{|\xi|} + \frac{1}{1+|\xi|} \right) \xi & \text{if } \xi \ne 0, \\ 0 & \text{if } \xi = 0. \end{cases}$$

The set of operators satisfying H($A$) is stable by addition, and by multiplication by Carathéodory maps $\theta : \Omega \times \mathbb{R} \to \mathbb{R}$ satisfying $\underset{(x,s) \in \Omega \times \mathbb{R}}{\text{ess inf}} \theta(x, s) > 0$. An example of a multivalued operator satisfying H($A$) can be obtained as the sum

$$\tilde{A}(x, s, \xi) = A(x, s, \xi) + \partial \psi(\xi),$$

where $A(x,s,\xi)$ satisfies H($A$) and where $\partial \psi$ denotes the subdifferential of $\psi$ : $\mathbb{R}^N \to \mathbb{R}$ convex and continuous, with $\psi(\xi) \geq \psi(0)$ for all $\xi \in \mathbb{R}^N$, and such that $|\xi^*| \leq \hat{c}(1 + |\xi|^{p-1})$ for all $\xi^* \in \partial \psi(\xi)$, all $\xi \in \mathbb{R}^N$, and some constant $\hat{c} > 0$. Then $\tilde{A}(x,s,\xi)$ is multivalued whenever $\psi$ is not Gâteaux differentiable (Proposition 3.12); the simplest example is $\psi(\xi) = |\xi|$.

Recall that by $\Gamma_0(\mathbb{R})$ we denote the cone of convex, lower semicontinuous functions $\vartheta : \mathbb{R} \to \mathbb{R} \cup \{+\infty\}$ with $\vartheta \not\equiv +\infty$, and by $\partial \vartheta$ we denote the subdifferential of the convex function $\vartheta$ (Definition 3.6). The hypotheses on $\beta$ are as follows.

H($\beta$)   $\beta = \partial \vartheta$ for some $\vartheta \in \Gamma_0(\mathbb{R})$ such that $\vartheta \geq 0$ and $\vartheta(0) = 0$.

Finally, the hypotheses on the multivalued term $G$ are as follows:

H($G$)$_1$   (i)  $G : \Omega \times \mathbb{R} \times \mathbb{R}^N \to 2^{\mathbb{R}}$ is a multifunction with nonempty, compact, convex values such that $x \mapsto G(x,s,\xi)$ has a measurable graph for all $(s,\xi) \in \mathbb{R} \times \mathbb{R}^N$ and $(s,\xi) \mapsto G(x,s,\xi)$ has a closed graph for a.a. $x \in \Omega$;
   (ii) There exists $c > 0$ such that

$$|z| \leq c(1 + |s|^{p-1} + |\xi|^{p-1})$$

   for a.a. $x \in \Omega$, all $(s,\xi) \in \mathbb{R} \times \mathbb{R}^N$, all $z \in G(x,s,\xi)$;
   (iii) We have

$$\liminf_{s \to \pm\infty} \left( \inf_{\xi \in \mathbb{R}^N} \frac{\min G(x,s,\xi)}{|s|^{p-2}s} \right) \geq -\eta_0(x) \text{ uniformly for a.a. } x \in \Omega,$$

   with $\eta_0 \in L^\infty(\Omega)_+$ such that

$$\eta_0(x) \leq c_1 \lambda_1 \text{ for a.a. } x \in \Omega \text{ and } \eta_0 \neq c_1 \lambda_1,$$

   where $\lambda_1 > 0$ denotes the first eigenvalue of the negative Dirichlet $p$-Laplacian and where $c_1 > 0$ is the same as in hypothesis H($A$) (iii).

*Remark 11.58.* Hypothesis H($G$)$_1$ (iii) is a version of the nonuniform nonresonance condition for the multivalued nonlinearity $s \mapsto G(x,s,\xi)$.

*Example 11.59.* A simple example of a multifunction satisfying H($G$)$_1$ is

$$G(x,s,\xi) = [\theta_1(x), \theta_2(x)](|s|^{r-2}s - \mu|s|^{p-2}s + \nu|\xi|^{p-1}),$$

where $r \in (1,p]$, $\mu \in \mathbb{R}$, $\nu \geq 0$ are constants and $\theta_1, \theta_2 \in L^\infty(\Omega)_+$ satisfy $\theta_1 \leq \theta_2$ a.e. in $\Omega$, $\mu\theta_2 \leq c_1\lambda_1$ a.e. in $\Omega$, and $\mu\theta_2 \neq c_1\lambda_1$.

The establishing of an existence result for solutions of problem (11.127) requires preliminary steps. Specifically, we need to point out properties of the different operators involved in (11.127). We start with the operator induced by the multifunction $A$. We will use the following variant of the Nemytskii operator $N_A$. For a fixed $u \in W_0^{1,p}(\Omega)$ we denote

$$N_A^u(y) = \{v \in L^{p'}(\Omega, \mathbb{R}^N) : v(x) \in A(x, u(x), \nabla y(x)) \text{ for a.a. } x \in \Omega\}$$

for all $y \in W_0^{1,p}(\Omega)$. Then we consider the multimap $E_u : W_0^{1,p}(\Omega) \to 2^{W^{-1,p'}(\Omega)}$ defined by

$$E_u(y) = \{-\operatorname{div} v : v \in N_A^u(y)\} \text{ for all } y \in W_0^{1,p}(\Omega).$$

**Lemma 11.60.** *Assume that* H(A) *holds. Then, for every* $u \in W_0^{1,p}(\Omega)$, *the multimap* $E_u$ *has values in* $\mathscr{P}_{wkc}(W^{-1,p'}(\Omega))$ *and is maximal monotone.*

*Proof.* For $y \in W_0^{1,p}(\Omega)$ the multimap $x \mapsto A(x, u(x), \nabla y(x))$ has a measurable graph, and its values are nonempty, closed, and convex [by H(A) (i)]. Also, using H(A) (ii), we may apply Proposition 11.56, which shows that the set $N_A^u(y)$ is nonempty and clearly closed, convex, and, by H(A) (ii), bounded in $L^{p'}(\Omega, \mathbb{R}^N)$. This implies that $E_u(y)$ is nonempty, closed, convex, and bounded in $W^{-1,p'}(\Omega)$ and so belongs to $\mathscr{P}_{wkc}(W^{-1,p'}(\Omega))$.

Hypothesis H(A) (i.b) implies that $E_u$ is monotone. By virtue of Proposition 2.40 and Remark 2.41, to prove the maximal monotonicity of $E_u$ (and so complete the proof of the lemma), it suffices to check the following technical claim.

*Claim 1:* Let $y, h \in W_0^{1,p}(\Omega)$, $\{t_n\}_{n \geq 1} \subset [0,1]$, $\{w_n\}_{n \geq 1} \subset W^{-1,p'}(\Omega)$ such that $w_n \in E_u(y + t_n h)$ for all $n \geq 1$ and $t_n \to 0$ as $n \to \infty$, and let $U \subset W^{-1,p'}(\Omega)$ be a weakly open neighborhood of $E_u(y)$. Then we have $w_n \in U$ for all $n \geq 1$ large enough.

It suffices to see that any relabeled subsequence $\{w_n\}_{n \geq 1}$ admits a subsequence weakly converging to an element of $E_u(y)$. By definition, $w_n = -\operatorname{div} v_n$, with $v_n \in N_A^u(y + t_n h)$ for all $n \geq 1$. Hypothesis H(A) (ii) implies that $\{v_n\}_{n \geq 1}$ is bounded in $L^{p'}(\Omega, \mathbb{R}^N)$, so we may assume that $v_n \overset{w}{\to} v$ in $L^{p'}(\Omega, \mathbb{R}^N)$ for some $v \in L^{p'}(\Omega, \mathbb{R}^N)$. Thus $w_n \overset{w}{\to} -\operatorname{div} v$ in $W^{-1,p'}(\Omega)$. Then it suffices to check that $-\operatorname{div} v \in E_u(y)$, i.e., that $v \in N_A^u(y)$. So we must prove

$$v(x) \in A(x, u(x), \nabla y(x)) \text{ for a.a. } x \in \Omega. \tag{11.128}$$

Using Mazur's theorem (e.g., Brezis [52, p. 61]), we find a sequence $\{\tilde{v}_n\}_{n \geq 1} \subset L^{p'}(\Omega, \mathbb{R}^N)$, where $\tilde{v}_n$ lies in the convex hull of $\{v_m : m \geq n\}$ for all $n \geq 1$ such that

$$\tilde{v}_n(x) \to v(x) \text{ as } n \to \infty \text{ for all } x \in \Omega \setminus S$$

for some Lebesgue-null subset $S \subset \Omega$. Up to enlarging $S$, we may also assume that

$$v_n(x) \in A(x, u(x), \nabla(y + t_n h)(x)) \text{ for all } n \geq 1, \text{ all } x \in \Omega \setminus S.$$

Fix $x \in \Omega \setminus S$ and $\varepsilon > 0$. By H(A) (i.a), (ii) [and in view of Remark 2.37(c)], the multifunction $(s, \xi) \mapsto A(x, s, \xi)$ is locally compact and u.s.c., hence [using Remark 2.37(b)] there is $n_0 \geq 1$ such that for all $n \geq n_0$ we have $d(v_n(x), A(x, u(x), \nabla y(x))) \leq \varepsilon$, whence $d(\bar{v}_n(x), A(x, u(x), \nabla y(x))) \leq \varepsilon$. Letting $n \to \infty$, we get

$$d(v(x), A(x, u(x), \nabla y(x))) \leq \varepsilon.$$

Finally, letting $\varepsilon \to 0$, we obtain (11.128). The proof is now complete. □

Next, we consider $E : W_0^{1,p}(\Omega) \to \mathscr{P}_{\text{wkc}}(W^{-1,p'}(\Omega))$, the differential operator of problem (11.127), defined by

$$E(u) = E_u(u) = \{-\operatorname{div} v : v \in N_A(u)\} \text{ for all } u \in W_0^{1,p}(\Omega). \tag{11.129}$$

**Proposition 11.61.** *If* H(A) *holds, then* $E$ *is bounded and pseudomonotone.*

*Proof.* The boundedness of $E$ is an easy consequence of hypothesis H(A) (ii), so it remains to check that $E$ is pseudomonotone. By virtue of Proposition 2.68, it suffices to show that $A$ is generalized pseudomonotone. To this end, let $\{(u_n, v_n^*)\}_{n \geq 1} \subset \operatorname{Gr} E$, and assume that

$$u_n \xrightarrow{w} u \text{ in } W_0^{1,p}(\Omega), \quad v_n^* \xrightarrow{w} v^* \text{ in } W^{-1,p'}(\Omega) \text{ and } \limsup_{n \to \infty} \langle v_n^*, u_n - u \rangle \leq 0. \tag{11.130}$$

We write $v_n^* = -\operatorname{div} v_n$, where $v_n \in N_A(u_n)$. We must show that

$$(u, v^*) \in \operatorname{Gr} E \text{ and } \langle v_n^*, u_n \rangle \to \langle v^*, u \rangle \text{ as } n \to \infty. \tag{11.131}$$

*Step 1:* $(u, v^*) \in \operatorname{Gr} E$.

We fix an arbitrary element $(y, y^*) \in \operatorname{Gr} E_u$, that is, $y \in W_0^{1,p}(\Omega)$ and $y^* = -\operatorname{div} h$, where $h \in L^{p'}(\Omega, \mathbb{R}^N)$ is such that $h(x) \in A(x, u(x), \nabla y(x))$ for a.a. $x \in \Omega$. For all $n \geq 1$ we consider the multifunction $H_n : \Omega \to 2^{\mathbb{R}^N}$ given by

$$H_n(x) = \{\xi \in A(x, u_n(x), \nabla y(x)) : |h(x) - \xi| = d(h(x), A(x, u_n(x), \nabla y(x)))\}$$

for a.a. $x \in \Omega$, where $d(h(x), A(x, u_n(x), \nabla y(x)))$ denotes the Euclidean distance from $h(x)$ to the set $A(x, u_n(x), \nabla y(x))$. Since the latter set is nonempty and compact [see H(A) (i)], we have $H_n(x) \neq \emptyset$ for all $x \in \Omega$.

*Claim 1:* For all $n \geq 1$ the multimap $H_n$ is graph measurable.

Since $A$ is graph measurable [see H(A) (i)], the graph of $A(\cdot, u_n(\cdot), \nabla y(\cdot))$ is a measurable set of $\Omega \times \mathbb{R}^N$. This also implies that $\gamma(x) := d(h(x), A(x, u_n(x), \nabla y(x)))$ is a Lebesgue measurable function on $\Omega$. Note that

$$\text{Gr}\,H_n = \text{Gr}\,A(\cdot, u_n(\cdot), \nabla y(\cdot)) \cap \{(x, \xi) \in \Omega \times \mathbb{R}^N : |h(x) - \xi| = \gamma(x)\}$$

is the intersection of two measurable sets. Therefore, $H_n$ is graph measurable.

Invoking Proposition 11.56 on the basis of Claim 1 and $H(A)$ (ii), we find an element $h_n \in L^{p'}(\Omega, \mathbb{R}^N)$ such that $h_n(x) \in H_n(x)$ for all $x \in \Omega$, all $n \geq 1$, that is, $h_n(x) \in A(x, u_n(x), \nabla y(x))$ for all $x \in \Omega$ and

$$|h(x) - h_n(x)| = d(h(x), A(x, u_n(x), \nabla y(x))) \quad \text{for a.a. } x \in \Omega. \tag{11.132}$$

*Claim 2:* We have $h_n \to h$ in $L^{p'}(\Omega, \mathbb{R}^N)$, and so $-\text{div}\,h_n \to -\text{div}\,h = y^*$ in $W^{-1,p'}(\Omega)$.

Because $u_n \xrightarrow{w} u$ in $W_0^{1,p}(\Omega)$ [see (11.130)], passing to a suitable subsequence if necessary, we may assume that

$$u_n \to u \text{ in } L^p(\Omega), \quad u_n(x) \to u(x) \text{ for a.a. } x \in \Omega \tag{11.133}$$

and that there is $\hat{k} \in L^p(\Omega)$ with

$$|u_n(x)| \leq \hat{k}(x) \text{ for a.a. } x \in \Omega, \text{ all } n \geq 1. \tag{11.134}$$

From (11.132), the second part of (11.133), and the fact that the multifunction $A(x, \cdot, \nabla y(x))$ has nonempty compact values and is l.s.c. [see $H(A)$ (i.c)], we infer that

$$h_n(x) \to h(x) \text{ for a.a. } x \in \Omega$$

[see Remark 2.37 (d)]. Invoking Lebesgue's dominated convergence theorem [see (11.134) and $H(A)$ (ii)], we deduce that $h_n \to h$ in $L^{p'}(\Omega, \mathbb{R}^N)$. This proves Claim 2.

Recall that $v_n(x) \in A(x, u_n(x), \nabla u_n(x))$, $h_n(x) \in A(x, u_n(x), \nabla y(x))$ for a.a. $x \in \Omega$, all $n \geq 1$. Exploiting the monotonicity of $A(x, u_n(x), \cdot)$ [see $H(A)$ (i.b)], we have

$$0 \leq \langle -\text{div}\,v_n + \text{div}\,h_n, u_n - y \rangle = \langle v_n^*, u_n - u \rangle + \langle v_n^*, u - y \rangle + \langle \text{div}\,h_n, u_n - y \rangle. \tag{11.135}$$

Since $v_n^* \xrightarrow{w} v^*$ in $W^{-1,p'}(\Omega)$ [see (11.130)], we note that

$$\langle v_n^*, u - y \rangle \to \langle v^*, u - y \rangle \text{ as } n \to \infty. \tag{11.136}$$

By Claim 2, we also have that

$$\langle \text{div}\,h_n, u_n - y \rangle \to \langle -y^*, u - y \rangle \text{ as } n \to \infty. \tag{11.137}$$

Passing to the limit as $n \to \infty$ in (11.135) by using (11.130), (11.136), and (11.137), we obtain

$$0 \le \langle v^* - y^*, u - y \rangle. \qquad (11.138)$$

Since $(y, y^*)$ is an arbitrary element in $\mathrm{Gr}\, E_u$, knowing (from Lemma 11.60) that $E_u$ is maximal monotone, from (11.138) we infer that $(u, v^*) \in \mathrm{Gr}\, E_u$, hence $v^* \in E_u(u) = E(u)$. This completes Step 1.

*Step 2:* $\langle v_n^*, u_n \rangle \to \langle v^*, u \rangle$ as $n \to \infty$.

By Step 1, there is $v \in N_A(u)$ such that $v^* = -\mathrm{div}\, v$. Arguing as in Step 1 with the multifunction $H_n$ constructed for $y = u$ and $y^* = v^* = -\mathrm{div}\, v$, we can find $h_n \in L^{p'}(\Omega, \mathbb{R}^N)$, with $h_n(x) \in A(x, u_n(x), \nabla u(x))$ for a.a. $x \in \Omega$, all $n \ge 1$, and $-\mathrm{div}\, h_n \to -\mathrm{div}\, v = v^*$ in $W^{-1, p'}(\Omega)$. Invoking H(A) (i.b), as in (11.135), we obtain

$$\langle v_n^*, u_n - u \rangle = \langle -\mathrm{div}\, v_n, u_n - u \rangle \ge \langle -\mathrm{div}\, h_n, u_n - u \rangle,$$

which implies that $\liminf\limits_{n \to \infty} \langle v_n^*, u_n - u \rangle \ge 0$. Combining the previous inequality with (11.130), we obtain that $\langle v_n^*, u_n - u \rangle \to 0$ as $n \to \infty$. Again invoking (11.130), this yields $\langle v_n^*, u_n \rangle \to \langle v^*, u \rangle$ as $n \to \infty$. This concludes Step 2.

Steps 1 and 2 yield (11.131). This completes the proof of the proposition. $\qquad \square$

Next, we focus on the term of (11.127) involving $\beta = \partial \vartheta$ [see H($\beta$)]. Recall that $\vartheta \in \Gamma_0(\mathbb{R})$ has nonnegative values. Let $\psi : L^2(\Omega) \to [0, +\infty]$ be the integral functional defined by

$$\psi(u) = \int_\Omega \vartheta(u(x))\, dx \text{ for all } u \in L^2(\Omega).$$

An easy application of Fatou's lemma shows that $\psi \in \Gamma_0(L^2(\Omega))$. Also, for every $n \ge 1$ let $\vartheta_n$ be the Moreau–Yosida regularization of $\vartheta$ corresponding to $\lambda = \frac{1}{n} > 0$ [see (3.4)].

*Remark 11.62.*

(a) In H($\beta$), we assume that $\vartheta \ge 0 = \vartheta(0)$. In this way, $0 \in \partial \vartheta(0) = \beta(0)$. Note that this implies that $J_{\frac{1}{n}}^{\partial \vartheta}(0) = 0$ [see (2.9)]; thus, by Theorem 3.18(b), we have $\vartheta_n \ge \vartheta_n(0) = 0$, and so $0 \in \partial \vartheta_n(0) = \{\vartheta_n'(0)\}$, i.e., $\vartheta_n'(0) = 0$ [Theorem 3.18(c)].

(b) Proposition 2.57(c) and Theorem 3.18(h) (with $z = 0$) imply that we have

$$|\vartheta_n'(s)| \le n|s| \text{ and } |\vartheta_n(s) - \vartheta_n(t)| \le n(|s| + |t|)|s - t| \text{ for all } s, t \in \mathbb{R}.$$

**Lemma 11.63.** *Assume that* H($\beta$) *holds. Then we have* $\vartheta_n(u(\cdot)) \in L^1(\Omega)$ *for all* $u \in L^2(\Omega)$, *all* $n \ge 1$. *Moreover, the functional* $\psi_n : L^2(\Omega) \to [0, +\infty)$ *defined by*

$$\psi_n(u) = \int_\Omega \vartheta_n(u(x))\,dx \ \ for\ all\ u \in L^2(\Omega)$$

*is the Moreau–Yosida regularization of $\psi$ corresponding to $\lambda = \frac{1}{n}$.*

*Proof.* The fact that we have $\vartheta(u(\cdot)) \in L^1(\Omega)$ whenever $u \in L^2(\Omega)$ is a consequence of Remark 11.62(b), which also implies that $\psi_n$ is continuous on $L^2(\Omega)$. Let $\overline{\psi}_n$ be the Moreau–Yosida regularization of $\psi$. We must check that $\psi_n(u) = \overline{\psi}_n(u)$ for all $u \in L^2(\Omega)$. Since $\overline{\psi}_n$ is also continuous on $L^2(\Omega)$ [by Theorem 3.18(a)], it is sufficient to check the previous relation for $u \in C(\Omega) \cap L^2(\Omega)$. By definition [see (3.4)], we have

$$\overline{\psi}_n(u) = \inf\left\{\psi(y) + \frac{n}{2}\|u - y\|_2^2 : y \in L^2(\Omega)\right\} \tag{11.139}$$

$$= \inf\left\{\int_\Omega \left(\vartheta(y(x)) + \frac{n}{2}|u(x) - y(x)|^2\right)dx : y \in L^2(\Omega)\right\}$$

$$\geq \int_\Omega \inf\left\{\vartheta(t) + \frac{n}{2}|t - u(x)|^2 : t \in \mathbb{R}\right\}dx = \int_\Omega \vartheta_n(u(x))\,dx = \psi_n(u).$$

It remains to show the reverse inequality. For all $x \in \Omega$ the map

$$t \mapsto \tilde{\vartheta}(t, u(x)) := \vartheta(t) + \frac{n}{2}|t - u(x)|^2$$

is strictly convex, lower semicontinuous, and coercive on $\mathbb{R}$ (since $\vartheta \geq 0$), hence there is $\gamma(x) \in \mathbb{R}$ unique such that

$$\tilde{\vartheta}(\gamma(x), u(x)) = \inf\{\tilde{\vartheta}(t, u(x)) : t \in \mathbb{R}\} = \vartheta_n(u(x)). \tag{11.140}$$

For all $x \in \Omega$, since $\vartheta(\gamma(x)) \geq \vartheta(0) = 0$, we have

$$\frac{n}{2}|\gamma(x) - u(x)|^2 \leq \vartheta(0) + \frac{n}{2}|u(x)|^2 - \vartheta(\gamma(x)) \leq \frac{n}{2}|u(x)|^2,$$

hence

$$|\gamma(x)| \leq 2|u(x)| \ \ for\ all\ x \in \Omega. \tag{11.141}$$

Let us check that $\gamma$ is continuous (hence measurable) on $\Omega$. Assume that $x_k \to x \in \Omega$ as $k \to \infty$. By (11.141), knowing that $u$ is continuous, we have that $\{\gamma(x_k)\}_{k \geq 1}$ is bounded in $\mathbb{R}$ and (along a relabeled subsequence) that $\gamma(x_k) \to s \in \mathbb{R}$ as $k \to \infty$. For all $t \in \mathbb{R}$, all $k \geq 1$, we have

$$\vartheta(\gamma(x_k)) + \frac{n}{2}|\gamma(x_k) - u(x_k)|^2 \leq \vartheta(t) + \frac{n}{2}|t - u(x_k)|^2.$$

Passing to the limit as $k \to \infty$ using the continuity of $u$ and the lower semicontinuity of $\vartheta$, we infer that $\tilde{\vartheta}(s, u(x)) = \inf\{\tilde{\vartheta}(t, u(x)) : t \in \mathbb{R}\}$, hence $s = \gamma(x)$. This establishes the continuity of $\gamma$. Finally, (11.141) and the fact that $u \in L^2(\Omega)$ imply that $\gamma \in L^2(\Omega)$. The latter fact, together with (11.139) and (11.140), yields

$$\overline{\psi}_n(u) \leq \int_\Omega \tilde{\vartheta}(\gamma(x), u(x))\,dx = \int_\Omega \vartheta_n(u(x))\,dx = \psi_n(u).$$

The proof is now complete.                                           □

Recall that $2 \leq p < +\infty$, and so $W_0^{1,p}(\Omega) \subset L^p(\Omega) \subset L^2(\Omega)$. Then we can set $\hat{\psi}_n = \psi_n|_{W_0^{1,p}(\Omega)}$. Recall that $\psi_n$ is convex and continuous [see Theorem 3.18(a)]. Thus, so is $\hat{\psi}_n$. Moreover, since $\psi_n$ is Fréchet differentiable [see Theorem 3.18(c)] and the embedding $W_0^{1,p}(\Omega) \hookrightarrow L^2(\Omega)$ is continuous, we obtain that $\hat{\psi}_n$ is Fréchet differentiable with

$$\hat{\psi}_n'(u) = \psi_n'(u) = \vartheta_n'(u) \in L^2(\Omega) \subset W^{-1,p'}(\Omega) \text{ for all } u \in W_0^{1,p}(\Omega). \quad (11.142)$$

Finally, we examine the term of problem (11.127) involving the multifunction $G$.

**Proposition 11.64.** *Assume that* $H(G)_1$ *holds. Then* $N_G : W_0^{1,p}(\Omega) \to 2^{L^{p'}(\Omega)}$ *has values in* $\mathscr{P}_{\text{wkc}}(L^{p'}(\Omega))$, *and it is u.s.c. from* $W_0^{1,p}(\Omega)$ *with the norm topology into* $L^{p'}(\Omega)$ *with the weak topology.*

*Proof.* The fact that $G$ has compact, convex values [by $H(G)_1$ (i)] easily implies that $N_G$ has closed, convex values in $L^{p'}(\Omega)$. From $H(G)_1$ (ii) it follows that $N_G$ has bounded values in $L^{p'}(\Omega)$, so $N_G(u)$ is weakly compact, convex for all $u \in W_0^{1,p}(\Omega)$.

Let us check that $N_G(u)$ is nonempty for all $u \in W_0^{1,p}(\Omega)$. Let $s_n : \Omega \to \mathbb{R}$ and $r_n : \Omega \to \mathbb{R}^N$ be step functions such that

$$s_n(x) \to u(x), \ r_n(x) \to \nabla u(x) \text{ as } n \to \infty \text{ for a.a. } x \in \Omega$$

and

$$|s_n(x)| \leq |u(x)|, \ |r_n(x)| \leq |\nabla u(x)| \text{ for a.a. } x \in \Omega, \text{ all } n > 1.$$

Hypothesis $H(G)_1$ (i) implies that for every $n \geq 1$, $x \mapsto G(x, s_n(x), r_n(x))$ is graph measurable, and so Proposition 11.56 implies that we can find functions $g_n \in L^{p'}(\Omega)$ such that $g_n(x) \in G(x, s_n(x), r_n(x))$ for a.a. $x \in \Omega$, all $n \geq 1$. Evidently, $\{g_n\}_{n \geq 1}$ is bounded in $L^{p'}(\Omega)$ [see $H(G)_1$ (ii)]. Thus, we may assume that

$$g_n \xrightarrow{w} g \text{ in } L^{p'}(\Omega) \text{ as } n \to \infty$$

for some $g \in L^{p'}(\Omega)$. Arguing as in Claim 1 of the proof of Lemma 11.60 on the basis of $H(G)_1$ (i), (ii) and Mazur's theorem and exploiting the fact that $G$ has convex values [see H$(G)_1$ (i)], we can see that, for all $x \in \Omega$ up to a Lebesgue-null set, $g(x)$ belongs to any $\varepsilon$-neighborhood of $G(x, u(x), \nabla u(x))$, whence

$$g(x) \in G(x, u(x), \nabla u(x)) \quad \text{for a.a. } x \in \Omega.$$

Hence $g \in N_G(u)$, and therefore $N_G(u)$ is nonempty.

It remains to show that $N_G$ is u.s.c. from $W_0^{1,p}(\Omega)$ to $L^{p'}(\Omega)$ furnished with the weak topology [denoted by $L^{p'}(\Omega)_w$]. Hypothesis H$(G)_1$ (ii) implies that $N_G$ is locally compact from $W_0^{1,p}(\Omega)$ to $L^{p'}(\Omega)_w$, so that it suffices to show that $N_G$ has a closed graph in $W_0^{1,p}(\Omega) \times L^{p'}(\Omega)_w$ [Remark 2.37(c)]. Thus, let $(u, g)$ belong to the closure of $\mathrm{Gr} N_G$ in $W_0^{1,p}(\Omega) \times L^{p'}(\Omega)_w$, and let us show that $(u, g) \in \mathrm{Gr} N_G$. Since $N_G$ is locally compact, we can find a $U \subset W_0^{1,p}(\Omega)$ bounded neighborhood of $u$ such that $M := (U \times L^{p'}(\Omega)) \cap \mathrm{Gr} N_G$ is bounded. Moreover, the closure of $M$ in $W_0^{1,p}(\Omega) \times L^{p'}(\Omega)_w$ contains $(u, g)$. Since the weak topology is metrizable on bounded sets of $L^{p'}(\Omega)$, we can find a sequence $\{(u_n, g_n)\}_{n \geq 1} \subset \mathrm{Gr} N_G$ such that

$$u_n \to u \text{ in } W_0^{1,p}(\Omega) \text{ and } g_n \xrightarrow{w} g \text{ in } L^{p'}(\Omega) \text{ as } n \to \infty. \qquad (11.143)$$

Thus, $g_n(x) \in G(x, u_n(x), \nabla u_n(x))$ for a.a. $x \in \Omega$, all $n \geq 1$. Again arguing as in Claim 1 of the proof of Lemma 11.60, we deduce that $g(x) \in G(x, u(x), \nabla u(x))$ for a.a. $x \in \Omega$. Hence $(u, g) \in \mathrm{Gr} N_G$. We conclude that $N_G$ is u.s.c. from $W_0^{1,p}(\Omega)$ to $L^{p'}(\Omega)_w$. The proof is then complete.  $\square$

Now we combine the operators $E$, $\psi_n$, and $N_G$ studied in the previous statements by considering, for every $n \geq 1$, the multivalued map $V_n : W_0^{1,p}(\Omega) \to \mathscr{P}_{\mathrm{wkc}}(W^{-1,p'}(\Omega))$ defined by

$$V_n(u) = E(u) + \psi_n'(u) + N_G(u) \quad \text{for all } u \in W_0^{1,p}(\Omega).$$

To determine the properties of $V_n$, we will use Lemma 11.3, which highlights the significance of hypothesis H$(G)_1$ (iii).

**Proposition 11.65.** *Assume that H(A), H($\beta$), and H$(G)_1$ hold. Then, for all $n \geq 1$, $V_n$ is bounded, pseudomonotone, and strongly coercive.*

*Proof.* We divide the proof into two steps.

*Step 1: $V_n$ is bounded and pseudomonotone.*

From Theorem 3.18(c) and Proposition 2.57(c) we know that $\psi_n'$ is bounded from $W_0^{1,p}(\Omega)$ to $W^{-1,p'}(\Omega)$. This fact, together with H(A) (ii) and H$(G)_1$ (ii), implies that $V_n$ is bounded. In view of Proposition 2.68, the property of $V_n$ being pseudomonotone will follow once we prove that $V_n$ is generalized pseudomonotone. To this end, let $\{(u_k, u_k^*)\}_{k \geq 1} \subset \mathrm{Gr} V_n$, and assume that

$$u_k \xrightarrow{w} u \text{ in } W_0^{1,p}(\Omega), \ u_k^* \xrightarrow{w} u^* \text{ in } W^{-1,p'}(\Omega) \text{ and } \limsup_{k \to \infty} \langle u_k^*, u_k - u \rangle \leq 0.$$

$$(11.144)$$

We can also assume that there is $\hat{k} \in L^p(\Omega)_+$ such that for a.a. $x \in \Omega$

$$|u_k(x)| \leq \hat{k}(x) \text{ for all } k \geq 1 \text{ and } u_k(x) \to u(x) \text{ as } k \to \infty. \qquad (11.145)$$

By definition, there are $v_k \in N_A(u_k)$ and $g_k \in N_G(u_k)$ such that we have

$$u_k^* = -\operatorname{div} v_k + \psi_n'(u_k) + g_k \text{ for all } k \geq 1. \qquad (11.146)$$

*Claim 1:* The sequences $\{v_k\}_{k\geq 1} \subset L^{p'}(\Omega, \mathbb{R}^N)$, $\{g_k\}_{k\geq 1} \subset L^{p'}(\Omega)$, $\{\psi_n'(u_k)\}_{k\geq 1} \subset L^2(\Omega)$ are bounded, so along relabeled subsequences we have

$$v_k \xrightarrow{w} v \text{ in } L^{p'}(\Omega, \mathbb{R}^N), \ g_k \xrightarrow{w} g \text{ in } L^{p'}(\Omega), \text{ and } \psi_n'(u_k) \xrightarrow{w} \eta \text{ in } L^2(\Omega)$$

as $k \to \infty$, with $v \in L^{p'}(\Omega, \mathbb{R}^N)$, $g \in L^{p'}(\Omega)$, and $\eta \in L^2(\Omega)$.

The boundedness of $\{v_k\}_{k\geq 1}$ and $\{g_k\}_{k\geq 1}$ is due to H(A) (ii) and H(G)$_1$ (ii), respectively. Also, since $\psi_n'$ is Lipschitz continuous with Lipschitz constant $n$ [see Theorem 3.18(c) and Proposition 2.57(c)] and $\psi_n'(0) = 0$ [because $\psi_n \geq \psi_n(0) = 0$], we have $\|\psi_n'(u_k)\|_2 \leq n\|u_k\|_2$ for all $k \geq 1$, which implies that $\{\psi_n'(u_k)\}_{k\geq 1}$ is bounded in $L^2(\Omega)$. This proves Claim 1.

*Claim 2:* We have $v \in N_A(u)$ [so $-\operatorname{div} v \in E(u)$], $\eta = \psi_n'(u)$, and

$$\langle u_k^*, u_k \rangle \to \langle -\operatorname{div} v + \eta + g, u \rangle \text{ as } k \to \infty.$$

First, since $g_k \xrightarrow{w} g$ in $L^{p'}(\Omega)$ whereas $u_k \to u$ in $L^p(\Omega)$, we have

$$\langle g_k, u_k - u \rangle = \int_\Omega g_k(x)(u_k(x) - u(x)) \, dx \to 0 \text{ as } k \to \infty. \qquad (11.147)$$

Since $\psi_n$ is convex and continuous, we have that $\psi_n' : L^2(\Omega) \to L^2(\Omega)$ is maximal monotone (Theorem 3.15), hence its graph is closed in $L^2(\Omega) \times L^2(\Omega)_w$ (Proposition 2.39). Then, from the fact that $u_k \to u$ in $L^2(\Omega)$ and $\psi_n'(u_k) \xrightarrow{w} \eta$ in $L^2(\Omega)$ as $k \to \infty$ we derive $\eta = \psi_n'(u)$. Thus,

$$\langle \psi_n'(u_k), u_k - u \rangle = \int_\Omega \psi_n'(u_k)(x)(u_k(x) - u(x)) \, dx \to 0 \text{ as } k \to \infty. \qquad (11.148)$$

If we act on (11.146) with $u_k - u$, pass to the limit as $k \to \infty$, and use (11.144), (11.147), and (11.148), then we obtain

$$\limsup_{k \to \infty} \langle -\operatorname{div} v_k, u_k - u \rangle \leq 0. \qquad (11.149)$$

From Proposition 11.61 we know that $E$ is pseudomonotone, hence generalized pseudomonotone (Proposition 2.67). Then, from (11.149) it follows that

$$- \operatorname{div} v \in E(u) \quad \text{and} \quad \langle -\operatorname{div} v_k, u_k \rangle \to \langle -\operatorname{div} v, u \rangle \quad \text{as } k \to \infty. \tag{11.150}$$

Combining (11.147), (11.148), and (11.150), we get $\langle u_k^*, u_k \rangle \to \langle -\operatorname{div} v + \eta + g, u \rangle$ as $k \to \infty$. This concludes the proof of Claim 2.

From Claim 2 we see that, to prove the generalized pseudomonotonicity of $V_n$, we need to show that $g \in N_G(u)$.

*Claim 3:* There is a Lebesgue-null set $N_0 \subset \Omega$ such that, for all $x \in \Omega \setminus N_0$, along a relabeled subsequence (in general depending on $x$) we have $u_k(x) \to u(x)$ and $\nabla u_k(x) \to \nabla u(x)$ as $k \to \infty$.

Reasoning as in Step 1 of the proof of Proposition 11.61 for $y = u$, $y^* = -\operatorname{div} v$, and $h = v$, for each $k \geq 1$ we can find $h_k \in L^p(\Omega, \mathbb{R}^N)$ such that, for a.a. $x \in \Omega$, all $k \geq 1$, we have

$$h_k(x) \in A(x, u_k(x), \nabla u(x)) \tag{11.151}$$

and

$$|v(x) - h_k(x)| = d(v(x), A(x, u_k(x), \nabla u(x))). \tag{11.152}$$

From Claim 2 we know that $-\operatorname{div} v \in E(u)$, i.e., $v(x) \in A(x, u(x), \nabla u(x))$ for a.a. $x \in \Omega$. This, together with (11.152), H(A) (i.c), and Remark 2.37(d), implies that

$$h_k(x) \to v(x) \quad \text{as } k \to \infty, \text{ for a.a. } x \in \Omega. \tag{11.153}$$

In view of (11.145), (11.151), and H(A) (ii), we can invoke Lebesgue's dominated convergence theorem, whence

$$h_k \to v \quad \text{in } L^{p'}(\Omega, \mathbb{R}^N) \quad \text{as } k \to \infty. \tag{11.154}$$

For every $k \geq 1$ we set

$$\mu_k(x) = (v_k(x) - h_k(x), \nabla u_k(x) - \nabla u(x))_{\mathbb{R}^N}.$$

The monotonicity of $A(x, u_k(x), \cdot)$ [see H(A) (i.b)] yields $\mu_k \geq 0$ a.e. in $\Omega$. By (11.150) and (11.154), we get $\mu_k \to 0$ in $L^1(\Omega)$ as $k \to \infty$, whence, along a subsequence,

$$\mu_k(x) \to 0 \quad \text{as } k \to \infty, \text{ for a.a. } x \in \Omega. \tag{11.155}$$

Hypotheses H($A$) (ii), (iii) imply that for all $x \in \Omega \setminus N_0$, where $N_0 \subset \Omega$ denotes a Lebesgue-null set, we have

$$\mu_k(x) \geq c_1(|\nabla u_k(x)|^p + |\nabla u(x)|^p) - 2c_2$$
$$-c_0|\nabla u(x)|(1 + |u_k(x)|^{p-1} + |\nabla u_k(x)|^{p-1})$$
$$-c_0|\nabla u_k(x)|(1 + |u_k(x)|^{p-1} + |\nabla u(x)|^{p-1}). \qquad (11.156)$$

Up to enlarging $N_0$, we may assume that for all $x \in \Omega \setminus N_0$ we have

$$u_k(x) \to u(x), \quad h_k(x) \to v(x), \quad \mu_k(x) \to 0 \text{ as } k \to \infty \qquad (11.157)$$

[see (11.145), (11.153), and (11.155)],

$$v(x) \in A(x, u(x), \nabla u(x)) \qquad (11.158)$$

(Claim 2), and

$$|v_k(x)| \leq c_0(1 + |\hat{k}(x)|^{p-1} + |\nabla u_k(x)|^{p-1}) \qquad (11.159)$$

[see (11.145) and H($A$) (ii)]. Now we fix $x \in \Omega \setminus N_0$. From (11.155) and (11.156) we see that $\{\nabla u_k(x)\}_{k \geq 1}$ is bounded in $\mathbb{R}^N$. Hence we can find a subsequence (depending on $x$), denoted for simplicity by the same symbol $\nabla u_k(x)$, such that

$$\nabla u_k(x) \to w(x) \text{ in } \mathbb{R}^N \text{ as } k \to \infty,$$

with $w(x) \in \mathbb{R}^N$. Also, because of (11.159), up to further extraction of a subsequence, we may assume that

$$v_k(x) \to \zeta(x) \text{ in } \mathbb{R}^N \text{ as } k \to \infty,$$

with $\zeta(x) \in \mathbb{R}^N$. Then hypothesis H($A$) (i.a) yields

$$\zeta(x) \in A(x, u(x), w(x)). \qquad (11.160)$$

On the other hand, since $\mu_k(x) \to 0$ [see (11.157)], in the limit we obtain

$$(\zeta(x) - v(x), w(x) - \nabla u(x))_{\mathbb{R}^N} = 0.$$

From (11.158), (11.160), and the strict monotonicity of $A(x, s, \cdot)$ [see H($A$) (i.b)] we finally deduce that $w(x) = \nabla u(x)$, hence $\nabla u_k(x) \to \nabla u(x)$ as $k \to \infty$. This proves Claim 3.

According to Mazur's theorem (e.g., Brezis [52, p. 61]), there exists a sequence $\{\tilde{g}_k\}_{k \geq 1} \subset L^{p'}(\Omega)$, with $\tilde{g}_k \in \text{conv}\{g_\ell : \ell \geq k\}$, such that $\tilde{g}_k \to g$ in $L^{p'}(\Omega)$ as $k \to \infty$.

Up to enlarging $N_0$ in Claim 3, we may assume that for all $x \in \Omega \setminus N_0$

$$\tilde{g}_k(x) \to g(x) \text{ as } k \to \infty, \quad g_k(x) \in G(x, u_k(x), \nabla u_k(x)) \text{ for all } k \geq 1,$$

and $(s, \xi) \mapsto G(x, s, \xi)$ is u.s.c. [see H$(G)_1$ (i), (ii) and Remark 2.37 (c)]. Fix $x \in \Omega \setminus N_0$, and let $\varepsilon > 0$. The fact that $G(x, \cdot, \cdot)$ is u.s.c., together with Claim 3, yields $k_0 \geq 1$ such that for all $k \geq k_0$ we have $d(g_k(x), G(x, u(x), \nabla u(x))) \leq \varepsilon$. Thus $d(\tilde{g}_k(x), G(x, u(x), \nabla u(x))) \leq \varepsilon$ for all $k \geq k_0$. Thus, $d(g(x), G(x, u(x), \nabla u(x))) \leq \varepsilon$. Finally, since $\varepsilon > 0$ is arbitrary, we conclude that $g(x) \in G(x, u(x), \nabla u(x))$, whence $g \in N_G(u)$. This, combined with Claim 2, proves the pseudomonotonicity of $V_n$.

*Step 2: $V_n$ is strongly coercive.*

Let $u \in W_0^{1,p}(\Omega)$ and $u^* \in V_n(u)$, thus,

$$\langle u^*, u \rangle = \langle -\operatorname{div} v, u \rangle + \langle \hat{\psi}_n'(u), u \rangle + \langle g, u \rangle, \tag{11.161}$$

with $v \in N_A(u)$, $g \in N_G(u)$. Since $\vartheta_n'$ is monotone and $\vartheta_n'(0) = 0$, we see that

$$\langle \hat{\psi}_n'(u), u \rangle = \int_\Omega \vartheta_n'(u) u \, dx \geq 0 \tag{11.162}$$

[see (11.142)]. Hypothesis H$(A)$ (iii) implies that

$$\langle -\operatorname{div} v, u \rangle = \int_\Omega (v, \nabla u)_{\mathbb{R}^N} \, dx \geq c_1 \|\nabla u\|_p^p - c_2 |\Omega|_N. \tag{11.163}$$

By H$(G)_1$ (iii), given $\varepsilon > 0$, there is $M = M(\varepsilon) \geq 1$ such that

$$ws \geq -(\eta_0(x) + \varepsilon) |s|^p \tag{11.164}$$

for a.a. $x \in \Omega$, all $|s| \geq M$, all $\xi \in \mathbb{R}^N$, all $w \in G(x, s, \xi)$. Combining (11.164) with hypothesis H$(G)_1$ (ii), we find a constant $\tilde{c} > 0$ such that

$$ws \geq -(\eta_0(x) + \varepsilon)|s|^p - \tilde{c}(1 + |\xi|^{p-1})$$

for a.a. $x \in \Omega$, all $s \in \mathbb{R}$, all $\xi \in \mathbb{R}^N$, all $w \in G(x, s, \xi)$. Hence we obtain

$$\langle g, u \rangle = \int_\Omega g(x) u(x) \, dx$$

$$\geq -\int_\Omega \eta_0(x) |u(x)|^p \, dx - \varepsilon \|u\|_p^p - c_3 \|\nabla u\|_p^{p-1} - c_4, \tag{11.165}$$

with $c_3 = \tilde{c} |\Omega|_N^{\frac{1}{p}}$ and $c_4 = \tilde{c} |\Omega|_N$. Returning to (11.161) and using (11.162), (11.163), (11.165), and Lemma 11.3, we have

$$\langle u^*, u \rangle \geq c_1 \|\nabla u\|_p^p - \int_\Omega \eta_0 |u|^p \, dx - \frac{\varepsilon}{\lambda_1} \|\nabla u\|_p^p - c_3 \|\nabla u\|_p^{p-1} - c_5$$

$$\geq \left( \tilde{c}_1 - \frac{\varepsilon}{\lambda_1} \right) \|\nabla u\|_p^p - c_3 \|\nabla u\|_p^{p-1} - c_5 \tag{11.166}$$

for some $\tilde{c}_1 > 0$ (independent of $\varepsilon$) and $c_5 > 0$. Choosing $\varepsilon \in (0, \lambda_1 \tilde{c}_1)$, from (11.166) we infer that $V_n$ is strongly coercive. □

Let $\beta_n = \vartheta_n'$ for $n \geq 1$, and consider the following approximation to problem (11.127):

$$\begin{cases} -\mathrm{div}\, A(x, u(x), \nabla u(x)) + \beta_n(u(x)) + G(x, u(x), \nabla u(x)) \ni e(x) & \text{in } \Omega, \\ u|_{\partial\Omega} = 0. \end{cases} \tag{11.167}$$

**Proposition 11.66.** *Assume that* H(A), H($\beta$), H(G)$_1$ *hold, and* $e \in L^2(\Omega)$. *Then, problem (11.167) has at least one solution* $u_n \in W_0^{1,p}(\Omega)$.

*Proof.* From Proposition 11.65 and Theorem 2.63 we have that $V_n$ is surjective. Thus, we can find $u_n \in W_0^{1,p}(\Omega)$ such that $V_n(u_n) = e$. Evidently, $u_n$ is a solution of problem (11.167). □

As an immediate consequence, we deduce the following corollary, which provides a solution of (11.127) in the case where $\beta = 0$.

**Corollary 11.67.** *Assume that* H(A) *and* H(G)$_1$ *hold,* $e \in L^2(\Omega)$, *and* $\beta = 0$. *Then problem (11.127) has at least one solution* $u \in W_0^{1,p}(\Omega)$.

In the general case, we will pass to the limit as $n \to \infty$ in (11.167) to obtain a solution to the original problem (11.127). To do this, we need to strengthen the assumption on $G$.

H(G)$_2$ Hypothesis H(G)$_1$ holds with (ii) replaced by

(ii)$'$ there is $c > 0$ such that

$$|z| \leq c(1 + |s| + |\xi|) \quad \text{for a.a. } x \in \Omega, \text{ all } (s, \xi) \in \mathbb{R} \times \mathbb{R}^N, \text{ all } z \in G(x, s, \xi).$$

**Theorem 11.68.** *Assume that* H(A), H($\beta$), H(G)$_2$ *hold, and* $e \in L^2(\Omega)$. *Then, problem (11.127) has at least one solution* $u \in W_0^{1,p}(\Omega)$.

*Proof.* By virtue of Proposition 11.66, for every $n \geq 1$, problem (11.167) has a solution $u_n \in W_0^{1,p}(\Omega)$. Thus we have

$$-\mathrm{div}\, v_n + \hat{\psi}_n'(u_n) + g_n = e \quad \text{in } W^{-1,p'}(\Omega), \tag{11.168}$$

with $v_n \in N_A(u_n)$ and $g_n \in N_G(u_n)$ for all $n \geq 1$. On (11.168) we act with $u_n$ and get

$$\langle -\mathrm{div}\, v_n, u_n \rangle + \langle \hat{\psi}_n'(u_n), u_n \rangle + \int_\Omega g_n u_n \, dx = \int_\Omega e u_n \, dx \quad \text{for all } n \geq 1. \tag{11.169}$$

By (11.142), the monotonicity of $\vartheta'_n$, and the fact that $\vartheta'_n(0) = 0$, we have

$$\langle \hat{\psi}'_n(u_n), u_n \rangle = \int_\Omega \vartheta'_n(u_n) u_n \, dx \geq 0 \quad \text{for all } n \geq 1. \tag{11.170}$$

Also, for every $\varepsilon > 0$, reasoning as we did in obtaining (11.165), from $H(G)_2$ (ii)$'$ we find constants $c_3, c_4 > 0$ such that

$$\int_\Omega g_n u_n \, dx \geq -\int_\Omega \eta_0 |u_n|^p \, dx - \frac{\varepsilon}{\lambda_1} \|\nabla u_n\|_p^p - c_3 \|\nabla u_n\|_p - c_4 \tag{11.171}$$

for all $n \geq 1$. Returning to (11.169) and using (11.170), (11.171), $H(A)$ (iii), and the continuity of the embedding $W_0^{1,p}(\Omega) \hookrightarrow L^2(\Omega)$, we obtain

$$c_1 \|\nabla u_n\|_p^p - c_2 - \int_\Omega \eta_0 |u_n|^p \, dx - \frac{\varepsilon}{\lambda_1} \|\nabla u_n\|_p^p \leq c_4 + c_5 \|\nabla u_n\|_p \quad \text{for all } n \geq 1$$

for some constant $c_5 > 0$. Using Lemma 11.3, we derive that

$$\left( \tilde{c}_1 - \frac{\varepsilon}{\lambda_1} \right) \|\nabla u_n\|_p^p \leq c_2 + c_4 + c_5 \|\nabla u_n\|_p \quad \text{for all } n \geq 1,$$

with $\tilde{c}_1 > 0$ independent of $\varepsilon$. Choosing $\varepsilon \in (0, \lambda_1 \tilde{c}_1)$, we infer that $\{u_n\}_{n \geq 1}$ is bounded in $W_0^{1,p}(\Omega)$. Hence we may assume that

$$u_n \xrightarrow{w} u \text{ in } W_0^{1,p}(\Omega) \quad \text{and} \quad u_n \to u \text{ in } L^p(\Omega) \quad \text{as } n \to \infty.$$

Note that $\hat{\psi}'_n(u_n) = \vartheta'_n(u_n) \in L^2(\Omega)$ [see (11.142)]. Actually, since $u_n \in W_0^{1,p}(\Omega)$ and $\vartheta'_n$ is Lipschitz continuous with Lipschitz constant $n$ [see Remark 11.62 (b)], we have $\vartheta'_n(u_n) \in W_0^{1,p}(\Omega)$ (Theorem 1.27 and Remark 1.28). Thus, acting on (11.168) with the test function $\vartheta'_n(u_n)$, for all $n \geq 1$, we obtain

$$\int_\Omega \vartheta''_n(u_n)(v_n, \nabla u_n)_{\mathbb{R}^N} \, dx + \|\vartheta'_n(u_n)\|_2^2 + \int_\Omega g_n \vartheta'_n(u_n) \, dx = \int_\Omega e \vartheta'_n(u_n) \, dx \tag{11.172}$$

(Theorem 1.27). Note that $\vartheta'_n$ is Lipschitz continuous and increasing [see Theorem 3.18(c) and Proposition 2.57(c)]. Hence $\vartheta''_n(u_n(x)) \geq 0$ a.e. in $\Omega$. Also, from the monotonicity of $A(x, u_n(x), \cdot)$ and because $0 \in A(x, s, 0)$ for a.a. $x \in \Omega$, all $s \in \mathbb{R}$, the first integral in (11.172) is nonnegative. Since $\{u_n\}_{n \geq 1}$ is bounded in $W_0^{1,p}(\Omega)$, by hypothesis $H(G)_2$ (ii)$'$ we have that $\{g_n\}_{n \geq 1}$ is bounded in $L^2(\Omega)$. Therefore, from (11.172) we infer that

$$\|\vartheta'_n(u_n)\|_2^2 \leq \|g_n\|_2 \|\vartheta'_n(u_n)\|_2 + \|e\|_2 \|\vartheta'_n(u_n)\|_2 \leq c_6 \|\vartheta'_n(u_n)\|_2 \quad \text{for all } n \geq 1,$$

for some $c_6 > 0$, which implies that $\{\vartheta_n'(u_n)\}_{n\geq1}$ is bounded in $L^2(\Omega)$. Therefore, we may assume that

$$\hat{\psi}_n'(u_n) = \vartheta_n'(u_n) \xrightarrow{w} \hat{\eta} \text{ in } L^2(\Omega) \text{ as } n \to \infty$$

for some $\hat{\eta} \in L^2(\Omega)$. Since $u_n \to u$ in $L^2(\Omega)$ (recall that $p \geq 2$), from Lemma 11.63, (11.142), and Theorem 3.18(e) we infer that $\hat{\eta} \in \partial\psi(u)$.

Hypothesis H(A) (ii) implies that we may assume that $v_n \xrightarrow{w} v$ in $L^{p'}(\Omega, \mathbb{R}^N)$, with $v \in L^{p'}(\Omega, \mathbb{R}^N)$, and we have $v \in N_A(u)$ (see the proof of Proposition 11.61). Since $\{g_n\}_{n\geq1}$ is bounded in $L^2(\Omega)$, we may also assume that $g_n \xrightarrow{w} g$ in $L^2(\Omega)$, with $g \in L^2(\Omega)$, and again $g \in N_G(u)$ [by Proposition 11.64; see also Remark 2.37(c)]. Thus, passing to the limit in (11.168) as $n \to \infty$, we obtain

$$-\text{div}\, v + \hat{\eta} + f = e.$$

Since $v \in N_A(u)$, $\hat{\eta} \in \partial\psi(u)$, and $g \in N_G(u)$, we conclude that $u \in W_0^{1,p}(\Omega)$ is a solution of problem (11.127). □

## 11.5 Remarks

**Section 11.1:** The existence and multiplicity of solutions of elliptic equations using degree theory can be found in Aizicovici et al. [3, 4], Ambrosetti and Arcoya [14], Ambrosetti and Malchiodi [15], Arcoya et al. [23], del Pino and Manásevich [110], Drábek [120], Hu and Papageorgiou [176], and Motreanu et al. [291, 292]. Usually the authors use the Leray–Schauder degree. Here, instead, we employ the degree theory for operators of monotone type (Sect. 4.3). Specifically, this section follows the work of Motreanu et al. [291]. The construction of the first solution in Theorem 11.13 is based on Proposition 11.9, a classic result due to Ôtani [312].

**Section 11.2:** The Ambrosetti–Rabinowitz condition, namely, there are $M > 0$ and $\mu > p$ such that

$$\underset{x\in\Omega}{\text{ess inf}}\, F(x,s) > 0 \quad \text{and} \quad \mu F(x,s) < f(x,s)s \quad \text{for a.a. } x \in \Omega, \text{ all } |s| > M$$

[where $F(x,s) = \int_0^s f(x,t)\,dt$], was first introduced by Ambrosetti and Rabinowitz [17] as a tool to study superlinear problems. It is a quite natural and useful condition to ensure the mountain pass geometry and the Palais–Smale condition. In this section, we use a slightly more general condition [see H($f$)$_8$ (iii)].

An important case of nonlinearities satisfying the Ambrosetti–Rabinowitz condition arises in problems with competing nonlinearities (concave and convex terms). Such problems were first investigated by Ambrosetti et al. [18] (for $p = 2$), who coined the term *concave–convex nonlinearities*. Their work was extended to

$p$-Laplacian equations by García Azorero et al. [147] and by Guo and Zhang [168] (for $p \geq 2$), who have a particular nonlinearity of the form

$$f(s) = \lambda |s|^{q-2}s + |s|^{r-2}s \text{ with } 1 < q < p < r < p^*,$$

and by Hu and Papageorgiou [178], who have a more general concave term. In these references, the authors establish the existence of two positive solutions and symmetrically two negative solutions of the problem (when the parameter $\lambda > 0$ goes to zero). Here, we point out the existence of a fifth nontrivial solution. This result is completed in Sect. 11.3, where we study a parametric problem with a concave–convex nonlinearity of a special form for which we show the existence of a fifth solution that is nodal.

This section follows the work of Motreanu et al. [289], which was itself inspired by Motreanu et al. [288] and Carl and Motreanu [69]. It relies on the so-called lower and upper solutions method developed by Dancer and Du [102] (in the case $p = 2$). In this section, we establish the existence of nodal solutions via the construction of extremal constant-sign solutions. For nodal solutions obtained via the same method, we refer to Carl and Motreanu [70, 71]. We mention that nodal solutions for superlinear equations were also produced by Bartsch and Liu [37], Bartsch et al. [38], Zhang et al. [394], and Zhang and Li [392] through a different approach based on the construction of a suitable pseudogradient vector field whose descent flow has appropriate invariance properties.

For other multiplicity results for Dirichlet boundary value problems involving the $p$-Laplacian, we refer to the works of Alves et al. [9], Averna et al. [26], Gasiński and Papageorgiou [153, 154], Jiu and Su [185], Kyritsi and Papageorgiou [213], Motreanu et al. [293], Motreanu and Tanaka [279], and Papageorgiou and Papageorgiou [319]. We mention the recent works of Motreanu [271] and Motreanu and Zhang [284] focusing on systems of quasilinear elliptic equations driven by the $p$-Laplacian. Finally, results on $(p, q)$-Laplacian equations have been obtained by Cherfils and Il'yasov [80], de Paiva et al. [116], Faria et al. [136], and Marano and Papageorgiou [244].

**Section 11.3:** The semilinear version (i.e., $p = 2$) of problem (11.72) with asymptotically linear nonlinearity was first studied by Amann and Zehnder [13]. They proved that if $f(x, s) = f(s)$ is of class $C^1$ and satisfies $\lim_{s \to \pm\infty} \frac{f(s)}{s} = \lambda \notin \sigma(\Delta)$ [where $\sigma(\Delta)$ is the spectrum of the negative Dirichlet Laplacian] and there exists at least one eigenvalue of $-\Delta^D$ between $\lambda$ and $\lambda + f'(0)$, then the problem has a nontrivial solution. Extensions of the Amann–Zehnder result can be found in Chang [76] and Lazer and Solimini [219]. Our analysis of problem (11.72) is based on the work of Motreanu et al. [297]. Theorem 11.44 partially extends to $p$-Laplacian equations the existence theorem of Amann and Zehnder [13]. In addition, we include in our setting the situation of resonance at $\pm\infty$ (with respect to the first eigenvalue of $-\Delta_p^D$ in Theorem 11.44 and with respect to any eigenvalue of $-\Delta^D$ in Theorem 11.46).

The second part of the section (parametric problems with concave term) is based on the work of Motreanu et al. [302] (although in Motreanu et al. [302] it is assumed that the nonlinearity has an asymmetric behavior in the positive and negative half-lines; see also Motreanu et al. [295]). Note that, although our assumptions indicate that the nonlinearity has a $(p-1)$-superlinear behavior at $\pm\infty$, we do not require the Ambrosetti–Rabinowitz condition, which is usual in such cases. Instead, we require hypothesis $H(g)_3$ (iv). Additional efforts to replace the Ambrosetti–Rabinowitz condition in order to deal with a wider class of nonlinearities were made in the works of Costa and Magalhães [90], Gasiński and Papageorgiou [155], Li et al. [226], Li and Yang [225], Liu and Wang [235], Miyagaki and Souto [260], Schechter and Zou [355].

**Section 11.4:** Problem (11.127) incorporates as a special case the so-called variational–hemivariational inequalities (Sect. 5.5). Such problems have been studied primarily for equations driven by the Laplacian or the $p$-Laplacian and with $\beta$ being the subdifferential of an indicator function: see Filippakis and Papageorgiou [138], Goeleven and Motreanu [161], Goeleven et al. [162], Kyritsi and Papageorgiou [210], and Liu and Motreanu [233]. We mention the work of Carl and Motreanu [68], who study a general nonlinear elliptic hemivariational inequality for a single-valued function $\beta$. For applications of the theory of monotone operators to variational inequalities, see also the work of Jebelean et al. [184]. Our presentation here is related to the work of Motreanu et al. [287], although here the multifunction $G$ depends also on the gradient of $u$, whereas $G(x,s,\xi) = G(x,s)$ in Motreanu et al. [287].

# Chapter 12
# Nonlinear Elliptic Equations with Neumann Boundary Conditions

**Abstract** This chapter aims to present relevant knowledge regarding recent progress on nonlinear elliptic equations with Neumann boundary conditions. In fact, all the results presented here bring novelties with respect to the available literature. We emphasize the specific functional setting and techniques involved in handling the Neumann problems, which are distinct in comparison with those for the Dirichlet problems. The first section of the chapter discusses the multiple solutions that arise at near resonance, from the left and from the right, in the Neumann problems depending on parameters. The second section focuses on nonlinear Neumann problems whose differential part is described by a general nonhomogeneous operator. The third section builds a common approach for both sublinear and superlinear cases of semilinear Neumann problems. Related comments and references are given in a remarks section.

## 12.1 Nonlinear Neumann Problems Using Variational Methods

Let $\Omega \subset \mathbb{R}^N$ be a bounded domain with a $C^2$-boundary $\partial\Omega$. In this section, we study the following nonlinear parametric Neumann problem:

$$\begin{cases} -\Delta_p u(x) = \lambda |u(x)|^{p-2} u(x) + f(x, u(x)) & \text{in } \Omega, \\ \dfrac{\partial u}{\partial n_p} = 0 & \text{on } \partial\Omega, \end{cases} \quad (12.1)$$

where $\lambda \in \mathbb{R}$ is a parameter, $1 < p < +\infty$, and $f : \Omega \times \mathbb{R} \to \mathbb{R}$ is a Carathéodory function. We recall that $\Delta_p$ denotes the $p$-Laplace differential operator defined by $\Delta_p u = \operatorname{div}(|\nabla u|^{p-2} \nabla u)$ for all $u \in W^{1,p}(\Omega)$ (where $|\cdot|$ stands for the Euclidean norm of $\mathbb{R}^N$) and $\frac{\partial u}{\partial n_p} := \gamma_n(|\nabla u|^{p-2} \nabla u) \in W^{-\frac{1}{p'}, p'}(\partial\Omega)$ denotes the generalized outward normal derivative (Theorem 1.39). We recall that the notion of solution to Neumann problems such as (12.1) is given in Definition 8.2 [see also Remark 8.3(b)].

D. Motreanu et al., *Topological and Variational Methods with Applications to Nonlinear Boundary Value Problems*, DOI 10.1007/978-1-4614-9323-5_12,
© Springer Science+Business Media, LLC 2014

We examine the existence and multiplicity of nontrivial solutions when the parameter $\lambda \in \mathbb{R}$ is near 0 (that is, when near resonance occurs with respect to the first eigenvalue $\lambda_0 = 0$ of the negative Neumann $p$-Laplacian). We consider two distinct cases depending on whether the parameter $\lambda$ approaches 0 from below or from above. In the first case (*near resonance from the left*) we establish the existence of three nontrivial, smooth solutions of problem (12.1); moreover, in the semilinear case (i.e., for $p = 2$), by strengthening the regularity conditions on $f(x, \cdot)$, we are able to produce four nontrivial, smooth solutions. In the second case (*near resonance from the right*), we produce two nontrivial, smooth solutions. In both cases, we use variational methods based on critical point theory (Chap. 5) and also on Morse theory (Chap. 6).

To start with, for later use, we state the following result on local $C^1(\overline{\Omega})$- versus local $W^{1,p}(\Omega)$-minimizers, similar to Proposition 11.4. The proof of this result will be given in wider generality in Sect. 12.2. Let $f_0 : \Omega \times \mathbb{R} \to \mathbb{R}$ be a Carathéodory function with a subcritical growth in the second variable, i.e.,

$$|f_0(x, s)| \leq c_0(1 + |s|^{r-1}) \quad \text{for a.a. } x \in \Omega, \text{ all } s \in \mathbb{R},$$

with $c_0 > 0$ and $1 < r < p^*$ (where $p^* \in (1, +\infty]$ denotes the Sobolev critical exponent; see Remark 1.50). We denote $F_0(x, s) = \int_0^s f_0(x, t)\, dt$ and consider the $C^1$-functional $\varphi_0 : W^{1,p}(\Omega) \to \mathbb{R}$ given by

$$\varphi_0(u) = \frac{1}{p}\|\nabla u\|_p^p - \int_\Omega F_0(x, u(x))\, dx \quad \text{for all } u \in W^{1,p}(\Omega).$$

**Proposition 12.1.** *If $u_0 \in W^{1,p}(\Omega)$ is a local $C^1(\overline{\Omega})$-minimizer of $\varphi_0$, i.e., there exists $\rho_0 > 0$ such that*

$$\varphi_0(u_0) \leq \varphi_0(u_0 + h) \quad \text{for all } h \in C^1(\overline{\Omega}), \ \|h\|_{C^1(\overline{\Omega})} \leq \rho_0,$$

*then $u_0$ is also a local $W^{1,p}(\Omega)$-minimizer of $\varphi_0$, i.e., there exists $\rho_1 > 0$ such that*

$$\varphi_0(u_0) \leq \varphi_0(u_0 + h) \quad \text{for all } h \in W^{1,p}(\Omega), \ \|h\| \leq \rho_1.$$

Here and in the sequel, by $\|\cdot\|$ we denote the usual Sobolev norm of $W^{1,p}(\Omega)$,

$$\|u\|^p = \|\nabla u\|_p^p + \|u\|_p^p \quad \text{for all } u \in W^{1,p}(\Omega).$$

### Near Resonance from Left: $\lambda < 0$

To deal with the case of near resonance from the left in problem (12.1) (i.e., when $\lambda$ approaches 0 while being negative), we impose the following conditions on the nonlinearity $f$ and its primitive $F(x, s) = \int_0^s f(x, t)\, dt$.

$H(f)_1$ (i) $f : \Omega \times \mathbb{R} \to \mathbb{R}$ is a Carathéodory function [i.e., $f(\cdot, s)$ is measurable for all $s \in \mathbb{R}$ and $f(x, \cdot)$ is continuous for a.a. $x \in \Omega$], with $f(x, 0) = 0$ a.e. in $\Omega$, and there are $c > 0$ and $r \in (p, p^*)$ such that

$$|f(x,s)| \leq c(1 + |s|^{r-1}) \text{ for a.a. } x \in \Omega, \text{ all } s \in \mathbb{R};$$

(ii) $\displaystyle\limsup_{s \to \pm\infty} \frac{F(x,s)}{|s|^p} \leq 0$ uniformly for a.a. $x \in \Omega$;

(iii) $\displaystyle\limsup_{s \to \pm\infty} \int_\Omega F(x,s)\,dx = +\infty$;

(iv) $\displaystyle\limsup_{s \to 0} \frac{F(x,s)}{|s|^p} \leq 0$ uniformly for a.a. $x \in \Omega$.

*Example 12.2.* The following function $f$, where for simplicity we drop the $x$-dependence, satisfies hypotheses $H(f)_1$:

$$f(s) = \begin{cases} c|s|^{\theta-2}s - \sin(-s-1) & \text{if } s < -1, \\ c|s|^{r-2}s & \text{if } -1 \leq s < 0, \\ 2cs^{r-1} & \text{if } 0 \leq s \leq 1, \\ 2(cs^{\theta-1} + \sin(s-1)) & \text{if } s > 1, \end{cases}$$

with constants $1 < \theta < p < r$ and $c > 0$. Note that $f$ has no symmetry properties.

**Theorem 12.3.** *Assume that $H(f)_1$ hold. Then there exists $\varepsilon_0 > 0$ such that for all $\lambda \in (-\varepsilon_0, 0)$ problem (12.1) has at least three nontrivial solutions in $C^1(\overline{\Omega})$.*

*Proof.* For any $\lambda < 0$ we consider the energy functional $\varphi_\lambda \in C^1(W^{1,p}(\Omega), \mathbb{R})$ for problem (12.1), defined by

$$\varphi_\lambda(u) = \frac{1}{p}\|\nabla u\|_p^p - \frac{\lambda}{p}\|u\|_p^p - \int_\Omega F(x, u(x))\,dx \quad \text{for all } u \in W^{1,p}(\Omega).$$

Evidently, the weak solutions of problem (12.1) coincide with the critical points of $\varphi_\lambda$. Moreover, by nonlinear regularity theory (Corollary 8.13), we know that any weak solution of (12.1) belongs to $C^1(\overline{\Omega})$. Therefore, the proof will be complete once we show the existence of three critical points of $\varphi_\lambda$ different from 0. This is done in several steps. The theorem will be implied by the claims in Steps 4 and 5.

*Step 1:* The functional $\varphi_\lambda$ is coercive, bounded below, and satisfies the (PS)-condition, and 0 is a local minimizer of $\varphi_\lambda$.

By virtue of hypotheses $H(f)_1$ (ii), (iv), given $\varepsilon \in (0, |\lambda|)$, we can find $\delta = \delta(\varepsilon) \in (0,1)$ and $M_1 = M_1(\varepsilon) > 1$ such that

$$F(x,s) \leq \frac{\varepsilon}{p}|s|^p \text{ for a.a. } x \in \Omega, \text{ all } s \in \mathbb{R} \text{ with } |s| \leq \delta \text{ or } |s| \geq M_1. \qquad (12.2)$$

From (12.2) we first deduce that $\varphi_\lambda$ is coercive. Indeed, combining (12.2) with $H(f)_1$ (i) yields $M_2 = M_2(\varepsilon) > 0$ such that

$$F(x,s) \leq \frac{\varepsilon}{p}|s|^p + M_2 \text{ for a.a. } x \in \Omega, \text{ all } s \in \mathbb{R}. \tag{12.3}$$

Then the fact that $\varphi_\lambda$ is coercive and bounded below is guaranteed by the relation

$$\varphi_\lambda(u) \geq \frac{1}{p}\|\nabla u\|_p^p + \frac{|\lambda| - \varepsilon}{p}\|u\|_p^p - M_2|\Omega|_N \text{ for all } u \in W^{1,p}(\Omega).$$

From (12.2) we also infer that

$$\varphi_\lambda(u) \geq \frac{1}{p}\|\nabla u\|_p^p + \frac{|\lambda| - \varepsilon}{p}\|u\|_p^p \text{ for all } u \in C^1(\overline{\Omega}) \text{ with } \|u\|_{C^1(\overline{\Omega})} \leq \delta.$$

Hence 0 is a local $C^1(\overline{\Omega})$-minimizer of $\varphi_\lambda$. Invoking Proposition 12.1, it is also a local $W^{1,p}(\Omega)$-minimizer of $\varphi_\lambda$.

It remains to check that $\varphi_\lambda$ satisfies the (PS)-condition. To do this, let $\{u_n\}_{n\geq 1} \subset W^{1,p}(\Omega)$ be a sequence such that

$$\{\varphi_\lambda(u_n)\}_{n\geq 1} \text{ is bounded and } \varphi_\lambda'(u_n) \to 0 \text{ in } W^{1,p}(\Omega)^*. \tag{12.4}$$

The boundedness in (12.4) and the coercivity of $\varphi_\lambda$ imply that $\{u_n\}_{n\geq 1}$ is bounded in $W^{1,p}(\Omega)$. Passing to a subsequence if necessary, we may assume that $u_n \overset{w}{\to} u$ in $W^{1,p}(\Omega)$ and $u_n \to u$ in $L^r(\Omega)$ as $n \to \infty$ [with $r$ as in $H(f)_1$ (i)]. We note that

$$\langle \varphi_\lambda'(u_n), h \rangle = \langle A(u_n), h \rangle - \int_\Omega (\lambda|u_n|^{p-2}u_n + f(x,u_n))h(x)\,dx$$

for all $h \in W^{1,p}(\Omega)$, all $n \geq 1$, where $A : W^{1,p}(\Omega) \to W^{1,p}(\Omega)^*$ is the operator defined by $\langle A(u), v \rangle = \int_\Omega (|\nabla u|^{p-2}\nabla u, \nabla v)_{\mathbb{R}^N}\,dx$ [see (2.28)]. The convergence in (12.4) leads to

$$\lim_{n\to\infty} \langle A(u_n), u_n - u \rangle = 0.$$

Since $A$ is an $(S)_+$-map (Proposition 2.72), we obtain $u_n \to u$ in $W^{1,p}(\Omega)$. So, $\varphi_\lambda$ satisfies the (PS)-condition. This completes Step 1.

We fix some notation. Let $\psi : W^{1,p}(\Omega) \to \mathbb{R}$ be the $C^1$-functional defined by

$$\psi(u) = \frac{1}{p}\|\nabla u\|_p^p - \int_\Omega F(x,u(x))\,dx \text{ for all } u \in W^{1,p}(\Omega).$$

Moreover, we consider the direct sum decomposition

$$W^{1,p}(\Omega) = \mathbb{R} \oplus V, \text{ with } V = \left\{ v \in W^{1,p}(\Omega) : \int_\Omega v(x)\,dx = 0 \right\}. \tag{12.5}$$

*Step 2:* $-\infty < m_V := \inf\limits_V \psi \leq \inf\limits_V \varphi_\lambda.$

Evidently (since $\lambda < 0$), we have that $\psi \leq \varphi_\lambda$. Thus, $\inf\limits_V \psi \leq \inf\limits_V \varphi_\lambda$. From the Poincaré–Wirtinger inequality (Theorem 1.44), there is $M_3 > 0$ such that

$$\|u\|_p^p \leq M_3\|\nabla u\|_p^p \text{ for all } u \in V.$$

Choosing $\varepsilon \in (0, |\lambda|)$ such that $\varepsilon M_3 \leq 1$ in (12.2), invoking (12.3), we obtain

$$\psi(u) \geq \frac{1 - \varepsilon M_3}{p}\|\nabla u\|_p^p - M_2|\Omega|_N \geq -M_2|\Omega|_N \text{ for all } u \in V.$$

Therefore, $m_V = \inf\limits_V \psi \geq -M_2|\Omega|_N > -\infty$. Step 2 is complete.

*Step 3:* There exist $\varepsilon_0, t > 0$ such that for all $\lambda \in (-\varepsilon_0, 0)$ we have $\varphi_\lambda(\pm t) < m_V$.
   For every $t \in (0, +\infty)$ we have

$$\varphi_\lambda(\pm t) = -\frac{\lambda}{p}t^p|\Omega|_N - \int_\Omega F(x, \pm t)\,dx. \tag{12.6}$$

By virtue of hypothesis $H(f)_1$ (iii), we can choose $t > 0$ large such that

$$-\int_\Omega F(x, \pm t)\,dx < m_V - 1. \tag{12.7}$$

Then, setting $\varepsilon_0 = \frac{p}{t^p|\Omega|_N}$, from (12.6) and (12.7) we get $\varphi_\lambda(\pm t) < m_V$ whenever $|\lambda| < \varepsilon_0$, which establishes Step 3.

Henceforth, we fix $\lambda \in (-\varepsilon_0, 0)$, with $\varepsilon_0 > 0$ provided by Step 3. We introduce the following two open subsets of $W^{1,p}(\Omega)$:

$$U_+ = \{u \in W^{1,p}(\Omega) : \int_\Omega u\,dx > 0\} \text{ and } U_- = \{u \in W^{1,p}(\Omega) : \int_\Omega u\,dx < 0\}.$$

Equivalently, $U_+$ (resp. $U_-$) is the subset of elements $u \in W^{1,p}(\Omega)$ of the form $u = \gamma + v$ with $v \in V$ and $\gamma \in (0, +\infty)$ [resp. $\gamma \in (-\infty, 0)$] [see (12.5)]. In particular, we have $\partial U_+ = \partial U_- = V$.

*Step 4:* We can find $u_0 \in U_+$ and $v_0 \in U_-$ such that $\varphi_\lambda(u_0) = \inf\limits_{U_+} \varphi_\lambda$ and $\varphi_\lambda(v_0) = \inf\limits_{U_-} \varphi_\lambda$. In particular, $u_0$ and $v_0$ are local minimizers (hence critical points) of $\varphi_\lambda$.

From Step 1 we know that $\varphi_\lambda$ is coercive. Also, using Theorem 1.49, it is easy to see that $\varphi_\lambda$ is sequentially weakly l.s.c. Moreover, the subsets $\overline{U_+}$ and $\overline{U_-}$ are closed and convex, and hence weakly closed. Thus, we can find $u_0 \in \overline{U_+}$ and $v_0 \in \overline{U_-}$ such that

$$\varphi_\lambda(u_0) = \inf\limits_{\overline{U_+}} \varphi_\lambda \text{ and } \varphi_\lambda(v_0) = \inf\limits_{\overline{U_-}} \varphi_\lambda. \tag{12.8}$$

Combining Steps 2 and 3, we note that

$$\varphi_\lambda(u_0) \le \varphi_\lambda(t) < m_V = \inf_V \psi \le \inf_V \varphi_\lambda,$$

whence $u_0 \notin V$. Since $V = \partial U_+$, this yields $u_0 \in U_+$. Similarly, we see that $v_0 \in U_-$. From this fact, combined with (12.8), we get Step 4.

*Step 5:* The functional $\varphi_\lambda$ admits another critical point $y_0 \in W^{1,p}(\Omega) \setminus \{0, u_0, v_0\}$. Moreover, in the case where $\varphi_\lambda$ has only a finite number of critical points, $y_0$ can be chosen such that $C_1(\varphi_\lambda, y_0) \ne 0$.

From Steps 1 and 4 we know that $0, u_0, v_0$ are local minimizers of $\varphi_\lambda$. Without any loss of generality, we may assume that each of them is an isolated critical point of $\varphi_\lambda$ and even that $\varphi_\lambda$ has only a finite number of critical points (otherwise we are done). In particular, we have

$$C_k(\varphi_\lambda, 0) = \delta_{k,0}\mathbb{F} \quad \text{for all } k \ge 0 \tag{12.9}$$

[Example 6.45(a)]. Moreover, we may assume that $\varphi_\lambda(v_0) \le \varphi_\lambda(u_0)$ (the analysis is similar if the opposite inequality holds). Reasoning as in the proof of Proposition 5.42, we can find $\rho > 0$ small such that

$$\varphi_\lambda(u_0) < \inf\{\varphi_\lambda(y) : \|y - u_0\| = \rho\} =: \eta_\rho \quad \text{and} \quad \|v_0 - u_0\| > \rho. \tag{12.10}$$

Since $\varphi_\lambda$ satisfies the (PS)-condition (by virtue of Step 1), because of (12.10), we can apply Corollary 6.81 and find $y_0 \in W^{1,p}(\Omega) \setminus \{u_0, v_0\}$ critical point of $\varphi_\lambda$ such that

$$C_1(\varphi_\lambda, y_0) \ne 0. \tag{12.11}$$

Comparing (12.11) and (12.9), we conclude that $y_0 \ne 0$. This completes Step 5 and the proof of the theorem.                                                                    □

In the semilinear case (i.e., for $p = 2$), by strengthening the regularity conditions on $f(x, \cdot)$, we can improve the conclusion of Theorem 12.3 by producing four nontrivial smooth solutions.

Now the parametric problem under consideration is as follows:

$$\begin{cases} -\Delta u(x) = \lambda u(x) + f(x, u(x)) & \text{in } \Omega, \\ \dfrac{\partial u}{\partial n} = 0 & \text{on } \partial\Omega, \end{cases} \tag{12.12}$$

with $\lambda \in \mathbb{R}$. The hypotheses on the nonlinearity $f$ are as follows.

$\mathrm{H}(f)_2$ (i) $f : \Omega \times \mathbb{R} \to \mathbb{R}$ is a function such that $f(\cdot, s)$ is measurable for all $s \in \mathbb{R}$, $f(x, \cdot)$ is of class $C^1$ for a.a. $x \in \Omega$, $f(x, 0) = 0$ a.e. in $\Omega$, and there are $c > 0$ and $r \in (2, 2^*)$ such that

$$|f_s'(x,s)| \le c(1+|s|^{r-2}) \text{ for a.a. } x \in \Omega, \text{ all } s \in \mathbb{R};$$

(ii) $\limsup\limits_{s\to\pm\infty} \dfrac{F(x,s)}{s^2} \le 0$ uniformly for a.a. $x \in \Omega$;

(iii) $\limsup\limits_{s\to\pm\infty} \int_\Omega F(x,s)\,dx = +\infty$;

(iv) $\limsup\limits_{s\to 0} \dfrac{F(x,s)}{s^2} \le 0$ uniformly for a.a. $x \in \Omega$.

*Example 12.4.* The following function $f$ satisfies hypotheses $H(f)_2$ (as before, for the sake of simplicity, we drop the $x$-dependence):

$$f(s) = \begin{cases} -c(4\sqrt{-s}-3) & \text{if } s < -1, \\ -cs^2 & \text{if } -1 \le s < 0, \\ 2cs^2 & \text{if } 0 \le s < 1, \\ 2c(4\sqrt{s}-3) & \text{if } s > 1, \end{cases}$$

with any constant $c > 0$. Note that $f$ has no symmetry properties.

We state the following multiplicity result concerning problem (12.12).

**Theorem 12.5.** *If hypotheses $H(f)_2$ hold, then there exists $\varepsilon_0 > 0$ such that for all $\lambda \in (-\varepsilon_0, 0)$ problem (12.12) has at least four nontrivial solutions in $C^1(\overline{\Omega})$.*

*Proof.* As in the proof of Theorem 12.3, let $\varphi_\lambda : H^1(\Omega) \to \mathbb{R}$ be the $C^1$-functional given by

$$\varphi_\lambda(u) = \frac{1}{2}\|\nabla u\|_2^2 - \frac{\lambda}{2}\|u\|_2^2 - \int_\Omega F(x,u(x))\,dx \text{ for all } u \in H^1(\Omega).$$

The solutions of (12.12) coincide with the critical points of $\varphi_\lambda$; moreover, regularity theory (Corollary 8.13) ensures that every solution belongs to $C^1(\overline{\Omega})$. Thus, we only have to show that $\varphi_\lambda$ admits at least four nontrivial critical points. To do this, we may assume without any loss of generality that $\varphi_\lambda$ has only a finite number of critical points. From Steps 1, 4, and 5 of the proof of Theorem 12.3 we know that 0 is a local minimizer of $\varphi_\lambda$ and that $\varphi_\lambda$ admits three other critical points $u_0, v_0, y_0$, where $u_0, v_0$ are local minimizers and $y_0$ satisfies

$$C_1(\varphi_\lambda, y_0) \ne 0.$$

By assumption, all these critical points are isolated. By Example 6.45(a), we have

$$C_k(\varphi_\lambda, 0) = C_k(\varphi_\lambda, u_0) = C_k(\varphi_\lambda, v_0) = \delta_{k,0}\mathbb{F} \text{ for all } k \ge 0. \tag{12.13}$$

It is also noted in Step 1 of the proof of Theorem 12.3 that $\varphi_\lambda$ is bounded below and satisfies the (PS)-condition. Thus, the critical groups of $\varphi_\lambda$ at infinity are well defined and, by Proposition 6.64(a), we have

$$C_k(\varphi_\lambda, \infty) = \delta_{k,0}\mathbb{F} \text{ for all } k \ge 0. \tag{12.14}$$

It remains to produce a fourth nontrivial critical point of $\varphi_\lambda$. Arguing indirectly, we assume that $0, u_0, v_0, y_0$ are the only critical points of $\varphi_\lambda$. We will derive a contradiction from the Morse relation (Theorem 6.62). To do this, we need to compute the critical groups of $\varphi_\lambda$ at $y_0$.

In view of H$(f)_2$ (i), we have that $\varphi_\lambda \in C^2(H^1(\Omega), \mathbb{R})$ and

$$\varphi_\lambda''(y)(u,v) = \int_\Omega (\nabla u, \nabla v)_{\mathbb{R}^N} \, dx - \lambda \int_\Omega uv \, dx - \int_\Omega f_s'(x,y) uv \, dx$$

for all $y, u, v \in H^1(\Omega)$. Recall that the nullity $v_0$ of $y_0$ is by definition the dimension of $\ker \varphi_\lambda''(y_0) = \{u \in H^1(\Omega) : \varphi_\lambda''(y_0)(u,v) = 0 \text{ for all } v \in H^1(\Omega)\}$, whereas the Morse index $m_0$ of $y_0$ is the supremum of the dimensions of the linear subspaces of $H^1(\Omega)$ on which $\varphi_\lambda''(y_0)$ is negative definite (Definition 6.46). We claim that

$$v_0, m_0 \text{ are finite and, if } m_0 = 0, \text{ then } v_0 \leq 1. \tag{12.15}$$

Under the assumption that (12.15) is satisfied, Proposition 6.101 can be applied to $\varphi_\lambda$ and $y_0$, and it yields

$$C_k(\varphi_\lambda, y_0) = \delta_{k,1} \mathbb{F} \text{ for all } k \in \mathbb{N}_0. \tag{12.16}$$

Then, from Theorem 6.62(b) (with $t = -1$), (12.13), (12.14), and (12.16), we get

$$3(-1)^0 + (-1)^1 = (-1)^0,$$

a contradiction, ensuring that there exists one more critical point $w_0 \in H^1(\Omega)$ of $\varphi_\lambda$ distinct from $0, u_0, v_0, y_0$. Therefore, all that remains is to establish (12.15). The rest of the proof is devoted to this purpose.

We need a preliminary construction. We know that $y_0 \in C^1(\overline{\Omega})$, hence H$(f)_2$ (i) yields $\beta := f_s'(\cdot, y_0(\cdot)) + \lambda \in L^\infty(\Omega)$. Let $\rho \in (\|\beta\|_\infty, +\infty)$. Thus $0 < \rho - \|\beta\|_\infty \leq \beta_\rho(x) := \rho - \beta(x) \leq \rho + \|\beta\|_\infty$ for a.a. $x \in \Omega$, so that

$$(h_1, h_2)_{\beta_\rho} := \int_\Omega (\nabla h_1(x), \nabla h_2(x))_{\mathbb{R}^N} \, dx + \int_\Omega \beta_\rho(x) h_1 h_2 \, dx$$

is a scalar product on $H^1(\Omega)$ leading to an equivalent structure of Hilbert space. Then, for every $h \in L^2(\Omega)$, the Riesz representation theorem (see, e.g., Brezis [52, p. 135]) yields a unique $S(h) \in H^1(\Omega) \subset L^2(\Omega)$ such that

$$(S(h), \cdot)_{\beta_\rho} = (h, \cdot)_{L^2(\Omega)} \text{ in } H^1(\Omega)^*. \tag{12.17}$$

Arguing as in the proof of Proposition 9.3, we can see that the map $S : L^2(\Omega) \to L^2(\Omega)$ is a compact self-adjoint linear operator. Relation (12.17) imposes that each eigenvalue of $S$ is positive. Then, Theorems 2.19 and 2.23 imply that the eigenvalues of $S$ consist of a decreasing sequence

$$\mu_1 \geq \mu_2 \geq \cdots \geq \mu_n \geq \cdots (> 0) \quad \text{with} \quad \lim_{n \to \infty} \mu_n = 0,$$

and there is an orthonormal basis $\{\hat{u}_n\}_{n \geq 1}$ of $L^2(\Omega)$, with $S(\hat{u}_n) = \mu_n \hat{u}_n$ for all $n \geq 1$. In particular, $E(S)_\mu := \ker(S - \mu \mathrm{id})$ has finite dimension for all $\mu \in \mathbb{R}$ and is nonzero only if $\mu = \mu_n$ for some $n \geq 1$. Also, note that $\hat{u}_n = \frac{1}{\mu_n} S(\hat{u}_n) \in H^1(\Omega)$ for all $n \geq 1$ and $\{\hat{u}_n\}_{n \geq 1}$ is an orthogonal basis of $(H^1(\Omega), (\cdot, \cdot)_{\beta_\rho})$.

Now we deal with the map $\varphi_\lambda''(y_0)$ in light of the properties of the operator $S$ constructed above. Observe that

$$\varphi_\lambda''(y_0)(u, v) = (u, v)_{\beta_\rho} - \rho(u, v)_{L^2(\Omega)} \quad \text{for all } u, v \in H^1(\Omega). \tag{12.18}$$

From (12.18) it clearly follows that

$$\ker \varphi_\lambda''(y_0) = E(S)_{\frac{1}{\rho}}. \tag{12.19}$$

Therefore, $\nu_0 = \dim \ker \varphi_\lambda''(y_0) < +\infty$. Moreover, for all $n \geq 1$ we have

$$\varphi_\lambda''(y_0)(\hat{u}_n, \hat{u}_n) = (\hat{u}_n, \hat{u}_n)_{\beta_\rho} - \rho(\hat{u}_n, \hat{u}_n)_{L^2(\Omega)} = (1 - \rho \mu_n)(\hat{u}_n, \hat{u}_n)_{\beta_\rho},$$

which implies that $\varphi_\lambda''(y_0)$ is nonnegative on $W := \overline{\mathrm{span}}\{\hat{u}_n : \rho \mu_n \leq 1\}$, where the notation $\overline{\mathrm{span}}$ indicates the closure in $H^1(\Omega)$. Thus,

$$m_0 = \mathrm{codim}\, W = |\{n \geq 1 : \rho \mu_n > 1\}| < +\infty.$$

We have shown the first part in (12.15).

To establish the second part in (12.15), we assume that $m_0 = 0$. This assumption reads as $\varphi_\lambda''(y_0)(u, u) \geq 0$ for all $u \in H^1(\Omega)$, that is,

$$\inf_{H^1(\Omega)} \psi \geq 0, \quad \text{where} \quad \psi(u) := \|\nabla u\|_2^2 - \int_\Omega \beta(x) u^2 \, dx. \tag{12.20}$$

We need to show that $\ker \varphi_\lambda''(y_0)$ has dimension $0$ or $1$. Note that $\ker \varphi_\lambda''(y_0)$ is equivalently the space of solutions of the linear problem

$$-\Delta u - \beta(x) u = 0 \text{ in } \Omega, \quad \frac{\partial u}{\partial n} = 0 \text{ on } \partial \Omega, \tag{12.21}$$

which also coincides with the set of critical points of the functional $\psi$. In particular, Corollary 8.13 yields $\ker \varphi_\lambda''(y_0) \subset C^1(\overline{\Omega})$.

We claim that every $u \in \ker \varphi_\lambda''(y_0) \setminus \{0\}$ satisfies

$$u > 0 \text{ in } \Omega \quad \text{or} \quad u < 0 \text{ in } \Omega. \tag{12.22}$$

Up to reasoning with $-u$ instead of $u$, we may assume that $u^+ \neq 0$. Acting on (12.21) with $u^+$, we get $\psi(u^+) = 0$. Then (12.20) implies that $u^+$ is a minimizer of the functional $\psi$, hence $u^+$ is also a solution of (12.21). Thus, Corollary 8.17 implies that $u^+ > 0$ in $\Omega$. This forces $u^- \equiv 0$, so $u = u^+ > 0$ in $\Omega$. We have checked (12.22).

Now, arguing by contradiction, we assume that $\dim \ker \varphi_\lambda''(y_0) \geq 2$. In view of (12.19), there are $k, \ell \geq 1$ with $k \neq \ell$ such that $\hat{u}_k, \hat{u}_\ell \in \ker \varphi_\lambda''(y_0)$. On the one hand, the orthonormality of the family $\{\hat{u}_n\}_{n \geq 1} \subset L^2(\Omega)$ yields $\int_\Omega \hat{u}_k \hat{u}_\ell \, dx = 0$. On the other hand, (12.22) implies that $\hat{u}_k$ and $\hat{u}_\ell$ must have a constant sign. This is contradictory. Therefore, we obtain that $v_0 \in \{0, 1\}$ whenever $m_0 = 0$. This establishes (12.15). The proof of the theorem is now complete. $\qquad\square$

So far, we have not provided sign information for solutions of problem (12.1). Next we look for nontrivial nonnegative solutions of (12.1), still in the case of near resonance from the left. For this purpose, we state the following hypotheses on $f$:

$\mathrm{H}(f)_3$ (i) $f : \Omega \times \mathbb{R} \to \mathbb{R}$ is a Carathéodory function with $f(x, 0) = 0$ for a.a. $x \in \Omega$ and there is $c > 0$ such that

$$|f(x, s)| \leq c(1 + s^{p-1}) \text{ for a.a. } x \in \Omega, \text{ all } s \geq 0;$$

(ii) There exists $\varepsilon_0 > 0$ such that

$$\liminf_{s \to +\infty} \frac{f(x, s)}{s^{p-1}} \geq \varepsilon_0 \text{ uniformly for a.a. } x \in \Omega;$$

(iii) $\limsup_{s \downarrow 0} \dfrac{f(x, s)}{s^{p-1}} \leq 0$ uniformly for a.a. $x \in \Omega$.

*Example 12.6.* The following function $f$, where for simplicity we drop the $x$-dependence, satisfies hypotheses $\mathrm{H}(f)_3$:

$$f(s) = \begin{cases} 0 & \text{if } s < 0 \\ cs^{r-1} & \text{if } 0 \leq s \leq 1 \\ cs^{p-1} & \text{if } s > 1, \end{cases}$$

with constants $c > 0$ and $r \in (p, +\infty)$.

**Theorem 12.7.** *If hypotheses* $\mathrm{H}(f)_3$ *hold, then for all* $\lambda \in (-\varepsilon_0, 0)$ *problem (12.1) has a nontrivial solution* $u_0 \in C^1(\overline{\Omega})$ *such that* $u_0 \geq 0$ *in* $\Omega$.

*Proof.* Fix $\lambda \in (-\varepsilon_0, 0)$. We deal with the functional $\varphi_{\lambda,+} : W^{1,p}(\Omega) \to \mathbb{R}$ defined for all $u \in W^{1,p}(\Omega)$ by

$$\varphi_{\lambda,+}(u) = \frac{1}{p}\|\nabla u\|_p^p + \frac{1}{p}\|u^-\|_p^p - \frac{\lambda}{p}\|u^+\|_p^p - \int_\Omega F(x, u^+(x)) \, dx.$$

Hypothesis H($f$)$_3$ (iii) clearly guarantees that $\varphi_{\lambda,+} \in C^1(W^{1,p}(\Omega), \mathbb{R})$. The proof of the theorem is based on the following claims.

*Claim 1:* $\varphi_{\lambda,+}$ satisfies the (PS)-condition.

To see this, let $\{u_n\}_{n\geq 1} \subset W^{1,p}(\Omega)$ be such that $\{\varphi_{\lambda,+}(u_n)\}_{n\geq 1}$ is bounded and

$$\varphi'_{\lambda,+}(u_n) \to 0 \text{ in } W^{1,p}(\Omega)^* \text{ as } n \to \infty. \tag{12.23}$$

From (12.23) we have

$$\left| \langle A(u_n), h \rangle + \int_\Omega (-(u_n^-)^{p-1} - \lambda(u_n^+)^{p-1})h\,dx - \int_\Omega f(x, u_n^+)h\,dx \right| \leq \varepsilon_n \|h\| \tag{12.24}$$

for all $h \in W^{1,p}(\Omega)$, all $n \geq 1$, with $\varepsilon_n \downarrow 0$ [where $A$ is defined in (2.28)]. Choosing $h = -u_n^- \in W^{1,p}(\Omega)$ in (12.24), we derive that $\|\nabla u_n^-\|_p^p + \|u_n^-\|_p^p \leq \varepsilon_n \|u_n^-\|$ for all $n \geq 1$, which shows that

$$u_n^- \to 0 \text{ in } W^{1,p}(\Omega) \text{ as } n \to \infty. \tag{12.25}$$

Let us show that $\{u_n^+\}_{n\geq 1}$ is bounded in $W^{1,p}(\Omega)$. Arguing by contradiction, suppose that, along a relabeled subsequence, $\|u_n^+\| \to +\infty$. Set $y_n = \frac{u_n^+}{\|u_n^+\|}$, $n \geq 1$. Then $\|y_n\| = 1$ and $y_n \geq 0$ for all $n \geq 1$. We may assume that

$$y_n \xrightarrow{w} y \text{ in } W^{1,p}(\Omega), \quad y_n \to y \text{ in } L^p(\Omega), \quad y_n(x) \to y(x) \text{ a.e. in } \Omega, \tag{12.26}$$

and $|y_n(x)| \leq k(x)$ a.e. in $\Omega$, with $k \in L^p(\Omega)_+$. From (12.24) and (12.25) we have

$$\left| \langle A(y_n), h \rangle - \int_\Omega \lambda y_n^{p-1} h\,dx - \int_\Omega \frac{f(x, u_n^+)}{\|u_n^+\|^{p-1}} h\,dx \right| \leq \varepsilon'_n \|h\| \tag{12.27}$$

for all $h \in W^{1,p}(\Omega)$, all $n \geq 1$, with $\varepsilon'_n \downarrow 0$. Hypothesis H($f$)$_3$ (i) implies that

$$\left\{ \frac{f(\cdot, u_n^+(\cdot))}{\|u_n^+\|^{p-1}} \right\}_{n\geq 1} \text{ is bounded in } L^{p'}(\Omega). \tag{12.28}$$

Choosing $h = y_n - y \in W^{1,p}(\Omega)$ in (12.27), passing to the limit as $n \to \infty$, and using (12.26) and (12.28), we obtain $\lim_{n\to\infty} \langle A(y_n), y_n - y \rangle = 0$. Since $A$ is an $(S)_+$-map (Proposition 2.72), we get

$$y_n \to y \text{ in } W^{1,p}(\Omega) \text{ as } n \to \infty, \tag{12.29}$$

so, in particular, $\|y\| = 1$, $y \geq 0$. Using hypothesis H($f$)$_3$ (ii) and reasoning as in Step 2 of the proof of Theorem 10.5, we can show that

$$\frac{f(\cdot, u_n^+(\cdot))}{\|u_n^+\|^{p-1}} \xrightarrow{w} \xi y^{p-1} \text{ in } L^{p'}(\Omega), \tag{12.30}$$

with $\xi \in L^\infty(\Omega)_+$, $\varepsilon_0 \le \xi(x) \le c$ for a.a. $x \in \Omega$. Thus, $\hat{\xi}(x) := \xi(x) + \lambda \ge \varepsilon_0 + \lambda > 0$ a.e. in $\Omega$. Passing to the limit as $n \to \infty$ in (12.27) and using (12.29) and (12.30), we deduce that

$$\langle A(y), h \rangle = \int_\Omega \hat{\xi} y^{p-1} h \, dx \quad \text{for all } h \in W^{1,p}(\Omega),$$

i.e., $y$ is an eigenfunction of the negative $p$-Laplacian under Neumann boundary conditions with respect to the weight $\hat{\xi}$, corresponding to the eigenvalue 1 [see Definition 9.24 (b)]. Invoking Proposition 9.33, we infer that $y$ must be sign changing, a contradiction of the fact that $y \ge 0$.

We have therefore shown that $\{u_n^+\}_{n \ge 1}$ is bounded in $W^{1,p}(\Omega)$. This fact and (12.25) imply that $\{u_n\}_{n \ge 1}$ is bounded in $W^{1,p}(\Omega)$. Consequently, we may assume that

$$u_n \xrightarrow{w} u \text{ in } W^{1,p}(\Omega) \text{ and } u_n \to u \text{ in } L^p(\Omega) \text{ as } n \to \infty \qquad (12.31)$$

for some $u \in W^{1,p}(\Omega)$. Choosing $h = u_n - u$ in (12.24), then passing to the limit and using (12.31), we obtain $\lim_{n \to \infty} \langle A(u_n), u_n - u \rangle = 0$, and so $u_n \to u$ in $W^{1,p}(\Omega)$ (Proposition 2.72). This establishes Claim 1.

*Claim 2:* There exists $\rho > 0$ such that $\eta_\rho := \inf\{\varphi_{\lambda,+}(u) : \|u\| = \rho\} > 0$.

Due to H($f$)$_3$ (iii), given $\varepsilon \in (0, |\lambda|)$, there is $\delta = \delta(\varepsilon) > 0$ such that

$$f(x,s) \le \varepsilon s^{p-1} \quad \text{for a.a. } x \in \Omega, \text{ all } s \in [0, \delta],$$

while hypothesis H($f$)$_3$ (i) implies that we can find $c_\varepsilon > 0$ and $r \in (p, p^*)$ such that

$$f(x,s) \le c_\varepsilon s^{r-1} \quad \text{for a.a. } x \in \Omega, \text{ all } s \ge \delta.$$

Thus, we have

$$F(x,s) \le \frac{\varepsilon}{p} s^p + \frac{c_\varepsilon}{r} s^r \quad \text{for a.a. } x \in \Omega, \text{ all } s \in [0, +\infty).$$

Using the continuity of the embedding $W^{1,p}(\Omega) \hookrightarrow L^r(\Omega)$, we infer that for some constant $\hat{c}_\varepsilon > 0$ we have

$$\varphi_{\lambda,+}(u) \ge \frac{1}{p} \|\nabla u\|_p^p + \frac{|\lambda| - \varepsilon}{p} \|u\|_p^p - \hat{c}_\varepsilon \|u\|^r \quad \text{for all } u \in W^{1,p}(\Omega). \qquad (12.32)$$

Since $\varepsilon < |\lambda|$ and $r > p$, from (12.32) it follows that Claim 2 holds whenever $\rho > 0$ is small enough.

*Claim 3:* $\varphi_{\lambda,+}(t) \to -\infty$ as $t \to +\infty, t \in \mathbb{R}$.

In view of $H(f)_3$ (i) and (ii), given $\varepsilon \in (0, \varepsilon_0 + \lambda)$, we can find $M_\varepsilon > 0$ such that

$$F(x,s) \geq \frac{\varepsilon_0 - \varepsilon}{p} s^p - s M_\varepsilon \text{ for a.a. } x \in \Omega, \text{ all } s \in [0, +\infty).$$

Hence, for every $t \in (0, +\infty)$ we have

$$\varphi_{\lambda,+}(t) \leq -t^p \frac{(\varepsilon_0 + \lambda) - \varepsilon}{p} |\Omega|_N + t M_\varepsilon |\Omega|_N,$$

and so, because $\varepsilon < \varepsilon_0 + \lambda$, Claim 3 holds.

Claims 1–3 permit the use of the mountain pass theorem (Theorem 5.40), which provides $u_0 \in W^{1,p}(\Omega)$ such that

$$\varphi_{\lambda,+}(0) = 0 < \eta_\rho \leq \varphi_{\lambda,+}(u_0) \tag{12.33}$$

and

$$\varphi'_{\lambda,+}(u_0) = 0. \tag{12.34}$$

From (12.33) we see that $u_0 \neq 0$. From (12.34) we have

$$A(u_0) - (u_0^-)^{p-1} = \lambda (u_0^+)^{p-1} + f(\cdot, u_0^+(\cdot)) \text{ in } W^{1,p}(\Omega)^*. \tag{12.35}$$

Acting on (12.35) with the test function $-u_0^- \in W^{1,p}(\Omega)$, we obtain $u_0^- = 0$, hence $u_0 \geq 0$. Thus, (12.35) becomes

$$A(u_0) = \lambda u_0^{p-1} + f(\cdot, u_0(\cdot)) \text{ in } W^{1,p}(\Omega)^*,$$

i.e., $u_0$ is a weak solution of (12.1). By nonlinear regularity theory (Corollary 8.13), we have $u_0 \in C^1(\overline{\Omega})$. The proof of the theorem is now complete.   □

## Near Resonance from Right: $0 < \lambda$

Next, we study the situation where $\lambda$ approaches 0 from the right. We first provide a simple existence result under the following assumptions on the nonlinearity $f$ and its primitive $F(x,s) = \int_0^s f(x,t)\,dt$.

$H(f)_4$ (i) $f : \Omega \times \mathbb{R} \to \mathbb{R}$ is a Carathéodory function with $f(x,0) = 0$ a.e. in $\Omega$ and such that there are $c > 0$ and $r \in (p, p^*)$ satisfying

$$|f(x,s)| \leq c(1 + |s|^{r-1}) \text{ for a.a. } x \in \Omega, \text{ all } s \in \mathbb{R};$$

(ii) There exists $\vartheta \in L^\infty(\Omega)$, with $\vartheta(x) \leq 0$ a.e. in $\Omega$, $\vartheta \neq 0$, such that

$$\limsup_{s \to \pm\infty} \frac{pF(x,s)}{|s|^p} \leq \vartheta(x) \text{ uniformly for a.a. } x \in \Omega;$$

(iii) There exists $s_0 \in \mathbb{R} \setminus \{0\}$ such that $\int_\Omega F(x,s_0)\,dx \geq 0$.

*Example 12.8.* The following function $f$ satisfies hypotheses H$(f)_4$ (for simplicity we drop the $x$-dependence):

$$f(s) = \begin{cases} -|s|^{p-2}s - 1 + \sin 1 & \text{if } s < -1, \\ \sin s^2 & \text{if } -1 \leq s \leq 1, \\ -s^{p-1} + 1 + \sin 1 & \text{if } s > 1. \end{cases}$$

**Theorem 12.9.** *If hypotheses H$(f)_4$ hold, then there exists $\varepsilon_0 > 0$ such that for all $\lambda \in (0, \varepsilon_0)$ problem (12.1) has a nontrivial solution $u_0 \in C^1(\overline{\Omega})$.*

*Proof.* By Lemma 9.29 (with $\xi = -\vartheta$), we find $\varepsilon_0 > 0$ such that

$$\|\nabla u\|_p^p - \int_\Omega \vartheta(x)|u(x)|^p\,dx \geq 2\varepsilon_0 \|u\|^p \quad \text{for all } u \in W^{1,p}(\Omega). \tag{12.36}$$

Assume that $\lambda \in (0, \varepsilon_0)$. As in the proof of Theorem 12.3, we consider the functional $\varphi_\lambda \in C^1(W^{1,p}(\Omega), \mathbb{R})$ such that

$$\varphi_\lambda(u) = \frac{1}{p}\|\nabla u\|_p^p - \frac{\lambda}{p}\|u\|_p^p - \int_\Omega F(x,u(x))\,dx \quad \text{for all } u \in W^{1,p}(\Omega).$$

By hypotheses H$(f)_4$ (i), (ii), we can find $c_\lambda > 0$ such that

$$F(x,s) \leq \frac{1}{p}(\vartheta(x)+\lambda)|s|^p + c_\lambda \quad \text{for a.a. } x \in \Omega, \text{ all } s \in \mathbb{R}. \tag{12.37}$$

Combining (12.36) and (12.37), for every $u \in W^{1,p}(\Omega)$, we obtain

$$\varphi_\lambda(u) \geq \frac{1}{p}\|\nabla u\|_p^p - \frac{1}{p}\int_\Omega \vartheta|u|^p\,dx - \frac{2\lambda}{p}\|u\|_p^p - c_\lambda|\Omega|_N$$

$$\geq \frac{2(\varepsilon_0-\lambda)}{p}\|u\|^p - c_\lambda|\Omega|_N, \tag{12.38}$$

which implies that $\varphi_\lambda$ is coercive. It is easy to see, using H$(f)_4$ (i), that $\varphi_\lambda$ is also sequentially weakly l.s.c. Therefore, we can find $u_0 \in W^{1,p}(\Omega)$ such that

$$\varphi_\lambda(u_0) = \inf\{\varphi_\lambda(u) : u \in W^{1,p}(\Omega)\}. \tag{12.39}$$

In particular, if $s_0 \in \mathbb{R} \setminus \{0\}$ is as in hypothesis $H(f)_4$ (iii), then

$$\varphi_\lambda(s_0) = -\frac{\lambda}{p}|s_0|^p|\Omega|_N - \int_\Omega F(x,s_0)\,dx < 0 = \varphi_\lambda(0),$$

hence $u_0 \neq 0$. From (12.39) we have $\varphi'_\lambda(u_0) = 0$, from which it follows that $u_0$ is a nontrivial solution of (12.1). As before, nonlinear regularity theory (Corollary 8.13) implies that $u_0 \in C^1(\overline{\Omega})$. $\square$

We now prove a multiplicity theorem for the case of near resonance from the right, involving the following hypotheses on the nonlinearity $f$. They coincide with $H(f)_4$, except the last assumption, which is now stronger. By $\lambda_1 > 0$ we denote the second eigenvalue of the negative Neumann $p$-Laplacian (with respect to the weight $\xi \equiv 1$; see Proposition 9.40).

$H(f)_5$ (i) $f : \Omega \times \mathbb{R} \to \mathbb{R}$ is a Carathéodory function with $f(x,0) = 0$ a.e. in $\Omega$ and such that there are $c > 0$ and $r \in (p, p^*)$ satisfying

$$|f(x,s)| \leq c(1+|s|^{r-1}) \quad \text{for a.a. } x \in \Omega, \text{ all } s \in \mathbb{R};$$

(ii) There exists $\vartheta \in L^\infty(\Omega)$, with $\vartheta(x) \leq 0$ a.e. in $\Omega$, $\vartheta \neq 0$, such that

$$\limsup_{s \to \pm\infty} \frac{pF(x,s)}{|s|^p} \leq \vartheta(x) \quad \text{uniformly for a.a. } x \in \Omega;$$

(iii) There exist $\delta > 0$ and $\eta \in (0, \lambda_1)$ such that

$$0 < F(x,s) \leq \frac{\eta}{p}|s|^p \quad \text{for a.a. } x \in \Omega, \text{ all } s \in [-\delta, \delta], s \neq 0.$$

*Example 12.10.* The following function $f$ satisfies hypotheses $H(f)_5$ (for simplicity we drop the $x$-dependence):

$$f(s) = \begin{cases} \eta|s|^{p-2}s & \text{if } |s| \leq 1, \\ 2\eta|s|^{\tau-2}s - \eta|s|^{p-2}s & \text{if } |s| > 1, \end{cases}$$

with constants $1 < \tau < p$ and $\eta \in (0, \lambda_1)$.

**Theorem 12.11.** *If hypotheses* $H(f)_5$ *hold and* $p \geq 2$, *then there is* $\varepsilon_0 > 0$ *such that for all* $\lambda \in (0, \varepsilon_0)$ *problem (12.1) has at least two nontrivial solutions* $u_0, v_0 \in C^1(\overline{\Omega})$.

*Proof.* From the proof of Theorem 12.9 we already know that there is $\varepsilon_1 > 0$ such that for $\lambda \in (0, \varepsilon_1)$ the functional $\varphi_\lambda$ is coercive and bounded below. Arguing as in Step 1 of the proof of Theorem 12.3, we deduce that $\varphi_\lambda$ satisfies the (PS)-condition.

We may assume that 0 is an isolated critical point of $\varphi_\lambda$ (otherwise we are done), so that the critical groups of $\varphi_\lambda$ at 0 are well defined. Set $\varepsilon_0 = \min\{\varepsilon_1, \lambda_1 - \eta\} > 0$ with $\eta$ from H$(f)_5$ (iii). We claim that for all $\lambda \in (0, \varepsilon_0)$ we have

$$\varphi_\lambda \text{ has a local } (1,1)\text{-linking at } 0 \tag{12.40}$$

(Definition 6.82). Once we prove (12.40), the theorem will follow from Corollary 6.94. Thus, it remains to prove (12.40).

Thus, let $\lambda \in (0, \varepsilon_0)$. In particular, $\lambda < \lambda_1 - \eta$. We denote

$$D = \{u \in W^{1,p}(\Omega): \int_\Omega |u|^{p-2} u \, dx = 0\}.$$

Since we assume that $p \geq 2$, Proposition 9.41(b) yields

$$\|u\|_p^p \leq \frac{1}{\lambda_1} \|\nabla u\|_p^p \quad \text{for all } u \in D. \tag{12.41}$$

Note that hypotheses H$(f)_5$ (i), (iii) imply the estimate

$$F(x,s) \leq \frac{\eta}{p} |s|^p + c_1 |s|^r \quad \text{for a.a. } x \in \Omega \text{ and } s \in \mathbb{R}, \tag{12.42}$$

with $c_1 > 0$. By (12.41), (12.42), and the continuity of the embedding $W^{1,p}(\Omega) \hookrightarrow L^r(\Omega)$, we have

$$\varphi_\lambda(u) \geq \frac{1}{p} \|\nabla u\|_p^p - \frac{\lambda + \eta}{p \lambda_1} \|\nabla u\|_p^p - c_2 \|\nabla u\|_p^r \quad \text{for all } u \in D$$

for some $c_2 > 0$. The choice of $\lambda < \lambda_1 - \eta$ and the fact that $r > p$ imply that we can find $\rho > 0$ small such that

$$\varphi_\lambda(u) > 0 \quad \text{for all } u \in D \text{ with } 0 < \|u\| < \rho. \tag{12.43}$$

On the other hand, by virtue of hypothesis H$(f)_5$ (iii) and by choosing $\rho > 0$ even smaller if necessary, we obtain

$$\varphi_\lambda(t) < 0 \quad \text{for all } t \in [-\rho_0, \rho_0] \setminus \{0\}, \text{ where } \rho_0 = \rho |\Omega|_N^{-\frac{1}{p}}. \tag{12.44}$$

Let $U = \{u \in W^{1,p}(\Omega): \|u\| \leq \rho\}$, $E_0 = \{-\rho_0, \rho_0\}$, and $E = [-\rho_0, \rho_0]$. Thus,

$$E_0 \subset E \subset U \quad \text{and} \quad E_0 \cap D = \emptyset. \tag{12.45}$$

Up to choosing $\rho > 0$ even smaller, we may assume that 0 is the only critical point of $\varphi_\lambda$ in $U$. Let $i: E_0 \to W^{1,p}(\Omega) \setminus D$ and $j: E_0 \to E$ be the inclusion maps, and let us consider the group homomorphisms $i_0 : H_0(E_0) \to H_0(W^{1,p}(\Omega) \setminus D)$ and $j_0 : H_0(E_0) \to H_0(E)$ induced between singular homology groups. Since $H_0(E, E_0) = 0$ [Example 6.42(a)], from Axiom 4 in Definition 6.9 we get that $j_0$ is surjective, hence

$$\dim \operatorname{im} j_0 = \dim H_0(E) = 1 \qquad (12.46)$$

[Example 6.42(b)]. Next, we may note that whenever $\gamma \in C([-1,1], W^{1,p}(\Omega))$ is a path from $\gamma(-1) = -\rho_0$ to $\gamma(1) = \rho_0$, because $\int_\Omega |\gamma(-1)|^{p-2}\gamma(-1)\,dx < 0 < \int_\Omega |\gamma(1)|^{p-2}\gamma(1)\,dx$, we can always find $t \in (-1,1)$ with $\gamma(t) \in D$. This shows that $-\rho_0, \rho_0$ belong to different connected components of $W^{1,p}(\Omega) \setminus D$. Thus there is a retraction $h : W^{1,p}(\Omega) \setminus D \to E_0$. As in the proof of Proposition 6.16, we deduce that $i_0$ is injective, so

$$\dim \operatorname{im} i_0 = \dim H_0(E_0) = 2 \qquad (12.47)$$

[Example 6.42(b)]. All together, relations (12.43)–(12.47) imply (12.40). The proof of the theorem is complete. □

## 12.2 Nonlinear Neumann Problems with Nonhomogeneous Differential Operators

In this section, we study the existence and multiplicity of nontrivial solutions for a Neumann problem

$$\begin{cases} -\operatorname{div} a(x, \nabla u) = f(x, u) & \text{in } \Omega, \\ \dfrac{\partial u}{\partial n_a} = 0 & \text{on } \partial\Omega, \end{cases} \qquad (12.48)$$

where $\Omega \subset \mathbb{R}^N$ ($N \geq 1$) is a bounded domain with $C^2$-boundary $\partial\Omega$, $f : \Omega \times \mathbb{R} \to \mathbb{R}$ is a Carathéodory function, and $\operatorname{div} a(x, \nabla u)$ is a differential operator patterned from the $p$-Laplacian for $p \in (1, +\infty)$, with the main difference that it is not assumed to be homogeneous with respect to the gradient of $u$. The boundary condition involves the generalized normal derivative $\frac{\partial u}{\partial n_a} = \gamma_n(a(x, \nabla u))$ (Theorem 1.38).

The purpose of this section is to illustrate, through the study of problem (12.48), the extent to which the variational method, which we used in the case of the $p$-Laplacian operator (in Chaps. 10 and 11 and Sect. 12.1), can be applied to the case of the more general operator $\operatorname{div} a(x, \nabla u)$.

The section is divided into two parts: first, a theoretical part, where we set forth the elementary properties of the operator $\operatorname{div} a(x, \nabla u)$ and necessary theoretical ingredients for dealing with problem (12.48). Then we apply these tools to obtain a multiplicity result for problem (12.48). Actually, our approach is not specific to the situation of the Neumann problem (12.48). In particular, the theoretical tools presented in what follows are also adapted to the study of the counterpart of (12.48) in the Dirichlet case:

$$\begin{cases} -\operatorname{div} a(x, \nabla u) = f(x, u) & \text{in } \Omega, \\ u = 0 & \text{on } \partial\Omega. \end{cases} \tag{12.49}$$

## Properties of Differential Operator

The assumptions on $a$ are as follows:

$H(a)_1$ (i) $a : \overline{\Omega} \times \mathbb{R}^N \to \mathbb{R}^N$ is a continuous map whose restriction to $\overline{\Omega} \times (\mathbb{R}^N \setminus \{0\})$ is of class $C^1$, and $a(x, 0) = 0$ for all $x \in \overline{\Omega}$. Moreover, $a$ is of the form

$$a(x, \xi) = \hat{a}(x, |\xi|)\xi \quad \text{for all } x \in \overline{\Omega}, \text{ all } \xi \in \mathbb{R}^N \setminus \{0\},$$

where $\hat{a} \in C^1(\overline{\Omega} \times (0, +\infty), (0, +\infty))$;

(ii) There is a constant $c_0 > 0$ such that

$$(a'_\xi(x, \xi)\eta, \eta)_{\mathbb{R}^N} \geq c_0 |\xi|^{p-2} |\eta|^2 \quad \text{for all } x \in \overline{\Omega}, \text{ all } \xi, \eta \in \mathbb{R}^N, \xi \neq 0;$$

(iii) There is a constant $c_1 > 0$ such that

$$\|a'_\xi(x, \xi)\| \leq c_1 |\xi|^{p-2} \quad \text{for all } x \in \overline{\Omega}, \text{ all } \xi \in \mathbb{R}^N \setminus \{0\};$$

(iv) There are constants $\alpha, c_2 > 0$ such that

$$|a(x, \xi) - a(y, \xi)| \leq c_2 |x - y|^\alpha (1 + |\xi|)^{p-2} |\xi| \quad \text{for all } x, y \in \overline{\Omega}, \text{ all } \xi \in \mathbb{R}^N.$$

*Example 12.12.* Examples 8.1(a) and (b) with $p \in (1, +\infty)$, (c) with $p \geq 2$, and (d) with $(p \in (1, 2]$ and $c \in (0, 4p(p - 1)))$ or $(p > 2$ and $c \in (0, 2p + 2\sqrt{2p}))$ satisfy $H(a)_1$. Moreover, it is readily seen that the class of maps $a : \overline{\Omega} \times \mathbb{R}^N \to \mathbb{R}^N$ satisfying $H(a)_1$ is stable by addition as well as by multiplication by any map $\theta \in C^1(\overline{\Omega}, (0, +\infty))$.

*Remark 12.13.*

(a) Clearly, $H(a)_1$ is stronger than the set of hypotheses $H(a)_2$ of Sect. 8.1. Moreover, from $H(a)_1$, for all $(x, \xi) \in \overline{\Omega} \times \mathbb{R}^N$, we obtain

$$(a(x, \xi), \xi)_{\mathbb{R}^N} = \int_0^1 (a'_\xi(x, t\xi)\xi, \xi)_{\mathbb{R}^N} \, dt \geq c_0 \int_0^1 t^{p-2} |\xi|^p \, dt = \frac{c_0}{p-1} |\xi|^p,$$

$$|a(x, \xi)| = \int_0^1 \frac{(a(x, t\xi), a'_\xi(x, t\xi)\xi)_{\mathbb{R}^N}}{|a(x, t\xi)|} \, dt \leq c_1 \int_0^1 t^{p-2} |\xi|^{p-1} \, dt = \frac{c_1}{p-1} |\xi|^{p-1}.$$

Thus, $H(a)_1$ is also stronger than the set of hypotheses $H(a)_1$ of Sect. 8.1.

(b) Hypothesis H$(a)_1$ (ii) implies that the operator $a$ is strictly monotone; more precisely, there is a constant $\tilde{c}_0 > 0$ such that for all $x \in \overline{\Omega}$, all $\eta, \xi \in \mathbb{R}^N$, we have

$$(a(x,\xi) - a(x,\eta), \xi - \eta)_{\mathbb{R}^N} \geq \tilde{c}_0 |\xi - \eta|^2 (|\xi| + |\eta|)^{p-2}$$

(Remark 8.18).

(c) The map $\hat{a}$ of H$(a)_1$ (i) is characterized by

$$\hat{a}(x,t)t = |a(x,t\xi)| = (a(x,t\xi), \xi)_{\mathbb{R}^N} \quad \text{for all } x \in \overline{\Omega}, \text{ all } t \in (0, +\infty),$$

whenever $\xi \in \mathbb{R}^N$, $|\xi| = 1$. In particular, the map $g : (x,t) \mapsto \hat{a}(x,t)t$ can be extended by continuity on $\overline{\Omega} \times [0, +\infty)$ by letting $g(x,0) = 0$.

The next result establishes a first important feature of the operator $\operatorname{div} a(x, \nabla u)$, which is in general an indispensable ingredient for checking the (PS)-condition for functionals associated to problems (12.48) and (12.49).

**Proposition 12.14.** *Assume that* H$(a)_1$ *holds. Then* $V : W^{1,p}(\Omega) \to W^{1,p}(\Omega)^*$ *defined by*

$$\langle V(u), v \rangle := \int_\Omega (a(x, \nabla u), \nabla v)_{\mathbb{R}^N} dx \quad \text{for all } u, v \in W^{1,p}(\Omega)$$

*is an* $(S)_+$*-map.*

*Proof.* This property is established in Proposition 2.72 in the case where $\operatorname{div} a(x, \nabla u)$ is the $p$-Laplacian. The proof in the present case is more involved. Let $\{u_n\}_{n \geq 1} \subset W^{1,p}(\Omega)$ be such that

$$u_n \xrightarrow{w} u \text{ in } W^{1,p}(\Omega) \quad \text{and} \quad \limsup_{n \to \infty} \langle V(u_n), u_n - u \rangle \leq 0. \tag{12.50}$$

The first part of (12.50) yields $M_1 > 0$ such that

$$\|\nabla u_n\|_p \leq M_1 \quad \text{for all } n \geq 1. \tag{12.51}$$

Note that $\langle V(u), u_n - u \rangle \to 0$ as $n \to \infty$. Then the monotonicity of $a$ [Remark 12.13(b)] and the second part of (12.50) imply

$$\lim_{n \to \infty} \int_\Omega (a(x, \nabla u_n) - a(x, \nabla u), \nabla u_n - \nabla u)_{\mathbb{R}^N} dx = 0.$$

Hence,

$$w_n \to 0 \text{ in } L^1(\Omega), \tag{12.52}$$

where $w_n(x) := (a(x, \nabla u_n(x)) - a(x, \nabla u(x)), \nabla u_n(x) - \nabla u(x))_{\mathbb{R}^N} \geq 0$.

*Claim 1:* $\nabla u_n \rightharpoonup \nabla u$ a.e. in $\Omega$.

We deal with any relabeled subsequence $\{\nabla u_n\}_{n\geq 1}$ so that it suffices to establish Claim 1 up to extraction of a new subsequence of $\{\nabla u_n\}_{n\geq 1}$. From (12.52), up to passing to a subsequence, we find $h \in L^1(\Omega)_+$ such that

$$w_n \to 0 \text{ a.e. in } \Omega \text{ and } 0 \leq w_n(x) \leq h(x) \text{ for a.a. } x \in \Omega, \text{ all } n \geq 1. \quad (12.53)$$

From (12.53) and Remark 12.13(b) we infer that there is a measurable subset $S \subset \Omega$ with $|S|_N = 0$ such that

$$\tilde{c}_0 |\nabla u_n(x) - \nabla u(x)|^2 (|\nabla u_n(x)| + |\nabla u(x)|)^{p-2} \leq h(x) \text{ for all } x \in \Omega \setminus S.$$

This readily implies that $\{\nabla u_n(x)\}_{n\geq 1}$ is bounded for all $x \in \Omega \setminus S$. We claim that

$$\nabla u_n(x) \to \nabla u(x) \text{ for all } x \in \Omega \setminus S. \quad (12.54)$$

To show (12.54), we fix $x \in \Omega \setminus S$ and we check that one can extract a subsequence converging to $\nabla u(x)$ from any relabeled subsequence $\{\nabla u_n(x)\}_{n\geq 1}$. Since $\{\nabla u_n(x)\}_{n\geq 1}$ is bounded in $\mathbb{R}^N$, along a subsequence (depending on $x$) we get $\nabla u_{n_k}(x) \to \xi(x) \in \mathbb{R}^N$ as $k \to \infty$. The first part of (12.53) then yields

$$(a(x, \xi(x)) - a(x, \nabla u(x)), \xi(x) - \nabla u(x))_{\mathbb{R}^N} = 0,$$

which, by the strict monotonicity of $a$ [Remark 12.13(b)], implies that $\nabla u(x) = \xi(x)$. We have shown (12.54), which establishes Claim 1.

*Claim 2:* For every $\varepsilon > 0$ we can find $m > 0$ such that for every $n \geq 1$

$$\int_{\Omega_{n,m}} |\nabla u_n(x)|^p \, dx \leq \varepsilon,$$

where $\Omega_{n,m} := \{x \in \Omega : |\nabla u_n(x)|^p > m\}$.

First, we note that (12.51) and the definition of $\Omega_{n,m}$ yield, for all $m \in (0, +\infty)$,

$$|\Omega_{n,m}|_N \leq \frac{1}{m^p} \int_{\Omega_{n,m}} |\nabla u_n(x)|^p \, dx \leq \frac{M_1^p}{m^p} \text{ for all } n \geq 1. \quad (12.55)$$

In view of (12.52), there is $n_0 \geq 1$ such that for all $n \geq n_0$ we have

$$\int_{\Omega} w_n(x) \, dx \leq \frac{c_0}{c_1} \frac{\varepsilon}{2}. \quad (12.56)$$

We can find $\delta > 0$ such that every measurable set $A \subset \Omega$ with $|A|_N \leq \delta^p$ satisfies

$$\int_A |\nabla u_n(x)|^p \, dx \leq \varepsilon \text{ for all } n \in \{1, \ldots n_0 - 1\} \quad (12.57)$$

and, letting $I_A(u) = \left( \int_A |\nabla u(x)|^p \, dx \right)^{\frac{1}{p}}$,

$$I_A(u)^p + I_A(u)^{p-1}M_1 + I_A(u)M_1^{p-1} \le \frac{c_0}{p-1} \frac{\varepsilon}{2}. \tag{12.58}$$

Set $m = \frac{M_1}{\delta} > 0$. Thus, we have $|\Omega_{n,m}|_N \le \delta^p$ [see (12.55)], so that the claimed relation holds for every $n \in \{1,\dots,n_0 - 1\}$ [by (12.57)]. For $n \ge n_0$, using Remark 12.13(a), (12.56), (12.51), (12.58), and Hölder's inequality, we compute

$$\int_{\Omega_{n,m}} |\nabla u_n(x)|^p \, dx \le \frac{p-1}{c_0} \int_{\Omega_{n,m}} (a(x,\nabla u_n), \nabla u_n)_{\mathbb{R}^N} \, dx$$

$$\le \frac{p-1}{c_0} \left( \int_{\Omega} w_n(x) \, dx + \int_{\Omega_{n,m}} ((a(x,\nabla u), \nabla u_n - \nabla u)_{\mathbb{R}^N} + (a(x,\nabla u_n), \nabla u)_{\mathbb{R}^N}) \, dx \right)$$

$$\le \frac{\varepsilon}{2} + \frac{c_1}{c_0} \int_{\Omega_{n,m}} (|\nabla u|^{p-1}(|\nabla u_n| + |\nabla u|) + |\nabla u_n|^{p-1}|\nabla u|) \, dx$$

$$\le \frac{\varepsilon}{2} + \frac{c_1}{c_0} \left( I_{\Omega_{n,m}}(u)^p + I_{\Omega_{n,m}}(u)^{p-1}M_1 + I_{\Omega_{n,m}}(u)M_1^{p-1} \right) \le \frac{\varepsilon}{2} + \frac{\varepsilon}{2} = \varepsilon.$$

This proves Claim 2.

The property shown in Claim 2 means that the family $\{|\nabla u_n|^p\}_{n\ge1} \subset L^1(\Omega)$ is uniformly integrable. Together with the fact that $|\nabla u_n(x)|^p \to |\nabla u(x)|^p$ for a.a. $x \in \Omega$ (Claim 1), this property permits the use of Vitali's theorem (e.g., Gasiński and Papageorgiou [151, p. 901]), which implies that

$$\lim_{n\to\infty} \int_{\Omega} |\nabla u_n|^p \, dx = \int_{\Omega} |\nabla u|^p \, dx. \tag{12.59}$$

Moreover, since $u_n \xrightarrow{w} u$ in $W^{1,p}(\Omega)$, we have $\|u_n\|_p \to \|u\|_p$ as $n \to \infty$, whence $\|u_n\| \to \|u\|$ [see (12.59)]. Since $W^{1,p}(\Omega)$ satisfies the Kadec–Klee property [Remark 2.47(a), (c)], the facts that $u_n \xrightarrow{w} u$ and $\|u_n\| \to \|u\|$ ensure that $u_n \to u$ in $W^{1,p}(\Omega)$ as $n \to \infty$. This shows that $V$ is an $(S)_+$-map. The proof is complete. □

We consider the following assumption on the nonlinearity $f$ involved in problems (12.48) and (12.49).

$H(f)_1$ The map $f:\Omega \times \mathbb{R} \to \mathbb{R}$ is a Carathéodory function, and there are $r \in (p,p^*)$ and $c > 0$ such that

$$|f(x,s)| \le c(1 + |s|^{r-1}) \quad \text{for a.a. } x \in \Omega, \text{ all } s \in \mathbb{R}.$$

Under hypotheses $H(a)_1$ and $H(f)_1$, we can introduce the notion of a weak solution of problems (12.48) and (12.49) as in Definition 8.2 as follows.

12 Nonlinear Elliptic Equations with Neumann Boundary Conditions

**Definition 12.15.** A (*weak*) *solution* of (12.48) [resp. of (12.49)] is an element $u \in W^{1,p}(\Omega)$ [resp. $u \in W_0^{1,p}(\Omega)$] such that the equality

$$\int_\Omega (a(x,\nabla u), \nabla v)_{\mathbb{R}^N} \, dx = \int_\Omega f(x, u(x)) v(x) \, dx$$

holds for all $v \in W^{1,p}(\Omega)$ [resp. $v \in W_0^{1,p}(\Omega)$].

In view of Corollary 8.12 and Remark 8.3(b), every weak solution $u$ of (12.48) belongs to $C^1(\overline{\Omega})$ and satisfies the boundary condition $\frac{\partial u}{\partial n_a} = 0$, whereas every solution of (12.49) belongs to $C_0^1(\overline{\Omega})$.

In our approach to problem (12.48), we use the variational method. This means that we need to associate to (12.48) a $C^1$-functional $\varphi : W^{1,p}(\Omega) \to \mathbb{R}$ whose critical points coincide with the solutions of (12.48). To do this, we rely on the particular form of the operator $a$ given in H(a)$_1$ (i). We define a map $G : \overline{\Omega} \times \mathbb{R}^N \to [0,+\infty)$ by letting

$$G(x,\xi) = \int_0^{|\xi|} \hat{a}(x,t) t \, dt \quad \text{for all } x \in \overline{\Omega}, \text{ all } \xi \in \mathbb{R}^N.$$

**Lemma 12.16.**

(a) *For every* $x \in \overline{\Omega}$ *the map* $\xi \mapsto G(x,\xi)$ *is of class* $C^1$, *and we have*

$$G'_\xi(x,\xi) = a(x,\xi) \quad \text{for all } x \in \overline{\Omega}, \text{ all } \xi \in \mathbb{R}^N.$$

(b) *For every* $x \in \overline{\Omega}$ *the map* $\xi \mapsto G(x,\xi)$ *is convex.*

(c) *We have*

$$(a(x,\xi),\xi)_{\mathbb{R}^N} \geq G(x,\xi) \geq \frac{c_0}{p(p-1)} |\xi|^p \quad \text{and} \quad G(x,\xi) \leq \frac{c_1}{p(p-1)} |\xi|^p$$

*for all* $x \in \overline{\Omega}$, *all* $\xi \in \mathbb{R}^N$, *with* $c_0, c_1$ *from* H(a)$_1$.

*Proof.* (a) The chain rule guarantees that $G(x,\cdot)$ is differentiable at every $\xi \neq 0$ and

$$G'_\xi(x,\xi) = \hat{a}(x,|\xi|)|\xi| \frac{\xi}{|\xi|} = a(x,\xi).$$

It follows from the relation $\hat{a}(x,t)t = |a(x,t\frac{\xi}{|\xi|})|$ and the estimate for $|a(x,\xi)|$ pointed out in Remark 12.13 that $G(x,\cdot)$ is also differentiable at 0 with $G'_\xi(x,0) = 0 = a(x,0)$. This proves (a).

(b) Note that $G(x,\xi) = G_0(x,|\xi|)$ with $G_0(x,s) = \int_0^s \hat{a}(x,t) t \, dt$ for all $s \geq 0$. It follows from the relation $\hat{a}(x,s)s = (a(x,s\xi),\xi)_{\mathbb{R}^N}$ for $|\xi| = 1$ and from the monotonicity of $a$ [Remark 12.13(b)] that $s \mapsto \hat{a}(x,s)s$ is nondecreasing, hence $G_0(x,\cdot)$ is convex. Therefore, $G(x,\cdot)$ is also convex.

(c) Using (a), we see that

$$G(x,\xi) = \int_0^1 (G'_\xi(x,t\xi),\xi)_{\mathbb{R}^N}\, dt = \int_0^1 (a(x,t\xi),\xi)_{\mathbb{R}^N}\, dt.$$

The claimed relations then follow from the estimates in Remark 12.13(a) and from the monotonicity of $a$ [Remark 12.13(b)].  $\square$

We denote $F(x,s) = \int_0^s f(x,t)\, dt$ for all $s \in \mathbb{R}$. The following proposition is a straightforward consequence of Lemma 12.16.

**Proposition 12.17.** *The functional* $\varphi : W^{1,p}(\Omega) \to \mathbb{R}$ *defined by*

$$\varphi(u) = \int_\Omega G(x,\nabla u)\, dx - \int_\Omega F(x,u)\, dx \text{ for all } u \in W^{1,p}(\Omega)$$

*is of class $C^1$, and we have*

$$\langle \varphi'(u),v \rangle = \int_\Omega (a(x,\nabla u),\nabla v)_{\mathbb{R}^N}\, dx - \int_\Omega f(x,u(x))v(x)\, dx \text{ for all } u,v \in W^{1,p}(\Omega).$$

*In particular, the critical points of $\varphi$ coincide with the weak solutions of (12.48), whereas the critical points of the restriction $\varphi_0 := \varphi|_{W_0^{1,p}(\Omega)}$ coincide with the weak solutions of (12.49).*

We conclude the theoretical part of this section with the following auxiliary result, which relates the local $C^1(\overline{\Omega})$- and $W^{1,p}(\Omega)$-minimizers of the functional $\varphi : W^{1,p}(\Omega) \to \mathbb{R}$ associated to problem (12.48). Actually, our statement addresses both the Neumann and the Dirichlet cases; in the latter it relates the local $C^1(\overline{\Omega})$- and $W^{1,p}(\Omega)$-minimizers of the functional $\varphi_0 : W_0^{1,p}(\Omega) \to \mathbb{R}$ associated to problem (12.49) (Proposition 12.17). We show the result simultaneously in both cases.

**Theorem 12.18.** *Assume that* H$(a)_1$ *and* H$(f)_1$ *hold. Let $(X,\psi)$ be any of the pairs $(W^{1,p}(\Omega),\varphi)$ or $(W_0^{1,p}(\Omega),\varphi_0)$, and let $u_0 \in X$. If $u_0$ is a local minimizer of $\psi$ with respect to the topology of $C^1(\overline{\Omega})$, i.e., there exists $\varepsilon > 0$ such that*

$$\psi(u_0) \leq \psi(u_0+h) \text{ for all } h \in X \cap C^1(\overline{\Omega}) \text{ with } \|h\|_{C^1(\overline{\Omega})} \leq \varepsilon,$$

*then $u_0$ is a local minimizer of $\psi$ with respect to the topology of $W^{1,p}(\Omega)$, i.e., there exists $\delta > 0$ such that*

$$\psi(u_0) \leq \psi(u_0+h) \text{ for all } h \in X \text{ with } \|\nabla h\|_p + \|h\|_p \leq \delta.$$

*Proof.* We start by pointing out a first consequence of the assumptions:

*Claim 1:* $u_0$ is a critical point of $\psi$ and $u_0 \in C^1(\overline{\Omega})$.

The assumption on $u_0$ clearly implies that $\langle \psi'(u_0), h \rangle = 0$ for all $h \in X \cap C^1(\overline{\Omega})$. Since $X \cap C^1(\overline{\Omega})$ is dense in $X$ (Definition 1.8 and Theorem 1.19), we deduce that $\psi'(u_0) = 0$, i.e., $u_0$ is a critical point of $\psi$. The fact that $u_0 \in C^1(\overline{\Omega})$ is then implied by Proposition 12.17 and Corollary 8.12. This proves Claim 1.

We prove the theorem by contradiction. Assume that $u_0$ is not a local $W^{1,p}(\Omega)$-minimizer of $\psi$. Then the continuity of the embedding $W^{1,p}(\Omega) \hookrightarrow L^r(\Omega)$ [with $r \in (p, p^*)$ as in H$(f)_1$] implies that for every $\delta > 0$ we have

$$m_\delta := \inf\{\psi(u_0 + h) : h \in X, \|h\|_r \leq \delta\} < \psi(u_0).$$

Note that $m_\delta > -\infty$ [see H$(f)_1$]. Moreover, Lemma 12.16(b), (c) and H$(f)_1$ easily ensure that $\psi(u_0 + \cdot)$ is sequentially weakly l.s.c. on $X$ and coercive on $\{h \in X : \|h\|_r \leq \delta\}$. Thus, we can find $h_\delta \in X$ such that

$$\|h_\delta\|_r \leq \delta \quad \text{and} \quad \psi(u_0 + h_\delta) = m_\delta < \psi(u_0). \tag{12.60}$$

Note that (12.60) and Lemma 12.16(c) yield $M_1 > 0$ such that

$$\|\nabla h_\delta\|_p + \|h_\delta\|_p \leq M_1 \quad \text{for all } \delta \in (0,1). \tag{12.61}$$

For the moment, we fix $\delta \in (0,1)$ and study $h_\delta$.

*Claim 2:* There is $\lambda_\delta \geq 0$ such that the equality

$$V(u_0 + h_\delta) = f(x, u_0 + h_\delta) - \lambda_\delta |h_\delta|^{r-2} h_\delta \tag{12.62}$$

holds in $X^*$.

Relation (12.60) implies that $h_\delta \neq 0$, so $\rho_\delta := \|h_\delta\|_r^r > 0$, and that

$$\psi(u_0 + h_\delta) = \inf\{\psi(u_0 + h) : h \in X, \|h\|_r^r = \rho_\delta\}.$$

Then the Lagrange multiplier rule yields $\lambda_\delta \in \mathbb{R}$ with

$$\psi'(u_0 + h_\delta) = -\lambda_\delta |h_\delta|^{r-2} h_\delta \quad \text{in } X^*, \tag{12.63}$$

which reads as (12.62) (Proposition 12.17). Finally, acting on (12.63) with the test function $h_\delta$ and invoking (12.60), we have

$$-\lambda_\delta \|h_\delta\|_r^r = \langle \psi'(u_0 + h_\delta), h_\delta \rangle = \lim_{t \downarrow 0} \frac{\psi(u_0 + h_\delta - t h_\delta) - \psi(u_0 + h_\delta)}{-t} \leq 0,$$

whence $\lambda_\delta \geq 0$. The proof of Claim 2 is complete.

*Claim 3:* We have $h_\delta \in L^\infty(\Omega)$, and there is $M_2 > 0$ independent of $\delta \in (0,1)$ such that $\|h_\delta\|_\infty \leq M_2$.

According to Claim 1, we have

$$V(u_0) = f(x, u_0) \text{ in } X^*. \tag{12.64}$$

Subtracting (12.64) from (12.62), we obtain

$$V_0(h_\delta) = f_0(x, h_\delta) - \lambda_\delta |h_\delta|^{p-2} h_\delta \text{ in } X^*, \tag{12.65}$$

where $V_0 : X \to X^*$ is defined by $\langle V_0(u), v \rangle = \int_\Omega (a_0(x, \nabla u), \nabla v)_{\mathbb{R}^N} \, dx$ for all $u, v \in X$, with $a_0(x, \xi) := a(x, \nabla u_0(x) + \xi) - a(x, \nabla u_0(x))$, and $f_0(x, s) := f(x, u_0(x) + s) - f(x, u_0(x))$. Evidently, $a_0(x, 0) = 0$ for all $x \in \overline{\Omega}$ and $f_0(x, 0) = 0$ for a.a. $x \in \Omega$. Moreover, by H$(f)_1$, Remark 12.13(a), (b), and the fact that $u_0 \in C^1(\overline{\Omega})$ (Claim 1), we see that

$$|f_0(x, s)| \leq c\big(2 + (\|u_0\|_\infty + |s|)^{r-1} + \|u_0\|_\infty^{r-1}\big) \leq \tilde{c}(1 + |s|^{r-1}) \tag{12.66}$$

for a.a. $x \in \Omega$, all $s \in \mathbb{R}$, for some $\tilde{c} > 0$ depending only on $c$, $r$, and $\|u_0\|_\infty$,

$$|a_0(x, \xi)| \leq \frac{c_1}{p-1}\big((\|\nabla u_0\|_\infty + |\xi|)^{p-1} + \|\nabla u_0\|_\infty^{p-1}\big)$$

$$\leq \tilde{c}_1(1 + |\xi|^{p-1}) \text{ for all } (x, \xi) \in \overline{\Omega} \times \mathbb{R}^N \tag{12.67}$$

for some $\tilde{c}_1 > 0$ depending only on $c_1$, $p$, and $\|\nabla u_0\|_\infty$, and

$$(a_0(x, \xi), \xi)_{\mathbb{R}^N} \geq \tilde{c}_0(|\nabla u_0(x) + \xi| + |\nabla u_0(x)|)^{p-2}|\xi|^2$$

$$\geq \tilde{c}_0(R + |\xi|)^{p-2}|\xi|^2 \text{ for all } (x, \xi) \in \overline{\Omega} \times \mathbb{R}^N, \tag{12.68}$$

with $\tilde{c}_0 > 0$ as in Remark 12.13(b) and where $R = 0$ (if $p \geq 2$) or $R = 2\|\nabla u_0\|_\infty$ (if $1 < p < 2$). Exploiting the fact that $\lambda_\delta \geq 0$ (Claim 2), from (12.65) we get

$$\int_\Omega (a_0(x, \nabla h_\delta), \nabla h)_{\mathbb{R}^N} \, dx \leq \int_\Omega f_0(x, h_\delta) h \, dx \tag{12.69}$$

whenever $h$ is of the form $h = \min\{h_\delta^+, \lambda\}^\alpha$ or $h = -\min\{h_\delta^-, \lambda\}^\alpha$ with $\lambda > 0$ and $\alpha \geq 1$. Applying Theorem 8.4 on the basis of (12.66)–(12.69) and taking (12.61) into account, we obtain that $h_\delta \in L^\infty(\Omega)$ and $\|h_\delta\|_\infty \leq M_2$ for some $M_2 > 0$ independent of $\delta \in (0,1)$. This establishes Claim 3.

*Claim 4:* There is $M_3 > 0$ independent of $\delta \in (0,1)$ such that $\lambda_\delta \|h_\delta\|_\infty^{r-1} \leq M_3$.

Recall that $h_\delta \neq 0$ [see (12.60)]. Set $\rho = \frac{1}{2}\|h_\delta\|_\infty > 0$. Note that we have $(h_\delta - \rho)^+ \neq 0$ or $(h_\delta + \rho)^- \neq 0$. Say $(h_\delta - \rho)^+ \neq 0$ (the argument is similar in the other situation). Acting on (12.65) with the test function $(h_\delta - \rho)^+ \in X$ and invoking (12.68) and (12.66), we get

$$0 \leq \int_{\{h_\delta > \rho\}} (a_0(x, \nabla h_\delta), \nabla h_\delta)_{\mathbb{R}^N} \, dx$$

$$= \int_{\{h_\delta > \rho\}} f_0(x, h_\delta(x))(h_\delta(x) - \rho) \, dx - \lambda_\delta \int_{\{h_\delta > \rho\}} h_\delta(x)^{r-1}(h_\delta(x) - \rho) \, dx$$

$$\leq (\tilde{c}(1 + \rho^{1-r}) - \lambda_\delta) \int_{\{h_\delta > \rho\}} h_\delta(x)^{r-1}(h_\delta(x) - \rho) \, dx,$$

whence $\lambda_\delta \leq \tilde{c}(1 + \rho^{1-r})$ (since the set $\{h_\delta > \rho\}$ is nonempty by assumption). Using Claim 3, we deduce

$$\lambda_\delta \|h_\delta\|_\infty^{r-1} \leq \tilde{c}(1 + 2^{r-1}\|h_\delta\|_\infty^{1-r})\|h_\delta\|_\infty^{r-1} \leq \tilde{c}(M_2^{r-1} + 2^{r-1}).$$

Thus, setting $M_3 = \tilde{c}(M_2^{r-1} + 2^{r-1})$, we have proven Claim 4.

According to Claims 1, 3, and 4, we find $M_4 > 0$ independent of $\delta \in (0,1)$ such that

$$-M_4 \leq \tilde{f}(x) := f(x, u_0(x) + h_\delta(x)) - \lambda_\delta |h_\delta(x)|^{r-2} h_\delta(x) \leq M_4$$

for a.a. $x \in \Omega$. Relation (12.62) reads as

$$V(u_0 + h_\delta) = \tilde{f}(x) \text{ in } X^*.$$

Claim 3 yields $\|u_0 + h_\delta\|_\infty \leq \|u_0\|_\infty + M_2$. Therefore, we can apply Theorem 8.10, which provides $\theta \in (0,1)$ and $M_5 > 0$, both independent of $\delta \in (0,1)$, such that $u_0 + h_\delta \in C^{1,\theta}(\overline{\Omega})$ and $\|u_0 + h_\delta\|_{C^{1,\theta}(\overline{\Omega})} \leq M_5$ for all $\delta \in (0,1)$. Due to the compactness of the embedding $C^{1,\theta}(\overline{\Omega}) \hookrightarrow C^1(\overline{\Omega})$ and to the first part of (12.60), we can find a sequence $\{\delta_n\}_{n \geq 1} \subset (0,1)$, with $\delta_n \to 0$ as $n \to \infty$ such that

$$\lim_{n \to \infty} \|h_{\delta_n}\|_{C^1(\overline{\Omega})} = 0.$$

This, combined with the second part of (12.60), contradicts the assumption that $u_0$ is a local $C^1(\overline{\Omega})$-minimizer of $\psi$. The proof of the theorem is complete. □

*Remark 12.19.* Let the pair $(X, \psi)$ be as in Theorem 12.18

(a) Claim 1 of the foregoing proof points out the noticeable fact that $u_0$ is a critical point of $\psi|_{X \cap C^1(\overline{\Omega})}$ if and only if $u_0$ is a critical point of $\psi$.

(b) Another property relating $C^1(\overline{\Omega})$- and $W^{1,p}(\Omega)$-topologies is that $u_0$ is an isolated critical point of $\psi|_{X \cap C^1(\overline{\Omega})}$ if and only if $u_0$ is an isolated critical point of $\psi$. Indeed, if there is a sequence $\{u_n\}_{n \geq 1} \subset W^{1,p}(\Omega)$ of critical points of $\psi$ converging to $u_0$ in $W^{1,p}(\Omega)$, then, by Theorems 8.4 and 8.10, we see that $\{u_n\}_{n \geq 1}$ is bounded in $C^{1,\alpha}(\overline{\Omega})$ for some $\alpha \in (0,1)$, so along a subsequence we have that $\{u_n\}_{n \geq 1}$ converges to $u_0$ in $C^1(\overline{\Omega})$, i.e., $u_0$ is not an isolated critical point of $\psi|_{X \cap C^1(\overline{\Omega})}$. The inverse implication is easy.

(c) In the case where $p = 2$ and $a(x, \xi) = |\xi|$ (corresponding to the Laplacian), Theorem 12.18 has the following Morse theoretical interpretation, due to Chang [79] and Liu and Wu [236]. Assume that $u_0$ is an isolated critical point of $\psi$, and so an isolated critical point of $\psi|_{X \cap C^1(\overline{\Omega})}$. Then we have the following equality of critical groups

$$C_k(\psi, u_0) = C_k(\psi|_{X \cap C^1(\overline{\Omega})}, u_0) \quad \text{for all } k \geq 1.$$

Invoking Proposition 6.95 [and under the assumption that $\psi(u_0)$ is isolated in the set of critical values of $\psi$], this equality yields that $u_0$ is a local minimizer of $\psi$ if and only if it is a local minimizer of $\psi|_{X \cap C^1(\overline{\Omega})}$. Actually the proof in Liu and Wu [236, Appendix] is formulated in the Dirichlet case [i.e., $X = H_0^1(\Omega)$], but the arguments also work in the Neumann case [i.e., $X = H^1(\Omega)$].

**Multiplicity Result**

We now apply the theoretical tools described in the first part of the section to establish a multiplicity result for problem (12.48). We assume that $a$ satisfies hypotheses H$(a)_1$, and we consider the following assumptions on the nonlinearity $f$ and its primitive $F(x, s) = \int_0^s f(x, t)\, dt$. Since the result will focus on nonnegative solutions of (12.48), the assumptions only concern the positive half-line (see, however, Remark 12.22).

H$(f)_2$ (i) The map $f : \Omega \times \mathbb{R} \to \mathbb{R}$ is a Carathéodory function with $f(x, 0) = 0$ a.e. in $\Omega$, and there are $r \in (p, p^*)$ and $c > 0$ such that

$$|f(x, s)| \leq c(1 + s^{r-1}) \quad \text{for a.a. } x \in \Omega, \text{ all } s \in [0, +\infty);$$

(ii) $\lim\limits_{s \to +\infty} \dfrac{F(x, s)}{s^p} = 0$ and $\lim\limits_{s \to +\infty} F(x, s) = -\infty$ uniformly for a.a. $x \in \Omega$;

(iii) There is a constant $c_+ > 0$ such that $\displaystyle\int_\Omega F(x, c_+) > 0$;

(iv) There is $\delta > 0$ such that $f(x, s) \leq 0$ for a.a. $x \in \Omega$, all $s \in [0, \delta]$.

*Example 12.20.* The following function $f$ satisfies hypotheses H$(f)_2$ (for simplicity we drop the $x$-dependence):

$$f(s) = \begin{cases} 0 & \text{if } s \leq 0, \\ -s^{\tau-1} + 2s^{\gamma-1} & \text{if } s \in (0, 1], \\ 2s^{\theta-1} - s^{q-1} & \text{if } s > 1, \end{cases}$$

where $1 < \tau < \gamma < 2\tau < +\infty$ and $1 < \theta < q < p$.

**Theorem 12.21.** *Assume that H$(a)_1$ and H$(f)_2$ hold. Then problem (12.48) admits at least two nontrivial solutions $u_0, v_0 \in C^1(\overline{\Omega})$ with $0 \leq v_0 \leq u_0$ in $\Omega$.*

*Proof.* We deal with the truncated functional $\varphi_+ \in C^1(W^{1,p}(\Omega), \mathbb{R})$ defined by

$$\varphi_+(u) = \int_\Omega G(x, \nabla u)\, dx + \frac{1}{p}\|u^-\|_p^p - \int_\Omega F(x, u^+)\, dx \quad \text{for all } u \in W^{1,p}(\Omega).$$

The proof splits into several steps.

*Step 1:* $\varphi_+$ is sequentially weakly l.s.c., bounded below, and coercive.

The sequential weak lower semicontinuity of $\varphi_+$ is a consequence of $\mathrm{H}(f)_2$ (i) and of the fact that $G(x, \cdot)$ is continuous and convex (Lemma 12.16).

We prove the coercivity of $\varphi_+$. Arguing by contradiction, assume that we can find a sequence $\{u_n\}_{n \geq 1} \subset W^{1,p}(\Omega)$ and a constant $M_1 > 0$ such that

$$\|u_n\| \to +\infty \text{ as } n \to \infty \text{ and } \varphi_+(u_n) \leq M_1 \text{ for all } n \geq 1. \tag{12.70}$$

Using Lemma 12.16(c), the second relation in (12.70) implies

$$\frac{c_0}{p(p-1)}\|\nabla u_n\|_p^p + \frac{1}{p}\|u_n^-\|_p^p - \int_\Omega F(x, u_n^+)\, dx \leq M_1 \text{ for all } n \geq 1. \tag{12.71}$$

From the first part of (12.70), (12.71), and the growth condition in $\mathrm{H}(f)_2$ (i), it follows that $\|u_n^+\| \to +\infty$ as $n \to \infty$. Thus, there is $y \in W^{1,p}(\Omega)$ such that, along a relabeled subsequence, we have

$$y_n := \frac{u_n^+}{\|u_n^+\|} \xrightarrow{\text{w}} y \text{ in } W^{1,p}(\Omega) \text{ and } y_n \to y \text{ in } L^\theta(\Omega) \text{ for each } \theta \in (1, p^*).$$

Arguing exactly as in Step 2 of the proof of Theorem 10.9, we can see that $y \equiv (\frac{1}{|\Omega|_N})^{\frac{1}{p}}$, whence, up to considering a subsequence, we may assume that $u_n^+(x) \to +\infty$ for a.a. $x \in \Omega$. Combining this with $\mathrm{H}(f)_2$ (i), (ii), (12.71), and Fatou's lemma, we get

$$M_1 \geq \varphi_+(u_n) \geq -\int_\Omega F(x, u_n^+)\, dx \to +\infty \text{ as } n \to \infty,$$

a contradiction. Thus, $\varphi_+$ is coercive.

The coercivity of $\varphi_+$, together with the assumption that $(x, s) \mapsto F(x, s^+)$ is bounded on bounded sets [see $\mathrm{H}(f)_2$ (i)], implies that $\varphi_+$ is bounded below. Step 1 is complete.

*Step 2:* $\varphi_+$ admits a global minimizer $u_0 \in W^{1,p}(\Omega)$. Moreover, we have $u_0 \in C^1(\overline{\Omega})$, $u_0 \geq 0$ in $\Omega$, $u_0 \neq 0$, and $u_0$ is a solution of problem (12.48).

In view of Step 1, there exists $u_0 \in W^{1,p}(\Omega)$, which is a global minimizer of $\varphi_+$. By $\mathrm{H}(f)_2$ (iii), we have

$$\varphi_+(u_0) \le \varphi_+(c_+) = -\int_\Omega F(x,c_+)\,dx < 0 = \varphi_+(0).$$

This ensures that $u_0 \ne 0$. The fact that $u_0$ is a critical point of $\varphi_+$ yields the relation

$$V(u_0) - (u_0^-)^{p-1} = f(x,u_0^+) \quad \text{in } W^{1,p}(\Omega)^*. \tag{12.72}$$

Acting on (12.72) with the test function $-u_0^-$, we obtain

$$\|u_0^-\|_p^p \le \int_{\{u_0<0\}} (a(x,\nabla u_0), \nabla u_0)_{\mathbb{R}^N}\,dx + \|u_0^-\|_p^p = 0,$$

hence $u_0^- = 0$, so $u_0 \ge 0$ a.e. in $\Omega$. Thus (12.72) reads as

$$V(u_0) = f(x,u_0) \quad \text{in } W^{1,p}(\Omega)^*,$$

i.e., $u_0$ is a weak solution of problem (12.48). Corollary 8.12 yields $u_0 \in C^1(\overline{\Omega})$. Therefore, $u_0$ satisfies all the claimed properties. This completes Step 2.

To look for an additional nonnegative solution to problem (12.48), we define a new truncated functional $\hat{\varphi}_+ \in C^1(W^{1,p}(\Omega), \mathbb{R})$ by

$$\hat{\varphi}_+(u) = \int_\Omega G(x,\nabla u)\,dx + \frac{1}{p}\|u^-\|_p^p + \frac{1}{p}\|(u-u_0)^+\|_p^p - \int_\Omega \hat{F}_+(x,u(x))\,dx$$

for all $u \in W^{1,p}(\Omega)$, with $\hat{F}_+(x,s) = \int_0^s f(x,\hat{\tau}_+(x,t))\,dt$, where we set

$$\hat{\tau}_+(x,s) = \begin{cases} 0 & \text{if } s \le 0, \\ s & \text{if } 0 < s < u_0(x), \\ u_0(x) & \text{if } s \ge u_0(x) \end{cases}$$

for all $(x,s) \in \Omega \times \mathbb{R}$.

*Step 3:* If $u$ is a critical point of $\hat{\varphi}_+$, then $u$ is a solution of (12.48), $u \in C^1(\overline{\Omega})$, and we have $0 \le u(x) \le u_0(x)$ for all $x \in \Omega$.

The assumption yields the following equality in $W^{1,p}(\Omega)^*$:

$$V(u) - (u^-)^{p-1} + ((u-u_0)^+)^{p-1} = f(x,\hat{\tau}_+(x,u(x))). \tag{12.73}$$

Acting on (12.73) with the test function $-u^-$, we obtain (as in Step 2) $u^- = 0$ a.e. in $\Omega$. On the one hand, acting on (12.73) with the test function $(u-u_0)^+$, we have

$$\int_\Omega (a(x,\nabla u), \nabla(u-u_0)^+)_{\mathbb{R}^N}\,dx + \|(u-u_0)^+\|_p^p = \int_\Omega f(x,u_0(x))(u-u_0)^+\,dx. \tag{12.74}$$

On the other hand, the fact that $u_0 \geq 0$ is a critical point of $\varphi_+$ yields

$$-\int_\Omega (a(x, \nabla u_0), \nabla(u - u_0)^+)_{\mathbb{R}^N}\, dx = -\int_\Omega f(x, u_0(x))(u - u_0)^+\, dx. \qquad (12.75)$$

Adding (12.74) and (12.75) and invoking the monotonicity of $a$ [Remark 12.13(b)], we obtain

$$\|(u - u_0)^+\|_p^p \leq \int_{\{u > u_0\}} (a(x, \nabla u) - a(x, \nabla u_0), \nabla(u - u_0))_{\mathbb{R}^N}\, dx + \|(u - u_0)^+\|_p^p = 0,$$

whence $(u - u_0)^+ = 0$. We therefore obtain that $0 \leq u \leq u_0$ a.e. in $\Omega$. In view of this relation, (12.73) becomes

$$V(u) = f(x, u(x)) \quad \text{in } W^{1,p}(\Omega)^*,$$

so $u$ is a solution of (12.48). By Corollary 8.12, we have $u \in C^1(\overline{\Omega})$, and the relation $0 \leq u \leq u_0$ holds everywhere in $\Omega$. This completes Step 3.

*Step 4:* The functional $\hat{\varphi}_+$ satisfies the (PS)-condition.
   We first check that $\hat{\varphi}_+$ is coercive. Since $u_0 \in C^1(\overline{\Omega})$ and using H$(f)_2$ (i), we find a constant $\hat{c} > 0$ such that

$$\left| \int_\Omega \hat{F}_+(x, u)\, dx \right| \leq \hat{c}\|u\|_p \quad \text{for all } u \in W^{1,p}(\Omega). \qquad (12.76)$$

Moreover, for every $u \in W^{1,p}(\Omega)$ we see that

$$\|(u - u_0)^+\|_p^p = \|(u^+ - u_0)^+\|_p^p = \|u^+ - u_0\|_p^p - \int_{\{u^+ \leq u_0\}} (u_0 - u^+)^p\, dx$$

$$\geq \frac{1}{2^{p-1}}\|u^+\|_p^p - \|u_0\|_p^p - \int_{\{u^+ \leq u_0\}} u_0(x)^p\, dx$$

$$\geq \frac{1}{2^{p-1}}\|u^+\|_p^p - 2\|u_0\|_p^p. \qquad (12.77)$$

From Lemma 12.16(c), (12.76), and (12.77) we derive

$$\hat{\varphi}_+(u) \geq \frac{c_0}{p(p-1)}\|\nabla u\|_p^p + \frac{1}{p}\|u^-\|_p^p + \frac{1}{p2^{p-1}}\|u^+\|_p^p - \frac{2}{p}\|u_0\|_p^p - \hat{c}\|u\|_p$$

for all $u \in W^{1,p}(\Omega)$. We easily deduce that $\hat{\varphi}_+$ is coercive.
   Now let $\{u_n\}_{n \geq 1} \subset W^{1,p}(\Omega)$ be a sequence such that

$$\{\hat{\varphi}_+(u_n)\}_{n \geq 1} \text{ is bounded and } \hat{\varphi}_+'(u_n) \to 0 \text{ in } W^{1,p}(\Omega)^* \text{ as } n \to \infty. \qquad (12.78)$$

The first part of (12.78) and the coercivity of $\hat{\varphi}_+$ imply that $\{u_n\}_{n\geq 1}$ is bounded in $W^{1,p}(\Omega)$. Hence, there is $u \in W^{1,p}(\Omega)$ such that, along a relabeled subsequence $\{u_n\}_{n\geq 1}$, we have

$$u_n \xrightarrow{w} u \text{ in } W^{1,p}(\Omega) \text{ and } u_n \to u \text{ in } L^{\theta}(\Omega) \text{ for each } \theta \in [1,p^*). \tag{12.79}$$

The second part of (12.78) yields

$$\int_{\Omega} (a(x,\nabla u_n), u_n - u)_{\mathbb{R}^N}\, dx - \int_{\Omega} \hat{g}_+(x, u_n(x))(u_n - u)\, dx \to 0 \text{ as } n \to \infty,$$

where $\hat{g}_+(x,s) = (s^-)^{p-1} - ((s - u_0(x))^+)^{p-1} + f(x, \hat{\tau}_+(x,s))$. The growth condition in $\mathrm{H}(f)_2$ (i) and the second part of (12.79) ensure that

$$\int_{\Omega} \hat{g}_+(x, u_n(x))(u_n(x) - u(x))\, dx \to 0 \text{ as } n \to \infty,$$

whence $\lim_{n\to\infty} \langle V(u_n), u_n - u \rangle = 0$. Since $V$ is an $(S)_+$-map (by Proposition 12.14), we conclude that $u_n \to u$ in $W^{1,p}(\Omega)$ as $n \to \infty$. This shows that $\hat{\varphi}_+$ satisfies the (PS)-condition.

*Step 5: $\hat{\varphi}_+$ admits a critical point $v_0 \in W^{1,p}(\Omega)$ different from $0, u_0$.*

Note that

$$\hat{\varphi}_+(u_0) = \varphi_+(u_0) < \varphi_+(0) = \hat{\varphi}_+(0) = 0. \tag{12.80}$$

We claim that

$$0 \text{ is a local minimizer of } \hat{\varphi}_+. \tag{12.81}$$

Using $\mathrm{H}(f)_2$ (iv) and the fact that $0 \leq \hat{\tau}_+(x,t) \leq |t|$ for a.a. $x \in \Omega$, all $t \in \mathbb{R}$, we have

$$\hat{F}_+(x,s) := \int_0^s f(x, \hat{\tau}_+(x,t))\, dt \leq 0 \text{ for a.a. } x \in \Omega, \text{ all } s \in [-\delta, \delta]. \tag{12.82}$$

Let $u \in C^1(\overline{\Omega})$, with $\|u\|_{C^1(\overline{\Omega})} \leq \delta$. Then (12.82) implies that $\hat{F}_+(x, u(x)) \leq 0$ for a.a. $x \in \Omega$, whence

$$\hat{\varphi}_+(u) \geq -\int_{\Omega} \hat{F}_+(x, u(x))\, dx \geq 0.$$

This shows that $0$ is a local minimizer of $\hat{\varphi}_+$ for the topology of $C^1(\overline{\Omega})$. Applying Theorem 12.18, we infer that $0$ is a local minimizer of $\hat{\varphi}_+$ for the topology of $W^{1,p}(\Omega)$. Hence, we obtain (12.81).

Note that we may assume that $0$ is a strict local minimizer of $\hat{\varphi}_+$, because otherwise, any neighborhood of $0$ in $W^{1,p}(\Omega)$ contains another critical point of $\hat{\varphi}_+$, and we are done. This fact, together with (12.80) and Step 4, permits the use of Proposition 5.42, which yields a critical point $v_0$ of $\hat{\varphi}_+$ different from $0$ and $u_0$. This completes Step 5.

The theorem now follows by combining Steps 2, 3, and 5.          □

*Remark 12.22.* If we assume, in addition, the counterparts of $H(f)_2$ (i)–(iv) on the negative half-line $(-\infty, 0]$, then the same reasoning as in the foregoing proof shows the existence of at least four nontrivial smooth solutions $u_0, v_0, y_0, z_0 \in C^1(\overline{\Omega})$ of the problem satisfying $y_0 \leq z_0 \leq 0 \leq v_0 \leq u_0$ in $\Omega$.


## 12.3   Sublinear and Superlinear Neumann Problems

In this section we focus on semilinear Neumann problems and provide a framework that permits a unified treatment of both superlinear and sublinear equations. We prove multiplicity results, providing sign information for the solutions.

Let $\Omega \subset \mathbb{R}^N$ $(N \geq 1)$ be a bounded domain with a $C^2$-boundary $\partial\Omega$. We deal with the Neumann problem

$$\begin{cases} -\Delta u = f(x, u(x)) & \text{in } \Omega, \\ \dfrac{\partial u}{\partial n} = 0 & \text{on } \partial\Omega, \end{cases} \tag{12.83}$$

driven by the Laplacian and involving a measurable nonlinearity $f(x, s)$ that will be assumed to be of class $C^1$ in the second variable. In the analysis of problem (12.83), we use the Sobolev space $H^1(\Omega)$, and the solutions will belong to the Banach space $C^1(\overline{\Omega})$. Recall that the negative Neumann Laplacian $u \mapsto -\Delta u$ can be identified with the linear operator $A : H^1(\Omega) \to H^1(\Omega)^*$ given by $\langle A(u), v \rangle = \int_\Omega (\nabla u, \nabla v)_{\mathbb{R}^N} \, dx$ [see (2.28)]. We consider the notion of (weak) solution of problem (12.83) given in Definition 8.2.


**Truncation Techniques and Preliminary Facts**

In this preliminary part, we explain how the truncation techniques described in Sects. 11.1 and 11.2 in the case of Dirichlet problems can be transposed here for Neumann problems. In what follows, we assume that the Carathéodory function $f : \Omega \times \mathbb{R} \to \mathbb{R}$ involved in problem (12.83) is subject to the following growth condition: there are constants $c > 0$ and $r \in (2, 2^*)$ such that

$$|f(x, s)| \leq c(1 + |s|^{r-1}) \quad \text{for a.a. } x \in \Omega, \text{ all } s \in \mathbb{R}. \tag{12.84}$$

Moreover, we suppose the following asymptotic behavior of $f$ near the origin: there is $\tilde{c} \in \mathbb{R}$ such that

$$\liminf_{s \to 0} \frac{f(x,s)}{s} \geq \tilde{c} \text{ uniformly for a.a. } x \in \Omega. \tag{12.85}$$

The sets of hypotheses $H(f)_1$, $H(f)_2$, and $H(f)_3$ that we will consider in this section will always imply (12.84) and (12.85). Relations (12.84) and (12.85) guarantee the following regularity properties of the (weak) solutions of (12.83):

**Proposition 12.23.** *Every solution* $u \in H^1(\Omega)$ *of problem (12.83) belongs to* $C^1(\overline{\Omega})$ *and satisfies* $\frac{\partial u}{\partial n}(x) := (\nabla u(x), n(x))_{\mathbb{R}^N} = 0$ *on* $\partial\Omega$ *[where $n(\cdot)$ stands for the outward unit normal]. Moreover:*

(a) *If* $u \geq 0$ *in* $\Omega$, $u \neq 0$, *then* $u > 0$ *on* $\overline{\Omega}$;
(b) *If* $u \leq 0$ *in* $\Omega$, $u \neq 0$, *then* $u < 0$ *on* $\overline{\Omega}$.

*Proof.* The fact that $u \in C^1(\overline{\Omega})$ is implied by Corollary 8.13. Then the equality $\frac{\partial u}{\partial n} = 0$ on $\partial\Omega$ is obtained by comparing Remarks 1.40 and 8.3(b). To prove part (a), suppose that $u \geq 0$ in $\Omega$, $u \neq 0$. Relations (12.84) and (12.85) imply that we can find $\hat{c} \in \mathbb{R}$ such that $f(x,s) \geq \hat{c}s$ for a.a. $x \in \Omega$, all $s \in [0, \|u\|_\infty]$, whence $-\Delta u \geq \hat{c}u$ in $H^1(\Omega)^*$. In view of this relation and the equality $\frac{\partial u}{\partial n} = 0$ on $\partial\Omega$, by the strong maximum principle (Theorem 8.27), we obtain that $u > 0$ on $\overline{\Omega}$. Part (b) can be proved similarly. □

We now outline the lower and upper solutions method for problem (12.83).

**Definition 12.24.** A *lower* (resp. *upper*) *solution* of problem (12.83) is a function $u \in H^1(\Omega)$ such that

$$\int_\Omega (\nabla u, \nabla v)_{\mathbb{R}^N} dx - \int_\Omega f(x, u(x)) v(x) dx \text{ is } \leq 0 \text{ (resp. } \geq 0\text{)}$$

for all $v \in H^1(\Omega)$ satisfying $v \geq 0$ a.e. in $\Omega$.

We adapt the truncation techniques from Sect. 11.1. Given measurable functions $\underline{u}: \Omega \to \mathbb{R} \cup \{-\infty\}$ and $\overline{u}: \Omega \to \mathbb{R} \cup \{+\infty\}$ such that $\underline{u} \leq \overline{u}$ a.e. in $\Omega$, we define the order interval

$$[\underline{u}, \overline{u}] = \{u \in H^1(\Omega): \underline{u}(x) \leq u(x) \leq \overline{u}(x) \text{ for a.a. } x \in \Omega\},$$

the truncated Carathéodory function

$$\hat{f}_{[\underline{u},\overline{u}]}(x,s) = \begin{cases} f(x, \underline{u}(x)) + \underline{u}(x) & \text{if } s \leq \underline{u}(x), \\ f(x,s) + s & \text{if } \underline{u}(x) < s < \overline{u}(x), \\ f(x, \overline{u}(x)) + \overline{u}(x) & \text{if } s \geq \overline{u}(x), \end{cases} \tag{12.86}$$

its primitive $\hat{F}_{[\underline{u},\overline{u}]}(x,s) = \int_0^s \hat{f}_{[\underline{u},\overline{u}]}(x,t)\,dt$, and the functional $\hat{\varphi}_{[\underline{u},\overline{u}]} \in C^1(H^1(\Omega),\mathbb{R})$ given by

$$\hat{\varphi}_{[\underline{u},\overline{u}]}(u) = \frac{1}{2}\|\nabla u\|_2^2 + \frac{1}{2}\|u\|_2^2 - \int_\Omega \hat{F}_{[\underline{u},\overline{u}]}(x,u(x))\,dx \tag{12.87}$$

for all $u \in H^1(\Omega)$. We will rely on the following lower and upper solutions principle.

**Proposition 12.25.**

(a) *Let $\underline{u}$ be either a lower solution of (12.83) or $-\infty$, and let $\overline{u}$ be either an upper solution of (12.83) or $+\infty$. Assume that $\underline{u} \leq \overline{u}$ a.e. in $\Omega$. Then any critical point of $\hat{\varphi}_{[\underline{u},\overline{u}]}$ is a solution of (12.83) belonging to $C^1(\overline{\Omega}) \cap [\underline{u},\overline{u}]$.*
(b) *Assume that $\underline{u},\overline{u} \in H^1(\Omega)$ are respectively a lower and an upper solution of (12.83) such that $\underline{u} \leq \overline{u}$ a.e. in $\Omega$. Then the functional $\hat{\varphi}_{[\underline{u},\overline{u}]}$ is coercive, sequentially weakly l.s.c., and bounded below and satisfies the (PS)-condition. Moreover, there is a solution $u \in C^1(\overline{\Omega}) \cap [\underline{u},\overline{u}]$ of (12.83) obtained as a global minimizer of $\hat{\varphi}_{[\underline{u},\overline{u}]}$.*

*Proof.* (a) A critical point $u \in H^1(\Omega)$ of $\hat{\varphi}_{[\underline{u},\overline{u}]}$ satisfies the relation

$$A(u) = -u + \hat{f}_{[\underline{u},\overline{u}]}(x,u) \text{ in } H^1(\Omega)^*. \tag{12.88}$$

From regularity theory (Corollary 8.13) we get $u \in C^1(\overline{\Omega})$. Let us check that $u \geq \underline{u}$ a.e. in $\Omega$. This property is clear in the case where $\underline{u} \equiv -\infty$, so we may assume that $\underline{u}$ is a lower solution of (12.83). Thus,

$$\langle A(\underline{u}), (\underline{u}-u)^+\rangle \leq \int_\Omega f(x,\underline{u})(\underline{u}-u)^+\,dx. \tag{12.89}$$

Moreover, acting on (12.88) with the test function $v = (\underline{u}-u)^+$, we have

$$\langle A(u), (\underline{u}-u)^+\rangle = \int_\Omega (-u+\underline{u}+f(x,\underline{u}))(\underline{u}-u)^+\,dx. \tag{12.90}$$

Subtracting (12.90) from (12.89), we deduce

$$\int_{\{\underline{u}>u\}} \|\nabla(\underline{u}-u)\|_2^2\,dx + \int_{\{\underline{u}>u\}} \|\underline{u}-u\|_2^2\,dx \leq 0,$$

which clearly implies that $u \geq \underline{u}$ a.e. in $\Omega$. Similarly, we can show that $u \leq \overline{u}$ a.e. in $\Omega$, thus $u \in [\underline{u},\overline{u}]$. In particular, this yields $\hat{f}_{[\underline{u},\overline{u}]}(x,u) = f(x,u)+u$, so in view of (12.88), we obtain that $u$ is a solution of (12.83).

(b) Relations (12.84) and (12.86) imply that for a.a. $x \in \Omega$, all $s \in \mathbb{R}$, we have

$$|\hat{f}_{[\underline{u},\overline{u}]}(x,s)| \leq \hat{k}(x) := c(1 + |\underline{u}(x)|^{r-1} + |\overline{u}(x)|^{r-1}) + |\underline{u}(x)| + |\overline{u}(x)|,$$

with $\hat{k} \in L^{r'}(\Omega)$. Estimating $\hat{\varphi}_{[\underline{u},\overline{u}]}(u)$ using this relation, we easily get that $\hat{\varphi}_{[\underline{u},\overline{u}]}$ is coercive, sequentially weakly l.s.c., and bounded below. These properties imply that there exists $u \in H^1(\Omega)$, which is a global minimizer of $\hat{\varphi}_{[\underline{u},\overline{u}]}$, and so, by part (a), $u$ is also a solution of (12.83) belonging to $C^1(\overline{\Omega}) \cap [\underline{u},\overline{u}]$. Finally, the fact that $\hat{\varphi}_{[\underline{u},\overline{u}]}$ satisfies the (PS)-condition can be deduced by using that $A$ is an $(S)_+$-map (Proposition 2.72). $\qquad\square$

*Remark 12.26.* The presence of the term $\frac{1}{2}\|u\|_2^2$ in $\hat{\varphi}_{[\underline{u},\overline{u}]}$ makes a slight difference with the definition of the truncated functional $\varphi_{[\underline{u},\overline{u}]}$ relative to the Dirichlet case [see (11.5)]. Since $u \mapsto \|\nabla u\|_2$ is not a norm on $H^1(\Omega)$ [whereas it is on $H_0^1(\Omega)$], this term is needed to guarantee the coercivity of $\hat{\varphi}_{[\underline{u},\overline{u}]}$ in Proposition 12.25(b).

We also mention the following property of lower and upper solutions, which can be obtained by reasoning as in Lemma 11.19.

**Lemma 12.27.**

(a) *If $\overline{u}_1, \overline{u}_2 \in H^1(\Omega)$ are upper solutions of problem (12.83), then so is $\overline{u} := \min\{\overline{u}_1, \overline{u}_2\}$.*
(b) *If $\underline{v}_1, \underline{v}_2 \in H^1(\Omega)$ are lower solutions of (12.83), then so is $\underline{v} := \max\{\underline{v}_1, \underline{v}_2\}$.*

**Two Extremal Constant-Sign and Two Nodal Solutions**

Let

$$\lambda_0 = 0 < \lambda_1 \leq \lambda_2 \leq \cdots \leq \lambda_m \leq \cdots, \quad \lim_{m\to\infty} \lambda_m = +\infty$$

be the nondecreasing sequence of eigenvalues of the negative Neumann Laplacian, repeated according to their (finite) multiplicities (Sect. 9.1). We state the following hypotheses on the nonlinearity $f$, under which we show the existence of two opposite constant-sign solutions and two nodal solutions for problem (12.83). Note that we do not ask for the moment for hypotheses at $\pm\infty$. Later we will add hypotheses on $f$ at $\pm\infty$ that distinguish between the superlinear and the sublinear cases, and under which we will show the existence of further constant-sign solutions.

$H(f)_1$ (i) $f : \Omega \times \mathbb{R} \to \mathbb{R}$ is a function such that $x \mapsto f(x,s)$ is measurable for all $s \in \mathbb{R}$, $s \mapsto f(x,s)$ is of class $C^1$ for a.a. $x \in \Omega$, $f(x,0) = 0$ a.e. in $\Omega$, and there exist $c > 0$ and $r \in (2, 2^*)$ such that

$$|f_s'(x,s)| \leq c(1 + |s|^{r-2}) \text{ for a.a. } x \in \Omega, \text{ all } s \in \mathbb{R};$$

(ii) There is an integer $m \geq 1$ such that

$$(f_s'(x,0) =) \lim_{s \to 0} \frac{f(x,s)}{s} \in [\lambda_m, \lambda_{m+1}] \text{ uniformly for a.a. } x \in \Omega$$

and $f_s'(\cdot,0) \not\equiv \lambda_m$, $f_s'(\cdot,0) \not\equiv \lambda_{m+1}$;
(iii) There exist real numbers $a_- < 0 < a_+$ such that

$$f(x,a_+) \leq 0 \leq f(x,a_-) \text{ for a.a. } x \in \Omega.$$

We denote $F(x,s) = \int_0^s f(x,t)\,dt$. Hypothesis H$(f)_1$ (i) implies that the energy functional $\varphi : H^1(\Omega) \to \mathbb{R}$ for problem (12.83) given by

$$\varphi(u) = \frac{1}{2} \|\nabla u\|_2^2 - \int_\Omega F(x,u(x))\,dx \text{ for all } u \in H^1(\Omega)$$

is well defined and of class $C^2$. The weak solutions of (12.83) coincide with the critical points of $\varphi$. Our first objective is to establish the existence of extremal constant-sign solutions of problem (12.83).

**Proposition 12.28.** *Assume that* H$(f)_1$ *holds. Then problem (12.83) admits a smallest positive solution* $u_+ \in C^1(\overline{\Omega})$ *and a biggest negative solution* $v_- \in C^1(\overline{\Omega})$. *Moreover, we have*

$$a_- \leq v_-(x) < 0 < u_+(x) \leq a_+ \text{ for all } x \in \overline{\Omega}.$$

*Proof.* We only show the existence of a smallest positive solution, the case of the biggest negative solution being similar. The proof splits into several steps.

*Step 1:* The function $\overline{u} := a_+$ is an upper solution of (12.83), and there is $\delta^* > 0$ such that for all $\varepsilon \in (0, \delta^*]$ the function $\underline{u}_\varepsilon := \varepsilon$ is a lower solution of (12.83).

The fact that $\overline{u} \equiv a_+$ is an upper solution of (12.83) is a consequence of H$(f)_1$ (iii). Hypothesis H$(f)_1$ (ii) allows us to find $\delta^* > 0$ such that $f(x,s) \geq 0$ for a.a. $x \in \Omega$, all $s \in [0, \delta^*]$, thus $\underline{u}_\varepsilon \equiv \varepsilon$ is a lower solution of (12.83) whenever $\varepsilon \in (0, \delta^*]$.

*Step 2:* For all $\varepsilon \in (0, \delta^*]$, there is a smallest solution $u_\varepsilon^*$ of (12.83) in the order interval $[\underline{u}_\varepsilon, \overline{u}] \subset H^1(\Omega)$.

This property can be obtained by arguing as in the proof of Proposition 11.20, on the basis of Lemma 12.27.

Now let $\{\varepsilon_n\}_{n \geq 1} \subset (0, \delta^*]$ be a decreasing sequence such that $\varepsilon_n \to 0$ as $n \to \infty$. According to Step 2, for all $n \geq 1$ there is $u_n^* \in C^1(\overline{\Omega})$ the smallest solution of (12.83) in the order interval $I_n^+ := [\underline{u}_{\varepsilon_n}, \overline{u}] \subset H^1(\Omega)$. In particular,

$$A(u_n^*) = f(x, u_n^*) \text{ in } H^1(\Omega)^* \text{ for all } n \geq 1. \tag{12.91}$$

From (12.91) and the fact that $u_n^* \in I_+^+$ it follows that $\{u_n^*\}_{n \geq 1}$ is bounded in $H^1(\Omega)$. Thus, we may assume that

$$u_n^* \xrightarrow{w} u_+ \text{ in } H^1(\Omega) \text{ and } u_n^* \to u_+ \text{ in } L^2(\Omega) \text{ as } n \to \infty \qquad (12.92)$$

for some $u_+ \in H^1(\Omega)$. Passing to the limit in (12.91) and using (12.92), we get that $u_+$ is a solution of (12.83) and $u_+ \in C^1(\overline{\Omega}) \cap [0, \overline{u}]$ (Proposition 12.23).

*Step 3:* $u_+ \neq 0$.

Arguing by contradiction, suppose that $u_+ = 0$. We directly deduce from (12.91) and (12.92) that $u_n^* \to 0$ in $H^1(\Omega)$ as $n \to \infty$. Set $y_n = \frac{u_n^*}{\|u_n^*\|}$ for $n \geq 1$. We may assume that

$$y_n \xrightarrow{w} y \text{ in } H^1(\Omega) \text{ and } y_n \to y \text{ in } L^2(\Omega) \text{ as } n \to \infty. \qquad (12.93)$$

From (12.91) we have

$$A(y_n) = \frac{f(x, u_n^*)}{\|u_n^*\|} \text{ in } H^1(\Omega)^* \text{ for all } n \geq 1. \qquad (12.94)$$

Note that, in view of H$(f)_1$ (i), (ii), the sequence $\left\{ \frac{f(x,u_n^*)}{\|u_n^*\|} \right\}_{n \geq 1}$ is bounded in $L^2(\Omega)$. Acting on (12.94) with $y_n - y \in H^1(\Omega)$ and passing to the limit as $n \to \infty$, we obtain that $\langle A(y_n), y_n - y \rangle \to 0$. Invoking Proposition 2.72, we infer that $y_n \to y$ in $H^1(\Omega)$. This implies that $\|y\| = 1$, hence $y \neq 0$. From the boundedness of the sequence $\left\{ \frac{f(x,u_n^*)}{\|u_n^*\|} \right\}_{n \geq 1}$ in $L^2(\Omega)$ we may also assume that

$$\frac{f(x, u_n^*)}{\|u_n^*\|} \xrightarrow{w} g \text{ in } L^2(\Omega) \text{ as } n \to \infty.$$

Using H$(f)_1$ (ii) and reasoning as in Claim 1 of the proof of Theorem 11.15, we show that $g = hy$ with $\lambda_m \leq h(x) \leq \lambda_{m+1}$ a.e. in $\Omega$, $h \neq \lambda_m$, $h \neq \lambda_{m+1}$, in particular $h \subset L^\infty(\Omega)_+ \setminus \{0\}$. Thus, passing to the limit as $n \to \infty$ in (12.94), we have

$$A(y) = hy \text{ in } H^1(\Omega)^*. \qquad (12.95)$$

On the one hand, from (12.95) and the fact that $y \neq 0$ we infer that $y$ is an eigenfunction of the negative Neumann Laplacian with respect to the weight $h$, corresponding to the eigenvalue $\lambda = 1$. On the other hand, we know that $y \geq 0$, whereas the only constant-sign eigenfunctions of $-\Delta^N$ are the nonzero constants, and they correspond to the first eigenvalue $0$ (Proposition 9.33). We reach a contradiction, so $u_+ \neq 0$.

*Step 4:* We have $0 < u_+(x) \leq a_+$ for all $x \in \overline{\Omega}$, and $u_+$ is the smallest positive solution of (12.83).

Since $u_n^* \in I_n^+$ for all $n \geq 1$, we have $0 \leq u_+(x) \leq a_+$ for all $x \in \overline{\Omega}$. In fact, since $u_+ \neq 0$, from Proposition 12.23(a), we obtain $u_+(x) > 0$ for all $x \in \overline{\Omega}$. Now let $u \in C^1(\overline{\Omega})$ with $u \geq 0$ in $\Omega$, $u \neq 0$, be another positive solution of (12.83). By Proposition 12.23(a), we have $u(x) > 0$ for all $x \in \overline{\Omega}$. Using Lemma 12.27, we see that the function $\tilde{u} := \min\{u, \overline{u}\}$ is an upper solution of (12.83). Choose $n_0 \geq 1$ such that $\underline{u}_{\varepsilon_n} \leq \tilde{u}$ in $\Omega$ for all $n \geq n_0$. Applying Proposition 12.25(b), we find a solution $\tilde{u}_n \in [\underline{u}_{\varepsilon_n}, \tilde{u}]$ of (12.83) for all $n \geq n_0$. Then, since $u_n^*$ denotes the smallest solution of (12.83) in the order interval $[\underline{u}_{\varepsilon_n}, \overline{u}]$, we get that $u_n^* \leq \tilde{u}_n \leq \tilde{u} \leq u$ in $\Omega$ for all $n \geq n_0$, whence $u_+ \leq u$ in $\Omega$. We have shown that $u_+$ is the smallest positive solution of (12.83). This completes the proof of the proposition.                           $\square$

Next we look for nodal (sign-changing) solutions of (12.83).

**Proposition 12.29.** *If* $H(f)_1$ *holds, then problem (12.83) has at least four distinct nontrivial solutions:* $u_0 \in C^1(\overline{\Omega})$ *positive,* $v_0 \in C^1(\overline{\Omega})$ *negative, satisfying*

$$a_- \leq v_0(x) < 0 < u_0(x) \leq a_+ \text{ for all } x \in \overline{\Omega},$$

*and* $y_0, \hat{y} \in C^1(\overline{\Omega}) \cap [v_0, u_0]$ *nodal.*

*Proof.* Let $u_0 = u_+ \in C^1(\overline{\Omega}) \cap [0, a_+]$ and $v_0 = v_- \in C^1(\overline{\Omega}) \cap [a_-, 0]$ be the extremal constant-sign solutions of (12.83) obtained in Proposition 12.28. It remains to construct the nodal solutions $y_0, \hat{y}$. To do this, we consider the truncated functional $\sigma \in C^1(H^1(\Omega), \mathbb{R})$ given by $\sigma = \hat{\varphi}_{[v_0, u_0]}$ [see (12.87)]. From Proposition 12.25(a) we know that the critical points of $\sigma$ are solutions of (12.83) belonging to $C^1(\overline{\Omega}) \cap [v_0, u_0]$. Due to the extremality property of $v_0$ and $u_0$, every critical point of $\sigma$ different from $0, v_0, u_0$ must be a nodal solution of (12.83). Therefore, our goal is to show that $\sigma$ admits at least two critical points different from $0, v_0, u_0$. We assume that $\sigma$ has only a finite number of critical points (otherwise we are done).

*Step 1:* $u_0$ *and* $v_0$ *are strict local minimizers of* $\sigma$.

We only give the proof for $u_0$ because the argument for $v_0$ is similar. By Proposition 12.25(b), the functional $\hat{\varphi}_{[0, u_0]}$ admits a global minimizer $u \in C^1(\overline{\Omega}) \cap [0, u_0]$ that is a solution of (12.83). The extremality property of $u_0$ imposes that $u \in \{0, u_0\}$. By $H(f)_1$ (ii), there is $\delta > 0$ such that $F(x, s) > 0$ for a.a. $x \in \Omega$, all $s \in (0, \delta)$. Choose $s \in (0, \delta)$ such that $s \leq u_0(x)$ for all $x \in \overline{\Omega}$. Then we have

$$\hat{\varphi}_{[0, u_0]}(u) \leq \hat{\varphi}_{[0, u_0]}(s) = -\int_\Omega F(x, s)\, dx < 0 = \hat{\varphi}_{[0, u_0]}(0).$$

Thus, $u \neq 0$. As a result, $u_0 = u$ is the unique global minimizer of $\hat{\varphi}_{[0, u_0]}$. Moreover, we know that $u_0$ belongs to the interior of $C^1(\overline{\Omega})_+$. Since the restrictions of the functionals $\hat{\varphi}_{[0, u_0]}$ and $\sigma$ to the set $C^1(\overline{\Omega})_+$ coincide, we obtain that $u_0$ is a local $C^1(\overline{\Omega})$-minimizer of $\sigma$. Invoking Theorem 12.18, we conclude that $u_0$ is a local minimizer of $\sigma$ with respect to the topology of $H^1(\Omega)$, which is strict because, by assumption, $\sigma$ has a finite number of critical points. This establishes Step 1.

*Step 2:* $C_k(\sigma,0) = \delta_{k,m+1}\mathbb{F}$ with $m \geq 1$ as in H($f$)$_1$ (ii).

The fact that $f(x,0) = 0$ a.e. in $\Omega$ guarantees that 0 is a critical point of $\sigma$, which is isolated (because we assume that $\sigma$ has a finite number of critical points). Note that the set $U = \{u \in C^1(\overline{\Omega}) : v_0(x) < u(x) < u_0(x)\}$ is an open neighborhood of 0 in $C^1(\overline{\Omega})$ such that $\sigma|_U = \varphi|_U$. By Remark 12.19, we deduce that 0 is an isolated critical point of $\varphi$ and

$$C_k(\sigma,0) = C_k(\sigma|_{C^1(\overline{\Omega})},0) = C_k(\varphi|_{C^1(\overline{\Omega})},0) = C_k(\varphi,0) \quad \text{for all } k \geq 0. \quad (12.96)$$

Recall that $\varphi \in C^2(H^1(\Omega),\mathbb{R})$. We claim that

$$\text{0 is a nondegenerate critical point of } \varphi \text{ with Morse index } m+1 \quad (12.97)$$

(Definition 6.46). Note that, once we have proven (12.97), we can apply Theorem 6.51, which yields $C_k(\varphi,0) = \delta_{k,m+1}\mathbb{F}$ for all $k \geq 0$. This equality, combined with (12.96), establishes Step 2. Therefore, it remains to check (12.97).

We can see that

$$\varphi''(0)(u,v) = \int_\Omega (\nabla u, \nabla v)_{\mathbb{R}^N}\,dx - \int_\Omega \xi(x)uv\,dx \quad \text{for all } u,v \in H^1(\Omega),$$

where $\xi(x) := f'_s(x,0)$ satisfies $\lambda_m \leq \xi(x) \leq \lambda_{m+1}$ for a.a. $x \in \Omega$, $\xi \neq \lambda_m$, and $\xi \neq \lambda_{m+1}$ [see H($f$)$_1$ (ii)]. Let $\{\hat{\lambda}_n(\eta)\}_{n\geq 1}$ denote the nondecreasing sequence of eigenvalues of $-\Delta^N$ with respect to a weight $\eta \in L^\infty(\Omega)_+ \setminus \{0\}$, repeated according to their multiplicity (Sect. 9.2). The monotonicity property of $\eta \mapsto \hat{\lambda}_n(\eta)$ (Proposition 9.53) yields

$$\hat{\lambda}_m(\xi) < \hat{\lambda}_m(\lambda_m) = 1 = \hat{\lambda}_{m+1}(\lambda_{m+1}) < \hat{\lambda}_{m+1}(\xi), \quad (12.98)$$

so $\lambda = 1$ is not an eigenvalue of $-\Delta^N$ with respect to $\xi$. This implies that each $u \in H^1(\Omega)$ such that $\varphi''(0)(u,\cdot) = 0$ is trivial. Thus, 0 is a nondegenerate critical point of $\varphi$. Finally, it follows from (12.98) and Proposition 9.52 that there exists an $(m+1)$-dimensional subspace $Y \subset H^1(\Omega)$ such that $\|\nabla u\|_2^2 < \int_\Omega \xi|u|^2\,dx$ for all $u \in Y \setminus \{0\}$, whereas for each $(m+2)$-dimensional subspace $\tilde{Y} \subset H^1(\Omega)$ there is $\tilde{u} \in \tilde{Y}$ such that $\|\nabla\tilde{u}\|_2^2 = \hat{\lambda}_{m+1}(\xi) > 1 = \int_\Omega \xi|\tilde{u}|^2\,dx$. This shows that $m+1$ is the Morse index of 0 as a critical point of $\varphi$. We have shown (12.97), and Step 2 is complete.

*Step 3:* There is a critical point $y_0$ of $\sigma$ different from $0, u_0, v_0$ and such that $C_k(\sigma,y_0) = \delta_{k,1}\mathbb{F}$.

Invoking Corollary 6.81 (by arguing as in the proof of Proposition 5.42, on the basis of Step 1, to check the condition of linking), we find $y_0 \in H^1(\Omega)$ a critical point of $\sigma$ different from $u_0, v_0$ such that

$$C_1(\sigma,y_0) \neq 0. \quad (12.99)$$

Step 2 yields $y_0 \neq 0$. Thus, to complete Step 3, it remains to compute the critical groups $C_k(\sigma, y_0)$. By $H(f)_1$ (i), there is $\hat{c} \in \mathbb{R}$ such that $f_s'(x,t) \geq \hat{c}$ for a.a. $x \in \Omega$, all $t \in [a_-, a_+]$. This implies that for a.a. $x \in \Omega$ we have

$$-\Delta(u_0 - y_0) = f(x, u_0(x)) - f(x, y_0(x)) = \int_{y_0(x)}^{u_0(x)} f_s'(x,t)\, dt \geq \hat{c}(u_0(x) - y_0(x)).$$

In addition, taking into account that $u_0 - y_0 \geq 0$ in $\Omega$, $y_0 \neq u_0$, and $\frac{\partial(u_0 - y_0)}{\partial n} = 0$ on $\partial\Omega$ (Proposition 12.23), by the strong maximum principle (Theorem 8.27), we obtain that $y_0(x) < u_0(x)$ for all $x \in \overline{\Omega}$. Arguing similarly, we get

$$v_0(x) < y_0(x) < u_0(x) \quad \text{for all } x \in \overline{\Omega}.$$

This relation implies that $V := C^1(\overline{\Omega}) \cap [v_0, u_0]$ is a neighborhood of $y_0$ in $C^1(\overline{\Omega})$. Since $\sigma|_V = \varphi|_V$, we deduce that $y_0$ is an isolated critical point of $\varphi$ and

$$C_k(\sigma, y_0) = C_k(\sigma|_{C^1(\overline{\Omega})}, y_0) = C_k(\varphi|_{C^1(\overline{\Omega})}, y_0) = C_k(\varphi, y_0) \quad \text{for all } k \geq 0 \quad (12.100)$$

(Remark 12.19). Combining this with (12.99), we infer that $C_1(\varphi, y_0) \neq 0$. Note that $\varphi \in C^2(H^1(\Omega), \mathbb{R})$. Moreover, arguing as in the proof of Theorem 12.5, we can see that each isolated critical point $y$ of $\varphi$ has finite nullity $\nu(y)$ and finite Morse index $m(y)$, with $\nu(y) \leq 1$ whenever $m(y) = 0$ (Definition 6.46). This fact allows us to apply Proposition 6.101, which yields $C_k(\varphi, y_0) = \delta_{k,1}\mathbb{F}$ for all $k \geq 0$. Combining this with (12.100), we obtain $C_k(\sigma, y_0) = \delta_{k,1}\mathbb{F}$ for all $k \geq 0$. This completes Step 3.

*Step 4:* There is a critical point $\hat{y}$ of $\sigma$ different from $0, u_0, v_0, y_0$.

Since $\sigma$ is bounded below and $u_0, v_0$ are local minimizers of $\sigma$, by Proposition 6.64(a) and Example 6.45(a), we obtain

$$C_k(\sigma, u_0) = C_k(\sigma, v_0) = C_k(\sigma, \infty) = \delta_{k,0}\mathbb{F} \quad \text{for all } k \geq 0. \quad (12.101)$$

Arguing by contradiction, suppose that $0, u_0, v_0, y_0$ are the only critical points of $\sigma$. Applying Theorem 6.62(b) with $t = -1$ on the basis of Steps 2 and 3 and relation (12.101), we have the equality

$$(-1)^0 + (-1)^0 + (-1)^1 + (-1)^{m+1} = (-1)^0,$$

which is impossible. This establishes Step 4.

By Steps 3 and 4, the functional $\sigma$ admits critical points $y_0, \hat{y}$ different from $0, u_0, v_0$. As explained at the beginning of the proof, $y_0, \hat{y}$ are nodal solutions of (12.83) belonging to $C^1(\overline{\Omega}) \cap [v_0, u_0]$. The proof is now complete. $\qquad\square$

**Two More Constant-Sign Solutions in Superlinear Case**

In the rest of this section, to the hypotheses in $H(f)_1$ we add hypotheses on the asymptotic behavior of the nonlinearity $f$ at $\pm\infty$, allowing us to show the existence of additional constant-sign solutions. First, we consider the case where $f$ is superlinear at $\pm\infty$. We also strengthen hypothesis $H(f)_1$ (iii).

$H(f)_2$ (i) $f: \Omega \times \mathbb{R} \to \mathbb{R}$ is a function such that $x \mapsto f(x,s)$ is measurable for all $s \in \mathbb{R}$, $s \mapsto f(x,s)$ is of class $C^1$ for a.a. $x \in \Omega$, $f(x,0) = 0$ a.e. in $\Omega$, and there exist $c > 0$ and $r \in (2, 2^*)$ such that

$$|f'_s(x,s)| \le c(1 + |s|^{r-2}) \quad \text{for a.a. } x \in \Omega, \text{ all } s \in \mathbb{R};$$

  (ii) There is an integer $m \ge 1$ such that

$$(f'_s(x,0) =) \lim_{s \to 0} \frac{f(x,s)}{s} \in [\lambda_m, \lambda_{m+1}] \quad \text{uniformly for a.a. } x \in \Omega$$

    and $f'_s(\cdot,0) \ne \lambda_m$, $f'_s(\cdot,0) \ne \lambda_{m+1}$;
  (iii) There exist real numbers $a_- < 0 < a_+$ such that

$$f(x,a_+) \le 0 \quad \text{and} \quad f(x,a_-) \ge 0 \quad \text{for a.a. } x \in \Omega,$$

    with strict inequalities on sets of positive measure;
  (iv) There exists $\gamma_0 > 0$ such that

$$\liminf_{s \to \pm\infty} \frac{F(x,s)}{s^2} \ge \gamma_0 \quad \text{uniformly for a.a. } x \in \Omega;$$

  (v) There exist $\tau \in ((r-2)\max\{1, \frac{N}{2}\}, 2^*)$ and $\beta_0 > 0$ such that

$$\liminf_{s \to \pm\infty} \frac{f(x,s)s - 2F(x,s)}{|s|^\tau} \ge \beta_0 \quad \text{uniformly for a.a. } x \in \Omega.$$

*Example 12.30.* The following function satisfies $H(f)_2$ (for the sake of simplicity we drop the $x$-dependence):

$$f(s) = \begin{cases} s\ln|s| - \mu_0 s + c_0 & \text{if } s < -1, \\ \lambda s - \mu |s|^{q-2}s & \text{if } |s| \le 1, \\ (s + \ln s)\ln s - \mu_0 s - c_0 & \text{if } s > 1, \end{cases}$$

with $\lambda \in (\lambda_m, \lambda_{m+1})$, $m \ge 1$, $q > 2$, $\mu > \lambda$, $\mu_0 = \mu(q-1) + 1 - \lambda$, and $c_0 = \mu - \mu_0 - \lambda$.

Our full multiplicity result in the superlinear case is the following theorem.

**Theorem 12.31.** *Assume that* H($f$)$_2$ *holds. Then problem (12.83) has at least six distinct, nontrivial solutions:* $u_0, \hat{u} \in C^1(\overline{\Omega})$ *positive,* $v_0, \hat{v} \in C^1(\overline{\Omega})$ *negative, satisfying*

$$\max\{\hat{v}(x), a_-\} < v_0(x) < 0 < u_0(x) < \min\{\hat{u}(x), a_+\} \ \textit{for all } x \in \overline{\Omega},$$

*and* $y_0, \hat{y} \in C^1(\overline{\Omega}) \cap [v_0, u_0]$ *nodal.*

*Proof.* From Proposition 12.29 we already have two nontrivial, smooth, constant-sign solutions $u_0, v_0 \in C^1(\overline{\Omega})$ such that $a_- \leq v_0(x) < 0 < u_0(x) \leq a_+$ for all $x \in \overline{\Omega}$ and two nodal solutions $y_0, \hat{y} \in C^1(\overline{\Omega}) \cap [v_0, u_0]$. Note that H($f$)$_2$ (iii) implies that $a_-, a_+$ are not solutions of (12.83). Therefore, $u_0 \leq a_+$ in $\Omega$, $u_0 \not\equiv a_+$, and $v_0 \geq a_-$ in $\Omega$, $v_0 \not\equiv a_-$. Arguing as in Step 3 of the proof of Proposition 12.29 by using the strong maximum principle (Theorem 8.27), we infer that

$$a_- < v_0(x) < 0 < u_0(x) < a_+ \ \text{ for all } x \in \overline{\Omega}. \tag{12.102}$$

We will only show the existence of a second positive solution $\hat{u}$ of (12.83) satisfying the claimed properties because the reasoning is the same for constructing the second negative solution $\hat{v}$. To do this, we consider the truncated functional $\hat{\varphi}_{[u_0, +\infty]} \in C^1(H^1(\Omega), \mathbb{R})$ [see (12.87)].

*Claim 1:* If $\hat{u} \in H^1(\Omega)$ is a critical point of $\hat{\varphi}_{[u_0, +\infty]}$ different from $u_0$, then $\hat{u} \in C^1(\overline{\Omega})$ is a solution of (12.83) satisfying $u_0(x) < \hat{u}(x)$ for all $x \in \overline{\Omega}$.

The facts that $\hat{u}$ is a solution of (12.83) and $\hat{u} \in C^1(\overline{\Omega}) \cap [u_0, +\infty]$ are implied by Proposition 12.25(a). By H($f$)$_2$ (i), there is $\hat{c} \in \mathbb{R}$ such that $f'_s(x, t) \geq \hat{c}$ for a.a. $x \in \Omega$, all $t \in [0, \|\hat{u}\|_\infty]$. This yields

$$-\Delta(\hat{u} - u_0) = f(x, \hat{u}(x)) - f(x, u_0(x)) \geq \hat{c}(\hat{u}(x) - u_0(x)) \ \text{ for a.a. } x \in \Omega.$$

Since $\hat{u} - u_0 \geq 0$ in $\Omega$, $\hat{u} \neq u_0$, and $\frac{\partial(\hat{u} - u_0)}{\partial n} = 0$ on $\partial\Omega$ (Proposition 12.23), applying the strong maximum principle (Theorem 8.27), we conclude that $u_0(x) < \hat{u}(x)$ for all $x \in \overline{\Omega}$. This proves Claim 1.

According to Claim 1, to complete the proof of the theorem, it remains to show that $\hat{\varphi}_{[u_0, +\infty]}$ admits a critical point different from $u_0$. Recall that $\overline{u} \equiv a_+$ is an upper solution of (12.83), and note that $u_0$ is in particular a lower solution of (12.83). By Proposition 12.25(b), the functional $\hat{\varphi}_{[u_0, a_+]}$ has a global minimizer $\hat{u}_0 \in C^1(\overline{\Omega}) \cap [u_0, a_+]$, which is also a solution of (12.83). If $\hat{u}_0 \neq u_0$, then $\hat{u}_0$ is a second critical point of $\hat{\varphi}_{[u_0, +\infty]}$, and we are done (by Claim 1). Thus, we may suppose that $u_0 = \hat{u}_0$ is a global minimizer of $\hat{\varphi}_{[u_0, a_+]}$. Note that the restrictions of the functionals $\hat{\varphi}_{[u_0, a_+]}$ and $\hat{\varphi}_{[u_0, +\infty]}$ to the set $W := \{u \in C^1(\overline{\Omega}) : \ u(x) < a_+ \text{ for all } x \in \overline{\Omega}\}$ coincide. Since $W$ is a neighborhood of $u_0$ in $C^1(\overline{\Omega})$, we conclude that $u_0$ is a local $C^1(\overline{\Omega})$-minimizer of $\hat{\varphi}_{[u_0, +\infty]}$ and, therefore, a local minimizer of $\hat{\varphi}_{[u_0, +\infty]}$ with respect to the topology of $H^1(\Omega)$. In fact, we may suppose that

$$u_0 \text{ is a strict local minimizer of } \hat{\varphi}_{[u_0, +\infty]} \tag{12.103}$$

(otherwise, $\hat{\varphi}_{[u_0,+\infty]}$ has infinitely many critical points, and again we are done).

*Claim 2:* $\hat{\varphi}_{[u_0,+\infty]}$ satisfies the (C)-condition.

Let $\{u_n\}_{n\geq 1} \subset H^1(\Omega)$ be a sequence such that

$$|\hat{\varphi}_{[u_0,+\infty]}(u_n)| \leq M_1 \text{ for all } n \geq 1, \tag{12.104}$$

for some $M_1 > 0$, and

$$(1 + \|u_n\|)\hat{\varphi}'_{[u_0,+\infty]}(u_n) \to 0 \text{ in } H^1(\Omega)^* \text{ as } n \to \infty. \tag{12.105}$$

From (12.105), for all $n \geq 1$, we have

$$|\langle \hat{\varphi}'_{[u_0,+\infty]}(u_n), h \rangle| \leq \frac{\varepsilon_n \|h\|}{1 + \|u_n\|} \text{ for all } h \in H^1(\Omega),$$

with $\{\varepsilon_n\}_{n\geq 1} \subset (0,+\infty)$ such that $\varepsilon_n \to 0$. This yields, for all $h \in H^1(\Omega)$,

$$\left| \langle A(u_n), h \rangle + \int_\Omega u_n h \, dx - \int_\Omega \hat{f}_{[u_0,+\infty]}(x, u_n) h \, dx \right| \leq \frac{\varepsilon_n \|h\|}{1 + \|u_n\|}. \tag{12.106}$$

Choosing $h = -u_n^- \in H^1(\Omega)$ in (12.106), by (12.86), we obtain

$$\|\nabla u_n^-\|_2^2 + \|u_n^-\|_2^2 \leq \varepsilon_n + \int_\Omega (f(x,u_0) + u_0)(-u_n^-) \, dx \leq c_1(1 + \|u_n^-\|_2)$$

for all $n \geq 1$, for some $c_1 > 0$. Therefore,

$$\{u_n^-\}_{n\geq 1} \text{ is bounded in } H^1(\Omega). \tag{12.107}$$

Now we check that

$$\{u_n^+\}_{n\geq 1} \text{ is bounded in } H^1(\Omega). \tag{12.108}$$

On the one hand, choosing $h = u_n^+ \in H^1(\Omega)$ in (12.106), we get

$$-\|\nabla u_n^+\|_2^2 - \|u_n^+\|_2^2 + \int_\Omega \hat{f}_{[u_0,+\infty]}(x, u_n) u_n^+ \, dx \leq \varepsilon_n \text{ for all } n \geq 1. \tag{12.109}$$

On the other hand, from (12.104) and (12.107) we have

$$\|\nabla u_n^+\|_2^2 + \|u_n^+\|_2^2 - \int_\Omega 2\hat{F}_{[u_0,+\infty]}(x, u_n) \, dx \leq M_2 \text{ for all } n \geq 1, \tag{12.110}$$

for some $M_2 > 0$. Adding (12.109) and (12.110), we obtain

$$\int_\Omega \left(\hat{f}_{[u_0,+\infty]}(x,u_n)u_n^+ - 2\hat{F}_{[u_0,+\infty]}(x,u_n)\right)dx \leq \varepsilon_n + M_2 \quad \text{for all } n \geq 1. \quad (12.111)$$

By (12.86), the inequality in (12.111) leads to

$$\int_{\{u_n > u_0\}} \left(f(x,u_n)u_n - 2F(x,u_n)\right)dx \leq M_3 \quad \text{for all } n \geq 1, \quad (12.112)$$

for some $M_3 > 0$. By virtue of hypotheses $H(f)_2$ (i), (v), we know that

$$f(x,s)s - 2F(x,s) \geq \beta_1 s^\tau - c_2 \quad \text{for a.a. } x \in \Omega, \text{ all } s \geq 0, \quad (12.113)$$

with $\beta_1 \in (0,\beta_0)$, $c_2 > 0$. Using (12.113) in (12.112), we obtain that

$$\{u_n^+\}_{n\geq 1} \text{ is bounded in } L^\tau(\Omega). \quad (12.114)$$

Since $u \mapsto \|\nabla u\|_2 + \|u\|_\tau$ is a norm on $H^1(\Omega)$ that is equivalent to the Sobolev norm (Proposition 1.53), in order to check (12.108), it remains to show that $\{\|\nabla u_n^+\|_2\}_{n\geq 1}$ is bounded. Taking $h = u_n^+$ in (12.106) and using $H(f)_2$ (i) and (12.86), we find a constant $M_4 > 0$ such that

$$\|\nabla u_n^+\|_2^2 \leq M_4(1 + \|u_n^+\|_r^r) \quad \text{for all } n \geq 1. \quad (12.115)$$

In the case where $\tau \geq r$, relations (12.114) and (12.115) imply that $\{\|\nabla u_n^+\|_2\}_{n\geq 1}$ is bounded, which ensures (12.108). Therefore, we may assume that $\tau < r$. The assumption that $\tau \in ((r-2)\max\{\frac{N}{2},1\},2^*)$ implies that we can always find $\ell \in (r,2^*)$ such that $\ell > \frac{2\tau}{2+\tau-r}$. The fact that $\tau < r < \ell$ yields $t \in (0,1)$ such that $\frac{1}{r} = \frac{1-t}{\tau} + \frac{t}{\ell}$. Moreover, the choice of $\ell$ implies that $tr < 2$. Using the interpolation inequality (e.g., Brezis [52, p. 93]), we have

$$\|u_n^+\|_r \leq \|u_n^+\|_\tau^{1-t} \|u_n^+\|_\ell^t.$$

By (12.114), this guarantees that $\|u_n^+\|_r^r \leq M_5 \|u_n^+\|_\ell^{tr}$ for all $n \geq 1$, for some $M_5 > 0$, whence, by (12.115),

$$\|\nabla u_n^+\|_2^2 \leq M_4(1 + M_5\|u_n^+\|_\ell^{tr}) \leq M_6(1 + (\|\nabla u_n^+\|_2 + \|u_n^+\|_\tau)^{tr}) \quad (12.116)$$

for all $n \geq 1$, for some $M_6 > 0$, where we also use the continuity of the embedding $H^1(\Omega) \hookrightarrow L^\ell(\Omega)$ and the equivalence between $u \mapsto \|\nabla u\|_2 + \|u\|_\tau$ and the Sobolev norm of $H^1(\Omega)$ (Proposition 1.53). Relations (12.114) and (12.116) and the fact that $tr < 2$ imply that $\{\|\nabla u_n^+\|_2\}_{n\geq 1}$ is bounded. We have therefore proven (12.108).

From (12.107) and (12.108) it follows that $\{u_n\}_{n\geq 1}$ is bounded in $H^1(\Omega)$. Thus, along a relabeled subsequence, we have

$$u_n \xrightarrow{w} u \text{ in } H^1(\Omega) \quad \text{and} \quad u_n \to u \text{ in } L^r(\Omega) \text{ as } n \to \infty \quad (12.117)$$

for some $u \in H^1(\Omega)$. Choosing $h = u_n - u \in H^1(\Omega)$ in (12.106) and passing therein to the limit as $n \to \infty$ using (12.117), we have $\lim_{n\to\infty} \langle A(u_n), u_n - u \rangle = 0$. By Proposition 2.72, we infer that $u_n \to u$ in $H^1(\Omega)$ as $n \to \infty$. We have proven Claim 2.

*Claim 3:* $\hat{\varphi}_{[u_0,+\infty]}(t) \to -\infty$ as $t \to +\infty$, $t \in \mathbb{R}$.

From (12.86) and (12.87) we have

$$\hat{\varphi}_{[u_0,+\infty]}(t) = \frac{t^2}{2}|\Omega|_N - \int_\Omega \hat{F}_{[u_0,+\infty]}(x,t)\,dx \leq M_7 - \int_\Omega F(x,t)\,dx \qquad (12.118)$$

for all $t > \|u_0\|_\infty$, for some $M_7 > 0$. On the other hand, by virtue of hypothesis H($f$)$_2$ (iv), there are $\gamma_1 \in (0,\gamma_0)$ and $M_8 > \|u_0\|_\infty$ such that

$$F(x,t) \geq \gamma_1 t^2 \text{ for a.a. } x \in \Omega, \text{ all } t \geq M_8. \qquad (12.119)$$

Claim 3 follows from (12.118) and (12.119).

By Claim 3, there is $t_0 > 0$ such that $\hat{\varphi}_{[u_0,+\infty]}(t_0) \leq \hat{\varphi}_{[u_0,+\infty]}(u_0)$. Relying on this fact, (12.103), and Claim 2, we can apply Proposition 5.42, which yields $\hat{u} \in H^1(\Omega)$, critical point of $\hat{\varphi}_{[u_0,+\infty]}$ different from $u_0$. Then, by Claim 1, $\hat{u}$ is a smooth solution of (12.83) such that $\hat{u} > u_0$ in $\overline{\Omega}$. The proof of the theorem is complete. $\square$

### Two More Constant-Sign Solutions in Sublinear Case

Finally, we investigate problem (12.83) when the nonlinearity $f$ is sublinear at $\pm\infty$. This means that the next set of hypotheses includes the situation where $f$ is resonant at $\pm\infty$ with respect to the first eigenvalue $\lambda_0 = 0$.

H($f$)$_3$ (i) $f:\Omega \times \mathbb{R} \to \mathbb{R}$ is a function such that $x \mapsto f(x,s)$ is measurable for all $s \in \mathbb{R}$, $s \mapsto f(x,s)$ is of class $C^1$ for a.a. $x \in \Omega$, $f(x,0) = 0$ a.e. in $\Omega$, and there exist $c > 0$ and $r \in (2,2^*)$ such that

$$|f'_s(x,s)| \leq c(1 + |s|^{r-2}) \text{ for a.a. } x \in \Omega, \text{ all } s \in \mathbb{R};$$

(ii) There is an integer $m \geq 1$ such that

$$(f'_s(x,0) =) \lim_{s\to 0} \frac{f(x,s)}{s} \in [\lambda_m, \lambda_{m+1}] \text{ uniformly for a.a. } x \in \Omega$$

and $f'_s(\cdot,0) \neq \lambda_m$, $f'_s(\cdot,0) \neq \lambda_{m+1}$;

(iii) There exist real numbers $a_- < 0 < a_+$ such that

$$f(x,a_+) \leq 0 \text{ and } f(x,a_-) \geq 0 \text{ for a.a. } x \in \Omega,$$

with strict inequalities on sets of positive measure;

(iv) There exists $\gamma_0 > 0$ such that

$$0 \le \liminf_{s \to \pm\infty} \frac{F(x,s)}{s^2} \le \limsup_{s \to \pm\infty} \frac{F(x,s)}{s^2} \le \gamma_0 \text{ uniformly for a.a. } x \in \Omega;$$

(v) There exist $\tau \in [1,2]$ and $\beta_0 > 0$ such that

$$\liminf_{s \to \pm\infty} \frac{2F(x,s) - f(x,s)s}{|s|^\tau} \ge \beta_0 \text{ uniformly for a.a. } x \in \Omega.$$

*Example 12.32.* The following function satisfies H($f$)$_3$ (for the sake of simplicity we drop the $x$-dependence):

$$f(s) = \begin{cases} \tilde{\lambda}_1 s + \mu_1 |s|^{\tau_1 - 2} s + v_1 \ln|s| + c_1 & \text{if } s < -1, \\ \lambda s - \mu |s|^{q-2} s & \text{if } |s| \le 1, \\ \tilde{\lambda}_2 s + \mu_2 s^{\tau_2 - 1} - v_2 \ln s - c_2 & \text{if } s > 1, \end{cases}$$

with $\lambda \in (\lambda_m, \lambda_{m+1})$, $m \ge 1$, $q > 2$, $\mu > \lambda$, and for $i \in \{1,2\}$: $\tilde{\lambda}_i \in [0, +\infty)$, $\tau_i \in (1,2)$, $\mu_i \in (0, +\infty)$, $c_i = \tilde{\lambda}_i + \mu_i + \mu - \lambda$, and $v_i = \tilde{\lambda}_i + \mu_i(\tau_i - 1) + \mu(q-1) - \lambda$.

*Remark 12.33.* As illustrated in Example 12.32, hypotheses H($f$)$_3$ allow resonance at infinity with respect to any eigenvalue of the negative Neumann Laplacian.

Our full multiplicity result in the sublinear case is as follows.

**Theorem 12.34.** *Assume that* H($f$)$_3$ *holds. Then problem* (12.83) *has at least six distinct, nontrivial solutions:* $u_0, \tilde{u} \in C^1(\overline{\Omega})$ *positive,* $v_0, \tilde{v} \in C^1(\overline{\Omega})$ *negative, satisfying*

$$\max\{\tilde{v}(x), a_-\} < v_0(x) < 0 < u_0(x) < \min\{\tilde{u}(x), a_+\} \text{ for all } x \in \overline{\Omega},$$

*and* $y_0, \hat{y} \in C^1(\overline{\Omega}) \cap [v_0, u_0]$ *nodal.*

*Proof.* Proposition 12.29 already provides two constant-sign solutions $u_0, v_0 \in C^1(\overline{\Omega})$, such that $a_- \le v_0(x) < 0 < u_0(x) \le a_+$ for all $x \in \overline{\Omega}$, and two nodal solutions $y_0, \hat{y} \in C^1(\overline{\Omega}) \cap [v_0, u_0]$. Actually, reasoning as in the beginning of the proof of Theorem 12.31, we have

$$a_- < v_0(x) < 0 < u_0(x) < a_+ \text{ for all } x \in \overline{\Omega}.$$

We only check the existence of the second positive solution $\tilde{u}$ (the argument for $\tilde{v}$ is the same). To do this, we consider the truncated functional $\hat{\varphi}_{[u_0, +\infty]}$ [see (12.87)]. The proof of this result is similar to the proof of Theorem 12.31. The only difference is the way to obtain the following two claims. The rest of the reasoning, based on these two claims, is unchanged:

*Claim 1:* $\hat{\varphi}_{[u_0,+\infty]}$ satisfies the (C)-condition.
*Claim 2:* $\hat{\varphi}_{[u_0,+\infty]}(t) \to -\infty$ as $t \to +\infty$, $t \in \mathbb{R}$.

The rest of the proof is devoted to checking Claims 1 and 2. First, we prove Claim 1. Let $\{u_n\}_{n\geq 1} \subset H^1(\Omega)$ be a sequence such that

$$|\hat{\varphi}_{[u_0,+\infty]}(u_n)| \leq M_1 \text{ for all } n \geq 1, \tag{12.120}$$

for some $M_1 > 0$, and $(1 + \|u_n\|)\hat{\varphi}'_{[u_0,+\infty]}(u_n) \to 0$ in $H^1(\Omega)^*$ as $n \to \infty$, that is,

$$\left|\langle A(u_n), h\rangle + \int_\Omega u_n h\, dx - \int_\Omega \hat{f}_{[u_0,+\infty]}(x, u_n) h\, dx\right| \leq \frac{\varepsilon_n \|h\|}{1 + \|u_n\|} \tag{12.121}$$

for all $h \in H^1(\Omega)$, all $n \geq 1$, where $\{\varepsilon_n\}_{n\geq 1} \subset (0, +\infty)$ is a sequence converging to 0. Choosing $h = u_n^-$ in (12.121) and taking (12.86) into account, we easily see that

$$\{u_n^-\}_{n\geq 1} \text{ is bounded in } H^1(\Omega). \tag{12.122}$$

Choosing $h = u_n^+ \in H^1(\Omega)$ in (12.121), we get

$$\|\nabla u_n^+\|_2^2 + \|u_n^+\|_2^2 - \int_\Omega \hat{f}_{[u_0,+\infty]}(x, u_n) u_n^+\, dx \leq \varepsilon_n \text{ for all } n \geq 1, \tag{12.123}$$

whereas from (12.120) and (12.122) we have

$$-\|\nabla u_n^+\|_2^2 - \|u_n^+\|_2^2 + \int_\Omega 2\hat{F}_{[u_0,+\infty]}(x, u_n)\, dx \leq M_2 \text{ for all } n \geq 1, \tag{12.124}$$

for some $M_2 > 0$. Adding (12.123) and (12.124), and using (12.86), we obtain

$$\int_\Omega \left(2F(x, u_n^+) - f(x, u_n^+)u_n^+\right) dx \leq M_3 \text{ for all } n \geq 1, \tag{12.125}$$

for some $M_3 > 0$. Combining H$(f)_3$ (i) and (v), we have

$$2F(x, s) - f(x, s)s \geq c_1 |s|^\tau - c_2 \text{ for a.a. } x \in \Omega, \text{ all } s \in \mathbb{R},$$

with constants $c_1, c_2 > 0$. Using this estimate in (12.125), we infer that

$$\{u_n^+\}_{n\geq 1} \text{ is bounded in } L^\tau(\Omega). \tag{12.126}$$

Fix $\ell \in (2, 2^*)$. Thus, $\tau \leq 2 < \ell$, so we can find $t \in [0, 1)$ such that $\frac{1}{2} = \frac{1-t}{\tau} + \frac{t}{\ell}$. Using the interpolation inequality, the continuity of the embedding $H^1(\Omega) \hookrightarrow L^\ell(\Omega)$, and (12.126), we obtain

$$\|u_n^+\|_2^2 \leq \|u_n^+\|_\tau^{2(1-t)} \|u_n^+\|_\ell^{2t} \leq M_4(\|u_n^+\|_2^{2t} + \|\nabla u_n^+\|_2^{2t}) \text{ for all } n \geq 1, \quad (12.127)$$

with $M_4 > 0$. Combining (12.127) with (12.120), (12.87), and H$(f)_3$ (iv), we infer that

$$\|\nabla u_n^+\|_2^2 + \|u_n^+\|_2^2 \leq M_5(1 + \|u_n^+\|_2^{2t} + \|\nabla u_n^+\|_2^{2t}) \text{ for all } n \geq 1, \quad (12.128)$$

for some $M_5 > 0$. Therefore, from (12.122) and (12.128) we obtain that

$$\{u_n\}_{n \geq 1} \text{ is bounded in } H^1(\Omega).$$

Then the verification of Claim 1 is completed by arguing as at the end of the proof of Claim 2 in the proof of Theorem 12.31.

Next, let us prove Claim 2. Hypothesis H$(f)_3$ (v) yields $c_3, M_6 > 0$ such that

$$\frac{d}{ds}\left(\frac{F(x,s)}{s^2}\right) = \frac{f(x,s)s - 2F(x,s)}{s^3} \leq -c_3 \frac{1}{s^{3-\tau}}$$

for a.a. $x \in \Omega$, all $s \geq M_6$. Clearly, in H$(f)_3$ (v), without any loss of generality, we may assume that $\tau < 2$. Integrating over $[t,s]$ with $s > t \geq M_6$, we see that

$$\frac{F(x,s)}{s^2} - \frac{F(x,t)}{t^2} \leq \frac{c_3}{2-\tau}\left(\frac{1}{s^{2-\tau}} - \frac{1}{t^{2-\tau}}\right) \text{ for a.a. } x \in \Omega, \text{ all } s > t \geq M_6.$$

Letting $s \to +\infty$, through hypothesis H$(f)_3$ (iv), we obtain

$$F(x,t) \geq \frac{c_3}{2-\tau} t^\tau \text{ for a.a. } x \in \Omega, \text{ all } s \geq M_6.$$

By (12.86) and (12.87), for every $t > \max\{\|u_0\|_\infty, M_6\}$ we have

$$\hat{\varphi}_{[u_0,+\infty]}(t) \leq -\int_\Omega F(x,t)dx + M_7 \leq -\frac{c_3|\Omega|_N}{2-\tau} t^\tau + M_7,$$

for some $M_7 > 0$. Claim 2 ensues. The proof of the theorem is then complete.  □


## 12.4 Remarks

**Section 12.1:** Multiplicity results for boundary value problems driven by the Neumann $p$-Laplacian were recently shown, for instance, in the works of Anello [21], Bonanno and Candito [48], Marano and Motreanu [241], Motreanu et al. [304], Motreanu and Papageorgiou [275], Motreanu and Perera [277], Ricceri [340], and Wu and Tan [383]. Anello [21], Marano and Motreanu [241], and Ricceri

[340] establish the existence of infinitely many solutions for a class of nonlinear elliptic problems by imposing certain oscillatory assumptions on the nonlinearity. In Marano and Motreanu [241] and Ricceri [340], it is assumed that $p > N$ (low-dimensional equations) and the authors exploit the fact that the space $W^{1,p}(\Omega)$ is compactly embedded into $C(\overline{\Omega})$. Nonlinear parametric problems subject to Neumann boundary conditions were studied by Bonanno and Candito [48] who establish a three-solution theorem (under the condition $p > N$). The assumption that $p > N$ is also present in the work of Wu and Tan [383], where the approach is variational based on critical point theory. In Motreanu and Papageorgiou [275], the Neumann problem under consideration involves a nonsmooth potential (hemi-variational inequalities), and the nonsmooth local linking theorem (Theorem 5.85) is used to produce two nontrivial smooth solutions. The work of Motreanu and Perera [277] deals with a Neumann $p$-Laplacian system for which the existence of two nontrivial solutions is shown through variational methods and Morse theory. This section focuses on the situation of near resonance with respect to the first eigenvalue of the negative Neumann $p$-Laplacian for a parametric problem and is based on the work of Motreanu et al. [294]. Its aim is to illustrate the variational techniques and the techniques of Morse theory that can be applied to the study of a parametric quasilinear Neumann problem, according to different situations of the parameter near resonance.

**Section 12.2:** Multiple solutions for Dirichlet problems driven by a nonhomogeneous differential operator were obtained by De Nápoli and Mariani [112], Kyritsi et al. [214], Motreanu and Tanaka [281], and Papageorgiou et al. [323].

Proposition 12.14 is due to Motreanu et al. [301]. Theorem 12.18 (in the Neumann case) is due to Motreanu and Papageorgiou [276] and Miyajima et al. [261], whose proof is essentially reproduced here. The first such result was proved for $H_0^1(\Omega)$, with $G(\xi) = \frac{1}{2}|\xi|^2$, for all $\xi \in \mathbb{R}^N$ by Brezis and Nirenberg [54]. It was extended to the space $W_0^{1,p}(\Omega)$ with $G(\xi) = \frac{1}{p}|\xi|^p$ for all $\xi \in \mathbb{R}^N$ by García Azorero et al. [147] and Guo and Zhang [168] (for $p \geq 2$). For the space $W^{1,p}(\Omega)$ with $G(\xi) = \frac{1}{p}|\xi|^p$ for all $\xi \in \mathbb{R}^N$, it was proved by Barletta and Papageorgiou [31] (for $2 \leq p < +\infty$) and by Motreanu et al. [294] (for $1 < p < +\infty$). Note that the argument of proof in Motreanu and Papageorgiou [276] and Miyajima et al. [261] is different and more direct than the proofs in all the aforementioned works. See also Khan and Motreanu [194].

Neumann boundary value problems driven by nonhomogeneous differential operators were recently studied by Miyajima et al. [261], Motreanu et al. [301], Motreanu and Papageorgiou [276], and Motreanu and Tanaka [280].

**Section 12.3:** Asymptotically linear Neumann problems driven by the Laplace differential operator were studied by Iannacci and Nkashama [179,180], Kuo [209], Mawhin [251], Mawhin et al. [254]. Iannacci and Nkashama [179] and Kuo [209] use variants of the well-known Landesman–Lazer conditions. Iannacci and Nkashama [180] use a sign condition, while Mawhin [251] and Mawhin et al. [254] assume a monotonicity condition on $f(x, \cdot)$. All the aforementioned works prove existence theorems. Multiplicity theorems can be found in Li [223], Motreanu et al. [303], Qian [333], and Tang and Wu [373].

Here, our approach for the study of problem (12.83), which is based on the work of Motreanu et al. [299], unifies the treatment of superlinear, asymptotically linear, and sublinear Neumann problems, allowing in addition resonance at any eigenvalue of the negative Neumann Laplacian. The hypotheses in this section are slightly weaker than in Motreanu et al. [299], where in counterpart the existence of a seventh nodal, nontrivial solution is established through flow invariance arguments.

# List of Symbols

| | |
|---|---|
| $\mathbb{N}$ | Natural numbers |
| $\mathbb{N}_0$ | $\mathbb{N} \cup \{0\}$ |
| $\mathbb{Z}$ | Integers |
| $\mathbb{R}$ | Real numbers |
| $\mathbb{R}_+$ | Nonnegative real numbers |
| $\mathbb{R}^N$ | $N$-dimensional Euclidean space |
| $\lvert \cdot \rvert$ | Euclidean norm of $\mathbb{R}^N$ |
| $E \subset F$ | $E$ is a subset of $F$ including $E = F$ |
| $f\rvert_E$ | Restriction of function $f$ to subset $E$ |
| $x^+$ | $\max\{x, 0\}$ |
| $x^-$ | $\max\{-x, 0\}$ |
| $\operatorname{sgn} x$ | $-1$, $0$, or $1$: sign of $x$ |
| $\lambda \downarrow 0$ | $\lambda > 0$ and $\lambda \to 0$ |
| $\overline{E}$ | Closure of a subset $E$ |
| $\operatorname{int}(E)$ | Interior of $E$ |
| $\partial E$ | Boundary of $E$ |
| $\partial^Y E$ | Boundary of $E \subset Y$ with respect to topology of $Y$ |
| $\operatorname{conv} E$ | Convex hull of $E$ |
| $\overline{\operatorname{conv}} E$ | Closure of $\operatorname{conv} E$ |
| $\operatorname{span} E$ | Linear subspace spanned by $E$ |
| $\overline{\operatorname{span}} E$ | Closure of $\operatorname{span} E$ |
| $2^X$ | Power set of set $X$, i.e., set of all subsets of $X$ |
| $\lambda^N$ | Lebesgue measure in $\mathbb{R}^N$ |
| $\lvert E \rvert_N$ | Lebesgue measure of a subset $E \subset \mathbb{R}^N$ |
| $\mu$ a.e. | Almost everywhere with respect to a measure $\mu$ |
| a.e. | Almost everywhere with respect to the Lebesgue measure |
| $\chi_E$ | Characteristic function of the set $E$ |
| $\Omega$ | Open domain in $\mathbb{R}^N$ |
| $\Omega' \subset\subset \Omega$ | $\Omega'$ is open, bounded and $\overline{\Omega'} \subset \Omega$ |
| id | Identity map |

D. Motreanu et al., *Topological and Variational Methods with Applications to Nonlinear Boundary Value Problems*, DOI 10.1007/978-1-4614-9323-5,
© Springer Science+Business Media, LLC 2014

| | |
|---|---|
| $(H, (\cdot, \cdot))$ | Real Hilbert space |
| $(X, \|\cdot\|)$ | Real Banach space |
| $(X^*, \|\cdot\|)$ | Dual space of $X$ |
| $\langle \cdot, \cdot \rangle$ | Duality brackets for pair $(X, X^*)$ |
| $X_w$ | $X$ endowed with weak topology |
| $X_{w^*}^*$ | $X^*$ endowed with weak$^*$ topology |
| $\xrightarrow{w}, \xrightarrow{w^*}$ | Weak (resp. weak$^*$) convergence |
| $\overline{E}^w, \overline{F}^{w^*}$ | Closure in weak (resp. weak$^*$) topology |
| $d(x, E)$ | Distance between $x$ and subset $E$ |
| $d(E, F)$ | $\inf\{\|x - y\| : x \in E, y \in F\}$ |
| $B_r(x)$ | Open ball |
| $\mathscr{L}(X, Y)$ | Space of bounded linear maps from $X$ to $Y$ |
| $\mathscr{L}_c(X, Y)$ | Space of compact linear maps from $X$ to $Y$ |
| $\mathscr{L}(X)$ | Abbreviation for $\mathscr{L}(X, X)$ |
| $\mathscr{L}_c(X)$ | Abbreviation for $\mathscr{L}_c(X, X)$ |
| $L^*$ | Adjoint map of $L \in \mathscr{L}(X, Y)$ |
| $\ker L$ | Kernel of $L$ |
| $\operatorname{im} L$ | Image of $L$ |
| $\sigma(L)$ | Spectrum of $L \in \mathscr{L}(X)$ |
| $\sigma_p(L)$ | Set of eigenvalues of $L$ |
| $\det L$ | Determinant of $L \in \mathscr{L}(\mathbb{R}^N)$ |
| $C(X, Y)$ | Space of continuous maps from $X$ to $Y$ |
| $C^k(X, Y)$ | Space of continuously $k$ times differentiable maps ($k \in \mathbb{N} \cup \{\infty\}$) |
| $\hookrightarrow, \overset{c}{\hookrightarrow}$ | Continuous (resp. compact) embedding |
| $D(A)$ | Domain of operator $A$ |
| $R(A)$ | Range of $A$ |
| $\operatorname{Gr} A$ | Graph of $A$ |
| $A^{-1}$ | Inverse of $A$ |
| $J_\lambda^A$ | Resolvent of $A$ |
| $A_\lambda$ | Yosida approximation of $A$ |
| $\mathscr{F}$ | Duality map |
| $\Gamma_0(X)$ | Set of lower semicontinuous, convex maps that are $\not\equiv +\infty$ |
| $\operatorname{dom} \varphi$ | Effective domain of map $\varphi : X \to \mathbb{R} \cup \{+\infty\}$ |
| $\varphi_\lambda$ | Moreau–Yosida regularization of $\varphi \in \Gamma_0(X)$ |
| $\varphi'(x; h)$ | Directional derivative |
| $\varphi'$ | Gâteaux or Fréchet differential |
| $\varphi^0(x; h)$ | Generalized directional derivative |
| $\partial \varphi$ | Convex or generalized subdifferential of $\varphi$ |
| $m_\varphi(x)$ | $\inf\{\|x^*\| : x^* \in \partial \varphi(x)\}$ for $\varphi$ locally Lipschitz |
| $J_\varphi(x)$ | Jacobian of $\varphi$ at $x$ |
| $K_\varphi$ | Set of critical points of $\varphi$ |
| $K_\varphi(U)$ | Set of critical points of $\varphi$ contained in $U$ |
| $K_\varphi^C$ | Set of critical points $x$ such that $\varphi(x) \in C$ |

| | |
|---|---|
| $K_\varphi^c$ | Abbreviation for $K_\varphi^{\{c\}}$ |
| $\varphi^a$ | Sublevel set $\{x : \varphi(x) \leq a\}$ |
| $d(\varphi, U, y)$ | Brouwer's degree of $\varphi$ |
| $d_{\mathrm{LS}}$ | Leray–Schauder degree map |
| $d_{(S)_+}, d_M$ | Degree map for $(S)_+$-maps (resp. for operators of monotone type) |
| gen | Krasnosel'skiǐ genus |
| $\frac{\partial}{\partial x_k}$ or $D_k$ | Partial derivative with respect to $k$th variable |
| $D^\alpha$ | Derivative with respect to multi-index $\alpha$ |
| $\nabla u$ | Gradient of $u$ |
| $\operatorname{div} G$ | Divergence of $G$ |
| $\Delta u$ | $\operatorname{div}(\nabla u)$ (Laplacian of $u$) |
| $\Delta_p u$ | $\operatorname{div}(\|\nabla u\|^{p-2}\nabla u)$ ($p$-Laplacian of $u$) |
| $C(\Omega)$ | Space of continuous functions $u : \Omega \to \mathbb{R}$ |
| $C^k(\Omega)$ | Space of continuously $k$ times differentiable functions $u : \Omega \to \mathbb{R}$ |
| $C_c^\infty(\Omega)$ | Space of infinitely differentiable functions with compact support in $\Omega$ |
| $C^k(\overline{\Omega})$ | Elements of $C^k(\Omega)$ admitting extension of class $C^k$ beyond $\Omega$ |
| $C_0^k(\overline{\Omega})$ | Elements of $C^k(\overline{\Omega})$ that are zero on $\partial\Omega$ |
| $C_0^k(\overline{\Omega})_+$ | Positive cone of $C_0^k(\overline{\Omega})$ |
| $C^{k,\lambda}(\overline{\Omega})$ | Elements of $C^k(\overline{\Omega})$ whose $k$th differential is $\lambda$-Hölder continuous |
| $L^p(\Omega)$ | Space of $p$ integrable functions $(1 \leq p \leq +\infty)$ |
| $L^p(\Omega)_+$ | A.e. nonnegative functions in $L^p(\Omega)$ |
| $\|u\|_p$ | Norm of $u$ in $L^p(\Omega)$ or in $L^p(\Omega, \mathbb{R}^N)$ |
| $W^{m,p}(\Omega)$ | Sobolev space |
| $W_0^{m,p}(\Omega)$ | $W^{m,p}(\Omega)$-functions with zero trace |
| $H^m(\Omega)$ | Hilbert space $W^{m,2}(\Omega)$ |
| $H_0^m(\Omega)$ | Hilbert space $W_0^{m,2}(\Omega)$ |
| $W_{\mathrm{loc}}^{1,p}(\Omega)$ | Space of locally $W^{1,p}$-functions |
| $W_{\mathrm{per}}^{1,p}((a,b))$ | Space of functions $u \in W^{1,p}((a,b))$ with $u(a) = u(b)$ |
| $\|u\|$ | Sobolev norm $(\|\nabla u\|_p^p + \|u\|_p^p)^{\frac{1}{p}}$ |
| $p'$ | $\frac{p}{p-1}$ |
| $p^*$ | Sobolev critical exponent, equal to $\frac{Np}{N-p}$ if $p < N$ or $+\infty$ if $p \geq N$ |
| $W^{-1,p'}(\Omega)$ | Dual space of $W_0^{1,p}(\Omega)$ |
| $H^{-1}(\Omega)$ | Dual space of $H_0^1(\Omega)$ |
| $\gamma(u)$ | Trace of $u$ or generalized boundary values of $u$ |
| $\frac{\partial u}{\partial n}, \gamma_n(u)$ | Classical and generalized normal derivatives of $u$ |
| $\frac{\partial u}{\partial n_p}$ | $\gamma_n(\|\nabla u\|^{p-2}\nabla u)$ |
| $\frac{\partial u}{\partial n_a}$ | $\gamma_n(a(x, \nabla u))$ for an operator $a : \Omega \times \mathbb{R}^N \to \mathbb{R}^N$ |
| $a_x'(x_0, \xi_0)$ | Differential of map $a(\cdot, \xi_0)$ evaluated at $x_0$ |
| $a_\xi'(x_0, \xi_0)$ | Differential of map $a(x_0, \cdot)$ evaluated at $\xi_0$ |
| $f_s'(x_0, s_0)$ | Differential of map $f(x_0, \cdot)$ evaluated at $s_0$ |
| $\Delta_p^{\mathrm{D}}, \Delta_p^{\mathrm{N}}$ | Dirichlet (resp. Neumann) $p$-Laplacian |
| $\sigma_p^{\mathrm{D}}(\xi)$ | Spectrum of $-\Delta_p^{\mathrm{D}}$ with respect to weight $\xi$ |

| | |
|---|---|
| $\sigma_p^N(\xi)$ | Spectrum of $-\Delta_p^N$ with respect to $\xi$ |
| $[\underline{u},\overline{u}]$ | Order interval between functions $\underline{u}$ and $\overline{u}$ |
| $f_{[\underline{u},\overline{u}]}, \hat{f}_{[\underline{u},\overline{u}]}$ | Truncated Carathéodory functions in Dirichlet and Neumann cases |
| $F_{[\underline{u},\overline{u}]}, \hat{F}_{[\underline{u},\overline{u}]}$ | Their primitives |
| $\varphi_{[\underline{u},\overline{u}]}, \hat{\varphi}_{[\underline{u},\overline{u}]}$ | Truncated energy functionals |
| (PS), (PS)$_c$ | Palais–Smale condition (resp. at level $c$) |
| (C), (C)$_c$ | Cerami condition (resp. at level $c$) |
| $\mathscr{R}$ | A ring with unit |
| $\mathbb{F}$ | A field of characteristic zero |
| $(X,A)$ | A topological pair |
| $H_k(X,A)$ | $k$th homology group of $(X,A)$ (usually singular homology) |
| $H_k(X)$ | $H_k(X,\emptyset)$ |
| $H_k(X,*)$ | Reduced homology group ($*$ stands for a point in $X$) |
| $f_*$ | Group homomorphism induced in homology by a map of pairs $f$ |
| $\partial$ or $\partial_k$ | Boundary homomorphism $H_k(X,A) \to H_{k-1}(A)$ |
| $C_k(\varphi,x)$ | $k$th critical group at $x$ of a functional $\varphi$ |
| $C_k(\varphi,\infty)$ | $k$th critical group of $\varphi$ at infinity |
| $M_k(a,b)$ | $k$th Morse type number between levels $a$ and $b$ |
| $\beta_k(a,b)$ | $\dim H_k(\varphi^b,\varphi^a)$ ($k$th Betti-type number) |

# References

1. Abramovich, Y.A., Aliprantis, C.D.: An Invitation to Operator Theory. American Mathematical Society, Providence, RI (2002)
2. Adams, R.A., Fournier, J.J.F.: Sobolev Spaces. Elsevier/Academic, Amsterdam (2003)
3. Aizicovici, S., Papageorgiou, N.S., Staicu, V.: Degree theory for operators of monotone type and nonlinear elliptic equations with inequality constraints. Mem. Amer. Math. Soc. **196**, 915 (2008)
4. Aizicovici, S., Papageorgiou, N.S., Staicu, V.: The spectrum and an index formula for the Neumann $p$-Laplacian and multiple solutions for problems with a crossing nonlinearity. Discrete Contin. Dyn. Syst. **25**, 431–456 (2009)
5. Aizicovici, S., Papageorgiou, N.S., Staicu, V.: Nonlinear resonant periodic problems with concave terms. J. Math. Anal. Appl. **375**, 342–364 (2011)
6. Aizicovici, S., Papageorgiou, N.S., Staicu, V.: Positive solutions for nonlinear periodic problems with concave terms. J. Math. Anal. Appl. **381**, 866–883 (2011)
7. Akhiezer, N.I., Glazman, I.M.: Theory of Linear Operators in Hilbert Space. Dover Publications Inc., New York (1993)
8. Allegretto, W., Huang, Y.X.: A Picone's identity for the $p$-Laplacian and applications. Nonlin. Anal. **32**, 819–830 (1998)
9. Alves, C.O., Carrião, P.C., Miyagaki, O.H.: Multiple solutions for a problem with resonance involving the $p$-Laplacian. Abstr. Appl. Anal. **3**, 191–201 (1998)
10. Amann, H.: Saddle points and multiple solutions of differential equations. Math. Z. **169**, 127–166 (1979)
11. Amann, H.: A note on degree theory for gradient mappings. Proc. Amer. Math. Soc. **85**, 591–595 (1982)
12. Amann, H., Weiss, S.A.: On the uniqueness of the topological degree. Math. Z. **130**, 39–54 (1973)
13. Amann, H., Zehnder, E.: Nontrivial solutions for a class of nonresonance problems and applications to nonlinear differential equations. Ann. Scuola Norm. Sup. Pisa Cl. Sci. **7**(4), 539–603 (1980)
14. Ambrosetti, A., Arcoya, D.: An Introduction to Nonlinear Functional Analysis and Elliptic Problems. Birkhäuser, Boston (2011)
15. Ambrosetti, A., Malchiodi, A.: Nonlinear Analysis and Semilinear Elliptic Problems. Cambridge University Press, Cambridge (2007)
16. Ambrosetti, A., Prodi, G.: A Primer of Nonlinear Analysis. Cambridge University Press, Cambridge (1993)
17. Ambrosetti, A., Rabinowitz, P.H.: Dual variational methods in critical point theory and applications. J. Funct. Anal. **14**, 349–381 (1973)

18. Ambrosetti, A., Brezis, H., Cerami, G.: Combined effects of concave and convex nonlinearities in some elliptic problems. J. Funct. Anal. **122**, 519–543 (1994)
19. Anane, A.: Simplicité et isolation de la première valeur propre du $p$-laplacien avec poids. C. R. Acad. Sci. Paris Sér. I Math. **305**, 725–728 (1987)
20. Anane, A., Tsouli, N.: On the second eigenvalue of the $p$-Laplacian. In: Nonlinear Partial Differential Equations (Fès, 1994). Longman, Harlow (1996)
21. Anello, G.: Existence of infinitely many weak solutions for a Neumann problem. Nonlin. Anal. **57**, 199–209 (2004)
22. Appell, J., Zabreĭko, P.P.: Nonlinear Superposition Operators. Cambridge University Press. Cambridge (1990)
23. Arcoya, D., Carmona, J., Pellacci, B.: Bifurcation for some quasilinear operators. Proc. Roy. Soc. Edinb. Sect. A **131**, 733–765 (2001)
24. Arias, M., Campos, J.: Radial Fučik spectrum of the Laplace operator. J. Math. Anal. Appl. **190**, 654–666 (1995)
25. Arias, M., Campos, J., Cuesta, M., Gossez, J.-P.: Asymmetric elliptic problems with indefinite weights. Ann. Inst. H. Poincaré Anal. Non Linéaire **19**, 581–616 (2002)
26. Averna, D., Marano, S.A., Motreanu, D.: Multiple solutions for a Dirichlet problem with $p$-Laplacian and set-valued nonlinearity. Bull. Aust. Math. Soc. **77**, 285–303 (2008)
27. Badiale, M., Serra, E.: Semilinear Elliptic Equations for Beginners: Existence Results via the Variational Approach. Springer, London (2011)
28. Balanov, Z., Krawcewicz, W., Steinlein, H.: Applied Equivariant Degree. American Institute of Mathematical Sciences (AIMS), Springfield (2006)
29. Barbu, V.: Nonlinear Semigroups and Differential Equations in Banach Spaces. Noordhoff International Publishing, Leiden (1976)
30. Barbu, V.: Analysis and Control of Nonlinear Infinite-Dimensional Systems. Academic Inc., Boston (1993)
31. Barletta, G., Papageorgiou, N.S.: A multiplicity theorem for the Neumann $p$-Laplacian with an asymmetric nonsmooth potential. J. Global Optim. **39**, 365–392 (2007)
32. Barletta, G., Papageorgiou, N.S.: Nonautonomous second order periodic systems: existence and multiplicity of solutions. J. Nonlin. Convex Anal. **8**, 373–390 (2007)
33. Bartolo, P., Benci, V., Fortunato, D.: Abstract critical point theorems and applications to some nonlinear problems with "strong" resonance at infinity. Nonlin. Anal. **7**, 981–1012 (1983)
34. Bartsch, T.: Topological Methods for Variational Problems with Symmetries. Lecture Notes in Mathematics, vol. 1560. Springer, Berlin (1993)
35. Bartsch, T.: Infinitely many solutions of a symmetric Dirichlet problem. Nonlin. Anal. **20**, 1205–1216 (1993)
36. Bartsch, T., Li, S.: Critical point theory for asymptotically quadratic functionals and applications to problems with resonance. Nonlin. Anal. **28**, 419–441 (1997)
37. Bartsch, T., Liu, Z.: On a superlinear elliptic $p$-Laplacian equation. J. Differ. Equat. **198**, 149–175 (2004)
38. Bartsch, T., Liu, Z., Weth, T.: Nodal solutions of a $p$-Laplacian equation. Proc. London Math. Soc. **91**(3), 129–152 (2005)
39. Benci, V.: On critical point theory for indefinite functionals in the presence of symmetries. Trans. Amer. Math. Soc. **274**, 533–572 (1982)
40. Benci, V., Rabinowitz, P.H.: Critical point theorems for indefinite functionals. Invent. Math. **52**, 241–273 (1979)
41. Ben-Naoum, A.K., De Coster, C.: On the existence and multiplicity of positive solutions of the $p$-Laplacian separated boundary value problem. Differ. Integr. Equat. **10**, 1093–1112 (1997)
42. Benyamini, Y., Sternfeld, Y.: Spheres in infinite-dimensional normed spaces are Lipschitz contractible. Proc. Amer. Math. Soc. **88**, 439–445 (1983)
43. Bessaga, C.: Every infinite-dimensional Hilbert space is diffeomorphic with its unit sphere. Bull. Acad. Polon. Sci. Sér. Sci. Math. Astronom. Phys. **14**, 27–31 (1966)
44. Beurling, A., Livingston, A.E.: A theorem on duality mappings in Banach spaces. Ark. Mat. **4**, 405–411 (1962)

45. Binding, P.A., Rynne, B.P.: The spectrum of the periodic $p$-Laplacian. J. Differ. Equat. **235**, 199–218 (2007)
46. Binding, P.A., Rynne, B.P.: Variational and non-variational eigenvalues of the $p$-Laplacian. J. Differ. Equat. **244**, 24–39 (2008)
47. Blanchard, P., Brüning, E.: Variational Methods in Mathematical Physics. Springer, Berlin (1992)
48. Bonanno, G., Candito, P.: Three solutions to a Neumann problem for elliptic equations involving the $p$-Laplacian. Arch. Math. (Basel) **80**, 424–429 (2003)
49. Borwein, J.M., Vanderwerff, J.D.: Convex Functions: Constructions, Characterizations and Counterexamples. Cambridge University Press, Cambridge (2010)
50. Brezis, H.: Équations et inéquations non linéaires dans les espaces vectoriels en dualité. Ann. Inst. Fourier (Grenoble) **18**, 115–175 (1968)
51. Brézis, H.: Opérateurs Maximaux Monotones et Semi-Groupes de Contractions dans les Espaces de Hilbert. North-Holland Publishing Co., Amsterdam (1973)
52. Brezis, H.: Functional Analysis, Sobolev Spaces and Partial Differential Equations. Springer, New York (2011)
53. Brezis, H., Nirenberg, L.: Remarks on finding critical points. Comm. Pure Appl. Math. **44**, 939–963 (1991)
54. Brezis, H., Nirenberg, L.: $H^1$ versus $C^1$ local minimizers. C. R. Acad. Sci. Paris Sér. I Math. **317**, 465–472 (1993)
55. Brøndsted, A., Rockafellar, R.T.: On the subdifferentiability of convex functions. Proc. Amer. Math. Soc. **16**, 605–611 (1965)
56. Brouwer, L.E.J.: Über Abbildung von Mannigfaltigkeiten. Math. Ann. **71**, 97–115 (1912)
57. Browder, F.E.: Nonlinear maximal monotone operators in Banach space. Math. Ann. **175**, 89–113 (1968)
58. Browder, F.E.: Nonlinear monotone and accretive operators in Banach spaces. Proc. Nat. Acad. Sci. U.S.A. **61**, 388–393 (1968)
59. Browder, F.E.: Nonlinear operators and nonlinear equations of evolution in Banach spaces. In: Nonlinear Functional Analysis (Proc. Sympos. Pure Math., vol. XVIII, Part 2, Chicago, Ill., 1968), pp. 1–308. American Mathematical Society, Providence (1976)
60. Browder, F.E.: Fixed point theory and nonlinear problems. Bull. Amer. Math. Soc. (N.S.) **9**, 1–39 (1983)
61. Browder, F.E.: Degree of mapping for nonlinear mappings of monotone type. Proc. Nat. Acad. Sci. USA **80**, 1771–1773 (1983)
62. Browder, F.E., Hess, P.: Nonlinear mappings of monotone type in Banach spaces. J. Funct. Anal. **11**, 251–294 (1972)
63. Các, N.P.: On nontrivial solutions of a Dirichlet problem whose jumping nonlinearity crosses a multiple eigenvalue. J. Differ. Equat. **80**, 379–404 (1989)
64. Čaklović, L., Li, S.J., Willem, M.: A note on Palais-Smale condition and coercivity. Differ. Integr. Equat. **3**, 799–800 (1990)
65. Cambini, A.: Sul lemma di M. Morse. Boll. Un. Mat. Ital. **7**(4), 87–93 (1973)
66. Candito, P., Livrea, R., Motreanu, D.: $\mathbb{Z}_2$-symmetric critical point theorems for non-differentiable functions. Glasg. Math. J. **50**, 447–466 (2008)
67. Candito, P., Livrea, R., Motreanu, D.: Bounded Palais-Smale sequences for non-differentiable functions. Nonlin. Anal.**74**, 5446–5454 (2011)
68. Carl, S., Motreanu, D.: Quasilinear elliptic inclusions of hemivariational type: extremality and compactness of the solution set. J. Math. Anal. Appl. **286**, 147–159 (2003)
69. Carl, S., Motreanu, D.: Constant-sign and sign-changing solutions of a nonlinear eigenvalue problem involving the $p$-Laplacian. Differ. Integr. Equat. **20**, 309–324 (2007)
70. Carl, S., Motreanu, D.: Constant-sign and sign-changing solutions for nonlinear eigenvalue problems. Nonlin. Anal. **68**, 2668–2676 (2008)
71. Carl, S., Motreanu, D.: Multiple and sign-changing solutions for the multivalued $p$-Laplacian equation. Math. Nachr. **283**, 965–981 (2010)

72. Carl, S., Le, V.K., Motreanu, D.: Nonsmooth Variational Problems and Their Inequalities. Comparison Principles and Applications. Springer, New York (2007)

73. Casas, E., Fernández, L.A.: A Green's formula for quasilinear elliptic operators. J. Math. Anal. Appl. **142**, 62–73 (1989)

74. Castro, A., Lazer, A.C.: Critical point theory and the number of solutions of a nonlinear Dirichlet problem. Ann. Mat. Pura Appl. **120**(4), 113–137 (1979)

75. Cerami, G.: An existence criterion for the critical points on unbounded manifolds. Istit. Lombardo Accad. Sci. Lett. Rend. A **112**, 332–336 (1978)

76. Chang, K.-C.: Solutions of asymptotically linear operator equations via Morse theory. Comm. Pure Appl. Math. **34**, 693–712 (1981)

77. Chang, K.-C.: Variational methods for nondifferentiable functionals and their applications to partial differential equations. J. Math. Anal. Appl. **80**, 102–129 (1981)

78. Chang, K.-C.: Infinite-Dimensional Morse Theory and Multiple Solution Problems. Birkhäuser, Boston (1993)

79. Chang, K.-C.: $H^1$ versus $C^1$ isolated critical points. C. R. Acad. Sci. Paris Sér. I Math. **319**, 441–446 (1994)

80. Cherfils, L., Il'yasov, Y.: On the stationary solutions of generalized reaction diffusion equations with $p\&q$-Laplacian. Commun. Pure Appl. Anal. **4**, 9–22 (2005)

81. Christensen, J.P.R.: Topology and Borel Structure. North-Holland Publishing Co., Amsterdam (1974)

82. Ciorănescu, I.: Geometry of Banach Spaces, Duality Mappings and Nonlinear Problems. Kluwer Academic Publishers, Dordrecht (1990)

83. Clark, D.C.: A variant of the Lusternik-Schnirelman theory. Indiana Univ. Math. J. **22**, 65–74 (1972/1973)

84. Clarke, F.H.: Generalized gradients and applications. Trans. Amer. Math. Soc. **205**, 247–262 (1975)

85. Clarke, F.H.: Optimization and Nonsmooth Analysis. A Wiley-Interscience Publication, Wiley Inc., New York (1983)

86. Coffman, C.V.: A minimum-maximum principle for a class of non-linear integral equations. J. Anal. Math. **22**, 391–419 (1969)

87. Cordaro, G.: Three periodic solutions to an eigenvalue problem for a class of second-order Hamiltonian systems. Abstr. Appl. Anal. **2003**, 1037–1045 (2003)

88. Corvellec, J.-N., Motreanu, V.V., Saccon, C.: Doubly resonant semilinear elliptic problems via nonsmooth critical point theory. J. Differ. Equat. **248**, 2064–2091 (2010)

89. Costa, D.G.: An Invitation to Variational Methods in Differential Equations. Birkhäuser, Boston (2007)

90. Costa, D.G., Magalhães, C.A.: Existence results for perturbations of the $p$-Laplacian. Nonlin. Anal. **24**, 409–418 (1995)

91. Costa, D.G., Silva, E.A.: The Palais-Smale condition versus coercivity. Nonlin. Anal. **16**, 371–381 (1991)

92. Courant, R., Hilbert, D.: Methods of Mathematical Physics, vol. I. Interscience Publishers, New York (1953)

93. Crandall, M.G., Rabinowitz, P.H.: Bifurcation from simple eigenvalues. J. Funct. Anal. **8**, 321–340 (1971)

94. Cuesta, M.: Minimax theorems on $C^1$ manifolds via Ekeland variational principle. Abstr. Appl. Anal. **2003**, 757–768 (2003)

95. Cuesta, M., Ramos Quoirin, H.: A weighted eigenvalue problem for the $p$-Laplacian plus a potential. NoDEA - Nonlin. Differ. Equat. Appl. **16**, 469–491 (2009)

96. Cuesta, M., Takáč, P.: A strong comparison principle for positive solutions of degenerate elliptic equations. Differ. Integr. Equat. **13**, 721–746 (2000)

97. Cuesta, M., Takáč, P.: Nonlinear eigenvalue problems for degenerate elliptic systems. Differ. Integr. Equat. **23**, 1117–1138 (2010)

98. Cuesta, M., de Figueiredo, D., Gossez, J.-P.: The beginning of the Fučik spectrum for the $p$-Laplacian. J. Differ. Equat. **159**, 212–238 (1999)

99. Damascelli, L.: Comparison theorems for some quasilinear degenerate elliptic operators and applications to symmetry and monotonicity results. Ann. Inst. H. Poincaré Anal. Non Linéaire **15**, 493–516 (1998)

100. Damascelli, L., Sciunzi, B.: Harnack inequalities, maximum and comparison principles, and regularity of positive solutions of $m$-Laplace equations. Calc. Var. Partial Differ. Equat. **25**, 139–159 (2006)

101. Dancer, E.N.: Remarks on jumping nonlinearities. In: Topics in Nonlinear Analysis, pp. 101–116. Birkhäuser, Basel (1999)

102. Dancer, E.N., Du, Y.: On sign-changing solutions of certain semilinear elliptic problems. Appl. Anal. **56**, 193–206 (1995)

103. Day, M.M.: Some more uniformly convex spaces. Bull. Amer. Math. Soc. **47**, 504–507 (1941)

104. de Figueiredo, D.G.: Positive solutions of semilinear elliptic problems. In: Differential Equations (Sao Paulo, 1981). Lecture Notes in Mathematics, vol. 957, pp. 34–87. Springer, Berlin (1982)

105. de Figueiredo, D.G.: Lectures on the Ekeland Variational Principle with Applications and Detours. Tata Institute of Fundamental Research, Bombay (1989)

106. de Figueiredo, D.G., Gossez, J.-P.: On the first curve of the Fučik spectrum of an elliptic operator. Differ. Integr. Equat. **7**, 1285–1302 (1994)

107. Degiovanni, M., Marzocchi, M.: A critical point theory for nonsmooth functionals. Ann. Mat. Pura Appl. **167**(4), 73–100 (1994)

108. Deimling, K.: Nonlinear Functional Analysis. Springer, Berlin (1985)

109. Del Pezzo, L.M., Fernández Bonder, J.: An optimization problem for the first weighted eigenvalue problem plus a potential. Proc. Amer. Math. Soc. **138**, 3551–3567 (2010)

110. del Pino, M.A., Manásevich, R.F.: Global bifurcation from the eigenvalues of the $p$-Laplacian. J. Differ. Equat. **92**, 226–251 (1991)

111. del Pino, M.A., Manásevich, R.F., Murúa, A.E.: Existence and multiplicity of solutions with prescribed period for a second order quasilinear ODE. Nonlin. Anal. **18**, 79–92 (1992)

112. De Nápoli, P., Mariani, M.C.: Mountain pass solutions to equations of $p$-Laplacian type. Nonlin. Anal. **54**, 1205–1219 (2003)

113. Denkowski, Z., Migórski, S., Papageorgiou, N.S.: An Introduction to Nonlinear Analysis: Theory. Kluwer Academic Publishers, Boston (2003)

114. Denkowski, Z., Migórski, S., Papageorgiou, N.S.: An Introduction to Nonlinear Analysis: Applications. Kluwer Academic Publishers, Boston (2003)

115. Deny, J., Lions, J. L.: Les espaces du type de Beppo Levi. Ann. Inst. Fourier Grenoble **5**, 305–370 (1953–54)

116. de Paiva, F.O., do Ó, J.M., de Medeiros, E.S.: Multiplicity results for some quasilinear elliptic problems. Topol. Methods Nonlin. Anal. **34**, 77–89 (2009)

117. Deville, R., Godefroy, G., Zizler, V.: Smoothness and Renormings in Banach Spaces. Longman Scientific and Technical, Harlow (1993)

118. DiBenedetto, E.: $C^{1+\alpha}$ local regularity of weak solutions of degenerate elliptic equations. Nonlin. Anal. **7**, 827–850 (1983)

119. Dold, A.: Lectures on Algebraic Topology. Springer, Berlin (1980)

120. Drábek, P.: On the global bifurcation for a class of degenerate equations. Ann. Mat. Pura Appl. **159**(4), 1–16 (1991)

121. Drábek, P., Manásevich, R.: On the closed solution to some nonhomogeneous eigenvalue problems with $p$-Laplacian. Differ. Integr. Equat. **12**, 773–788 (1999)

122. Dugundji, J.: An extension of Tietze's theorem. Pacific J. Math. **1**, 353–367 (1951)

123. Dugundji, J.: Topology. Allyn and Bacon Inc., Boston (1966)

124. Dunford, N., Schwartz, J.T.: Linear Operators I: General Theory. Interscience Publishers, Inc., New York (1958)

125. Eilenberg, S., Steenrod, N.: Foundations of Algebraic Topology. Princeton University Press, Princeton (1952)

126. Ekeland, I.: On the variational principle. J. Math. Anal. Appl. **47**, 324–353 (1974)

127. Ekeland, I.: Nonconvex minimization problems. Bull. Amer. Math. Soc. (N.S.) **1**, 443–474 (1979)
128. Ekeland, I.: Convexity Methods in Hamiltonian Mechanics. Springer, Berlin (1990)
129. Ekeland, I., Temam, R.: Convex Analysis and Variational Problems. North-Holland Publishing Co., Amsterdam (1976)
130. Evans, L.C.: Partial Differential Equations. American Mathematical Society, Providence (1998)
131. Evans, L.C., Gariepy, R.F.: Measure Theory and Fine Properties of Functions. CRC, Boca Raton (1992)
132. Fabry, C., Fonda, A.: Periodic solutions of nonlinear differential equations with double resonance. Ann. Mat. Pura Appl. **157**(4), 99–116 (1990)
133. Fadell, E.R., Rabinowitz, P.H.: Generalized cohomological index theories for Lie group actions with an application to bifurcation questions for Hamiltonian systems. Invent. Math. **45**, 139–174 (1978)
134. Fadell, E.R., Husseini, S.Y., Rabinowitz, P.H.: Borsuk-Ulam theorems for arbitrary $S^1$ actions and applications. Trans. Amer. Math. Soc. **274**, 345–360 (1982)
135. Faraci, F.: Three periodic solutions for a second order nonautonomous system. J. Nonlin. Convex Anal. **3**, 393–399 (2002)
136. Faria, L.F.O., Miyagaki, O.H., Motreanu, D.: Comparisonand positive solutions for problems with $(p, q)$-Laplacian and convection term. Proc. Edinb. Math. Soc. (2) (to appear)
137. Fernández Bonder, J., Del Pezzo, L.M.: An optimization problem for the first eigenvalue of the $p$-Laplacian plus a potential. Commun. Pure Appl. Anal. **5**, 675–690 (2006)
138. Filippakis, M.E., Papageorgiou, N.S.: Solutions for nonlinear variational inequalities with a nonsmooth potential. Abstr. Appl. Anal. **8**, 635–649 (2004)
139. Finn, R., Gilbarg, D.: Asymptotic behavior and uniqueness of plane subsonic flows. Comm. Pure Appl. Math. **10**, 23–63 (1957)
140. Floret, K.: Weakly Compact Set. Lecture Notes in Mathematics, vol. 801. Springer, Berlin (1980)
141. Fonseca, I., Gangbo, W.: Degree Theory in Analysis and Applications. The Clarendon Press and Oxford University Press, New York (1995)
142. Fredholm, J.: Sur une classe d'équations fonctionnelles. Acta Math. **27**, 365–390 (1903)
143. Fučík, S.: Boundary value problems with jumping nonlinearities. Časopis Pěst. Mat. **101**, 69–87 (1976)
144. Führer, L.: Ein elementarer analytischer Beweis zur Eindeutigkeit des Abbildungsgrades im $R^n$. Math. Nachr. **54**, 259–267 (1972)
145. Gagliardo, E.: Proprietà di alcune classi di funzioni in più variabili. Ricerche Mat. **7**, 102–137 (1958)
146. Gallouët, T., Kavian, O.: Résultats d'existence et de non-existence pour certains problèmes demi-linéaires à l'infini. Ann. Fac. Sci. Toulouse Math. **3**(5), 201–246 (1981)
147. García Azorero, J.P., Manfredi, J.J., Peral Alonso, I.: Sobolev versus Hölder local minimizers and global multiplicity for some quasilinear elliptic equations. Commun. Contemp. Math. **2**, 385–404 (2000)
148. Garofalo, N., Lin, F.-H.: Unique continuation for elliptic operators: a geometric-variational approach. Comm. Pure Appl. Math. **40**, 347–366 (1987)
149. Gasiński, L.: Positive solutions for resonant boundary value problems with the scalar $p$-Laplacian and nonsmooth potential. Discrete Contin. Dyn. Syst. **17**, 143–158 (2007)
150. Gasiński, L., Papageorgiou, N.S.: Nonsmooth Critical Point Theory and Nonlinear Boundary Value Problems. Chapman and Hall/CRC, Boca Raton (2005)
151. Gasiński, L., Papageorgiou, N.S.: Nonlinear Analysis. Chapman and Hall/CRC, Boca Raton, FL (2006)
152. Gasiński, L., Papageorgiou, N.S.: Three nontrivial solutions for periodic problems with the $p$-Laplacian and a $p$-superlinear nonlinearity. Commun. Pure Appl. Anal. **8**, 1421–1437 (2009)

153. Gasiński, L., Papageorgiou, N.S.: Nodal and multiple constant sign solutions for resonant $p$-Laplacian equations with a nonsmooth potential. Nonlin. Anal. **71**, 5747–5772 (2009)
154. Gasiński, L., Papageorgiou, N.S.: Multiple solutions for asymptotically $(p-1)$-homogeneous $p$-Laplacian equations. J. Funct. Anal. **262**, 2403–2435 (2012)
155. Gasiński, L., Papageorgiou, N.S.: Bifurcation-type results for nonlinear parametric elliptic equations. Proc. Roy. Soc. Edinb. Sect. A **142**, 595–623 (2012)
156. Ghoussoub, N.: Duality and Perturbation Methods in Critical Point Theory. Cambridge University Press, Cambridge (1993)
157. Ghoussoub, N., Preiss, D.: A general mountain pass principle for locating and classifying critical points. Ann. Inst. H. Poincaré Anal. Non Linéaire **6**, 321–330 (1989)
158. Gilbarg, D., Trudinger, N.S.: Elliptic Partial Differential Equations of Second Order. Springer, Berlin (2001)
159. Giles, J.R.: Convex Analysis with Application in the Differentiation of Convex Functions. Pitman, Boston (1982)
160. Godoy, T., Gossez, J.-P., Paczka, S.: On the antimaximum principle for the $p$-Laplacian with indefinite weight. Nonlin. Anal. **51**, 449–467 (2002)
161. Goeleven, D., Motreanu, D.: A degree-theoretic approach for the study of eigenvalue problems in variational-hemivariational inequalities. Differ. Integr. Equat. **10**, 893–904 (1997)
162. Goeleven, D., Motreanu, D., Panagiotopoulos, P.D.: Eigenvalue problems for variational-hemivariational inequalities at resonance. Nonlin. Anal. **33**, 161–180 (1998)
163. Gossez, J.-P., Omari, P.: Periodic solutions of a second order ordinary differential equation: a necessary and sufficient condition for nonresonance. J. Differ. Equat. **94**, 67–82 (1991)
164. Gossez, J.-P., Omari, P.: A necessary and sufficient condition of nonresonance for a semilinear Neumann problem. Proc. Amer. Math. Soc. **114**, 433–442 (1992)
165. Granas, A., Dugundji, J.: Fixed Point Theory. Springer, New York (2003)
166. Gromoll, D., Meyer, W.: On differentiable functions with isolated critical points. Topology **8**, 361–369 (1969)
167. Guedda, M., Véron, L.: Quasilinear elliptic equations involving critical Sobolev exponents. Nonlin. Anal. **13**, 879–902 (1989)
168. Guo, Z., Zhang, Z.: $W^{1,p}$ versus $C^1$ local minimizers and multiplicity results for quasilinear elliptic equations. J. Math. Anal. Appl. **286**, 32–50 (2003)
169. Halmos, P.R.: Introduction to Hilbert Space and the Theory of Spectral Multiplicity. AMS Chelsea Publishing, Providence (1998)
170. Heinz, E.: An elementary analytic theory of the degree of mapping in $n$-dimensional space. J. Math. Mech. **8**, 231–247 (1959)
171. Hilbert, D.: Grundzüge einer allgemeinen Theorie der linearen Integralgleichungen. IV. Gött. Nachr. **1906**, 157–227 (1906)
172. Hofer, H.: Variational and topological methods in partially ordered Hilbert spaces. Math. Ann. **261**, 493–514 (1982)
173. Hofer, H.: A geometric description of the neighbourhood of a critical point given by the mountain-pass theorem. J. London Math. Soc. **31**(2), 566–570 (1985)
174. Hu, S., Papageorgiou, N.S.: Generalizations of Browder's degree theory. Trans. Amer. Math. Soc. **347**, 233–259 (1995)
175. Hu, S., Papageorgiou, N.S.: Handbook of Multivalued Analysis. vol. I. Theory. Kluwer Academic Publishers, Dordrecht (1997)
176. Hu, S., Papageorgiou, N.S.: Multiple positive solutions for nonlinear eigenvalue problems with the $p$-Laplacian. Nonlin. Anal. **69**, 4286–4300 (2008)
177. Hu, S., Papageorgiou, N.S.: Nontrivial solutions for superquadratic nonautonomous periodic systems. Topol. Methods Nonlin. Anal. **34**, 327–338 (2009)
178. Hu, S., Papageorgiou, N.S.: Multiplicity of solutions for parametric $p$-Laplacian equations with nonlinearity concave near the origin. Tohoku Math. J. **62**(2), 137–162 (2010)
179. Iannacci, R., Nkashama, M.N.: Nonlinear boundary value problems at resonance. Nonlin. Anal. **11**, 455–473 (1987)

448                                                                    References

88

180. Iannacci, R., Nkashama, M.N.: Nonlinear two-point boundary value problems at resonance without Landesman-Lazer condition. Proc. Amer. Math. Soc. **106**, 943–952 (1989)
181. Ioffe, A.D., Tihomirov, V.M.: Theory of Extremal Problems. North-Holland Publishing Co., Amsterdam (1979)
182. Ize, J., Vignoli, A.: Equivariant Degree Theory. Walter de Gruyter and Co., Berlin (2003)
183. Jabri, Y.: The Mountain Pass Theorem: Variants, Generalizations and Some Applications. Cambridge University Press, Cambridge (2003)
184. Jebelean, P., Motreanu, D., Motreanu, V.V.: A unified approach for a class of problems involving a pseudo-monotone operator. Math. Nachr. **281**, 1283–1293 (2008)
185. Jiu, Q., Su, J.: Existence and multiplicity results for Dirichlet problems with $p$-Laplacian. J. Math. Anal. Appl. **281**, 587–601 (2003)
186. Jost, J.: Partial Differential Equations. Springer, New York (2002)
187. Kačurovskiĭ, R.I.: Monotone operators and convex functionals. Uspekhi Mat. Nauk **154**(94), 213–215 (1960)
188. Kačurovskiĭ, R.I.: Nonlinear monotone operators in Banach spaces. Uspekhi Mat. Nauk **232**(140), 121–168 (1968)
189. Kartsatos, A.G., Skrypnik, I.V.: Topological degree theories for densely defined mappings involving operators of type $(S_+)$. Adv. Differ. Equat. **4**, 413–456 (1999)
190. Kato, T.: Perturbation Theory for Linear Operators. Springer, Berlin (1995)
191. Kavian, O.: Introduction à la Théorie des Points Critiques et Applications aux Problèmes Elliptiques. Springer, Paris (1993)
192. Kenmochi, N.: Nonlinear operators of monotone type in reflexive Banach spaces and nonlinear perturbations. Hiroshima Math. J. **4**, 229–263 (1974)
193. Kenmochi, N.: Pseudomonotone operators and nonlinear elliptic boundary value problems. J. Math. Soc. Japan **27**, 121–149 (1975)
194. Khan, A.A., Motreanu, D.: Local minimizers versus $X$-local minimizers. Optim. Lett. **7**, 1027–1033 (2013)
195. Kielhöfer, H.: Bifurcation Theory: An Introduction with Applications to PDEs. Springer, New York (2004)
196. Kien, B.T., Wong, M.M., Wong, N.-C.: On the degree theory for general mappings of monotone type. J. Math. Anal. Appl. **340**, 707–720 (2008)
197. Kobayashi, J., Ôtani, M.: Topological degree for $(S)_+$-mappings with maximal monotone perturbations and its applications to variational inequalities. Nonlin. Anal. **59**, 147–172 (2004)
198. Kobayashi, J., Ôtani, M.: An index formula for the degree of $(S)_+$-mappings associated with one-dimensional $p$-Laplacian. Abstr. Appl. Anal. **2004**, 981–995 (2004)
199. Kobayashi, J., Ôtani, M.: Degree for subdifferential operators in Hilbert spaces. Adv. Math. Sci. Appl. **14**, 307–325 (2004)
200. Kobayashi, J., Ôtani, M.: The principle of symmetric criticality for non-differentiable mappings. J. Funct. Anal. **214**, 428–449 (2004)
201. Kondrachov, W.: Sur certaines propriétés des fonctions dans l'espace. C. R. (Doklady) Acad. Sci. URSS (N.S.) **48**, 535–538 (1945)
202. Krasnosel'skiĭ, M.A.: Topological Methods in the Theory of Nonlinear Integral Equations. The Macmillan Co., New York (1964)
203. Krasnosel'skiĭ, M.A., Zabreĭko, P.P., Pustyl'nik, E.I., Sobolevskiĭ, P.E.: Integral Operators in Spaces of Summable Functions. Noordhoff International Publishing, Leiden (1976)
204. Krawcewicz, W., Marzantowicz, W.: Some remarks on the Lusternik-Schnirel'man method for nondifferentiable functionals invariant with respect to a finite group action. Rocky Mountain J. Math. **20**, 1041–1049 (1990)
205. Krawcewicz, W., Wu, J.: Theory of Degrees with Applications to Bifurcations and Differential Equations. Wiley Inc., New York (1997)
206. Kufner, A.: Weighted Sobolev Spaces. Wiley Inc., New York (1985)
207. Kufner, A., John, O., Fučík, S.: Function Spaces. Noordhoff International Publishing, Leyden (1977)

208. Kuiper, N.H.: $C^1$-equivalence of functions near isolated critical points. In: Symposium on Infinite-Dimensional Topology (Louisiana State University, Baton Rouge, LA, 1967). Annals of Mathematics Studies, vol. 69, pp. 199–218. Princeton University Press, Princeton (1972)
209. Kuo, C.C.: On the solvability of a nonlinear second-order elliptic equation at resonance. Proc. Amer. Math. Soc. **124**, 83–87 (1996)
210. Kyritsi, S.T., Papageorgiou, N.S.: Nonsmooth critical point theory on closed convex sets and nonlinear hemivariational inequalities. Nonlin. Anal. **61**, 373–403 (2005)
211. Kyritsi, S.T., Papageorgiou, N.S.: Solutions for doubly resonant nonlinear non-smooth periodic problems. Proc. Edinb. Math. Soc. **48**(2), 199–211 (2005)
212. Kyritsi, S.T., Papageorgiou, N.S.: Positive solutions for the periodic scalar $p$-Laplacian: existence and uniqueness. Taiwanese J. Math. **16**, 1345–1361 (2012)
213. Kyritsi, S.T., Papageorgiou, N.S.: Multiple solutions for nonlinear elliptic equations with an asymmetric reaction term. Discrete Contin. Dyn. Syst. Ser. A **33**, 2469–2494 (2013)
214. Kyritsi, S.T., O'Regan, D., Papageorgiou, N.S.: Existence of multiple solutions for nonlinear Dirichlet problems with a nonhomogeneous differential operator. Adv. Nonlin. Stud. **10**, 631–657 (2010)
215. Ladyzhenskaya, O.A., Ural'tseva, N.N.: Linear and Quasilinear Elliptic Equations. Academic, New York (1968)
216. Landesman, E.M., Lazer, A.C.: Nonlinear perturbations of linear elliptic boundary value problems at resonance. J. Math. Mech. **19**, 609–623 (1969/1970)
217. Laurent, P.-J.: Approximation et Optimisation. Hermann, Paris (1972)
218. Lax, P.D., Milgram, A.N.: Parabolic equations. In: Contributions to the Theory of Partial Differential Equations, pp. 167–190. Princeton University Press, Princeton (1954)
219. Lazer, A.C., Solimini, S.: Nontrivial solutions of operator equations and Morse indices of critical points of min-max type. Nonlin. Anal. **12**, 761–775 (1988)
220. Lê, A.: Eigenvalue problems for the $p$-Laplacian. Nonlin. Anal. **64**, 1057–1099 (2006)
221. Lebourg, G.: Valeur moyenne pour gradient généralisé. C. R. Acad. Sci. Paris Sér. A **281**, 795–797 (1975)
222. Leray, J., Schauder, J.: Topologie et équations fonctionnelles. Ann. Sci. École Norm. Sup. **51**(3), 45–78 (1934)
223. Li, C.: The existence of infinitely many solutions of a class of nonlinear elliptic equations with Neumann boundary condition for both resonance and oscillation problems. Nonlin. Anal. **54**, 431–443 (2003)
224. Li, S.J., Willem, M.: Applications of local linking to critical point theory. J. Math. Anal. Appl. **189**, 6–32 (1995)
225. Li, G., Yang, C.: The existence of a nontrivial solution to a nonlinear elliptic boundary value problem of $p$-Laplacian type without the Ambrosetti-Rabinowitz condition. Nonlin. Anal. **72**, 4602–4613 (2010)
226. Li, S., Wu, S., Zhou, H.-S.: Solutions to semilinear elliptic problems with combined nonlinearities. J. Differ. Equat. **185**, 200–224 (2002)
227. Liapounoff, A.: Problème général de la stabilité du mouvement. Ann. Fac. Sci. Toulouse Sci. Math. Sci. Phys. **9**(2), 203–474 (1907)
228. Lieberman, G.M.: Boundary regularity for solutions of degenerate elliptic equations. Nonlin. Anal. **12**, 1203–1219 (1988)
229. Lindqvist, P.: On the equation $\operatorname{div}(|\nabla u|^{p-2}\nabla u) + \lambda |u|^{p-2}u = 0$. Proc. Amer. Math. Soc. **109**, 157–164 (1990)
230. Ling, J.: Unique continuation for a class of degenerate elliptic operators. J. Math. Anal. Appl. **168**, 511–517 (1992)
231. Liu, J.: The Morse index of a saddle point. Syst. Sci. Math. Sci. **2**, 32–39 (1989)
232. Liu, J., Li, S.: An existence theorem for multiple critical points and its application. Kexue Tongbao (Chinese) **29**, 1025–1027 (1984)
233. Liu, Z., Motreanu, D.: A class of variational-hemivariational inequalities of elliptic type. Nonlinearity **23**, 1741–1752 (2010)

234. Liu, J., Su, J.: Remarks on multiple nontrivial solutions for quasi-linear resonant problems. J. Math. Anal. Appl. **258** 209–222 (2001)

235. Liu, Z., Wang, Z.-Q.: Sign-changing solutions of nonlinear elliptic equations. Front. Math. China **3**, 221–238 (2008)

236. Liu, J., Wu, S.: Calculating critical groups of solutions for elliptic problem with jumping nonlinearity. Nonlin. Anal. **49**,779–797 (2002)

237. Lloyd, N.G.: Degree Theory. Cambridge University Press, Cambridge (1978)

238. Lucia, M., Prashanth, S.: Strong comparison principle for solutions of quasilinear equations. Proc. Amer. Math. Soc. **132**, 1005–1011 (2004)

239. Manásevich, R., Mawhin, J.: Periodic solutions for nonlinear systems with $p$-Laplacian-like operators. J. Differ. Equat. **145**, 367–393 (1998)

240. Manes, A., Micheletti, A.M.: Un'estensione della teoria variazionale classica degli autovalori per operatori ellittici del secondo ordine. Boll. Un. Mat. Ital. **7**(4), 285–301 (1973)

241. Marano, S.A., Motreanu, D.: Infinitely many critical points of non-differentiable functions and applications to a Neumann-type problem involving the $p$-Laplacian. J. Differ. Equat. **182**, 108–120 (2002)

242. Marano, S.A., Motreanu, D.: A deformation theorem and some critical point results for non-differentiable functions. Topol. Methods Nonlin. Anal. **22**, 139–158 (2003)

243. Marano, S.A., Motreanu, D.: Critical points of non-smooth functions with a weak compactness condition. J. Math. Anal. Appl. **358**, 189–201 (2009)

244. Marano, S.A., Papageorgiou, N.S.: Constant-sign and nodal solutions of coercive $(p,q)$-Laplacian problems. Nonlin. Anal. **77**, 118–129 (2013)

245. Marcus, M., Mizel, V.J.: Absolute continuity on tracks and mappings of Sobolev spaces. Arch. Rational Mech. Anal. **45**, 294–320 (1972)

246. Marcus, M., Mizel, V.J.: Continuity of certain Nemitsky operators on Sobolev spaces and the chain rule. J. Anal. Math. **28**, 303–334 (1975)

247. Marcus, M., Mizel, V.J.: Every superposition operator mapping one Sobolev space into another is continuous. J. Funct. Anal. **33**, 217–229 (1979)

248. Margulies, C.A., Margulies, W.: An example of the Fučik spectrum. Nonlin. Anal. **29**, 1373–1378 (1997)

249. Marino, A., Prodi, G.: Metodi perturbativi nella teoria di Morse. Boll. Un. Mat. Ital. **11**(4, suppl. 3), 1–32 (1975)

250. Martio, O.: Counterexamples for unique continuation. Manuscripta Math. **60**, 21–47 (1988)

251. Mawhin, J.: Semicoercive monotone variational problems. Acad. Roy. Belg. Bull. Cl. Sci. **73**(5), 118–130 (1987)

252. Mawhin, J.: Forced second order conservative systems with periodic nonlinearity. Ann. Inst. H. Poincaré Anal. Non Linéaire **6**, 415–434 (1989)

253. Mawhin, J., Willem, M.: Critical point theory and Hamiltonian systems. Springer, New York (1989)

254. Mawhin, J., Ward, J.R., Willem, M.: Variational methods and semilinear elliptic equations. Arch. Rational Mech. Anal. **95**, 269–277 (1986)

255. Maz'ja, V.G.: Sobolev Spaces. Springer, Berlin (1985)

256. Megginson, R.E.: An Introduction to Banach Space Theory. Springer, New York (1998)

257. Meyers, N.G., Serrin, J.: $H = W$. Proc. Nat. Acad. Sci. USA **51**, 1055–1056 (1964)

258. Minty, G.J.: Monotone (nonlinear) operators in Hilbert space. Duke Math. J. **29**, 341–346 (1962)

259. Minty, G.J.: On a "monotonicity" method for the solution of non-linear equations in Banach spaces. Proc. Nat. Acad. Sci. USA **50**, 1038–1041 (1963)

260. Miyagaki, O.H., Souto, M.A.S.: Superlinear problems without Ambrosetti and Rabinowitz growth condition. J. Differ. Equat. **245**, 3628–3638 (2008)

261. Miyajima, S., Motreanu, D., Tanaka, M.: Multiple existence results of solutions for the Neumann problems via super- and sub-solutions. J. Funct. Anal. **262**, 1921–1953 (2012)

262. Montenegro, M.: Strong maximum principles for supersolutions of quasilinear elliptic equations. Nonlin. Anal. **37**, 431–448 (1999)

263. Mordukhovich, B.S.: Variational Analysis and Generalized Differentiation I: Basic Theory. Springer, Berlin (2006)
264. Moroz, V.: Solutions of superlinear at zero elliptic equations via Morse theory. Topol. Methods Nonlin. Anal. **10**, 387–397 (1997)
265. Morrey, C.B.: Functions of several variables and absolute continuity II. Duke Math. J. **6**, 187–215 (1940)
266. Morrey, C.B.: Multiple Integrals in the Calculus of Variations. Springer, New York (1966)
267. Morse, M.: Relations between the critical points of a real function of $n$ independent variables. Trans. Amer. Math. Soc. **27**, 345–396 (1925)
268. Morse, M.: The Calculus of Variations in the Large, vol. 18, p. IX+368. American Mathematical Society Colloquium Publications, New York (1934)
269. Moser, J.: A new proof of De Giorgi's theorem concerning the regularity problem for elliptic differential equations. Comm. Pure Appl. Math. **13**, 457–468 (1960)
270. Motreanu, V.V.: Multiplicity of solutions for variable exponent Dirichlet problem with concave term. Discrete Contin. Dyn. Syst. **5**, 845–855 (2012)
271. Motreanu, D.: Three solutions with precise sign properties for systems of quasilinear elliptic equations. Discrete Contin. Dyn. Syst. Ser. S **5**, 831–843 (2012)
272. Motreanu, D., Motreanu, V.V.: Coerciveness property for a class of non-smooth functionals. Z. Anal. Anwend. **19**, 1087–1093 (2000)
273. Motreanu, D., Motreanu, V.V.: Nonsmooth variational problems in the limit case and duality. J. Global Optim. **29**, 439–453 (2004)
274. Motreanu, D., Panagiotopoulos, P.D.: Minimax Theorems and Qualitative Properties of the Solutions of Hemivariational Inequalities. Kluwer Academic Publishers, Dordrecht (1999)
275. Motreanu, D., Papageorgiou, N.S.: Existence and multiplicity of solutions for Neumann problems. J. Differ. Equat. **232**, 1–35 (2007)
276. Motreanu, D., Papageorgiou, N.S.: Multiple solutions for nonlinear Neumann problems driven by a nonhomogeneous differential operator. Proc. Am. Math. Soc. **139**, 3527–3535 (2011)
277. Motreanu, D., Perera, K.: Multiple nontrivial solutions of Neumann $p$-Laplacian systems. Topol. Methods Nonlin. Anal. **34**, 41–48 (2009)
278. Motreanu, D., Rădulescu, V.: Variational and Non-Variational Methods in Nonlinear Analysis and Boundary Value Problems. Kluwer Academic Publishers, Dordrecht (2003)
279. Motreanu, D., Tanaka, M.: Sign-changing and constant-sign solutions for $p$-Laplacian problems with jumping nonlinearities. J. Differ. Equat. **249**, 3352–3376 (2010)
280. Motreanu, D., Tanaka, M.: Existence of solutions for quasilinear elliptic equations with jumping nonlinearities under the Neumann boundary condition. Calc. Var. Partial Differ. Equat. **43**, 231–264 (2012)
281. Motreanu, D., Tanaka, M.: Generalized eigenvalue problems of nonhomogencous elliptic operators and their application. Pacific J. Math. **265**, 151–184 (2013)
282. Motreanu, D., Winkert, P.: On the Fučik spectrum for the $p$-Laplacian with Robin boundary condition. Nonlin. Anal. **74**, 4671–4681 (2011)
283. Motreanu, D., Winkert, P.: The Fučík spectrum for the negative $p$-Laplacian with different boundary conditions. In: Nonlinear Analysis, pp. 471–485. Springer, New York (2012)
284. Motreanu, D., Zhang, Z.: Constant sign and sign changing solutions for systems of quasilinear elliptic equations. Set-Valued Var. Anal. **19**, 255–269 (2011)
285. Motreanu, D., Motreanu, V.V., Paşca, D.: A version of Zhong's coercivity result for a general class of nonsmooth functionals. Abstr. Appl. Anal. **7**, 601–612 (2002)
286. Motreanu, D., Motreanu, V.V., Papageorgiou, N.S.: Periodic solutions for nonautonomous systems with nonsmooth quadratic or superquadratic potential. Topol. Methods Nonlin. Anal. **24**, 269–296 (2004)
287. Motreanu, D., Motreanu, V.V., Papageorgiou, N.S.: Existence of solutions for strongly nonlinear elliptic differential inclusions with unilateral constraints. Adv. Differ. Equat. **10**, 961–982 (2005)

288. Motreanu, D., Motreanu, V.V., Papageorgiou, N.S.: Multiple nontrivial solutions for nonlinear eigenvalue problems. Proc. Amer. Math. Soc. **135**, 3649–3658 (2007)
289. Motreanu, D., Motreanu, V.V., Papageorgiou, N.S.: A unified approach for multiple constant sign and nodal solutions. Adv. Differ. Equat. **12**, 1363–1392 (2007)
290. Motreanu, D., Motreanu, V.V., Papageorgiou, N.S.: Two nontrivial solutions for periodic systems with indefinite linear part. Discrete Contin. Dyn. Syst. **19**, 197–210 (2007)
291. Motreanu, D., Motreanu, V.V., Papageorgiou, N.S.: A degree theoretic approach for multiple solutions of constant sign for nonlinear elliptic equations. Manuscripta Math. **124**, 507–531 (2007)
292. Motreanu, D., Motreanu, V.V., Papageorgiou, N.S.: A multiplicity theorem for problems with the $p$-Laplacian. Nonlin. Anal. **68**, 1016–1027 (2008)
293. Motreanu, D., Motreanu, V.V., Papageorgiou, N.S.: Positive solutions and multiple solutions at non-resonance, resonance and near resonance for hemivariational inequalities with $p$-Laplacian. Trans. Amer. Math. Soc. **360**, 2527–2545 (2008)
294. Motreanu, D., Motreanu, V.V., Papageorgiou, N.S.: Nonlinear Neumann problems near resonance. Indiana Univ. Math. J. **58**, 1257–1279 (2009)
295. Motreanu, D., Motreanu, V.V., Papageorgiou, N.S.: Multiple solutions for Dirichlet problems which are superlinear at $+\infty$ and (sub-)linear at $-\infty$. Commun. Appl. Anal. **13**, 341–357 (2009)
296. Motreanu, D., Motreanu, V.V., Papageorgiou, N.S.: Multiple solutions for resonant nonlinear periodic equations. NoDEA - Nonlin. Differ. Equat. Appl. **17**, 535–557 (2010)
297. Motreanu, D., Motreanu, V.V., Papageorgiou, N.S.: Existence and multiplicity of solutions for asympotically linear, noncoercive elliptic equations. Monatsh. Math. **159**, 59–80 (2010)
298. Motreanu, D., Motreanu, V.V., Turinici, M.: Coerciveness property for conical nonsmooth functionals. J. Optim. Theory Appl. **145**, 148–163 (2010)
299. Motreanu, D., O'Regan, D., Papageorgiou, N.S.: A unified treatment using critical point methods of the existence of multiple solutions for superlinear and sublinear Neumann problems. Commun. Pure Appl. Anal. **10**, 1791–1816 (2011)
300. Motreanu, D., Motreanu, V.V., Papageorgiou, N.S.: Nonautonomous resonant periodic systems with indefinite linear part and a nonsmooth potential. Commun. Pure Appl. Anal. **10**, 1401–1414 (2011)
301. Motreanu, D., Motreanu, V.V., Papageorgiou, N.S.: Multiple constant sign and nodal solutions for nonlinear Neumann eigenvalue problems. Ann. Scuola Norm. Sup. Pisa Cl. Sci. **10**(5), 729–755 (2011)
302. Motreanu, D., Motreanu, V.V., Papageorgiou, N.S.: On $p$-Laplace equations with concave terms and asymmetric perturbations. Proc. Roy. Soc. Edinb. Sect. A **141**, 171–192 (2011)
303. Motreanu, D., Motreanu, V.V., Papageorgiou, N.S.: On resonant Neumann problems. Math. Ann. **354**, 1117–1145 (2012)
304. Motreanu, D., Motreanu, V.V., Papageorgiou, N.S.: Existence and nonexistence of positive solutions for parametric Neumann problems with $p$-Laplacian. Tohoku Math. J. (to appear)
305. Mugnai, D., Papageorgiou, N.S.: Resonant nonlinear Neumann problems with indefinite weight. Ann. Sc. Norm. Super. Pisa Cl. Sci. **11**(5), 729–788 (2012)
306. Nagumo, M.: A theory of degree of mapping based on infinitesimal analysis. Amer. J. Math. **73**, 485–496 (1951)
307. Naniewicz, Z., Panagiotopoulos, P.D.: Mathematical Theory of Hemivariational Inequalities and Applications. Dekker Inc., New York (1995)
308. Ni, W.-M., Wang, X.: On the first positive Neumann eigenvalue. Discrete Contin. Dyn. Syst. **17**, 1–19 (2007)
309. Nirenberg, L.: Topics in Nonlinear Functional Analysis. New York University Courant Institute of Mathematical Sciences, New York (2001)
310. Njoku, F.I., Zanolin, F.: Positive solutions for two-point BVPs: existence and multiplicity results. Nonlin. Anal. **13**, 1329–1338 (1989)
311. Nussbaum, R.D.: The fixed point index for local condensing maps. Ann. Mat. Pura Appl. **89**(4), 217–258 (1971)

312. Ôtani, M.: Existence and nonexistence of nontrivial solutions of some nonlinear degenerate elliptic equations. J. Funct. Anal. **76**, 140–159 (1988)
313. Palais, R.S.: Morse theory on Hilbert manifolds. Topology **2**, 299–340 (1963)
314. Palais, R.S.: Lusternik-Schnirelman theory on Banach manifolds. Topology **5**, 115–132 (1966)
315. Palais, R.S.: The principle of symmetric criticality. Comm. Math. Phys. **69**, 19–30 (1979)
316. Palais, R.S., Smale, S.: A generalized Morse theory. Bull. Amer. Math. Soc. **70**, 165–172 (1964)
317. Panagiotopoulos, P.D.: Hemivariational Inequalities. Applications in Mechanics and Engineering. Springer, Berlin (1993)
318. Papageorgiou, N.S., Kyritsi-Yiallourou, S.T.: Handbook of Applied Analysis. Springer, New York (2009)
319. Papageorgiou, E.H., Papageorgiou, N.S.: A multiplicity theorem for problems with the $p$-Laplacian. J. Funct. Anal. **244**, 63–77 (2007)
320. Papageorgiou, N.S., Papalini, F.: On the existence of three nontrivial solutions for periodic problems driven by the scalar $p$-Laplacian. Adv. Nonlin. Stud. **11**, 455–471 (2011)
321. Papageorgiou, N.S., Papalini, F.: Multiple solutions for nonlinear periodic systems with combined nonlinearities and a nonsmooth potential. J. Nonlin. Convex Anal. **13**, 681–693 (2012)
322. Papageorgiou, N.S., Staicu, V.: Multiple nontrivial solutions for doubly resonant periodic problems. Canad. Math. Bull. **53**, 347–359 (2010)
323. Papageorgiou, N.S., Rocha, E.M., Staicu, V.: A multiplicity theorem for hemivariational inequalities with a $p$-Laplacian-like differential operator. Nonlin. Anal. **69**, 1150–1163 (2008)
324. Pascali, D., Sburlan, S.: Nonlinear Mappings of Monotone Type. Martinus Nijhoff Publishers, The Hague (1978)
325. Pauli, W.: Theory of Relativity. Pergamon, New York (1958)
326. Perera, K.: Homological local linking. Abstr. Appl. Anal. **3**, 181–189 (1998)
327. Perera, K.: Nontrivial critical groups in $p$-Laplacian problems via the Yang index. Topol. Methods Nonlin. Anal. **21**, 301–309 (2003)
328. Phelps, R.R.: Convex Functions, Monotone Operators and Differentiability. Lecture Notes in Mathematics, vol. 1364. Springer, Berlin (1993)
329. Pigola, S., Rigoli, M., Setti, A.G.: Maximum principles on Riemannian manifolds and applications. Mem. Amer. Math. Soc. **174**, 822 (2005)
330. Pucci, P., Serrin, J.: A mountain pass theorem. J. Differ. Equat. **60**, 142–149 (1985)
331. Pucci, P., Serrin, J.: The structure of the critical set in the mountain pass theorem. Trans. Amer. Math. Soc. **299**, 115–132 (1987)
332. Pucci, P., Serrin, J.: The Maximum Principle. Birkhäuser, Basel (2007)
333. Qian, A.: Existence of infinitely many nodal solutions for a superlinear Neumann boundary value problem. Bound. Value Probl. **2005**, 329–335 (2005)
334. Rabinowitz, P.H.: Some global results for nonlinear eigenvalue problems. J. Funct. Anal. **7**, 487–513 (1971)
335. Rabinowitz, P.H.: A note on topological degree for potential operators. J. Math. Anal. Appl. **51**, 483–492 (1975)
336. Rabinowitz, P.H.: Some minimax theorems and applications to nonlinear partial differential equations. In: Nonlinear Analysis (collection of papers in honor of Erich H. Rothe), pp. 161–177. Academic, New York (1978)
337. Rabinowitz, P.H.: Periodic solutions of Hamiltonian systems. Comm. Pure Appl. Math. **31**, 157–184 (1978)
338. Rabinowitz, P.H.: Minimax Methods in Critical Point Theory with Applications to Differential Equations. American Mathematical Society, Washington (1986)
339. Rellich, F.: Ein Satz über mittlere Konvergenz. Nachrichten Göttingen **1930**, 30–35 (1930)
340. Ricceri, B.: Infinitely many solutions of the Neumann problem for elliptic equations involving the $p$-Laplacian. Bull. London Math. Soc. **33**, 331–340 (2001)

341. Riesz, F.: Über lineare Funktionalgleichungen. Acta Math. **41**, 71–98 (1916)
342. Roberts, A.W., Varberg, D.E.: Convex Functions. Academic, New York (1973)
343. Rockafellar, R.T.: Convex Analysis. Princeton University Press, Princeton (1970)
344. Rockafellar, R.T.: On the maximality of sums of nonlinear monotone operators. Trans. Amer. Math. Soc. **149**, 75–88 (1970)
345. Rockafellar, R.T.: On the maximal monotonicity of subdifferential mappings. Pacific J. Math. **33**, 209–216 (1970)
346. Rockafellar, R.T., Wets, R.J.-B.: Variational Analysis. Springer, Berlin (1998)
347. Roselli, P., Sciunzi, B.: A strong comparison principle for the $p$-Laplacian. Proc. Amer. Math. Soc. **135**, 3217–3224 (2007)
348. Rothe, E.H.: Morse theory in Hilbert space. Rocky Mountain J. Math. **3**, 251–274 (1973)
349. Rynne, B.P.: Spectral properties of $p$-Laplacian problems with Neumann and mixed-type multi-point boundary conditions. Nonlin. Anal. **74**, 1471–1484 (2011)
350. Schauder, J.: Der Fixpunktsatz in Funktionalräumen. Studia **2**, 171–180 (1930)
351. Schechter, M.: The Fučík spectrum. Indiana Univ. Math. J. **43**, 1139–1157 (1994)
352. Schechter, M.: Infinite-dimensional linking. Duke Math. J. **94**, 573–595 (1998)
353. Schechter, M.: Linking Methods in Critical Point Theory. Birkhäuser, Boston (1999)
354. Schechter, M.: Principles of Functional Analysis. American Mathematical Society, Providence (2002)
355. Schechter, M., Zou, W.: Superlinear problems. Pacific J. Math. **214**, 145–160 (2004)
356. Schmidt, E.: Zur Theorie der linearen und nichtlinearen Integralgleichungen III, Teil: Über die Auflösung der nichtlinearen Integralgleichungen und die Verzweigung ihrer Lösungen. Math. Ann. **65**, 370–399 (1908)
357. Schwartz, L.: Théorie des Distributions: Tome I. Hermann and Cie., Paris (1950)
358. Schwartz, L.: Théorie des Distributions: Tome II. Hermann and Cie., Paris (1951)
359. Serrin, J.: Local behavior of solutions of quasi-linear equations. Acta Math. **111**, 247–302 (1964)
360. Skrypnik, I.V.: Nonlinear Elliptic Boundary Value Problems. Teubner Verlagsgesellschaft, Leipzig (1986)
361. Silva, E.A., Teixeira, M.A.: A version of Rolle's theorem and applications. Bol. Soc. Brasil. Mat. (N.S.) **29**, 301–327 (1998)
362. Smale, S.: Morse theory and a non-linear generalization of the Dirichlet problem. Ann. of Math. **80**(2), 382–396 (1964)
363. Sobolev, S.L.: Applications of Functional Analysis in Mathematical Physics. Amer. Math. Soc., Providence (1963)
364. Sobolev, S.L.: On a theorem of functional analysis. Amer. Math. Soc. Transl. **34**, 39–68 (1963)
365. Spanier, E.H.: Algebraic Topology. McGraw-Hill Book Co., New York (1966)
366. Stampacchia, G.: Équations Elliptiques du Second Ordre à Coefficients Discontinus. Les Presses de l'Université de Montréal, Montreal (1966)
367. Struwe, M.: Variational Methods. Springer, Berlin (1996)
368. Su, J., Zhao, L.: An elliptic resonance problem with multiple solutions. J. Math. Anal. Appl. **319**, 604–616 (2006)
369. Szulkin, A.: Minimax principles for lower semicontinuous functions and applications to nonlinear boundary value problems. Ann. Inst. H. Poincaré Anal. Non Linéaire **3**, 77–109 (1986)
370. Tang, C.-L.: Periodic solutions for nonautonomous second order systems with sublinear nonlinearity. Proc. Amer. Math. Soc. **126**, 3263–3270 (1998)
371. Tang, C.-L., Wu, X.-P.: Periodic solutions for second order systems with not uniformly coercive potential. J. Math. Anal. Appl. **259**, 386–397 (2001)
372. Tang, C.-L., Wu, X.-P.: Periodic solutions for a class of nonautonomous subquadratic second order Hamiltonian systems. J. Math. Anal. Appl. **275**, 870–882 (2002)
373. Tang, C.-L., Wu, X.-P.: Existence and multiplicity for solutions of Neumann problem for semilinear elliptic equations. J. Math. Anal. Appl. **288**, 660–670 (2003)

374. Tartar, L.: An Introduction to Sobolev Spaces and Interpolation Spaces. Springer, Berlin (2007)
375. Thews, K.: Nontrivial solutions of elliptic equations at resonance. Proc. Roy. Soc. Edinb. Sect. A **85**, 119–129 (1980)
376. Tolksdorf, P.: Regularity for a more general class of quasilinear elliptic equations. J. Differ. Equat. **51**, 126–150 (1984)
377. Troyanski, S.L.: On locally uniformly convex and differentiable norms in certain non-separable Banach spaces. Studia Math. **37**, 173–180 (1971)
378. Trudinger, N.S.: On Harnack type inequalities and their application to quasilinear elliptic equations. Comm. Pure Appl. Math. **20**, 721–747 (1967)
379. Vázquez, J. L.: A strong maximum principle for some quasilinear elliptic equations. Appl. Math. Optim. **12**, 191–202 (1984)
380. Wang, J.: The existence of positive solutions for the one-dimensional $p$-Laplacian. Proc. Amer. Math. Soc. **125**, 2275–2283 (1997)
381. Weyl, H.: Space, Time, Matter. Dover Publications, New York (1951)
382. Willem, M.: Minimax Theorems. Birkhäuser, Boston (1996)
383. Wu, X., Tan, K.-K.: On existence and multiplicity of solutions of Neumann boundary value problems for quasi-linear elliptic equations. Nonlin. Anal. **65**, 1334–1347 (2006)
384. Yang, C.-T.: On theorems of Borsuk-Ulam, Kakutani-Yamabe-Yujobô and Dyson. I. Ann. Math. **60**(2), 262–282 (1954)
385. Yang, X.: Multiple periodic solutions of a class of $p$-Laplacian. J. Math. Anal. Appl. **314**, 17–29 (2006)
386. Zeidler, E.: The Ljusternik-Schnirelman theory for indefinite and not necessarily odd nonlinear operators and its applications. Nonlin. Anal. **4**, 451–489 (1980)
387. Zeidler, E.: Nonlinear Functional Analysis and Its Applications III: Variational Methods and Optimization. Springer, New York (1985)
388. Zeidler, E.: Nonlinear Functional Analysis and Its Applications II/A: Linear Monotone Operators. Springer, New York (1990)
389. Zeidler, E.: Nonlinear Functional Analysis and Its Applications II/B: Nonlinear Monotone Operators. Springer-Verlag, New York (1990)
390. Zhang, M.: The rotation number approach to eigenvalues of the one-dimensional $p$-Laplacian with periodic potentials. J. London Math. Soc. **64**(2), 125–143 (2001)
391. Zhang, Q.: A strong maximum principle for differential equations with nonstandard $p(x)$-growth conditions. J. Math. Anal. Appl. **312**, 24–32 (2005)
392. Zhang, Z., Li, S.: On sign-changing and multiple solutions of the $p$-Laplacian. J. Funct. Anal. **197**, 447–468 (2003)
393. Zhang, Q., Li, G.: On a class of second order differential inclusions driven by the scalar $p$-Laplacian. Nonlin. Anal. **72**, 151–163 (2010)
394. Zhang, Z., Chen, J., Li, S.: Construction of pseudo-gradient vector field and sign-changing multiple solutions involving $p$-Laplacian. J. Differ. Equat. **201**, 287–303 (2004)
395. Zhong, C.-K.: A generalization of Ekeland's variational principle and application to the study of the relation between the weak P.S. condition and coercivity. Nonlin. Anal. **29**, 1421–1431 (1997)
396. Ziemer, W.P.: Weakly Differentiable Functions. Sobolev Spaces and Functions of Bounded Variation. Springer, New York (1989)
397. Zou, W.: Variant fountain theorems and their applications. Manuscripta Math. **104**, 343–358 (2001)

# Index

**A**

Ambrosetti–Rabinowitz condition, 341, 383

antimaximum principle, 262

**B**

Banach submanifold, 133

Betti-type number, 157

bifurcation point, 182

bilinear form
  coercive, 98
  continuous, 98
  symmetric, 98

Borsuk's theorem, 65

Borsuk–Ulam theorem, 67

boundary values, 8

Brouwer index, 165

Brouwer's fixed point theorem, 64

**C**

Carathéodory map, 41

Cerami condition, 102

chain rule, 6, 53

characteristic function, 12

compact map, 15

completely continuous map, 16

concave–convex nonlinearity, 346

contractible, 147

convex map, 45

convex subdifferential, 25, 46

Courant nodal domain theorem, 232, 242

Courant–Fischer theorem, 230, 252

critical group, 153
  at infinity, 159

critical point, 62, 133, 135
  constrained, 134

nondegenerate, 154

of a locally Lipschitz map, 55

of mountain pass type, 175

**D**

deformation, 110

deformation theorem
  first, 106
  second, 110

degree
  Brouwer's degree, 61
  Leray–Schauder degree, 70
  for $(S)_+$-maps, 76
  for operators of monotone type, 84

demicontinuous, 26

distributional derivative, 2

duality map, 29

**E**

effective domain, 45

eigenfunction, 224, 232, 235

eigenspace, 18

eigenvalue, 18, 255
  (LS)-eigenvalue, 241, 249
  of the $p$-Laplacian, 235, 253
  of the Laplacian, 224, 232

eigenvector, 18

Ekeland variational principle, 99

epigraph, 46

exact sequence, 142

**F**

finite-rank map, 16

fountain theorem, 132

D. Motreanu et al., *Topological and Variational Methods with Applications to Nonlinear Boundary Value Problems*, DOI 10.1007/978-1-4614-9323-5,
© Springer Science+Business Media, LLC 2014

Printed in the United States
By Bookmasters

Printed in the United States
By Bookmasters